The Anthropocene: Politik—Economics—Society—Science

Volume 35

Series Editor

Hans Günter Brauch, Peace Research and European Security Studies (AFES-PRESS), Mosbach, Baden-Württemberg, Germany

The *Anthropocene* represents a major shift in Earth history as a result of the direct intervention by humankind into the earth system and nature. The term, coined by Stoermer, was popularized by Nobel Laureate, Paul J. Crutzen, when he declared in Cuernavaca (Mexico) in February 2000: "We are in the Anthropocene". Crutzen argued that the result of the massive intervention by human economic activities into the earth system, scientifically documented in the rise of GHG (parts per million [ppm] of CO2) in the atmosphere, is that earth history has transitioned from the Holocene, the latest period of earth history following the end of the glacial period 12,000 years ago, to the Anthropocene.
When the Anthropocene started is disputed. It may have started at any time between the start of the industrial revolution around 1750 and the end of World War II in 1945, or in 1958 when the scientific measurement of the concentration (ppm) of CO2 in the atmosphere began and since when two-thirds of the increase has occurred. The societal and political impacts of this silent transition from the 'Holocene' to the 'Anthropocene' may be more severe and long-lasting than the impacts of a) the long-term major *technical transitions* (the Neolithic or agricultural and industrial revolutions), b) *major changes* in the political order (Vienna in 1815, Versailles in 1919, Yalta and San Francisco in 1945, the fall of the Berlin Wall in 1989), c) medium-term transitions of governments and major office holders, and d) short-term structure-creating events (e.g. the Russian revolution of 1917).

This peer-reviewed scientific book series addresses this very long-term and severe transition process, where the cause of the change and its potential violent societal consequences are no longer the ' *other* ' nation, ethnic, religious or political group, but ' *we* ': "we are the threat" and only *we* as part of *humankind* can offer a remedy by containing the causes and addressing, facing and coping with the consequences.

This book series reviews and analyses selective societal consequences of this fundamental change in earth history and the political, economic, societal and scientific discourses and policy-oriented societal debates on i) achieving the goal of 'sustainable development', ii) creating processes of 'sustainability transition', iii) the need for 'a new contract for sustainability', iv) the need for a paradigmatic change in world view (a scientific revolution) towards sustainability, or even v) the need for a 'sustainability revolution'.

The key actors for bringing about such a change are: 1) *Politik* in its three distinct meanings of 'politics' (process), 'policy' (field, area) and 'polity' (legal and institutional framework); 2) *Economics* (as the field, actor and process), 3) *Society* (as the innovative groups and processes), and 4) *Science* as the source of technical innovation and societal and philosophical reflection.

All books in this series range from 55,000 to 250,000 words, and are published as printed books in softcover and as eBooks, and are also available as individual chapters. All members of academic institutions globally that subscribe to the relevant book package of SpringerLink have free access to these eBooks and may obtain *MyCopy* at a highly discounted price.

More details on the website of the Series Editor.

Hans Günter Brauch
Editor

Towards Rethinking Politics, Policy and Polity in the Anthropocene

Multidisciplinary Perspectives

With Forewords by Ulrich Pöschl, Director at the Max-Planck-Institute for Chemistry, Mainz
Ernst Ulrich von Weizsäcker, Honorary President of the Club of Rome

Hans Günter Brauch Foundation
for Peace and Ecology in the Anthropocene

Springer

Editor
Hans Günter Brauch
Peace Research and European Security Studies (AFES-PRESS)
Hans Günter Brauch Foundation for Peace and Ecology in the Anthropocene (HGBS)
Mosbach, Baden-Württemberg, Germany

Acknowledgement: The first cover photo on the first test of a nuclear weapon on 16 July 1945 in Alamogordo in New Mexico was taken from: First Nuclear weapon test_trinity_ copyright: US-DoE, Manhattan District project that is in the public domain. The second photo on Hurricane Otis of 25 October 2023 was originally provided by the US government NoA and is as such in the public domain. It was used by CIRA, Reuters and Science and the US press media. The US government institutions [US-DoE; NoA] © with an exclusive license for Springer Nature Switzerland. More on this book is at: http://www.afes-press-books.de/html/APESS_35.htm.

ISSN 2367-4024 ISSN 2367-4032 (electronic)
The Anthropocene: Politik—Economics—Society—Science
ISBN 978-3-031-71806-9 ISBN 978-3-031-71807-6 (eBook)
https://doi.org/10.1007/978-3-031-71807-6

This work was supported by Max Planck Institute for Chemistry, Mainz, Germany.

Copyediting: PD Dr. Hans Günter Brauch, AFES-PRESS e.V., Mosbach, Germany
Copy and Language Editor: Ms Elizabeth Norton-Amor.
Proofreading: By the authors.

This Springer imprint is published by the registered company Springer Nature Switzerland AG
The registered company address is: Gewerbestrasse 11, 6330 Cham, Switzerland

To my parents
To my father who shared with me his memories and stories of the fighting during World War II in North Africa and to my mother who sensitised me about the human impact and the aftermath of this war with the burning of Dresden she observed from the farm of my grandfather in Germany's mineral mountains

To Úrsula Oswald Spring who carried the torch of peace research further as President (1998–2000) and Secretary General of IPRA (2016–2018) and as Secretary General of the Latin American Council of Peace Research and to her grandchildren Charlotte (2018), Luca Kai (2018), Gianpaolo Nalu (2023) and Olivia (2023)

Foreword: Climate, Health, and Open Scholarship in a Prosperous Anthropocene

Ulrich Pöschl[1]

The Anthropocene as the current epoch in Earth history is characterised by a globally pervasive and rapidly increasing influence of human activities on the entire planet, except its interior—from the equator to the poles and from the land surface,

[1] Ulrich Pöschl, Prof. Dr., Director, Max Planck Institute for Chemistry, Mainz, Hahn-Meitner-Weg 1, D-55128 Mainz, Germany; Phone/Fax: +49-6131-305-7000/7111; e-mail: u.poschl@mpic.de, Websites: www.mpic.de; www.mpic.de/forschung/multiphasenchemie/profil-ulrich-poeschl.html; www.webofscience.com/wos/author/record/849640; scholar.google.com/citations?user=czxG87cAAAAJ.

atmosphere and biosphere to the oceans and deep sea. The intensive use of land and water, as well as the emission of greenhouse gases and pollutants, has led to climate change and has put pressure on ecosystems, biodiversity, and human health. Related phenomena, such as global warming and sea level rise resulting from the anthropogenically enhanced greenhouse effect, have long been known, scientifically understood, and predicted.

During the development of modern humans (*Homo sapiens*) in recent Earth history—i.e. in the Pleistocene, beginning about a million years ago—the concentration of carbon dioxide (CO_2) in the Earth's atmosphere varied with a periodicity of about 100 thousand years between minimum values around 200 ppm in ice ages and maximum values around 300 ppm in the warm periods, accompanied by fluctuations in the global mean surface temperature of about 5 °C and changes in sea level of more than a hundred metres. These ice age cycles can be explained by natural variations in the Earth's orbital motion around the sun (Milankovic cycles), associated changes in solar energy input, and coupled feedback loops in the climate system such as the global material cycling (biogeochemical cycles) of carbon, nitrogen, aerosols, and water.

In the geological epoch of the Holocene, which began around 10,000 years ago, the atmospheric CO_2 concentration was fairly stable at around 260–280 ppm, and the fluctuations of global mean surface temperature and sea level were relatively small. Under these favourable and stable conditions, human civilisation developed from the Neolithic Age to modern times. Since global industrialisation in the 19th century, however, there has been a sharp increase in CO_2 and other greenhouse gases such as methane (CH_4) and nitrous oxide (N_2O). The current atmospheric greenhouse gas concentrations of more than 420 ppm CO_2 and 1900 ppb CH_4 are not only far above the stable values during the last 10,000 years in the Holocene (more than + 50% for CO_2 and +130% for CH_4). They are also far outside the range of natural fluctuations in recent Earth history, i.e. far above the approximately 200–300 ppm CO_2 and 400–800 ppb CH_4 that were characteristic for the regular cycling between ice ages and warm periods during the last million years in the Pleistocene.

The massive changes in the concentration of greenhouse gases and other trace gases in the atmosphere are a clear indicator of the extent to which humans are interfering with the global biogeochemical cycles that can be regarded as the metabolism of planet Earth. Particularly far-reaching are the human interventions in the carbon cycle through the release of CO_2 from the combustion of fossil fuels (coal, crude oil, natural gas) and the release of CH_4 from industrial and agricultural processes. Similarly, the nitrogen cycle has been massively altered and enhanced by the industrial production and agricultural use of ammonia and nitrogen fertiliser, as well as the release of nitrogen oxides such as N_2O from fertilised soils and NO or NO_2 from combustion and other high-temperature processes. The observed steep increase in greenhouse gas concentrations far outside the range of natural fluctuations is clearly attributable to anthropogenic emissions, and there is no plausible alternative explanation. The same applies to the rapid global warming of our time, i.e. the increase in global mean surface temperature by more than one degree Celsius in less than two centuries, from 1850 until today.

At a scientific conference of the International Geosphere-Biosphere Programme (IGBP) in the year 2000, the atmospheric researcher and Chemistry Nobel laureate Paul Crutzen realised and stated that the Earth system had transitioned from the Holocene to a new geological epoch shaped by humanity, for which he coined the term Anthropocene. Scientists had already recognised an increasing influence of humans on their environment and planet Earth since the 19th century (Leclerc, Marsh, Stoppani, Arrhenius, Vernadsky, Teilhard de Chardin, Stoermer, Markl et al.), but had not yet identified a clear transition to a new geological epoch.

Since the discovery and conceptualisation by Paul Crutzen and colleagues, there have been various proposals for the formal beginning and dating of the Anthropocene. In addition to global industrialisation in the 18th/19th century, which led to the above-mentioned increase in CO_2 and CH_4 in the global atmosphere and polar ice, the beginning of agriculture in the Neolithic period has also been proposed as a starting point; this led to regional changes but is hardly detectable on a global scale and coincides with the beginning of the climatically very stable Holocene. Equating the beginning of the Anthropocene with the Holocene would be neither obvious nor useful from a scientific point of view, and although equating it with the beginning of industrialisation would be possible based on the increase in greenhouse gas concentrations in ice cores and similar parameters, it would not meet the formal requirements for the official definition of a geological epoch, which should be traceable in sediments worldwide.

Accordingly, the Anthropocene Working Group of the International Commission on Stratigraphy (ICS/IUGS) proposes to define the beginning of the Anthropocene on the basis of plutonium concentration peaks, which originate from radioactive fallout from nuclear weapons tests in the middle of the 20th century and are detectable worldwide. This period of the "Great Acceleration" of human activities and influence on the Earth system also saw particularly high growth rates and turning points in the characteristic S-shaped growth curves for the world population, global primary energy consumption, and other indicators of human expansion.

Due to human activities and emissions, the CO_2 concentration in the Earth's atmosphere is now as high as it was around 3 million years ago, when the average global temperature and sea level were around 3 °C and 10–20 metres higher, respectively. If the CO_2 concentration remains this high or continues to rise, a correspondingly strong global warming and sea level rise can be expected—in accordance with modern Earth system and climate models, as well as with earlier estimates of the greenhouse effect by pioneers of natural and Earth system sciences in the 19th and 20th centuries (Fourier, Arrhenius, Revelle, Bolin, Keeling, Manabe, Hasselmann et al.). This will result in an increase in the frequency and intensity of extreme weather events (heat waves, droughts, heavy rainfall, landslides, etc.), the flooding of coastal areas—where a large part of the world's population lives—and corresponding migration movements, as explained in detail in the reports of the Intergovernmental Panel for Climate Change (IPCC, www.ipcc.ch). If the anthropogenic release of CO_2 and other greenhouse gases is not curbed, a transition from the Holocene warm period to an Anthropocene hot period is to be expected, tipping the stable environmental conditions under which human civilisation developed.

In addition to climate change, anthropogenic emissions have also led to a massive increase in the atmospheric concentration of aerosols and photooxidants on a local, regional, and global scale. For example, the average mixing ratios of ozone in continental background air have increased by a factor of ~2–4 from about 10–20 ppb at the beginning of the 19th century to 30–40 ppb in the 21st century. The amount of particulate matter in polluted urban air is typically ~10 times higher than in clean air of remote continental regions (~10–100 $\mu g\ m^{-3}$ compared to ~1–10 $\mu g\ m^{-3}$). Exposure to air pollutants leads to oxidative stress in human, animal, and plant organisms and to an increase in morbidity and mortality due to respiratory, cardiovascular and metabolic diseases. Epidemiological and toxicological studies show that pollutants from combustion processes and other anthropogenic emission sources are among the greatest risk factors for human health and life expectancy worldwide (World Health Organisation, www.who.int; Global Burden of Disease, www.healthdata.org/research-analysis/gbd).

Environmental pollution severely affects climate and public health in the Anthropocene. Thus, we need to understand how humanity can best deal with the sources and effects of environmental pollutants in relation to economic development and social welfare. Recent advances in scientific research provide deep insights into the underlying physical, chemical, and biological processes and reveal the relative importance of different pollutants and their sources, including natural and anthropogenic contributions. This knowledge enables the development of efficient strategies and policies to mitigate and counteract the adverse effects of environmental pollutants on the Earth system, climate, and health ("planetary health"). With regard to the development, societal communication, and political implementation of appropriate policies for environmental protection, it seems worthwhile to emphasise that climate and health effects are two facets of global change that can—and need to be—handled together.

Further scientific, technical, and societal progress towards a climate-neutral, environmentally friendly and sustainable energy and circular economy appears urgently needed in order to achieve a lasting and stable Anthropocene, ensure the health, prosperity and continued existence of human civilisation, and avoid a deterioration in, or even the loss of, the basis of our livelihood. The Anthropocene concept may help humanity to recognise and communicate, both cognitively and emotionally, the message: We are shaping the planet, so let's try to get it right.

To avoid the perils and grasp the chances related to the power of shaping the planet, it seems critically important to explore and follow potentially suitable pathways according to the basic principles of critical rationalism, i.e. according to the best state of knowledge as validated by observations, experiments, and theoretical considerations, but without claiming absolute certainty.

Multi- and transdisciplinary exchange will be required for proper consideration and communication of all relevant perspectives, for societal decision making, and for the practical implementation of suitable directives, regulations, and measures—with regard to climate protection and human health as well as economic prosperity, social justice, and global equity. Key issues are outlined in the 17 Sustainable Development

Goals of the United Nations and related statements from political and scholarly institutions around the world (e.g., sdgs.un.org/goals; www.usaid.gov/sdgs; international-partnerships.ec.europa.eu/policies/sustainable-development-goals_en; www.pas.va/en/publications/scripta-varia/sv139_springer.html).

Recent experiences at the science-policy interface related to the COVID-19 pandemic, air quality, climate, and human health confirmed that open scholarship (including open access publications, open data, open source codes, open peer review, interactive public discussion, and the principles of critical rationalism) are essential for efficient multi- and transdisciplinary exchange among scholars, politics, and society at large. New and improved forms of scholarly communication and quality assurance need to be developed while converting the existing corpus of scholarly journals from subscription to open access (OA) as outlined in the OA2020 Expression of Interest and the latest Berlin Open Access Conference (oa2020.org; oa2020.org/b16-conference).

To gain the full benefits of open scholarship, a dualism of complementary approaches is required ("as-well-as", not "either-or"): We need top-down approaches to achieve large-scale progress, as well as bottom-up initiatives to maintain scholarly diversity. We need appropriate transformative agreements with traditional publishers to achieve an efficient conversion of subscription journals that are still important for some communities, and we also need to liberate and provide funds for proper OA publishers and new publishing platforms to maintain competition and innovation (scilog.fwf.ac.at/en/magazine/its-hard-to-explain-why-this-is-taking-so-long). Most importantly, we have to uphold and improve scholarly quality assurance, evaluation, and self-regulation. For this purpose, we need to maintain functioning elements of traditional peer review and add advances such as open review, interactive discussion, and article level metrics that are traceable, transparent, and inclusive for the global scientific community and society at large. Indeed, transcending hidden peer review and citation counting oligopolies by transparent forms of quality assurance that can be followed, reproduced, and refined by everybody around the world is one of the main reasons why we need full and immediate open access to scholarly research publications.

In times of growing irrationality, "alternative facts", postfactual discussions, and fake news, it seems ever more important to practise and promote open scholarship and critical rationalism, i.e. openly show and share the basis, methods, and rigour of the scientific discourse as a role model and reference for societal exchange, decision making, and evolution. Accordingly, open peer review and interactive public discussion are important steps on the way to an epistemic web that documents the scientific discourse in accordance with critical rationalism—openly telling the world not only what we know, but also how well we know it, and where the limitations are. A global commons of knowledge will help to augment, communicate, and utilise scholarly insights and understanding for rational and transparent approaches to resolving complex questions and problems in the Anthropocene.

January 2024

Prof. Dr. Ulrich Pöschl
Max Planck Institute for Chemistry
Mainz, Germany

Acknowledgement This article is dedicated to Paul J. Crutzen, whose curiosity, ingenuity, and wisdom have helped to prevent ozone layer destruction and nuclear winter, and will help humankind to avoid the perils and grasp the opportunities of the Anthropocene.

Prof. Dr. Ulrich Pöschl is the director at the Max Planck Institute for Chemistry and professor at the Johannes Gutenberg University (JGU) in Mainz, Germany. He studied Chemistry at the Technical University of Graz and was postdoctoral researcher, group leader, and university lecturer at the Massachusetts Institute of Technology (MIT), Departments of Chemistry and of Earth, Atmospheric, and Planetary Sciences; the Max Planck Institute for Chemistry (MPIC), Atmospheric Chemistry and Biogeochemistry Departments; and the Technical University of Munich (TUM), Institute of Hydrochemistry. Among other internationally leading scientists, Pöschl worked with the Nobel laureates Paul Crutzen and Mario Molina. His current research and teaching are focused on the effects of multiphase processes, i.e. interactions between gases, liquids, and solids, in the Earth system, climate, life, and health. He is engaged in numerous international scientific collaborations, has been honoured with prestigious scientific awards, and is among the world's most highly cited researchers (Web of Science; Google Scholar). Pöschl has advised and mentored numerous students and postdoctoral researchers, many of whom are now faculty members or senior researchers in scientific leadership positions around the world. In accordance with Pöschl's open and equitable approach to scientific teaching and mentoring, his students, postdocs, and alumni successfully pursuing an academic career come from a wide range of ethnic and societal backgrounds (~50% international) with a fairly even gender balance (~50% female). Pöschl actively promotes open science, and he initiated the innovative and successful approach of interactive open access publishing with public peer review and online discussion (multi-stage open peer review) as established in the leading international scientific journal Atmospheric Chemistry and Physics (ACP, www.atmospheric-chemistry-and-physics.net) and the European Geosciences Union (EGU, www.egu.eu). Moreover, Pöschl initiated and co-chairs the global open access initiative OA2020 (oa2020.org).

Preface: Towards a New Enlightenment for the Anthropocene

Ernst Ulrich von Weizsäcker[2]

[2] Prof. Dr. Ernst Ulrich von Weizsäcker, Founding President of the Wuppertal Institute and Honorary President Club of Rome; Emmendingen, Germany; e-mail: ernst@weizsaecker.de.

The New Global Context of the Anthropocene

The "Anthropocene" is the new name of today's face of our planet Earth. The name was coined by biologist Eugene F. Stoermer and by atmospheric chemist and Nobel Prize winner Paul J. Crutzen. The Anthropocene concept is a combination of *anthropo-* from *anthropos* (ancient Greek: ἄνθρωπος), meaning "human", and *-cene* from *kainos* (ancient Greek: καινός), meaning "new" or "recent". The plural *Anthropoi* is the ancient Greek name for humans that should usually be understood in this inclusive sense if the context suggests it, and the singular *anthropos* is often used as a collective term (like the English "mankind") which obviously is meant to include both males and females. So "the Anthropocene" means the planet's appearance at the time when the humans are dominating the face of the Earth.

Alas, we humans are underway to transforming the planet in a way that is very unpleasant for millions of animal and plant species. Thousands, if not millions, of species simply disappear because they can't survive. One reason is global warming. Polar bears are accustomed to living in an icy environment. But global warming makes icy places disappear. Another reason is human agriculture for feeding eight billion humans. Such agriculture requires hundreds of millions of hectares of soil, which then is no longer available for thousands of specific "wild" animals and plants. A third reason is human fishery, again for feeding humans. Thousands of fish species have vanished already because they were simply killed.

Hans Günter Brauch is describing the political need for handling the Anthropocene, including strategies for reducing global warming. A great approach, no doubt, but terribly difficult to follow. People tend to continue their usual way of life—which has pushed us into the global warming disasters. But what else can we do?

I am afraid that we have no easy strategy at hand. Most of us simply prefer to continue our usual habits. We consider forests and meadows and rivers and lakes as our legitimate property. We dig out minerals and coal and oil. We catch fish, and we raise and kill pigs and cows and fowls. We construct endless buildings and streets and rails. Humanity rose from 2.7 billion humans in 1950 to 8.1 billion today—three times more than 74 years before. This jump created the Anthropocene.

Need for a New Enlightenment in the Anthropocene

What we absolutely need to create is a new civilisation that reduces and ultimately stops the increase of world population and global warming. Moreover, it must halt the relentless extermination of animals and plants.

How can this be achieved? I suggest that we develop *a new kind of Enlightenment*. The Enlightenment was the historical era, roughly between 1500 and 1900, based on ancient civilisations, such as Greek and Latin, but also Egyptian, Chinese, Indian, and other languages and cultures. The Enlightenment opened a universe of natural sciences, of legal rules, of artistic painting and literature, and of medicine. But it also developed brutal military tools and the conquering of other regions and cultures, arrogantly called "colonisation".

A new Enlightenment must take care of climate and biodiversity; this means stabilising and later shrinking human population. Similarly, our civilisations ought to develop technologies and habits allowing for human prosperity, but stabilising the climate and allowing the recovery of rare animals and plants.

A new Enlightenment must be peaceful. It must overcome the follies of wars and of conquering other parts of the world. It must build on modern sciences and medicine. It should have an appreciation of *balance* instead of dogmatism.

Some twenty or more cases of balance can be formulated, such as female and male, parents and children, humans and nature, short term and long term, monthly and daily, the state and the markets, rigid law and freedom of action, religion and freedom of action, innovation and customary tradition, economic equity and incentives for specific achievement. All such balanced achievements and attitudes should develop with a view of peace and friendship.

The neoliberal economy of making the strongest the winner and the weaker the loser is almost the opposite of "peace and friendship". Neoliberal economics is closely related to what is called "Social Darwinism" The term was coined by Herbert Spencer and chiefly developed in the USA. It is not what Charles Darwin meant by referring to "survival of the fittest". For him, the fittest could just as well be a mouse or a small bird quickly fleeing from a predator.

It is evident that Hans Günter Brauch and his eight co-authors come with an extremely ambitious agenda. Readers will be surprised by that ambition. But that ambition can lead hundreds or thousands of readers to becoming the pioneers of a new face of the Anthropocene and of a new Enlightenment.[3]

Our World Is in a State of Disarray, Confusion and Uncertainty

No sense of direction is really visible. Deep social discrepancies, failed states, wars and civil wars, unemployment, and mass migrations have left hundreds of millions of people in a state of despair. The United Nations has unanimously adopted the *2030 Agenda*, which is seen by many as the epitome of a sense of direction. But when looking at it more closely, a reader cannot escape the impression that a successful implementation of the Agenda's eleven socio-economic goals could more than likely

[3] The following text is based on key arguments developed by the author and Anders Wijkman as co-Presidents in Part II of their report to the Club of Rome, *Come On!* (2018), published by Springer in New York and translated into ten languages. Permission was granted by co-author Anders Wijkman and by Ms. Alice Essenprice from Springer's Permissions Department on 5 January 2024.

destroy any hopes of it being able to realise its three ecological goals of stabilising climate, stopping the dangerous overexploitation of the oceans, and halting the massive losses of living species.[4]

The First Enlightenment in the Holocene

The Enlightenment of the 17th and 18th centuries prompted the rise of science, technology, and a subsequent adoration of technological "miracles". Korten proposed a new social narrative and a new cosmology, the Sacred Life and Living Earth story. He used self-governing local communities, the 2010 Cochabamba Declaration on the Rights of Mother Earth, and emerging movements worldwide towards a Living Economy as examples of how the Earth and all the life on it can be preserved by wide adherence to a different kind of story.

The first part of a Report to the Club of Rome (2018), *Come On!*, analysed the Cold War and the multiple discussions and assessments in the post-Cold War period, including the security, economic, ideological and philosophical discussions and the new environmental challenges since 1988, concluding that: "Historians know that the new (capitalistic) mind-set consolidated its influence after the end of the Cold War".

The authors deduced that the new era of "globalisation" "meant essentially that small and medium states had to surrender much of their governing powers to the markets" and that "[t]he period since 1989 was also the period during which the finance sector extended its domination over the worldwide economy". Graeme Maxton and Jorgen Randers (2016) noted

---

[4] See a shortened version of the summary: "Part I of the book offers a descriptive diagnosis of the non-sustainable trends of our time. This current era is being termed the 'Anthropocene', a reference to the fact that humans are dominant, changing and shaping the bio-geochemistry of the planet. The Earth's 'planetary boundaries', established by physics, are in danger of being destroyed by humanity's selfish priority of feeding human 'needs'. … Part II will offer a deeper analysis, describing society's fundamental philosophical crisis at this juncture, starting with the message of the Encyclical Letter *Laudato Sí* by Pope Francis. The foundations of today's economics, stemming from a time of a (relatively) 'empty world' … are inadequate for dealing with our current 'full world'. The currently dominant scientific 'analytical philosophy' and materialist thinking are shallow and inadequate for understanding and dealing with a living world. Instead, a systems approach will be needed to address and understand the complex and interactive realities of our time. … Part II suggests a radical change: developing a new Enlightenment, one that is more based on balance than on doctrine, and one that is fitting for the 'full world' and for *sustainable* development. Part II can therefore be seen as the most revolutionary part of the book. Can the planet's beleaguered natural systems wait until all of human civilization has gone through the long process of a new Enlightenment? No, explains Part III, which lists an optimistic collection of opportunities that exist already, and of practical policies that support beneficial solutions and open a wide horizon into a better, sustainable world. It is *people* who can and should move our civilization into the new direction—as they simultaneously create and live a new, enlightened philosophy. … The book closes with an invitation to readers and other discussants to engage themselves in the many possible ways of leading to a sustainable world society".

> dangerous failures: climate change, pollution (especially of the oceans), resource depletion and biodiversity loss; poverty, inequality, and social friction; and unemployment, especially for the young. Civil, religious and territorial wars are also increasing along with terrorism, partly as a result of these endemic problems. These multiple crises are driving mass migrations of refugees. Finally, geopolitical friction is also rising. Many of the wars in recent decades have implicitly been about access to the resources needed to fuel economic growth, notably oil and water. Such wars, ironically, both push social and environmental problems further down the agenda and exacerbate them. Most of the migrants across Asia, as well as those from the Middle East and North Africa, are leaving areas riven by conflict and ravaged by poverty and resource destruction. ... All of them have the same basic cause: "the desire for endless consumption growth without due concern for the impacts on the environment and inequality." The capitalist incentive structure rewards cost cutting and short-term profits. ... Extreme free-market thinking is at the root of the damage humanity is inflicting on the planet (Maxton/Randers).

Such views are shared by a broad range of analysts and experts. Similarly, Anders Wijkman and Johan Rockström (2012), in their *Report to the Club of Rome—Bankrupting Nature*, have illustrated the fact that the destruction of nature and the emergence of financial collapses.[5] Following essentially the same logic of greed, impatience and short-termism, by the 1980s, neoliberal thinking had become a mainstream view in US academic circles and a broadly accepted alternative to the European social market economy.

Today the collapse of communism is history, and free-market fundamentalism, in combination with the steeply rising influence of the finance sector, is our new reality. An understanding of the negative effects of extreme market thinking and practice has not reached average citizens. Nonetheless, increasing numbers of voices can be heard recognising that neoliberalism has been "oversold" and that fiscal austerity policies can actually be detrimental to a country's economy. So far, despite protectionist challenges to compulsory free trade, the free market doctrine still prevails, and the transnational financiers it has favoured still control much of the global economy. However, as far as the planet and the vast majority of its people are concerned, this doctrine has failed us all.

Three major tenets in terms of relevance deserve further investigation and clarification:

- Adam Smith's concept of the *invisible hand* and the related conviction (chiefly from the Chicago School of Economics) that markets by definition are superior to states or to lawmakers in finding the optimum path of development;
- David Ricardo's discovery of *comparative advantages*, which in theory makes trade a win-win operation for both sides of the transaction; and
- Charles Darwin's conceptual edifice, which has been wrongly interpreted as postulating that competition, the fiercer the better, leads to *continued progress and evolution*.

---

[5] Wijkman, Anders; Rockström, Johan, 2012: *Bankrupting Nature: Denying Our Planetary Boundaries. A Report to the Club of Rome* (London: Earthscan).

All three tenets contain valid points, but all three have to be better understood and put into an historical perspective. The *Report to the Club of Rome* (2018) noted "Gaps between Theory, Education and Social Reality":

> A reductionist philosophy as described in the last chapter and our growing tendency to divide reality in order to apprehend it, led to a divorce between our knowledge, our educational system and the social reality that we live in. Not only has the ivory tower of academia been far away from social reality for many centuries but also has the increasing reductionist tendency led to ever more specialised disciplines. Human endeavour has moved away from looking at reality as a whole to cutting it up in small pieces—a major source of the challenges we are facing.
>
> The division between economy and ecology that persisted for nearly two centuries is a dramatic example of a general problem. Fragmentation of knowledge results in a loss of perspective regarding the interrelationships and interdependencies between the parts and the wider whole of which they are components. This disconnect has shaped the organisation of universities and research institutes on narrow disciplinary lines. It has shaped the organisation of government of policy and administrative functions in increasingly specialised areas of expertise divorced from the wider social reality of which they are parts. It has resulted in laws and policies that seek to address specialised issues, unmindful of their impact on other fields. It is the reason why until the 1970s, environmental impact studies were never considered important in planning for new commercial or public projects.

Fragmentation and Specialisation of Academic Perspectives, Disciplines, Economic Theories and Practice

The impact of this divorce on policy and practice has been equally apparent and influential at the theoretical level. It led to a fragmentation of economic theory into countless sub-disciplines and the formulation of countless theories and models confined to describing the internal operations of a specific type of activity, while marginalising important factors as mere externalities. The same tendency facilitated economics in developing separately and independently of the other social sciences. The physicists who founded neoclassical economics in the late 19th century also played a part, determined as they were to elevate economics to the status of a true science based on Physics. These trends resulted in the prevalence of economic theories that show mathematical virtuosity while partly ignoring political, legal, social, cultural, and psychological factors of relevance to an understanding of human economic behaviour and systems. The same fragmentation of disciplines has occurred in all the natural and social sciences, with similar consequences. Specialisation triumphs over integral views.

Governance: Tolerance and Long-Term Perspectives

The philosophical crisis has massive effects on the way the world is governed. One important aspect has so far been completely neglected. It is the relation between national and global governance. The United Nations was created at a time when the nation state was the only entity entitled to adopt legally binding laws. Indeed, there were also legally binding international treaties, but they were meaningless unless ratified by all nation states involved.

When the world takes the sustainability challenges seriously, many other areas of sovereignty have to be put in question. This requires the establishment of a truly new mind-set. The climate discussions are the most familiar case in point. It is becoming increasingly immoral to continue blasting greenhouse gases into the atmosphere. And yet national parliaments and governments, not to speak of nationalistic citizens' movements, often consider it unacceptable to be forced into national legislation by international institutions or conferences.

The European Union, on the other hand, is a shining example of the fact that surrendering national sovereign rights to a larger authority can actually be beneficial for the countries doing so. In fields like commerce, consumer protection, agriculture, and the environment, some 80 per cent of the legislation of an EU member state is prescribed by European directives, or even directly binding regulations. Of course, there are procedures in which national governments negotiate about the contents of directives and regulations. But increasingly, qualified majorities of countries can outvote minorities of countries, still making the EU law binding for all; there are exceptions where unanimity rules, such as issues related to taxation. All economic analyses have shown that the EU countries greatly benefitted from this reduction in their national sovereignty.

On the global scale, nothing of that kind is visible so far. Of course, there is the Law of the Sea, the UN Conventions on climate and biodiversity, and many other international legal instruments. But enforcement is hardly possible—with one major exception: the rulings of the WTO. It is time to take a new initiative for a more stringent international legal system, adequate for the "Full World". The very idea of national sovereignty was a product of the "Empty World". Some religions and cultural traditions stemming from the empty world had a basically intolerant credo legitimising aggression, expansion, and victimisation of the people of other creeds, colours, or cultures. This must definitely also be overcome.

From the Past Enlightenment in the Holocene Towards a New Enlightenment in the Anthropocene

The strongest figures of the European Enlightenment of the 18th century were David Hume, Jean-Jacques Rousseau, Voltaire, Adam Smith, and Immanuel Kant but it was built on powerful precursor philosophers such as René Descartes, Blaise Pascal, Francis Bacon, Erasmus, John Locke, Baruch de Spinoza, Montesquieu, G. W. Leibniz, and Isaac Newton, to name but a few. Together, they caused and created a revolutionary change in European civilisation.

One of the most revolutionary features was the separation of the nation state from the church(es). While the then existing Church had little sympathy for independent intellectualism, the enlightened state saw free thinking and acting of citizens as a great hope for the future. They also saw an enlightened citizenry as the prime source of scientific endeavour, technological inventiveness, and the entrepreneurial spirit. And, in fact, the 18th century witnessed an explosive development of science and technology. Antoine de Lavoisier and James Watt were among the first, but after them an avalanche of technological innovations led to the Industrial Revolution.

The Enlightenment is also credited for freeing human individuals from the suffocating pressures of churches and the absolutist state of the 17th and 18th centuries. But this new individualism also led to the gradual decay of earlier communities. The commons (like grazing land, forests or fishing grounds) were the foundation of earlier human survival. However, in parallel with growing private wealth and with the new appreciation of individual achievements, these were eroded, privatised and in some cases destroyed.

For civilisations outside Europe, the downsides of the Enlightenment had even worse effects. European armies, colonists, and missionaries had already conquered and colonised much of the world during the 16th and 17th century, and the Industrial Revolution that followed made Europe, notably the British Empire, essentially invincible. European supremacism and missionary war went as far as justifying the subjection and killing of peoples living in conquered territories. Many alternative traditions and cultures were destroyed, cultures that had existed and developed for thousands of years.

It has become fashionable to call for new Enlightenments, but motives and contents differ widely. In many cases, the word is used simply to revive or modernise the old Enlightenment concepts of rationalism and freedom, and an outlook that is anti-normative, anti-regulation, and anti-state dominance. One example out of many comes from the British Libertarian Alliance. The (Austrian) European Forum Alpbach, widely known for showcasing new intellectual fashions, devoted its entire 2016 programme to questions around a New Enlightenment; but this was mostly a call for a renewal of rationalism.

The new enlightenment, "Enlightenment 2.0", is unlikely to be Europe-centred. It should look at the great traditions of other civilisations. To give two very different examples:

The Hopi tradition in North America remained essentially stable and sustainable for 3000 years. They are one of the oldest living cultures in documented history, with a past stretching back thousands of years. They developed a sustainable kind of agriculture, maintained an essentially stable population size, avoided wars, and became masters in building stone houses. Under almost every definition of sustainability, they would be among the champions. Their very complex religion is based on the idea of balance—between resources like water and light, between differing peoples, between day and night and the various seasons, even between humour and sobriety.

In most of the Asian traditions, there is a strong sense of balance, as opposed to the monotheistic dogmatic view where only one side is right. Balance is sought between rational thinking (the brain) and emotional feeling (the heart). Yin and Yang are symbols of balanced contrast. Mark Cartwright (2012)[6] offers a simplified definition of what is also an essential part of Confucian cosmology:

> The principle of Yin and Yang is a fundamental concept in Chinese philosophy and culture in general dating from the third century BCE or even earlier. This principle is that all things exist as inseparable and contradictory opposites, for example female-male, dark-light and old-young. The two opposites attract and complement each other and, as their symbol illustrates, each side has at its core an element of the other. Neither pole is superior to the other and, as an increase in one brings a corresponding decrease in the other, a correct balance between the two poles must be reached in order to achieve harmony.
>
> Yin is feminine, black, dark, north, water (transformation), passive, moon (weakness and the goddess Changxi), earth, cold, old, even numbers, valleys, poor, soft, and provides spirit to all things. Yin reaches its height of influence with the winter solstice. Yin may also be represented by the tiger, the colour orange and a broken line in the trigrams of the I Ching (or Book of Changes).
>
> Yang is masculine, white, light, south, fire (creativity), active, sun (strength and the god Xihe), heaven, warm, young, odd numbers, mountains, rich, hard, and provides form to all things. Yang reaches its height of influence with the summer solstice. Yang may also be represented by the dragon, the colour blue and a solid line trigram (Wijkman/von Weizsäcker, 2018; Cartwright. 2012).[7]

This brief description cannot explain the whole richness of the Yin and Yang philosophy, which can also be criticised for typified and hence unfair gender roles, or having the static features of zero-sum games (positive sum games are much preferable). But it reflects a wisdom in the comprehension that contrasts can be creative. This wisdom differs from prevailing Western and Islamic habits that see contrasts as an invitation to decide which one is right (or benign) and which one is wrong (or evil), often leading to bitter and violent feuds.

Towards a Philosophy of Balance, Not of Exclusion

The wisdom of the synergies to be found between contrasts can also help overcome deficits of the analytical philosophy of science, making room for a more future-oriented philosophy. To be sure, technical and scientific measurements must be done correctly; facts must be treated as facts. But modern physics has shown that the precise measurement of one feature can destroy the measurability of its contrasting (complementary) feature. For example, Heisenberg's Uncertainty Principle established that the momentum and position of a particle cannot be measured simultaneously with unlimited precision. The physical basis of this astounding finding lies in the fact that

[6] Cartwright, Mark, 2012: *Yin and Yang. Definition. Ancient History Encyclopedia.*

[7] Wijkman, Anders; von Weizsäcker, Ernst Ulrich von, 2018: *Come On! Capitalism, Short-termism, Population and the Destruction of the Planet* (New York: Springer).

the particle also has wave properties that interfere with the waves (e.g. light waves) of the measuring instrument. Also, particle properties and wave properties are mutually "complementary".

Complementarity can be a door-opener to perceiving parallels between modern physics and Eastern wisdom and religions. In his bestselling *The Tao of Physics* Frithjof Capra,[8] who was once an academic assistant of Heisenberg's, showed that Buddhism, Hinduism and Taoism had the power of dealing with unexplainable realities that people call mysticism. At the end of this book, Capra stated that, "science does not need mysticism and mysticism does not need science, but man needs both".

Complementarity, balance, and the wisdom of synergies between contrasts should be milestones on the way to a New Enlightenment. Surely there will be more philosophical steps for overcoming the deficits of analytical philosophy, of selfishness, individualism, short-termism, and other traits mentioned by Pope Francis in *Laudato Sí* as destructive and suicidal in terms of "our common home". But there is clearly a short list of topics where a renewed appreciation of balance is necessary. Most of the topics listed are not new at all, but all of them suffer from a lack of balance in our time.

The New Enlightenment Should Work on a Balance

- *Between humans and nature*: this is one of the core messages. In the "empty world" it was a given. In the "full world" it is a huge challenge. Using remaining natural landscapes, waterbodies and minerals chiefly as resources for an ever-growing human population and the fulfilment of ever-growing consumption is not balance, but destruction.
- *Between short term and long term*: Humans appreciate quick gratification, such as something to drink when thirsty, and quarterly financial reports by publicly traded companies. But there is a need for a counterbalance to ensure long-term action such as policies to re-stabilise the Earth's climate. In addition to long-term ethics, short-term incentives will be needed to reward long term action.
- *Between speed and stability:* Technological and cultural progress benefits from competition for temporal priority. That is the most important criterion for scientific careers and is essential for commercial success: "disruptive" innovations carry tremendous appreciation. But speed by itself can be a horror for the slow ones, for most of the elderly, for babies, and for village communities (think of the Hopis!). What is worse is that the current civilisational addiction to speed is destructive to structures, habits, and cultures that have emerged under the sustainability criterion. Sustainability, after all, basically means stability.

---

[8] Capra, Frithjof; Luisi, Pier Luigi, 2014: *The Systems View of Life* (Cambridge: Cambridge University Press).

- *Between private and public:* The discovery of the human values of individualism, private property, and protection against state intrusion has been among the most valuable achievements of the European Enlightenment. But in our times, public goods are much more endangered than private goods. There are dangers to the commons, the public infrastructure, the system of justice and reliable order. Under international competition for lowest taxes (attracting investors), public goods tend to be neglected and under-financed. The state (public) should set the rules for the market (private), not the other way round. Paul de Grauwe and Anna Asbury have lucidly described how history has produced pendulum swings between private dominance and state dominance. However, history has not brought about anything like a lasting balance between the two.
- *Between women and men:* Many early cultures developed through wars, during which women were chiefly entrusted with caring for the family and men for defence (or aggression). This model is outdated. Riane Eisler in *The Chalice and the Blade* has offered archaeological insights into cultures thriving in partnership models and has also called for "The Real Wealth of Nations", claiming that the conventional (male-dominated) "wealth of nations" is almost a caricature of real well-being. Balance may not be well achieved by pushing and pulling as many women as possible into positions that have been typified as "male". Balance may be better attained by changing the typology of the job functions.
- *Between equity and awards for achievements:* Without awards for achievement, societies can get sleepy and lose out in the competition with other societies. But there must be a publicly guaranteed system of justice and equity. Inequity, according to Wilkinson and Pickett,[9] tends to be correlated with undesirable social parameters, poor education, high criminality, infant mortality, etc.
- *Between state and religion:* It was a great achievement of the European Enlightenment to separate public from religious leadership, fully respecting religious values and communities. This has to be maintained in a balanced manner. Religions dominating the public sector are in high danger of destroying human rights and the great civilisational achievement of an independent legal system with independent High Courts. Dominance of religion tends to be intolerant to persons working outside the religious community. On the other hand, states that are intolerant against religious communities tend to lose touch with ethical (and long-term) needs.

Towards a New Enlightenment for the Anthropocene

This is a modest and only an indicative list of the principle of balance. Part II of the Report to the Club of Rome consisted of a philosophical critique of mainstream thinking, and ended with the quest for a new Enlightenment. But the pragmatic

[9] Wilkinson, Richard; Pickett, Kate, 2009: *The Spirit Level. Why greater equality makes societies stronger.*

intentions of *Come On!* cannot be to work on a new Enlightenment and then patiently and comfortably wait until all people on Earth, or nearly all, have adopted that new worldview. That process could take a hundred years, similar to the process of the European Enlightenment from the late 16th century until the French Revolution.

Instead of waiting, it is better to act now. And there is reason for optimism. Actually, many things are already moving in the right direction. There was a decisive reduction in global poverty from 1820 (over 90% of people then living in poverty) through 1970 (60% in poverty) to 2011 (only 14% in poverty). Even the rapid increase in the world's population did not prevent the absolute number of poor people from decreasing.

The entire series of good news is anthropocentric. Ecological achievements are restricted to very few successes of local pollution control and some successes in repatriation of rare animals. Individuals can be catalysts for social change, innovation, development, and evolution. This fact is a source of hope. Indeed, every significant change in the history of civilisation has first occurred as an inspired idea or aspiration in the mind and heart of an individual who then transmitted it to others and awakened the same fire in them. What is true of the individual is also true of small groups and larger organisations, many of which are the seeds of what has grown over the decades into large movements striving to change the course of human development. Individuals, small groups and organisations can serve as the prime catalysts of change.

This book of nine authors offers in ten chapters a new framework for a "second *Sattelzeit*" with new enlightenment in the Anthropocene Epoch of Earth history that started with the end of World War II, the first test of a nuclear weapon on 16 July 1945 and the first use of nuclear bombs against Hiroshima and Nagasaki on 6 and 9 August 1945.

Hans Günter Brauch argues in his texts for this volume on the coincidence of two changes of Earth history in the Anthropocene and the Nuclear Era that unleashed two anthropogenic existential threats for the survival of humanity:

- *Nuclear war* with side effects in a *nuclear winter* for human food security;
- *Global environmental change* with climate change and biodiversity loss that has already resulted in more intensive and frequent extreme weather events and that may result in the coming decades in chaotic interactions in the Earth system that may trigger tipping points with existential geopolitical consequences.

Brauch claims that the two anthropogenic existential threats for the survival of humanity differ: while the first is triggered by a threat of the "other" (leader, country, superpower, alliance, camp), in the second we as a species (humanity) have become the existential threat to our own survival as a result of our burning of fossil energy sources and of our way of life.

Paul J. Crutzen pointed to these multiple existential complex threats on 23 February 2000 in a scientific workshop of the International Geosphere-Biosphere Programme when he stated: "We are not living any longer in the Holocene but in the Anthropocene".

Hans Günter Brauch combines both existential threats in his texts and calls for the new enlightenment which the Club of Rome stressed in its report *Come On!* in 2018. He shares the critique of fragmentation and specialisation of academia and calls for a multidisciplinary, transformative holistic approach in science and practice as goals of a socio-ecological transformation. For moving "towards rethinking politics, polity, and polity in the Anthropocene" he calls for a New Enlightenment in a "New *Sattelzeit*", which Reinhart Koselleck suggested in the 1970s and a few younger historians have taken up without any framing in the Anthropocene, the new epoch of Earth history.

Thinking on a new Enlightenment can bring different strands of innovative thinking on the Anthropocene together: a) in the natural sciences and Earth System Science, as the German Academy of Science Leopoldina has called for in its report in 2022; and b) in the thinking on the History of Science and Geoanthropology as seen in the work by Jürgen Renn (2020) and his colleagues.

Brauch argues that his own discipline of Political Science and International Relations should no longer ignore these multiple efforts of thinking together and integrate the thinking on the Anthropocene and on a New Enlightenment where he has focused on the linkages between peace, security, and ecology—or a peace ecology and an ecological peace policy.

Emmendingen, Baden-Württemberg, Germany
January 2024

Prof. Dr. Ernst Ulrich von Weizsäcker

Prof. Dr. Ernst Ulrich von Weizsäcker was born 25 June 1939 in Zürich, Switzerland. 1965 M.A. Physics, Hamburg; 1969 Ph.D. Biology Freiburg, Germany. 1972 Professor of Biology, Essen University. 1975 Founding President, University of Kassel. 1981 Director, UN Centre for Science and Technology for Development, New York. 1984 Director, Institute for European Environmental Policy, Bonn, London, Paris. 1991 Founding President, Wuppertal Institute for Climate, Environment, Energy. 1998 Member of Parliament, SPD (serving as Chair, consecutively, of the Bundestag Study Commission on Economic Globalisation, and of the Environment Committee.) 2006 Dean, Bren School for Environmental Science and Management, UC Santa Barbara, California. 2007–2009 He returned to Germany in 2009 where he is an honorary professor at Freiburg University. Co-Chair China Council (CCICED) Taskforce on Economic Instruments for Energy Efficiency and the Environment. 2009 retired; *pro bono* Co-Chair, since 2007, International Resource Panel, and Co-President, Club of Rome with Anders Wijkman 2011–2019.

Publications: 1989: *Erdpolitik*; (With: Jochen Jesinghaus, 1992): *Ecological Tax Reform* (London: Zed Books); 1994: *Earth Politics* (London: Zed Books); (With: Amory and Hunter Lovins, 1997): *Factor Four. Doubling Wealth—Halving Resource Use* (London: Earthscan); (With: Oran Young and Matthias Finger, 2005): *Limits to Privatization* (London: Earthscan); (With: Charlie Hargroves et al., 2009): *Factor Five* (London, Earthscan) [The last three books are also available in Chinese]; (With Anders Wijkman, 2018): *Come On! Capitalism, Short-termism, Population and the Destruction of the Planet*; 2022: *So reicht das nicht: Außenpolitik, neue Ökonomie, neue Aufklärung—Was wir in der Klimakrise jetzt wirklich brauchen. Awards: 1989:* Premio de Natura; 1996: Duke of Edinburgh Gold Medal of WWF International; 2001: Takeda Award, shared with. F. Schmidt-Bleek; 2008: German Environment Prize; 2009: Commander's Cross of the Order of

Merit of the Federal Republic of Germany; 2010: Federal High Cross of Merit; 2011: Theodor Heuss Prize: 2012: Order of Merit of Baden-Württemberg; 2014: Sustainability Prize of the Saxonian Hans-Carl-von-Carlowitz-Society; 2016: Honorary Doctorate of Nagoya University; 2021: Order of Merit of the State of North Rhine-Westphalia; Honorary Councillor of the World Future Council.

Address: P.O. Box 1125, 79325 Teningen, Germany
e-mail: ernst@weizsaecker.de
Homepage: www.ernst.weizsaecker.de

Acknowledgement

On the eve of my 70th birthday on 31 May 2017, I organised a day's brainstorming in my private library in Mosbach on "'Politik' in and for the Anthropocene: Obstacles, Challenges, Opportunities and Tasks for the Social Sciences in the 21st Century"[10] to which I had invited a few former students, friends, authors, and members of the AFES-PRESS Board, which the Dutch Nobel Laureate in Chemistry, Paul J. Crutzen, attended (see the photo gallery in the back matter).

Most of the photos were taken by Ph.D. Candidate Mr. A. S. M. Mostafizur Rahman, Bangladesh (Heidelberg University) as part of a tax-exempt media project of AFES-PRESS, an international scientific non-profit organisation, with the Editors photo and film equipment. Hans Günter Brauch retains the copyright on behalf of AFES-PRESS for all photos and videos. The idea was to prepare a book on this theme that took several years to realise. Dr. Brauch is grateful to then Ph.D. Candidate A. S. M. Mostafizur Rahman who received his Ph.D. in March 2024, Bangladesh; Ms. Astrid Kaltenbach, Kaltenbach, personal secretary of Prof. Crutzen of the Max Planck Institute for Chemistry in Mainz, and to Michael Headon of North Wales, UK for making the photos in the photo gallery available free of charge.

For different reasons, this book has been long in the making. Most participants were then busy with other projects. I gave top priority to the completion of the first volume of this book series devoted to the theme of our brainstorming on "The Anthropocene: Politik-Economics-Society-Science" (APESS) that was published in December 2021 (Benner/Lax/Crutzen/ Pöschl/Lelieveld/Brauch) on *Paul Crutzen and the Anthopocene: A New Epoch in Earth's History*.[11] Several participants have published in the meantime their own monographs (Arizpe Schlosser 2019) and edited books (Serrano 2021). Since two participants from the Netherlands have passed away,

---

[10] See: "Structure of the Brainstorming on 31 May 2017: 'Politik' in and for the Anthropocene: Obstacles, Challenges, Opportunities and Tasks for the Social Sciences in the 21st Century"; at: http://www.afes-press-books.de/html/events.htm please add the password [anthropocene2017] and http://www.afes-press-books.de/html/PDFs/2017/From%20the%20Brainstorming%20to%20the%20Book%20Schedule_22-6-2017.pdf.

[11] Benner; Lax; Crutzen; Pöschl; Lelievel; Brauch (Eds.), 2021: *Paul Crutzen and the Anthopocene: A New Epoch in Earth's History* (Cham: Springer Nature Switzerland).

Paul J. Crutzen[12] on 28 January 2021 and Frederick Arends[13] on 3 June 2020, with whom I shared a room at the Collegium Academicum of Heidelberg University (1967/1968) and who has been a friend ever since.

In the meantime, several monographs[14] and other major readers were published in Political Science and International Relations[15] primarily by young scholars from the perspective of distinct theories or methodological approaches.

The editor would like to thank those authors who finally delivered the chapters that are included in this volume.

With Springer Nature I would like to thank my counterpart, Dr. Christian Witschel and the two producers of the book Ms. Birke Dalia in Heidelberg (Germany) and Mr. Arulmurugan Venkatasalam in Chennai, Tamil Nadu (India) as well as the manager for the production of this book and the team of typesetters, designers working on this book in different divisions and in different parts of the world.

Mosbach, Germany and Cuernavaca, Mexico
January 2024

Hans Günter Brauch
Peace Research and European Security Studies (AFES-PRESS)
Hans Günter Brauch Foundation for Peace and Ecology in the Anthopocene (HGBS)

[12] Obituary on Paul J. Crutzen: http://www.afes-press-books.de/html/SpringerBriefs_PSP_Crutzen.htm.

[13] Obituary on Frederik Arends: http://www.afes-press-books.de/html/PDFs/2020/Obituary_Jacob_Frederik_Martinus_Arends.pdf.

[14] See the references in my Chapter 1 on: Kelly, D., 2019: *Politics and the Anthropocene* (Cambridge—Medford, MA: Polity) and Dryzek, J. S.; Pickering, J., 2019: *The Politics of the Anthropocene* (Oxford: Oxford University Press) in this volume.

[15] See the references in Table 1.1 in my Chapter 1 below.

Introduction

Hans Günter Brauch[16]

The press conference took place as a special event of the international Max Planck Conference for a Sustainable Anthropocene,[17] where a background report was released on "The Anthropocene's Signal: What the Geology of the Present Beckons for the Future of Research" by Christoph Rosol and Georg Schäfer (7 June 2023).[18] At this press conference:

> the *Anthropocene Working Group* (AWG) announced Crawford Lake, Canada as the proposed *Global* boundary *Stratotype Section and Point* (GSSP) candidate site of the Anthropocene series. Selecting Crawford Lake marks a significant step in the AWG's effort to test the geological validity of the Anthropocene hypothesis and towards the scientific, and thereby also political and social recognition of the scale and severity of planetary transformation processes unleashed by industrialized humanity.[19] …

This news was widely covered globally marking a major step by the geological community and its responsible institutions to initiate the formal decision making process to officially end the Holocene Epoch of Earth history and to start with the Anthropocene Epoch.[20] "The AWG must now submit three proposals for a vote by the *Subcommission on Quaternary Stratigraphy* (SQS): … The first names Crawford Lake as the golden spike, and the second offers supporting evidence from several other sites. … The third and most contentious proposes formal recognition of the Anthropocene as an epoch".

---

[16] Hans Günter Brauch has been chairman of the non-profit scientific society *Peace Research and European Security Studies* (AFES-PRESS) since 1987 and since May 2020 chairman of the board of *the Hans Günter Brauch Foundation for Peace and Ecology in the Anthropocene* (HGB Foundation); e-mail: hg.brauch@onlinehome.de.

[17] See at: https://climate-conference.mpg.de/.

[18] See at: https://www.mpiwg-berlin.mpg.de/feature-story/anthropocene-signal.

[19] See Anthropocene Working Group proposes Crawford Lake as GSSP candidate site of the Anthropocene series; at: https://www.shh.mpg.de/2347073/anthropocene-working-group-crawford-lake-candidate-anthropo-cene-site.

[20] See the *Anthropocene Working Group Newsletter*.

Negative Vote on the Anthropocene as a 'New Epoch of Earth History' by the *Subcommission on Quaternary Stratigraphy* (SQS)

After submission of these three proposals, the Anthropocene was to be voted on by the *International Congress on Stratigraphy* (ICS) and finally "they would be submitted for final approval from the *International Union of Geological Sciences*" (IUGS). If the vote fails it cannot be submitted again for a decade. However, many geologists and other natural scientists argue that the Anthropocene will stay an important scientific concept marking a fundamental change in earth history, even after a failed vote.

On 16 March 2024, the chairman of the Anthropocene Working Group, Colin Waters, released Volume 13 of the AWG Newsletter on its activities in 2023:

> … After three rounds of voting, core from Crawford Lake received supermajority support, as required by ICS statutes. Consequently, the AWG proposal was for an Anthropocene epoch/series and Crawfordian age/stage. Having resolved this task, voting members turned to selecting candidate Standard Auxiliary Boundary Stratotypes (SABSs). Newly introduced in 2023 by ICS as a requirement, we were fortunate that Martin Head was the lead author on the significance of, and procedures for, proposing SABSs and was able to guide us through selection of three candidate sites, Beppu Bay, Sihailongwan Lake and Śnieżka Peatland, with the remaining eight sites as reference sections. Following the binding vote on selection of Crawford Lake as the proposed site for the GSSP, a new core had been collected in order to carry out annual resolution analysis of the primary marker (239+240Pu) and some secondary markers. This dataset providing exceptional resolution of the onset of the Pu bomb-spike and further binding votes were required to select the new core and propose the preferred precise position of the GSSP. There was very strong support to select the base of the organic lamina in the varve level equivalent to 1952 CE (predominantly deposited in the autumn), the varve in which the initial major increase in 239+240Pu is recorded. Subsequently, it was formally agreed in a final vote that in addition to the GSSP, it was recommended that if required, a notional start of the Crawfordian age and Anthropocene epoch could be aligned with the detonation of the first thermonuclear device (the Ivy Mike detonation) which occurred on 1 November 1952 at 07:15 local time (19:15 on 31 October, GMT) and hence broadly coincident with the elevated Pu signal in the GSSP. Full details of all of the votes carried out from late 2022 to beginning of 2024 are recorded later in this Newsletter. I wish to acknowledge the great dedication of the voting members during such a demanding process to address these series of complex votes with the great consideration required given the magnitude of their decisions.
>
> Armed with the results of the selection of Crawford Lake it was possible to announce the AWG decision at the 4th International Congress on Stratigraphy STRATI 2023 at Lille on 11 July.
>
> In a session on "The Anthropocene: stratigraphical concepts and evidence", there were four presentations by: Colin Waters et al. on the stratigraphic enrichment of proxies during the Anthropocene; Francine McCarthy et al. on the Crawford Lake GSSP candidate; Michael Wagreich et al. on the urban Anthropocene of Karlsplatz, Vienna; and Colin Waters et al. on the understanding of 'event stratigraphy' in the context of the Anthropocene. Having agreed to requests from ICS not to announce the results of our voting to the media in advance of STRATI 2023, a joint online press conference with the Max Planck Society (MPS) was held independently of the congress on the evening of the 11th July as part of the MPS Sustainable Anthropocene conference, with presentations by Jürgen Renn (Director MPI for the History of Science and Founding Director MPI of Geoanthropology), Bernd Scherer

(Former Director of Haus der Kulturen der Welt (HKW), Katrin Klingan (Former Head of the Department Literature and Humanities, HKW), Colin Waters, Simon Turner, Francine McCarthy and Martin Head. It was an opportunity to celebrate the role of HKW in securing the funding that made the analysis of the 12 sites possible. Georg Schäfer is particularly thanked for his patience in coordinating the event given the very fluid nature of the plans.

Following the Congress, ICS President imposed without discussion a deadline to get the AWG submission completed by the 29 September 2023, which on complaint by AWG was extended to 31 October 2023. AWG members then worked tirelessly to ensure compilation of the report, which had to be written in parallel with continued rounds of voting and data analysis. Ultimately, the submission went to SQS at 23.45 pm on the 31st October, and comprised an executive summary, part 1 report (and appendices) detailing the evidence for the Anthropocene

AWG Newsletter 2023 4 as a candidate series/epoch and part 2 report on the GSSP and SABSs proposals (and appendices for the reference sections). It is our intention to make this complete submission available online in the near future, but at time of writing the Executive Summary is accessible at EarthArXiv at: https://eartharxiv.github.io/moderation.html. As has been the case throughout the life of the AWG, we value the importance of the evidence of our work being openly available to scrutiny.

As one now expects from AWG members, this Newsletter records the phenomenal output of publications and media outreach over the year. Plaudits to Francine for working tirelessly to promote the cause of Crawford Lake as the candidate GSSP and the amazing efforts of Team Crawford to further refine the data for the site. It is a worthy candidate. The compilation of the results of the analytical work of the 12 initial candidate sites, as published in a special thematic set of papers in Anthropocene Review, volume 10, provides a comprehensive assessment of the evidence for the Anthropocene across five continents and eight distinct environmental settings.

As we come towards the end of the life of the AWG (as a working group of the SQS at least), it is time to reflect on the early days of inception of the group in 2009. Only four members of the original group persisted through to the very end: An Zhisheng, Davor Vidas, Mark Williams, and Jan Zalasiewicz. I would like to show my appreciation for their determination over the years. Mark did a wonderful job in being the original Secretary, helped coordinate the very influential 2011 Geological Society of London (GSL) meeting and has been a leading light in developing the group's biostratigraphical analysis and formidable presenter at conferences. But, without Jan none of this would have happened. He led that first tentative exploration of the Anthropocene by the GSL Stratigraphy Commission that ultimately inspired the inception of the group and during his time as Chair (from 2009 to 2020), and his phenomenal output of publications at the forefront of investigations into the concept have truly shaped how we view our impact upon the planet.

At the time of writing, it would appear that our worthy submission to SQS was rejected, though we only found out via a leak in the New York Times and the SQS vote moderators have still not contacted us with the results or provided explanation of the basis by which it was rejected. This is an unacceptable way for the AWG members, who had worked for 15 years on behalf of SQS, to be treated. A catalogue of complaints about unethical behaviour by those in high-office over the past decade who have aimed at preventing fair assessment of any proposal had been compiled and submitted to an independent IUGS Geoethics Commission at the beginning of this year before voting commenced. Their reasonable assessment of a suitable schedule for ensuring a fair vote was largely ignored by IUGS in the haste of seeing the vote completed, aided by ICS clearly breaching their own statutes and rules. The full story will eventually be told but is not for here and now. Certainly, the media have been greatly interested in understanding how SQS could possibly have come to their conclusion. The number of interviews by AWG members over the last few weeks are not included in the list provided later in this Newsletter.

Box 1 Voting record of the Anthropocene Working Group from late 2022 to early 2024. *Source* Anthropocene Working Group Newsletter, Vol. 13, 16 February 2024, p. 7–9:

2022 binding vote on selection of rank for proposed Anthropocene chronostratigraphic unit

No. of potential voting members: 23
No. required to be quorate (60%): 14
No. of votes received: 23 (100% of voting membership).

Round 1 (concluded 17 December 2022):

Epoch/series and corresponding stage: 22
Votes (95.7%)
Stage alone: 1 vote (4.3%)
Other rank: 0 votes
Abstentions: 0

2022/2023 binding vote on selection of GSSP candidate

No. of potential voting members: 23
No. required to be quorate (60%): 14
No. of votes received: 23 (100% of voting membership).

Round 1 (concluded 17 December 2022):

Crawford Lake: 11 votes (47.8%)
Sihailongwan Lake: 5 votes (21.7%)
Beppu Bay: 4 votes (17.4%)
Searsville Lake: 3 votes (13.0%)
All other sites: 0
Abstentions: 0

Round 2 (concluded 20 February 2023):

Crawford Lake: 11 votes (47.8%)
Sihailongwan Lake: 8 votes (34.8%)
Beppu Bay: 3 votes (13.0%)
Abstentions: 1 (4.3%)

Round 3 (concluded 19 April 2023):

Crawford Lake: 14 votes (60.9%)
Sihailongwan Lake: 7 votes (30.4%)
Abstentions: 2 (8.7%)

Hence, the proposal is for an Anthropocene epoch/series and Crawfordian age/stage.

2023 binding vote on selection of candidate Standard Auxiliary Boundary Stratotypes (SABSs)

No. of potential voting members: 22
No. required to be quorate (60%): 13
No. of votes received: 21 (95% of voting membership).

Peer-review of the published data for the 12 candidate sites determined three sites should be considered reference sections: Ernesto Cave (Italy), Karlsplatz (Vienna) and San Francisco Estuary (USA). A non-binding vote was undertaken (completed on 4 July 2023) with 19 members voting, to determine relative support for remaining sites. The candidates with least support, East Gotland Basin (Baltic Sea), Flinders Reef (Australia), and Palmer ice core (Antarctica) were excluded from the SABSs vote and were considered to represent reference sections. The remaining five sites then were included in a formal vote.

Round 1 (concluded on 6 August 2023)

Beppu Bay, Japan: 20 votes (95.2%)
Sihailongwan Lake, China: 19 votes (90.5%)
Śnieżka Peatland, Poland: 16 votes (76.2%)
Searsville Lake, USA: 12 votes (57.1%)
West Flower Garden Bank, USA: 9 votes 42.9%)
No vote: 1
Abstentions: 0

Hence, Beppu Bay, Sihailongwan Lake and Śnieżka Peatland all achieved greater than 60% support to be proposed as SABSs for the GSSP. Searsville Lake and West Flower Garden Bank Reef are included as reference sites.

2023 binding vote on the level (year) of the GSSP in the Crawford Lake core

No. of potential voting members: 22
No. required to be quorate (60%): 13
No. of votes received: 22 (100% of voting membership).

Following the binding vote on selection of Crawford Lake as the proposed site for the GSSP, a new core had been collected in order to carry out annual resolution analysis of the primary marker 239+240Pu) and some secondary markers. Given that the Crawford Lake submission voted on by AWG members was on an earlier core and proposing 1950 CE as the level of the GSSP in that core as outlined by McCarthy et al., 2023), it was necessary to confirm this in a binding vote. The results of the vote were concluded on 18th October 2023.

Q1. *Can the preferred core from Crawford Lake be revised from CRA22-1FRA-3, as published in the original proposal McCarthy et al. (2023) in the Anthropocene Review Special Publication and selected by vote by the AWG, to the new core CRA23-BC-1F-A with annual-resolution Pu data?*

Yes: 21 votes (95%)
No: 0 votes (0%)
Abstentions: 1 (5%)

Q2. *What is your preferred level for the proposed Global boundary Stratotype Section and Point (GSSP) in the Crawford Lake core?*

1950 CE1: 4 votes (18%)
1952 CE2: 18 votes (82%)

1. Specifically, at the base of the calcite lamina in the varve level equivalent to 1950 CE, fixed at a convenient date close to a lithological boundary evident in the core and consistent with the common understanding of the onset of the Great Acceleration. For practical purposes to be taken as 1 January 1950 CE.
2. Specifically to be taken at the base of the organic lamina in the varve level equivalent to 1952 CE, the varve in which the initial major increase in 239+240Pu is recorded. For practical purposes to be taken midpoint between the two adjacent annual samples that show the biggest proportional increase within the run of samples that all have detectable Pu. At Crawford Lake this definition coincides in core CRA23-BC-1F-A with the Fall (October–December) 1952 CE closely coincident with the first thermonuclear detonation on 1 November 1952 CE. This option is conditional on selected 'YES' for question 1.

Hence, the Crawford Lake core CRA23-BC-1F-A has been selected as the host for the proposed GSSP, which is specifically taken in this core at the initial major increase in 239+240Pu coincident with the boundary between the lower calcite and upper organic component of the varve level equivalent to 1952 CE at a depth of 16.2 cm.

2023/2024 binding vote on nominal start time for the Anthropocene epoch/ Crawfordian age

No. of potential voting members: 22
No. required to be quorate (60%): 13
No. of votes received: 22 (100% of voting membership).

Q1. *Can the preferred core from Crawford Lake be revised from CRA23-BC-1F-A (for which all material had been consumed during analysis) to the alternative CRA23-BC-1F-B which contributed Pu data to the annual analysis and for which intact core is preserved, as published in McCarthy et al. (submitted) in the Anthropocene Review? The boundary will continue to be taken at the base of the organic lamina in the varve level equivalent to 1952 CE, the varve in which the initial major increase in 239+240Pu is recorded.*

Yes: 21 votes (95%)
No: 0 votes (0%)
Abstentions: 1 (5%)

Q2. *Can a notional start of the proposed Crawfordian age/Anthropocene epoch be proposed at a specific date and time. Options are: (a) 00.00 GMT 1 January 1952; (b) 1 November 1952 coincident with detonation of first thermonuclear device, augmenting the proposed Global boundary Stratotype Section and Point (GSSP) in the Crawford Lake core present at the boundary between summer and fall 1952 CE varves? The Ivy Mike detonation occurred on 1 November 1952 at 07:15 local time (19:15 on 31 October, GMT); or (c) 00.00 GMT 1 January 1953. Please select only one option.*

(a) 1 January 1952: 1 vote (5%)
(b) 1 November 1952: 19 votes (86%)
(c) 1 January 1953: 1 vote (5%)
(d) Abstentions: 1 (5%)

On 5 March 2024, an alleged negative vote of the SQS taken on 4 March 2024 was leaked to the New York Times[21] and CNN[22]: "after the SQS had concluded its month-long voting process, but before committee leaders had finalised discussions and made an official announcement".[23] On 6 March 2024, Alexandra Witze reported in *Nature*: "Geologists reject the Anthropocene as Earth's new epoch":

> The ruling … is meant to be final, but is being challenged by two leading members of the committee that ran the vote. Twelve members of the international Subcommission on Quaternary Stratigraphy (SQS) voted against the proposal to create an Anthropocene epoch, and only four voted for it. That would normally constitute an unqualified defeat, but a dramatic challenge has arisen from the chair of the SQS, palaeontologist Jan Zalasiewicz at the University of Leicester, UK, and one of the group's vice-chairs, stratigrapher Martin Head at Brock University in St Catharines, Canada.[24]
>
> In a 6 March press statement, they said that they are asking for the vote to be annulled. They added that "the alleged voting has been performed in contravention of the statutes of the International Commission on Stratigraphy" (ICS), including statutes governing the eligibility to vote. Zalasiewicz told *Nature* that he couldn't comment further just yet, but that neither he nor Head had "instigated the vote or agreed to it, so we are not responsible for procedural irregularities". … Normally, there would be no appeals process for a losing vote. ICS chair David Harper, a palaeontologist at Durham University, UK, had confirmed to *Nature* before the 6 March [2024] press statement that the proposal "cannot be progressed further". Proponents could put forward a similar idea in the future.
>
> If successful, the proposal would have codified the end of the current Holocene epoch, … and set the start of the Anthropocene in the year 1952. This is when plutonium from hydrogen-bomb tests showed up in the sediment of Crawford Lake near Toronto, Canada, a site chosen by some geologists to be designated as a 'golden spike' as capturing a pristine record of humans' impact on Earth. Other signs of human influence in the geological record include microplastics, pesticides and ash from fossil-fuel combustion.

Witze described the Anthropocene as a "cultural concept":

> Although the Anthropocene probably will not be added to the geological timescale, it remains a broad cultural concept already used by many to describe the era of accelerating human impacts, such as climate change and biodiversity loss. … For now, the SQS and the ICS will sort out how to handle Zalasiewicz and Head's request for a vote annulment. Meanwhile, scientific and public discussions about how best to describe the Anthropocene continue.

---

[21] "Are We in the 'Anthropocene', the Human Age? Nope, Scientists Say", in: *New York Times*, 5 March 2024.

[22] Katie Hunt: "Proposal to mark a new chapter in Earth's history rejected, scientists say", in: *CNN*, 5 March 2024.

[23] Comment by Google as top news on 13 March 2024; at: https://www.google.com/search?client=firefox-b-d&q=rejection+of+the+Anthropocene+on+5+March+2024. See additional reports: Paul Voosen: "The Anthropocene is dead. Long live the Anthropocene Panel—rejects a proposed geologic time division reflecting human influence, but the concept is here to stay", in: *Science*, 5 March 2024: Chen Li: "Surprise decision not to define the Anthropocene shocks scientists. A proposal to define the Anthropocene, a geologic epoch defined by human activity, has been rejected—surprising even scientists who consulted the voting group", in: *New Scientist*, 5 March 2024; Alexander Damianos: "What was the Anthropocene", in: *Critical Legal Thinking*, 8 March 2024; at: https://criticallegalthinking.com/2024/03/08/what-was-the-anthropocene/.

[24] Alexandra Witze: "Geologists reject the Anthropocene as Earth's new epoch—after 15 years of debate—But some are now challenging the vote, saying there were 'procedural irregularities'", in: *Nature*: 627 (14 March): 249–250 (2024); at: https://doi.org/10.1038/d41586-024-00675-8.

With regard to the future use of the Anthropocene concept Paul Voosen concluded in *Science*:

> Even if it is not formally recognized by geologists, the Anthropocene is here to stay. It is used in art exhibits, journal titles, and endless books. And Gibbard, Ellis, and others have advanced the view that it can remain an informal geologic term, calling it the "Anthropocene event." ... the Anthropocene marks a huge transition, but one without an exact date. "Let us work together to ensure the creation of a far deeper and more inclusive Anthropocene event," Ellis says.
>
> Waters and his colleagues will continue to press that the Anthropocene is worthy of recognition in the geologic timescale, even if that advocacy has to continue in an informal capacity, he says. Although small in size, Anthropocene strata such as the 10 centimeters of lake mud are distinct and can be traced using more than 100 durable geochemical signals, he says. And there is no going back to where the planet was 100 years ago, he says. "The Earth system changes that mark the Anthropocene are collectively irreversible."[25]

On 6 March, in a German article for *Spektrum der Wissenschaft*,[26] Christian Schwägerl offered additional background information from among geologists and members of the AWG and of the SQS. On 6 March 2024 at 10.00 GMT the AWG chairman, Colin Walters, published this press release:

> The results of voting on the Anthropocene within SQS were announced 05 March 2024 without the authorization of the Chair of the SQS or overall authorization of its executive. Issues remain that question the validity of the vote and the circumstances surrounding it, in part concerning relevant information just received today from the President, International Union of Geological Sciences. When these are resolved and clarified, we hope shortly, we will be positioned to comment further, but are unable to talk directly on this matter at present.
>
> Irrespective of the vote, the AWG stands fully behind its proposal, which demonstrated beyond reasonable doubt the following:
>
> - The Earth System now clearly lies outside of the relatively stable interglacial conditions that characterized the Holocene Epoch beginning ~11,700 years ago.
> - The Earth System changes that mark the Anthropocene are collectively irreversible, meaning that a return to the stable conditions of the Holocene is no longer possible.
> - Anthropocene strata are distinct from Holocene strata. They can be characterized and traced using>100 durable sedimentary signals including anthropogenic radionuclides, microplastics, fly ash and pesticide residues, most of which show sharp increases in the mid-20th century, concurrent with the "Great Acceleration" of population, industrialization and globalization.
> - The base of the Anthropocene is clearly identified in the proposed stratotype section at Crawford Lake, Canada, by a sharp upturn in plutonium concentrations in annually laminated sediments deposited in 1952 CE, coincident with the beginning of thermonuclear bomb testing. This marker level has been traced with great precision in strata around the

---

[25] Paul Voosen: "Dramatic shift in ice age rhythm pinned to carbon dioxide: Southern Ocean 'pump' sparked deep freezes, new temperature record shows", in: *Science*, 24 February 2024.

[26] See C. Schwägerl: "Anthropozän: Keine Epoche für die Menschheit—Verändern wir Menschen die Erde so massiv, dass eine neue Erdepoche angebrochen ist? Darüber streiten Geologen seit Jahren. Doch der Vorschlag ist nun überraschend gescheitert"; Christian von Eichhorn: Leben wir jetzt im Anthropozän oder nicht? Bereits zum zweiten Mal haben Geologen den Vorschlag abgelehnt, ein neues 'Erdzeitalter des Menschen' auszurufen. Forscher machen einen ungewöhnlichen Vorschlag—und warnen vor einem verheerenden Signal, in: *Süddeutsche Zeitung*, 20 March 2024.

> world including at the three proposed Standard Auxiliary Boundary Stratotypes (SABS) and other reference sections.
>
> All these lines of evidence indicate that the Anthropocene, though currently brief, is—we emphasise—of sufficient scale and importance to be represented on the Geological Time Scale and terminating the Holocene.
>
> We, as a group of many eminent researchers in our field of expertise, wish to carry on, in an informal capacity if necessary, and will continue to argue the case that the evidence for the Anthropocene as an epoch should be formalised, as consistent with the scientific data presented in the submission. If the above vote is confirmed by ICS then the current proposal cannot progress. Given significant issues with the procedure and circumstances of the vote, though, currently being pursued, we cannot confirm that the vote will be upheld.[27]

On 21 March 2024, *Nature* asked: "Are we in the Anthropocene yet?"[28] pointing to the confusion and concern:

> among those currently working on the Anthropocene science. This confusion arises because the term is understood and widely used by scientists, as well as people outside research, to mean a time in Earth history when humans are having severe biophysical impacts on the planet. … In 2023, we launched called 'Nature Portfolio: Anthropocene', highlighting research about humanity's footprint on Earth. … The difficulty is that the concept has taken off while geologists have been locked in discussion about how the Anthropocene should be measured, and when it started. ... The current lack of agreement on a start date and which marker to use should not distract from the Anthropocene as a concept. … Measurement matters. It is need not least so that the world is confident that the Anthropocene's start date and marker are grounded in the broadest consensus of scholarly knowledge. Geologists must quickly resolve their disagreements. At the same time there is little doubt that the world is in an Anthropocene, as understood by researchers who use the term, and that course correction is needed.

T. P. Roland, G. T. Swindles; A. Ruffell[29] claimed in *Nature* that

> the rejection of formalization of the Anthropocene represents an opportunity. Released from the constraints of strict geological definition, it will remain open to interpretation and critique, and can flourish as an informal and inclusive concept that crosses disciplines, cultures and world view.

In March 2024, when this book went to press, the future of the Anthropocene was still uncertain, and it will remain so for many years to come. Between 2000 and 2009, when the AWG was set up, it had been discussed in the international global change and Earth Systems Science networks and communities before it gradually spread to geology, the Social Sciences, the Humanities, Law and many other practical disciplines (medicine, engineering) but also religious groups (Brauch 2021). Thus, this author's key thesis in his two chapters for this book remains valid: that

---

[27] Updates will be accessible after publication at: http://www.afes-press-books.de/html/APESS_35.htm. See also: https://scilogs.spektrum.de/der-anthropozaeniker/anthropozaen-keine-epoche-fuer-die-menschheit/ and: "The 'Anthropocene Epoch' bites the dust?", in: Earth-logs, 8 March 2024; at: https://earthlogs.org/2024/03/08/the-anthropocene-epoch-bites-the-dust/.

[28] "Are we in the Anthropocene yet?", in: *Nature*, 627, 8004 (21 March 2024): 466.

[29] Thomas P. Roland; Graeme T. Swindles; Alastair Ruffell: "Correspondence: Opportunity in 'Anthropocne' rejection", in: *Nature*, 627/8005 (28 March 2024): 735.

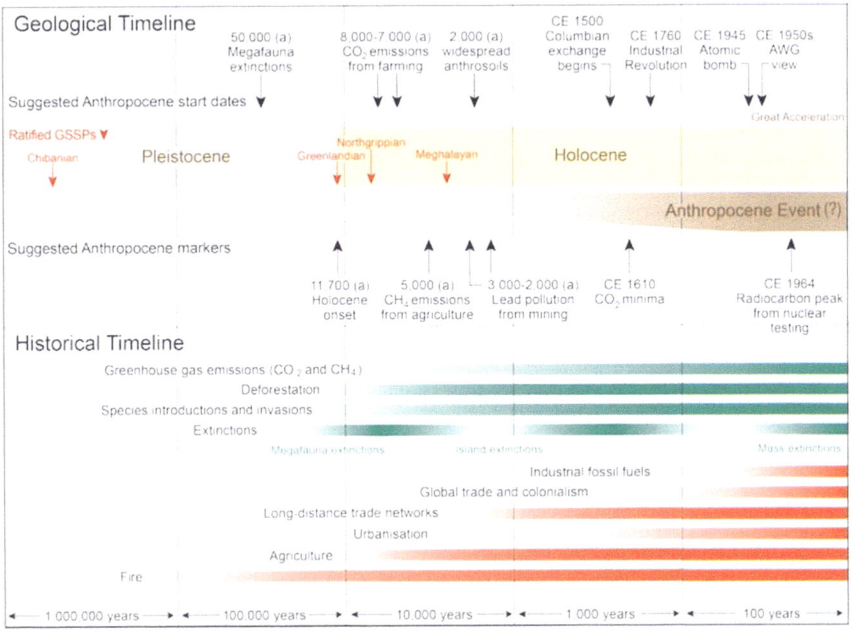

Fig. 1 A chart reflecting timing of the 'Anthropocene Event' shows how various human activities have affected the planet over millennia in the recent geologic time scale. *Source* Used by Erle C. Ellis based on Philip Gibbard, et al., 2022

the Anthropocene marks a transition in the Earth system that coincides with major changes in human and political history and in political order. Therefore, rethinking politics, policy and polity in the Anthropocene remains a challenge, irrespective of whether it started on 16 July 1945 or on 1 November 1952 to mark a change in 'Earth epoch' or an 'event'.

Erle C. Ellis, who left the AWG in 2023, argued on 5 March 2024 in *The Conversation*[30] that "the Anthropocene is not an epoch—but the age of humans is most definitely underway" and that "such massive planetary changes did not begin all at once at any single place or time.... starting precisely in 1952" [and that] "the marker was radioactive fallout from hydrogen bomb tests". Ellis stated that "despite the 'no' vote on the Anthropocene Epoch, the Anthropocene will continue to be as useful as it has been for more than 20 years in stimulating discussions and research into the nature of human transformation of this planet" (Fig. 1).

[30] Erle Ellis: "The Anthropocene is not an epoch—but the age of humans is most definitely underway", in: *The Conversation*, 5 March 2024; at: https://theconversation.com/the-anthropocene-is-not-an-epoch-but-the-age-of-humans-is-most-definitely-underway-224495.

Richard Fisher,[31] in a text for the BBC on 6 March 2024, discussed "Why it's proving difficult to define the official dawn of the Anthropocene". Noting that it is not clear-cut, he asked, "When exactly did human beings begin influencing nature on a planetary scale?". For Fisher:

> The term "Anthropocene"—popularised in the 2000s—has become a widely-used concept to describe our current moment in time: when human beings are having profound impacts on the living and physical world. We have transformed Earth's chemistry and atmospheric composition, triggered extinctions and specifications, and even created new rock types, such as "plastiglomerate"—a conglomeration of rock fragments, plastic debris and organic material.

Fisher quoted Colin Waters, the present chairman of the AWG, who argued that

> the group acknowledged that people had been altering nature for millennia, but from the 1950s onwards many of the metrics of planetary change suddenly shot up, from CO_2 to ocean acidification. … The pushback appears to be over the AWG's choice of the mid-20th Century as a beginning: other researchers have argued, for instance, that a methane rise from the first farming could equally serve as a marker, or the uptick in lead pollution from early mining and smelting. Not to mention the massive increase in fossil fuel burning during the Industrial Revolution. …
>
> For starters, the results are not official unless authorised by senior members of the SQS, including its chair, says Waters. "We note at this stage that there remain several issues that need to be resolved about the validity of the vote and the circumstances surrounding it." In a statement on 6 March, the SQS suggested that the vote may not be final. "The alleged voting has been performed in contravention of the statutes of the International Commission on Stratigraphy," it said. The SQS chair, Jan Zalasiewicz of the University of Leicester, has requested "an inquiry including instituting a procedure to annul the putative vote".

On 6 March 2024, Brad Johnson commented on the issue in a *Hill Heat* article entitled "Scientific Leaders: 'No Anthropocene' Vote Was a Sham. The fraudulent decision made the front page of the New York Times today":

> Zalasiewicz and Head note that "a large majority of SQS members who took part in the alleged voting" (11 out of 16) were "not eligible as voting members at the time they cast their votes," as they had been members of the subcommission for more than 12 years. The five eligible members do not represent a needed quorum for a vote to take place.
>
> Moreover, the sham vote was held even as a Geoethics Commission report on the workings of the subcommission's *Anthropocene Working Group* (AWG) was being reviewed by the president of the IUGS, geologist ***John Ludden***.
>
> Zalasiewicz requested the report be distributed to the subcommission members before any vote was held, but Zhou and Bertini ignored him, pushed forward the sham vote, with members notifying the *New York Times* about the fraudulent result. Ludden released the report to the SQS and the AWG on March 5th, after the *Times* story was published. According to Zalasiewicz,
>
> *The findings of that report included: that the AWG, in preparing its proposal, was unfairly treated, via conflicts of interest, application of different standards than to other working groups, and unreasonable requests and restrictions, while insufficient time was allowed for*

[31] Richard Fisher: "Why it's proving difficult to define the official dawn of the Anthropocene", in: BBC, 6 March 2024; at. https://www.bbc.com/future/article/20240306-why-its-proving-difficult-to-define-the-official-dawn-of-the-anthropocene.

comment on the proposal, and the AWG were not asked to provide feedback on the discussions as would be normal practice. The Geoethics Commission further observed that the process as a whole between AWG/SQS/ICS/IUGS was dysfunctional; it thus recommended the urgent suspension of any voting procedures (though not examining their validity).

Panel member ***Naomi Oreskes***, the noted historian of climate science, responds:

The irregularities in the SQS voting procedures strongly suggest that the SQS did not make its decision on scientific grounds. The argument put forward by the AWG—and overwhelmingly endorsed by the AWG membership—was never given a fair hearing.

What's particularly sad about this to me—as a person who cut my teeth in field geology—is that by rejecting the Anthropocene proposal, *the SQS suggests to the world that they are unwilling or unable to recognize what we all can now see: that we do indeed live in the Anthropocene. By denying the obvious, the stratigraphers threaten to undermine the credibility of the science that they claim to be protecting.*

It will come as no surprise to *Hill Heat* readers that the online version of the *Times* story is running with fossil-fuel propaganda from the American Petroleum Institute.

On 13 March 2024, in *The Nation*, Leigh Phillips advised: "keep calm and carry on with the Anthropocene".[32]

The failure of the Anthropocene to be formally recognized as an epoch, the analogy of the current moment to the Great Oxygenation Event, and the continued utility of the Great Acceleration as a geological concept but outside the GTS thus remind us that we should be thinking about how the profit incentive undermines our ability to correct environmental problems in a timely fashion. It is this that we should emphasize rather than thinking that humanity is an interloper in nature, a scourge on the planet. Recent developments also remind us that we don't need the imprimatur of geology to be able to recognize that human actions are undermining ecosystem services. Even without a formal geological epoch designation, we already know, and all too well, that the problems we face here are unprecedented in our species' history.

On 13 March 2024, Simon Turner (UCL), Colin Waters (University of Leicester), Jan Zalasiewicz (University of Leicester), Martin J. Head (Brock University), all four members of the AWG and two of them members of the SQS, argued in *The Conversation* and in *Climate and Capitalism* that "the Anthropocene's critics are missing the point".[33]

The speed and magnitude of recent change jumps out at you. The sharp upturns are essentially Crutzen's Anthropocene, representing the last 72 years of what has been called the "great acceleration" of population, consumption, industrialization, technical innovation and globalization (a more detailed way of expressing the "big zoom").

Similar graphs can be drawn for species extinction and invasion rates, or the production and spread of fly ash, concrete, plastics, and a host of other things. They show that Crutzen's Anthropocene is real, evidence based, and represents an epoch-scale change (at least). The significance for us all, of course, is that the near-vertical recent trends in these graphs are still, for the most part, rising, zooming us into a new kind of planet. The repercussions cannot fail to last for many thousands of years—and some will change the Earth for ever.

---

[32] See at: https://www.thenation.com/article/environment/anthropocene-voted-down-formal-recognition/.

[33] See at: https://climateandcapitalism.com/2024/03/13/why-the-anthropocenes-critics-are-wrong/.

> So the Anthropocene as an epoch is very different from the "event" of Erle Ellis and others, which encapsulates all human influence on the planet (and so is about a thousand times longer than the epoch, and differs in many other ways). They're both valid concepts of course, and have some overlap, just like a mouse in some ways overlaps with a blue whale (they're both mammals, and share a good deal of their genetic code). But they're different.
>
> It's absurd, therefore, to give them the same name: to take Crutzen's term and appropriate it for a wholly different purpose, and in doing so obscuring the real meaning of his insight and its significance. Under a different name (the Anthropolithic, perhaps?), it could perfectly well complement an Anthropocene epoch.
>
> Humans have had a long and complex impact on the planet, true. For almost all that time, they left their marks on Earth—but did not utterly overwhelm it. Less than a century ago, processes that began during the Industrial Revolution swung into overdrive. That's the Anthropocene as an epoch. It's real, it's already made geology, and it won't go away. Best to acknowledge it, to help us cope with its consequences.

Arjun Sengupta explained on 8 March 2024 in *The Indian Express* "Why scientists voted down [a] proposal to declare start of the Anthropocene—and what it means". He asked, "What is the proposed Anthropocene epoch? Why, despite the downvote, does the concept of the Anthropocene as a transformative time for Earth as a result of human activity remain relevant?"[34] and concluded that "The concept of the Anthropocene as a transformative time for the planet due to human activity, remains very much relevant".

On 26 February 2024, just prior to the vote of the SQS on the Anthropocene, Alessio Giacometti and Paolo Giardullo,[35] writing in *Science and Culture*, reviewed the "Trajectories of the Anthropocene as a boundary concept bridging debates about climate change and ecological collapse (years 2000–2019)" and recognised

> four main uses: a "proving use" employed for collecting evidence for the human epoch's official recognition; a "questioning use" for criticising its epistemological and political shortcomings; a "mentioning use" when taking it for granted as a catchword; and a "metaphorical use" for conveying the rising problem of global warming and ecological collapse. Regarding this last metaphorical use, the Anthropocene appears to act as a boundary concept bridging debates about climate change and aligning apocalyptic imaginaries of the future. The success of the Anthropocene concept might therefore be explained not by virtue of its intrinsic properties, unspecified virality and predetermined trajectory, but rather by the crystallisation and frequency of its use as a boundary concept. As such, it plays an evoking function in discursively introducing the whole semantic domain of the climate crisis condition and establishing the basis for talking about its apocalyptic consequences. Being used for unifying these narratives about the past and future of humankind and for capturing the variety of changes caused to the planet, the Anthropocene's fields of circulation have thus enlarged vastly outside its geological debate.

---

[34] See at: https://indianexpress.com/article/explained/explained-sci-tech/scientists-voted-down-proposal-anthropocene-9201316/.

[35] See at: https://www.tandfonline.com/doi/full/10.1080/09505431.2024.2321366?src=exp-la.

The debate within the geological community has remained controversial—even more so after the negative and contested vote of the SQS on 4 March 2024. The debate within the Global Environmental Change and Earth Systems Science communities, as well as the discussion among social and policy scientists, those in the humanities, and among lawyers will not be affected by the vote of the SQS of 4 March 2024 and subsequent statements by other groups of the geological community.

On 16 March 2024, the chairman of the AWG, Colin Waters, and Simon Turner, as the AWG's Secretary, offered the following programme and outlook for its future work for 2024 in the Group's 13th Newsletter:

> With submission of the proposal to SQS, and completion of the voting process, the AWG will stand down as an SQS working group and will have no official function as part of ICS. The last formal function is for several members of AWG to attend the 37th International Geological Congress 2024 in Busan, Korea from 25 to 31 August.
>
> It is envisioned that many members will be keen to continue working together as a loose group investigating the evolving details of the Anthropocene as a concept that encapsulates overwhelming human impact on the geology and Earth System state of the planet. However, this will be the final official AWG Newsletter. From its inception as a working group in 2009, this has been an incredible journey from early exploration of Paul Crutzen's inspired improvisation of the term, through appreciation of the significance of the Great Acceleration championed by Will Steffen and to our current understanding based upon rigorous multi-proxy analysis across diverse environments. The series of newsletters records the vast number of publications, conference presentations and media interviews undertaken by AWG membership. Against significant pressures to the contrary, we have always aspired to ensure that our work has been made open to all see and comment upon, not the work of a closed group making decisions in isolation and without justification. We hope this will be some legacy!

On 20 March 2024, "the International Union of Geological Sciences (IUGS) … announced it is upholding a decision made earlier … by a group of geoscientists. … The IUGS decision effectively terminates a dramatic challenge to that earlier vote. … The results of the vote [of 4 March 2024] were then approved by the full ICS and, as of today, the IUGS. … Zalasiewicz … and other Anthropocene-epoch advocates are likely to continue their campaign in other venues, he says: 'Another means will have to be found' to codify the Anthropocene as a concept outside of the official geological timescale".[36]

A joint statement by the IUGS and ICS on the vote by the ICS Subcommission on Quaternary Stratigraphy announced that, with the delegated authority of the IUGS President and Secretary General and on behalf of the International Commission on Stratigraphy (ICS) … the vote by the ICS Subcommission on Quaternary Stratigraphy (SQS) to reject the proposal for an Anthropocene Epoch as a formal unit of the Geologic Time Scale is approved".[37] …

---

[36] A. Witze: "It's final: the Anthropocene is not an epoch, despite protest over vote. Governing body upholds earlier decision by geoscientists amid drama", in: *Nature*, 20 March 2024

[37] International Commission on Stratigraphy, "Joint statement by the IUGS and ICS on the vote by the ICS Subcommission on Quaternary Stratigraphy", at: https://stratigraphy.org/news/152 (21 March 2024).

Zalasiewicz says he has been asked to step down as SQS chair and does not expect to be part of any ICS group going forward. He and other Anthropocene-epoch advocates are likely to continue their campaign in other venues, he says: "Another means will have to be found" to codify the Anthropocene as a concept outside of the official geological timescale. Regardless of there being no formal Anthropocene epoch, the term will continue to be used in broad popular and scientific usage as the era of human-induced change. "As such, it will remain an invaluable descriptor in human-environment interactions", the IUGS says [emphasising]

> Despite its rejection as a formal unit of the Geologic Time Scale, the Anthropocene will nevertheless continue to be used not only by Earth and environmental scientists, but also by social scientists, politicians and economists, as well as by the public at large. It will remain an invaluable descriptor of human impact on the Earth system.

In the social and policy sciences, the 'Anthropocene' concept marks a fundamental change in human action that has become an existential threat to humanity since 16 July 1945 when the atomic age (or the nuclear era) began with the first test of a nuclear device in the atmosphere—or since 1952 when the nuclear debris of the first hydrogen bombs was deposited in Crawford Lake, near Toronto in Canada, which the AWG described as a 'golden spike' for the change of an environmental epoch in Earth history.

The Anthropocene concept coined by Paul J. Crutzen on 23 February 2000 during a discussion at an IGBP meeting in Cuernavaca (Crutzen/Stoermeer 2000) will remain a new and contested concept in the Natural and Social Sciences, in the Humanities and Law, and in many other scientific disciplines; it has triggered a new narrative that was not yet socially constructed by scientists and policy makers in 1945.

The vote that was taken 24 years later, on 4 March 2024, by a majority of the SQS, remains contested within the geological community and does not bind the Global Environmental Change and the Earth Systems Science communities, which had both used and promoted the validity of the Anthropocene concept. For the Social and Policy Sciences and the Humanities, geologists cannot prevent many disciplines from continuing to use the Anthropocene concept as a metaphorical change in political and human history that marks an existential threat to human survival along with that of nuclear weapons.

Thus, a key argument of this book and its goal of contributing towards *Rethinking Politics, Policy and Polity in the Anthropocene* also remains relevant and possibly becomes even more urgent. Social and policy scientists designed, approved and implemented a new global political order whose institutions have survived many violent conflicts and wars during the Cold War, the post-Cold War era since 1990, and the system's collapse after the aggression of Russia against the Ukraine since 2014 and especially since 24 February 2022.

Three Networks Using and Promoting the Anthropocene Concept: Global Environmental Change, Earth Systems Science and the Anthropocene Working Group

During the first decade (2000–2010), the concept had spread primarily in the Natural Sciences, especially in the 'Global Environmental Change' and the 'Earth Systems Science' communities and since 2008 (Brauch 2021: 328–333) through the *Anthropocene Working Group* (AWG). The members of the AWG voted on 21 May 2019 with a large majority that the first test of a nuclear bomb on 16 July 1945 (and the so far first and only use of a nuclear weapon against Hiroshima and Nagasaki on 6 and 9 August 1945) as the *Turning Point* when the Holocene Epoch of Earth history ended and the Anthropocene Epoch has started. According to the AWG's website:

> Following guidance from the Subcommission on Quaternary Stratigraphy and the International Commission on Stratigraphy, the AWG have completed a binding vote to affirm some of the key questions that were voted on and agreed at the IGC Cape Town meeting in 2016. The details are as follows: No. of potential voting members: 34; No. required to be quorate (60%): 21 No. of votes received: 33 (97% of voting membership)
>
> Q1. Should the Anthropocene be treated as a formal chrono-stratigraphic unit defined by a GSSP? 29 voted in favour (88% of votes cast); 4 voted against; no abstentions
> Q2. Should the primary guide for the base of the Anthropocene be one of the stratigraphic signals around the mid-twentieth century of the Common Era? 29 voted in favour (88% of votes cast); 4 voted against; no abstentions
>
> Both votes exceed the 60% supermajority of cast votes required to be agreed by the Anthropocene Working Group as the official stance of the group and will guide their subsequent analysis.

The AWG defined the 'Anthropocene' on its official website:

- The 'Anthropocene' is a term widely used since its coining by Paul Crutzen and Eugene Stoermer in 2000 to denote the present geological time interval, in which many conditions and processes on Earth are profoundly altered by human impact. This impact has intensified significantly since the onset of industrialization, taking us out of the Earth System state typical of the Holocene Epoch that post-dates the last glaciation.
- The 'Anthropocene' has developed a range of meanings among vastly different scholarly communities. Here we examine the Anthropocene as a geological time (chronostratigraphic) unit and potential addition to the Geological Time Scale, consistent with Crutzen and Stoermer's original proposal. The *Anthropocene Working Group* (AWG) is charged with this task as a component body of the *Subcommission on Quaternary Stratigraphy* (SQS) which is itself a constituent body of the *International Commission on Stratigraphy* (ICS).
- Phenomena associated with the '**Anthropocene**' include: an order-of-magnitude increase in erosion and sediment transport associated with urbanization and agriculture; marked and abrupt anthropogenic perturbations of the cycles of elements such as carbon, nitrogen, phosphorus and various metals together with new chemical compounds; environmental changes generated by these perturbations, including global warming, sea-level rise, ocean acidification and spreading oceanic 'dead zones'; rapid changes in the biosphere both on land and in the sea, as a result of habitat loss, predation, explosion of domestic animal populations and species invasions; and the proliferation and global dispersion of

many new 'minerals' and 'rocks' including concrete, fly ash and plastics, and the myriad 'technofossils' produced from these and other materials.

- Many of these changes will persist for millennia or longer, and are altering the trajectory of the Earth System, some with permanent effect. They are being reflected in a distinctive body of geological strata now accumulating, with potential to be preserved into the far future.
- The '**Anthropocene**' is not currently a formally defined geological unit within the Geological Time Scale; officially we still live within the Meghalayan Age of the Holocene Epoch. A proposal to formalise the '**Anthropocene**' is being developed by the **AWG**. Based on preliminary recommendations made by the **AWG** in 2016, this proposal is being developed on the following basis:
    1. It is being considered at series/epoch level (and so its base/beginning would terminate the Holocene Series/Epoch as well as Meghalayan Stage/Age);
    2. It would be defined by the standard means for a unit of the Geological Time Scale, via a Global boundary Stratotype Section and Point (GSSP), colloquially known as a 'golden spike';
    3. Its beginning would be optimally placed in the mid-20th century, coinciding with the array of geological proxy signals preserved within recently accumulated strata and resulting from the 'Great Acceleration' of population growth, industrialization and globalization;
    4. The sharpest and most globally synchronous of these signals, that may form a primary marker, is made by the artificial radionuclides spread worldwide by the thermonuclear bomb tests from the early 1950s.

Between 2019 and July 2023, the 'analyses of potential 'golden spike' locations [were] underway. The proposal that was announced in July 2023 still needs a

- supermajority (> 60%) agreement by the **AWG** and its parent bodies (successively the **SQS** and **ICS**) and ratification by the Executive Committee of the ***International Union of Geological Sciences***. The success of any such proposal is not guaranteed.
- Broadly, to be accepted as a formal geological time term the '**Anthropocene'** needs to be (a) scientifically justified, i.e. the 'geological signal' currently being produced in strata now forming must be significantly large, clear and distinctive; sufficient evidence has now been gathered to demonstrate this phenomenon (b) useful as a formal term to the scientific community. In terms of (b), the currently informal term '**Anthropocene**' has already proven highly useful to the global change and Earth System science research communities and thus will continue to be used. Its value as a formal geological time term to other communities continues to be discussed.
- The '**Anthropocene**' has emerged as a popular scientific term used by scientists, the scientifically engaged public and the media to designate the period of Earth's history during which humans have a decisive influence on the state, dynamics and future of the Earth System. It is widely agreed that the Earth is currently in such a state. The term has also been used in a non-chronostratigraphic context to be an informal term to denote a broader interpretation of anthropogenic impact on the planet that is markedly diachronous, reaching back many millennia. In geology, such an interpretation is already encompassed by lithostratigraphy, in which the character of stratified rocks is based solely on their physical features and not by age. Such an interpretation represents a concept sharply distinct from the Anthropocene as a chronostratigraphic unit, though it can be complementary with it.

Since 2010 and even more since 2015, different scientific discourses on the Anthropocene concept evolved and widely spread across scientific disciplines according to data in the *Web of Science* and *Scopus* data bases in early 2021 (Brauch 2021: 334–343).

Two historians have shaped since 1945 a 'rethinking' on time (F. Braudel 1969, 1972) and on a 'reconceptualization of concepts' due to changes in the context (R. Kosseleck 1979, 2006). However, both were still unaware of the specific dual turning point of 16 July 1945 that was coined by P. J. Crutzen as the 'Anthropocene' in February 2000 and the first nuclear weapons test to which the AWG referred to in its statement of May 2019.

The AWG's stated in 2023 that the Anthropocene started in 1952 with "other markers [that] offer the 'golden spike' … Principal among these is the distinctive fingerprint of radioactive isotopes, particularly plutonium, produced by cold war H-bomb tests, the first of which was carried out by the US on 1 November 1952 on the Eniwetok atoll in the Pacific Marshall Islands.[38]

Many ecologists have reacted critically against the vote of the SQS of 4 March 2024 to reject the Anthopocene as the starting point of a new Epoch of the Earth system.[39]

> But whether or not it's formally approved as a stratigraphic term, the idea of the Anthropocene is now firmly rooted in research. So, how are scientists using the term, and what does it mean to them and their fields? … [For] Chris Thomas, … the word Anthropocene neatly summarizes the sense hat humans are part of Earth's system and integral to its processes—what he calls indivisible connectedness. "That helps move us away from the notion that somehow humanity is apart from the rest of nature and natural systems," he says. "It's undoable—the change is everywhere."
>
> The concept of an era of human-driven change also provides convenient common ground for him to collaborate with researchers from other disciplines. "This is something that people in the arts and humanities and the social sciences have picked up as well ... "It is a means of enabling communication about the extent to which we are living in a truly unprecedented and human-altered world."
>
> Jacquelyn Gill, a palaeoecologist at the University of Maine … agrees. "It's a term that belongs to everyone. To people working in philosophy and literary criticism, in the arts, in the humanities, the sciences," she says. "I think it's far more meaningful in the way that it is currently being used, than in any attempts that stratigraphers could have made to restrict or define it in some narrow sense." She adds: "It serves humanity best as a loose concept that we can use to define something that we all widely understand, which is that we live in an era where humans are the dominant force on ecological and geological processes."

The concept of the Anthropocene is owned by many fields, Gill says. "This reiterates the importance of understanding that the role of people on our planet requires many different ways of knowing and many different disciplines". … In a world in which the threat of climate change dominates environmental debates, the term

---

[38] D. Carrington, 2023: "H-bombs or chicken bones: the race to define the start of the Anthropocene", in *The Guardian*, 6 January 2023.

[39] D. Adam, 2024: "Ditching 'Anthropocene': why ecologists say the term still matters—Beyond stratigraphic definitions, the name has broader significance for understanding humans' place on Earth", in *Nature*, 24 March.

Anthropocene can help to broaden the discussion, says Yadvinder Malhi, a biodiversity researcher [in] the ... UK. ... Although the word 'Anthropocene' makes some researchers think of processes that began 10,000 years ago, others consider it to mean those of the past century. This author's key hypthesis is that 16 July 1945 or 1 November 1952 mark a simultaneous change in earth history, in the history of humankind and in political history of *long durée* (international order) that have affected changes in the conscious actions of policy makers (*politics*), in political fields (*policies*) and in treaties and institutions (*polities*). It is so far the only overlapping transition in Earth and human history where human action in a specific global policy framework has directly interfered into the earth system. I propose to think since this joint transition changes in Earth and political history together and to analyse the changes on international order that expedited the human impacts on the chaotic earth system. For this approach, an understanding and analysis of human and earth interactions and a cooperation between earth systems scientists and social scientists and historians and anthropologists is essential.

The French social historian of the *Annales* school, Braudel (1946, 1969, 1972), distinguished in his book on the *Mediterranean and Philipp II* between history of events, conjunctures and structures of long duration and thus offered different levels of time, to which this author (Brauch 2016: 14–26) proposed to add even much longer time spans with the concepts of *cosmic*, *geological* and *technological* time.

1. *Cosmic time*, used in models of *physical cosmology* (since Max Planck), and referring to time since the Big Bang about 13.8 billion years ago; this is beyond the scope of any collective human impact.
2. *Geological time*, which describes the timing of and relationships between events throughout the earth's history of about 4.54 billion years; its scales are adopted by geologists and earth scientists and defined by members of the *International Commission on Stratigraphy*. Its most recent accepted epoch is the *Holocene*, the period since the retreat of the Quaternary Ice Age some 12,000 years ago that made the rise of human civilizations possible.
3. The time of the *technical revolutions* (the 'Neolithic' or 'agricultural' revolution of 10,000 to 6,000 BCE, and the 'Industrial Revolution' from about 1750 CE and its different phases of innovation).

Two Pioneers on Human History in the Anthropocene: Fernand Braudel and Reinhart Koselleck

In human history, the French social historian F. Braudel (1946, 1969, 1972) distinguished between three historical times: a) long duration (*la longue durée*), b) repeating historical cycles (*histoire de conjuncture*), and c) events (*l'histoire événementielle*). Braudel's periodisation has been used in history and in the social sciences (political science and economics). But he was not aware of the changes in geological time, nor did he address the technical revolutions and deal with the environment (ecology).

The first refers to "a history in which all change is slow, a history of constant repetition, ever-recurring cycles" (Braudel 1972, vol. 1: 20–21). The second historical time has "slow but perceptible rhythms ... One could call it *social history*, the history of groups and social groupings" with the goal of showing "how all these deep-seated forces were at work in the complex arena of warfare". The third historical time deals with "traditional history ...what [was] called '*l'histoire événementielle*', that is the history of events. ... a history of brief rapid, nervous fluctuations, by definition ultra-sensitive ... But as such it is the most exciting of all, the richest in human interest, and also the most dangerous".

Braudel's temporal trilogy was discussed in a different context with structural time (*histoire de longue durée*) which is used here to refer to the *time of changes in national and international order* due to revolutions (American, 1776; French, 1789; Soviet, 1917; Chinese, 1949) and major wars resulting in the international orders of Vienna (1815), Versailles (1919), Yalta and San Francisco (1945) (Brauch 2008; Holsti 1991, 1992, 2015, 2015a).

Since 1945, this author distinguishes among three phases of international order since the start of the Anthropocene on 16 July 1945 that is addressed in the first two chapters by the editor in the framework of 'peace' and 'security narratives':

(a) The *bipolar world* of the Cold War (1945/1947–1989/1990) in which nuclear deterrence prevailed in a system of collective self-defence (Art. 51 UNCH) while the collective security system envisaged in the UN Charter was paralysed.
(b) The *unipolar world* of globalisation (since 1990) was dominated by the sole remaining military hegemon: the United States of America (USA) that has both initiated the process of economic globalisation and promoted neoliberal reforms (deregulation) but simultaneously implemented in the Trump Administration neo-isolationist and protectionist policies.
(c) The unipolar international order has gradually eroded since the attack of non-state actors (terrorists) against the US territory on 11 September 2001. Since the Russian annexation of Crimea Peninsula in 2014 and the its aggression against Ukraine on 22 February 2022 the previous European peace and security order collapsed. The unipolar world dominated by the sole military superpower (USA) is being challenged by an emerging bilateral alliance between China and Russia and by an expanded group of BRICS+6 states of Argentina, Egypt, Ethiopia, Iran, Saudi Arabia, and the United Arab Emirates (since 24 August 2023).
(d) During the first Trump (2017–2021) and the Biden Administration, the US has perceived China as a political rival and as an emergent new major military threat to its hegemony. Within the institutional framework set up since 1945, a *multipolar world* is emerging where the US and the West tries to move from the globalisation it has promoted since the 1990s towards shifting to a diversification

of trade and financial streams and a partial de-globalisation with increases in autarky, protectionism, sanctions, and barriers for the free flow of finance, goods and peoples and access to new technologies.

(e) This erosion of the old international order may result in multiple new tensions, instabilities and conflicts including military confrontations and wars. This new phase makes cooperative agreements on international environmental and climate change more difficult and may thus increase the societal impacts of environmental threats.

Since 1945, this author distinguishes among different phases of the change in global CO_2 emissions since the start of the Anthropocene on 16 July 1945 that are addressed in the context of the Anthropocene narrative and the interaction between international order and global environmental and climate change:

(a) The first *Takeoff* period since 1750–1950 with a relatively *slow acceleration* from 279 ppm of CO_2(at the beginning of the Industrial Revolution until 1950 with the End of World War II and the start of the Cold War when global CO_2concentrations in the atmosphere had reached 315 ppm) during a time period when environmental issues were not yet socially constructed and thus did no yet exist for policy makers, researchers and human beings.

(b) The second period of the *Great Acceleration* (Mc Neill/Engelke 2014) in global CO_2 emissions started with the adoption of a global US military strategy with a quadrupling of US military expenditures that were suggested and discussed in the NSC-68 paper in 1949/1950 by the *National Security Council* (NSC) of the Truman Administration and approved after the start of the Korean War. The US security umbrella fostered high growth rates during the reconstruction period and expansion of global trade.

(c) The third period of an *Exponential Acceleration* of global trends since the fall of the Berlin Wall (9 November1989) after the adoption of the Charter of Paris (19–21 November 1990) that ended of the Cold War, resulted in the decline and dissolution of the USSR, the new framework of economic globalisation, deregulation and the Washington consensus on the one hand and the rapid economic rise of China since the implementation of the economic reforms of Deng Xiaoping. As a result in 2007, China replaced the USA as the largest emitter of CO_2 and has become the second largest global economy (in GDP) and even the biggest producer based on ppp parity.

(d) Since the Russian aggression of 2022, the overall change during the fourth phase of the rise of CO_2emissions and their concentration in the atmosphere is uncertain as the changes in international growth, trade, and GDP are at present unpredictable given the uncertainty of the future global political and security order, economic relations, trade, and the CO_2 emissions regime. This uncertainty will further increase during the second Trump Administration since January 2025.

Since the 1970s, R. Koselleck[40] (1923–2006) coedited with W. Conze and O. Brunner an eight volume *Encyclopaedia* on *Geschichtliche Grundbegriffe* (basic historical concepts) and gradually established a discourse on a 'history of concepts' which

> focuses on the change in the meaning of expressions in order to highlight the experience of reality in past epochs. Because this change was particularly great around 1800 as a result of political and industrial revolutions, Koselleck coined the term 'saddle time' for the period from around 1750 to 1850. Old words have thus gained new meaning … The term threshold period is used synonymously. More recent work, however, assumes that political-social language once again underwent a profound structural change in the course of the 20th century, which requires new categories of analysis that go beyond Koselleck's.[41]

The fundamental change that was triggered by the first test of an Atomic Bomb on 16 July 1946 in Alamogordo in New Mexico, to which the AWG referred to as a marker for the transition from the Holocene to the Anthropocene in earth history, the end of World War II (8 May 1945 in Europe and in August 1945 for Asia), the start of the Nuclear or Atomic Age (use of Nuclear Weapons against Hiroshima and Nagasaki), the creation of the United Nations Systems with the signing of the UN Charter in San Francisco on 26 June 1945 and the Establishment of the Bretton Woods Institutions between 1944 and 1948 (World Bank, IMF, GATT) may have resulted in a new 'threshold period' or *Sattelzeit II*[42] that may require both a 'rethinking' of the substantive object of the wider German concept of *Politik* and in English a 'reconceptualisation' of its three components of 'politics', 'policy' and 'polity'.

Traditionally, the concept of 'Politik' was closely linked with national sovereignty. However, in the Anthropocene, we have a multi-level system with a divided sovereignty where Politik as politics, policy and polity occurs on each level. G. Simonis has argued that in the Anthropocene the concept of sovereignty has fundamentally changed.[43]

The 'threshold period' since 16 July 1945 requires a 'rethinking' of humanity as a geological actor whose actions have caused dual consequences. The 'radionucleids' in the sediments geologists have proposed as a marker for the change in earth's history. Both narratives that were caused by this 'single event' require a distinct analysis of the different causes, impacts and countermeasures. The nuclear age and the age of environment were closely linked by the scientific and technological progress that was fostered during World War II and in the framework of globalisation during the Cold War and thereafter.

---

[40] Brunner, O.; Conze, W.; Koselleck, R. (1972–1997); Koselleck (1979, 2006); see on "*Conceptual History*", at: https://en.wikipedia.org/wiki/Conceptual_history#Descrip-tion_and_history; see also "Concepta—International Research School in Conceptual History and Political Thought", at: https://concepta-net.org/.

[41] On "Reinhart Koselleck", see at: https://de.wikipedia.org/wiki/Reinhart_Koselleck (25 August 2023); (Müller/Schmieder 2016): 383–392.

[42] See Jordan: (2005–2012); http://dx.doi.org/10.1163/2352-0272_emho_COM_026960 (27 August 2023); Kirchberger (2010); Décultot/Fulda (2016).

[43] I am grateful to Georg Simonis who has raised this point in his comment to the author on an earlier draft of this text in October 2023.

The German *Encyclopaedia* on basic historical concepts coedited by Koselleck, Conze and Brunner in eight volumes contains a long chapter on 'Politik' by V. Sellin (vol. 8: 789–874) that analysed the concept of Politik since ancient times (from Plato, Aristoteles and Cicero until modern times in philosophy) until the French Revolution and the enlightenment period until the 1970s in German history, sociology until Carl Schmitt and Max Weber based on literature published. A possible impact of the nuclear age since the emergence of the Atomic bomb in 1945 was not yet addressed. While the concepts on 'peace' (by W. Janssen,1972, in vol. 2: 543–591) and 'security' (by W. Conze,1972, in vol. 5: 831–862) but also on 'pacifism' (by K. Holl in vol. 4: 767–787) and 'militarism' (by W. Conze, 1972, in vol. 4: 1–47) are covered there is no entry on 'environment' (in vol. 6) and 'ecology' (in. vol. 4). This indicates that until in the 1970s these two concepts that are used on the Anthropocene narrative were not yet recognised.

The concept of a second *Sattelzeit* has not been used for the transition period since 1940 or 1945 for the atomic age until now representing the different phases of the Anthropocene. While the 'enlightenment period' emerged during the so-called Sattelzeit (1750–1850), the need for a 'second enlightenment era' has been suggested by several organisations such as the Club of Rome and several authors like E.U. v. Weizsäcker and A. Wijkman (2017, 2018) and C. Pelluchon (2021).

Two Narratives on the Emerging Anthropocene Epoch: The Peace and Security Versus the Ecology Narrative

For the 'security and peace narrative', the threat and enemy is the *other* side by its actions, intentions and employment of its military and weapons, while for the 'Anthropocene narrative', 'we as humanity' are the major threat against which other countermeasures are needed: a cooperative socio-economic transition of our activity (production and consumption), of the economy, society and political realm that require peaceful cooperation among us as members of humanity, of society, of the economy and of the state.

The change that has occurred since 1945 in both Earth and human history is unique, as for the first and so far only time the nuclear weapons test on 16 July 1945 has triggered manifold, different, and diverse changes. Nobody (policymakeres and scientists) in 1945 was aware of this coincidence and the environment had not yet been socially constructed until the 1960s and 1970s and the concept of the Anthropocene was not yet coined until 2000.

The proposed second 'threshold period' (Sattelzeit II) was much more severe as the new political, economic and environmental processes made it for the first time possible that humanity has intervened as a geological force into the earth system and may trigger consequences that undermine the livelihood of millions of people and in the worst case the survival of humanity as a species.

This raises many fundamental research questions whether the unleashed consequences and multiple potential 'tipping points' of two existential threats triggered by the military use of the splitting of the atom (nuclear weapons, arms race, deterrence) and the 'chaotic impacts' of human actions in the earth system and the lack of sufficient countermeasures to face and cope with global environmental and climate change may unleash consequences that threaten the survival of humanity.

However, the causes, the nature of these two existential fundamental threats and their potential impacts differ fundamentally:

- The *Nuclear era* ('Atomzeitalter') has been analysed in the political realm from different theoretical perspectives that are triggered according to the 'English school' (Wight 2005; Bull 1977, 1995; Buzan 2002)[44] by three ideal types influenced by the ideas of *Thomas Hobbes* (power prevails), *Hugo Grotius* (legal agreements and international institutions matters) and *Immanuel Kant* (conceptual ideas, approaches and theories are relevant) in three different approaches of the *peace* and *security narratives* that have become again important and are in 2023 in the centre of political action and reflection.
- *Global environmental and climate change* (*Anthropocene narrative*). While Svante Arrhennius had hypothesised already in 1896 a linkage between the burning of hydrocarbon energy sources and the concentration of CO_2in the atmosphere, it took another 75 years until 1971 when anthropogenic climate change has gradually become a scientific issue and until 1988 when climate change was finally put on the policy agenda.

This book on "towards rethinking politics, polity and polity in the Anthropocene" on the dual agenda of a political, economic, societal, and individual' world of action, analysis, and reflection in the social sciences cannot answer the impact these multiple changes had for many have on the substance of analysis and reflections in political science, international relations as the concept is still being discussed only on the peripheries of social science and in the humanities and has so far still been ignored by the most cited mainstream authors.

This book with ten substantive chapters cannot answer to these questions yet but it tries to put the 'Anthropocene narrative' on the agenda of 'Politik', to foster intellectual curiosity, interest and public support for systematic efforts to 'rethink' 'politics', 'policy', and 'politiy' in the Anthropocene and to 'reconceptualise the

[44] See "English school of international relations theory"; at: https://en.wikipedia.org/wiki/English_school_of_international_relations_theory; (Wight 2005; Wight/Porter 1991; Bull 1977/1995; Buzan 2002).

needed key terms'. While the first period of enlightenment (1750–1860)[45] has triggered many changes analysed in the framework of the *Sattelzeit* (1750–1850), a new enligthenment is needed as a framework for the tasks ahead.[46]

While many books on political theory and history ideas (von Beyme 2003) covered primarily the period of an extended *Sattelzeit* from 1789 to 1945, new theories on Politik are needed that address and analyse the new 'threshold period' or a new *Sattelzeit* II for the Anthropocene since the Anthropocene Epoch has started in 1945.[47] However, the proposal by Geulen (2010) and the comment by Nolte did not discuss the 'Anthropocene concept' and its impact on and relevance for a 'reconceptualisation' of new basic concepts since 1945.

But these discussions go far beyond the goal of this book whose focus is limited to 10 chapters on moving "Towards rethinking politics, polity and polity" which try to put the Anthropozene agenda and narrative on the political and scientific era for the present and future.

Structure of the Book

This book emerged from a workshop on "'Politik' in and for the Anthropocene: Obstacles, Challenges, Opportunities and Tasks for the Social Sciences in the 21st Century" that took place on 31 May 2017 on the eve of the editor's 70th birthday in the presence of Nobel Prize Laureate Paul J. Crutzen in the editor's private library in Mosbach (Germany).[48] Two participants at this workshop from the Netherlands

---

[45] See for an overview of key thinkers during and on the enlightenment at: "Age on Enlightenment", at: https://en.wikipedia.org/wiki/Age_of_Enlightenment; on the need for a new enlightenment, see Bivona (2017); Gabriel, Horn, Katsman, Krull, Lippold, Pelluchon, Venzke (2022); at: https://www.transcript-verlag.de/shopMedia/openaccess/pdf/oa9783839465707.pdf; Deutsche Gesellschaft Club of Rome (2017); Weizsäcker, Wijkman (2017): *Wir sind dran. Club of Rome: Der große Bericht: Was wir ändern müssen, wenn wir bleiben wollen. Eine neue Aufklärung für eine volle Welt* (Gütersloh: Gütersloher Verlagshaus).

[46] On the need for a new enlightenment: Bivona (2017); Gabriel, Horn, Katsman, Krull, Lippold, Pelluchon, Venzke (2022); at: https://www.transcript-verlag.de/shopMedia/openaccess/pdf/oa9783839465707.pdf; Deutsche Gesellschaft Club of Rome, 2017; E. U. von Weizsäcker and A. Wijkman (2018).

[47] See: C. Geulen, 2010: "Plädoyer für eine Geschichte der Grundbegriffe des 20. Jahrhunderts", in: *Zeit-historische Forschungen/Studies in Contemporary History*, 7,1 (2010): 79–97; Online ed., at: https://zeithistorische-forschungen.de/1-2010/4488; DOI at: https://doi.org/10.14765/zzf.dok-1790; see for a comment on Geulen; Nolte, 2010: "Vom Fortschreiben und Umschreiben der Begriffe. Kommentar zu Christian Geulen", in: *Zeithistorische Forschungen/Studies in Contemporary History*, 7,1 (2010): 98–103; online at: https://zeithistorische-forschungen.de/1-2010/4616; DOI: https://doi.org/10.14765/zzf.dok-1789.

[48] See the programme of the workshop on 31 May 2017 at: http://www.afes-press-books.de/html/PDFs/2017/From%20the%20Brainstorming%20to%20the%20Book%20Schedule_22-6-2017.pdf; see also on the website on 2 books by P. J. Crutzen at: http://www.afes-press-books.de/html/SpringerBriefs_PSP_Crutzen.htm; see also on the book with Crutzen's selected texts on the Anthropocene at: https://link.springer.com/book/10.1007/978-3-030-82202-6.

have since passed away: F. Arends on 3 June 2020 and Paul J. Crutzen on 28 January 2021.[49] This workshop with Nobel Laureate P. J. Crutzen is documented in an appendix with photos.

The book is structured in three parts addressing in Part I: the context for international order as a challenge for human survival, in Part II: the three concepts which the German term 'politics', 'policy' and 'polity' in the Anthropocene comprises, and dealing in Part III with integrative geography, governance, and gender.[50]

In Chapter 1 *Hans Günter Brauch* reviews the controversial discussion on the start of the Anthropocene claiming that the change in Earth history coincided with the emergence of a new international order. During the Cold War the challenge of the liberal order contributed to the collapse of the Soviet Union and the intensification of globalisation resulted in a rapid increase of GHG emissions. The collapse of the European security order (2022) upgraded Hobbesian geopolitics and postponed in many countries the realisation of the goal of a climate neutral society and economy. The Anthropocene has not yet affected the traditional definitions of concepts of *Politik* prevailing in text books written by authors representing the mainstream of the discipline. The debate on the Anthropocene within Political Science and International Relations is addressed in ten anthologies and a few monographs with texts written by innovative social scientists. Their contributions differ in the assessment of the Anthropocene and on its relevance, definitions, and purpose.

Being no actor the Anthropocene by itself could not foster a new international environmenttal order to counter Global Environmental and Climate Change. However, 'Politik' as politics, policy and polity as the result of decisions taken by individual and collective actors and humankind as a representation of collective behaviour within the temporal framework of the 'Anthropocene Epoch' in Earth History should be reconceptualised detached from the concept of national sovereignty. This proposal has so far not been taken up and considered within the discipline of political science and the sub discipline of International Relations.

In Chapter 2 *Hans Günter Brauch* discusses "The Anthropocene and International Political Order: Towards an Integrated Analysis" so far as "the only epoch in Earth history in which humankind as a 'geological actor' has directly intervened into and thus impacted on the earth system". I will argue that "in the early 21st century two totally different narratives co-exist: the *peace and security* or *war and peace narrative* and the *ecological and environmental narrative*". I argue "that an integrated analysis of both narratives is needed in the social sciences and humanities on peace and security and in the natural and social sciences on environmental issue areas and ecological concerns". I further emphasise that in the Anthropocene the start of "a new epoch of Earth history and of a new global international order since 1945" have coincided. Therefore I suggest that "both peace and security problems

---

[49] See the obituaries by Hans Günter Brauch on Paul J. Crutzen at: http://www.afes-press-books.de/html/Crutzen_obituary.htm and on Frederik Arends at: http://www.afes-press-books.de/html/PDFs/2020/Obituary_Jacob_Frederik_Martinus_Arends.pdf (15 August 2022).

[50] This section is based exclusively on the abstracts written by the authors themselves in which they summarise their major arguments, reflections and results.

and environmental issues and ecological concerns have been interrelated and must thus be analysed and addressed from a common political strategy to face and cope with fundamental threats to human history that are associated with the atomic bomb and the Anthropocene". In Chapter 2 I will "address two threats for the survival of humankind: (a) the Nuclear Threat, the Security Dilemma and the logic of Nuclear Deterrence and (b) the impacts of global climate change that create human survival threats". However, "the response to the Russian war in Europe has already resulted in a downgrading of the effects of Global Environmental Change and to a fragmentation and uncertainty during the third phase of the Anthropocene". I will propose "a *rethinking* of politics, policy and polity in the Anthropocene, *thinking* geological and geographic time together, *combining* the peace and security narrative with the ecological and Anthropocene narrative and *integrating* Earth Systems Science and history of events". But I am fully aware that this ambitious, broad and holistic agenda setting can only partially be realised in this book.

Part II on *Politics, Policy and Polity in the Anthropocene* contains five chapters. Chapter 3, "The Anthropocene and Complexity", prepared by *Czesław Mesjasz* and based on his professional background in physics and management, offered "a Survey of Ideas" linking the Anthropocene with various interpretations of complexity. The aim of this chapter was to identify, order, and evaluate preliminarily the applications of broadly defined complexity-related ideas in the development of the concept of the Anthropocene and in the Anthropocene-related areas of environmental studies. The main conclusion of the chapter is that complexity-related ideas constitute an essential support for enhancing the scientific value of policy-oriented considerations on the Anthropocene.[51]

In Chapter 4 *Simon Dalby* argues in "Politics and Geology in Anthropocene Geopolitics" that "The Anthropocene concept requires an extension of thinking well beyond traditional environmental formulations to recontexualise humanity as a major earth system force that is transforming the biosphere. This new context requires moving beyond conventional social science assumptions that the planet is a stable stage for human rivalries and engaging with the world making implications of understanding humanity as a geological actor. If humanity is to shape the future so that life operates in conditions loosely analogous to the last dozen millennia, then a 'planet politics' that focuses on production processes that transcends the systems of fossil fuelled capitalism of the last couple of centuries is essential. The task for social scientists is to facilitate such transitions, and the Anthropocene concept, if it is used carefully is a key formulate to facilitate transitions to more sustainable futures".

In Chapter 5 *Richard Ned Lebow* is "Thinking Politically about the Anthropocene". He examines "possible interactions between the Anthropocene and the study of international relations". He asks "how the concept of the Anthropocene

---

[51] The author added a remark (3.1.1) on the impact of the rejection of the Anthropocene as a new epoch in Earth History by several official geological institutions in March 2024, reviewed above in this Introduction.

encourage[s] new and productive frameworks for realism, liberalism, and constructivism". He suggests "that suitably revised paradigms can help us reduce and cope with the effects of climate change".

In Chapter 6, Dahlia Simangan offers "An International Relations Perspective" on "Peace and Security Discourses on the Anthropocene". Her review of relevant publications shows that the IR discipline's critical engagement with the Anthropocene discourse enables the conceptual expansion of the traditional notions of peace and security. She further argues that the plurality of perspectives on the value of the Anthropocene concept within IR is vital for further theorising politics and enhancing interdisciplinary conversations on peace and security. Simangan concludes the chapter with the following questions for future research. What are the conditions that will support peace in the Anthropocene, and how can IR advance the conversations surrounding these conditions? And what are the implications of intersecting the Anthropocene concept with peace and security for different modes of peace formation amid global environmental challenges?

In a case study, "The Anthropocene and German Political Science", *Georg Simonis* addresses in Chapter 7, the question of why the concept and hypothesis of the Anthropocene has elicited little response in German political science. From the margins of the discipline, the notion of a new age in which anthropogenically generated technologies and their forms of use threaten the conditions of human life is only slowly diffusing into the core of political science. The explanatory approach refers to Niklas Luhmann's systems theory. However, there are individual political scientists, especially from the field of environmental politics, who have made important contributions to the political science discussion of the Anthropocene. Oriented towards policy fields, the chapter provides an overview of relevant contributions of German Anthropocene research, mainly in the fields of governance research and international relations theory. As a result of its thematically structured analysis, the article provides an overview of focal points of Anthropocene research in political science.

In Part III on *Integrative Geography, Governance, and Gender* in Chapter 8 the trained physicist and geographer Jürgen Scheffran who is a retired professor of integrative geography and was the founder and head of the *Research Group Climate Change and Security* (CLISEC) discussed in his long chapter key themes of his career: "Planetary Boundaries, Polycrisis and Politics in the Anthropocene: Climate Pathways, Tipping Cascades and Transition to Sustainable Peace in Integrative Geography". The Dutch physicist *John Grin* became a professor of political science and was twice head of the Amsterdam School of Social Research. He writes on "Governance and Politics in the Anthropocene" and *Ursula Oswald Spring* a trained ecologist from CRIM UNAM in Cuernavaca (Mexico) argues that "Patriarchy has Produced the Patriacene" and asks "Can Gender Challenge Its Persistence?"

Based on Integrative Geography conceptions of landscapes, interspaces, and pathways of complex human-nature interactions, Scheffean investigates the impacts of limits to growth and planetary boundaries on the polycrisis in the Anthropocene. Geopolitical conflicts are challenging the existing world order, leading to interconnected instabilities, compounding risks and tipping cascades, connecting the climate-migration-conflict nexus, disasters and extreme events, biodiversity loss and ecocides, the Corona crisis, the Russia–Ukraine war and other violent conflicts. Future challenges and opportunities for multipolar world orders include the transformation of fossil capitalism, North-South power shifts, and the role of civil society in democracy.

Instead of a fundamental shift (*Zeitenwende*) towards military force and geopolitical power struggles to defend the existing world order, continuing the fatal triumvirate of growth, power and violence driving the Earth into a vicious circle between conflict, environmental degradation and underdevelopment, a different 'Zeitenwende' is necessary towards a positive virtuous circle between sustainability, development and peacebuilding. To balance human needs and available natural resources, an efficient, sufficient and fair use and distribution of resources is required. Possible futures are shaped by critical thresholds between pathways of disruption and construction, conflict and cooperation, war logic and peace logic, risk and resilience, exclusion and coexistence, identity and diversity, trade-offs, and synergies. Adaptive and anticipative governance integrates social-ecological transformation and conflict transformation, sustainable and environmental peacebuilding, disarmament and common security, climate mitigation and adaptation, conflict-sensitive and resilient climate policies, climate justice and climate matching in North-South climate cooperation and the global energy transition.

In Chapter 9 entitled: "Governance and Politics in the Anthropocene" *John Grin* contends:

> that the key factor that has driven humankind and its planet into the Anthropocene is the modernist idea to achieve human progress through *controlling* nature and society on basis of universal principles, without paying attention to or even suppressing local conditions of both 'nature' and 'society'. An alternative orientation for development is to (i) replace the instrumentalisation of nature to serve human needs by mutual alignment of human and social needs with nature through (ii) re-appreciating opportunities for regeneration of energy, resources and the atmosphere. The chapter elaborates this building on two examples: modern, productivist agriculture and associate water management and food consumption practices in Europe and the US; and energy, water and waste disposal systems in American and European cities. These will show how place-based, regenerative development, creating synergy with rather than exploiting nature offers a promising way to remedy society's adverse impacts on the planet during the Anthropocene, and may help improve and maintain a fair and sustainable level of well-being of current and future generations globally. Guiding principles for associate place-based governance include new constitutional understanding of notions like equity and accountability; new finance arrangements and standards that allow for synergy and adaptive governance.

Finally in Chapter 10 on "Patriarchy has Produced the Patriacene: Can Gender Challenge Its Persistence?" *Ursula Oswald Spring* reviews the evolution of the Patriacene

> over the past 6000 years, when the 'system of rule' changed from maternal clans to fratrilineal kinships, god-kings, *pater familias*, colonisation, capitalism, and multinational corporations. Patriarchy is based on violence, wars, and exploitation of humans and nature. Facing these imperialist powers, people in the Global South are helpless to limit the extraction and mistreatments. Thus, women experimented with radical alternatives such as gift economy, the Aymara's Cosmovision of 'living well' and the Zapatista's 'leading by obeying'. They reversed violent accumulation by understanding that "there is no better life", promoting

subsistence production, inclusive education, traditional health care, and diverse productive activities. Their bottom-up cultural options against the Patriacene are improving well-being, social care, local cohesion, sorority, and ecosystem restoration.

It has not been possible to present and summarise these ten chapters within one narrowly focused research design in the social sciences. Therefore, I will propose in the back matter in my "Summary and Outlook" on "The Anthropocene and the Social Sciences" to move the discourse on Crutzen's initial concept in the natural sciences in the social sciences "from 'Agenda-Setting' towards a 'Rethinking' of the 'Anthropocene' and a 'Reconceptualisation' of Concepts in Earth and World History and Planetary Politics Thought Together".

References

Adam, D., 2024: "Ditching 'Anthropocene': why ecologists say the term still matters - Beyond stratigraphic definitions, the name has broader significance for understanding humans' place on Earth", in *Nature*, 24 March.

Benner, S.; Lax, G.; Crutzen, P. J.; Pöschl, U.; Lelieveld, J.; Brauch, H. G. (Eds.): *Paul J. Crutzen and the Anthropocene: A New Epoch in Earth's History* (Cham: Springer Nature Switzerland).

Bivona, R., 2017: *The New Enlightenment: A Twenty First Century Peaceful American Revolution* (North Fort Myers, FL: New Enlightenment Citizen's Union).

Brauch, H. G., 2016: "Historical Times and Turning Points in a Turbulent Century: 1914, 1945, 1989 and 2014?", in: Brauch, H. G.; Oswald Spring, Ú.; Bennett, J.; Serrano Oswald, S. E. (Eds.): *Addressing Global Environmental Challenges from a Peace Ecology Perspective* (Cham—Heidelberg—New York—Dordrecht—London: Springer International Publishers): 11–54.

Brauch, H. G., 2021: "The Anthropocene concept in the natural and social sciences, the humanities and law—a bibliometric analysis and a qualitative interpretation (2000-2020), in: Benner, S.; Lax, G.; Crutzen, P. J.; Pöschl, U.; Lelieveld, J.; Brauch, H. G. (Eds.): *Paul J. Crutzen and the Anthropocene: A New Epoch in Earth's History* (Cham: Springer Nature Switzerland): 289–438.

Brauch, H. G., 2021a: "Peace Ecology in the Anthropocene", in: Oswald Spring, Ú.; Brauch, H. G (Eds.): *Decolonising Conflicts, Security, Peace, Gender, Environment and Development in the Anthropocene* (Cham: Springer Nature Switzerland): 51–186.

Brauch, H. G., 2008: "Introduction: Globalization and Environmental Challenges: Reconceptualizing Security in the 21st Century", in: Brauch, H. G.; Oswald Spring, Ú.; Mesjasz, C.; Grin, J.; Dunay, P; Behera, N. B.; Chourou, B.; Kameri-Mbote, P.; Liotta, P. H. (Eds.): *Globalization and Environmental Challenges—Reconceptualizing Security in the 21st Century* (Berlin: Springer-Verlag): 27 43.

Brauch, H. G.; Oswald Spring, Ú.; Grin, J.; Scheffran, J. (Eds.), 2016: *Handbook on Sustainability Transition and Sustainable Peace (Cham:* Springer International Publishing).

Braudel, F., 1946, 1966: *La Méditerranée et le Monde Méditerranéen à l'époque de Philippe II* (Paris: Armand Colin).

Braudel, F., 1969: "Histoire et Science Sociales: La Longue Durée", in: *Écrits Sur l'Histoire* (Paris: Flammarion): 41–84.

Braudel, F., 1972: *The Mediterranean and the Mediterranean World in the Age of Philip II*, 2 volumes (New York: Harper & Row).

Brunner, O.; Conze, W.; Koselleck, R. (Eds.) 1972 - 1997: *Geschichtliche Grundbegriffe. Historisches Lexikon zur politisch-sozialen Sprache in Deutschland*, vol. 1–8 (Stuttgart: Klett—Darmstadt: Wissenschaftliche Buchgemeinschaft, 2023).

Bull, H., 1977/1995: *The Anarchical Society* (New York: Columbia University Press; London: Red Globe Press).

Buzan, B., 2002, 2004: *From International to World Society? English School Theory and the Social Structure of Globalisation* (Cambridge: Cambridge University Press).

Carrington, D., 2023: "H-bombs or chicken bones: the race to define the start of the Anthropocene", in *The Guardian*, 6 January.

Conze, W., 1972: "Militarismus", in: Brunner, O.; Conze, W.; Koselleck, R. (Eds.) 1972—1979, 2023: *Geschichtliche Grundbegriffe. Historisches Lexikon zur politisch-sozialen Sprache in Deutschland*, vol. 1–8 (Stuttgart: Klett—Darmstadt: Wissenschaftliche Buchgemeinschaft, 2023), vol. 4: 1–47.

Conze, W., 1972: "Sicherheit", in: Brunner, O.; Conze, W.; Koselleck, R. (Eds.) 1972—1979, 2023: *Geschichtliche Grundbegriffe. Historisches Lexikon zur politisch-sozialen Sprache in Deutschland*, vol. 1–8 (Stuttgart: Klett—Darmstadt: Wissenschaftliche Buchgemeinschaft, 2023), vol. 5: 831-862.

Crutzen, P. J., 2002: "Geology of mankind—The Anthropocene", in: *Nature*, 415: 23.

Crutzen, P. J.; Brauch, H. G. (Eds.), 2016: *Paul J. Crutzen: A Pioneer on Atmospheric Chemistry and Climate Change in the Anthropocene* (Cham—Heidelberg—New York—Dordrecht—London: Springer International Publishing).

Crutzen, P. J.; Stoermer, E. F., 2000: "The Anthropocene", in: *IGBP Newsletter*, 41 (May): 17–18; at: http://www.igbp.net/download/18.316f18321323470177580001401/1376383088452/NL41.pdf.

Damianos, A., 2024: "What was the Anthropocene", in: *Critical Legal Thinking*, 8 March; at: https://criticallegalthinking.com/2024/03/08/what-was-the-anthropocene/.

Décultot, E.; Fulda, D. (Eds.), 2016: *Die Vielfalt der Sattelzeit. Strukturen und Tendenzen des Historischen Erzählens um 1800 im Deutsch-Französischen Vergleich Veranstaltung 2013 Halle (Saale)* (Berlin-Boston: Walter de Gruyter).

Deutsche Gesellschaft Club of Rome, 2017: "Wir brauchen eine Neue Aufklärung"; at: https://clubofrome.de/neue-aufklarung.

Fisher, R., 2024: "Why it's proving difficult to define the official dawn of the Anthropocene", in: *BBC*, 6 March; at: https://www.bbc.com/future/article/20240306-why-its-proving-difficult-to-define-the-official-dawn-of-the-anthropocene.

Gabriel, M.; Horn, C.; Katsman, A.; Krull, W.; Lippold, A.L.; Pelluchon, C.; Venzke, I., 2022: *Towards a New Enlightenment—The Case for Future-Oriented Humanities* (Bielefeld: transcript); at: https://www.transcript-verlag.de/shopMedia/openaccess/pdf/oa9783839465707.pdf

Geulen, G., 2010: "Plädoyer für eine Geschichte der Grundbegriffe des 20. Jahrhunderts", in: *Zeithistorische Forschungen/Studies in Contemporary History*, 7,1 (2010): 79–97; Online ed., at: https://zeithistorische-forschungen.de/1-2010/4488; DOI at: https://doi.org/10.14765/zzf.dok-1790.

Gibbard, P.; Walker, M.; et al., 2022 Bauer, A.: Edgeworths, M.; Edwards, E.E.; Finney, S.; Gill, J.; Maslin, M.; Merritts, D.; Ruddiman, W., 2022: "Rapid communication: The Anthropocene as an Event, not an Epoch", in: *Journal of Quaternary Science*, 16,3.

Holl, K., 1972: "Pazifismus", in: Brunner, O.; Conze, W.; Koselleck, R. (Eds.) 1972–1979, 2023: *Geschichtliche Grundbegriffe. Historisches Lexikon zur politisch-sozialen Sprache in Deutschland*, vol. 1–8 (Stuttgart: Klett—Darmstadt: Wissenschaftliche Buchgemein-schaft, 2023), vol. 4: 767–787.

Holsti, K. J., 1991, 1992: *Peace and War: Arms Conflicts and International Order 1648–1989* (Cambridge: Cambridge University Press).

Holsti, K., 2015: *Kalevi Holsti: A Pioneer in International Relations Theory, Foreign Policy Analysis, History of International Order, and Security Studies* (Cham—Heidelberg—New York—Dordrecht—London: Springer);

Holsti, K., 2015a: *Kalevi Holsti: Major Texts on War, Peace, and International Order* (Cham—Heidelberg—New York—Dordrecht—London: Springer).

Hunt, K., 2024: "Proposal to mark a new chapter in Earth's history rejected, scientists say", in: *CNN*, 5 March.

Janssen, W., 1972: "Frieden", in: Brunner, O.; Conze, W.; Koselleck, R. (Eds.) 1972–1979, 2023: *Geschichtliche Grundbegriffe. Historisches Lexikon zur politisch-sozialen Sprache in Deutschland*, vol. 1–8 (Stuttgart: Klett—Darmstadt: Wissenschaftliche Buchgemeinschaft), vol. 2: 543–591.

Jordan, S., 2005–2012: "Sattelzeit", in: *Enzyklopädie der Neuzeit* (J.B. Metzlersche Verlagsbuchhandlung and Carl Ernst Poeschel Verlag); at: http://dx.doi.org/10.1163/2352-0272_emho_COM_026960 (27 August 2023).

Jordan, S.: "Sattelzeit", in: *Encyclopedia of Early Modern History Online*, Editors of the English edition: Graeme Dunphy, Andrew Gow.

Kirchberger, U., 2010: "Eine Globalgeschichte der Sattelzeit, 1760-1830", in: *H-Soz-Kult* (07.10.2010); at: www.hsozkult.de/event/id/event-64855.

Koselleck, R. (Ed.): *1979: Historische Semantik und Begriffsgeschichte* (Stuttgart: Klett-Cotta).

Koselleck, R., 2006: *Begriffsgeschichten* (Frankfurt: Suhrkamp).

Li, C., 2024: "Surprise decision not to define the Anthropocene shocks scientists. A proposal to define the Anthropocene, a geologic epoch defined by human activity, has been rejected—surprising even scientists who consulted the voting group", in: *New Scientist*, 5 March.

McNeill, J.R.; Engelke, P., 2014: *The Great acceleration, an environmental History of the Anthropocene since 1945* (Cambridge, MA—London: The Belknap Press of Harvard University Press).

Müller, E.; Schmieder, F., 2016: *Begriffsgeschichte und historische Semantik* (Frankfurt: Suhrkamp).

Nolte, P. 2010: "Vom Fortschreiben und Umschreiben der Begriffe. Kommentar zu Christian Geulen", in: *Zeithistorische Forschungen/Studies in Contemporary History*, 7,1 (2010): 98–103; online at: https://zeithistorische-forschungen.de/1-2010/4616; DOI: https://doi.org/10.14765/zzf.dok-1789.

Pelluchon, C., 2021: *Das Zeitalter des Lebendigen—Eine neue Philosophie der Aufklärung* (Darmstadt: wbg Academic).

Phillips, L. 2024: "Keep calm and carry on with the Anthropocene", in: *The Nation*, 13 March.

Schwägerl, C., 2024: "Anthropozän: Keine Epoche für die Menschheit - Verändern wir Menschen die Erde so massiv, dass eine neue Erdepoche angebrochen ist? Darüber streiten Geologen seit Jahren. Doch der Vorschlag ist nun überraschend gescheitert", in Spektrum.de, on 6 March; at: https://www.spektrum.de/news/geologen-lehnen-neues-erdzeitalter-anthropozaen-ueberraschend-ab/2210153.

Sellin, V., 1972: "Politik", in: Brunner, O.; Conze, W.; Koselleck, R. (Eds.) 1972–1979, 2023: *Geschichtliche Grundbegriffe. Historisches Lexikon zur politisch-sozialen Sprache in Deutschland*, vol. 1–8 (Stuttgart: Klett—Darmstadt: Wissenschaftliche Buchgemeinschaft), vol. 8: 789–854.

Simon, Z. B.; Thomas. J. A., 2022: "Earth System Science, Anthropocene Historiography, and Three Forms of Human Agency", in: *Isis*, 113, 2 (June): 396–406.

SIPRI, 2023: *SIPRI Yearbook 2003: Armaments, Disarmament and International Security* (Oxford: Oxford University Press).

Turner, S.; Waters, C.; Zalasiewicz, J.; Head, M.J., 2024: "The Anthropocene's critics are missing the point", in: *Climate and Capitalism*, 13 March.

Voosen, P., 2024: "The Anthropocene is dead. Long live the Anthropocene -Panel rejects a proposed geologic time division reflecting human influence, but the concept is here to stay", in: *Science*, 5 March.

Weizsäcker, E. U. v.; Wijkman, A., 2017: *Wir sind dran. Club of Rome: Der große Bericht: Was wir ändern müssen, wenn wir bleiben wollen. Eine neue Aufklärung für eine volle Welt* (Gütersloh: Gütersloher Verlagshaus).

Weizsäcker, E. U. v.; Wijkman, A., 2018: *Come On! Capitalism, Short-termism, Population and the Destruction of the Planet—A Report to the Club of Rome* (New York: Springer).

Wight, G.; Porter, B. (Eds.), 1991: *International Theory—The Three Traditions Martin Wight* (Leicester—London: Leicester University Press for The Toyal Institute of International Affairs).

Wight, M. 2005: *Four Seminal Thinkers in International Theory: Machiavelli, Grotius, Kant, and Mazzini* (Oxford: Oxford Academic).

Witze, A., 2023: "This quiet lake could mark the start of a new Anthropocene epoch—The dawn of a new geological epoch is recorded in the contaminated sediment at the bottom of Crawford Lake in Canada", in: *Nature*, 12 July 2023.

Witze, A., 2024: "Geologists reject the Anthropocene as Earth's new epoch— after 15 years of debate—But some are now challenging the vote, saying there were 'procedural irregularities'", in: *Nature*: **627**, 249–250; at: https://doi.org/10.1038/d41586-024-00675-8.

Zalasiewicz, J. 2017: "The Extraordinary Strata of the Anthropocene", in: Oppermann, S.; Iovino, S. (Eds.): *Environmental Humanities. Voices from the Anthropocene* (London: Rowman & Littlefield

Zhong, R.: "Are We in the 'Anthropocene,' the Human Age? Nope, Scientists Say", in: *New York Times*, 5 March 2024.

Contents

Abbreviations

°C	Degree of Celsius
ABM	Antiballistic Missile Treaty (1972)
ACLED	Armed Conflict Location and Event Data Project, University of Sussex
ACP	Atmospheric Chemistry and Physics
AFES-PRESS	Peace Research and European Security Studies, Mosbach, Germany
AL	Arab League
ANZUS	Australia, New Zealand, USA Military Alliance
APESS	The Anthropocene: Politik—Economics—Society—Science
AR1	IPCC First Assessment Report
AR5	IPCC Fifth Assessment Report
AR6	IPCC Sixth Assessment Report
AU	African Union
AWAG	Anthropocene Working Group
BAU	Business As Usual
BCE	Before Christ
BICC	Bonn International Centre on Conversion
BMZ	Germany, Federal Ministry on Economic Cooperation
BRICS	Brazil, Russia, India, China, South Africa
BRICS+6	Argentina, Egypt, Ethiopia, Iran, Saudi Arabia, and the United Arab Emirates (since 24 August 2023)
BWC	Biological Weapons Convention (1971)
CAS	Complex Adaptive Systems
CBD	Convention on Biological Diversity
CC	Anthropogenic Climate Change
CE	After Christ
CEN	Center for Earth System Research and Sustainability
CEPEAS	Center for Peaceful and Sustainable Futures, Hiroshima University

CFCs-11	Trichlorofluoromethane
CFE	Conventional Armed Forces in Europe Treaty (1990)
CFR	Council on Foreign Relations, NewYork, USA
CH_4	Methane
CIGI	Chair in Political Economy of Climate Change at the Balsillie School of International Affairs, Waterloo, Ontario, Canada
CLAIP	Latin-American Council for Peace Research
CLISAP	The "Integrated Climate System Analysis and Prediction" (CliSAP) Cluster of Excellence is a research association of the Universität Hamburg, the Max Planck Institute for Meteorology (MPI-M), the Helmholtz-Zentrum Geesthacht (Centre for Materials and Coastal Research, HZG), and the German Climate Computing Center (DKRZ)
CliSAP/CLICCS	Climate Excellence Clusters, Hamburg University
CLISEC	Research Group Climate Change and Security
CNHS	Climate change, natural resources, human security, and societal stability
CO_2	Carbon dioxide
CO_2-LULUCF	Land Use, Land-Use Change and Forestry
COMECON	Council for Mutual Economic Assistance (1950–1990)
COP 27	UNFCCC Conference of Parties in Sharm el Sheik (Egypt)
COP 28	UNFCCC Conference of Parties in Dubai (United Arab Emirates, UAE)
COP	Conference of the Parties
COVID-19	Pandemic
CRIM	Regional Multidisciplinary Centre
CRIM	Regional Multidisciplinary Research Center. UNAM, Cuernavaca, Mexico
CSBM	Confidence and Security Building Measures
CSCE	Conference on Security and Cooperation in Europe
CTBT	Comprehensive Test Ban Treaty
CTBT	The Comprehensive Test Ban Treaty
CWC	Chemical Weapons Convention (1993)
DRC	Democratic Republic of Congo
DVPW	Deutsche Vereinigung für Politikwissenschaft
ECLAC	The United Nations Economic Commission for Latin America and the Caribbean
EEA	European Environment Agency
EGU	European Geosciences Union
EM-DAT	The Emergency Events Database
EMICs	Earth System Models of Intermediate Complexity
EnDev	Energising Development

ENMOD	Convention on the Prohibition of Military or Any Other Hostile Use of Environmental Modification Techniques (1972)
EnPAX	The Environmental Peacebuilding Association
ESA	Earth System Analysis
ESD	Education for Sustainable Development
ESDP	SpringerBriefs on Environment, Security, Development, and Peace
ESG	Environment, Society, Governance
ESR	Earth System Research
ESS	Earth System Science
ESS	Ecosystem Services
ESSP	Earth System Science Partnership
ETP	South East Asia Energy Transition Partnership
EU	European Union
EZLN	Zapatistas
FAO	Food and Agriculture Organisation of UN, Rome
FAR	IPCC First Assessment Report
FIIA	Finnish Institute for International Affairs
FU	Freie University of Berlin
GATT	General Agreement on Tariff and Trade
GCMs	General Circulation Models
GDP	Gross Domestic Product
GDR	German Democratic Republic
GEC	Global Environmental Change
GERD	Gross Domestic Expenditure on Research and Development
GHG emissions	Greenhouse Gas Emissions
GQL	Good Quality of Life
GRPS	Global Risks Perception Survey
GSSP	Global boundary Stratotype Section and Point
HDR	Human Development Report
HEXAGON Series	Hexagon Book Series on Human and Environmental Security and Peace
HGBS	Hans Günter Brauch Foundation for Peace and Ecology in the Anthropocene, Mosbach, Germany
HIIK	Conflict Monitor of the Heidelberg Institute for International Conflict Research
HSFK	Hessen (Frankfurt) Foundation on Peace and Conflict Research
IAMs	Integrated Assessment Models
IASC	Inter-Agency Standing Committee
ICC	International Criminal Court
ICE	International Court for the Environment
ICI	International Climate Initiative

ICJ	International Court of Justice
ICS	International Commission on Stratigraphy
IDMC	Internal Monitoring Displacement Centre
IEA	International Energy Agency
IFSH	Institute for Peace and Security Research at Hamburg University
IGBP	International Geosphere-Biosphere Programme
IGC Cape Town	International Geological Conference in Cape Town in 2016
IGY	International Geophysical Year
IHDP	International Human Dimensions Programme
IIASA	International Institute for Applied Systems Analysis
IISS	International Institute for Strategic Studies, London
ILC	International Law Commission
IMF	International Monetary Fund, Washington D.C.
INEF	The Institute for Development and Peace (Duisburg, Essen, Germany)
INF	Intermediate Nuclear Force Agreement (1987)
IOM	International Organization for Migration
IPBES	Intergovernmental Science-Policy Platform on Biodiversity and Ecosystem Services
IPCC	Intergovernmental Panel for Climate Change
IPRA	International Peace Research Association
IR	International Relations
IRENA	International Renewable Energy Agency
ISA	International Studies Association
ISA-Peace	International Studies Association, Peace Section
IUGS	International Union of Geological Sciences
JGU	Johannes Gutenberg University
JSPS	Japan Society for the Promotion of Science
LTBT	Limited Test Ban Treaty
MA	Assessment
MAD	Mutual Assured Destruction
MIT	Massachusetts Institute of Technology
MNE	Multinational Enterprise
MPI Geoanthropology	MPI for Geoanthropology, Jena (renamed in 2022)
MPI	Max Planck Institute
MPIC	Max Planck Institute for Chemistry, Mainz, Germany
N_2O	nitrous oxide
NASA	US National Aerospace Agency
NATO	North Atlantic Treaty Organization
NAZCA platform	Global Climate Action (GCAP) Platform was adopted by COP20 in Lima in December 2014
NCP	Nature's Contributions to People

NDCs	Nationally Determined Contributions (Paris Climate Treaty, 2015)
NERPS	Network for Education and Research on Peace and Sustainability
NGO	Non-governmental Organisation
NoAA	The National Oceanic and Atmospheric Administration, a Washington, D.C.-based scientific agency within the US Department of Commerce
NOAA/ESRL	Earth Systems Research Laboratories
NPT	Nonproliferation Treaty
NRW	Northrhine-Westfalia
NSC	National Security Council
NSC-68	National Security Council Paper 69 (1950)
OAS	Organization of American States
OECD	Organization of Economic Cooperation and Development
OPEC	The Organization of the Petroleum Exporting Countries
OPEP	Organización de Países Exportadores de Petróleo
OPIC	Overseas Private Investment Corporation
OSCE	Organization for Security and Co-operation in Europe (OSCE) in 1994
OWD	Open Web Design
PA	Paris Agreement
PAHSEP	Pioneers in Arts, Humanities, Science, Engineering, Practice
PIK	Potsdam Institute on Climate Change Impact Research
Ppm	Parts Per Million
ProKnow-C	An instrument for a structured selection of bibliographic materials that ensures completeness and the relevance of the bibliographic portfolio to the review purpose
PSP	SpringerBriefs on Pioneers in Science and Practice
R&D	Research and Development
SALT I	Strategic Ams Limiation Treaty (1972)
Sars-CoV-2	Corona 19 Pandemic
SCAD	Social Conflict Analysis Database; University of Texas
Scopus	Indexing Company Owned by Elsevier
SDGs	Sustainable Development Goals
SEATO	South East Asia Treay Organization
SIPRI	Stockholm International Peace Research Institute
SPD	Social DemocraticParty of Germany
SQS	Subcommission on Quaternary Stratigraphy
TAB	Office of Technology Assessment at the German Bundestag
TUM	Technical University of Munich
UCDP	Uppsala Conflict Data Program
UCL	University College London

UNAM	National Autonomous University of Mexico
UNCCD	UN Convention to Combat Desertification and Drought
UNCh	United Nations Charter
UNDP	United Nations Development Programme
UNEP	United Nations Environment Programme
UNESCO	United Nations Education, Science, Culture Organization
UNFCCC	United Nations Framework Agreement on Climate Change
UNHCR	High Commissioner for Refugees
UNSC	United Nations Security Council
UNSG	United Nations Secreary General
UNU-CPR	United Nations University—Centre for Policy Research
US EPA	US Environment Protection Agency
US, NRC	United States, National Research Council
US, NSC	United States, National Security Council
USA	United States of America
US-DoE	US Department of Energy
USNSF	US National Science Foundation
USSR	Union of Soviet Socialist Republics (dissolved in 1990)
UVA	Amsterdam University
VSM	Viable System Model
WBGU	Wissenschaftlicher Beirat der Bundesregierung Globale Umweltverän-derungen (German Advisory Council on Global Change)
WCED	World Commission on the Environment and Development
Web of Science	US Indexing Company Owned by Clarivate
WEF	World Economic Forum
WHO	World Health Oganisation
WorldCat	World Catalogue
WoS	Web of Science
WSUD	Water-Sensitive Urban Design
WTO	Warsaw Treaty Organisation
WTO	World Trade Organization
WWF	Worldwide Fund for Nature

List of Figures

List of Photos

List of Tables

Part I
Context for International Order—A Challenge for Human Survival

Chapter 1
Towards Rethinking Politics, Policy and Polity in the Anthropocene: Context and Threat to Survival

Hans Günter Brauch

Abstract This chapter reviews the controversial discussion on the start of the Anthropocene claiming that this change in Earth history coincided with the emergence of a new international order. During the Cold War this liberal order contributed to the collapse of the Soviet Union, and the intensification of globalisation resulted in a rapid increase in GHG emissions. The collapse of the European security order (2022) upgraded Hobbesian geopolitics and may postpone the realisation of a climate-neutral society and economy. During the first phase of the Anthropocene (bipolarity in the Cold War) the security discourse clearly dominated. In the second phase, since the end of the Cold War (unipolarity, 1990–2022), the planetary ecological debate emerged within new forms of "governance without government" extending beyond national sovereignty. During the third phase of increased uncertainty since 24 February 2022, both discourses coexist, while in policymaking the Hobbesian logic on national sovereignty clearly dominates, and the Anthropocene narrative has been downgraded. Within the three phases of international order since 1945 the Anthropocene has not yet affected the definitions of concepts of *Politik*. It has been addressed in ten edited collections in Political Science and in a few monographs, without a consensus on its relevance, definitions and purpose. During the Anthropocene the creation of a new planetary environmental order to counter global environmental and climate change has not yet emerged to prevent a "hothouse Earth" and to counter possible geopolitical tipping points resulting from chaotic interactions in the Earth system.

Dr. Hans Günter Brauch, Chairman of Peace Research and European Security Studies (APES-PRESS) and Chairman of the Boatd of the Hans Günter Brauch Foundation (HGBS) on Peace and Ecology in the Anthropocene, Alte Bergsteige 47, 74821 Mosbach, Germany; Email: hg.brauch@onlinehome.de.

H. G. Brauch (✉)
Hans Günter Brauch Foundation on Peace and Ecology in the Anthropocene (HGBS), Peace Research and European Security Studies (AFES-PRESS), Mosbach, Germany
e-mail: hg.brauch@onlinehome.de

H. G. Brauch (ed.), *Towards Rethinking Politics, Policy and Polity in the Anthropocene*, The Anthropocene: Politik—Economics—Society—Science 35,
https://doi.org/10.1007/978-3-031-71807-6_1

Keywords Anthropocene · Climate change · Cold War · Europe · Hobbesian perspective · Peace · Politics · Policy · Polity · Security · Security order · Socio-ecological transformation · War against Ukraine

1.1 Introduction[1]

1.1.1 Paul J. Crutzen and the 'Anthropocene' Concept

The Anthropocene is a new scientific concept in the natural and social sciences and the humanities that was coined on 23 February 2000 by the Dutch atmospheric chemist and Nobel laureate Paul J. Crutzen during a scientific committee meeting of the International Geosphere-Biosphere Programme (IGBP) in Cuernavaca, Mexico. Crutzen disagreed with a colleague who referred to the Holocene when discussing recent developments in global change dominated by human actions, and exclaimed (Brauch 2021), "We are not in the Holocene any more … We are in the … the Anthropocene!" He recalled this debate when speaking to a BBC journalist (Falcon-Lang 2011): "I was at a conference where someone said something about the Holocene. I suddenly thought this was wrong. The world has changed too much. No, we are in the Anthropocene. I just made up the word on the spur of the moment. Everyone was shocked." (cited by Lewis/Maslin 2018: 21).Helmuth Tritschler (2014) of the German Technological Museum in Munich noted later:

> Crutzen has repeated this story of a sudden flash of insight, a 'eureka' moment of our day, on multiple occasions (Schwägerl [2015]), and Will Steffen has confirmed it, thus codifying an attractive founding myth about the origins of the term (Steffen 2004, 2006, 2010, 2013: 486). Crutzen tells an additional story as part of this founding myth: after colleagues suggested that he should claim ownership of the term, he discovered that Stoermer had come up with the term independently. The joint authorship of the article in the *IGBP Newsletter* [Crutzen/ Stoermer 2020] that initiated the entire current debate was a reflection of this co-creation. … To him, the new term offers a powerful tool to validate the ongoing process of anthropogenic climate change, which has impacted the Earth in such an enduring manner that it needs a new geological era to depict it properly. (Tritschler 2016)

In 1995, Crutzen received the Nobel Prize for Chemistry, with his American and Mexican colleagues F. Sherwood Rowland and Mario J. Molina, "for their work in atmospheric chemistry, particularly concerning the formation and decomposition of ozone". As early as 1982, Crutzen and Birks had pointed to the danger of a nuclear winter as an outcome of a massive use of nuclear weapons in a war.

[1] The author is grateful to Simon Dalby (emeritus, Wilfrid Laurier Univ., Waterloo, Canada); Czeslaw Mesjasz (ret. from Economic Univ. Cracow); Dahlia Simangan (Hiroshima Univ.); Jürgen Scheffran (emeritus, Univ. Hamburg) and Georg Simonis (emeritus, Distant Univ. Hagen, Germany) for comments and suggestions.

In 2021 Steffen claimed that Crutzen's proposal "was an insight that was to change the course of scientific research and to challenge humanity's perception of itself and its role on the planet". Steffen also noted that Crutzen's proposal "for the Anthropocene included a direct challenge to the geological research community by specifically proposing that the Anthropocene became a new epoch in the Geologic Time Scale, coming after the Holocene" (Steffen 2021: 21).

Will Steffen had worked closely with Paul Crutzen as executive director (1998–2004) of the *International Geosphere-Biosphere Programme* (IGBP), a global research programme studying changes to the Earth system as a whole; he was present at the meeting in Cuernavaca and has since referred to this episode in many lectures.[2]

1.1.2 The Anthropocene Concept: Scientific Context and Spread

Based on Greek origins, the Anthropocene concept is a combination of *anthropo-* from *anthropos* (ancient Greek: ἄνθρωπος), meaning "human", and *-cene* from *kainos* (ancient Greek: καινός), meaning "new" or "recent".[3] Various concepts analogous to the Anthropocene were suggested earlier in the literature (Crutzen/Stoermer 2000), and many other concepts were subsequently proposed (Brauch 2021) in many disciplines (including in the social sciences), among them a "Capitalocene" (Moore 2016). However, all these proposals failed to replace the Anthropocene in cross-disciplinary discourses.

During the first decade (2000–2010) the debate on the Anthropocene emerged primarily within three networks in the natural sciences: the global environmental change (GEC) and the Earth System Science (ESS) communities and, since 2008, the Anthropocene Working Group (AWG), which has used the concept and whose members have spread the term globally across disciplines.

---

[2] In lectures, oral interventions, and discussions W. Steffen has referred to Crutzen's intervention during a discussion at an IGBP meeting in Cuernavaca which he attended on 23 February 2000; see the hyperlinks to 5 lectures (2010–2019) by W. Steffen in Benner, S.; Lax, G.; Crutzen, P. J.; Crutzen, P. J.; Pöschl, U.: Lelieveld, J.; Brauch, H. G. (2021), Appendix VI, part 3, p. 546. The *IGBP Newsletter* with Crutzen and Stoermer (May 2000: 17–18) is at: http://www.igbp.net/download/18.316f18321323470177580001401/1376383088452/NL41.pdf. A synthesis of the IGBP Series by Steffen, W.; Sanderson A.; Tyson, P. D.; et al. (2004: 81–142) can be downloaded here: https://link.springer.com/chapter/10.1007/3-540-26607-0_3 and the whole book for free here: http://www.igbp.net/download/18.56b5e28e137d8d8c09380001694/1376383141875/SpringerIGBPSynthesisSteffenetal2004_web.pdf.

[3] On the Greek origins and meanings of the term Anthropos, see F. Arends' (1943–2020) oral contribution on "Conceptual Origins: 'Anthropos' and 'Politik' as seen by the Greeks"; at: http://www.afes-press-books.de/html/Brainstorming_Mosbach.html [use the password: 'anthropocene2017' to listen]. An obituary of Dr J. F. M. Arends is at: http://www.afes-press-books.de/html/PDFs/2020/Obituary_Jacob_Frederik_Martinus_Arends.pdf; see also: http://www.afes-press-books.de/html/PDFs/2020/Brauch_Personal_Memories_on_Frederik_Arends.pdf.

- The global environmental change community comprises climate change, desertification, water and disaster specialists, who were represented by the International Geosphere-Biosphere Programme (IGBP, Noone et al. 2011), the International Human Dimensions Programme (IHDP, von Falkenhayn et al. 2011; Arizpe et al. 2016; Ehlers 2016), and since the Amsterdam Conference of 2001 (Steffen et al. 2002) also in the framework of the Earth System Science Partnership (ESSP, Leemans et al. 2011).
- The Earth Systems Science (ESS) or the Earth System Analysis (ESA) Community[4] met during the Amsterdam Conference in 2001 and at the Dahlem Workshop on Earth System Analysis for Sustainability (May 2003) in Berlin. Scientists at several large research institutes had a significant impact on this community, among them the Potsdam Institute on Climate Change Impact Research (PIK), which was founded in 1991, and the Stockholm Resilience Centre, which was formed in 2007 by Stockholm University, and the Beijer Institute of Ecological Economics[5] (see Steffen et al. 2020) who offered different frameworks.
- The geological community that is represented by the Anthropocene Working Group (AWG)[6] was set up in 2009 as a working group of the *Subcommission on Quaternary Stratigraphy* (SQS)[7] a constituent body of the *International Commission on Stratigraphy* (ICS), the largest scientific organisation within the *International Union of Geological Sciences* (IUGS).[8]

The meaning of the Anthropocene has differed in these three scientific communities and networks and many different definitions were offered by authors in the natural sciences, social sciences and humanities. Only the AWG and its stratigraphic bodies can officially decide whether the Anthropocene will become an official epoch in Earth history by simultaneously ending the Holocene (see Introduction). However, as the Anthropocene has been widely accepted by both the global environmental change and the Earth Systems Science communities, the Anthropocene model will remain a key concept in the natural and social sciences, as well as in the humanities,

[4] On *Earth Systems Science* (ESS), see at: https://en.wikipedia.org/wiki/Earth_system_science; Steffen et al. (2020); On *Earth System Analysis* (ESA), see: Schellnhuber and Wenzel (1998); Schellnhuber et al. (2004).

[5] On the *Potsdam Institute on Climate Change Impact Research* (PIK), see at: https://www.pik-potsdam.de/en/home. On the *Stockholm Resilience Centre* see at: https://www.stockholmresilience.org/ (13 February 2021).

[6] The AWG's first newsletter was published in December 2009. Its editorial referred to Crutzen's (2002) suggestion of the term "proposing that we are now living in a new geological interval of time — the Anthropocene — that is dominated by human activities. Since then, the term (currently informal) has been widely used by a range of Earth and environmental scientists, and the concept has been further examined (e.g. Steffen et al. 2008). Recent assessment by the Stratigraphy Commission of the Geological Society of London (Zalasiewicz et al. 2008) suggested a case for formally incorporating the term into the Geological Time Scale."

[7] See at: http://quaternary.stratigraphy.org/.

[8] On the voting within the AWG between 2022 and 2024 and by the *Subcommission on Quaternary Stratigraphy* (SQS) of the body of the *International Commission on Stratigraphy* (ICS) in March 2023 see the update by the editor in the frontmatter.

irrespective of the decision by the *IUGS*. During the second decade after the coining of the term (2011–2020) this concept spread to the social sciences, the humanities, law, the applied sciences (e.g. medicine), engineering and beyond (Brauch 2021).

1.1.3 A Global Survey of the Increase in Publications on the Anthropocene from 2000 to 2023

Table 1.1 on the quantitative increase in publications on the Anthropocene since 2000 relies on three sources: the "World Catalogue" and two US-based indexing companies, Clarivate (the owner of *World of Science*) and *Scopus* (owned by Elsevier, one of the largest scientific publishers globally). While the "World Catalogue" (WorldCat.org) "is supported by OCLC, a non-profit global library organization"[9] headquartered in the US that primarily covers library products (books, reports, pamphlets etc.) the two commercial U.S. companies have primarily included peer-reviewed journal articles in English.

A wider take-off in the general literature on the Anthropocene has occurred since 2015 in the World Catalogue (*WorldCat*) and in the peer-reviewed publications on the Anthropocene examined in the *Web of Science* and by *Scopus* (Table 1.1). In a bibliometric analysis on the use of the Anthropocene concept this author (Brauch 2021) listed these disciplines among the social sciences: Human Geography, Political Science, Sociology, Economics, Psychology and Education. The *World Catalogue* includes all references associated with this term that were listed in the catalogues of its member institutions, while the two indexing companies the *Web of Science* (Clarivate) and *Scopus* (Elsevier) primarily include peer-reviewed literature published in English in journals and in a few books and book chapters, which favoured English-speaking countries. Table 1.1 distinguishes five phases:

(a) In the starting phase (2000–2005) only very few publications on the Anthropocene were listed in the *WorldCat* based on the titles (73) and for all results (89), while the *Web of Science* (WoS) listed only 15 and *Scopus* 16 results.
(b) Over the next five years (2006–2010) a slow increase of texts on the Anthropocene occurred in the *WorldCat* as at 1 September 2023 based on 'titles' (with 262) to reach 332 for the first decade (2000–2010), and for 'all results' (with 429 texts) to reach 515 publications (2000–2010), while on 30 August 2023 the *Web of Science* listed an increase of 100 to reach 115 results, and on 31 August 2023 *Scopus* recognised 131 texts (2006–2010) to reach 142 publications for 'all results' from 2000 to 2010.

During the first decade the publications were primarily contributed by authors from the natural sciences. This changed during the third five-year period (2011–2015), when increasingly publications on the Anthropocene were contributed from the social sciences, the humanities and other disciplines (Brauch 2021: 313).

[9] See 'Worldcat", at: https://www.worldcat.org/ (3 September 2023).

Table 1.1 Anthropocene texts by year in the World Catalogue, the Web of Science, and Scopus (2000–2023)

Year	World Catalogue all results (1 February 2021)		Web of Science	Scopus	Year	World Catalogue (31 August 2023)		Web of Science	Scopus
			(31 August 2023)					(30 August 2023)	
	All results	Title	Topic	Theme		All results	Title	Topic	Theme
2000	3	2							
2001	6	3	1		2011	302	209	58	108
2002	10	9	4	3	2012	337	219	69	154
2003	12	9	3	3	2013	710	472	115	252
2004	21	19	2	4	2014	1379	1056	302	494
2005	42	34	5	5	2015	1866	1517	494	789
Subtotal [2001–2005]	[89]	[73]	[15]	[16]	Subtotal [2011–2015]	[4835]	[3717]	[1038]	[1103]
2006	60	49	6	15	2016	2531	1975	625	524
2007	61	43	13	31	2017	3614	2905	1060	810
2008	91	48	20	39	2018	4291	3195	1147	832
2009	78	31	20	50	2019	4333	3195	1740	1187
2010	152	95	41	72	2020	5003	3576	1853	1.363
Subtotal [2006–2010] [2001–2010]	[429] [515]	[262] [332]	[100] [115]	[131] [142]	Total [2016–2020] [2011–2020] [2000–2020] own total	[20,188] [25,236] [25,538]	[15,639] [19,955] [19,688]	[6425] [7578]	[5162] [5952]
					Own HGB total (2000–2020)	[25,541]	19,691	[7578]	[6412]
					2021	4322	3074	1848	1539

(continued)

Table 1.1 (continued)

Year	World Catalogue all results (1 February 2021)		Web of Science	Scopus	Year	World Catalogue (31 August 2023)		Web of Science	Scopus
			(31 August 2023)					(30 August 2023)	
	All results	Title	Topic	Theme		All results	Title	Topic	Theme
					2022	4211	2850	1762	1581
					2023	2797	1975	884	937
Total [2000–2023]	[40,941]	[31,069]	[12,072]	[10,010	Total [2021–2023] [2000–2023]	[11,547] [40,941]	[8260] [31,069]	[4494] [12,072]	[4057] [10,469]
HGB own total [2000–2 Sep. 2023]	[37,288]	[27,951]	[12,072]	[10,469]	HGB own total [2000–2 Sep. 2023]	[37,288]	[27,951]	[12,072]	[10,469]

Source Websites of the WorldCat, Web of Science and Scopus (30 and 31 August 2023), published first by Brauch (2021: 313). The data were all updated on 30 August 2023 for (Scopus and Web of Science) and on 1 September 2023 for the World Catalogue

(c) In the third phase (2011–2015) a much larger increase of titles on the Anthropocene listed in the *WorldCat* occurred, of 3.717 publications by title and 4.835 for 'all results', while the *Web of Science* listed an increase of 1038 and for *Scopus* the texts rose by 1.103.
(d) A rapid spread of titles on the Anthropocene occurred during the fourth phase (2016–2020) in the *WorldCat* for 'titles' by 15,639 publications and for 'all results' to reach 19,955 texts (2011–2020), and for all publications (for 2016–2020) by 20,188 to reach a total of 25.236 texts for the second decade (2011–2020). The *Web of Science* listed an increase (2016–2020) of 6425 publications to reach a total of 7578 texts for the second decade (2011–2020). For *Scopus* the texts rose by 5162 to reach 5952 texts.
(e) Finally, for the emerging fifth phase from 1 January 2021 until 1 September 2023, in the WorldCat the texts increased by an additional 8260 titles to reach a total of 31,069 and for all results by an additional 11,547 texts to reach 40,941 texts on the Anthropocene. The *Web of Science* listed an increase (from 2021 to 1 September 2023) of 4494 publications to reach a total of 12,072 texts for the whole time span (from 2000 to 1 September 2023). For *Scopus* the texts rose by 4057 to reach 10,010 texts.

Box 1.1 Detailed bibliometric analysis on literature in political science and international relations on the Anthropocene in Web of Science and Scopus between 2021 and 30 August to 2 September 2023. *Source* Compiled by the authors in two surveys in 2021 and 2023 based on information in databases of both *Web of Science* and *Scopus*

On 2 September 2023 the *Web of Science* listed the Anthropocene among the 12,113 results from its Core Collection for the whole time period (2000–2 September 2023) under Political Science (417), Social Sciences: Interdisciplinary (436), International Relations (363), Area Studies (158), Development Studies (191), Social Issues (115), Urban Studies (76), Asian Studies (56), Women's Studies (59), and Public Administration (56).

Of the major subcategories of the 417 texts that were indexed under Political Science 166 counts (39.8%) were in Environmental Studies, 107 entries (25.7%) in International Relations, 47 in Sociology (11.3%), 30 in Public Administration (7.2%), 26 in History (6.2%), 21 in Geography (5.0%), 21 in Philosophy (5.0%), 18 in Economics (4.3%), as well as in Development Studies (3.8%), 16 in Women's Studies (3.8%), 14 in Ethics (3.4%), 7 in Law (1.7%), and 3 in Area Studies (0.7%)

The Political Science publications on the Anthropocene started with 3 texts in 2011 and gradually grew to 5 in 2013, 18 in 2014, 57 in 2016, climaxed with 74 entries in 2019 and 2020 and then declined to 41 in 2021, 47 in 2022 and to 20 in 2023 up until 1 September.

Of the 136 texts written in Political Science on the Anthropocene by authors with more than 3 entries, the most prolific authors were, according the *Web of Science*: F. Biermann (20), O. Young with 14 texts, S. Dalby with 8 results, W. Gabardi and R. E. Kim with 7, L. J. Kotze and J. Gupta with 6 publications, and P. Wapner, T. W. Luke and C. Inoue with 5 texts each; there were 9 authors with 4 results and 15 authors with 3 publications each on the Anthropocene.

The 15 countries whose authors contributed most texts on the Anthropocene were from the USA (109), United Kingdom (61), Germany (44), Australia (42), the Netherlands (42), Sweden (33), Canada (31), Spain (15), Uganda (15), Italy (12), Brazil (11), South Africa (10), France (9), India (8) and Mexico (8).

On 2 September 2023 the *Web of Science* listed 363 results under International Relations among the 12,113 results from its Core Collection on the Anthropocene. Among the 115 texts whose authors had more than 4 texts included were S. Steiner (12), L. Arizpe Schlosser (11), T. E. Kim (10), D. Curran (8), E. Emkic (8), Z. Savasan (8), D. S. Kotze (6), D. Simangan (5), S. Dalby (5), C. Inoue (5), A. Burke (4), J. C. Pereira (4), F. Biermann (4), V. Lamb (4), D. Rothe (4), C. Middleton (4), K. N. Chris (4), L. Jourdan (4) and U. Oswald Spring (4). The listed authors of the 363 texts on the Anthropocene in International Relations came from these countries: USA (56), United Kingdom (52), Australia (38), Canada (36), Germany (25), the Netherlands (22), Italy (20), Sweden (16), Mexico (16), Spain (12), Brazil (11), Japan (11), Turkey (11), South Africa (9) and Thailand (8).

Could major changes be observed between this author's earlier bibliometric review (Brauch 2021: 313), done in early February 2021, and this new assessment of late August and early September 2023? The climax for all publications listed on the *Web of Science* on the Anthropocene was during the Covid 19 pandemic in 2020 and 2021, while the results slightly declined in 2022 to the level of 2019. The number of publications in Political Science and in International Relations rose significantly above the modest levels for Political Science (186) and in International Relations (109) in early February 2021, reaching 417 for Political Science, and 363 for International Relations (363) as at early September 2023.

There are 31 months (February 2021–September 2023) between these two bibliometric surveys by this author on texts on the Anthropocene that were categorised as Political Science by the *Web of Science*, and the number of authors increased significantly after early February 2021. In the first survey on publications on the Anthropocene (Brauch 2023: 336) "the *Web of Science* listed the following authors of 6 or more texts: J. Zalasiewicz (18), M. Williams (15), W. Steffen (12), A. D. Barnosky (11), E. C. Ellis (9), C. Waters (9), F. Biermann (8), S. Dalby (7), A. Cearreta (6), J. Rockström (6), J. P. M. Syvitski (6), C. D. Thomas (6), and O. R. Young (6)". In September 2023 the number of texts under Political Science had increased for F. Biermann from 8 to 20, for O. Young from 6 to 14 texts, and for S. Dalby from 7 to 8 results; L. J.

Kotze and J. Gupta had 6 publications, P. Wapner, T. W. Luke and C. Inoue had 5 texts each, and there were 9 authors with 4 results and 15 authors with 3 publications each on the Anthropocene.

On 27 July 2022 *Scopus* categorised none of its 7691 results on the Anthropocene in the subject areas of Political Science and International Relations. If one searches directly for Political Science, *Scopus* showed 2456 document results for this field and 1483 texts for International Relations. The authors who, according to *Scopus*, had between 7 and 20 texts were: S. Dalby (20), M. Arias-Maldonado, D. Chandler and L. J. Kotze (13 titles each), W. Steffen (12), N. Castree (11), N. Clark and S. J. Cooke (11 titles each), M. Di Paola (10), R. E. Kim, P. Pattberg and J. J. Schmidt (9 titles), F. Biermann, E. C. Ellis, T. W. Luke, J. Rockström, M. Williams and J. Zalasiewicz (8 titles), and at least 7 texts were listed that day by V. Galaz, M. Glaser, A. Hornborg, T. Karlsson, J. Lorimer, E. Lövbrand, Z. B. Simon, J. Syvitski, S. Wakefield, and K. Yusoff. All these authors together had a total of 220 document results on Political Science. The eleven countries whose authors contributed—according to Scopus—most texts on the Anthropocene were the USA (731), the UK (497), Australia (298), Canada (227), Germany (217), Sweden (161), the Netherlands (128), France (99), Spain 82), South Africa (80), Brazil (68), Norway (67), Italy (63) and Denmark (57). Due to different methods of categorisation, the data listed in *Web of Science* and in *Scopus* are not directly comparable.

While these bibliometric analyses in both the *Web of Science* and *Scopus* offer some general trends for Political Science and International Relations, they do not provide a systematically representative sample of the key texts on *Politik* as *politics*, *policy* and *polity* in the Anthropocene. A representative analysis of the authors of selected key English edited collections on the Anthropocene by Political Science and International Relations (IR) specialists may be more representative of whether these texts have contributed to a rethinking of *Politik* in the new epoch of Earth history.

1.1.4 Goal of the Chapter and Research Question

Key goals of this introductory chapter are whether the Anthropocene debate within political science and IR has so far resulted in a rethinking of *Politik* in German, *La Politique* in French, *La Politica* in Italian or Spanish, its process (*politics*), fields (*policies*) and legal texts and institutions (*polity*), and whether the Anthropocene concept has impacted on the research programmes on peace, security, development, environment and gender studies. This chapter will address these two key research questions:

- How and in which theoretical frameworks has the Anthropocene been analysed so far in Political Science and International Relations and in the peace and security research programmes, and in development and environment as well as in gender studies? (See especially Chap. 6 by Simangan in this volume).
- Has the Anthropocene affected our understanding of *Politik* (*la politique, la politica*)? Did the debate on the Anthropocene trigger a reconceptualisation of *Politik, la politique* and *la politica* and of its components of *politics* (the process), *policy* (the field or issue area) and *polity* (the legal framework and the institutions)?

This chapter argues that the Anthropocene has not yet reached the mainstream of the disciplines of Political Science and International Relations and that the most-cited authors in Political Science have not yet addressed the conceptual and historical challenge posed by the Anthropocene concept.

The second chapter will address the relationships between the proposed transition in Earth history from the Holocene to the Anthropocene (beginning with the start of the nuclear era on 16 July 1945, as was approved by the AWG in a majority vote in May 2019) discussed in the natural sciences and in human and international political, economic and societal history.

This chapter will first review the proposed start of the Anthropocene, primarily in the global environmental change, Earth Systems Science and geological (AWG) communities since Crutzen proposed the concept to illustrate the deep turn in human-Earth relations that has made human beings into "geological" factors that have not only changed nature (increasingly since the Neolithic revolution), but have interfered directly in the Earth System. This has been especially evident in the atmosphere, causing an ozone layer depletion and an exponential change in the concentration of greenhouse gases in the atmosphere that has resulted in global warming and in four major physical effects of anthropogenic global climate change: temperature increases, precipitation variability, sea-level rise and an increase in the number and intensity of extreme weather events that has led to more human-induced natural disasters.

1.2 Start of the Anthropocene: Epoch of Earth History

Crutzen and Stoermer (2000) initially suggested that the Anthropocene began with the Industrial Revolution, and especially with Watt's invention of the steam engine in 1782. Zalasiewicz (2008, 2008a), who chaired the Anthropocene Working Group (AWG) from 2009 to 2020, shared Crutzen's initial proposal when he wrote: "The Anthropocene does seem geologically real and could be judged to have begun in 1800." Steffen et al. (2011) argued a decade later that "the term Anthropocene … suggests that the Earth has left its natural geological epoch … called the Holocene. Human activities have become so pervasive and profound that they rival the great forces of Nature and are pushing the Earth into planetary *terra incognita*" (cited by

Angus 2016). During the first decade at least four different starting points for the Anthropocene were discussed in the scientific literature:

- The Neolithic or Agricultural Revolution (12,000 BP)
- The Impact of the Columbian Exchange (1492–1600)
- The start of the Industrial Revolution (1750–1800)
- The beginning of the Nuclear Era and the fallout from nuclear testing (since 16 July 1945) that coincided with the end of World War II (Subramanian 2019)
- The "Great Acceleration" (since 1950).

Within the geological community, in 2022 the AWG reduced its list of 12 of their so-called Global Stratotype Section and Point (GSSP) sites to 9 remaining sites and it was planned to further reduce their number at the end of 2022.

> The Anthropocene Working Group (AWG), …, [has] been working to decide precisely when the Anthropocene began, with a focus around the mid-20th century. The definition will need to identify specific physical properties in sediment layers, or strata, that capture the effects of recent increases in human population; unprecedented industrialization and globalization; and changes imposed on the landscape, climate, and biosphere. (Waters/Turner 2022: 706–708)

McKenzie Prillaman (*Nature*, 613, 7942) (5 January 2023) indicated: "if no clear winner emerges, more voting will be needed to narrow down the candidate list, delaying the decision possibly until May 2023".

A minority of natural scientists, especially Ruddiman (2003, 2005, 2016), argued that humans have had an impact on the climate since the Neolithic or Agricultural Revolution (Lewis/Maslin 2015, 2018: 141–142; Renn 2020: 348–354). Since the sixteenth century the successive scientific revolutions (Renn 2020: 118–142) in astronomy triggered by Copernicus, Kepler and Galileo, and the many new discoveries in the natural sciences by Newton, Bacon, Leibnitz et al. did trigger many turning points in the epochs of Earth's history, but they also fundamentally changed the worldview of scientists—despite the orthodoxy of the Catholic Church.

Lewis and Maslin (2015, 2018) identified four major transitions related to (a) energy use and its environmental impacts on the Earth System; (b) the gradual start of colonialism since Columbus reached the Americas in 1492, which resulted in a "Columbian Exchange" of species (animals and plants) and the emergence of a global capitalist economy; (c) the availability and use of coal as a result of mining; and (d) the Great Acceleration after the end of World War II, partly based on the easy access to cheap fossil fuel energy sources.

A decade later, influenced by the discussions during the Dahlem workshop (Steffen et al. 2004), the results of the Millennium Ecosystem Assessment (MEA 2005), and the projects of the IGBP, Steffen et al. (2007) agreed that "the 20th century can be characterised by global change processes of a magnitude which have never occurred in human history" (Steffen et al. 2015). They further agreed that human-environment interactions increased after 1950 due to the "amplified human impact on the environment" in the period they termed the "Great Acceleration" (Angus 2016: 42).

In its discussions, the AWG later rejected the Agricultural Revolution, the Columbian Exchange, and the start of the Industrial Revolution as starting points for the Anthropocene. The AWG opted for two more recent turning points that were instrumental in triggering more intrusive global environmental and an anthropogenic climate change: the start of the Nuclear Era that caused a fundamental change in military strategy and politics during the Cold War and created, for the first time, the ability to destroy a significant part of humankind during a major nuclear war; and the Great Acceleration (since 1950) that was made possible by the US-planned new political, security and economic world order. This was facilitated by the active multilateral global leadership role of the US and the economic globalisation process the US-dominated political and security order gradually made possible during the Cold War and further facilitated after its end in 1990 (see my Chap. 2 below).

In December 2022, the members of the Anthropocene Working Group (AWG) had narrowed their choice of the transition from the Holocene to the Anthropocene to nine geological sites. After the AWG agreed on a final candidate in July 2023 (see Introduction), seven other committees of geologists also had to approve this choice. But first, 60% of the AWG's voting members "must agree on a final candidate", which occurred in July 2023.

The *Guardian*'s science writer, Robert McKie, concluded on 1 January 2023:

> "Another indicator is provided by plastics which became widespread in use in the 1950s – which most of us point to as the dawn of the Anthropocene," added Waters. "The main thing we now have to decide is which of the sites we have shortlisted provides the clearest signal among the measures we have selected – plutonium, aluminium, plastics, and other variables – that demonstrate a fundamental epoch change."[10]

This was done in July 2023 with the announcement of Crawford Lake in Canada (see Introduction).

1.2.1 The Anthropocene and Change in International Order

The period from 1940 to 1950 was also a major turning point in international politics. It has been intensively analysed in international political, military and security history, and also in economic, social and human history. It heralded a fundamentally new international order that was intensively discussed and planned within the US government, most particularly by the US State Department, as well as by the War and Navy Departments, after 1939 (Notter 1949; Brauch 1976, 1977).

---

[10] McKie, R.: "Hunt on for spot to mark the start of the Anthropocene epoch - Scientists are to pick a location that sums up the current epoch when Homo sapiens made its mark", in: *The Guardian*, 1 January 2023.

1.2.1.1 Prehistory: Nuclear Fission, the Manhattan Project and First Nuclear Weapons Test and Employment of Atomic Weapons in 1945

After the first nuclear fission by Otto Hahn and Lise Meitner in Berlin (Hahn/Strasssmann 1939, cited in Renn 2020: 231), Albert Einstein and Leo Szilard[11] persuaded President F. D. Roosevelt to launch the top secret "Manhattan District" Project (1942–1945),[12] which "produced the atomic bomb" that "created not only a new technology, but also a new kind of socio-economic complex of technological, political, economic, and knowledge structures that seems impossible to abolish" (Renn 2020: 373). For the first time, humankind had the military tools to destroy the human species in a major nuclear war. This new "super weapon" was first tested secretly on 16 July 1945 in Alamogordo in New Mexico and was then employed on 6 August 1945 against Hiroshima and on 9 August 1945 against Nagasaki, thus forcing Japan to accept unconditional defeat in World War II.

1.2.1.2 Secrecy of Nuclear Research and Lack of Public Awareness of Environmental Policy Issues of Decision-Makers and of Scientists

In 1945 there was no scientific awareness and no research on global environmental change (GEC) and anthropogenic climate change (CC). Policy makers were similarly unaware of "the environment" and "development" as policy issues to be addressed by the United Nations Charter that was adopted in June 1945 in San Francisco as the framework for the post-World War II global political and security order. The social construction of reality (Berger/Luckmann 1966, 1967), science (Hacking 1999) and scientific knowledge (Mendelsohn 1977) of GEC, CC and the Anthropocene had not yet occurred and had no impact on the framing and development of the Post-World War II order. These ecological problems had not yet been "socially constructed" and thus did not then exist for either policy makers or scientists.

1.2.1.3 Start of Scientific Debate on Anthropogenic Climate Change

Although in 1896 Svante Arrhenius (Rodhe et al. 1997)—and other natural scientists—had already hypothesised a link between the burning of fossil fuels and the concentration of greenhouse gases, especially carbon dioxide (CO_2), in the atmosphere, the scientific debate on climate change did not take off until the 1970s (v.

---

[11] "The Einstein-Szilard Letter – 1939", in: "Atomic Heritage Foundation" website; at: https://www.atomic-heritage.org/history/einstein-szilard-letter-1939.

[12] "Manhattan District History Project: The Los Alamos Project, Vol. I: Inception until August 1945"; at: https://digital.library.unt.edu/ark:/67531/metadc841449/; "Manhattan Project"; at: https://en.wikipedia.org/wiki/Manhattan_Project.

Weizsäcker 1989; Brauch 1996), and the argument for a transition to the Anthropocene Epoch of Earth's history was not suggested until 2000, when Crutzen coined the term.

1.2.1.4 First Nuclear Weapons Test on 16 July 1945: Start of the Anthropocene

In 2015, Jan Zalasiewicz[13] et al. linked the Anthropocene with the start of the nuclear age when they suggested:

> The entry into the nuclear age (signalled by the detonation of the Trinity A-bomb in New Mexico on July 16, 1945) could be used to mark the beginning of the Anthropocene because of the stratographic evidence from artificial radionuclides but also from the sediment of global industrialisation such as fly ash, concrete and plastic. This was a unique event, comparable to the meteorite impact causing the extinction of the dinosaurs and marking the end of the Mesozoic era. (Renn 2020: 360)

In 2023 the AWG stated that the Anthropocene started in 1952 with "other markers [that] offer the 'golden spike' … Principal among these is the distinctive fingerprint of radioactive isotopes, particularly plutonium, produced by cold war H-bomb tests, the first of which was carried out by the US on 1 November 1952 on the Eniwetok atoll in the Pacific Marshall Islands."[14] See the details in Box 1 in the Introduction cited above. Thus, according to two votes by the AWG between 2019 and 2023, the transition between the Holocene and Anthropocene has occurred between two US nuclear weapon tests: (a) of the first Atomic bomb on 16 July 1945 in Alamogordo in New Mexico and (b) the first thermonuclear or hydrogen bomb rest on 1 November 1952 in the atoll Eniwetok in the Pacific Marshall Islands.

1.2.1.5 The Hypothesis: Dual Turning Point in Earth and Human History

Thus, in 1945 two turning points—in geological time, with the transition from the Holocene to the Anthropocene, and in political time of the *longue durée* (Braudel 1966, 1969, 1972)—coincided with the start of the new "glocal era" (Sassen 1991, 2001, 2008) and the end of World War II, which resulted in a fundamental change in the international diplomatic, peace and security order (Brauch 2016). In the twentieth century, three turning points in international order occurred after World War I (1914–1918); after World War II (1939–1945), with the adoption of the United Nations

[13] Zalasiewicz et al. (2015); Waters et al. (2018): 379–429.

[14] D. Carrington, 2023: "H-bombs or chicken bones: the race to define the start of the Anthropocene", in *The Guardian*, 6 January 2023. A detailed description of a 'nuclear weapon' and of a 'Thermonuclear weapon' are at: https://en.wikipedia.org/wiki/Thermonuclear_weapon.

Charter (1945); and after the fall of the Berlin Wall with the end of the Cold War (1989–1991) that resulted in the reunification of Germany and several expansions of the European Union and NATO. These linkages will be addressed in Chap. 2.

1.2.1.6 US Planning for an International Order of Peace and Security After the End of World War II and International Negotiations

During World War II, President Roosevelt was already discussing the new global order for peace and security with British Prime Minister Winston Churchill (in 1941 and 1943) and with Joseph Stalin, the Chairman of the *Council of Ministers of the Soviet Union* (at Yalta in February 1945). After meetings in Dumbarton Oaks (1944) and Chapultepec (January 1945) this resulted in the UN Charter (1945), which was weakened before it was signed in June 1945 by the veto power Roosevelt granted to Stalin at the Yalta Conference in February 1945.

Thus, during the Cold War, the system of the "collective security" of the UN was paralysed from the outset by this veto power of the five permanent members of its Security Council. Instead, the US's military dominance and security influence was achieved by a series of systems of "collective self-defence" (Art. 51, UN Ch), among them the North Atlantic Treaty Organization (NATO). The US's intention of adopting a leadership role and playing this leading role in the post-war global economy resulted in the creation of several international institutions as an outcome of the Bretton Woods Conference (1944): the World Bank (WB), the International Monetary Fund (IMF) and, in 1948, the General Agreement on Tariff and Trade (GATT).

This multilateral institutional framework, based on US political, military, and economic power, created a stable global order aiming for a free trade policy that fostered a process of economic and financial globalisation. As part of this process, the "Great Acceleration" has developed since 1950, when, with the acceptance of the security planning paper NSC-68 in June 1950, the US launched its first global military order, which was instrumental in tripling the US defence expenditure at the outset of the Korean War (Brauch 1976, 1977).

1.2.2 Start of the Anthropocene as an Epoch of Earth History

Crutzen initially suggested that the Anthropocene began with the Industrial Revolution, and especially with Watt's invention of the steam engine in 1782. Steffen et al. (2011) argued a decade later that "the term Anthropocene … suggests that the Earth has left its natural geological epoch … called the Holocene". Influenced by the discussions during the Dahlem workshop (Steffen et al. 2004), the results of the *Millennium Ecosystem Assessment* (MEA 2005), and the projects of the IGBP, Steffen et al. (2005) agreed that "the 20th century can be characterised by global change processes of a magnitude which have never occurred in human history" (Steffen et al. 2015).

They further agreed that human-environment interactions increased after 1950 due to the "amplified human impact on the environment" in the period they termed the "Great Acceleration" (Angus 2016: 42).

1.2.2.1 Debate on the Anthropocene (2000–2010) in the Natural Sciences

In 2009 the Anthropocene Working Group (AWG) was established as

> an interdisciplinary research group dedicated to the study of the Anthropocene as a geological time unit. It is part of the Subcommission on Quaternary Stratigraphy (SQS), a constituent body of the *International Commission on Stratigraphy* (ICS). … The main goal of the AWG is providing scientific evidence robust enough for the Anthropocene to be formally ratified by the *International Union of Geological Sciences* (IUGS) as an Epoch within the Geologic time scale.[15]

1.2.2.2 Scientific Agenda-Setting by the Anthropocene Working Group (2009–2023)

The AWG was tasked by the SQS to "examine the status, hierarchical level and definition of the Anthropocene as a potential new formal division of the Geological Time Scale", which was interpreted by some authors as a "stratigraphic turn" or "geological turn", where the AWG acknowledged the Anthropocene as an object of geological interest. The AWG's key research goal is

> establishing *when*, *where*, and *how* to locate the lower boundary of the Anthropocene. This means assigning a starting date to the Anthropocene …, locating primary as well as auxiliary markers defining Anthropocene geologic record, and determining the proper methodology to implement in the overall process of formalization (…, what proxies to use as markers, etc.). … The AWG has been considering the Anthropocene to best fit … to be taxonomically recognized as an Epoch. … In May, 2019, the AWG completed a binding vote determining two major research questions:
>
> - "Should the Anthropocene be treated as a formal chronostratigraphic unit defined by a GSSP?"
> - "Should the primary guide for the base of the Anthropocene be one of the stratigraphic signals around the mid-twentieth century of the Common Era?"
>
> Both questions received a positive response, with 29 votes in favour, 4 votes against, and no abstention (33 votes received out of 34 potential voting members).[16]

On 21 September 2021, the AWG held an Anthropogenic Markers Workshop, and an additional meeting was held on 18–19 May 2022 (both in Berlin); it planned

[15] "Anthropocene Working Group", at: https://en.wikipedia.org/wiki/Anthropocene_Working_Group (9 August 2022).

[16] "Anthropocene Working Group", at: https://en.wikipedia.org/wiki/Anthropocene_Working_Group.

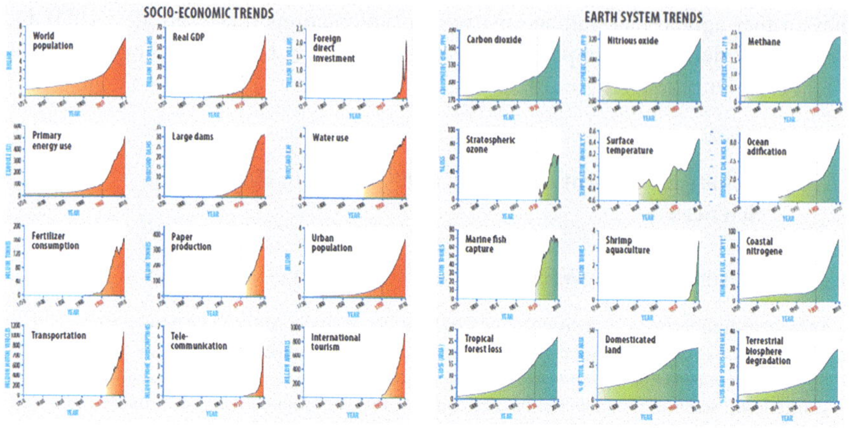

Fig. 1.1 The "Anthropocene" began around 1950. *Source Globaia* 2011; at: https://globaia.org/great-acceleration; based on Steffen et al. (2007)[18]

a formal AWG voting process to take place over October and November 2022.[17] Among the most recent publications of AWG members are two texts by Steffen (2021), who stated that (Fig. 1.1)

> the evidence for a mid-twentieth century start date for the Anthropocene is overwhelming. Potential markers for the beginning of the Anthropocene abound. They include radioactive nuclides from atomic bomb tests; hydrogen and oxygen isotopes in ice cores; nitrates in coastal ocean sediments; plastics; synthetic fibres; technofossils; extinction rates; city strata including metro systems; concrete along the Spanish coast; and even the change in size of chicken bones from 1950 to the present. The Anthropocene has also triggered a large and creative response from the arts community, becoming the theme of several films, cartoons and songs, as well as lending its name to cafés and pubs, and their products, around the world. (Steffen 2021, 2021a)

The meaning of the Anthropocene differs in these three scientific communities of AWG, GEC and ESS networks. Only the AWG and its stratigraphic bodies can officially decide whether the Anthropocene will become an official epoch in Earth history by simultaneously ending the Holocene (Fig. 1.2).

However, as the Anthropocene has been widely accepted by both the global environmental change and the Earth Systems Science communities, the Anthropocene concept is to remain a key concept in the natural sciences, irrespective of the decision by the IUGS. During the second decade after the coining of the term (2011–2020) this concept spread to the social sciences, the humanities, law, the applied sciences (e.g. medicine) and engineering, as has been shown above.

---

[17] "Anthropocene GSSP project Update", in: *Newsletter of the Anthropocene Working Group*, Volume 11: Report of activities 2021, February 2022".

[18] Source: *Globaia*, based on Steffen et al. (2007, 2015). Design: © Félix Pharand-Deschênes, *Globaïa*.

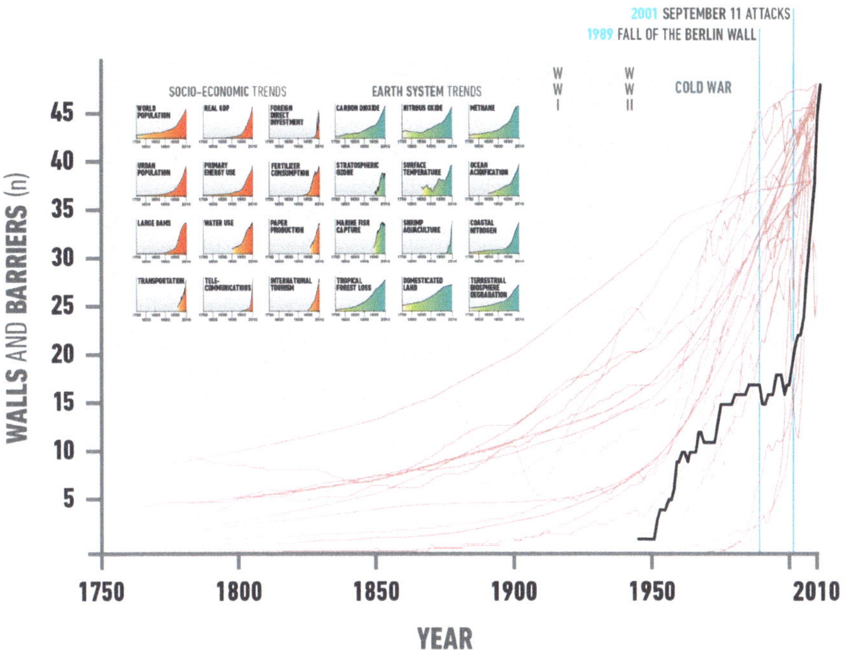

Fig. 1.2 An increasingly walled world and the broader context of the Great Acceleration. *Source Globaia* 2011

1.2.2.3 Impacts of New International Order Since 1945 on GHG Emissions

According to *Future Earth*, the following observation is apparent: "the synchronous acceleration of trends from the 1950s to the present day … with little sign of abatement".[19] The emergence of the Anthropocene reflects the impacts since 1945 of the US-dominated world order of the nuclear era, a world trade system based on "open doors", cheap fossil energies and the "American way of life", or a "Western lifestyle" based on abundance and a high consumption of resources.

1.2.2.4 Dual Existential Threat for Humanity Since 1945: Atomic Bombs, Thermonuclear Weapon, and Global Environmental Change in the Anthropocene

Since 1945 humankind has, for the first time, threatened its own survival through the fallout from atmospheric nuclear tests and a dominant economic system that has

[19] See the figures on the twenty-four IGBP categories of the "Great Acceleration" on the website of Future Earth; at: https://futureearth.org/2015/01/16/the-great-acceleration/ (1 December 2019); Steffen et al. (2004, 2015).

resulted in a massive increase in the burning of cheap fossil energy sources (coal, oil and gas). However, there is a major difference between the two existential threats (which were connected by Scheffran 2011; Scheffran et al. 2016). While the first threat is influenced by rival superpowers or geopolitical camps, the second is the result of our own individual and joint economic behaviour through the individual and collective burning of fossil energy resources.

The nuclear era and the Anthropocene gradually emerged between 1940 and 1950 and are two sides of the same coin (Scheffran 2019). However, they are addressed by different disciplines and research programmes: the first has been primarily tackled in History and in the social sciences via security and peace research, while the second has been primarily developed in the natural sciences via environmental research studies focusing on global environmental change. A new peace ecology approach in the social sciences could combine both research programmes by addressing the manifold interlinkages (Brauch 2020).

1.2.3 Rise in CO_2 Emissions and Interactions with International Order and Security Policy During the Cold War

Carbon dioxide (CO_2) as the key indicator increased from about 279 ppm in 1750 to 310 ppm in 1950, and soared to 354 ppm by July 1990, increasing by 44 ppm in just forty years. Over the next thirty years, CO_2 rose to 416 ppm in April 2020—i.e. by another 62 ppm (Figs. 1.3 and 1.4). In December 2021 it reached an average of 416.71 ppm, by December 2022 an average of 418.95 ppm and in June 2023 an average of 419.51 ppm (see at: https://gml.noaa.gov/ccgg/trends/global.html#global (5 January 2023).

There were also close interactions between the contextual changes in international order with regard to international peace and security and international economic relations. The fundamental structural change in the USA during World War II, its rise to the status of world power, and its acceptance of a world leadership role (Brauch 1977) acted as multiple political triggers that made the launch of the "Great Acceleration" between 1945 and 1950 possible.

1.2.3.1 Humanity Has Become an Existential Threat for the First Time

In May 2019 the Anthropocene Working Group proposed that the Anthropocene started in the mid-twentieth century, and 22 of its members suggested that the first test of a nuclear weapon on 16 July 1945 was the starting point of the new epoch. This implies that the transition from the Holocene to the Anthropocene coincides with the start of the nuclear era and the new post-World War II security and economic order. This refers to a fundamental change: that humankind (humanity, *Menschheit*)

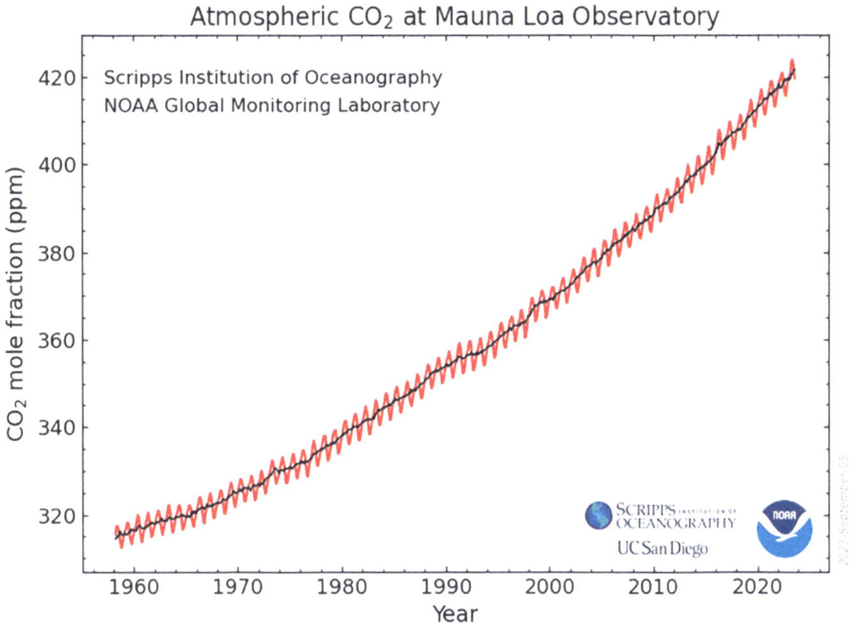

Fig. 1.3 Atmospheric CO_2 at Mauna Loa Observatory (1958–2023). *Source* At: https://gml.noaa.gov/ccgg/trends/. This figure is in the public domain

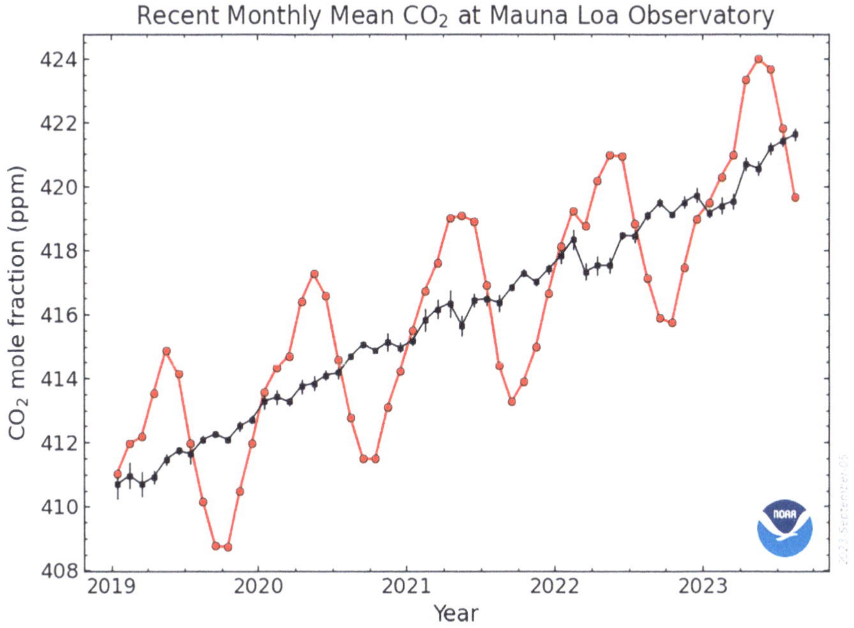

Fig. 1.4 Atmospheric CO_2 at Mauna Loa Observatory (2019–2023). *Source* See at: https://gml.noaa.gov/ccgg/trends/ (6 September 2023). This figure is in the public domain

has, for the first time, become an anthropogenic threat to our survival as a species. But the threat and impact of the first atomic weapon test has differed[20] from that caused by global environmental change:

- With a massive use of thermonuclear weapons (atomic and hydrogen warheads: A or H-bombs) in an all-out nuclear war, humankind would not only become a victim of a nuclear winter (Crutzen/Birks 1982), but its very survival as a species might be at risk. A nuclear war may be the result of policy decisions of the top political and military elite in the two major nuclear powers (USA or Russia) or the result of technical errors and human misperceptions. This debate has occurred in the framework of nuclear strategy in international security or strategic studies where the "other", the rival bloc or system, poses the threat to our survival.
- In the Anthropocene, humankind has become a geological actor (Crutzen 2002; Antweiler 2022) that has directly interfered in the Earth system by causing ozone layer depletion and anthropogenic climate change, with an exponential increase in six greenhouse gases (especially CO_2) in the atmosphere. In the Anthropocene, we as human beings and as humankind have become the threat with our burning of fossil fuels, and through our lifestyles, production and consumption, which have caused the exponential increase in GHG since the Industrial Revolution and triggered four physical effects: temperature increase, sea-level rise, precipitation variability and extreme weather events.

1.2.3.2 Both Anthropogenic Threats to Human Survival Fundamentally Differ and Are Analysed in Two Unrelated Scientific Worlds

Both threats to our survival have been addressed by different disciplines and research programmes. Nuclear weapons have been addressed since 1945 by International Relations and security studies and strategic research programmes but also—from a critical, idealist perspective—by peace researchers.

Since 2000 the Anthropocene has been discussed initially and primarily by the global environmental change, Earth systems science, and geological communities, but by 2021 the concept had spread across disciplines and cultures and was used in the natural and social sciences, in the humanities and arts, and in many other professional contexts (Brauch 2021).

However, so far, no integrated analysis has been proposed in the framework of a "peace [and] ecology" approach (Brauch 2020). In Chap. 2 the different developments in the dominant peace and security realm and in the still-not-recognised developments in global environmental change will be presented from a dual peace and ecology approach.

The institutional framework in the US and the bipartisan foreign policy consensus in the US Congress was the backbone of the US leadership position from 1941 until 2017, when President Donald Trump started to gradually tear it down and tried

[20] Besides on the nuclear winter, there is little research on connecting the nuclear and climate threat, mostly by Scheffran (2011, 2019); Scheffran et al. (2016).

to replace the dominant multilateral trade system and liberal global order with his self-centred motto of "making America great again".[21]

1.2.4 The Great Acceleration (1950)

In the United States of America (USA) a new domestic structure for the creation of a national security state was institutionalised via the National Security Act of 1947, which offered a new rationale for developing and declaring US national interests. The US's National Security Council (NSC) justified the Act on the grounds that its purpose was to achieve "national security" goals. The contest between rival ideologies, systems of rule and visions of world order resulted in a Cold War that combined militarisation in a bipolar world with the establishment of an open door policy based on free trade that superseded the British imperial preference system. The system of Bretton Woods (World Bank, IMF), NATO and the first global security strategy devised by Paul Nitze via NSC-68 in January 1950 was gradually implemented after the start of the Korean War on 25 June 1950.

1.2.4.1 The Year 1950 as Another Dual Turning Point

Thus, in 1950, NSC-68 became the turning point towards an internationalist, globalist and partly also interventionist US foreign, foreign economic (open door) and military (national security) policy (Brauch 1976, 1977). This multiple turning point in US global policy coincided with what was later called the "Great Acceleration", marking the silent transition in Earth's history from the Holocene to the Anthropocene. Between 1945 and 1950, several major turning points and transitions occurred in the realms of international politics and national policies that fundamentally changed the context for the study of international peace and security in the social sciences.

For the first time in Earth's history, the manifold human economic and military activities directly interfered with the Earth System, which, according to Crutzen (2000), triggered a transition in Earth's history from the Holocene to the Anthropocene Epoch. Thus, the emergence of the U.S. national security state had created the preconditions for the Great Acceleration. Several natural and social scientists have stated that this silent transition to the Anthropocene was triggered by a Great Acceleration in twelve exponential Earth systems and twelve socio-economic trends that began around 1950 and further intensified after the global "turn" of 1990 (when the Cold War ended) with an "Exponential Acceleration". Neoliberal globalisation trends have benefited from the rapid economic and scientific innovation and expansion of the People's Republic of China since Deng Xiao Ping's economic reforms (1979–1992).

---

[21] See at: https://en.wikipedia.org/wiki/Enlargement_of_NATO (29 July 2022).

1.2.4.2 Global Environmental Regimes: Governance without Government

In International Relations, with the rapid political and economic decolonisation since the 1950s and during the 1960s, development policy was set up as a new field of global and national policymaking, partly managed by the World Bank, the IMF and the United Nations Development Programme (UNDP). In the 1960s and 1970s, national and international environment policies were added, and in the late 1980s and the 1990s several new regimes of "governance without government" (Rosenau/ Czempiel 1992) were set up.

Despite these global contextual changes since the Great Acceleration of the 1950s and the intensified or exponential acceleration since 1990, only a few industrialised countries have implemented the quantitative reductions in GHGs and CO_2 emissions in accordance with the framework of the Kyoto Protocol (1997–2012). While the societal impact of the physical consequences of anthropogenic climate change (temperature rise, change in precipitation, sea-level rise, and increase in the number, intensity and cost of natural hazards—storms, floods, forest fires, drought, soil erosion and desertification) is apparent, the sense of urgency among top policy makers concerning the need to address them has been inadequate.

A window of opportunity opened with the co-existence of global, regional and national challenges in the realms of peace, international security, and the international political economy. As a result of negotiations, global environmental governance emerged towards the end of the Cold War in the autumn of 1988. However, momentum stalled somewhat in 1998, when the US Senate refused to ratify the Kyoto Protocol (1997) in case it damaged the US economy.

The conceptual innovation in the discussions on human and environmental security since the 1990s and the policy debate on the possible impacts of climate change on security since 2004 reflected efforts to highlight the conceptual link between the two policy areas. Unfortunately, most peace researchers, peace research foundations and many peace research institutes on the one hand, and environmental specialists on the other, did not address these contextual factors that directly affected their research and analyses. Instead, disciplinary specialisation has prevailed over transformative and multi-, inter-, or transdisciplinary approaches, while broader holistic perspectives which address the links and interactions between different policy fields have declined. During the brief period of conceptual innovation there was hardly any discussion on the need for a "peace ecology" approach that could conceptually and politically link both issue areas.

1.3 Three Phases of International Order in the Anthropocene

Steffen et al. (2007: 616–618) initially distinguished between two stages of the Anthropocene: (a) the industrial era (1800–1945), and (b) the Great Acceleration (1946–2015) (documented in Fig. 1.1):

> From 1950 to 2000 the percentage of the world's population living in urban areas grew from 30% to 50% and continues to grow strongly. ... The pressure on the global environment from this burgeoning human enterprise is intensifying sharply. Over the past 50 years, humans have changed the world's ecosystems more rapidly and extensively than in any other comparable period in human history ... The Great Acceleration took place in an intellectual, cultural, political, and legal context in which the growing impacts upon the Earth System counted for very little in the calculations and decisions made in the world's ministries, boardrooms, laboratories ... The exponential character of the Great Acceleration is obvious from our quantification of the human imprint on the Earth System, using atmospheric CO_2 concentration as the indicator ... Although by the Second World War the CO_2 concentration had clearly risen above the upper limit of the Holocene, its growth rate hit a take-off point around 1950. Nearly three-quarters of the anthropogenically driven rise in CO_2 concentration has occurred since 1950 (from about 310 to 380 ppm), and about half of the total rise (48 ppm) has occurred in just the last 30 years. (Steffen et al. 2007: 616–618)

Steffen et al. (2007: 619–620) identified three philosophical responses—(a) business as usual, (b) mitigation, and (c) geoengineering—any of which "can raise serious ethical questions and intense debate", given "the possibility for unintended and unanticipated side-effects that could have severe consequences. The cure could be worse than the disease." They concluded that "The Great Acceleration is reaching criticality", but pointed to "evidence for radically different directions built around innovative, knowledge-based solutions".

McNeill and Engelke (2014: blurb) claimed that in the Anthropocene

> humans are the most powerful influence on global ecology. Since the mid-twentieth century, the accelerating pace of energy use, greenhouse gas emissions, and population growth has thrust the planet into a massive uncontrolled experiment. *The Great Acceleration* explains its causes and consequences, highlighting the role of energy systems, as well as trends in climate change, urbanization, and environmentalism. More than any other factor, human dependence on fossil fuels inaugurated the Anthropocene. Before 1700, people used little in the way of fossil fuels, but over the next two hundred years coal became the most important energy source. When oil entered the picture, coal and oil soon accounted for seventy-five percent of human energy use. This allowed far more economic activity and produced a higher standard of living than people had ever known—but it created far more ecological disruption. We are now living in the Anthropocene. The period from 1945 to the present represents the most anomalous period in the history of humanity's relationship with the biosphere. Three-quarters of the carbon dioxide humans have contributed to the atmosphere has accumulated since World War II ended, and the number of people on Earth has nearly tripled. So far, humans have dramatically altered the planet's biogeochemical systems without consciously managing them. If we try to control these systems through geoengineering, we will inaugurate another stage of the Anthropocene. Where it might lead, no one can say for sure. (Engelke/McNeill 2016)

Engelke and McNeill (2016) stated in a text for the Washington Post that:

> since about 1945, human interference in the Earth's ecology has reached a new intensity, the result of more humans, more energy use and more transformative technologies, among other things. According to this new way of thinking, the *Great Acceleration* (of human impact on the global environment) helped launch the Anthropocene (a new interval in Earth's history). Both began around the middle of the 20th century. … The Great Acceleration's ecological results are all around us: biodiversity loss, soil erosion, ocean acidification, overfishing, aquifer depletion, rapid carbon loading of the atmosphere and oceans, deforestation, altered global geo-chemistry (the carbon, nitrogen, sulphur and phosphorous cycles, among others). …

The International Geosphere-Biosphere Programme (IGBP)[22] analysed data for 1750–2010 by distinguishing between twelve socio-economic trends (population, real GDP, foreign direct investment, urban population, primary energy use, fertiliser consumption, large dams, water use, paper production, transportation, telecommunications, and international tourism) and twelve Earth systems trends (carbon dioxide, nitrous oxide, methane, stratospheric ozone, surface temperature, ocean acidification, marine fish capture, shrimp aquaculture, nitrogen to coastal zone, tropical forest loss, domesticated land and terrestrial biosphere degradation) (Fig. 1.1).

1.3.1 The Anthropocene as a New Epoch of Earth History in the Natural Sciences

A previous text (Brauch 2021) addressed the research questions of how the scientific literature on this concept developed within natural sciences over two decades, and how this concept spread from a debate in the natural sciences during the first decade (2000–2010) to the social sciences, the humanities, law, the applied sciences (including medicine) and engineering during the second decade (2011–2020). It specifically reviewed the role that two multidisciplinary scientific networks, the International Geosphere-Biosphere Programme (IGBP) and the Anthropocene Working Group (AWG), played in spreading the Anthropocene concept globally and across disciplinary boundaries, and it discussed the role of key scientific authors who were instrumental in initiating the scientific debate and whose work was highly cited in the peer-reviewed scientific literature. In the natural sciences, three scientific groups or networks were instrumental in developing, using and promoting the Anthropocene concept (see above).

The meaning of the Anthropocene differs within these three scientific communities and networks in the natural sciences. Only the geological community and its stratigraphic bodies can officially decide whether the Anthropocene will become an official epoch in Earth history by simultaneously ending the Holocene. However, as the Anthropocene has been widely accepted by both the global environmental change and the Earth systems science communities, the Anthropocene concept is to remain a

[22] See at: http://www.igbp.net/ (1 December 2019).

key concept in the natural sciences, irrespective of the eventual decision by the International Union of Geological Sciences. Authors working on global environmental change discussed the Anthropocene, arguing that

> *Homo sapiens* has had such a significant impact on Earth and its inhabitants that we will have a lasting – and potentially irreversible – influence on its systems, environment, processes and biodiversity. … These human impacts have unfolded at an unprecedented rate and scale. This period is sometimes known as the Great Acceleration. Carbon dioxide emissions, global warming, ocean acidification, habitat destruction, extinction and widescale natural resource extraction are all signs that we have significantly modified our planet. …Scientists now agree that human activity, rather than any natural progress, is the primary cause of the accelerated global warming. Agriculture, urbanisation, deforestation and pollution have caused extraordinary changes on Earth. … Our awareness of both the current state of the planet and the effects of our actions is a key factor in the Anthropocene.[23]

From Earth systems science, Steffen et al. (2018) analysed "Trajectories of the Earth System in the Anthropocene" by exploring

> the risk that self-reinforcing feedbacks could push the Earth System toward a planetary threshold that, if crossed, could prevent stabilisation of the climate at intermediate temperature rises and cause continued warming on a "Hothouse Earth" pathway even as human emissions are reduced. … If the threshold is crossed, the resulting trajectory would likely cause serious disruptions to ecosystems, society, and economies. Collective human action is required to steer the Earth System away from a potential threshold and stabilise it in a habitable interglacial-like state. Such action entails stewardship of the entire Earth System—biosphere, climate, and societies—and could include decarbonisation of the global economy, enhancement of biosphere carbon sinks, behavioural changes, technological innovations, new governance arrangements, and transformed social values. (Steffen et al. 2018)

The Anthropocene Working Group (AWG) offered this definition of the Anthropocene on its website that was cited in the introduction to this book (see above). Analyses of potential "golden spike" locations are now completed (see Introduction above) and require supermajority (> 60%) agreement by the AWG and its parent bodies, where a success of any such proposal is not guaranteed.[24]

While most natural scientists agree that the Anthropocene is a new epoch of Earth history, the scientific definitions, contexts, and frameworks differ. In the social sciences and in the humanities, no institutional body can offer authoritative definitions of its key concepts. In Political Science and International Relations, no integrated analysis exists of the two related dangers posed by anthropogenic human activities: the threat of a nuclear war and of the dangerous impacts of human interventions into the Earth system in the Anthropocene.

---

[23] See at: https://www.nhm.ac.uk/discover/what-is-the-anthropocene.html.

[24] This long quote was taken from the AWG website at: http://quaternary.stratigraphy.org/working-groups/anthropocene/.

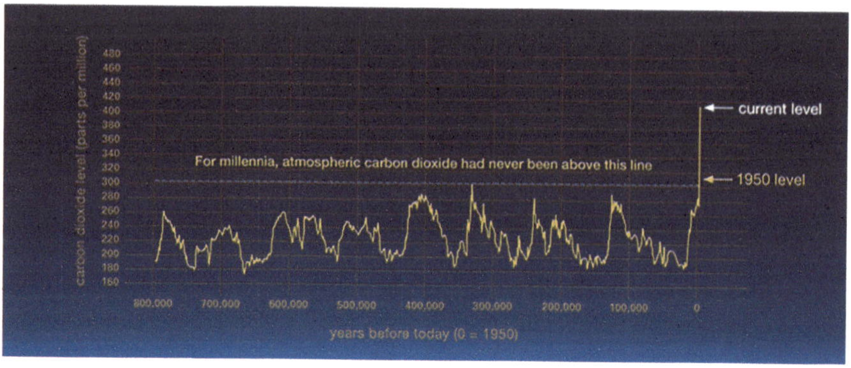

Fig. 1.5 Development of atmospheric CO_2 for the past 800,000 years. *Source* https://climate.nasa.gov/climate_resources/24/graphic-the-relentless-rise-of-carbon-dioxide/

1.3.2 Carbon Dioxide Emissions into the Atmosphere

During the Holocene epoch of Earth history a complex interaction between nature, the Earth system, humankind and individual human beings has occurred. The end of the last glacial period resulted in the transition from the Pleistocene to the Holocene. In human prehistory, humankind made a series of major transitions from nomadic hunters and collectors to settled farmers during the agricultural revolution, and these transitions were accompanied by increasing human interventions in nature which contributed to the emergence of villages, towns, kingdoms and states that interacted both peacefully and in violent conflicts.[25]

Crutzen (2002) claimed in "The geology of mankind" that the intervention of humankind in the Earth system differs from the multiple interferences in nature during the Holocene. Through the impact of GHG in the atmosphere humankind has for the first time become a geological (f)actor whose economic activities have resulted in an exponential increase in atmospheric CO_2 since 1950, and may also trigger tipping points in the climate system. During the past 800,000 years and during the Holocene epoch (until 1950), atmospheric carbon dioxide (CO_2) was below 300 ppm (Figs. 1.4, 1.5 and 1.6).

Figure 1.7 documents the increase of global carbon dioxide emissions between 1850 and 2020 and gives projections until 2040, while Fig. 1.8 illustrates the changes in the six major economies between 2000 and 2019 and provides projections until 2040 based on the background of global negotiations on climate change since 1992 (UNFCCC), 1997 (Kyoto Protocol) and 2015 (Paris Climate Change Agreement),

[25] Zalasiewicz (2008); Wolfe, Birks, Briner, Holmgren, Ingólfsson, Kaushal, Miller, Pagani, Saros, Vinebrooke (2013): "Stratigraphic expressions of the Holocene–Anthropocene transition revealed in sediments from remote lakes", in: *Earth-Science Reviews*, 116 (January): 17–34; Smith, Zeder, 2013: "The onset of the Anthropocene", in: *Anthropocene*, 4 (December): 8–13.

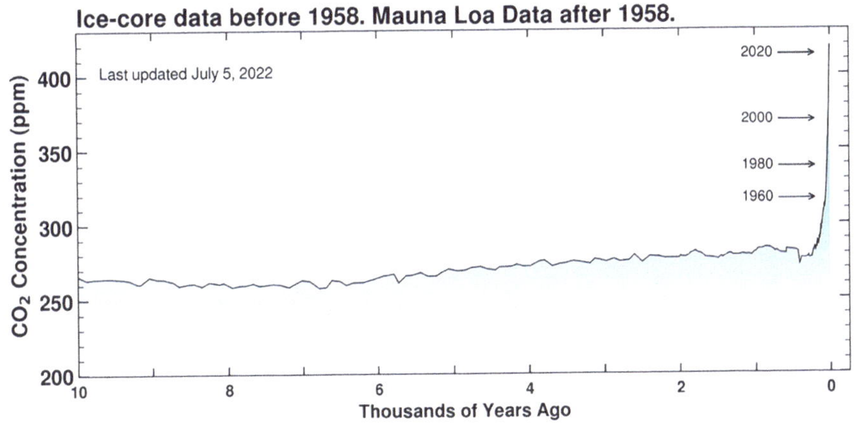

Fig. 1.6 Ice-core data before 1958 (during most of the Holocene, over the past 10,000 years). Mauna Loa data since 1958. *Source* https://keelingcurve.ucsd.edu/

when international climate policy started and most states signed and ratified these three climate agreements. The so-called Kurling curve (Fig. 1.9) has measured CO_2 emissions since 1958 to the present (reaching 418.90 ppm in July 2022, and 419.68 in August 2023).

From 1900 until 2017 extreme weather events (storms, floods, droughts) have significantly increased (Fig. 1.10)

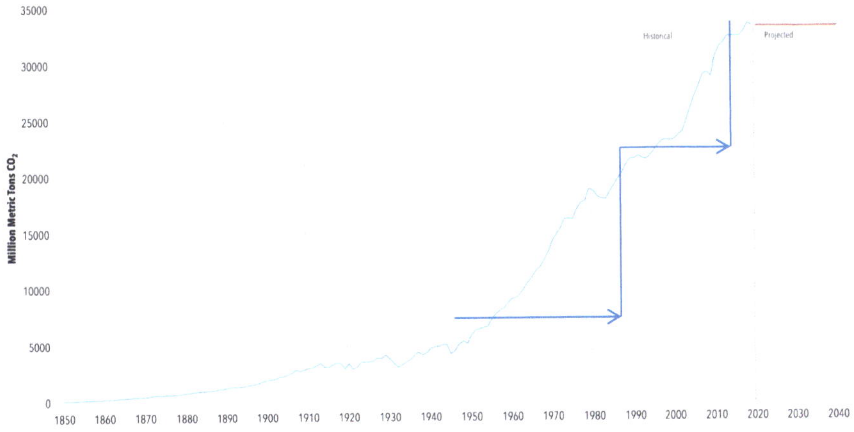

Fig. 1.7 Global carbon dioxide emissions, 1850–2040. *Source* Center for Climate and Energy Solutions: "Climate Basics: Energy/Emissions Data Global Emissions: Global Carbon Dioxide Emissions, 1850–2040"; at: https://www.c2es.org/content/international-emissions/; based on: IEA (2020); Oak Ridge National Laboratory (2017)

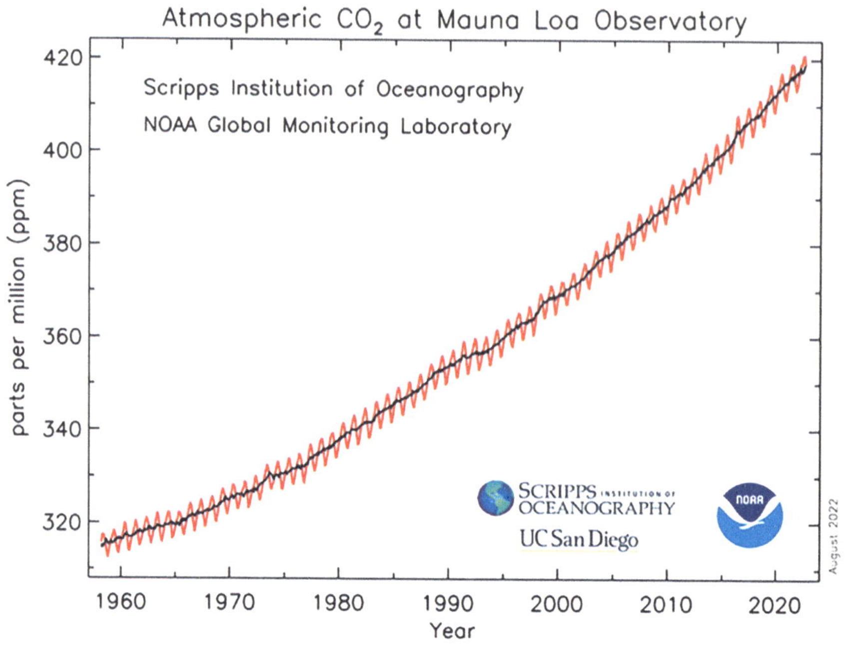

Fig. 1.8 Atmospheric CO_2 at Mauna Loa Observatory (1958–August 2023). *Source* https://gml.noaa.gov/ccgg/trends/. Dr Pieter Tans, NOAA/ESRL, at: www.esrl.noaa.gov/gmd/ccgg/trends/ and Dr Ralph Keeling, Scripps Institution of Oceanography, at: scrippsco2.ucsd.edu/. This figure is in the public domain

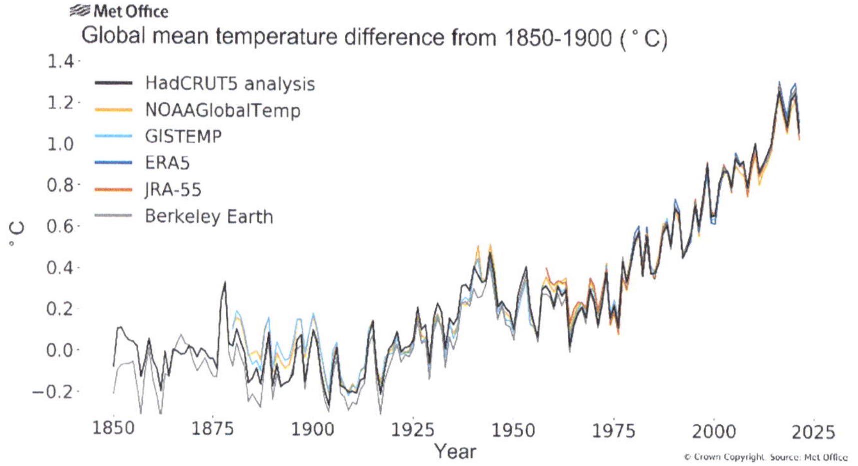

Fig. 1.9 Met Office: global mean temperature difference from 1850 to 1900–present. *Source* Met Office. Crown copyright

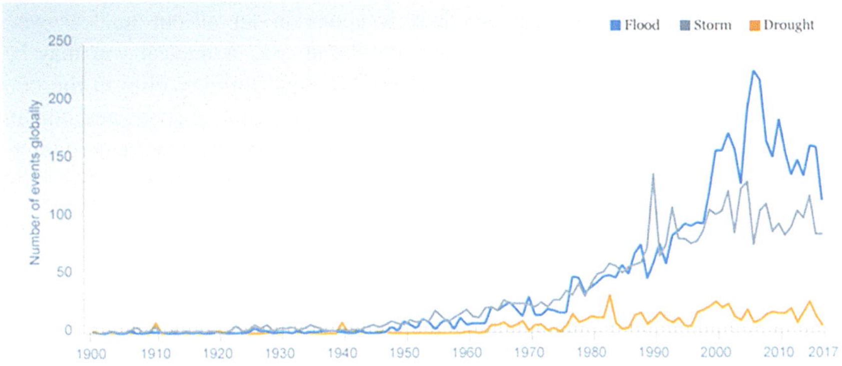

Fig. 1.10 Extreme weather on the rise: floods, storms and droughts (1900–2017). *Source* Deutsche Welle based on The Emergency Events Database (EM-DAT)

From 1850 to the present, global mean temperature has increased by more than 1.2 °C (Fig. 1.8). Based on data collected in the Emergency Events Database (EM-DAT) for 1900–2017, extreme weather (floods, storms and droughts) was rising (Fig. 1.10). A combined analysis and projection of global carbon dioxide emissions (2000–2040) by the Center for Climate and Energy Solutions (in Arlington, VA) illustrates the rapid increase in CO_2 emissions for China, that overtook the USA in 2004, and in India, that overtook Brazil in 2021 and may overtake the USA by 2050 (Fig. 1.11).

When the Anthropocene started in the mid-twentieth century, humankind became, for the first time, a dual anthropogenic threat to our survival. But the threat and impact of the first atomic weapons test and of global climate change differ fundamentally:

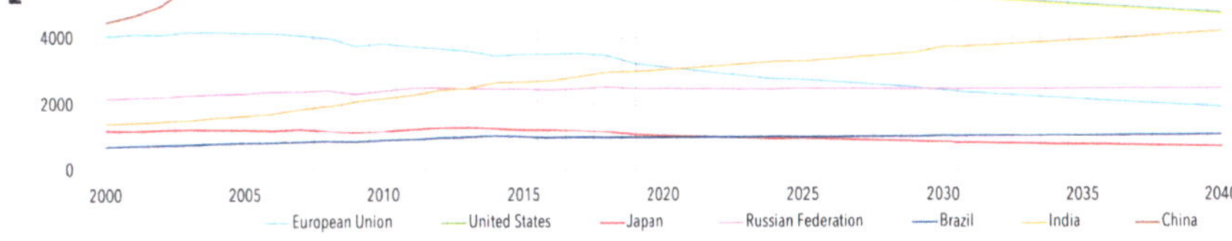

Fig. 1.11 Global carbon dioxide emissions, 2000–2040. *Source* Center for Climate and Energy Solutions: "Climate Basics: Energy/Emissions Data Global Emissions: Greenhouse Gas Emissions for Major Economies, 1990–2030", at: https://www.c2es.org/content/international-emissions/; based on: IEA (2020, 2021), U.S. EPA (2019)

- With a massive use of thermonuclear weapons in an all-out nuclear war, humankind's very survival as a species may be at risk. A nuclear war may be the result of policy decisions of the top political and military elite in the two major nuclear powers (USA, Russia) or the result of technical errors and human misperceptions. This debate occurs in the framework of nuclear strategy in international security or strategic studies, where the "other", the rival bloc or system, poses the threat to our survival.
- In the Anthropocene, humankind has become a geological actor (Crutzen 2002) that has directly interfered in the Earth system. In the Anthropocene, we have become the threat, as human beings and as humankind, with the burning of fossil fuels through our lifestyles, production and consumption, which have caused the rapid increase in GHG.

Nuclear weapons have been addressed since 1945 by International Relations and security studies and strategic research programmes, but also by "critical" peace researchers. The Anthropocene has been discussed since 2000, initially primarily by natural scientists, but by 2021 the concept had spread across disciplines and cultures and was used in the natural and social sciences, in the humanities and arts and in many other professional contexts (Brauch 2021). So far, no integrated analysis from peace and security as well as ecological perspectives exists that has been proposed in the framework of a "peace [and] ecology" approach (Brauch 2021).

1.4 The Anthropocene in the Social Sciences

Crutzen's (2000, 2002) 'Anthropocene concept' and his thesis of the emergence of a 'geology of mankind' has become a fundamental challenge for the social sciences in general, and political science in particular, its major sections (political theory, comparative government, international relations and policy studies). Influenced by the emergence of the debate in Earth Systems Science, for the first time human beings and humankind have directly impacted on the processes of earth systems through their consumption of fossil energy sources (Arrhennius 1896) and their lifestyles and economic behaviour (Steffen 2020).

Since the French Revolution (1789) until end of World War II (1945) political thinking and theory was influenced by three major schools of political thought: conservatism, liberalism and socialism, including anarchism and communism (v. Beyme 2002, 2013, 2013a, b, 2019). During the Cold War the world was divided among two political camps: the so-called 'free world' where forces of a moderate liberalism, conservatism and social democracy were dominating the mind-set of policy makers in the West, while Stalinist communism prevailed in the Soviet Union and in the countries of the rival 'socialist' political camp.

The 'ecology' concept was coined in 1866 by the German biologist E. Häckel and was initially defined by natural scientists with it has gradually become since the 1960s and 1970s a term for a fourth major political school of thought representing green and ecological thinking, movements and parties with a special focus on issues of environment policies and global environmental change (climate change, biodiversity, renewable energy sources etc.).[26]

In his last 'memorandum' on the emergence of an ecological class Bruno Latour in cooperation with Schultz (2022: 9–13) reflected on how an 'ecological class' could become a powerful political actor for transforming the conditions of life that are being destroyed by the insistence on production' to outline a common horizon for collective action towards a socio-ecological transformation in implementing global ecological concerns from the bottom-up.

On the other end of the political spectrum right wing populism and extremism (von Beyme 2019) has denied or downgraded the issues of global environmental change, opposes the rule of law and the freedom of the media and calls for a revisionist worldview that opposes and challenges modern lifestyles and family preferences. Rightwing populism is an obstacle for the development and implementation of socio-ecological transformations and reforms.[27]

The 'Anthropocene' has become in the social sciences an indication of a fundamental contextual change that has generated a policy narrative that differs fundamentally from the Hobbesian geopolitics where changes are occasionally implemented by revisionist powers in violation of international law with the unilateral use of military power to revise the territorial and political status quo that was previously mutually accepted by political principles and contractual obligations (treaties).

The Anthropocene narrative or discourse stands for ecological political concerns, actors and action programmes to transform the production system that evolved in the framework of liberal, conservative and social-democratic and socialist political thought and that resulted since the 1950s in the Western way of life with an improved standard of living and higher income aiming at societal reforms to redistribute social wealth based on criteria of social justice. The Anthropocene discourse

---

[26] A school of ecological political thought, of ecological and green political thinking, has developed since the 1960s: Dobson (2007); Biro (2006); Tokar (1988); Blühdorn and Wels (2007); Humphrey (2007); Dobson and Eckersley (2006); Morton (2010); Code (2006).

[27] "Right-wing populism", at: https://en.wikipedia.org/wiki/Right-wing_populism; "Right-Wing Populism"; at: https://www.populismstudies.org/Vocabulary/right-wing-populism/; "Understanding right-wing populism and what to do about it"; at: https://blogs.lse.ac.uk/europpblog/2022/06/01/understanding-right-wing-populism-and-what-to-do-about-it/; "Right-wing populism and climate policy in Europe", at: https://www.adelphi.de/en/in-focus/right-wing-populism-and-climate-policy-europe (11 January 2022); Lockwood and Lockwood (2022).

aims at an ecological agenda that reduces the intervention of hidden societal activities into the earth systems that may trigger chaotic interactions among the forces of the earth system through the increase in the number and intensity of the physical consequences of global environmental change and their potential conflictual societal outcomes. The Anthropocene stands for a determined societal and political effort for a long-term structural contextual change. Dalby (2016) has distinguished among three alternative perspectives of the Anthropocene representing a good, a bad and an ugly Anthropocene.

1.4.1 Reviews of the Anthropocene in Political Science

What does the Anthropocene discourse imply for political science as a scientific discipline and for political action aiming at socio-economic transformation and climate neutrality within the next two to four decades to stabilise the increase of GHG emissions with an increase of global average temperature with only 1.5–2 °C above the average temperature level since the industrial revolution in the eighteenth and nineteenth centuries. Since 2011, the Anthropocene has gradually become an object of analysis and a theme of reflection in political science in political theory, in environmental policy studies, in international relations and in specialised policy programmes of peace, security, development studies and ecology (Brauch 2021).

1.4.1.1 Definitions of *Politik, la Politique, la Politica*

Political Science, political theory, comparative government and International Relations (IR) have gradually developed as academic disciplines and sub-disciplines since the late nineteenth and early twentieth centuries in the United States and in selected European countries and in countries of the Global South. Political Science was established after World War II in several Western European countries and in some post-communist countries with the end of the Cold War in the 1990s (Smith 2003).

Many reviews on the history of the discipline were published by American scholars (Almond 1996: 3–49; Goodin/Klingemann 1996: 50ff.) based on publications by American and Anglo-Saxon authors. In political theory, or in the history of political ideas, many reviews started with Greek (Aristotle, Plato) and Roman philosophers and their impact on Machiavelli, Hobbes, Locke, Montesquieu et al., and the nineteenth century political theory that was influenced by Hegel, Comte, Marx, Saint-Simon, Spencer, Tönnies, Pareto, Durkheim, Weber et al.

In his review of the professionalisation of political science in the twentieth century, Almond (1996: 65–78) referred to the Chicago School (1920–1940) focusing on empirical studies (Merriam, Gosnell, Lasswell) up to World War II; the post-behavioural revolution (including the emergence of security studies, international political economy, public opinion, political culture studies); and political refugees from Europe who helped to spread political science in the USA and in Europe. Brunner (1996: 622) claimed—based on Lerner and Lasswell (1951)—that "the policy sciences are concerned with understanding the decision processes of public and private institutions and with assessing the significance of all knowledge for purposes of decision".

More comprehensive and wider concepts of the political prevail in German (*Politik*),[28] French (*la Politique*), Italian and Spanish (*la Politica*) while in English at least three different concepts have been used: "politics", which refers to the political process or the processual aspects; "policy", which applies to fields or issue areas of political activities, the administration and content of political decisions; and "polity", which refers to the legal frameworks (laws) and institutions, its form and political order (Schmidt 2004: 538).

The *Brockhaus Encyclopedia* (21st ed., vol. 21, 2006: 655) states that the German concept of *Politik* comprises "the entirety of procedures and actions by individuals and collectives aimed at the binding regulation of public concerns". On this basis *Politik* is defined as "the institutional, procedural and decision-making dimensions of the quest for power or for influencing the balance of power that affect the totality of public concerns".

With his proposal of the Anthropocene concept, Crutzen (2002) changed the context for human-Earth systems since 1945, but also for Political Science and politics, policy and polity. For the first time in Earth and human history, humankind has been considered as a geological factor or geological actor. When the post-World War II order was planned and gradually implemented, the focus was primarily on the political and military dimension (the peace and security system in the framework of the UN, and later on systems of collective self-defence in the framework of military alliances, such as NATO).

1.4.1.2 Political Science Perspectives on "Politics" in the Anthropocene

What impact does the focus on the Anthropocene concept from a political science perspective have on the thinking on politics, policy and polity? From the vantage point of modern political theory, Kelly (2019: 1) of Cambridge University discusses *Politics and the Anthropocene* as "a concept for a world 'after nature', to the extent that it presupposes no distinction between human action and the natural world". As there is not yet a scientific consensus "whether or not a new geological epoch exists… [,] the Anthropocene remains a conceptual work in progress."

Kelly (2019: 2) points to two challenges of the Anthropocene for modern politics that forces "politics to be more overtly ecological" by addressing "the continued 'greening' of business, politics and policy" and a "more radically 'Anthropocened'

[28] For historical reviews of the German concepts of 'Politik' see: Sellin (1978, vol. 4: 789–854); Meier (1989, vol. 7: 1038–1047); Weinacht (1989, vol. 7: 1047–1055); Vollrath (1989, vol. 7: 1055–1071).

sense of politics [that] challenges the way in which we frame the terms of our engagement", which implies that "we need to rewrite its terminology, its concepts, and thus its history" and that we must "critique our conventional thinking in relation to political values, economic limits, population growth and the nature of unevenly distributed ecological indebtedness". For Kelly (2019: 3), the Anthropocene poses a triple challenge, "a radically humanistic one, that is to say, as it is scientific and geological one". For Kelly, the Anthropocene "opens up a new space for thinking about the connections between long-term geological and planetary time-scapes and human or earthly time-frames". In the subsequent chapters Kelly reviews the timings (2019: 8–18), the ecological inequalities (29–42), the discussion on limiting growth (43–65), economic debts (66–81), population futures (82–98) and values (99–111), while in his epilogue he addresses "historical possibilities for an Anthropocene Politics" (112–122).

In their co-authored book *The Politics of the Anthropocene*, Dryzek and Pickering (2019: v–vi) see the challenge the Anthropocene "presents to humanity '…' that in future all politics should be first and foremost a politics of the Anthropocene" where the purpose of their book is to make "all politics a politics of the Anthropocene" in a modern world "in which systems once categorized as non-human can be joined in a more productive relationship". They start with a discussion on the "good, the bad, and the inescapable of the Anthropocene" (1–19), contrast the governance in the Holocene (20–33) with that in the Anthropocene (34–57), review the concept in the framework of planetary justice (58–81) and sustainability (82–102) by discussing "who will form the Anthropocene?" (103–127) and provide questions related to a "democratic Anthropocene" (128–150). In their conclusions they address "a practical politics of the Anthropocene" (151–162):

> The core problem is pathological path dependency in institutions, practices, and ideas that developed under Holocene conditions. In ways they did not recognize the possibility that the Earth system could be anything but benign. … The solution is easily summarized, but much less easily achieved: to establish ecological reflexivity as a core priority of social, political and economic institutions. (Dryzek/Pickering 2019: 151)

Dryzek and Pickering (2019: 155–161) stress "that an effective response to the Anthropocene requires institutions to cultivate unprecedented capacities for ecological reflexivity, [and] development of those capacities has to start where we are now." They consider "social movements and individual action [as] vital for transition to more reflexive societies" and recommend "pathways towards transforming states and markets" comprising:

- Dismantling barriers to reflexivity;
- Avoiding renewed lock-in;
- Introducing elements of reflexivity through public inquiries, independent review bodies, procedural environmental rights, mandated attention to ecological concerns, requirements for periodic review of legislation and institutions in light of performance, sunset clauses for legislation and policies, custodians for non-humans and future generations, linking citizens with Earth systems expertise,

learning from experimentation, a global dissent channel, electoral system reform, reflexivity unity and educating for reflexivity.

Hickmann et al. (2019: 1–12) outline "a political science perspective on the Anthropocene" in which they address two fundamental questions:

- What is the contribution of political science to the Anthropocene debate, e.g. in terms of identified problems, answers and solutions?
- What are the conceptual and practical implications of the Anthropocene debate for the discipline of political science?

While on the first question Crutzen and Schwägerl (2011) argued: "It's no longer us against 'nature'. It's we who decide what nature is and what it will be", Hickmann et al. stress that "we have to understand the Anthropocene as a threatening scenario and as a call for action". In the context of the second question the following aspects will be discussed: the contribution of political science to the Anthropocene discourse, and the implications of the Anthropocene debate for political science.

In a commentary, Hickmann et al. (2020: 659–670) called for more engagement in the political science debate on the Anthropocene in Germany. In a debate among political scientists on the evolving "Planetary Political Thinking" (Pauls et al. 2022: 703–728) the Anthropocene was also peripherally considered.

As of January 2023, no systematic study that addresses the two questions by Hickmann et al. (2019, 2020) had been published and there has also been no rethinking of politics, policy and polity in the Anthropocene since 1945. However, besides the three classical political theories influenced by liberalism, conservatism and socialism that have evolved since the French Revolution (1789) and the Industrial Revolution (1750–1900), a fourth school of ecological thinking has gradually been developing since the 1980s, influenced by environmental movements and the protests against nuclear energy and weapons during the 1970s and 1980s.

In political practice, during the Cold War the dominant Hobbesian approach to security and peace in international and national security and war studies was challenged by scholars in the social sciences from the 1950s onwards who focused on alternative pathways aiming at détente, arms control and disarmament. However, since the Russian aggression against Ukraine these approaches have been discredited by some hawks, who have called for a major rearmament and for exporting heavy battle weapons to the victim of the Russian aggression.

In the late 1980s and in the early 1990s two discourses stressed cooperation among nations on both arms control and disarmament, and the establishment of a cooperative security order based on the concept of mutual, common or human security. In this framework, environmental regimes on ozone layer depletion, climate change, biodiversity, desertification and water were established.

The latter discourse can be summarised under a wider Anthropocene discourse: Since 24 February 2022 Hobbesian geopolitics has again prevailed while the Anthropocene agenda has—at least temporarily—lost its urgency in many countries in the remaining old "West". Countries in the Global South condemned the Russian aggression but did not support Western countermeasures, such as economic sanctions, and refrained from supporting Ukraine with weapons supplies.

1.4.2 The Anthropocene as a Theme of Eight Edited Collections in Political Science

A decade after the discussion and reflections on the Anthropocene had started in the natural sciences (Brauch 2021) the *Web of Science* and *Scopus* included the first publications by political scientists in 2011, and the inclusion of political science texts started in 2012. There was no detailed discussion about why Crutzen had triggered this debate in February 2000 in Cuernavaca, on the discourse within the Global Environmental Change (Steffen et al. 2004) and the Earth Systems Science communities after the Amsterdam Conference in 2001 (Steffen et al. 2002) and the Dahlem Workshop in 2003 (Schellnhuber et al. 2004), or after the Anthropocene Working Group (AWG) was established in 2009.

Since 2012 eight edited collections have been published with the Anthropocene as a theme in Political Science, International Relations, Political Theory and in Environment and Earth Politics:

(1) Glaser, M.; Krause, G.; Ratter, B. M. W.; Welp, M. (Eds.), 2012: *Human-Nature Interactions in the Anthropocene: Potentials of Social-Ecological Systems Analysis* (London: Routledge).
(2) Hamilton, C.; Bonneuil, C.; Gemenne, F., 2015: *The Anthropocene and the Global Environmental Crisis: Rethinking Modernity in a New Epoch* (London – New York: Routledge).
(3) Nicholson, S.; Jinnah, S., 2016: *New Earth Politics. Essays from the Anthropocene* (Cambridge: MIT Press).
(4) Pattberg, P., Zelli, F. (Eds.), 2016: *Environmental Politics and Governance in the Anthropocene: Institutions and Legitimacy in a Complex World* (Abingdon: Routledge).
(5) Biermann, F.; Lövbrand, E. (Eds.), 2019: *Anthropocene Encounters: New Directions in Green Political Thinking* (Cambridge – New York – Melbourne: Cambridge University Press).
(6) Arias-Maldonado, M.; Trachtenberg, Z. (Ed.), 2019c: *Rethinking the Environment for the Anthropocene: Political Theory and Socio-natural Relations in the New Geological Epoch* (Abingdon: Routledge).
(7) Hickmann, T.; Partsch, L.; Pattberg, P.; Weiland, S. (Eds.), 2019: *The Anthropocene Debate and Political Science* (London – New York: Routledge).
(8) Chandler, D.; Müller, F.; Rothe, D. (Eds.), 2021: *International Relations in the Anthropocene: New Agendas, New Agencies and New Approaches* (Cham: Springer Nature Switzerland).

Seven of these eight readers in Political Science on the Anthropocene (Table 1.2) were published by British publishers—Routledge (5), Cambridge University Press (1) and Palgrave Macmillan (1)—and one by MIT Press in the USA, between 2012 and 2021. Three volumes were exclusively edited (1, 3, 7) and three were coedited (2, 4, 7) volumes; among these five volumes are two by German scholars working in the Netherlands (Pattberg [2, 7] and Biermann [5]) and one volume by a German scholar in Sweden (Zelli [1]). Only one book was edited by two American researchers (Nicholson, Jinnah [3]), one was coedited by a Spanish (Arias-Maldonado) and an American (Trachtenberg) expert, while one was edited by a senior British professor (Chandler) and by two junior German researchers (Müller, Rothe [8]). None of the eight edited collections included a natural scientist, and none of the authors was from the Global South and working there as well.

The main goal of most contributions of the 117 chapters in these eight readers has been to discuss the Anthropocene from a Political Science (Hickmann et al. 2019), an International Relations (Biermann/Lövbrand 2019; Chandler et al. 2021) and a Political Theory (Arias-Maldonado/Trachtenberg 2019) perspective as well as from the point of view of Environmental Studies (Pattberg/Zelli 2016; Arias-Maldonado/Trachtenberg 2019), the thinking on Earth Politics (Nicholson/Jinnah 2016) and on Green Political thinking (Biermann/Lövbrand 2019) in the context of the Anthropocene.

None of these edited collections addresses how the start of the Anthropocene between 1945 and 1950 coincided with the multiple changes in the international order and in political history, how humankind and the totality of the human species has interfered as a geological force in the Earth system, or the impact of the physical consequences of global climate change on the survival of the human species due to linear projected trends and unpredictable chaotic interactions within the Earth system as possible "tipping points".

More recently, research in environmental and sustainability studies and on human security and peacebuilding by scholars from the Global South (from Africa: Kiyala/Harris 2022; Hussein/Cocodia 2023; Kiyala/Chivasi 2024; from Asia: Nayak 2019, 2024; and from Latin America: Padilla 2021; Prado de Sousa/Serrano Oswald 2021) have been contextualised in the framework of the Anthropocene.

The thesis is that the silent transition to the Anthropocene in 1945–1950 was not yet socially constructed until the year 2000, when Crutzen coined the term "Anthropocene" and it began to be used in global environmental studies, in Earth systems science, and in the geological communities.

Between 1939 and 1945, policy planners in the US Department of State and policy makers designed a new international political, security and peace order that would make major violent conflicts and wars impossible by cooperation among the five permanent members of the Security Council (SC). Their planning assumptions and goals were affected by the outcome of the top secret Manhattan Project (with the test of the first nuclear weapon in July 1945 and the first use of nuclear weapons against Hiroshima and Nagasaki in August 1945), but also by the compromise reached by Roosevelt and Stalin in Yalta in February 1945 on a veto right in the global UN collective security system. But environmental issues and global challenges posed by

Table 1.2 Eight edited volumes on the Anthropocene in Political Science, International Relations, Political Theory, and Environment and Earth Studies

Title	*Human-Nature Interactions in the Anthropocene: Potentials of Social-Ecological Systems Analysis* [1]	*The Anthropocene and the Global Environmental Crisis: Rethinking Modernity in a New Epoch* [2]	*New Earth Politics. Essays from the Anthropocene* [3]	*Environmental Politics and Governance in the Anthropocene: Institutions and Legitimacy in a Complex World* [4]	*Anthropocene Encounters: New Directions in Green Political Thinking* [5]	*Rethinking the Environment for the Anthropocene: Political Theory and Socionatural Relations in the New Geological Epoch* [6]	*The Anthropocene Debate and Political Science* [7]	*International Relations in the Anthropocene: New Agendas, New Agencies and New Approaches* [8]
Editors	Glaser, Krause, Ratter, Welp (4 German eds.)	Hamilton, Bonneuil, Gemenne (1 Australian, 2 French eds.)	Nicholson, Jinnah (2 American eds.)	Patttberg, Zelli (2 German eds., in Netherlands, Sweden)	Biermann, Lövbrand (1 German ed., in Netherlands)	Arias-Maldonado/ Trachtenberg (1 Spanish, 1 American ed.)	Hickmann/Partsch/ Pattberg/Weiland (4 German eds. (NL/ F))	Chandler/ Müller/Rothe (1 British, 2 German eds.)
Publisher	Routledge, UK	Routledge, UK	MIT, US/UK	Routledge, UK	Cambridge UP, UK	Routledge, UK	Routledge, UK	Palgrave Macmillan, UK
Publ. year	2012	2016	2016	2016	2019	2019	2019	2021
Authors	E. Becker, D K. Eisenack, D B. Glaeser, D M. Glaser, D A. Halliday, UK D. Hummel, D G. Krause, D R. Leemans, NL B. Ratter, D S. J. Singh, Aust. F. Tretter, D M. Welp, D A. Wildenberg, Austria	C. Hamilton, Australia C. Bonneuil, F F. Gemenne, F D. Chakrabarty, US/India A. Hornborg, Sweden J. B. Fressoz, F L. Semal, F M., Northcot Y. Cochet, F V. Maris, F. I. Stengers, B. B. Latour, F	Assadourian, US F. Biermann, NL/ D W. Burns, US K. Conca, US P. Dauvergne, Can. D. Deudney, US N. K. Dubash, India R. Falk, US J. Gupta, NL/ India M. Ivanova, US P. J. Jacques, US	E. Arias-Maldonado, Spain W. F. Barber, UK R. Bailis, UK R. V. Bartlett, UK S. Hailwood, UK M. Isailovic, NL M. Kühner, NL S. Meisch, D C. Moser, D P. Pattberg, NL/D C. Prokopf, D./NL J. van Leuwen, NL L. Walbott, D O. Widerberg, NL	E. Arias-Maldonado, E J. Barry, UK J. Baskin, Australia S. Beck, D F. Biermann, NL, D A. Burke, Australia N. Castree, UK S. Fishel, Australia A. Fremaux, UK V. Galaz, Sweden E. Lövbrand, Sweden J. Meadowscroft, Canada A. Mert, Sweden	Y. Ariffin, CH S. Baker, UK Barak, USA/Israel J. Batty, UK A. Fremaux, UK S. Hailwood, UK M. Hammond, UK J. M. Meyer, USA L. Pellizzoni, I M. Wissenburg, NL H. Ward, UK D. White, USA	B. Bornemann, D A. Brunnengräber, D J. N. Hardt, D T. Hermanns, D L. Hermwille, D R. Hickmann, D C. Höhne, D. Qiuri Li, D J. Lundershausen, D J. Marquardt, D F. Müller, D. L. Partzsch, D P. Pattberg, NL/D S. Swindenhammer, D	S. Alt, NL A.-K. Benner, D H. Bhat, UK B. Bornemann, D D. Chandler, UK O. Corry, UK E. Cudworth, UK S. Dalby, Canada J. N. Hardt, D C. Harrington, UK I. Hermann, D

(continued)

Table 1.2 (continued)

Title	*Human-Nature Interactions in the Anthropocene: Potentials of Social-Ecological Systems Analysis* [1]	*The Anthropocene and the Global Environmental Crisis: Rethinking Modernity in a New Epoch* [2]	*New Earth Politics. Essays from the Anthropocene* [3]	*Environmental Politics and Governance in the Anthropocene: Institutions and Legitimacy in a Complex World* [4]	*Anthropocene Encounters: New Directions in Green Political Thinking* [5]	*Rethinking the Environment for the Anthropocene: Political Theory and Socionatural Relations in the New Geological Epoch* [6]	*The Anthropocene Debate and Political Science* [7]	*International Relations in the Anthropocene: New Agendas, New Agencies and New Approaches* [8]
		I. Blühdorn, UK B. Szerszynski, UK	S. Jinnah, US K. T. Liftin, US M. F. Maniates, US, Singapore E. Mendenhall, US S. Nicholson, US K. O'Neill, US J. Shapiro, US P. Wapener, US O. A. Young, US	M. Wissenburg, NL F. Zelli, Sweden	A. Nikoleris, Sweden J. Stripple, Sweden P. Tenngart, Sweden P. Wapner, USA	E. Arias-Maldonado, Spain Z. Trachtenberg, USA	D. Themann, D J. Tremmel, D M. Weißpflug, D S. Weiland, F/D	S. Hobden, UK N. Kornblech, Dk A. Leander, CH M. McDonald, Australia C. McEavan, UK A. Mert, Sweden F. Müller, D J. C. Pereita, Portugal J. Reddekop, US D. Rothe, D. C. Ryan, NL D. Simangan, Japan M. J. Trombetta, China T. Trownwell, Ecuador R. V. Munster, Dk S. Wakefield, USA

(continued)

Table 1.2 (continued)

Title	*Human-Nature Interactions in the Anthropocene: Potentials of Social-Ecological Systems Analysis* [1]	*The Anthropocene and the Global Environmental Crisis: Rethinking Modernity in a New Epoch* [2]	*New Earth Politics. Essays from the Anthropocene* [3]	*Environmental Politics and Governance in the Anthropocene: Institutions and Legitimacy in a Complex World* [4]	*Anthropocene Encounters: New Directions in Green Political Thinking* [5]	*Rethinking the Environment for the Anthropocene: Political Theory and Socionatural Relations in the New Geological Epoch* [6]	*The Anthropocene Debate and Political Science* [7]	*International Relations in the Anthropocene: New Agendas, New Agencies and New Approaches* [8]
Title	*Human-Nature Interactions in the Anthropocene: Potentials of Social-Ecological Systems Analysis* [1]	*The Anthropocene and the Global Environmental Crisis: Rethinking Modernity in a New Epoch* [2]	*New Earth Politics. Essays from the Anthropocene* [3]	*Environmental Politics and Governance in the Anthropocene: Institutions and Legitimacy in a Complex World* [4]	*Anthropocene Encounters: New Directions in Green Political Thinking* [5]	*Rethinking the Environment for the Anthropocene: Political Theory and Socionatural Relations in the New Geological Epoch* [6]	*The Anthropocene Debate and Political Science* [7]	*International Relationsin the Anthropocene: New Agendas, New Agencies and New Approaches* [8]
Countries of Authors	Germany (9) Austria (2) Netherlands (1) UK (1) [13 authors, 4 countries]	Australia (1) Belgium (1) France (7) Sweden (1) UK (3) USA (1) India (1) [14 authors, 7 countries]	US (16) India (2) Netherlands (2) Canada (1) [20 authors, 4 countries]	Netherlands (5) Germany (5) UK (4) Spain (1) Sweden (1) [16 authors, 5 countries]	Sweden (6) UK (3) Australia (3) Canada (1) Netherlands (1) Germany (1) Spain (1) USA (1) [17 authors, 8 countries]	UK (6) US (4) Netherlands (1) Italy (1) Israel (1) Switzerland (1) Spain (1) [14 authors, 7 countries]	Germany (18) France (1) Netherlands (1) [18 authors, 3 countries]	UK (7) Germany (5) Netherlands (2) Denmark (2) US (2) Switzerland (2) Canada (1) Australia (1) Sweden (1) Portugal (1) Japan (1) Philippines (1) Italy (1) [27 authors, 13 countries]

(continued)

Table 1.2 (continued)

Title	*Human-Nature Interactions in the Anthropocene: Potentials of Social-Ecological Systems Analysis* [1]	*The Anthropocene and the Global Environmental Crisis: Rethinking Modernity in a New Epoch* [2]	*New Earth Politics. Essays from the Anthropocene* [3]	*Environmental Politics and Governance in the Anthropocene: Institutions and Legitimacy in a Complex World* [4]	*Anthropocene Encounters: New Directions in Green Political Thinking* [5]	*Rethinking the Environment for the Anthropocene: Political Theory and Socionatural Relations in the New Geological Epoch* [6]	*The Anthropocene Debate and Political Science* [7]	*International Relations in the Anthropocene: New Agendas, New Agencies and New Approaches* [8]
Disciplines	Sociology Economics Geography		Political Science International Relations	Environmental Politics	International Relations	Political Theory Environment	Political Science	International Relations
Themes	*Social-Ecological Systems Analysis*		*New Earth Politics*	*Environmental Governance*	*Green Political Thinking*	*Political Theory and Socionatural Relations*		*New Agendas, Agencies and Approaches*
Multidiscipl. interdiscipl. transformative	Multidisciplinary research programme			Subdisciplinary research programme	Disciplinary	Multidisciplinary	Disciplinary	Disciplinary
Chapters	10	15	16	14	12	12	14	24

Source Compiled by the author
Conclusions: No natural scientists, primarily mono-disciplinary, no scholar from the Global South (northern, Eurocentric approaches prevail)

environmental change and climate change in the Anthropocene were political *terrae incognita* for scientists and policymakers. Neither yet existed in 1945.

1.4.3 *Rethinking* Politik *in Political Science Anthropocene Texts*

Key goals of this introductory chapter are to review whether the Anthropocene debate within Political Science and International Relations (IR) has resulted in a rethinking of the wider concepts of *Politik* (*la Politique*, *la Politica*), or the three narrower concepts in English that refer its processes (*politics*), fields (*policies*) and legal texts and institutions (*polity*), and whether the Anthropocene concept has impacted on research programmes in peace, security, development, environment and gender studies.

By September 2023, the most cited authors in Political Science and International Relations had not yet addressed the Anthropocene concept, and there had been no systematic rethinking of the key concepts of politics, policy and polity due to the Anthropocene. After the first decade of the discussion within Political Science and International Relations (2012–2023) on the Anthropocene, the impact of the Anthropocene concept in the framework of the ecological and environmental narrative and on the major discourse on *Politik* has been rather peripheral.

While the debate on the utility of nuclear weapons in deterrence theories of mutually assured destruction (MAD) has reinforced the Hobbesian logic of the enemy as the major threat, in the Anthropocene humankind itself has become the major challenge to its own survival by its energy- and carbon-intensive way of life, for which this logic and its means do not apply.

Box 1.2 Detailed searches in the databases of *Web of Science* and *Scopus* on authors and texts on the Anthropocene in Political Science and in International Relations in August 2022, and in January 2023. *Source* Compiled by the author based on both databases

In the *Web of Science* on 14 August 2022 in a combined search for texts on the Anthropocene from a political science and an international relations perspective, 560 entries were listed. Among them were 18 texts by F. Biermann, a German political scientist at Utrecht University, 15 by O. Young, an American environmental politics specialist, 11 by L. Arizpe Schlosser, an anthropologist from Mexico, 9 by S. Dalby, a Canadian geographer from W. Laurier University in Waterloo, 5 by L. J. Kotzé, an international lawyer from South Africa, 5 by D. Chandler, an International Relations specialist from the UK, 4 by U. Oswald Spring, an ecologist from UNAM in Mexico, 4 by D. Simangan, an IR specialist teaching at Hiroshima University, 4 by M. Arias-Maldonado, a political scientist from Spain, 4 by J. S. Dryzek from the University of Canberra in Australia and 4 from A. Burke in Australia.

A second *Web of Science* search on 27 January 2023 for texts on the Anthropocene had 10,837 documents listed, whereof 380 texts were noted under Political Science and 307 under International Relations—or a total of 594 texts. Among them were 18 texts by F. Biermann, 15 by O. Young, 11 by L. Arizpe Schlosser, 9 by S. Dalby, 6 by L. J. Kotzé and by D. Chandler, 5 by A. Burke, P. Wapner, and J. Gupta, and 4 texts each by D. Simangan, E. Lövbrand, J. J. Schmitt, J. S. Dryzek, J. Castro Pereira, M. Arias-Maldonado, J. Pickering, U. Oswald Spring, D. Chandler and S. Nicholson.

On 27 January 2023 the Scopus database did not list the Anthropocene texts on Political Science and International Relations as subject areas. Among authors' names it listed texts for the following social scientists: L. J. Kotzé (27), S. Dalby (27), M. Arias-Maldonado, (13), F. Biermann (12), M. Glaser (11), J. Gupta (11), A. Hornborg (10), C. Hamilton (9), P. Heikkurinen (9), B. Latour (9), A. Haines (8), E. Lövbrand (7) and A. Burke (7). Both lists of authors referred to in the *Web of Science* and *Scopus* as (co)authors of texts on the Anthropocene point to a relatively small, limited—but not identical—group of experts, many of them having a publication record on environmental policy studies.

1.4.4 **Politik** ***in the Anthropocene by Political Scientists***

Since 2011, only a few political scientists have offered definitions of *Politik* in the Anthropocene. Biermann (2016: 407, 412–413) in his text: "Epilogue Politics for a New Earth: Governing in the 'Anthropocene'"argues that

> an Anthropocene lens emphasizes an integrated perspective of evolving social systems that require not only active management of human influences but also the adaptation of human societies to inescapable changes. The governance of such societal adaptation processes, from local to global levels, poses formidable challenges for policy … [Biermann states that] the emergence of the Anthropocene changes the context for both politics and political science … [and that] the Anthropocene is not only a major challenge for political science and the traditional field of environmental studies. … [He concludes that this requires] an integrated understanding of socioecological systems. … Global change research and the merging field of integrated Earth systems science require a sophisticated perspective from an interdisciplinary Earth systems governance research program that cuts across social science disciplines and levels of analysis. … seeking a holistic understanding of the Anthropocene.

Biermann (2016: 414, 416–418) proposes a transition "from disciplinary science to transdisciplinary knowledge generation", moving "from incrementalism toward transformative thinking". He concludes that "it is up to our generation to consciously embark on a political and ethical discourse on how best to navigate the Anthropocene, with a view toward effective, equitable, and legitimate systems of Earth systems governance that range from local politics to planetary-scale institutions and governance".

Barber and Bartlett (2016: 167–181) note that

> the interpenetration of the environmental problematique, the impact severity of some crucial environmental trajectories, and the unfathomable diversity of humans and human cultures combine to make governing the interaction with earth's natural systems the most daunting challenge humans will ever face. … [They conclude that] deliberative democratic techniques can contribute to norm mapping and building, to choosing among policy models, and in developing and implementing policy action plans. By establishing new democratic responsibilities and the expectation that policies and actions must be consistent with democratically embraced normative principles, appropriately institutionalized deliberative democracy in the Anthropocene has the potential to add significantly to the accountability, and therefore political legitimacy, of global environmental governance.

Crutzen's (2002) thesis that in the Anthropocene human beings have become a "geological factor", which implies a fundamental change in human-environment-relations where humans dominate "over biological, chemical and geological processes on Earth" provoked Crutzen and Schwägerl (2011) to argue that "it's no longer us against 'Nature'. It is we who decide what nature is and what it will be".

For Hickmann et al. (2019: 2–11) the Anthropocene points to an urgency to act "in order to fight climate change, species extinction and other global problems", where the Anthropocene must be understood both "as a threatening scenario and as a call for action". They argue that the implications of the Anthropocene debate for Political Science "is not about the new geological epoch, but rather about the political and social dimensions of global environmental change". For them,

> the main objective of political science research should be to identify a diversity of understandings, problem descriptions and future ideas about the design of environmental and sustainability politics in the Anthropocene. … [They claim that] the implications of human-induced global system changes for political science research are diverse. … We strive for more political science engagement to foster a broader and deeper debate on the Anthropocene between scholars from the natural sciences and the social sciences. … [In conclusion, they argue that] the overarching goal of this book is to complement the Anthropocene debate with a well-grounded reflection on how the planetary scale crisis alters the ways in which humans respond to the most pressing environmental challenges of the 21st century … [In their view] the concept of the Anthropocene raises numerous questions with regard to environmental and development policy-making and provides a unique opportunity for rethinking core concepts of the scholarship on global sustainability politics. (Hickmann et al. 2019: 2–11)

In their concluding chapter Hickmann et al. (2019: 246–250) identify

> three broader challenges and opportunities from adopting a political science perspective on the Anthropocene debate. First, the question of *governance* is paramount, as both the protagonists of a 'crisis' reading of the Anthropocene and those that regard the Anthropocene as a chance, develop ideas about how to 'steer' the coupled nature-social system in the right direction. Second, in determining this 'right' direction, *political theory* might offer important normative insights. And third, while disciplinary depth is an important prerequisite for meaningfully engaging in the Anthropocene debate, the complexity of the challenge requires to move beyond disciplinary silos and to practice genuine *interdisciplinarity*.

Arias-Maldonado and Trachtenberg (2019) review "the most current thinking about the Anthropocene in the field of *Environmental Political Theory* (EPT)" and try to develop "the idea of 'socio-natural relations'" focusing on the concern how "human beings transform their environments" and "where human transformative powers have brought about a qualitative change in natural systems on a planetary level". They further argue that "the Anthropocene demands a blurring of the boundaries between disciplines, becoming itself a new framework for interdisciplinary research, moral reflection, and political action" (Arias-Maldonado/Trachtenberg 2019: 1, 8).

Rothe et al. (2021: 2–11) refer to the Anthropocene

> as a *condition that we are in* rather than as an *external set of problems which we are confronted with*. Thus, the Anthropocene is not merely a question of new or more pressing problems, such as climate change and extreme weather events, but also a matter of the tools and understanding that are available to us: in other words, it is a matter of how we know—of epistemology—and also of what we understand the world to consist of—that is, questions of ontology. …
>
> [For International Relations] the Anthropocene as a new epoch brings into question the traditional modes of conceptualising International Relations. … Firstly, International Relations was understood as an elite or 'top-down' discipline, narrowly preoccupied with 'high politics'. … Secondly, the disciplinary concerns revolved around seemingly timeless questions of war and peace. … [They also note that] the slowness of International Relations to react to the seismic shifts heralded by the Anthropocene has been highlighted by a number of authors in recent years. (Harrington 2016; Mitchell 2017; Fagan 2017; Simangan 2020)

The edited book by Chandler et al. (2021) is divided into four parts, focusing in Part I on "The Anthropocene: From the Global to the Planetary", then on "The Challenge of Security" (Part II), on "Governance and Agency" (Part III) and on "Methods and Approaches Beyond the Human/Nature Divide" (Part IV).

Biermann's (2014: ix) book on *Earth Systems Governance: World Politics in the Anthropocene* was one of the first IR monographs to promise "a new perspective on the governance of human-nature coevolution at a planetary scale". Purdy (2015: 3, 17) notes that the Anthropocene marks the last of three revolutions in politics, economics and nature, arguing that "the world we inhabit will henceforth be the world we inhabit" and that "the Anthropocene begins amid a threefold crisis of ecology, economics, and politics".

For Kelly (2019), "the Anthropocene is a concept for a world 'after nature', to the extent that it presupposes no distinction between human action and the natural world". In the Anthropocene humans are the first species with the power "in having transformed the earth". Kelly approaches the theme from "the perspective of the history of modern political theory". He sees a twofold challenge that forces "politics to be more openly ecological", which requires that we must rewrite "its concepts, and thus its history, to show the interconnections between 'nature' and the 'artificial' world of politics". In his 'Epilogue' on "Historical Possibilities for an Anthropocened Politics", Kelly (2019: 113, 122) argues that "Anthropocened politics is a set of administrative forms of 'solutionism', because neither the Anthropocene, nor politics, is a problem with a solution". He concludes, "if representative politics is to do more than merely survive or just settle for adaptation, it must meet the challenge to the established narratives of a liberal world order, which to my mind remains the most realistic future for representative political thinking to pursue into the new times of the Anthropocene."

Dryzek and Pickering (2019) argue in *The Politics of the Anthropocene* that "the Anthropocene is an emerging epoch of human-induced instability in the Earth system. The challenge the Anthropocene presents to humanity is profound, meaning in future all politics should be first and foremost a politics of the Anthropocene." The book is structured in eight chapters, addressing the Anthropocene as the good, bad and inescapable (this chapter), governance in the Holocene (Chap. 2), governance in the Anthropocene (Chap. 3), planetary justice (Chap. 4), and sustainability (Chap. 5). It discusses who will form the Anthropocene (Chap. 6) and the question of a democratic Anthropocene (Chap. 7), and in the concluding chapter (Chap. 8) the authors outline a practical politics of the Anthropocene.

In this book both authors "focus on the politics and governance of an emerging epoch of potentially catastrophic impact on the character of the Earth system that is coming to be known as the Anthropocene". Inspired by Dalby (2016), who distinguished three versions of the Anthropocene ("the good, bad and ugly"), they discuss the bad (impacts of dangerous climate change), the good (vision of the eco-modernists) and the "inescapable Anthropocene" as a "game changer". They argue that "the broad-ranging requirements to be able to rethink everything means that although environmental concerns are prominent, the Anthropocene has implications well beyond what has traditionally been classified as environmental affairs. Accordingly this book is not confined to environmental politics, but it is about politics in general in the Anthropocene." They conclude that the

> Anthropocene now represents an inescapable condition with dramatic consequences for human institutions, values, principles, and practices. If that condition is not recognized or (even worse) if it is explicitly denied, humanity will continue to stumble into a succession of catastrophes. If it is recognized, then there remains much work to be done to figure out how to respond most effectively, for it is not at all obvious how to do so. (Dryzek/Pickering 2019: 19)

Harrington and Shearing (2017: 1, 17) shifted their focus from the prevailing interest in environmental issues to "security in the Anthropocene", especially "for our safety and our very survival", as we have become "powerful 'geological agents' (Charakbarty 2009) who are actively shaping the Earth system in ways that seriously undermines our safe spaces". In conclusion, Harrington and Shearing (2017: 154–155) believe that

> the Anthropocene can find its way into traditional security analyses that are less interested in examining meta-theoretical questions in two principle ways. First, the Anthropocene pushes open even further the security black box by signalling the diversity of social impacts that arise from a rapidly shifting planet. … It could mean the more forceful incorporations of ignored (or previously absent) threats. … Or, the Anthropocene could lead to the inclusion of so-called 'new security actors' as the energy sector, wildlife poachers, urban planners, and the insurance industry. One might also envision a growing fascination with how environmental norms, regimes, and institutions seep into security discussions. Finally, the increasing role of sciences and technologies, both new and inherited, could be an important component of situating Anthropocene security.

> Second, the Anthropocene alerts scholars and policymakers to the necessity of building cooperative security relationships in advance of future changes that are now locked-in. ... The Anthropocene troubles certain dominant security logics, like the threat/defence and friend/enemy distinctions. As exposure to environmental hazards grows and spread, it should become clear that shared goals and vulnerabilities might produce impactful cooperative and multilateral agreements.

The eight readers on the Anthropocene and selected monographs from a disciplinary perspective in Political Science and International Relations have so far only partially addressed the impact of the survival challenges posed by the policy agendas associated with UN Security Council for the five permanent SC members which has paralysed the adopted Anthropocene on the disciplines' key concepts. The co-editors of one of the Anthropocene readers, Hickmann et al. (2020: 659–670) urge

> the discipline of political science to enhance its engagement in the debate on the so-called human age and its far-reaching implications. The Anthropocene is characterized first and foremost by complex interactions, non-linear dynamics and tipping points that entail abrupt changes. Against this backdrop, it becomes clear that there can be no simple management of human impacts on the earth system in the Anthropocene. To explore the root causes and potential solutions of human-induced environmental problems, we need to raise questions on key topics, such as prevailing power relations, clashes of interests and norm conflicts. Political theory is moreover highly relevant to address the question of how the existing political institutions could be reformed in the Anthropocene. The discipline of political science plays hence a particular role in the Anthropocene debate. Yet, the Anthropocene should not only remain a critical topic for those political scientists concerned with the environment. Instead, what is needed is a political science contribution that reflects the entire breadth of the discipline to develop adequate response strategies for system-threatening global environmental changes in an interdisciplinary dialogue.

Their self-critical assessment of the discourses on the Anthropocene in political science points to valuable comments and suggestions disciplinary authors have made during the first decade (2012–2022) that have been briefly reviewed. From the perspective of International Relations, Simangan (2019, 2020, 2020a, 2021) noted in a review on the Anthropocene:

> Several disciplines outside the natural sciences, including International Relations (IR), have engaged with the Anthropocene discourse in order to theorize their relevance and translate their practical value in this new phase of the Earth's history. Some IR scholars have called for a post-humanist IR, planet politics, a cosmopolitan view, and ecological security, among other approaches, to recalibrate the theoretical foundations of the discipline, making it more attuned to the realities of the Anthropocene. Existing discussions, however, tend to universalize human experience and gravitate towards western ontologies and epistemologies of living in the Anthropocene. Within this burgeoning scholarship, how is the IR discipline engaging with the Anthropocene discourse? Although the Anthropocene has become a new theoretical landscape for the conceptual broadening of conventional IR subjects, this review reveals the need for sustained discussion that highlights the differentiated human experiences in the Anthropocene. The existing IR publications on the Anthropocene locate the non-spatial narratives of vulnerability and historical injustice, the non-modernist understanding of nature, the agency of the vulnerable, and the amplification of security issues in the Anthropocene. It is in amplifying these narratives that the IR discipline can broaden and diversify the discourse on the Anthropocene and, therefore, affirm its relevance in this new geological age.

Both reviews and assessments of the Anthropocene discourses in Political Science and International Relations have been highly diverse and after a decade of complex efforts for research and analysis, so far no systematic assessment exists for future research. In Political Science and International Relations the Anthropocene stands for a transition in Earth history from the Holocene to an Anthropocene epoch where human beings have directly interfered in the Earth system, causing both physical effects and multiple societal consequences that have already been put on both the global and national political agendas.

These new dangers for human survival in the Anthropocene differ fundamentally from the other human survival threat posed by weapons of mass destruction, most particularly thermonuclear and hydrogen weapons. Both the logic causing and driving each survival threat and also the coping strategies and tools differ profoundly. Tropical storms, droughts, forest fires, food insecurity and hunger cannot be solved by a logic driven by Hobbesian fear. Each survival threat requires different worldviews of societies and mind-sets of policy makers. the Anthropocene and the global environmental agenda is not only a theme for scientific reflection and analysis, but also a political priority that requires both precautionary steps and active policy responses.

The aggression of the Russian Federation against Ukraine on 24 February 2022 has resulted in a return of geopolitical and "realist" or "Hobbesian" thinking, where security issues are given utmost primacy and the environmental narrative in the framework of the Anthropocene is put on hold. Climate change negotiations that require multilateral cooperation based on mutual trust instead of exclusion and punishment are at least temporarily frozen.

A combination of the effects of a severe drought in parts of the Global South (especially in East Africa) and the blockade of export of grain and maize from both Ukraine and Russia has increased global food prices and has already resulted in food insecurity, undernourishment and hunger. The two narratives have again been separated, and the security agenda has regained top priority—at least in many countries in the Global North. Progress on the climate agenda will require the active participation of China (as the most populous country with the highest GHG emissions and the largest exports), but also of Russia (as the country with the biggest territory and a major exporter of carbon energy sources of coal, oil and natural gas).

The Political Science and International Relations literature on the Anthropocene has thus far only partially addressed the following fundamental research questions:

1. *Does the Anthropocene as the only turning point in political history require a fundamental rethinking of the key concept of Politik, la Politique or la Politica or in English of "Politics" (process), "Policy" (field) and "Polity" (laws and institutions)?*

The Anthropocene as a proposed new Epoch of the Earth system started, based on the recommendation of the Anthropocene Working Group (2019),[29] with the

[29] The AWG website with links to the AWG Newsletters at: http://quaternary.stratigraphy.org/working-groups/anthropocene/ (16 May 2021). Click at: "Anthropocene" … "https://quaternary.stratigraphy.org/working-groups/anthropocene"

first atomic bomb test in New Mexico and with the "Great Acceleration" (McNeill/Engelke 2014) in 1950:

> We are now living in the Anthropocene. The period from 1945 to the present represents the most anomalous period in the history of humanity's relationship with the biosphere. Three-quarters of the carbon dioxide humans have contributed to the atmosphere has accumulated since World War II ended, and the number of people on Earth has nearly tripled. So far, humans have dramatically altered the planet's biogeochemical systems without consciously managing them.

2. *Does the start of the Anthropocene coincide with a fundamental change in world politics, in human history and in international relations?*

There is no agreement in the social sciences and in history on the major turning points in human history, in world history, in world politics (the object of analysis) and in *International Relations* (discipline). Polanyi (1944) argued in his book *The Great Transformation* "that the modern *market economy* and the modern *nation-state* should be understood not as discrete elements but as the single human invention he calls the 'Market Society'". In *Peace and War: Armed Conflicts and International Order 1648–1989* Holsti (1991) analysed the development of new orders after major wars as a result of peace treaties and the "Problem of Change in International Relations Theory" (Holsti 2016: 37–56).

Buzan and Lawson (2015) claim that the global transformation during the "long nineteenth century" (1776–1914) "was a period of political, economic, military and cultural revolutions that re-forged both domestic and international societies" and that "the global transformation also changed the basic sources, or *mode of power*, stimulating the emergence of global modernity". They argue "that the global transformation generated four basic, but linked, types of change in international relations": 1. "Industrialization and the extension of the market to a global scale" ...; 2. "The reconstitution of power associated with the emergence of modernity was sustained by processes of rational state-formation" ...; 3. "The new ideologies that rose to prominence during the nineteenth century" ...; 4. "The tripartite configuration that lay behind the global transformation" (Buzan/Lawson 2015: 1–4).

Polanyi, Holsti, and Buzan and Lawson do not discuss whether there has been a fundamental turning point in world politics since the end of World War II, nor do they consider the impact of environmental challenges. In my view, the outcome of the Second World War marks a dual fundamental change in human history and in international order that may undermine human survival:

- The danger of a *major nuclear war* that may put the survival of humankind at risk due to the irradiation of major parts of the globe and of the territory of the combatants, and the consequences of a nuclear winter (Crutzen/Birks 1982) for the territories of non-combatant countries also. This theme has been discussed from two opposite perspectives, those of strategic studies and peace research. Since the use of nuclear weapons against Hiroshima and Nagasaki in August 1945, and after the Soviet Union, China and other countries obtained these weapons of mass destruction, this threat became obvious and has remained with us.

- The threat of a *sixth extinction* based on the impact of *global environmental and climate change* and a *major environmental crisis* (Buzan and Lawson mention this for the future as a possible wild card) did not exist in the perception of decision makers until the late 1980s. This threat has been socially constructed by natural scientists since the early 1970s and by policy makers since 1988, when they launched policies that resulted in new environmental regimes (UNFCCC, CBD, UNCCD) and new forms of global environmental governance (Biermann et al.) and has been considered as a "security threat" since the early twenty-first century.

Since Crutzen coined the term "Anthropocene" both threats to the survival of humankind have coexisted in the perception of many analysts, but their legitimisations have differed fundamentally. The nuclear threat is posed by the rival political camp or nation and has been responded to by efforts to achieve a nuclear superiority and by obtaining a dual-strike nuclear capability. The risk of a severe environmental crisis is the result of the cumulative burning of hydrocarbon energy sources (coal, oil, gas) and the rising greenhouse gas emissions as a result of our way of life, production and consumption. Each threat points to different intellectual traditions: with the atomic weapon the threat is posed by the "other" (Hobbesian realist tradition), whereas with the environmental crisis we are the threat and only we can solve the challenge with cooperative strategies, with peaceful means, with technical innovation and with behavioural changes.

The policy outcome in the Anthropocene depends on the collective future behaviour of the large emitters of greenhouse gases and on whether the goal to reduce the growth of average global temperature on earth to 2 °C—or even to 1.5 °C above the level that existed in the pre-industrial era (before 1750–1800)—can be achieved.

3. *In the best case the largest emitters of GHG develop and implement a development path that aims at and achieves a process of a sustainable transition.*

To achieve this goal the largest GHG polluters must adopt and strictly implement the goal of a carbon neutral economy, to be achieved by a decarbonisation of the energy, mobility, production, consumption, transportation, agricultural and housing sectors. However, since 1992 a "climate paradox" (Brauch 2012) has developed. While most governments have endorsed the first six Assessment Reports of the IPCC, many governments have so far failed to implement their declared climate change goals and commitments, as well as their Nationally Determined Contributions (NDCs) under the Paris Climate Change Treaty of 2015.

4. *In the worst case, if the largest emitters of GHG adhere to strategies of business as usual or if they even increase their GHG emissions further, the societal consequences of the physical outcomes of climate change may be major climate-induced distress migrations, domestic crises and domestic, regional or interregional violent conflicts.*

The linear projections of the climate change models may trigger chaotic interactions within the climate system that could cause major tipping points (Lenton et al. 2008: 1786–1793; 2019), with significant geopolitical consequences.

In the Anthropocene epoch of Earth history, which started between 1945 and 1950, it has become obvious, both in global, international and national politics as well as in Political Science and International Relations, that an integrated approach for addressing, facing and coping with the threats for the survival of humankind is needed, linking the peace and security narrative with the ecological perspective in the social sciences by a Peace Ecology approach and in politics with strategies that address both survival threats in an integrated way. This requires linking the diplomatic and military agenda with environmental tasks and ecological concerns in an ecological peace policy, which has been postponed into the distant future since the Russian aggression against Ukraine.

1.5 Three Phases of International Political Order During the Anthropocene Since 1945

Since 1945 to 1950, when the Anthropocene started, three phases of international order have developed. In the direct aftermath of the Second World War the United Nations organisation and the Bretton Woods Institutions (the World Bank, the International Monetary Fund [IMF] and GATT—that become the World Trade Organization [WTO]) were formed. The institutional framework ("polity") of the multilateral international order survived the three phases of international order in the Anthropocene and adapted to the new emerging circumstances.

1.5.1 The Cold War as the First Phase of International Order in a Bipolar World in the Anthropocene

The Cold War (1947–1989) relied on a system of collective security and its regional arrangements and agencies (Chapter VII UN Ch), such as the Organization of American States (OAS), the Organisation of African Unity (OAU), and the Arab League, but the major decision-making moved to several systems of collective self-defence (Art. 51, UN Ch), including the Southeast Asia Treaty Organization (SEATO) and the Australia, New Zealand and United States Security Treaty (ANZUS) that have been dissolved (SEATO) or are not used any longer (ANZUS).

Only NATO, whose membership has increased from 12 to 31 members, has survived. The enlargement of the Alliance has been an ongoing and dynamic process. Since NATO was created in 1949, its membership has grown from the 12 founding members to 31 members (in Spring 2023) through eight rounds of enlargement in 1952, 1955, 1982, 1999, 2004, 2009, 2017 and 2020–2023. Turkey and Hungary have temporarily blocked the membership of Sweden, while Bosnia Herzegovina, Georgia, Ukraine and Moldova have indicated an interest in joining NATO (Sarotte 2021).

In 1945 environment policies and environmental concerns were not yet socially constructed and thus did not yet exist, either for natural and social scientists or in the political realm at the national and international level. Partly triggered by Rachel Carson's book *Silent Spring* (1962), the first environmental concerns (April 1970 saw the First "Earth Day" in the USA) and movements emerged, and the first environment agencies (in December 1970 the US Congress set up the Environment Protection Agency [EPA]) and environment concerns emerged in North America and in Europe. In 1972 international environment policy emerged during and in the aftermath of the first UN Conference on the Environment in Stockholm in June 1972. As an outcome the United Nations Environment Programme (UNEP) was established in Nairobi in the Autumn of 1972.

After 1970, arms control and disarmament policy and détente between the two blocs gradually emerged between the USA and the Soviet Union (on the bilateral level through SALT I in 1972), in Europe through the international relaxation of tensions (multilateral: CSCE with its Helsinki Agreement, 1975) and between Germany and its eastern neighbours through its *Ostpolitik* with the Moscow and Warsaw Treaty of 1970 and the German-German Agreement of 1972.

With the progress of détente between the military, economic, societal and ideological blocs, new issue areas emerged in the multilateral cooperative context of the United Nations; for example, international climate policy and its institutions were set up between 1988 and 1992 with mandates from the UN General Assembly to establish the Intergovernmental Panel on Climate Change (IPCC) and to launch international negotiations (in December 1988) that resulted in the first Earth Summit, the UN Conference on Development and Environment (UNCED) in Rio de Janeiro in June 1992. Here the United Nations Framework Agreement on Climate Change (UNFCCC) treaty, the Convention on Biological Diversity (CBD), the Earth Charter and several other documents were adopted; these were instrumental in the creation of several functional global secretariats that formed a new environmental regime for "governing without government" (Rosenau/Czempiel 2002).

1.5.2 Globalisation as the Second Phase of International Order in a Unipolar World in the Anthropocene

During the Post-Cold War phase (1990–2022), the Conference on Security and Cooperation in Europe (CSCE) was converted into the Organization for Security and Co-operation in Europe (OSCE) in 1994 with 57 member states:

> in North America, Europe and Asia, the OSCE … works for stability, peace and democracy for more than a billion people, through political dialogue about shared values and through practical work. … The OSCE is a forum for political dialogue on a wide range of security issues and a platform for joint action to improve the lives of individuals and communities.

> The organization uses a comprehensive approach to security that encompasses the politico-military, economic and environmental, and human dimensions. Through this approach ... the OSCE helps bridge differences and build trust between states by co-operating on conflict prevention, crisis management and post-conflict rehabilitation. With its institutions, expert units and network of field operations, the OSCE addresses issues that have an impact on our common security, including arms control, terrorism, good governance, energy security, human trafficking, democratization, media freedom and national minorities.[30]

As a result of the aggression of the Russian Federation against Ukraine, the role of the OSCE has been downgraded while that of NATO has been further upgraded. In Europe, the attempts since 1990 to build a collective security system in the framework of the OSCE have declined, while NATO as the only remaining system of collective self-defence has been strengthened as the anchor of European security, especially for the former Warsaw Pact countries and the former member states of the Soviet Union that have become NATO members (Estonia, Latvia and Lithuania) or have applied for membership (Ukraine, Georgia, Moldova).

1.5.3 Third Phase of International Order: Erosion, Collapse and Uncertainty in a Fragmented World in the Anthropocene

The emerging new international order that may evolve after the end of the Russian war against Ukraine is still uncertain in November 2024. The implication of this uncertainty and lack of trust will be higher armament spending and a postponement of the necessary decisions on global environmental change. This period of postponement and inability to act will increase the costs for coping with the socio-political outcomes of anthropogenic climate change.

The erosion of the European security order in 2022 was preceded by a gradual dismantlement of the rule-based security order and the violation of key legal principles by both the sole remaining hegemon and its challengers. The lack of leadership of the G8 in implementing the national commitments to international climate policy has emerged since 1997 with the failure of the US to ratify the Kyoto Protocol and the announcement of its withdrawal from the Paris Climate Treaty on 1 June 2017 and again in November 2025 after the re-election of President Trump. This lack of leadership became evident in 2012 during Rio II, when the governments failed to agree on new legally binding commitments. This author has called this a "climate paradox" (Brauch 2012) between recognising the consequences of anthropogenic climate change and the lack of willingness and commitment to fully implement legal commitments and unilateral promises.

[30] OSCE; at: https://www.osce.org/who-we-are (30 March 2023).

1.5.3.1 Gradual Dismantlement of Rule-Based Security Order and Violation of International Legal Principles

Since the George W. Bush Administration in the US and the Presidency of Vladimir Putin in the Russian Federation, the arms control regime that had evolved during détente since 1972 was partly dismantled by both the US and Russia (Congressional Research Service 2022).

The US withdrew from several arms control and disarmament agreements, especially during the George W. Bush, the Obama and the Trump administrations, such as the

- Antiballistic Missile (ABM) Treaty (of 1972) during the G. W. Bush administration (2002);
- Intermediate Nuclear Force (INF) Agreement (of 1987) by the Trump administration on 20 October 2018 announced its withdrawal and on 2 August 2019 it formally withdrew; and the
- Conventional Armed Forces in Europe Treaty (CFE) of 1990 that was unilaterally modified by both the USA and Russia to accord with the changed geostrategic conditions when both suspended their cooperation because the CFE limited their military flexibility.
- In March 2015 Russia suspended its participation in the Joint Consultative Group of the CFE Treaty, objected to any on-site inspections and no longer provided data mandated by this treaty. In November 2011 the US announced that it would stop implementing its data exchange obligations under the CFE Treaty with respect to Russia.
- The *Open Skies Treaty* was signed in 1992 but it did not enter into force until 2002. But in late 2020 the US withdrew from this treaty during the Trump Administration.

The *Comprehensive Test Ban Treaty* (CTBT) was first signed in 1996 by the Clinton administration, but it failed to be ratified by the Senate with 48–51 on 13 October 1999 due to the opposition of the Republican Party. This institutional framework in the US and the bipartisan foreign policy consensus in the US Congress had been the backbone of the US leadership position since 1941 until 2017, when President Donald Trump started to gradually tear it down and tried to replace the dominant multilateral trade system and liberal global order with a partly protectionist trade system based on high import tariffs.

Key elements were the Arms Control and Disarmament Agreements that were negotiated, signed and ratified by both Democratic and Republican Administrations during the Cold War by the following US Presidents: J. F. Kennedy (Limited Test Ban Treaty [LTBT], 1963), L. B. Johnson (Nonproliferation Treaty [NPT], 1968), R. Nixon (Biological Weapons Convention [BWC], 1971; Strategic Arms Limitation Treaty [SALT I] including the Antiballistic Missile Treaty [ABM], 1972), G. Ford (the Conference on Security and Cooperation in Europe [CSCE] Helsinki Accord, 1975), R. Reagan (the Intermediate Nuclear Forces Treaty [INF], 1987), G. Bush

(Treaty on Conventional Armed Forces In Europe [CFE], 1990, and the Chemical Weapons Convention [CWC], 1993). In the post-Cold War era this positive trend suddenly changed during the Clinton Administration, when the Republican-controlled US Senate did not ratify the CTBT in 1999. Since the administrations of G. W. Bush (ABM), B. Obama (CFE) and D. Trump (INF, Open Sky), the US has unilaterally modified or withdrawn from several key arms control and disarmament treaties.

A neo-isolationist mind-set has gradually grown within the US Republican party since 1997, when Rep. N. L. Gingrich became its leader in the US House of Representatives and J. Helms (Rep., NC), as chairman of the US Senate's Foreign Relations Committee, succeeded in defeating both the ratification of the Kyoto Protocol on Climate Change (1998–2001) and the Comprehensive Test Ban Treaty (CTBT).

The Clinton Administration, with the support of the Republicans, increased US military spending and approved the first round of the eastern enlargement of NATO (1999), led by Secretary of State M. Albright, when Poland, Hungary and the Czech Republic joined, followed by Bulgaria, Estonia, Latvia, Lithuania, Romania, Slovakia, and Slovenia in 2004. Albania and Croatia joined on 1 April 2009, Montenegro on 5 June 2017, North Macedonia on 27 March 2020 and Finland on 4 April 2023.[31] The NATO membership application of Sweden has temporally been blocked by Turkey and Hungary (Sarotte 2021).

With the Trump administration's goal "to make America Great Again", the United States of America shifted towards a neo-isolationist foreign economic, foreign and security policy, threatening and implementing unilateral fees on imports against both global competitors and allies, and by questioning international trade agreements and even the US's continued membership in NATO.

The Biden administration returned to a conciliatory policy on NATO with its European allies, but partly continued the 'China-bashing' of the Trump Administration, using pressure and persuasion on their allies with regard to economic cooperation with China, and protectionism through massively supporting US industry with regard to investments for a sustainable transformation of its energy and industrial sectors.

The President of the Russian Federation, Putin, announced on 21 February 2023 that Russia would no longer participate in the START II Treaty, the last remaining major nuclear arms control treaty with the US (Roth/Borger 2021). On 3 November 2023 Putin "signed into law Russia withdrawing its ratification of the Comprehensive Nuclear Test Ban Treaty (CTBT)".[32]

[31] See at: https://en.wikipedia.org/wiki/Enlargement_of_NATO (29 July 2022).

[32] "Putin criticised as he withdraws Russia's ratification of nuclear test ban treaty", in: *The Guardian*, 3 November 2023; Blinken (2023).

1.5.4 *The Changing Military Framework Since the Russian Aggression Against Ukraine: Rising Military Expenditure and Arms Exports and Imports in Europe*

After Russia's aggression against Ukraine, the US Secretary of Defense, Lloyd Austin, described the war goals of the US:

> The US defense secretary's declaration that Washington wanted to see Russia weakened militarily and unable to recover quickly, marks a shift in Washington's declared aims underlying its military support for Ukraine. At a press conference in Poland …, Lloyd Austin was asked if he would now define US goals differently from those set out soon after the Russian invasion. In response, he started out with the established administration line about helping Ukraine retain its sovereignty and defend its territory. Then Austin added a second goal: "*We want to see Russia weakened* to the degree that it can't do the kinds of things that it has done in invading Ukraine." That meant Russia should "not have the capability to very quickly reproduce" the forces and equipment that had been lost in Ukraine. (Borger 2022)

This goal was repeated in a second report in the *Guardian* on 25 April 2022. Austin stressed:

> "We want to see Ukraine remain a sovereign country, a democratic country able to protect its sovereign territory. *We want to see Russia weakened to the point where it can't do things like invade Ukraine*". … The western military alliance NATO has said repeatedly it does not intend to fight directly in the conflict, but Austin's comments also make clear that the US has become increasingly drawn into a proxy war—in which it has become the leading supplier of military aid. (Sabbagh/Livingstone 2022)

The US Department of State summarised the US arms support on 22 August 2023 in a fact sheet on "US Security cooperation with Ukraine":

> Since January 2021, the United States has invested more than $43.8 billion in security assistance to demonstrate our enduring and steadfast commitment to Ukraine's sovereignty and territorial integrity. This includes more than $43.1 billion since Russia launched its premeditated, unprovoked, and brutal war against Ukraine on February 24, 2022. Since 2014, the United States has provided more than $45.9 billion in security assistance for training and equipment to help Ukraine preserve its territorial integrity, secure its borders, and improve interoperability with NATO.[33]

In an update of global arms exports of 22 March 2023, the Stockholm International Peace Research Institute (SIPRI) summarised the global arms export trends:

> The new data revealed that imports of major arms by European states increased by 47 per cent between 2013–17 and 2018–22, while the global level of international arms transfers decreased by 5.1 per cent. Arms imports fell overall in Africa (−40 per cent), the Americas (−21 per cent), Asia and Oceania (−7.5 per cent) and the Middle East (−8.8 per cent)—but imports to East Asia and certain states in other areas of high geopolitical tension rose sharply. The United States' share of global arms exports increased from 33 to 40 per cent while Russia's fell from 22 to 16 per cent.[34]

---

[33] U.S. Department of State, Fact Sheet: "US Security cooperation with Ukraine, 22 August 2023"; at: https://www.state.gov/u-s-security-cooperation-with-ukraine/ (5 September 2023).

[34] SIPRI Email, 22 March 2023, 12.08: "SIPRI Update March 2023: New data on international arms transfers, invasion of Iraq 20 years on, Annual Review 2022, nuclear security during Ukraine war and more".

The SIPRI Fact Sheet of March 2023 on "Trends in International Arms Transfers, 2022" noted that

> [t]he war in Ukraine had only a limited impact on the total volume of arms transfers in 2018–22, but Ukraine did become a major importer of arms in 2022. In addition, most European states substantially increased their arms import orders and the war will have significant ramifications for future supplier-recipient arms trade relations globally. …

Between 2018 and 2022, the three largest arms suppliers to Ukraine were, according to SIPRI, the US with 34%, Poland with 17% and Germany with 11%.

> Arms imports by European states were 47 per cent higher in 2018–22 than in 2013–17. The biggest European arms importer in 2018–22 was the UK, which was the 13th largest arms importer in the world, followed by Ukraine … The USA accounted for 56 per cent of the region's arms imports in 2018–22, Russia for 5.8 per cent (mainly to Belarus) and Germany for 5.1 per cent. (Wezeman et al. 2023)

In an expert discussion organised by SIPRI on 8 November 2022 on "Global military spending and the war in Ukraine" its report noted:

> After Russia invaded Ukraine, several European countries announced increases to coming military budgets. The war is likely to accelerate an ongoing upward trend in global military spending; in April, SIPRI estimated global military spending to reach $2.1 trillion in 2021, an all-time high.
>
> The war in Ukraine also casts a shadow over recent efforts to revive discussions on reducing military spending. …

By the end of 2023 Russia, China, the USA, and NATO and EU member states had all announced major increases in their military expenditure, which may be an indicator that Europe and the global North will end up in a new Cold War, with unknown consequences for a ceasefire and a settlement of the ongoing war of aggression between Russia and its victim Ukraine.

This has already resulted in a downgrading of policies and expenditure for dealing with the consequences of global environmental change and climate change and with addressing the European "green deal" to achieve climate neutrality for the EU countries by 2050.[35]

1.5.5 *Emergence and Weakening of the International Order of Peace and Security in Europe (1972–2022)*

After the détente era and during the Cold War, a relatively stable international order of peace and security emerged in a divided Europe based on the spheres of influence that had been discussed between the United States, the United Kingdom and the Soviet Union during World War II, primarily for Eastern and South Eastern Europe.

[35] The European Think Tanks Group (ETTG): "The European Green Deal and the war in Ukraine Addressing crises in the short and long term" (Brussels: ETTG, Jul 2022); at: https://ettg.eu/wp-content/uploads/2022/07/The-European-Green-Deal-and-the-war-in-Ukraine.pdf; Dennison (2022).

Although these spheres of influence were never certified by a treaty, they were nevertheless respected by both superpowers during the Cold War to avoid any escalation into a major war that could have intensified into a nuclear inferno.

During the upheaval of workers in East Berlin and in the German Democratic Republic (GDR) in June 1953, during the violent protests in Hungary and the protests in Poland in 1956, during the invasion of Soviet and allied forces of the Warsaw Pact into Czechoslovakia in August 1968, and during the protests of *Solidarnosc* in the 1980s in Poland, NATO countries refrained from the use of force in support of the people that were struggling for their own democratic rights and freedom.

During the Cold War, in 1954 the USSR aimed at the Geneva Conference at a "formal recognition of the political boundaries in Eastern Europe that had been established after World War II". While no progress was made during the 1950s and 1960s, in the early 1970s, in the framework of the US-Soviet détente and the German policy towards Eastern Europe (*Ostpolitik*), the 1972 Helsinki Consultations resulted on 1 August 1975 in the framework of the formal Conference on Security and Cooperation in Europe (CSCE), with the signing of the so-called Helsinki Final Act by all European countries (except Albania), and the United States and Canada.[36] The Helsinki Final Act (1 August 1975) consisted of four baskets, where the

> first basket included ten principles covering political and military issues, territorial integrity, the definition of borders, peaceful settlement of disputes and the implementation of confidence building measures between opposing militaries. The second basket focused on economic issues like trade and scientific cooperation. The third basket emphasized human rights, including freedom of emigration and reunification of families divided by international borders, cultural exchanges and freedom of the press. Finally, the fourth basket formalized the details for follow-up meetings and implementation procedures.[37]

The U.S. State Department later concluded that

> the Helsinki Process ... led to greater cooperation between Eastern and Western Europe, where representatives from non-aligned countries acted as intermediaries, helping to broker deals between members of the North Atlantic Treaty Organization and the Warsaw Pact. The Vienna Review Meeting [1986–1989] introduced recognition of the rights of emigration and religious freedom. ... These shifts helped bring an end to Soviet dominance in Eastern Europe and the end of the Cold War.[38]

The core of the first basket of the CSCE Final Act of Helsinki was the so-called Decalogue with the *Declaration on Principles Guiding Relations between Participating States*

I. Sovereign equality, respect for the rights inherent in sovereignty
II. Refraining from the threat or use of force
III. Inviolability of frontiers
IV. Territorial integrity of States

---

[36] Department of State, Office of the Historian, "Helsinki Final Act, 1975", at: https://history.state.gov/milestones/1969-1976/helsinki (8 January 2023).

[37] *Ibid.*

[38] *Ibid.*

V. Peaceful settlement of disputes
VI. Non-intervention in internal affairs
VII. Respect for human rights and fundamental freedoms, including the freedom of thought, conscience, religion or belief
VIII. Equal rights and self-determination of peoples
IX. Co-operation among States
X. Fulfilment in good faith of obligations under international law.

With the symbolic end of the Cold War, through the sudden opening of the Berlin Wall on 9 November 1989, a process of peaceful change was launched in all former Warsaw Pact countries that resulted on 26 December 1991 in the peaceful dissolution of the Union of 15 Soviet Socialist Republics into 15 independent sovereign countries, and also of the Warsaw Treaty Organisation (WTO). The Council for Mutual Economic Assistance (COMECON) had already been dissolved on 28 June 1991.

On 21 November 1990 the heads of state and government of all CSCE member countries adopted the "Charter of Paris for a New Europe" in which they called for a "New Era of Democracy, Peace and Unity" and stated that "the era of confrontation and division of Europe has ended. We declare that henceforth our relations will be founded on respect and co-operation."[39] On security the CSCE member countries stated that: "the changing political and military environment in Europe opens new possibilities for common efforts in the field of military security" based on existing treaties and agreements (CFE and, CSBM [Confidence and Security Building Measures] Negotiations). … "Following a period for national preparations, we look forward to a more structured co-operation among all participating States on security matters, and to discussions and consultations among the thirty-four participating States". They stressed

> the duty of States to refrain from the threat or use of force [through] the peaceful settlement of disputes, both being essential factors for the maintenance and consolidation of international peace and security, we will not only seek effective ways of preventing, through political means, conflicts which may yet emerge, but also define, in conformity with international law, appropriate mechanisms for the peaceful resolution of any disputes which may arise. Accordingly, we undertake to seek new forms of co-operation in this area, in particular a range of methods for the peaceful settlement of disputes, including mandatory third-party involvement.[40]

In December 1991 in the framework of the dissolution of the Soviet Union, Russia fully recognised the sovereignty of Ukraine and Belarus, and on 5 December 1994 the Budapest Memorandum on Security Assurances was signed when Belarus, Kazakhstan and Ukraine acceded to the Treaty on the Non-Proliferation of Nuclear Weapons (NPT).

> The *Budapest Memorandum on Security Assurances* comprises three substantially identical political agreements … to provide security assurances by its signatories relating to the accession of Belarus, Kazakhstan and Ukraine to the Treaty on the Non-Proliferation of Nuclear

[39] OSCE, Charter of Paris for a New Europe, Paris 1990; at: https://www.osce.org/files/f/documents/0/6/39516.pdf (8 January 2023).

[40] *Ibid*.

> Weapons (NPT). The three memoranda were originally signed by three nuclear powers: the Russian Federation, the United Kingdom and the United States. China and France gave somewhat weaker individual assurances in separate documents. The memoranda … prohibited the Russian Federation, the United Kingdom and the United States from threatening or using military force or economic coercion against Ukraine, Belarus, and Kazakhstan, "except in self-defence or otherwise in accordance with the Charter of the United Nations."[41]

In the security guarantees, the three nuclear powers (Russia, the USA and the UK)[42] confirmed:

1. to respect the independence and sovereignty and the existing borders of Ukraine …
2. to refrain from the threat or use of force against the territorial integrity or political independence of Ukraine, and that none of their weapons will ever be used against Ukraine except in self-defence or otherwise in accordance with the Charter of the United Nations …
3. to refrain from economic coercion designed to subordinate to their own interest the exercise by Ukraine of the rights inherent in its sovereignty and thus to secure advantages of any kind …
4. to seek immediate United Security Council action to provide assistance to Ukraine, as a non-nuclear weapon state party to the Treaty on the Non-Proliferation of Nuclear Weapons, if Ukraine should become a victim of an act of aggression in which nuclear weapons are used …
5. in the case of Ukraine, their commitment not to use nuclear weapons against any non-nuclear weapon state party to the Treaty on the Non-proliferation of Nuclear Weapons, except in the case of an attack on themselves, their territories or dependent territories, their armed forces, or their allies, by such a state in association or alliance with a nuclear weapon state.
6. Ukraine, the Russian Federation, the UK and the USA will consult in the event a situation arises which raises a question concerning these commitments.

On 24 February 2022, the President of the Russian Federation, Vladimir Putin, through launching a war of aggression against Ukraine, violated multiple legal treaties and agreements, among them the Charter of the United Nations of 26 June 1945, the Helsinki Final Act of 1 August 1976, the Charter of Paris for a New Europe of 21 November 1990 and the Budapest Memorandum on Security Assurances of 5 December 1994. With this unilateral act of aggression by the Russian Federation, a permanent member of the United Nations Security Council (UNSC) violated the UN Charter and prevented the UNSC from acting by using its veto right in defence of its own aggression and thus thwarted the UN from acting in the framework of Chapter

---

[41] Wikipedia: "Budapest Memorandum", at: https://en.wikipedia.org/wiki/Budapest_Memorandum (8 January 2023).

[42] See the full text of the Budapest Memorandum with the Security Guarantees for Belarus, Kazakhstan and Ukraine at: https://treaties.un.org/doc/Publication/UNTS/Volume%203007/Part/volume-3007-I-52241.pdf.

VI on the Pacific Settlement of Disputes (Art. 33–38) and Chapter VII on Action with Respect to Threats to the Peace, Breaches of the Peace, and Acts of Aggression (Art. 39–51).

1.5.6 Violent Conflicts and Wars by Members of the UN Security Council in Violation of the UN Charter Since 1990

During the Cold War, the five permanent members of the United Nations Security Council were involved in many violent activities (interventions, anticolonial conflicts) in violation of different principles of the United Nations Charter, especially of Art. 2.4. on the non-use of force and of Art. 2.7 on the principle of non-intervention in the internal activities of sovereign states, with the clandestine support of *coups d'etat* (Iran, 1954; Guatemala, 1954; Chile 1972), involvements in wars of independence (Korea, 1950–1953; Vietnam, 1961–1976), in violent repression of national upheavals in the GDR (1953) and Poland and Hungary (1956), and in wars against liberation from colonial rule in Africa and Asia, notably in Algeria (1954–1961). The superpowers have repeatedly used their veto against condemnations of their military force abroad.[43]

Since the end of the Cold War the unilateral violent aggression and war by the Russian Federation against Ukraine has not been the only violation of the UN Charter by a Permanent Member of the Security Council, using its veto to prevent a condemnation of its aggression and collective countermeasures under Chapters VI and VII of the UN Charter. Among the major unilateral violations by the US and its allies without the support of a UN Security Council Resolution were the bombing of Yugoslavia in 1999 by the Clinton administration, the invasion of Afghanistan in 2001 and the war against Iraq in 2003 by the G. W. Bush administration, and the bombing of Libya in 2011 by the Obama Administration (Cohn 2022).

Russia violated the UN Charter with the invasion and annexation of the Crimean peninsula in 2014 and with a major war of aggression against Ukraine in 2022, despite the explicit security guarantees it had granted in the Budapest Memorandum of 1994 (Merezhko 2015). The UN failed to act in Kosovo, in Iraq and in several cases in Africa (Allain 2004).

[43] See: "List of vetoed United Nations Security Council resolutions", at: https://en.wikipedia.org/wiki/List_of_vetoed_United_Nations_Security_Council_resolutions#cite_note-1 (6 September 2023).

1.6 Uncertain Global Trends

Since the military aggression of Russia against Ukraine the security order in Europe has fallen apart, while globalisation is in retreat. The aggressor Russia and its political, military, economic and religious leadership have been objects of Western (NATO, EU, US) sanctions. Russian exports of fossil energy sources to Western (and especially EU) countries have stopped and have been partly redirected to China and India and to other countries, especially in the Global South.

Since 1990 the US has been the sole hegemon, while the President of the Russian Federation aimed at regaining the position as a co-equal partner with the US. China has benefited from both Deng Xiaoping's economic reforms and US-led economic globalisation in trade and finance to become the largest producer of goods and the second-largest economic power. China has also expanded its science base and significantly increased its military expenditure.

The US military interest in Russia is to weaken Russia's military potential; this began with its Lend and Lease Policy first introduced in 1941 during World War II when both countries were fighting against the German aggressor. US policy also tries to contain China's military potential and its economic and scientific goals to become a second hegemon competing with or replacing the USA. This new global trend may lead to another Cold War between democracies (of OECD, NATO, and EU countries) and autocracies (Russia and China et al.). Many states in the Global South (Fig. 1.14) did not participate in Western sanctions against Russia and military aid to Ukraine. These developments may result in a new multipolar world system where the prospects for new arms control and disarmament treaties and prospects for global climate agreements are—at least during the ongoing war—declining (Menzel 2023).

A change is observable towards a massive rearmament of NATO, the Russian Federation and China while the spending for cooperative climate change policies is declining. This may also have negative impacts on policies for global poverty reduction and in addressing this goal at the national level, thus leading to progressing economic inequality.

Since February 2022 geopolitical concerns and analyses have moved to the top of the global, regional and national policy agendas. Since then, national military spending has significantly increased, not only in Russia and Ukraine, but also in most NATO, EU and many OECD countries, while there was hardly any progress in the climate change negotiations in COP 27 in Sharm el Sheik (Egypt)[44] in November–December 2022 and during the Bonn climate talks[45] in June 2023.[46]

What are the key points of the political summary of the Synthesis Report of the Sixth Assessment of the IPCC and the projections until 2050 and 2100, and what

[44] COP 27 of IPCC in Sharm el Shaik, in November/December 2022; see: at: https://unfccc.int/cop27.

[45] See at: https://unfccc.int/sb58.

[46] COP 28 of IPCC in the United Arab Emirates, from 30 November to 12 December 2023; at: https://unfccc.int/cop28.

are the key messages of the risk report of the World Economic Forum published in 2023? The security agenda triggered by the war in Ukraine has, in many Western countries, replaced climate change as a major threat or challenge—both in political urgency and in public spending.

1.6.1 Increased Urgency of Global Climate Change: Synthesis of the IPCC's Sixth Assessment Report, 20 March 2023

The Intergovernmental Panel on Climate Change (IPCC) was established based on a Resolution of the UN's General Assembly (48/58) that was adopted on 6 December 1988 with a mandate

> to prepare a comprehensive review and recommendations with respect to the state of knowledge of the science of climate change; the social and economic impact of climate change, and potential response strategies and elements for inclusion in a possible future international convention on climate. Since 1988, the IPCC has had [six] assessment cycles and delivered [six] Assessment Reports, the most comprehensive scientific reports about climate change produced worldwide. It has also produced a range of Methodology Reports, Special Reports and Technical Papers, in response to requests for information on specific scientific and technical matters from the United Nations Framework Convention on Climate Change (UNFCCC), governments and international organizations consequences of climate change. … The Fifth Assessment Report (AR5) was finalized between 2013 and 2014. It provided the scientific input into the Paris Agreement.[47]

During its Sixth Assessment cycle the IPCC prepared several Special Reports, a Methodology Report and the Sixth Assessment Report on *Global Warming of 1.5 °C* (SR15), a *2019 Refinement* to the 2006 IPCC Guidelines on National Greenhouse Gas Inventories, a *Special Report on Climate Change and Land* (SRCCL, 2019) and the *Special Report on the Ocean and Cryosphere in a Changing Climate* (SROCC, 2019). The Sixth Assessment Report (AR6)[48] resulted in three reports of its three working groups between 2021 and 2022 and the Synthesis Report that was released on 20 March 2023 that noted in its summary for policy makers:

> A.1 Human activities, principally through emissions of greenhouse gases, have unequivocally caused global warming, with global surface temperature reaching 1.1 °C above 1850–1900 in 2011–2020. Global greenhouse gas emissions have continued to increase, with unequal historical and ongoing contributions arising from unsustainable energy use, land use and land-use change, lifestyles and patterns of consumption and production across regions, between and within countries, and among individuals (high confidence). …

[47] "The *Intergovernmental Panel on Climate Change* (IPCC) was established by the *United Nations Environment Programme* (UNEP) and the *World Meteorological Organization* (WMO) in 1988"; at: https://www.ipcc.ch/about/history/ (28 March 2023).

[48] The three Working Group contributions to AR6 are: AR6 Climate Change 2021: The Physical Science Basis; AR6 Climate Change; 2022: Impacts, Adaptation and Vulnerability; and AR6 Climate Change 2022: Mitigation of Climate Change.

> A.1.3 Observed increases in well-mixed GHG concentrations since around 1750 are unequivocally caused by GHG emissions from human activities over this period. Historical cumulative net CO_2 emissions from 1850 to 2019 were 2400 ± 240 $GtCO_2$ of which more than half (58%) occurred between 1850 and 1989, and about 42% occurred between 1990 and 2019 (high confidence). In 2019, atmospheric CO_2 concentrations (410 parts per million) were higher than at any time in at least 2 million years (high confidence), and concentrations of methane (1866 parts per billion) and nitrous oxide (332 parts per billion) were higher than at any time in at least 800,000 years (very high confidence). (IPCC 2023)

This report was hardly noted and underreported in the quality media (in daily and weekly newspapers, by radio and TV stations) in the spring of 2023 and did not create public controversy. The summary for policy makers of the synthetic report of the Sixth IPCC Assessment Report noted in March 2023 (Fig. 1.12):

> A.1.4 Global net anthropogenic GHG emissions have been estimated to be 59 ± 6.6 $GtCO_2$-eq in 2019, about 12% (6.5 $GtCO_2$-eq) higher than in 2010 and 54% (21 $GtCO_2$-eq) higher than in 1990, with the largest share and growth in gross GHG emissions occurring in CO_2 from fossil fuels combustion and industrial processes (CO_2-FFI) followed by methane, whereas the highest relative growth occurred in fluorinated gases (F-gases), starting from low levels in 1990. Average annual GHG emissions during 2010–2019 were higher than in any previous decade on record, while the rate of growth between 2010 and 2019 (1.3% year-1) was lower than that between 2000 and 2009 (2.1% year-1). In 2019, approximately 79% of global GHG emissions came from the sectors of energy, industry, transport and buildings together and 22% from agriculture, forestry and other land use (AFOLU). Emissions reductions in CO_2-FFI due to improvements in energy intensity of GDP and carbon intensity of energy, have been less than emissions increases from rising global activity levels in industry, energy supply, transport, agriculture and buildings (high confidence).

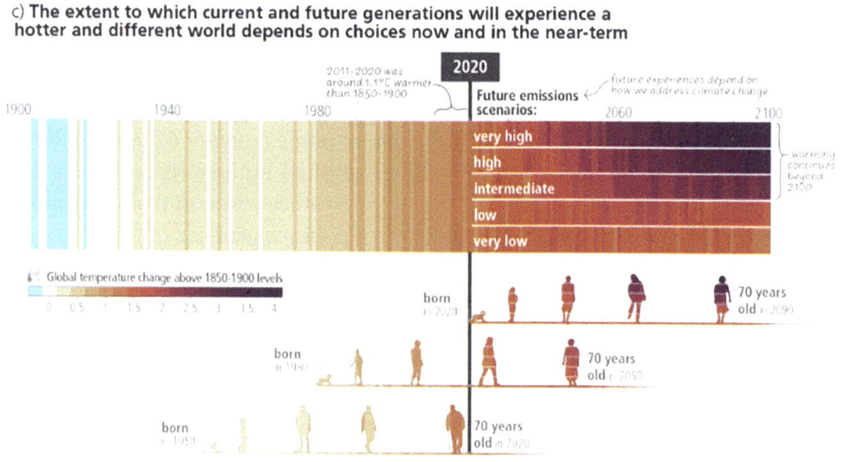

Fig. 1.12 The extent to which current and future generations will experience a hotter and different world depends on choices now and in the near-term. *Source* IPCC: *Synthesis Report of the IPCC Sixth Assessment Report (AR6), Summary for Policymakers* (Geneva: IPCC, 2023); Sig. SPM 1: IPCC_AR6_SYR_SPM.pdf. The figure is in the public domain

The summary for policy makers of the IPCC's Synthesis Report of March 2023 clearly distinguishes the different levels of historical contributions of GHG emissions, noting that the Small Island Developing States and least developed countries—with lower emissions—experience higher impacts:

> A.1.5 Historical contributions of CO_2 emissions vary substantially across regions in terms of total magnitude, but also in terms of contributions to CO_2-FFI and net CO_2 emissions from land use, land-use change and forestry (CO_2-LULUCF). In 2019, around 35% of the global population live in countries emitting more than 9 tCO_2-eq per capita[11] (excluding CO_2-LULUCF) while 41% live in countries emitting less than 3 tCO_2-eq per capita; of the latter a substantial share lacks access to modern energy services. Least developed countries (LDCs) and Small Island Developing States (SIDS) have much lower per capita emissions (1.7 tCO_2-eq and 4.6 tCO_2-eq, respectively) than the global average (6.9 tCO_2-eq), excluding CO_2-LULUCF. The 10% of households with the highest per capita emissions contribute 34–45% of global consumption-based household GHG emissions, while the bottom 50% contribute 13–15%. …

The IPCC Report also noted that:

> A.2 Widespread and rapid changes in the atmosphere, ocean, cryosphere and biosphere have occurred. Human-caused climate change is already affecting many weather and climate extremes in every region across the globe. This has led to widespread adverse impacts and related losses and damages to nature and people (high confidence). Vulnerable communities who have historically contributed the least to current climate change are disproportionately affected (high confidence).

The IPCC's Sixth Synthesis Report addresses the gap between national commitments and policy implementation and the lack of financial resources "to meet climate goals":

> A.4 Policies and laws addressing mitigation have consistently expanded since AR5. Global GHG emissions in 2030 implied by nationally determined contributions (NDCs) announced by October 2021 make it likely that warming will exceed 1.5 °C during the 21st century and make it harder to limit warming below 2 °C. There are gaps between projected emissions from implemented policies and those from NDCs and finance flows fall short of the levels needed to meet climate goals across all sectors and regions (high confidence).

According to the *Summary for Policymakers of the Synthesis Report of the IPCC Sixth Assessment Report* (2023), the impact of these trends will differ significantly for the older and the younger generations (IPCC 2023). The report describes future trends of climate change:

> B.1 Continued greenhouse gas emissions will lead to increasing global warming, with the best estimate of reaching 1.5 °C in the near term in considered scenarios and modelled pathways. Every increment of global warming will intensify multiple and concurrent hazards (high confidence). Deep, rapid, and sustained reductions in greenhouse gas emissions would lead to a discernible slowdown in global warming within around two decades, and also to discernible changes in atmospheric composition within a few years (high confidence).
>
> B.2 For any given future warming level, many climate-related risks are higher than assessed in AR5, and projected long-term impacts are up to multiple times higher than currently observed (high confidence). Risks and projected adverse impacts and related losses and damages from climate change escalate with every increment of global warming (very high

confidence). Climatic and non-climatic risks will increasingly interact, creating compound and cascading risks that are more complex and difficult to manage (high confidence).

B.2.1 In the near term, every region in the world is projected to face further increases in climate hazards (medium to high confidence, depending on region and hazard), increasing multiple risks to ecosystems and humans (very high confidence). Hazards and associated risks expected in the near-term include an increase in heat-related human mortality and morbidity (high confidence), food-borne, water-borne, and vector-borne diseases (high confidence), and mental health challenges (very high confidence), flooding in coastal and other low-lying cities and regions (high confidence), biodiversity loss in land, freshwater and ocean ecosystems (medium to very high confidence, depending on ecosystem), and a decrease in food production in some regions (high confidence). Cryosphere-related changes in floods, landslides, and water availability have the potential to lead to severe consequences for people, infrastructure and the economy in most mountain regions (high confidence). The projected increase in frequency and intensity of heavy precipitation (high confidence) will increase rain-generated local flooding (medium confidence).

C.6 Effective climate action is enabled by political commitment, well-aligned multilevel governance, institutional frameworks, laws, policies and strategies and enhanced access to finance and technology. Clear goals, coordination across multiple policy domains, and inclusive governance processes facilitate effective climate action. Regulatory and economic instruments can support deep emissions reductions and climate resilience if scaled up and applied widely. Climate resilient development benefits from drawing on diverse knowledge.

However, the *Summary for Policymakers* of the *Synthesis Report of the IPCC Sixth Assessment Report* does not give a critical analysis of why governments have again failed to implement their national commitments. Thus, the "climate paradox" (Brauch 2012) has again not been addressed, possibly because the governments have to approve this text before it can be released to the public, the media and citizens.

C.7 Finance, technology and international cooperation are critical enablers for accelerated climate action. If climate goals are to be achieved, both adaptation and mitigation financing would need to increase many-fold. There is sufficient global capital to close the global investment gaps but there are barriers to redirect capital to climate action. Enhancing technology innovation systems is key to accelerate the widespread adoption of technologies and practices. Enhancing international cooperation is possible through multiple channels (high confidence).

Again, the *Synthesis Report of the IPCC Sixth Assessment Report* neither discusses the reasons why the financial commitments have again not been met nor which governments have again failed to do so since the Fifth Assessment Report was released (2013/2014–2021/2023). It is unlikely that the goals for the seventh Assessment Cycle (2023–2027/2030) will be met if the world enters a new Cold War between democracies and autocracies during the next decade, with high military spending and an economic containment of the largest global economy and emitter of GHG (especially of CO_2). The ambitious climate policy goals cannot be met without the active involvement of China and Russia and the BRICs countries.

However, due to the costs of the war in Ukraine, and the rebuilding of Ukraine once the war has ended, it is highly unlikely "that adaptation and mitigation financing" will increase "manyfold" during the period before the seventh IPCC Assessment Report is scheduled to be released at end of this decade. Successful climate negotiations are required to regain "trust" in facing and coping with the impacts of multiple climate-induced hazards. This requires a willingness to achieve compromises with regard to the terms, by both parties: both those in the security realm and those in the area of global environmental change. So far the interlinkages of both issue areas have been ignored by policy makers and have hardly been addressed by social scientists in peace and security studies on the one side, and global environmental research on the other, due to the prevailing education and thinking in two different 'silos' that has made conceptual bridge-building difficult.

1.6.2 Increasing Risks as Addressed by the Global Economic Forum in Switzerland in January 2022 and 2023

The 17th edition of the World Economic Forum (WEF)'s *Global Risks Report 2022–Insight Report* was written by Marsh McLennan, the SK Group and the Zurich Insurance Group, and reflected contributions of its three Academic Advisers: the National University of Singapore, the Oxford Martin School and the Wharton Risk Management and Decision Presses Center, University of Pennsylvania. Its risk perception survey for 2022 noted that:

- Global risk perceptions highlight societal and environmental concerns
- A divergent economic recovery threatens collaboration on global challenges
- *A disorderly climate transition will exacerbate inequalities*
- Growing digital dependency will intensify cyber threats
- Barriers to mobility risk compounding global insecurity
- Opportunities in space could be constrained by frictions
- Year two of the pandemic yields insights on resilience

The Executive Summary and the full *Global Risks Report 2022* of the WEF was published before Russia's aggression against Ukraine and discusses "a global disorderly climate transition" that will exacerbate inequalities:

> Respondents to the GRPS rank 'climate action failure' as the number one long-term threat to the world and the risk with potentially the most severe impacts over the next decade. Climate change is already manifesting rapidly in the form of droughts, fires, floods, resource scarcity and species loss, among other impacts. ... Yet a disorderly climate transition characterized by divergent trajectories worldwide and across sectors will further drive apart countries and bifurcate societies, creating barriers to cooperation. Given the complexities of technological, economic and societal change at this scale, and the insufficient nature of current commitments, it is likely that any transition that achieves the net zero goal by 2050 will be disorderly.
>
> Countries continuing down the path of reliance on carbon-intensive sectors risk losing competitive advantage through a higher cost of carbon, reduced resilience, failure to keep up with technological innovation and limited leverage in trade agreements. Yet shifting away from carbon-intense industries, which currently employ millions of workers, will trigger economic volatility, deepen unemployment and increase societal and geopolitical tensions. Adopting hasty environmental policies will also have unintended consequences for nature

> … while lack of public support for land use transitions or new pricing schemes will create political complications that further slow action. A transition that fails to account for societal implications will exacerbate inequalities within and between countries, heightening geopolitical frictions.

The 18th edition of the WEF's *Global Risks Report 2023–Insight Report*, published after Russia's aggression against Ukraine, noted in its Executive Summary:

> As 2023 begins, the world is facing a set of risks that feel both wholly new and eerily familiar. We have seen a return of 'older' risks—inflation, cost-of-living crises, trade wars, capital outflows from emerging markets, widespread social unrest, geopolitical confrontation and the spectre of nuclear warfare—which few of this generation's business leaders and public policy-makers have experienced. These are being amplified by comparatively new developments in the global risks landscape, including unsustainable levels of debt, a new era of low growth, low global investment and de-globalization, a decline in human development after decades of progress, rapid and unconstrained development of dual-use (civilian and military) technologies, and the growing pressure of climate change impacts and ambitions in an ever-shrinking window for transition to a 1.5 °C world. Together, these are converging to shape a unique, uncertain and turbulent decade to come. …
>
> The *Global Risks Report 2023* presents the results of the latest Global Risks Perception Survey (GRPS). We use three time frames for understanding global risks. Chapter 1 considers the mounting impact of current crises (i.e. global risks which are already unfolding) on the most severe global risks that many expect to play out over the short term (two years). Chapter 2 considers a selection of risks that are likely to be most severe in the long term (10 years), exploring newly emerging or rapidly accelerating economic, environmental, societal, geopolitical and technological risks that could become tomorrow's crises. Chapter 3 imagines mid-term futures, exploring how connections between the emerging risks outlined in previous sections may collectively evolve into a 'polycrisis' centred around natural resource shortages by 2030. The report concludes by considering perceptions of the comparative state of preparedness for these risks and highlighting enablers to charting a course to a more resilient world. Below are key findings of the report. …
>
> The report describes four potential futures centred around food, water and metals and mineral shortages, all of which could spark a humanitarian as well as an ecological crisis—from water wars and famines to continued overexploitation of ecological resources and a slowdown in climate mitigation and adaption. Given uncertain relationships between global risks, similar foresight exercises can help anticipate potential connections, directing preparedness measures towards minimizing the scale and scope of polycrises before they arise.

Among the 10 largest global risks ranked by severity over the short-term (2023–2025) the WEF 2023 lists 5 climate-related risks: natural disasters and extreme weather events (2nd most severe); failure to mitigate climate change (4th); large-scale environmental damage incidents (6th); failure of climate change adaptation (7th); and natural rescue crises (9th). These constitute three societal risks, one geopolitical risk and one technological risk. For government officials, three of ten risks are climate change-related, while business expects five of ten risks to be climate-related. With regard to the strategies to be considered for the future, the WEF 2023 Report concludes:

> In this context, defensive, fragmented and crisis-oriented approaches are short-sighted and often perpetuate vicious cycles. Lack of preparedness for longer-term risks will destabilize the global risks landscape further, bringing ever tougher trade-offs for policy-makers and business leaders scrambling to address simultaneous crises. A rigorous approach to foresight and preparedness is called for, as we aim to bolster our resilience to longer-term risks and chart a path forward to a more prosperous world.
>
> Each risk requires concerted, specific and customized efforts but several cross-cutting principles can support preparedness across themes. In this concluding section, we outline four principles for preparedness in this new era of concurrent shocks: 1) strengthening risk identification and foresight, 2) recalibrating the present value of 'future' risks, 3) investing in multi-domain risk preparedness, and 4) strengthening preparedness and response cooperation.

The WEF Report addresses neither the short-term consequence of the Russian war against Ukraine on climate diplomacy nor the strategic goals of some policy advisers and decision makers to contain both Russia and China in the geopolitical effort to uphold US hegemony in a multipolar world, given a neutral Global South that has resisted being co-opted by the West or by China and Russia by shifting public expenditure from climate change to a new longer-term arms race with the consequences of these policies on the prospects of parties and government for being re-elected.

1.6.3 A Dual Challenge: Anthropocene Versus Hobbesian Geopolitics

Two existential discourses have co-existed since the end of the Cold War (1989–2022):

- The emerging debates on the national and international environment and on global environmental and climate change have been present since the 1960s, 1970s and 1980s. Crutzen's proposed concept of the Anthropocene and his thesis of a "geology of [hu]mankind" linked human action and its impact on the Earth system and the complex interactions of anthropogenic interventions, the physical effects of climate change and potentially violent societal outcomes that may trigger chaotic interactions in the climate system, resulting in geopolitical tipping points with (at present) unforeseeable existential implications for different regions of the planet. In this new evolving existential Anthropocene discourse, humankind as a species has become a threat to our own survival, which only we can counter with human ingenuity and multiple innovations.
- The international order of 1945, at the end of World War II, resulted in a Cold War with a costly global arms competition, or nuclear arms race, which made a nuclear war among the two major nuclear armed superpowers impossible.
- Since Russia, as a permanent member of the UN Security Council with a veto power, has become a revisionist power—at least since 2014 with its annexation of the Crimean Peninsula from Ukraine and 24 February 2022 with its aggression

against Ukraine—the liberal international order that emerged after 1945 and was consolidated after 1990, and the prevailing legal, political, security and military order among the big powers has collapsed in Europe.
- With the Russian aggression of 2022 the Anthropocene discourse and its political agenda has been pushed to the back burner, while Hobbesian pessimism and military geopolitics have prevailed and the global cooperative liberal order and its proponents have nearly disappeared from the political discourse.
- COP 27 of the UNFCCC at Sharm el-Sheikh (6–18 November 2022) produced only minimal agreements with a compromise on "loss and damage"'. The war in Ukraine has clearly replaced climate change as an existential threat in public perceptions and on the key global political agenda.
- Since February 2022, the co-existence of two major survival threats and discourses—environment versus war—has resulted in a clear priority on the political security agenda compared with global environmental challenges. However, the embargo against Russian oil purchases by EU countries, the US, Canada and Australia, and the Russian cut-off of gas supplies to selected EU member states may expedite the transition to renewable energies (solar energy, wind power and green hydrogen). However, the geoeconomics of "energy insecurity", not climate change, has become a key economic driver of innovation.

1.7 Emerging New Cold War in Anthropocene's Third Phase: Relevance for Political Science and *Politik*

The Anthropocene will be discussed in Chap. 2 as a new context for Political Science and for the political dimension of human activities since the end of World War II. This recontextualisation of politics, policy and polity requires a rethinking of the basic goals and means of *Politik* since the turning point in world history that evolved during the 1940s in the Second World War in Europe (1939–1945) and in Asia (1941–1945), with the start of the nuclear era and the build-up of the post-World War II international order.

Preliminary proposals of the AWG of May 2019 place this turning point as coinciding with the end of the Holocene epoch of Earth history and the start of a new epoch, the Anthropocene. In May 2019 a majority of the voting members of the AWG concluded that the first nuclear weapons test in Alamogordo in New Mexico on 16 July 1945 was a highly visible event whose impacts could already be observed decades later in sediments in the neighbourhood of the many atmospheric nuclear weapons tests between 1945 and 1963, when the Limited Nuclear Weapons Test Treaty prohibited nuclear weapons tests in the atmosphere. This Treaty was also a major turning point in the first phase of the Cold War and the limited détente that encouraged arms control agreements, economic cooperation between the two rival military powers and alliances and diplomatic consultations on human rights protection in the framework of the Final Act of Helsinki of the CSCE on 1 August 1975.

Key questions that will be discussed below are:

- What impacts have been triggered in the turn from the Holocene to the Anthropocene since the decade of a dual structural change in Earth and human history for the definition of *Politik* (1940–1950)? Have these changes in the definition of the concept been reflected in the definition of *Politik* from 1940 until the early 2020s by policy makers and social scientists? So far, no evidence for a textual adaptation of the concept of *Politik* and of its three related terms of "politics", "policy" and "polity" can be observed.
- The structural changes in international and national *Politik* significantly changed the context of factors that influenced policy making, often widening the scope but limiting the impact of individual policy makers on persons and arenas.
- Did the timely coincidence of these dual changes in political and Earth history influence the perceptions of threats, vulnerabilities and risks in policies on security with respect to peace, conflicts and wars?

Since the decade of multiple changes (1940–1950) two different policy perspectives and narratives have developed, based on different sources of knowledge and evidence, that offer different answers to the fundamental question: which existential dangers, threats, vulnerabilities and risks do we as human beings, as members of nation states and as a human species face?

In classical thinking since the French Revolution (1789), three schools of thought or political ideologies or macro-theories have co-existed—conservatism, liberalism and socialism (von Beyme 2003)—which have excluded environmental developments and changes in Earth systems and in human and nature connections. These worldviews have influenced the mindset of policy makers and scientists.

Between 1940 and 1950 the environment as a social and political issue did not yet exist, and issues of global environmental change were not even addressed by the mainstream natural sciences until the 1970s, and for policy makers and policy institutions until the late 1980s, with the first legal frameworks (regimes) being approved in June 1992 at the Rio Earth summit. Ecology as a fourth major school of political thought and ideology gradually evolved between the 1960s and 1970s and influenced environmental networks and movements, and since the 1980s has been instrumental in the setting up of green political parties.

In the three classical schools of thought the major threat was posed by the "other", at the level of the family, ethical religious group, country or alliance, that might destroy human existence—in the worst case, through a major nuclear war. This form of threat dominated during the Cold War (1947–1990) and was reflected in the nuclear deterrence doctrine of a mutually assured destruction (MAD).

Some ecologists have challenged this view by claiming that we have become the major threat to our own survival as human beings and as a species through our lifestyles, economic production, and consumption and especially through our use of fossil energies in production (of food and goods), consumption, and for heating, electricity generation and transportation. If humankind and governments fail to make timely decisions, future generations will have to pay the costs and for the damage caused by an increase in the number and intensity of extreme weather events (Stern

2006, 2007). These consequences cannot be attributed to the "other" and cannot be solved by any type of military weapon. As we, as human beings and as humankind, are the source of this fundamentally different threat, *we must become the solution together* and bear the consequences that will not distinguish between rich and poor people and countries. The poor will face a higher degree of social and environmental vulnerability (Bohle et al. 1994; Bohle 2009).

However, with Russia's war of aggression against Ukraine since 2014 (with the annexation of the Crimean Peninsula) and since 24 February 2022 (with an invasion into the territory of Ukraine) Russia violated its own guarantees for Ukraine's territorial integrity outlined in the agreement of the dissolution of the Soviet Union in December 1991 and in the Budapest Memorandum of December 1994. The liberal international security order in Europe has been destroyed by a revisionist power that has, with its veto, also totally paralysed the UN Security Council from implementing its international legal duties and obligations under the United Nations Charter of 26 June 1945.

While a large majority of UN members condemned Russia's act of aggression in the UN Security Council, and especially in the UN General Assembly, the policy response of the members of the United Nations has been divided:

- A new Cold War of the West (world of democratically organised countries) in Europe (all countries except Serbia and the two remaining neutral countries of Austria and Switzerland), North America (USA, Canada), and the Asia Pacific (Australia, New Zealand, Japan, South Korea, Taiwan) against the world of autocracies (Russia, China, Iran, North Korea et al.) has emerged (Menzel 2023).
- However, the majority of countries and of the world population has remained neutral and did not participate in implementing economic sanctions against Russia, among them especially the so-called BRICS countries (Brazil, Russia, India, China, South Africa) plus the five new BRICS members (Egypt, Ethiopia, Iran, Saudi Arabia, United Arab Emirates) that joined on 24 August 2023. Argentina had also joined in August 2023 but its new government withdrew in December 2023.
- Including a few democratic states from the OECD world (Chile, Israel, Mexico, Turkey, etc.).

A large number of countries representing the Global South from Africa, Asia and Latin America did not support Ukraine with weapons and ammunition supplies and continued to purchase Russian coal, gas and oil despite Western economic sanctions by NATO and EU countries.

In November 2022, about two months after the G-20 meeting in Bali (Indonesia), the International Crisis Group (Fig. 1.13) documented the divided attitude of the G-20 countries between three groups of states:

(a) those that had condemned Russia at the UN,
(b) those that had condemned Russia at the UN but opposed sanctions and
(c) those that did not condemn Russia (China, India, South Africa) and remained neutral.

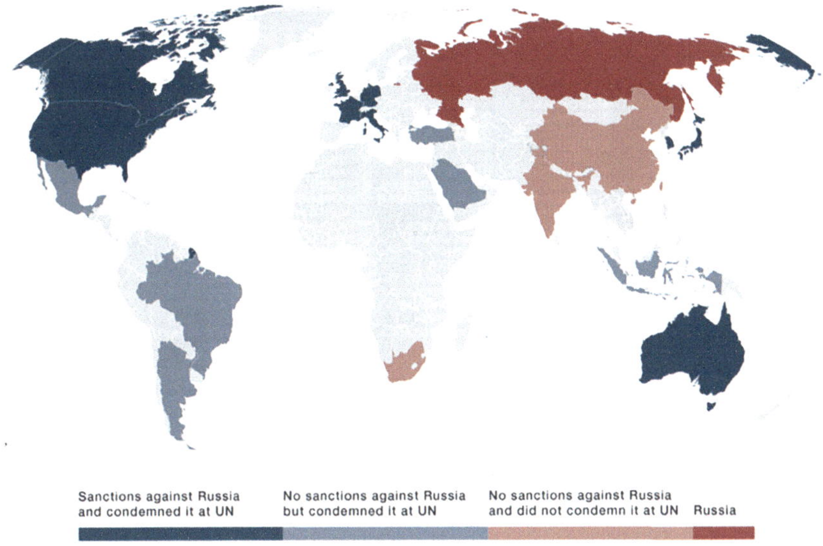

Fig. 1.13 The International Crisis Group (Nov. 2022), documented after the G20 meeting in Bali (September 2022). *Source* See at: https://www.crisisgroup.org/global/toward-common-set-signals-g20-about-russias-war-ukraine (20 February 2023)

The global environmental agenda, most particularly the global negotiations on anthropogenic climate change in the framework of the UNFCC (1992) and the Paris Climate Change Treaty (2015), has been downgraded and delayed. The publication of the Sixth Synthesis Report of the IPCC[49] has repeatedly been delayed, COP 27 of the UNFCCC in Sharm-El-Shaik[50] from 6 to 20 November 2022 failed to achieve its previously announced goals, and many experts have become doubtful whether the 1.5 °C goal of the parties to the Paris climate change treaty can still be realised.

Both NATO and EU countries have, in the meantime, announced and stressed a need to significantly enhance the military spending of their member states and to supply more heavy weapons and ammunition to Ukraine—and even to establish a "war economy"—although they are not direct parties to the conflict but are trying to avoid becoming one and striving to reduce the likelihood of an escalation of an all-out conventional war into a nuclear war (Fig. 1.14).

In September 2023 a small growth in countries in the Global South that do not support Western policies against Russia and China can be observed. Two analysts of the Washington think tank CSIS assessed the possible impact of the BRICS' enlargement from 5 to 11 countries:

> With the addition of Saudi Arabia, the UAE, and Iran, this expanded group would include three of the world's largest oil exporters and would constitute 42 percent of global oil supply.

---

[49] See IPCC; at: https://www.ipcc.ch/report/sixth-assessment-report-cycle/ (20 February 2023).

[50] See UNFCC; at: https://unfccc.int/cop27 (20 February 2023).

Fig. 1.14 Vote of the General Assembly of the United Nations (UNGA Res. No. A/ES-11/L.7, 23 February 2023) "calling for a comprehensive, just and lasting peace in Ukraine in line with the principles of the Charter of the United Nations". *Source* See at: https://news.un.org/en/story/2023/02/1133847. This figure is in the public domain

> … But over the long term, an expanded BRICS grouping could be significant for energy markets. … More recently, the EU embargoes on Russian seaborne crude oil and petroleum products and the EU-G7 price caps have created a new sanctions mechanism that targets revenues rather than export volumes. … An enlarged BRICS would include both oil and gas exporters and two of the largest importers, China and India—both of which refused to join the 'price cap coalition' targeting Russia. Producers and consumers in this group have a shared interest in creating mechanisms to trade commodities outside the reach of the G7 financial sector. … Dollar-denominated energy trade persists for many reasons: the dollar is liquid and freely convertible. … As energy sanctions have become more prevalent, a growing number of bilateral energy deals are being settled in other currencies such as the Chinese renminbi or Indian rupees. … For energy markets, enlarging BRICS is largely symbolic for now—but it is another sign that countries are exploring ways to skirt the U.S. financial system and the reach of the dollar. (Baskaran/Cahill 2023)

All initial five BRICS states and two new members (Argentina and Saudi Arabia) are among the G-20 countries.[51] However, bilateral security and political problems between two BRICSs members (India and China) became obvious at the G-20 meeting in New Delhi on 9 and 10 September 2023. The twenty G-20 member states—including all five original BRICS states—account "for 80 percent of the global gross domestic product (GDP), 75 percent of international trade and about two-thirds of the world population".

Thus, in September 2023 it is impossible to foresee and project the future of the global international economic or peace and security order, or the importance

[51] See: Aljazeera: "Explainer: G20 in India: When will 2023 New Delhi summit be held and who is attending?"; at: https://www.aljazeera.com/news/2023/9/5/g20-in-india-when-will-2023-new-delhi-summit-be-held-and-who-is-attending (6 September 2023).

attributed to an Anthropocene policy agenda within the existing global environmental regimes that were established in June 1992 during the UN Earth Summit. As long as this uncertainty dominates, the peace and security agenda from a realist or Hobbesian perspective is likely to prevail, while the Anthropocene agenda will be further downgraded by this policy world view of researchers and the scientific community, and the respective mindset of policy makers. However, the renewed postponement of courageous strategies and decisions towards a sustainability transition as part of a socio-ecological transformation will increase the costs of non-acting, as Stern (2006, 2007) stressed in his review 15 years ago.

References

Allain, J., 2004: "The True Challenge to the United Nations System of the Use of Force: The Failures of Kosovo and Iraq and the Emergence of the African Union", in: *Max Planck UNYB 8*: 237–289.

Almond, G., 1996: "Political Science: The History of the Discipline", in: Goodin, R. E.; Klingemann, H.-D. (Eds.): *A New Handbook of Political Science* (Oxford – New York: Oxford University Press): 50–96.

Angus, I., 2016: *Facing the Anthropocene: Fossil Capitalism and the Crisis of the Earth System* (New York: Monthly Review Press).

Antweiler, C., 2022: *Anthropologie im Anthropozän: Theoriebausteine für das 21. Jahrhundert* (Darmstadt: Wissenschaftliche Buchgemeinschaft Academic).

Arias-Maldonado, M.; Trachtenberg, Z. (Eds.), 2019: *Rethinking the Environment for the Anthropocene: Political Theory and Socionatural Relations in the New Geological Epoch* (Abingdon: Routledge).

Arias-Maldonado, M.; Trachtenberg, Z., 2019: "Introduction", in: Arias-Maldonado, M.; Trachtenberg, Z, (Eds.), 2019: *Rethinking the Environment for the Anthropocene: Political Theory and Socio-natural Relations in the New Geological Epoch* (Abingdon: Routledge): 1–16

Arizpe, L.; Price, M. F.; Worcester, R., 2016: "The First Decade of Initiatives for the Research on the Human dimensions of Global (environmental) change (1986–1995)", in: Brauch, H. G.; Oswald Spring, Ú.; Grin, J.; Scheffran, J. (Eds.): *Handbook on Sustainability Transition and Sustainable Peace* (Cham: Springer International Publishing Switzerland): 349–358.

Arrhennius, S., 1896: "On the Influence of Carbonic Acid in the Air upon the Temperature of the Ground", in: *Philosophical Magazine and Journal of Science*, Series 5, 41 (April): 237–276.

Barber, W. F.; Bartlett, R. V., 2016: "Democratic Accountability in the Anthropocene: Toward a Non-legislative Model", in: Pattberg, P.; Zelli, F. (Eds.), 2016: *Environmental Politics and Governance in the Anthropocene – Institutions and Legitimacy in a Complex World* (London – New York: Routledge): 167–183.

Baskaran, G.; Cahill, B., 2023: "Six New BRICS: Implications for Energy Trade" *(Washington, D.C.: Center for Strategic and International Studies (CSIS))*; at: https://www.csis.org/analysis/six-new-brics-implications-energy-trade (6 September 2023).

Berger, P. L.; Luckmann, T., 1966, 1967: *The Social Construction of Reality: A Treatise in the Sociology of Knowledge* (Garden City, NY: Anchor).

Bevir, M., 2022: *A History of Political Science* (Cambridge: Cambridge University Press).

Beyme, K. von, 2002: *Politische Theorien im Zeitalter der Ideologien 1789–1945* (Wiesbaden: Westdeutscher Verlag).

Beyme, K. von, 2013: *Konservatismus: Theorien des Konservatismus und Rechtsextremismus im Zeitalter der Ideologien 1789–1945* (Wiesbaden: Springer VS).

Beyme, K. von, 2013a: *Liberalismus: Theorien des Liberalismus und Radikalismus im Zeitalter der Ideologien 1789–1945* (Wiesbaden: Springer VS)
Beyme, K. von, 2013b: *Sozialismus: Theorien des Sozialismus, Anarchismus und Kommunismus im Zeitalter der Ideologien 1789–1945* (Wiesbaden: Springer VS).
Beyme, K., von, 2019: *Rightwing Populism. An Element of Neodemocracy* (Cham: Springer International Publishing, 2019).
Biermann, F., 2014: *Earth Systems Governance. World Politics in the Anthropocene* (Cambridge, MA – London: MIT Press).
Biermann, F., 2016: "Epilogue Politics for a New Earth: Governing in the 'Anthropocene'", in: Nicholson, S.; Jinnah, S. (Eds.): *New Earth Politics: Essays From the Anthropocene* (Cambridge, MA: The MIT Press): 405–420.
Biermann, F.; Lövbrand, E. (Eds.), 2019: *Anthropocene Encounters: New Directions in Green Political Thinking* (Cambridge – New York – Melbourne: Cambridge University Press).
Biro, A., 2006: *Denaturalizing Ecological Politics. Alienation from Nature from Rousseau to the Frankfurt School and Beyond* (Toronto: Toronto University Press).
Blinken, A. J., Secretary of State: "Russia's Planned Withdrawal of Its CTBT Ratification"; at: https://www.state.gov/russias-planned-withdrawal-of-its-ctbt-ratification/ (2 Nov. 2023).
Blühdorn, I.; Welsh, I., 2007: "Eco-politics beyond the paradigm of sustainability: A conceptual framework and research agenda", in: *Environmental Politics*, 16, 2: 185–205.
Bohle, H.-G., 2009: "Sustainable Livelihood Security. Evolution and Application", in: Brauch, H. G.; Oswald Spring, Ú.; Grin, J.; Mesjasz, C.; Kameri-Mbote, P.; Behera, N. Chadha; Chourou, B.; Krummenacher, H. (Eds.): *Facing Global Environmental Change: Environmental, Human, Energy, Food, Health and Water Security Concepts* (Berlin – Heidelberg – New York: Springer-Verlag): 521–528.
Bohle, H. G.; Downing, T. E.; Watts, M. J., 1994: "Climate change and social vulnerability: Toward a sociology and geography of food insecurity", in: *Global Environmental Change*, 4, 1 (March): 37–48.
Borger, J. 2022: "Pentagon's chief's remarks show shift in US declare aims in Ukraine", in: *The Guardian*, 25 April 2022; at: https://www.theguardian.com/world/2022/apr/25/russia-weaked end-lloyd-austin-ukraine (22 March 2023).
Brauch, H. G., 1976: *Struktureller Wandel and Rüstungspolitik der USA (1940–1950): Zur Weltführungsrolle and ihren innenpolitischen Bedingungen* [Structural Change and Armament Policy of the USA (1940–1950): Towards its World Leadership Role and its Domestic Conditions] (Inaugural doctoral dissertation, Ruprecht-Karl University of Heidelberg, Faculty of Philosophy and History, 30 June).
Brauch, H. G., 1977: *Struktureller Wandel and Rüstungspolitik der USA (1940–1950): Zur Weltführungsrolle and ihren innenpolitischen Bedingungen* [Structural Change and Armament Policy of the USA (1940–1950): Towards its World Leadership Role and its Domestic Conditions] (Ann Arbor – London: University Microfilms).
Brauch, H. G., 1996: *Klimapolitik: Naturwissenschaftliche Grundlagen, internationale Regimebildung and Konflikte, ökonomische Analysen sowie nationale Problemerkennung and Politikimplementation* (Berlin – Heidelberg: Springer-Verlag).
Brauch, H. G., 2012: "Climate paradox of the G-8: legal obligations, policy declarations and implementation gap", in: *Revista Brasileira de Política Internacional*, 55 (spe); at: https://doi.org/10.1590/S0034-73292012000300003.
Brauch, H. G., 2020: "Peace Ecology in the Anthropocene", in: Oswald Spring, Ú.; Brauch, H. G. (Eds.): *Decolonising Conflicts, Security, Peace, Gender, Environment and Development in the Anthropocene* (Cham: Springer Nature Switzerland): 51–186.
Brauch, H. G., 2021: "The Anthropocene concept in the natural and social sciences, the humanities and law – a bibliometric analysis and a qualitative interpretation (2000–2020), in: Benner, S.; Lax, G.; Crutzen, P. J.; Pöschl, U.; Lelieveld, J.; Brauch, H. G. (Eds.): *Paul J. Crutzen and the Anthropocene: A New Epoch in Earth's History* (Cham: Springer Nature Switzerland): 289–438.

Brauch, H. G.; Oswald Spring, Ú.; Grin, J.; Scheffran, J. (Eds.), 2016: *Handbook on Sustainability Transition and Sustainable Peace (Cham:* Springer International Publishing).

Braudel, Fernand, 1966: *La Méditerranée et le Monde Méditerranéen à l'époque de Philippe II* (Paris: Armand Colin).

Braudel, Fernand, 1969: "Histoire et Science Sociales: La Longue Durée", in: *Écrits Sur l'Histoire* (Paris: Flammarion): 41–84.

Braudel, Fernand, 1972: *The Mediterranean and the Mediterranean World in the Age of Philip II,* 2 volumes (New York: Harper & Row).

Brockhaus Enzyklopädie [21st ed.], 2006: "Politik", in: *Brockhaus Enzyklopädie,* vol. 21 (Leipzig, Mannheim: Brockhaus Enzyklopädie): 655–656.

Buzan, B.; Lawson, G., 2015: *The Global Transformation: History, Modernity and the Making of International Relations* (Cambridge: Cambridge University Press).

Chandler, D.; Müller, F.; Rothe, D. (Eds.), 2021: *International Relations in the Anthropocene - New Agendas, New Agencies and New Approaches* (Cham: Springer Nature Switzerland).

Charakbarty, D., 2009: "The Climate of History: Four theses", in: *Critical Inquiry*, 35, 2: 197–222.

Charakbarty, D., 2021: *The Climate of History in a Planetary Age* (Chicago: The University of Chicago Press).

Charakbarty, D., 2022: *Das Klima der Geschichte im planetarischen Zeitalter* (Berlin: Suhrkamp).

Code, L., 2006: *Ecological Thinking: The Politics of Epistemic Location* (Oxford: Oxford University Press).

Cohn, M., 2022: "Serial US Violations of the UN Charter", in: *IADL* [International Association of Democratic Lawyers] (1 June 2022); at: https://iadllaw.org/2022/06/serial-us-violations-of-the-un-charter-marjorie-cohn/ (21 March 2023).

Congressional Research Service, 2022: *Arms Control and Nonproliferation: A Catalogue of Treaties and Agreements* (Washington D.C.: CRS, 25 April).

Council of Councils, 2023: "Global Perspectives - The BRICS Summit 2023: Seeking an Alternate World Order?"; at: https://www.cfr.org/councilofcouncils/global-memos/brics-summit-2023-seeking-alternate-world-order (6 September 2023).

Crutzen, P. J., 2002: "Geology of mankind – The Anthropocene", in: *Nature*, 415: 23.

Crutzen, P. J.; Birks, J. W., 1982: "The Atmosphere After a Nuclear War: Twilight at Noon", in: *Ambio*, 11, 2, 3: 114–125.

Crutzen, P. J.; Schwägerl, C., 2011: "Living in the Anthropocene: Toward a New Global Ethos", in: *Yale Environment*, 360.

Crutzen, P. J.; Stoermer, E. F., 2000: "The Anthropocene", in: *IGBP Newsletter*, 41 (May): 17–18; at: http://www.igbp.net/download/18.316f18321323470177580001401/1376383088452/NL41.pdf.

Dalby, S., 2016: "Framing the Anthropocene: The Good, the Bad and the Ugly", in: *Anthropocene Review*, 3, 1: 33–51.

Dennison, S.; 2022: "Green peace: How Europe's climate policy can survive the war in Ukraine" (Brussels: European Council on Foreign Relations, 27 June 2022; at: https://ecfr.eu/publication/green-peace-how-europes-climate-policy-can-survive-the-war-in-ukraine/ (30 March 2023).

Department of State, Office of the Historian: "Helsinki Final Act, 1975", at: https://history.state.gov/milestones/1969-1976/helsinki (8 January 2023).

Dobson, A., 2007: *Green Political Thought* (London: Routledge).

Dobson, A.; Eckersley, R., 2006: *Political theory and the ecological challenge* (Cambridge: Cambridge University Press).

Dryzek, J. S.; Pickering, J., 2019: *The Politics of the Anthropocene* (Oxford: Oxford University Press).

Ehlers, E., 2016: "From HDP to IHDP: "Evolution of the International Humans Dimensions of Global Environmental change Programme", in: Brauch, H. G.; Oswald Spring, Ú.; Grin, J.; Scheffran, J. (Eds.): *Handbook on Sustainability Transition and Sustainable Peace* (Cham: Springer International Publishing Switzerland): 359–376.

Engelke, P.; McNeill, J. R., 2016: "Earth Day: Are we at the beginning of a new geological era?", in: *Washington Post* (21 April).
Fagan, M., 2016: "Security in the Anthropocene: Environment, ecology, escape", in: *European Journal of International Relations*, 1–23.
Fagan, M., 2017: "Security in the Anthropocene: environment, ecology, escape", in: *European Journal of International Relations*, 23, 2: 292–314.
Falcon-Lang, H.: "Anthropocene: How humans created a new geological age"; at: *BBC website*, 11 May.
Falkenhayn, L. von; Rechkemmer, A.; Young, O. T., 2011"The International Human Dimensions Programme on Global Environmental Change – Taking Stock and Moving Forward", in: Brauch, H. G.; Oswald Spring, Ú; Mesjasz, C., et al. (Eds.): *Coping with Global Environmental Change, Disasters and Security – Threats, Challenges, Vulnerabilities and Risks* (Berlin – Heidelberg – New York: Springer-Verlag): 1221–1234.
Glaser, M.; Krause, G.; Ratter, B. M. W.; Welp, M. (Eds.), 2012: *Human-Nature Interactions in the Anthropocene: Potentials of Social-Ecological Systems Analysis* (London: Routledge).
Goodin, R. E.; Klingemann, H.-D.: "Political Science; The Discipline", in: Goodin, R. E.; Klingemann, H.-D. (Eds.): *A New Handbook of Political Science* (Oxford – New York: Oxford University Press): 3–49.
Häckel, E., 1866: *Generelle Morphologie der Organismen*, vol. 1: *Allgemeine Anatomie der Organismen*; vol.2: *Allgemeine Entwicklungsgeschichte der Organismen* (Georg Reimer).
Hacking, Ian, 1999: *The Social Construction of What?* (Cambridge, MA: Harvard University Press).
Hallé, C.; Milon, A-S., 2020: "The infinity of the Anthropocene: A (Hi)story with a thousand names", in: Latour, B.; Weibel, P. (Eds.): *Critical Zones: The Science and Politics of Landing on Earth* (pp.). (Cambridge, MA: MIT Press): 44–49.
Hamilton, C.; Bonneuil, C.; Gemenne, F., 2015: *The Anthropocene and the Global Environmental Crisis: Rethinking modernity in a new epoch* (London – New York: Routledge).
Harrington, C., 2016: "The Ends of the World: International Relations and the Anthropocene", in: *Millennium*, 44, 3: 478–498.
Harrington, C.; Shearing C., 2017: *Security in the Anthropocene. Reflections on Safety and Care* (Bielefeld: Transcript).
Hickmann, T.; Partsch, L.; Pattberg, P.; Weiland, S. (Eds.), 2019: *The Anthropocene Debate and Political Science* (London – New York: Routledge).
Hickmann, T.; Partzsch, L.: Pattberg, P.; Weiland, S., 2019a: "Introduction: A political science perspective on the anthropocene", in: Hickmann, T.; Partsch, L.; Pattberg, P.; Weiland, S. (Eds.): *The Anthropocene Debate and Political Science* (London - New York: Routledge): 1–12.
Hickmann, T.; Partzsch, L.; Pattberg, P.; Weiland, S. (Eds.), 2019b: "Conclusion: towatds a 'deep debate' on the Anthropocene", in: Hickmann, T.; Partzsch, L.; Pattberg, P.; Weiland, S. (Eds.): *The Anthropocene Debate and Political Science* (London – New York: Routledge): 237–251.
Hickmann, T.; Partzsch, L.; Pattberg, P.; Weiland, S., 2020: "Mehr Engagement der Politikwissenschaft in der Anthropozän-Debatte", in: *Politische Vierteljahresschrift, 61*, 4: 659–670.
Holsti, K. J., 1991, 1992: *Peace and War: Arms Conflicts and International Order 1648–1989* (Cambridge: Cambridge University Press).
Holsti, K. J., 2015: *Kalevi Holsti: A Pioneer in International Relations Theory, Foreign Policy Analysis, History of International Order, and Security Studies* (Cham – Heidelberg – New York – Dordrecht – London: Springer).
Holsti, K. J., 2016: *Kalevi Holsti: Major Texts on War, Peace, and International Order* (Cham – Heidelberg – New York – Dordrecht – London: Springer, 2015).
Humphrey, M., 2007: *Ecological Politics and Democratic Theory. The Challenge to the Deliberative Ideal* (London: Routledge).
Hussein, S.; Cocodia, J., 2023: *African Security in the Anthropocene* (Cham: Springer Nature Switzerland).
IEA, 2020: *World Energy Outlook* (Paris: OECD).
IEA, 2021: *CO_2 Highlight* (Paris: OECD).

IPCC, 2023: *Synthesis Report of the IPCC Sixth Assessment Report (AR6), Summary for Policymakers* (Geneva; 2023).

Kelly, D., 2019: *Politics and the Anthropocene* (Cambridge – Medford, MA: Polity).

Kiyala, J. C. K.; Chivasa, N. (Eds.), 2023: *Peace, Security and Sustainable Development: Climate Change and Socio-political Violence in Sub-Saharan Africa* (Cham: Springer Nature Switzerland).

Kiyala, J. C. K.; Harris, G. T., 2022: Civil Society and Peacebuilding in Sub-Saharan Africa in the Anthropocene: An Overview (Cham: Springer Nature Switzerland).

Latour, B.; Schultz, N., 2022: *Zur Entstehung einer ökologischen Klasse: Ein Memorandum* (Berlin: Suhrkamp).

Latour, B.; Schultz, N., 2022a: *On the Emergence of an Ecological Class: A Memo* (Cambridge: Polity).

Leemans, R.; Rice, M.; Henderson-Sellers, A.; Noone, K., 2011: Research Agenda and Policy Input of the Earth Systems Science Partnership for Coping with global Environmental Change", in: Brauch, H. G.; Oswald Spring, Ú; Mesjasz, C., et al. (Eds.): *Coping with Global Environmental Change, Disasters and Security – Threats, Challenges, Vulnerabilities and Risks* (Berlin – Heidelberg – New York: Springer-Verlag): 1205–1220.

Lenton, T.; Held, H.; Kriegler, E.; Hall, J. W.; Lucht, W.; Ramstorf, S.; Schellnhuber, H. J., 2008: "Tipping Elements in the Earth's Climate System", in: *Proceedings of the National Academy of Science* (PNAS), 105, 6 (12 February): 1, 786–1, 793.

Lenton, T. M.; Rockström, J.; Gaffney, O.; Rahmstorf, S.; Richardson, K.; Steffen, W.; Schellnhuber, H. J., 2019: "Climate Tipping Points – Too Risky to Bet Against: The Growing Threat of Abrupt and Irreversible Climate Changes Must Compel Political and Economic Action on Emissions", in: *Nature*, 575 (27 November): 592–595.

Lewis, S. L.; Maslin, M. A., 2015: "Defining the Anthropocene", in: *Nature*, 519, 7542: 171–180.

Lewis, S. L.; Maslin, M. A., 2018: *The Human Planet: How We Created the Anthropocene* (New Haven: Yale University Press).

Lockwood, B.; Lockwood, M., 2022: "How Do Right-Wing Populist Parties Influence Climate and Renewable Energy Policies? Evidence from OECD Countries", in: *Global Environmental Politics*, 22, 3: 12–37.

McKenzie Prillaman, 2022: "Are we in the Anthropocene? Geologists could define new epoch for Earth - Researchers have zeroed in on nine sites that could describe a new geological time, marked by pollution and other signs of human activity", in: *Nature*, (13 December).

McKenzie Prillaman: "Geologists seek to define the Anthropocene", in: *Nature*, 613, 7942: 14–15.

McKenzie Prillaman: "There's been a fundamental change in our planet'; *Nature*, 613, 7942 (5 January): 14–15.

McKie, R., 2023: "Hunt on for spot to mark the start of the Anthropocene epoch - Scientists are to pick a location that sums up the current epoch when Homo sapiens made its mark", in: *The Guardian*, 1 January.

McNeill, J. R.; Engelke, P., 2014: *The Great acceleration, an environmental History of the Anthropocene since 1945* (Cambridge, MA – London: The Belknap Press of Harvard University Press).

Meier, C., 1989: "Politik", in: Ritter, J.; Gründer, K. (Eds.): *Historisches Wörterbuch der Philosophie* (Darmstadt: Wissenschaftliche Buchgesellschaft), vol. 7: 1038–1047.

Mendelsohn, E., 1977: "The Social Construction of Scientific Knowledge", in: Mendelsohn, E.; Weingart, P.; Whitley, R. (Eds.): *The Social Production of Scientific Knowledge: Sociology of the Sciences: A Yearbook*, vol. 1 (Dordrecht: Springer).

Menzel, U., 2023: *Wendepunkte zum autoritären Jahrhundert* (Berlin: Suhrkamp).

Merezhko, O., 2015: "Crimea's Annexation by Russia – Contradictions of the New Russian Doctrine of International Law", in: *ZaöRV*, 75: 167–194.

Millennium Ecosystem Assessment [MEA], 2005: Ecosystems and Human Well-Being: Synthesis-Report (Washington, DC: Island Press).

Mitchell, A., 2017: "Is IR going extinct", in: *European Journal of International Relations*, 23, 1: 3–25.
Moore, J. M. (Ed.), 2016: *Anthropocene or Capitalocene? Nature, History, and the Crisis of Capitalism* (Oakland, CA: PM Press).
Morton, T., 2010: *The Ecological Thought* (Cambridge, Mass: Harvard University Press).
Nayak, A. K. J. R, (Ed.). 2019: *Transition Strategies for Sustainable Community Systems. Design and Systems Perspectives* (Cham: Springer International Publishing, 2019).
Nicholson, S.; Jinnah, S. (Eds.), 2016: *New Earth Politics. Essays from the Anthropocene* (Cambridge, MA: MIT Press).
Noone, K.; Nobre, C.; Seitzinger, S., 2011: "The International Geosphere and Biosphere Programme's (IGBP Scientific Research Agenda for Coping with Global Environmental Change", in: Brauch, H. G.; Oswald Spring, Ú; Mesjasz, C., et al. (Eds.): *Coping with Global Environmental Change, Disasters and Security – Threats, Challenges, Vulnerabilities and Risks* (Berlin – Heidelberg – New York: Springer-Verlag): 1249–1256.
Notter, H., 1949: *Postwar Foreign Policy Preparation, 1939–1945* (Washington, D.C.: U.S. Government Printing Office).
Oak Ridge National Laboratory, 2017: *Carbon Dioxide Information Analysis Center* (Oak Ridge, Tennessee: Oak Ridge National Laboratory).
OSCE, Charter of Paris for a New Europe, Paris 1990; at: https://www.osce.org/mc/39516.
Padilla, L.-A., 2021: *Sustainable Development in the Anthropocene. Towards a New Holistic and Cosmopolitan Paradigm* (Cham: Springer International Publishing).
Pattberg, P.; Zelli, F. (Eds.), 2016: *Environmental Politics and Governance in the Anthropocene: Institutions and legitimacy in a complex world* (Abingdon: Routledge).
Pauls, C.; Oppelt, M.; Weber, N. K., et al. 2022: "Das Planetarische Politische Denken [Planetary Political Thinking]", in: *Politische Vierteljahresschrift*, 4, 2022: 703–728.
Polanyi, K., 1944: *The Great Transformation: The Political and Economic Origins of Our Time* (Boston: Beacon Press).
Prado de Sousa, C.; Serrano Oswald, S. E. (Eds.), 2021: *Social Representations for the Anthropocene: Latin American Perspectives* (Cham: Springer International Publishing, 2021).
Purdy, J. (2015): *After Nature: A Politics for the Anthropocene* (Cambridge, MA – London: Harvard University Press).
Renn, J., 2020: *Evolution of Knowledge: Rethinking Science for the Anthropocene*, (Princeton - Oxford: Princeton University Press).
Rodhe, H.; Charlson, R.; Crawford, E., 1997: "Svante Arrhenius and the Greenhouse Effect", in: *Ambio*, 26, 1 (February): 2–5; at: http://www.jstor.org/stable/4314542.
Rosenau, J.; Czempiel, E. O., 1992: *Governance without Government: Order and Change in World Politics* (London: Cambridge University Press).
Roth, A.; Borger, J., 2021: "Putin says Russia will halt participation in New Start nuclear arms treaty", in: *The Guardian* (21 February 2021); at: https://www.theguardian.com/world/2023/feb/21/putin-russia-halt-participation-new-start-nuclear-arms-treaty (30 March 2023).
Rothe, D.; Müller, F.; Chandler, D., 2019: "Introduction: International Relations in the Anthropocene", in: Chandler, D.; Müller, F.; Rothe, D. (Eds.), 2021: *International Relations in the Anthropocene - New Agendas, New Agencies and New Approaches* (Cham: Springer Nature Switzerland): 1–16.
Ruddiman, W. F., 2003: "The anthropogenic greenhouse era began thousands of years ago", in: Climatic Change, 61, 3: 261–293; at: https://doi.org/10.1023/B:CLIM.0000004577.17928.fa.
Ruddiman, W. F., 2005: Plows, Plagues, and Petroleum: How Humans Took Control of Climate (Princeton, NJ: Princeton University Press).
Ruddiman, W. F., 2016: "Late Holocene climate: natural and anthropogenic", in: Review of Geophysics, 54: 93–118.

Sabbagh, D.; Livingstone, H., 2022: "US pledges extra $713m for Ukraine war effort and to weaken Russia", in: *The Guardian*, 25 April 2; at: https://www.theguardian.com/world/2022/apr/25/us-diplomats-to-return-to-ukraine-and-fresh-military-aid-unveiled-after-blinken-visit (22 March 2023).

Sarotte, M. E., 2021: *Not One Inch: America, Russia, and the Making of Post-Cold War Stalemate* (New Haven – London: Yale University Press).

Sassen, S., 1991, 2001: The *Global City* (Princeton: Princeton University Press).

Sassen, S., 2008: "The Global City", in: Nugent, David; Vincent, Joan (Eds.): *A Companion to the Anthropology of Politics* (Hoboken, NJ: Wiley-Blackwell, March): 168–178; at: https://doi.org/10.1002/9780470693681.ch11.

Scheffran, J., 2011: *Climate Change, Nuclear Risks and Nuclear Disarmament: From Security Threats to Sustainable Peace. Report World Future Council* (Hamburg: World Future Council).

Scheffran, J. A., 2019: "The entwined Cold War roots of missile defense and climate geoengineering", in: *Bulletin of the Atomic Scientists*, 75, 5: 222–228; https://doi.org/10.1080/00963402.2019.1654256.

Scheffran, J.; Burroughs, J.; Leidreiter, R.; van Riet, R.; Ware, A., 2016: *The Climate-Nuclear Nexus* (London: World Future Council).

Schellnhuber, H. J.; Wenzel, V. (Eds.). 1998: *Earth System Analysis – Integrating Science for Sustainability* (Berlin – Heidelberg: Springer-Verlag).

Schellnhuber, H. J.; Crutzen, P. J.; Clark, W. C.; Claussen, M.; Held, H. (Eds.), 2004: *Earth Systems Analysis for Sustainability* (Cambridge, Mass.: MIT Press).

Schmidt, M. G., 2004: "Politik", in: *Wörterbuch zur Politik* (Stuttgart: Kröner): 538.

Schwägerl, C., 2015: "A Concept with a Past", in: Möllers, N.; Schwägerl, C.; Trischler, H. (Eds.), 2014: *Welcome to the Anthropocene: The Earth in Our Hands* (Munich: Deutsches Museum): 128–129.

Sellin, V., 1978: "Politik", in: Brunner, O.; Conze, W.; Koselleck, R. (Eds.): *Geschichtliche Grundbegriffe – Historisches Lexikon zur politisch-sozialen Sprache in Deutschland* (Stuttgart: Klett-Cotta), vol. 4: 789–854.

Simangan, D., 2019: "Situating the Asia Pacific in the age of the Anthropocene", in: *Australian Journal of International Affairs*, 73, 6: 564–584.

Simangan, D., 2020: "Literature review Where is the Anthropocene? IR in a new geological epoch", in: *International Affairs*, 96, 1: 211–227.

Simangan, Dahlia, 2020a: "Where is the Anthropocene? IR in a new geological epoch", in: *International Affairs*, 96, 1 (January): 211–224.

Simangan, D., 2021: "Where is the Asia Pacific in mainstream international relations scholarship on the Anthropocene?", in: *Pacific Review*, 34, 2021, 5: 724–746.

Smith, S., 2003: "The United States and the Discipline of International Relations: 'Hegemonic Country, Hegemonic Discipline'", in: *International Studies Review* 4, 2 (February): 67–85.

Smith, B. D.; Zeder, M. A., 2013: "The onset of the Anthropocene", in: *Anthropocene*, 4 (December): 8–13.

Steffen, W., 2004: "Group Report: Earth Systems Dynamics in the Anthrpocene", in: Schellnhuber, H. J.; Crutzen; Clark, W. C.; et al. (Eds.): *Earth System Analysis for Sustainability* (Cambridge, MA - Berlin: MIT Press – Freie Universität Berlin): 313–340.

Steffen, W., 2006: "The anthropocene, global change and sleeping giants: Where on Earth are we going?", in: *Carbon Balance and Management*, 1, 1: 3.

Steffen, W., 2010: "Observed trends in Earth System behavior", in: *Wiley Interdisciplinary Reviews: Climate Change*, 1, 3: 428–449.

Steffen, W., 2013: "Commentary: Paul J. Crutzen and Eugene F. Stoermer", in: Robbin, L.; Sörlin, S.; Warde, P. (Eds.): *The Future of Nature. Documents of Global Change* (New Haven: Yale University Press): 486–490.

Steffen, W., 2021: "The Earth System, the Great Acceleration, and the Anthropocene", in: Williams, S. J.; Taylor, R. (Eds.): *Sustainability and the New Economics: Synthesizing ecological economics and modern monetary theory* (Cham: Springer Nature Switzerland AG): 21.

Steffen, W., 2021a: "Introducing the Anthropocene: The human epoch", in: *Ambio*, https://link.springer.com/article/10.1007/s13280-020-01489-4 (9 August 2022).
Steffen, W.; Jäger, J.; Carson, D. J.; Bradshaw, C. (Eds.), 2002: *Challenges of a Changing Earth - Proceedings of the Global Change Open Science Conference, Amsterdam, The Netherlands, 10–13 July 2001* (Berlin Heidelberg: Springer-Verlag).
Steffen, W.; Andreae, M. O.; Bolin, B.; Cox, P. M.; *Crutzen*, P. J.; Cubasch, U.; Held, H.; Nakicenovic, N.; Talaue-McManus, L.; Turner II, B. L., 2004: "Group Report: Earth Systems Dynamics in the Anthropocene", in: Schellnhuber, H. J.; Crutzen, P. J.; Clark, W. C.; Claussen, M.; Held, H. (Eds.): *Earth System Analysis for Sustainability* (Cambridge, MA; London: MIT Press): 313–340.
Steffen, W.; Sanderson, R. A.; Tyson, P. D.; Jäger, J.; Matson, P. A.; Moore III, B.; Oldfield, F.; Richardson, K.; Schellnhuber, H. J.; Turner, B. L.; Wasson, R. J., 2004: *Global Change and the Earth System: A planet under pressure* (Berlin - Heidelberg: Springer-Verlag).
Steffen, W.; Crutzen, P. J.; McNeill, J. R., 2007: "The Anthropocene: Are Humans Now Overwhelming the Great Forces of Nature?", in: *Ambio*, 36: 614–621.
Steffen, W.; Grinevald, J.; Crutzen, P.; McNeill, J., 2011: "The anthropocene: Conceptual and historical perspectives", in: *Philosophical Transactions of the Royal Society A: Mathematical, Physical and Engineering Sciences*, 369, 1938; 842–867.
Steffen, W.; Broadgate, W.; Deutsch, L.; Gaffney, O.; Ludwig, C., 2015: "The trajectory of the anthropocene: The great acceleration", in: *Anthropocene Review*, 2, 1: 81–98.
Steffen, W.; Rockström, J.; Richardson, K.; et al., 2018: "Trajectories of the Earth System in the Anthropocene", in: *PNAS*, 115, 33: 8252–8259 (6 August 2023); at: https://www.pnas.org/doi/10.1073/pnas.1810141115.
Steffen, W.; Richardson, K.; Rockström, J.; Schellnhuber, H. J.; Dube, O. P.; Dutreuil, S.; Lenton, T. M.; Lubchenco, J., 2020: "The emergence and evolution of Earth System Science", in: *Nat Rev Earth Environ*, 1: 54–63; at: https://doi.org/10.1038/s43017-019-0005-6.
Stern, Nicholas, 2007: *The Economics of Climate Change* (Cambridge: Cambridge University Press).
Subramanian, M., 2019: "Anthropocene now: influential panel votes to recognize Earth's new epoch 'Atomic Age' would mark the start of the current geologic time unit, if proposal receives final approval", in: *Nature* (21 May).
Tokar, B., 1988: "Social Ecology, Deep Ecology and the Future of Green Political Thought", in: *Ecologist*, 18, 4–5: 132–141.
Trischler, H. (Eds.), 2015: *Welcome to the Anthropocene: The Earth in Our Hands* (Munich: Deutsches Museum): 128–129.
Tritschler, H., 2016: "The Anthropocene: A Challenge for the History of Science, Technology, and the Environment", in: *NTM Zeitschrift für Geschichte der Wissenschaften, Technik und Medizin*, 24: 309–335; at: https://doi.org/10.1007/s00048-016-0146-3.
U.S. EPA, 2019: *Global Non CO_2 Emission Projections (Washington, D.C.: EPA).*
Vollrath, E., 1989: "Politik", in: Ritter, J.; Gründer, K. (Eds.): *Historisches Wörterbuch der Philosophie* (Darmstadt: Wissenschaftliche Buchgesellschaft), vol. 7: 1055–1071.
Waters, C. N.; Turner, S. D., 2022: "Defining the onset of the Anthropocene - Twelve sites are considered for defining the Anthropocene geological epoch", in: *Science* (17 Nov.), 378, 6621: 706–708; at: https://doi.org/10.1126/science.ade2310.
Waters, C. N.; Zalasiewicz, J.; Summerhayes, C.; et al. 2018: "Global Boundary Stratotype Section and Point (GSSP) for the Anthropocene Series: Where and how to look for potential candidates", in: *Earth-Science Reviews*, 178: 379–429.
Weinacht P. L., 1989: "Politik", in: Ritter, J.; Gründer, K. (Eds.): *Historisches Wörterbuch der Philosophie* (Darmstadt: Wissenschaftliche Buchgesellschaft), vol. 7: 1047–1055.
Weizsäcker, Ernst-Ulrich von, 1989, ^{5}1997: *Erdpolitik. Ökologische Realpolitik als Antwort auf die Globalisierung* (Darmstadt: Wissenschaftliche Buchgemeinschaft).

Wezeman, P. D.; Gadon, J.; Wezeman, S. R.: *SIPRI Fact Sheet, March 2023: Trends in International Arms Transfers, 2022*, at: https://www.sipri.org/sites/default/files/2023-03/2303_at_fact_sheet_2022_v2.pdf (22 March 2023).

Williams, S. J.; Taylor, R. (Eds.): *Sustainability and the New Economics: Synthesizing ecological economics and modern monetary theory* (Cham: Springer Nature Switzerland AG).

Williams, M.; Zalasiewicz, J.; Haff, P., et al., 2015: "The Anthropocene biosphere", in: *The Anthropocene Review*, 2, 3: 196–219.

Wolfe, A. P.; Birks, H. H.; Briner, J. P.; Holmgren, S. U.; Ingólfsson, Ó.; Kaushal, S. S.; Miller, G. H.; Pagani, M.; Saros, J. E.; Vinebrooke, R. D., 2013: "Stratigraphic expressions of the Holocene–Anthropocene transition revealed in sediments from remote lakes", in: *Earth-Science Reviews*, 116 (January): 17–34.

Zalasiewicz, J., 2008: "Our Geological footprint", in: Wilkinson, C. (Ed.): *The Observer Book of the Earth* (London: Observer Books).

Zalasiewicz, J., 2008a: "Are we now living in the Anthropocene?", in: *Geological Society of America Today*, 18: 4–8.

Zalasiewicz, J., 2015: "Epochs. Disputed start dates for Anthropocene", in: *Nature*, 520: 436.

Zalasiewicz, M.; Williams, A.; Smith, et al., 2008: "Are we now living in the Anthropocene?", in: *GSA Today*, 18, 2 (4 February): 4–8; https://doi.org/10.1130/GSAT01802A.1.

Zalasiewicz, J.; Waters, C. N.; Barnosky, A. D.; …; Williams, M.; Wolfe, A. P., 2015: "Colonization of the Americas, 'little ice age' climate, and bomb produced carbon: Their role in defining the anthropocene", in: *Anthropocene Review*, 2, 2: 117–127.

Chapter 2
The Anthropocene and International Political Order: Towards an Integrated Analysis

Hans Günter Brauch

Abstract The Anthropocene is thus far the only epoch in Earth history in which humankind as a "geological actor" has directly intervened in and thus impacted on the Earth system. In the early twenty-first century two totally different narratives co-exist: The peace and security or war and peace narrative and the ecological and environmental narrative. This author argues that an integrated analysis of both narratives is needed in the social sciences and humanities on peace and security, and in the natural and social sciences on environmental issue areas and ecological concerns in the Anthropocene, the start of which represents both a new epoch in Earth history and a new global international order since 1945. Peace and security problems and environmental issues and ecological concerns are interrelated and must thus be analysed and addressed from a common political strategy if we are to face and cope with fundamental threats to human history that are associated with the atomic bomb and the Anthropocene. This chapter addresses two threats for the survival of humankind: (a) the nuclear threat, the security dilemma and the logic of nuclear deterrence and (b) the impacts of global climate change that create threats to human survival. When the third phase of post-World War II order began is disputed, and the assessments have differed among policy analysts. For the USA, the 9/11 attack by terrorists on its territory was perceived as the key turning point by which its "war on terror" (especially the War against Iraq and Libya) was justified. The US and Western use of force in the Balkans (in Bosnia in 1995, and Serbia in 1999) and in the extended Middle East conflict (in Afghanistan in 2001, Iraq in 2003, Libya in 2011) set a precedent of a unilateral use of force that was not endorsed by the UNSC or the UN Charter. The

Hans Günter Brauch has been chairman of the non-profit scientific society *Peace Research and European Security Studies* (AFES-PRESS) since 1987 and since May 2020 chairman of the board of *the Hans Günter Brauch Foundation for Peace and Ecology in the Anthropocene* (HGB Foundation); Email: hg.brauch@onlinehome.de.

The author is grateful to Czeslaw Mesjasz (ret. from Economic Univ. Cracow), and Georg Simonis (ret. from FU Hagen) for very thoughtful comments and suggestions.

H. G. Brauch (✉)
Hans Günter Brauch Foundation on Peace and Ecology in the Anthropocene (HGBS), Peace Research and European Security Studies (AFES-PRESS), Mosbach, Germany
e-mail: hg.brauch@onlinehome.de

H. G. Brauch (ed.), *Towards Rethinking Politics, Policy and Polity in the Anthropocene*, The Anthropocene: Politik—Economics—Society—Science 35,
https://doi.org/10.1007/978-3-031-71807-6_2

Russian invasion of Ukraine in 2014 and the war against Ukraine which began on 24 February 2022 were also severe violations of international law. The response to the Russian war in Europe has already resulted in a downgrading of the response to the effects of global environmental change to fragmentation and uncertainty during the third phase of the Anthropocene. This author calls for a rethinking of 'politics', 'policy' and 'polity' in the Anthropocene, "thinking geological and geographic time together", "combining the peace and security narrative with the ecological and Anthropocene narrative" and "integrating Earth Systems Science" (ESS) with the "history of structures and events" (Braudel).

Keywords Anthropocene · Atomic bomb · Climate change · Earth history epoch · Earth Systems Science · Ecological and environmental narrative · Fragmentation · Geological and geographic time · Geological actor · Global international order · Humankind · Integrated analysis · Nuclear threat · Nuclear deterrence · Peace and security narrative · Political order · Russian war · Security dilemma · Survival dilemma · Ukraine uncertainty

2.1 Simultaneous Change in Earth and Political History: Need for a Joint and Integrated Perspective

In the Anthropocene epoch of Earth history humankind has, for the first time, directly intervened in and thus impacted on the Earth system as a "geological actor" (Crutzen 2002; Antweiler 2022, 2023, 2023a). In Earth history the transition from the Holocene epoch—that started with the end of the last glacial period 11,700 years ago—to the Anthropocene occurred in the mid-twentieth century (between 1940 and 1950).

The Anthropocene Working Group (AWG) agreed in a majority vote in 2019 that the first test of a nuclear bomb on 16 July 1945 in Alamogordo, New Mexico ended the Holocene and marks the start of the Anthropocene.[1] This unique event was a direct outcome of the top secret Manhattan Project in Los Alamos, where the first nuclear weapons were developed by top natural scientists from the US with émigrés—among them several Nobel laureates—from the UK, Germany, Italy, Hungary, Poland and other European countries.

[1] See: Subcommission on Quaternary Stratigraphy, Working Group on the 'Anthropocene': "Results of binding vote by AWG, Released 21st May 2019", at: http://quaternary.stratigraphy.org/working-groups/anthropocene/. See the detailed story of the decision-making process of the Anthropocene Working Group up to July 2023 cited in the 'Introduction' in this book's front matter. See also a documentary on Joseph Rotblat: "The Strangest Dream".

The fundamental change in the post-war order—where the consequences of the nuclear bomb became a major "game changer"—has been analysed in the different scientific disciplines of History, International Law, Philosophy, and International Relations, as well as in Peace and Security Studies. In the early twenty-first century two totally different narratives co-exist:

- The *peace and security* or *war and peace narrative* still prevails. This was planned, negotiated and implemented by top decision-makers and government officials—primarily in the United States—and was later analysed in the social sciences, especially in History, International Law, International Relations, and Peace and Security Studies, but also in Philosophy.
- The *ecological and environmental narrative* has evolved gradually since the 1960s in the global environmental change and the Earth Systems Science community, primarily in the natural sciences. The narrative has been increasingly framed by the Anthropocene concept which the Dutch Nobel laureate P. J. Crutzen suggested during a discussion in the International Geophysical and Biological Programme (IGBP) on 23 February 2000 in Cuernavaca in Mexico (see Chap. 1; Crutzen/Stoermer 2000: 17–18; Benner et al. 2021).

In both narratives the development of the nuclear bomb, its first test, and its first (and so far, only) use, against Hiroshima and Nagasaki on 6 and 9 August 1945, became a multiple "game changer". The nuclear bombs may have ended the war against Japan, and they made the US the single most powerful military, economic and political 'superpower' or hegemon, that framed and implemented the post-war order between 1945 and 1950.

Within the geological scientific community, the Anthropocene Working Group (AWG) voted in May 2019, with a large majority, that "the primary guide for the base of the Anthropocene [should] be one of the stratigraphic signals *around the mid-twentieth century of the Common Era*".[2] The AWG also noted that "The 'Anthropocene' has developed a range of meanings among vastly different scholarly communities". This is reflected in the scientific consensus that argues that the Anthropocene began with the first nuclear test in 1945 and with 12 socio-economic and 12 ecological trends during the period of a "great acceleration" (McNeill/Engelke 2014) since 1950.

This author argues that an integrated analysis of both narratives is needed: in the social sciences and humanities on the one hand (on peace and security) and in the natural and social sciences (on environmental issue areas and ecological concerns) on the other. Both peace and security problems and environmental issues and ecological concerns are interrelated and must thus be analysed within an integrated scientific approach and addressed from a common political strategy to face and cope with the fundamental threats to human history that are associated with the atomic bomb and the Anthropocene.

[2] See at: http://quaternary.stratigraphy.org/working-groups/anthropocene/.

Earlier, this author suggested an integrated review of developments between "peace research" and "ecological concerns", which have so far been assessed separately by both communities, in the framework of a proposed "peace ecology" approach (Brauch 2020), addressing both issue areas in the temporal framework of the Anthropocene, a concept that has increasingly been employed in the natural and social sciences, the humanities, law and even in the arts (Brauch 2021).

The temporal context for this proposed integrated "peace ecology approach", starting with the developments between 1940 and 1950, was addressed by this author in his Ph.D. thesis (Brauch 1976, 1977) on *Structural Change and Armament Policy of the USA: America's World Power Role and Its Domestic Preconditions* (1940–1950). At this time, global environmental change and anthropogenic climate change were not yet socially and politically constructed, and thus did not exist in the minds of either policymakers or social scientists, or in that of this author, who was provoked to consider them for the first time in 1978 by a text of the British diplomat Tickell (2003) on climate change, security and conflicts.

However, this situation has gradually changed since the UN Stockholm Conference of 5–16 June 1972 and the UN Earth Summit in Rio de Janeiro of June 1992. In the aftermath of the Stockholm Conference the UN Environment Programme (UNEP) was set up in Nairobi, with regional offices in Europe, North America, Latin America and the Caribbean, West Asia and the Asia-Pacific. After the Rio Earth summit, secretariats for global environmental change regimes were established in Montreal (CBD) and Bonn (UNFCC, UNCCD). Subsequently, international environment policies (Caldwell 1996; Porter/Brown 1991; Stevenson 2017) and global environmental regimes (Breitmeier et al. 2006; Young 2011: 19853–19860) became research areas for political science, international relations and environmental studies.

2.2 The Peace and Security Narrative: Planning, Framing and Establishing the Global Political Order After 1945

The end of World War II on 8 May 1945 in Europe and on 2 September 1945 in East Asia resulted in a multifaceted turning point in international order and in Earth History.

A successful scientific experiment by the chemists O. Hahn and F. Straßmann in Berlin on 17 December 1938 and the explanation by the physicist L. Meitner and her nephew O. Frisch of the splitting of a uranium atom (fission), that was published on 6 January 1939 in *Naturwissenschaft* and on 11 February 1939 by *Nature*, initiated the dual development of nuclear weapons and later of nuclear energy (Hahn/Straßmann 1939; Meitner/Fritsch 1939). O. Hahn received the Nobel Prize in chemistry in 1944 for this revolutionary breakthrough in the natural sciences: "his discovery of the fission of heavy nuclei".

Hahn and Straßmann's successful experiment caused much concern amongst the elite physicists and chemists in the United States, and resulted in a letter of 2 August 1939, drafted by L. Szilard after consultation with fellow Hungarian physicists and political émigrés E. Teller and E. Wigner, and signed by Nobel laureate A. Einstein. The letter was addressed to President F. D. Roosevelt and warned of the danger of the development of a German nuclear bomb[3]:

> In the course of the last four months it has been made probable—through the work of Joliot in France as well as Fermi and Szilard in America—that it may become possible to set up a nuclear chain reaction in a large mass of uranium, by which vast amounts of power and large quantities of new radium-like elements would be generated. Now it appears almost certain that this could be achieved in the immediate future.
>
> This new phenomenon would also lead to the construction of bombs, and it is conceivable—though much less certain—that extremely powerful bombs of a new type may thus be constructed. A single bomb of this type, carried by boat and exploded in a port, might very well destroy the whole port together with some of the surrounding territory. However, such bombs might very well prove to be too heavy for transportation by air.

Einstein specifically warned about Germany:

> I understand that Germany has actually stopped the sale of uranium from the Czechoslovakian mines which she has taken over. That she should have taken such early action might perhaps be understood on the ground that the son of the German Under-Secretary of State, [Carl Friedrich] von Weizsäcker, is attached to the Kaiser-Wilhelm-Institute in Berlin where some of the American work on uranium is now being repeated.

Einstein's letter—a month prior to the aggression of Germany against Poland on 1 September 1939—triggered a series of research, technological, military and political developments that initiated a change in the U.S. worldview from isolationism (1939) towards an active leadership position during World War II and as a framer of the post-World War II international order. Between 1941 and 1945, the USA underwent a fundamental structural change of its military system and its foreign and military policy, with its new industrial policy resulting in a war economy, the build-up of a huge governmental sector with a global intelligence system, the emergence of a government-sponsored and military-controlled research effort and a Congress that supported both the war effort and the new post-World War II political (United Nations), economic (the Bretton Woods institutions: the World Bank, the IMF and GATT) and military order based on NATO and other military alliances (Brauch 1976, 1977).

This initiative resulted in the top-secret military-controlled and -supervised Manhattan Project in Los Alamos in New Mexico, where between 1942 and July 1945 several thousand natural scientists and engineers developed the first nuclear bomb, which was successfully tested on the eve of the Potsdam Conference (17 July–2 August 1945).[4]

---

[3] Letter by Albert Einstein to President Franklin Delano Roosevelt, 2 August 1939. *Source*: https://en.wikipedia.org/wiki/Einstein%E2%80%93Szil%C3%A1rd_letter.

[4] See: Homeland Security Digital Library: "The Manhattan District History: Full-Text Now Available Online", at: https://www.hsdl.org/c/the-manhattan-district-history-full-text-now-available-online/.

During the Potsdam Conference U.S. President H. S. Truman, Soviet Communist leader J. V. Stalin, and British Prime Minister W. Churchill (who was succeeded as Prime Minister by C. Attlee during this conference) met to adopt the post-war order in Europe after the end of World War II in Europe (VE Day), President Truman decided to use this new weapon against Japan before the Soviet Union could declare war against Japan.[5] On 26 July 1945,

> the Potsdam Declaration gave Japan an ultimatum to surrender unconditionally or meet "prompt and utter destruction", which did not mention the new bomb but promised that "it was not intended to enslave Japan". The Soviet Union was not involved in that declaration since it was still neutral in the war against Japan. Japanese Prime Minister Kantarō Suzuki did not respond, which was interpreted as a sign that the Japanese had ignored the ultimatum. As a result, the United States dropped atomic bombs on Hiroshima on 6 August and on Nagasaki on 9 August 1945.[6]

The first nuclear test in Alamogordo and the first military use of US nuclear bombs that were detonated above Hiroshima and over Nagasaki triggered the end of World War II in Asia "with the surrender of Imperial Japan [that] was announced by Japanese Emperor Hirohito on August 15 and formally signed on September 2, 1945, bringing the hostilities of World War II to a close".[7]

Alamogordo was the first step towards a nuclear arms race after World War II by two competing superpowers and political camps. The existence of nuclear weapons made it technically possible for the first time for humankind to destroy itself by damaging its living conditions (Jungk 1956, 1958).However, in 1945 most policy-makers that had framed the post-World War II order were not aware of the consequences of the subsequent thousands of nuclear tests in the atmosphere and below ground by the US, the Soviet Union, the UK, France, the People's Republic of China, India and Pakistan (Table 2.1).[8]

This single event of the first test of a nuclear device in 1945 started a new scientific and political era: the Nuclear Era or the *Atomzeitalter* in which political and military leaders could, for the first time, threaten not only the survival of the enemy nation, but also the lives of millions other people, including those living in the country to be protected in the framework of the nuclear doctrine of Mutually Assured Destruction. This could result, with its nuclear fallout and debris, in a "Nuclear Winter" (Crutzen/Birks 1982) and in health and food security impacts on millions of people in all parts of the world.

---

[5] See the historical reports and different interpretations of US policies that resulted in the first use of nuclear weapons in war against Japan on 6 August 1945 (against Hiroshima) and on 9 August 1945 (against Nagasaki).

[6] "Potsdam Conference", at: https://en.wikipedia.org/wiki/Potsdam_Conference.

[7] "Surrender of Japan", at: https://en.wikipedia.org/wiki/Surrender_of_Japan; The Committee for the Compilation of materials and Damage Cause by the Atomic Bomb in Hiroshima and Nagasaki (1981).

[8] US EPA: "Radioactive Fallout From Nuclear Weapons Testing", at: https://www.epa.gov/radtown/radioactive-fallout-nuclear-weapons-testing. US EPA: "Health implications of fallout from nuclear weapons testing through 1961", Report of the Federal Radiation Council (May 1962); at: https://www.epa.gov/sites/default/files/2015-05/documents/frc_rpt3.pdf.

Table 2.1 Towards a peace ecology approach: linking the peace and security and the ecological and environmental perspectives in the Anthropocene

Perspective	Integrated peace (and) ecology approach			
	Peace and security narrative		Ecological and environmental narrative	
Key terms	Security	Peace	Ecology approach	Environment (policy)
Terms, concepts	'Securitas'	'Pax'	'Anthropocene' was coined by Crutzen in 2000	
Authors, theories	Thukidides, *Hobbes*, Machiavelli, v. Clausewitz, Morgenthau'realism' (geopolitics) "power matters"	*Grotius*, *Kant* Galtung idealist approach "idealism and values matter"	E. Haeckel coined the 'ecology' concept (1866)	Crutzen/Stoermer (2000); Crutzen (2002) coined the 'Anthropocene' concept
Science	Social sciences and humanities		Natural sciences and engineering	
Discipline	History, state science, international law, *political science*	Theology, philosophy, history, law, international law, *political science*, *international relations*	Biology (botany, zoology)	Pollution (air, water, soil), global environmental and climate change, Earth Systems Science and analysis
Subdiscipline (dominance for policy actors and scientists)	• Political theory • Comparative government • Policy studies • *International relations*		Environmental concerns and movements started in the 1960s, *International Environment Policy* in 1972 (Stockholm Conference) and *Climate Change Policy* in 1988 (UN, IPCC)	
Research programme	*Diplomacy and military studies*		*Environment policy studies*	
	• *Security studies*	• *Peace studies*	• *Ecology* (political ecology, social ecology, human ecology, *peace ecology*) • *Sustainable development* was coined in 1987 by the Brundtland Report	• *Global environmental change* • *Earth Systems Science* • These research programmes started after 1972, 1992
	• *Development studies* (started in 1950s)			
Networks	• *International Studies Association* (ISA) • ISA (ISSS; ESS)	• *International Peace Research Association* (IPRA) • ISA (ISA-PEACE)	• *Environmental Peacemaking Association (EPax)* • *Ecopeace Middle East*	• *Geology (stratigraphy)* The AWG was set up in 2009 • ISA (ESS)

(continued)

Table 2.1 (continued)

Perspective	Integrated peace (and) ecology approach			
	Peace and security narrative		Ecological and environmental narrative	
Key terms	Security	Peace	Ecology approach	Environment (policy)
Goals	• *Collective security* • *Collective self-defence* (Art. 51, UNCh)	• *Eternal peace* (Kant 1795)	• Key indicators: accumulation of GHG (especially of CO_2) in the atmosphere • 'Keeling curve' since 1958 on GHG emissions	
Concepts, dilemmas (in terms of security: state- vs. human-centred security)	*State security dilemma* (Herz 1951; Butterfield 1951; Jervis 1976)	*Human security* (UNDP 1994) survival dilemma (Brauch 2004, 2009)	*Ecological security* (Mische 1990)	*Environmental security* (Brown 1977; Ullman 1983; Westing 1986; Mathews 1989)
Approach (collective security)	Shift from Order of Versailles (League of Nations) to that of San Francisco (UN)			
Transition (1) after World War I	System of Versailles (1919) ended with aggression and territorial revisionism by Japan (1931), Italy (1935) and Germany (1939)		Not yet a human concern and no political issue	
Transition (2) after World War II	End of World War II on 8 May 1945 (in Europe) and on 2 September 1945 (in East Asia)		End of the Earth epoch of the *Holocene* and start of the *Anthropocene* Earth epoch (AWG)	

(continued)

Table 2.1 (continued)

Perspective	Integrated peace (and) ecology approach			
	Peace and security narrative		Ecological and environmental narrative	
Key terms	Security	Peace	Ecology approach	Environment (policy)
Order (history of long duration) *Polity*: UN Charter, Bretton Woods Agreements (1945)	*Political*: United Nations system (UN Charter) *Economic*: system of Bretton Woods (World Bank, International Monetary Fund (IMF), GATT and World Trade Organization) *Military*: system of military alliances (NATO)		The Anthropocene and its political phases • Cold War (1946/1947–1989/1990) • Post-Cold War (1990–2022) • End of rule-based security order (24 February 2022) Challenge for rule-based trade and international environment regime order (2017–)	
1945 (end of World War II, start of nuclear era)	Start of the *Nuclear Era*: first nuclear weapons test and on 6 and 9 August: first use of atomic bombs against Hiroshima and Nagasaki		Start of the *Anthropocene* Earth epoch due to radioactive nucleoids (plutonium) in earth sediments	
1945–1950: end of WW II, establishment of US military order	Build-up of the US National Security state (NSA of 1947) and adoption of US global strategy and military build-up (NSC 68) in 1950		*Great acceleration* (30 socio-economic and environmental trends, since 1950)	
	Post-1945 an integrated peace ecology perspective or approach in the Anthropocene should be developed in the social sciences addressing the manifold linkages			

Source The author

This event had a direct impact on the emerging international political and economic order that had been planned both by the US government since 1939 (Notter 1949) and discussed by planning groups of elite organisations such as the Council on Foreign Relations (CFR) and by business organisations that monitored the establishment of the UN system and the Bretton Woods Institutions (the World Bank, the International Monetary Fund, GATT, the WTO).

During the war, the post-war peace and security order was discussed from 1941 by top US and British leaders. On 14 August 1941, U.S. President F. D. Roosevelt and UK's Prime Minister W. Churchill proposed a set of principles for international collaboration in maintaining peace and security which they adopted in the Atlantic Charter. On 1 January 1942, representatives of thirty nations—led by the "Big Four" powers of China, the Soviet Union, the U.K., and the U.S.—signed the Declaration of the United Nations, which reaffirmed the purposes and principles of the Atlantic Charter. In the Moscow Declaration of *30 October 1943*, the Governments of the Soviet Union, the UK, the US and China called for an early establishment of an international organisation to maintain peace and security. During the Dumbarton Oaks Conference in Washington, D.C. (21 September–7 October 1944), the US, the UK, the USSR and China agreed on the aims, structure and functioning of a world organisation. After the meeting on 11 February 1945 at Yalta, President Roosevelt, Prime Minister Churchill and Premier Joseph Stalin agreed to establish: "a general international organisation to maintain peace and security" after Roosevelt had conceded to Stalin a veto right for the permanent members of its Security Council. On 26 June 1945, in San Francisco, 50 countries signed the Charter of the United Nations (UN); these countries were determined to "save succeeding generations from the scourge of war …, to reaffirm faith in fundamental human rights …, and to establish conditions under which justice and respect for the obligations arising from treaties and other sources of international law could be maintained, and to promote progress and the better standards of life in larger freedom". The Charter stated as its purposes in Article 1:

1. To *maintain international peace and security*, and to that end: to take effective collective measures for the prevention and removal of threats to the peace, and for the suppression of acts of aggression or other breaches of peace, …;
2. To *develop friendly relations among nations* based on respect for the principle of equal rights and self-determination of peoples …;
3. To *achieve international co-operation in solving international problems* of an economic, social, cultural, or humanitarian character, …;
4. To be a *centre for harmonising the actions of nations* in the attainment of these common ends.

The goal of the United Nations, as the framework for the new international political order for the post-World War II world, was strictly limited to threats to "international peace and security". Both in the post-war planning and in the negotiations for the UN Charter, references to policy issues related to development, the environment and sustainable development were lacking, as these issues did not exist in the minds of policy planners, government officials or top policymakers.

For the system of military alliances based on Article 51 of the UN Charter (collective self-defence) the North Atlantic Treaty Organisation (NATO) became the most important and longest-lasting institution. It complemented the global (Chap. VII of the UN charter, Art. 39–50) and regional collective security system (Chap. VIII of UNCh, Art. 52–54 on regional arrangements).[9]

The US government rapidly demobilised its personnel in the US military by almost 90% between mid-1945 and mid-1947, from more than 12 million to 1,566,000 by 30 June 1947.[10] In the same year the U.S. Congress passed the US National Security Act that offered the framework for its postwar military force and for the establishment of the CIA as its global intelligence organisation.[11]

2.2.1 United Nations Charter: From Collective Security to Collective Self-defence

During the Yalta conference in February 1945, the Soviet political leader Joseph Stalin insisted on three votes for the USSR (for Russia, Ukraine and Belarus) and on a veto right for the five permanent members of the UN Security Council. Article 51 was a compromise that President Roosevelt had negotiated with the Republican Senator Arthur H. Vandenberg to achieve a bipartisan support for the ratification of the UN Treaty in the US Senate. During the Cold War, the UN's "collective security" system was paralysed from the outset by this veto power. Instead, the US's military dominance and security influence was achieved by a series of systems of "collective self-defence" (Art. 51, UNCh), among them the North Atlantic Treaty Organization (NATO).

In 1945 both compromises fundamentally transformed the UN global and regional collective security system that had been planned for since 1941, where the "individual and collective self-defence of an armed attack occurs against a Member of the United Nations" (UNCh, Art. 51) became the rule and collective efforts became an exception that was often blocked due to a veto by permanent members of the UN Security Council.

[9] See: Russell (1958); Simma (1994); Volger (2002); Czempiel (1966).

[10] The National World War II Museum, New Orleans: "The Points Were All That Mattered: The US Army's Demobilization After World War II"; at: https://www.nationalww2museum.org/war/articles/points-system-us-armys-demobilization; "Demobilization of United States armed forces after World War II", at: https://en.wikipedia.org/wiki/Demobilization_of_United_States_armed_forces_after_World_War_II (7 August 2022).

[11] "Ref Book - 1947 National Security Act", at: https://www.dni.gov/index.php/ic-legal-reference-book/national-security-act-of-1947 (7 August 2022); "National Security Act of 1947"; at: https://en.wikipedia.org/wiki/National_Security_Act_of_1947; Czempiel (1966); Brauch (1976, 1977).

In 1950, as chairman of the US National Security Council (NSC), Paul H. Nitze drafted the first U.S. global post-World War II military strategy in the National Security Council Paper 68 (NSC-68),[12] that suggested at least a tripling of US military spending prior to the start of the Korean War (1950–1953).

2.2.2 *Bretton Woods: Precondition for the Great Acceleration*

The US intention of playing the leading role in the post-war global economy resulted in the creation of several international institutions as an outcome of the Bretton Woods Conference (1944): the World Bank (WB), the International Monetary Fund (IMF) and, in 1948, the General Agreement on Tariffs and Trade (GATT). This multilateral institutional framework, based on US political, military and economic power, created a stable global order aiming at a free trade policy that fostered a process of economic and financial globalisation which resulted in a "great acceleration" of greenhouse gas (GHG) emissions during the Cold War that was further intensified after the end of the Cold War through an exponential increase in CO_2 emissions.

Within the United Nations, development policy had become a major new problem area since the 1950s in the context of the decolonisation process and the struggle of many countries in Africa, Asia and Latin America for national independence based on the principle of self-determination. Many international economic development assistance programmes and negotiations have taken place in the framework of the WB and the IMF, since 1948 in the GATT, and since 1 January 1995 in the World Trade Organisation (WTO).

While the system of Bretton Woods realised the US strategy of an "Open Door" global free trade system (at least among the US and its allies), during the Cold War trade with the USSR and its COMECON members and hostile countries remained strictly controlled by the USA and its allies. The US military dominance guaranteed the free flow of goods and of cheap coal, oil and gas as a precondition for the great acceleration during the Cold War. After the end of the Cold War, neoliberal goals and strategies led to deregulation of national economies, and international trade constraints against the former USSR and its successor states and the COMECON (which were both dissolved in 1991) were deleted.

---

[12] See the extensive debate on NSC 68: USA Department of State, Office of the Historian: "NSC-68, 1950"; at: https://history.state.gov/milestones/1945-1952/NSC68 (4 August 2022); "NSC 68: United States Objectives and Programs for National Security"; at: https://irp.fas.org/offdocs/nsc-hst/nsc-68-cr.htm; "NSC-68: A Report to the National Security Council", in: *Naval War College Review*, 27, 6 (May–June 1975): 51–108 (U.S. Naval War College Press); at: http://www.jstor.com/stable/44641594 (4 August 2022); May (1950); Hammond (1962); for an overview of the extensive literature: "NSC 68"; at: https://en.wikipedia.org/wiki/NSC_68; Cardwell (2011); Casey (2005): 655–690; Gaddis (1980); Guerrier (1988); Hogan (1998): 12; Nitze (1980): 170–176; (1989); Rosenberg (1983): 3–72; Wells (1979): 116–158; Young (2013).

Table 2.2 China's carbon (CO_2) emissions: historical data

Year	Kilotons of CO_2	Metric tons per capita
2020	10,944,686.20	7.76
2015	9,859,281.20	7.15
2010	8,474,922.70	6.34
2005	5,824,625.10	4.47
2001	3,529,081.30	2.77
2000	3,346,525.80	2.65
1995	3,088,620.20	2.56
1991	2,302,185.20	2.00
1990	2,173,364.20	1.91

Source "China Carbon (CO_2) Emissions 1990–2023"; at: https://www.macrotrends.net/countries/CHN/china/carbon-co2-emissions (20 September 2023)

After 2001 China's WTO membership[13] contributed to the rapid increase of its CO_2 emissions with the result that in 2007 China overtook the USA as the major emitter. Early working papers concluded in 2005 (Vennemo et al. 2005) that "China's accession to the WTO in 2001 completed the country's entry into the global economy". A South Asian study by Shahbaz et al. (2015) concluded that "[c]oal consumption increases CO_2 emissions significantly." An additional study by Zheng et al. (2019) argued that "China's CO_2 emissions have plateaued under its commitment to reaching peak carbon emissions before 2030 in order to mitigate global climate change."

These three analyses are not representative of the impact of China's trade, and the increase in consumption due to lifestyle changes of its population, on the increase in its CO_2 emissions between 1990 and 2022. After 2001 the CO_2 emissions for China tripled from 3,529,081.30 kilotons of CO_2 (or 2.77 metric tons per capita) to 10,944,686.20 kilotons of CO_2 (or 7.76 metric tons per capita) in 2020 (Table 2.2).

This snapshot is not representative of other countries in the Global South. However, it indicates the role of geopolitical and geo-economic trade factors. China's WTO membership fostered its integration into the globalised world economy and resulted between 2001 and 2020 in a significant rise of CO_2 emissions, reflected both in total kilotons of CO_2 and a near tripling of its CO_2 in metric tons per capita.

2.2.3 *Nuclear Weapons Testing and Nuclear Arms Control*

Between the first nuclear weapons test on 16 July 1945 and the signing of the Limited Test Ban Treaty (LTBT) in 1963, 499 nuclear tests were conducted in the atmosphere

[13] See: "China and the World Trade Organization"; at: https://en.wikipedia.org/wiki/China_and_the_World_Trade_Organization (20 September 2023).

and by the time the Comprehensive Nuclear-Test-Ban Treaty (CTBT) was signed in 1996, over 2000 nuclear tests had been carried out globally by the United States (1032, 1945–1992), the Soviet Union (715, 1949–1990), the UK (45, 1952–1991), *France* (210, 1960–1996), China (45, 1964–1996), and India (1, 1974). Since September 1996, 10 nuclear tests have been conducted by India and Pakistan (2, 1998) and the Democratic People's Republic of Korea. Before the LTBT was signed and entered into force in 1963 nuclear testing had risen significantly, while with the start of the limited détente nuclear weapons testing in the atmosphere (and also underground tests) declined and continued to fall after the Cold War ended.

During the late 1950s nuclear weapons testing triggered the first multinational peace movement in North America, Europe, East Asia and Australia, which pointed to the negative implications of nuclear radiation for human beings generally and in the milk of mothers for their children.[14] Wittner (1993, 1997, 2003, 2009) analysed "the struggle against the bomb". This nuclear disarmament and peace movement built up public pressure to which the J. F. Kennedy administration responded with the first arms control agreement it negotiated with the Soviet Union during the Cold War, the LTBT that was signed in Moscow on 5 August 1963 and entered into force on 10 October 1963 (Badash 1995).

In addition, the economic reforms launched by Deng Xiaoping (1978–2005) transformed China from a planned to a politically controlled socialist market economy and opened it up to foreign investment and technology. This introduced its huge labour force to the global market, thus turning China into one of the world's fastest-growing economies. Thus, China became, within three decades, the largest producer and exporter of goods, and the second-largest global economy, which had overtaken the United States by 2007 as the largest emitter of greenhouse gases—and especially of CO_2. Not surprisingly, since 1988 (when international climate policy started) the GHG and especially CO_2 concentration in the atmosphere increased exponentially.

After the end of the Cold War—33 years after the ratification of the LTBT (1963)—the Comprehensive Nuclear Test Ban Treaty (CTBT) was signed and adopted by the United Nations General Assembly on 10 September 1996; "as at August 2022, 174 states ratified the CTBT while another 12 states have signed but not ratified it. … As of 2016 … China, Egypt, Iran, Israel and the United States have signed but not ratified the Treaty; India, North Korea and Pakistan have not signed it".[15] On 13 October 1997, the US Congress failed to ratify this last important arms control treaty "by a vote of 51–48, marking the first time that it [had] defeated a security-related treaty since the Treaty of Versailles nearly 80 years [before]."[16] Nuclear weapons testing in the atmosphere and the build-up of nuclear arsenals increasingly turned public opinion in

[14] "Nuclear Testing 1945 – today", at: https://www.ctbto.org/nuclear-testing/history-of-nuclear-testing/nuclear-testing-1945-today/ (1 August 2022).

[15] See: "Comprehensive Nuclear-Test-Ban Treaty"; at: https://en.wikipedia.org/wiki/Comprehensive_Nuclear-Test-Ban-Treaty (8 August 2022).

[16] "Senate Rejects Comprehensive Test Ban Treaty; Clinton Vows to Continue Moratorium", at: https://www.armscontrol.org/act/1999-09/press-releases/senate-rejects-comprehensive-test-ban-treaty-clinton-vows-continue (8 August 2022).

the US and in Western Europe against the testing of nuclear weapons and the intensifying competition among a rapidly increasing number of nuclear weapon states, with an increasingly critical philosophical debate on the nuclear predicament (Fig. 2.1; Churchill 1983; Anders 1956, 1972, 1980). Both R. Jungk and G. Anders later joined the antinuclear movement of the late 1970s and 1980s and offered critiques of nuclear energy as well, thus contributing intellectually to the environmental and anti-nuclear energy debates.

Of the over 2000 nuclear test explosions (1945–1996), "over 500 bombs were exploded in the atmosphere: over 200 by the United States, over 200 by the Soviet Union, about 20 by Britain, about 50 by France and over 20 by China".[17] The nuclear fallout from these atomic tests fell on the Earth's surface and could be identified and measured by geologists at some of the selected GSSP Sites. In July 2023, Crawford Lake in Canada was finally selected by the AWG as the so-called golden spike (see Fig. 2.2 and Introduction).

2.2.4 *Development of Social and Political Sciences, International Relations, and Security and Peace Studies During the Cold War*

After World War II the social sciences spread, both in the liberal democratic states in the West with market economies influenced by the discourses in philosophy, history, sociology and law, and in the East within the framework of Marxist-Leninist ideology and thinking on the state. In many Western countries Political Science and its subdisciplines of Political Theory, Comparative Government and International Relations emerged as disciplines for teaching and research in institutions of higher education, primarily at universities.

Within International Relations the primary focus shifted from idealist approaches in the Kantian tradition (which prevailed in the USA between the two world wars), where 'ideas' mattered, to realist frameworks in a Hobbesian tradition where 'power' and 'geopolitics' became relevant. This theoretical competition resulted in two rival research programmes of global, international and national security studies and peace and conflict studies. In the state socialist East, Marxist-Leninism offered a 'realist' geopolitical framework for the analysis of the state and its international relations.

Based on the UN Charter, peace and international security issues prevailed, and nuclear weapons and deterrence strategies were analysed from the vantage point of security studies, and later of peace research. With the progressing decolonisation process, "development policy" became a task from the 1950s onwards, followed by "environment policy" in the 1960s as challenges to national state policies—and

[17] United Nations: "International Day against Nuclear Tests, 29 August", at: https://www.un.org/en/observances/end-nuclear-tests-day/history (10 August 2022).

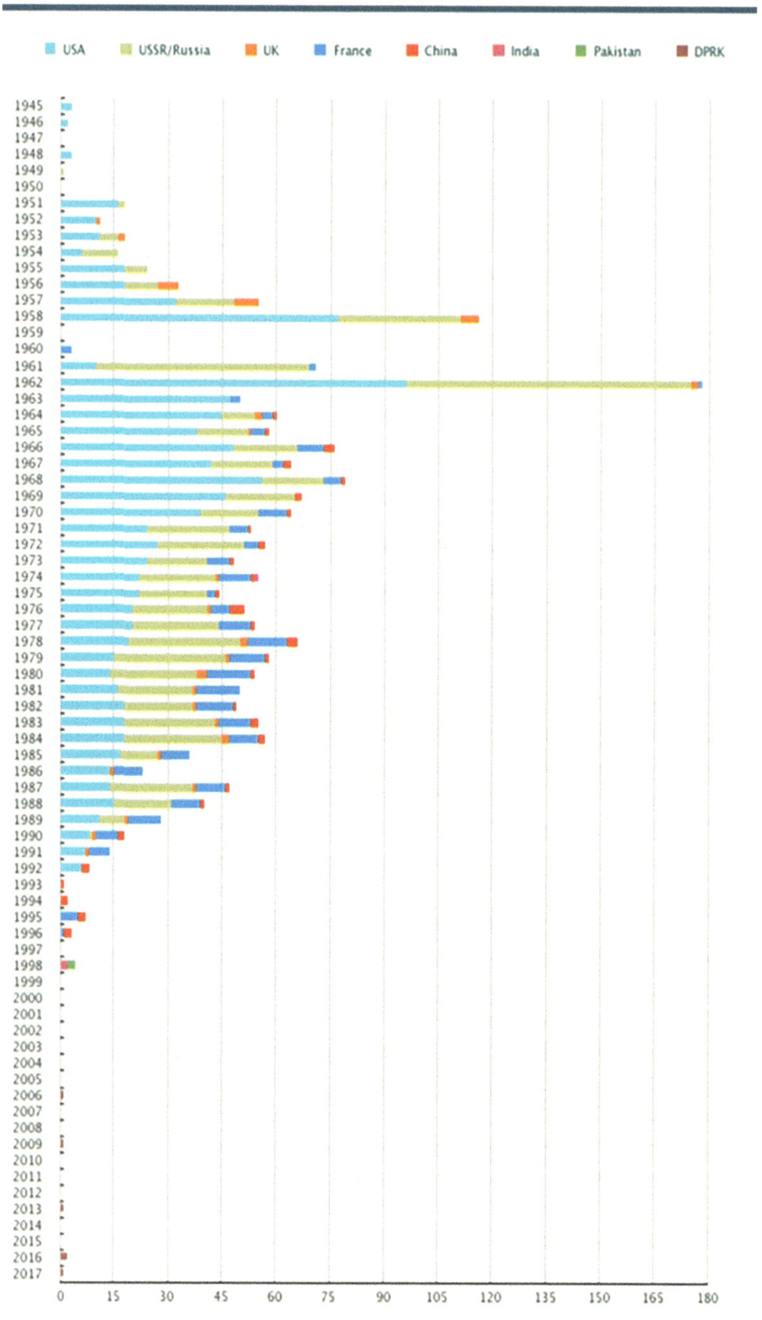

Fig. 2.1 Comprehensive Nuclear Test Ban Treaty Organization: "Worldwide Nuclear Testing" (1945–2017); in: CTBTO, Vienna; at: https://www.ctbto.org/nuclear-testing/history-of-nuclear-testing/nuclear-testing-1945-today/ (1 August 2022)

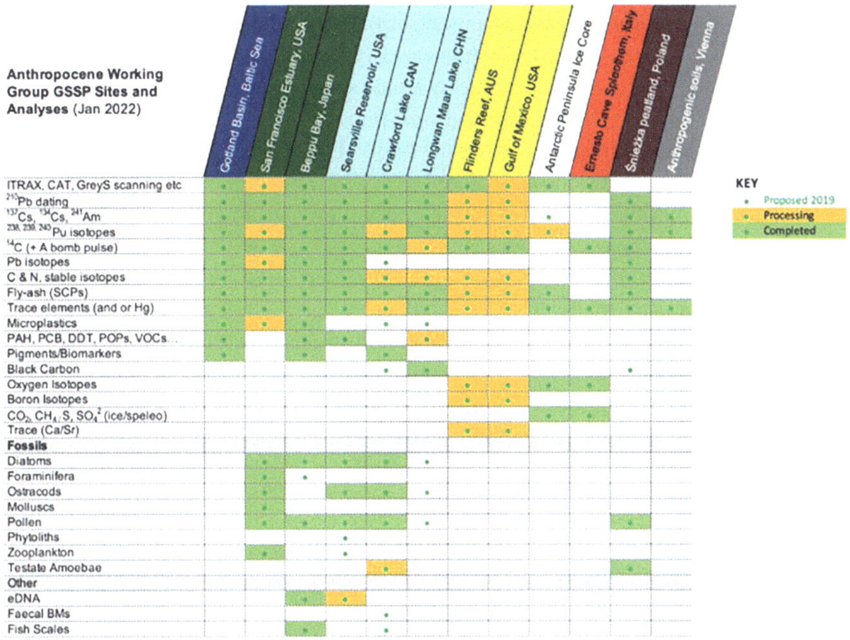

Fig. 2.2 Matrix providing a summary of analyses being processed and completed as of January 2022. The majority of analyses proposed in 2019 during the initial planning of the analytical work have been completed. *Source AWG Newsletter, 2021*, vol. 11 (February 2022): 5; at: http://quaternary.stratigraphy.org/wp-content/uploads/2022/03/AWG-Newsletter-2021_final.pdf (10 August 2022)

after the Stockholm Conference in 1972 as agenda items for international affairs. Based on this gradually expanding policy agenda, nuclear weapons and strategies were addressed as peace and security issues, while issues of pollution and global environmental change were studied in the framework of environment policies.

The segmentalisation of the political agendas was replicated in the social sciences by a further specialisation of topics and of theoretical approaches. Thus, peace research and global, political, social and human ecology developed in different research groups with hardly any substantial or theoretical debates between them (Brauch 2021).

2.3 Emergence of the Ecological Narrative: Global Environmental Change and Climate Change Issues

While ecology concerns have been addressed by many natural scientists since the nineteenth century, the environment did not become a goal of policy until the 1960s, when Rachel Carson's book *Silent Spring* (1962) generated public awareness and strengthened the environmental movement in the US, where the first environment agency, the Environmental Protection Agency (EPA), was set up on 2 December 1970. In June 1972, the UN Conference on the International Environment in Stockholm put the environment on the agenda of international policy making by preparing the establishment of the United Nations Programme on the Environment (UNEP) in Nairobi. In 1987, the World Commission on Environment and Development (WCED), chaired by Gro Harlan Brundtland, introduced the concept of "sustainable development" in its Report on *Our Common Future*.

Two decades after the Stockholm Conference, in June 1992, during the UN Conference on the Environment and Development (UNCED), the legal framework for several global environmental regimes and institutions was set up with the signing of the United Nations Framework Convention on Climate Change (UNFCCC) and the Convention on Biological Diversity (CBD), and in 1994 the United Nations Convention to Combat Desertification (UNCCD) followed.

2.3.1 Start of Climate Policy and of Research on Climate Change

The history of scientific research on climate change began in the early nineteenth century when the greenhouse effect was first recognised. In 1896 the Swedish Nobel laureate and physicist S. Arrhenius claimed that the burning of fossil fuels (coal, oil and natural gas) would result in an increase of the concentration of carbon dioxide and of other greenhouse gases (GHG) in the atmosphere. In 1899, Chrowder Chamberlin (1899) claimed "that climate changes could result from changes in the concentration of atmospheric carbon dioxide".

It took some 75 years until natural scientists took S. Arrhenius' thesis (1896) seriously. The scientific debate slowly emerged during the 1960s in the USA and spread to Europe during the early 1970s, where issues of anthropogenic climate change were increasingly discussed.

As early as 1965, a report of President Lyndon B. Johnson's Science Advisory Committee declared "the rise of atmospheric carbon dioxide levels to be a direct result of fossil fuel burning" (White House 1965; Sellers 1969). Since the early 1970s natural scientists started to analyse the effects of anthropogenic global environmental change and climate change, which resulted in a rapid scientisation of a new research field with global political relevance. In 1972, John Sawyer summarised the available knowledge "on the anthropogenic attribution of the carbon dioxide greenhouse gas, distribution and exponential rise, findings which still hold today" ... and "he accurately predicted the rate of global warming for the period between 1972 and 2000".[18]

[18] See: "History of climate change science", at: https://en.wikipedia.org/wiki/History_of_climate_change_science; See: Sawyer (1972): 23–26 and Nichols (1972): 992.

These new challenges were at first only discussed by natural scientists at their scientific conferences, in their journals and in reports and assessments addressed to natural scientists and government advisers. In 1985, at an International Conference on Assessment of the Role of Carbon Dioxide and of Other Greenhouse Gases in Climate Variations and Associated Impacts, held in Villach, Austria, scientists called on politicians to collaborate in the exploration of policies to mitigate human-induced climate change. This conference noted a consensus among experts "that some global warming seems inevitable, [and called] on governments to consider international agreements to restrict emissions". In 1987, the Montreal Protocol of the Vienna Convention imposed international restrictions on emission of ozone-destroying gases. In 1986, the German Physical Society warned in a report of a pending climate catastrophe (Kords 1996: 204).

In 1988 record temperatures and drought in the US generated political concern about climate change in the U.S. Congress. In September 1988, the Reagan administration added climate change to the agenda of the first G-7 meeting in Toronto. With the adoption of the UN General Assembly Resolution 43/53 of 6 December 1988, the Intergovernmental Panel on Climate Change (IPCC)[19] was established that was

> to prepare a comprehensive review and recommendations with respect to the state of knowledge of the science of climate change; the social and economic impact of climate change, and possible response strategies and elements for inclusion in a possible future international convention on climate. Today the IPCC's role is as defined in Principles Governing IPCC Work, '… to assess on a comprehensive, objective, open and transparent basis the scientific, technical and socio-economic information relevant to understanding the scientific basis of risk of human-induced climate change, its potential impacts and options for adaptation and mitigation'. IPCC reports should be neutral with respect to policy, although they may need to deal objectively with scientific, technical and socio-economic factors relevant to the application of particular policies.[20]

On 21 December 1990, GA resolution 45/212 approved a mandate for starting international negotiations in the framework of the Intergovernmental Negotiating Committee (INC) for a UN Framework Convention on Climate Change (UNFCCC).

> In 1990, at the Second World Climate Conference in Geneva, it became clear that there was a 'North-South' divide on how developed and developing countries viewed climate change. While for the former it was, at that time, primarily a scientific and environmental issue, the latter emphasised the implications for poverty and development of any future regime. These seemingly joint positions of the 'North' and 'South' were based on different and often diverging underlying interests with respect to obligations. In the first group, the majority of [OECD] countries favoured the adoption of an agreement that would reduce global GHG emissions, though there was no agreement by how much these should be reduced. In contrast, the United States, agreeing in principle to the need for an environmental agreement, did not want to be subject to any obligation to reduce emissions. The second group was united in the argument that the new legal instrument must not obstruct their economic development. Apart from this position, interests were divided: while the oil exporting countries feared for

[19] "The Discovery of Global Warming – Timeline (Milestones)", https://history.aip.org/climate/timeline.htm. So far there have been six IPCC assessment reports: I (1990), II (1995), III (2001), IV (2007), V (2013/2014) and VI (2021/2023).

[20] "IPCC History", at: https://archive.ipcc.ch/organization/organization_history.shtml.

> their revenue streams under a new instrument that might restrict the use of fossil fuels, and charcoal consuming countries were concerned about the future use of their primary source of energy, the Small Island States and States with low-lying coastal areas created an alliance to protect themselves from the threat of rising oceans.[21]

Thus, between 1988 and 1990, a process of politicisation of global environmental change and climate change had begun and resulted in the adoption of the UN Framework Convention on Climate Change (UNFCCC) and of the Convention on Biological Diversity (CBD) during the first UN Summit on Environment and Development (UN Earth Summit) in June 1992, and two years later in the UN Convention to Combat Desertification and Drought (UNCDD). On 11 December 1997 the Kyoto Protocol (1997) was signed; this entered into force on 15 February 2005 and expired in 2012.

On 15 December 2015 the Paris Agreement on Climate Change was adopted by 196 Parties at the Conference of the Parties (COP 21) as a legally binding international treaty on climate change; it entered into force on 4 November 2016 with the aim of limiting global warming to well below 2 °C, preferably to 1.5 °C, compared to pre-industrial levels based on nationally determined contributions (NDCs).

In 2019, an increase in the number of disasters (cyclones, wildfires etc.) spurred scientific warnings and public demonstrations. In the same year, mean global average temperature increased to 14.8 °C, the warmest in tens of thousands of years, while the level of CO_2 concentration in the atmosphere reached 415 ppm, the highest in millions of years. Due to COVID 19, COP 21, originally planned for 2020 in Glasgow (UK), was postponed to 31 October to 13 November 2021, and reached an agreement on three pillars of collective climate action.[22]

---

[21] Boisson de Chazournes, L.: *"The Human Impact on Climate Change", at: "Audiovisual Library of International Law"*; at: https://legal.un.org/avl/ha/ccc/ccc.html.

[22] See: UN Climate, "COP26 Reaches Consensus on Key Actions to Address Climate Change" (13 Nov. 2021), at: https://unfccc.int/news/cop26-reaches-consensus-on-key-actions-to-address-climate-change.

[On] "*Adaptation* … parties established a work programme to define the global goal on adaptation, which will identify collective needs and solutions to the climate crisis already affecting many countries. The Santiago Network was further strengthened by elaborating its functions in support of countries to address and manage loss and damage. …

[On] *Finance* … there was consensus in the need to continue increasing support to developing countries. The call to at least double finance for adaptation was welcomed by the Parties. The duty to fulfil the pledge of providing 100 billion dollars annually from developed to developing countries was also reaffirmed. And a process to define the new global goal on finance was launched.

On *mitigation*, the persistent gap in emissions has been clearly identified and Parties collectively agreed to work to reduce that gap and to ensure that the world continues to advance during the present decade, so that the rise in the average temperature is limited to 1.5°. Parties are encouraged to strengthen their emissions reductions and to align their national climate action pledges with the Paris Agreement.

In addition, a key outcome is the conclusion of the so-called Paris rulebook. An agreement was reached on the fundamental norms related to Article 6 on carbon markets, which will make the Paris Agreement fully operational. This will give certainty and predictability to both market and non-market approaches in support of mitigation as well as adaptation. And the negotiations on the

After 30 years of global climate change policy (1992–2021) the assessment is mixed:

- Countries have debated how to combat climate change since the early 1990s. These negotiations have produced several important accords, including the Kyoto Protocol (1997) and the Paris Agreement (2015).
- Governments generally agree on the science behind climate change but have diverged on who is most responsible and how to set emissions-reduction goals.
- Experts say the Paris Agreement is not enough to prevent the global average temperature from rising by 1.5 °C. When that happens, the world will suffer devastating consequences, such as heat waves and floods.[23]

While there has been a general agreement on the goals and on the science of climate change among most parties to the Paris Agreement, the implementation of the quantitative CO_2 reduction has been limited, and during these three decades the GHG emissions have exponentially increased from 1990 to July 2022, which has been analysed as a "climate paradox" for the G-8 countries (Brauch 2012: 30–52).

As with the environment, anthropogenic climate change and global warming were not an issue of public concern among natural scientists until the early 1970s, when several conferences and summer schools were held in Austria (v. Weizsäcker et al. 1995; Von Weizsäcker [4]1994) It had taken more than 75 years from the thesis of S. Arrhenius, who had in 1896 hypothesised a link between the burning of fossil fuels (coal, oil, gas) and the concentration of CO_2 (Lanchberry/Victor 1995: 30ff.; Brauch 1996: xxiii).

The global process of the scientisation of climate change emerged rapidly after the early 1970s, the process of politicisation started in 1988 and the debate on the securitisation of climate change has occurred since the start of the new millennium (Oberthür 1993: 25; Brauch 1996, 2002).

2.4 Two Threats to the Survival of Humankind

Since 1945, humankind has for the first time created the military means (nuclear and hydrogen bombs and their launching systems as the most dangerous weapons of mass destruction, intercontinental ballistic missiles, submarine-launched ballistic missiles and bomber aircraft) to cause major damage to military targets, human infrastructure and human beings (the military and, increasingly, civilians). But humankind has also become a *geological actor* directly interfering in the Earth system through its lifestyle, production, consumption, and mobility patterns.

Enhanced Transparency Framework were also concluded, providing for agreed tables and formats to account and report for targets and emissions."

[23] L. Maizland, 2023: "Global Climate Agreements: Successes and Failures" (17 Nov.), at: https://www.cfr.org/backgrounder/paris-global-climate-change-agreements.

These two threats are analysed separately by different disciplines and research programmes and addressed by different government agencies and international organisations and regimes. The security agenda, its theories and its approaches still dominate, while environmental issues have become and remain a secondary priority, especially since 24 February 2022, when policy issues of global climate change were put on hold and progress in negotiations was paralysed. The Hobbesian security thinking and agenda has since prevailed again in a permanent short-term mode of crisis management where the prospects are dim, as progress on the environmental realm cannot be achieved with (heavy) weapons but only by cooperation based on trust. This delay due to two ongoing major wars in Europe (of Russia against Ukraine) and in the Near East may further increase the ecological cost for future generations.

2.4.1 The Nuclear Threat, the Security Dilemma and the Logic of Nuclear Deterrence

During the Cold War, the worst military threat in the nuclear age was posed by the nuclear arsenals of the competing military alliances and superpowers that had the military capability to retaliate against any nuclear or other attack by a military, political, or economic opponent or enemy. The persistent fear and uncertainty of the military potential of the rival created a Hobbesian security dilemma where the doctrine of a mutually assured destruction (MAD), based on a second strike capability, was to deter the other side from launching a nuclear first strike.

While the number of countries controlling nuclear weapons (US, 1945; Soviet Union, 1949; UK, 1952; France, 1960; People's Republic of China, 1964; India, 1974; and Pakistan, 1998) has increased and there are several unconfirmed aspiring atomic weapons states (Israel, North Korea and Iran), nuclear bombs reached about 100,000 warheads during the 1980s, and have since declined to "an estimated 12,705 warheads at the start of 2022, [of which] about 9440 were in military stockpiles for potential use. Of those, an estimated 3732 warheads were deployed with missiles and aircraft, and around 2000—nearly all of which belonged to Russia or the USA—were kept in a state of high operational alert."[24] During the past 77 years, their military damage potential has so far succeeded in deterring their use during the Cold War and thereafter due to the nuclear doctrines and policies of a mutually assured destruction (MAD).

[24] SIPRI, 2022: "Global nuclear arsenals are expected to grow as states continue to modernize–New SIPRI Yearbook out now", at: https://sipri.org/media/press-release/2022/global-nuclear-arsenals-are-expected-grow-states-continue-modernize-new-sipri-yearbook-out-now (10 August 2022).

During Russia's aggression against Ukraine, Russian President Vladimir Putin put his country's nuclear weapons units on alert.[25] By pointing to the Russian nuclear arsenal, President Putin and foreign minister Lavrov repeatedly warned NATO countries not to interfere directly in this ongoing war against Ukraine, threatening them with a possible use of these weapons.

In International Relations the concept of a "security dilemma" was first developed by H. Butterfield and J. Herz and developed further by R. Jervis in 1976 and 1978 and has been widely discussed since 1951 (Butterfield 1951; Herz 1951; Jervis 1976, Chap. 3; Jervis 1978: 167–214; Tang 2009: 587–623). Herz (1951: 157) originally used the term "security dilemma" arguing:

> Groups and individuals who live alongside each other without being organised into a higher unity … must be … concerned about their security from being attacked, subjected, dominated, or annihilated by other groups and individuals. Striving to attain security from such attacks, they are driven to acquire more and more power in order to escape the effects of the power of others. This, in turn, renders the others more insecure and compels them to prepare for the worst. Because no state can ever feel entirely secure in such a world of competing units, power competition ensues, and the vicious circle of security and power accumulation is on.

The fear of the behaviour, intentions and goal of the 'other' (the nuclear rival and the military, political, economic enemy) was a driver of the nuclear and conventional arms race during the Cold War and may again become a legitimation for an upgrading of military spending and for an arms build-up after the collapse of the post-Cold War order in 2022. In Herz's (1951: 3–4) analysis it is not the question,

> [w]hether man is 'by nature' peaceful and cooperative, or aggressive and domineering … It is his uncertainty and anxiety as to his neighbours' intentions that places man in this basic [security] dilemma, and makes the 'homo homini lupus' a primary fact of the social life of man. Basically it is the mere instinct of self-preservation which, in the vicious circle [of the security dilemma], leads to competition for ever more power. … It is one of the tragic implications of the security dilemma that mutual fear of what initially may never have existed may subsequently bring about exactly that which is feared most. (Herz 1951: 234, 1961: 241)

---

[25] Harding, Luke, 2022: "Putin signals escalation as he puts Russia's nuclear force on high alert", in: *The Guardian*, 27 February; at: https://www.theguardian.com/world/2022/feb/27/vladimir-putin-puts-russia-nuclear-deterrence-forces-on-high-alert-ukraine.

Tang (2009: 591–592) has taken up the original discussions of Herz[26] and Jervis[27] arguing that they denied "that the security dilemma is at the heart of all conflicts". In his reconceptualisation, Tang (2009: 394–395) points to eight features of a security dilemma. Many liberal scholars used the "security dilemma" in interpreting international relations and politics during the Cold War, as well as the arms competition between both super-powers and alliances within the framework of national and international security. Since the post-Cold War order fell apart after Russia's aggression against Ukraine in February 2022, thinking in terms of a "security dilemma" may experience a revival in International Relations and in security and strategic studies. However, it may not help in interpreting the challenges posed by anthropogenic climate change: its physical impacts through an increase in number and intensity of extreme weather events and possible societal outcomes.

2.4.2 *In Global Climate Change "We Are the Threat" and "We Must Also Be the Solution": Towards a Survival Dilemma*

Global environmental change and climate change and its physical effects have already caused multiple societal outcomes (Brauch 2002; WBGU 2007). The framework of analysis is a much wider human security concept based on a broadening of its dimensions from political and military threats to environmental security challenges in terms of the costs posed by storms, floods, droughts, forest fires, food and health insecurity, and undernourishment and hunger (Brauch 2009, 2009a). The referent object (Møller 2003) of environmental security (Dalby 2022) is not primarily the nation state, but is individual human beings or humankind. Thus, for global environmental

[26] According to Tang "Herz also spelled out six aspects of the security dilemma: (1) the ultimate source of the security dilemma is anarchy—the lack of 'a higher unity'; (2) an immediate cause of the security dilemma is states' uncertainty and fears about each other's intentions to do harm under anarchy; (3) states' means of self-help—trying to escape from the security dilemma by accumulating more and more power—generates a cycle of power competition; (4) states' attempt to escape from the security dilemma by accumulating more and more power may not increase their security at all, becoming self-defeating and even tragic; (5) the security dilemma can cause war, but is not the cause of all wars; and (6) the dynamic of the security dilemma is a self-reinforcing 'vicious cycle'."

[27] For Tang, Jervis stressed seven aspects of the security dilemma: "(1) the security dilemma is structural in origin; (2) states' uncertainty and fears about each other's present and future intentions is crucial for forming and maintaining the security dilemma; (3) it is caused by defensive actions, thus unintentional (Jervis 1976: 36); (4) it tends to produce unintended and self-defeating results—that is, decreases in one's own security; (5) it tends to produce unintended and tragic results—that is, war; (6) the security dilemma can cause war, but is not the cause of all wars; and (7) the dynamic of the security dilemma is self-reinforcing and resembles a spiral."

and climate change, it is not national or international security that primarily applies, but rather "human security" (UNDP 1994)[28] or "planetary security".[29]

While the "security dilemma" has been an approach for analysing key security problems during the Cold War, this author suggested a "survival dilemma" (Brauch 2009) for interpreting the human and planetary security consequences of global environmental and climate change. The "survival dilemma" concept has gradually emerged during the past two decades, focusing on the need for cooperation in 'facing' (responding) and "coping with" (adapting, mitigating) manifold new environmental and societal challenges that do not respect state boundaries (Brauch et al. 2007, 2011).

This concept tries to overcome the "Hobbesian fear" (Butterfield 1950, 1950a, 1952) that arose after the Cold War ended, but also the increase of non-military soft security dangers that require primarily non-military, economic, societal, and environmental mitigation strategies. As a human-centred concept it focuses on the causes of global environmental change (GEC) and its impacts on humankind and human beings, especially on the poor, the marginalised and the environmentally and socially highly vulnerable people whose personal survival and that of their families and communities has been put at risk and who have been confronted with several unpleasant choices. The "survival dilemma" links a widened and deepened security concept with issues that are caused by unsustainable development and human-induced GEC, and it offers an attempt to conceptualise a "survival dilemma" including ideas for coping with this dilemma using survival strategies (Oswald Spring 1991).

The focus of the "survival dilemma" is neither a "survival of the fittest" as Darwin has often been misquoted,[30] nor a "survival society" (Hillmann 1998), but rather the dilemma that often confronts the poor, and the environmentally and socially highly vulnerable people who face several unpleasant choices in response to war, hazards, disasters, and complex emergencies that fundamentally challenge the survival of their traditional livelihoods. The key questions are, "who is to survive?" (states or human beings?) and "against what?" (wars, hazards, and disasters as well as complex emergencies) against the background of emerging new security concerns in the Anthropocene (Crutzen/Stoermer 2000; Crutzen 2002; Clark et al. 2005).

---

[28] UNDP (1994); there exists a voluminous literature on the human security debate.

[29] On "planetary security": The *Planetary Security Initiative* (PSI) was launched in 2015 by the Netherlands Ministry of Foreign Affairs aiming at "best practice, strategic entry points and new approaches to reducing climate-related risks to conflict and stability, thus promoting sustainable peace in a changing climate"; see at: https://www.planetarysecurityinitiative.org/about (4 August 2022); PSI (2016); Rademaker et al. (2016).

[30] "According to Darwin's Origin of Species, it is not the most intellectual of the species that survives; it is not the strongest that survives; but the species that survives is the one that is able best to adapt and adjust to the changing environment in which it finds itself." See: Megginson (1963: 4).

Since the early twenty-first century, climate change has increasingly been perceived as a threat to national, international, and human security. Climate change has gradually been securitised in government reports and in statements of government officials in several countries (Brauch 2002, 2004, 2006; WBGU 2007; CNA 2007) and has repeatedly been addressed by the UN Security Council (Hardt et al. 2023).

While the last three global orders of Vienna, Versailles and San Francisco (1815–1989) were primarily based on power legitimised in terms of a *security dilemma*, new non-military global challenges of the twenty-first century may require a new international order based on a *survival dilemma* (Brauch 1996, 2000, 2009). This may necessitate additional multilateral cooperation on international security (arms control, terrorism) and environmental regimes (climate, desertification, water), in international and supranational organisations. Thus, the zero-sum games of many realist approaches in the Hobbesian tradition of the nineteenth and twentieth centuries must be replaced by non-zero-sum games where all major players should aim for the survival of humankind (Axelrod 1984). However, this optimistic outlook has fundamentally changed since the Russian aggression against Ukraine in 2022, and the future in the emerging third phase of the Anthropocene has become highly uncertain and unpredictable, where a security dilemma and a survival dilemma may coexist.

Brauch (2000: 281–318) argued earlier that the root causes of global environmental change could become "severe challenges for the survival of governments", and he stated that the environmental conditions for human life may be fundamentally challenged as a result of a complex process of incremental change caused by soil erosion and desertification leading to more frequent and intensive droughts, water scarcity and a lack of food that will force people to migrate, which may sometimes lead to violent conflicts (Brauch 2002; IPCC 2014a, b).

If such a "*survival dilemma* is more than an idealist construct of good intentions, effective mechanisms for an efficient implementation of adopted norms and goals and a comprehensive verification regime for effective sanctions against violators are needed" (Brauch 2000: 286, 2008a). This poses the question of a legitimate use of force for these non-military global challenges. Thus, Earth policy requires an increase in the effectiveness of international organisations, which requires a fundamental reform of the United Nations system (von Weizsäcker 1993; Brauch 1996). However, with the many unresolved national, ethnic, and religious conflicts around the globe, distrust and arms competition still prevail in many regions. Thus, the security dilemma has not been overcome and it has re-emerged since 24 February 2022.

The most likely implication of the threats, challenges, vulnerabilities, and risks posed by global environmental and climate change for small island states may be a survival dilemma (Brauch 2000, 2004, 2008, 2009) confronting their poor and highly vulnerable population with unattractive alternatives: to stay at home on their threatened islands and be exposed to an increasing number of, and more intense, tropical hurricanes and cyclones, or to be forced to migrate from the Caribbean to

North America, and from the small islands in the Indian and Pacific Ocean to countries that offer their families better prospects for survival and economic well-being. In other regions and circumstances the survival option may be further reduced:

- To stay at home, to suffer from hunger, and, as the worst alternative, to die;
- To wait for governmental help and to survive in bad conditions;
- To migrate and to face conflicts elsewhere for water, land, food and jobs;
- To migrate from the villages to urban slums with a low quality of life in a dangerous environment with a lack of employment;
- To separate, with the young and the men emigrating to where there are jobs and prospects for a better life, and with women, children, and the old often staying at home and living on the remittances of emigrated family members;
- Or to develop local resilience, adaptation, and coping mechanisms, as well as survival strategies (Oswald Spring 1991).

In contemporary English, the concept of a "survival dilemma" has so far not rigorously been defined as a social science concept for the analysis of global environmental change, its extreme and fatal outcomes and its potential violent societal repercussions. What is the dilemma about, and what are the choices for whom? Whose survival is at stake? That of humankind, the state, an ethnic group, a family or of an individual? What is the referent of such a "survival dilemma"? International anarchy, the nation state, society, one's own ethnic or religious group, a clan, a village, a family or the individual? What are the reasons that necessitate a choice between survival or forced migration or even death (decline, disintegration)? Are they socially or environmentally driven, or both?

A dilemma requires a choice between two often unattractive options, e.g. between survival or death of an individual, or loss of his/her home and livelihood. Thus, first of all, the survival of the individual human being is at stake, but also of the dwellers on marginal land or in vulnerable housing. Often the poor, as well as minority ethnic or religious groups or clans with limited adaptive and mitigation capabilities, are among the first victims of hazards and disasters partly caused by human- and nature-induced global environmental change.

But in ecologically sensitive arid and semi-arid regions, as in the Sahel and in other deserts in the global sunbelt, and also in earthquake-prone regions, the survival of individuals, tribes, regional or national governments in failed or disintegrated states is at stake (Somalia, Afghanistan). Global environmental change, and especially climate change, can also become an additional challenge to humankind, but its impacts will be unequal due to the different degrees of societal (poor vs. rich) and environmental (tropical, subtropical vs. moderate regions) vulnerability. Rapid population growth increases the demand for water, food and housing, and contributes to rapid urbanisation, with negative repercussions on the environment (pollution of soil, water, and air).

A *survival dilemma* exists if the livelihood of human beings (individuals, families, clans, tribes or ethnic or religious groups) is severely challenged by the extreme or fatal outcomes of global environmental change (hydro-meteorological hazards due to extreme weather events) and by extreme poverty. The survival dilemma is the highest where poverty is high and the adaptive and mitigation capacity and resilience are low (high societal vulnerability), and where the probability and impact of natural disasters is severe (high environmental vulnerability).

Initially, the referent for the survival dilemma is the family and the ethnic and religious group (e.g. tribe or clan), but also the state. If the impact of a severe drought results in a major famine (as was the case in the Sahel during the 1980s), then the states and the sub-regional, regional, and global organisations, as well as nongovernmental aid organisations, become a major referent. The referent for major causes of global environmental change—climate change, desertification, and either water scarcity or flooding—is the international community (states, regional and universal organisations, and regimes).

The survival dilemma confronts its victims with a vicious circle (stay and starve, or leave and lose their own identity) which requires fundamental and difficult choices by the people (individuals, families, tribes, clans) in the affected regions, primarily in the developing countries. These choices often imply leaving their home, tribe, or clan, losing their livelihood and identity, and giving up their inherited customs and culture. The survival dilemma poses a major challenge to the individual and society in states and regions where there is a high personal and societal vulnerability due to poverty, and a high environmental vulnerability due to extreme weather events and natural hazards, where the impact is high, but the resilience is low.

Until recently, this survival dilemma was hardly a concern for strategists in industrialised countries, because in their states the societal vulnerability is relatively low and even in cases of a high environmental vulnerability, the resources and coping capacities exist to reduce human fatalities. However, on 29 August 2005, Hurricane Katrina caused 1833 reported deaths in the USA, affected about 500,000 people, and caused an estimated economic damage of 125 billion US dollars.

While the security dilemma confronts the state and its national security elites and decision-makers with tough strategic choices, the survival dilemma initially confronts the individual victim and his/her family, community, ethnic or religious group with difficult decisions that affect the future of their community and society. While military means are crucial to cope with the security dilemma, they are in most cases irrelevant for coping with the survival dilemma.

While the security dilemma is closely linked with a narrow political and military security concept, and is often associated with Hobbesian or Grotian security perspectives, the survival dilemma is primarily linked with *human* and *environmental* security concepts, and with Grotian and Kantian security worldviews, as well as with equity-oriented distributionist environmental standpoints, and less with optimistic Cornucopian perspectives.

While the security dilemma has been in the centre of the classical security agenda since ancient times and certainly since the modern territorial state emerged in the Westphalian order, the survival dilemma addresses a new agenda where solutions have become more urgent during the twenty-first century and where the old (and still dominant) Machiavellian and Hobbesian mindset of strategists is bound to fail. The survival dilemma requires a combined mitigation strategy that aims at "peace with the environment" and "development with security", or a combination of sustainable peace with sustainable development.

The political strategies for coping with the security and the survival dilemmas differ in the post-Cold War era, taking the specific geopolitical context and the status of socio-economic development of states and regions into account. Efforts to cope with the security dilemma of states require national, regional or international multilateral diplomatic initiatives to deal with the perceived threats by other states, through cooperative efforts to address the structural root causes that may cause, contribute to or trigger conflicts within and between states, and to resolve conflicts and to prevent them from escalating into violence as well as to prevent natural hazards from turning into complex emergencies.

Initiatives to cope with the manifold political, societal, economic, and environmental threats, challenges, vulnerabilities, and risks—most of a non-military nature—that cause, contribute to, intensify or trigger a survival dilemma for states (e.g. drought-stricken regions) and human beings forced to leave their homes and livelihoods, require different coping strategies. Where the enemy is *us*, the consumers of fossil fuels, it is impossible to use the military to deter or to combat that threat; rather, the tools provided by international development and environment policies aiming at sustainable development and peace are needed. As these new security dangers are of a global or regional transboundary nature, only inter- and transnational coping strategies involving not only the states but also the societies and the national and transnational business communities will enable humankind to adapt to and mitigate against these security concerns.

While top-down initiatives to relieve or overcome the survival dilemma of the most-affected poor, marginalised, socially and environmentally vulnerable, with the least ability to adapt and mitigate, are necessary, they nevertheless have often failed. They must be complemented by bottom-up survival strategies that involve and empower these people in actively creating new livelihoods instead of waiting solely for relief aid in refugee camps. Such survival strategies of individuals and of the marginalised poor in urban centres have been successfully developed. These initiatives that rely on the dignity and ingenuity of the affected victims need encouragement, and national, trans- and international support (Oswald Spring 1991, 2007).

During the post-Cold War years (1990–2022) the existence of a Hobbesian security dilemma was less discussed, and the proposed survival dilemma was not taken up by mainstream International Relations. Which policy goals and means for addressing both existential human survival threats of a nuclear war and of serious impacts of global climate change were considered?

2.5 Means for Addressing Both Human Survival Threats

2.5.1 Response to Military Attack and Armament Build-Up

From a Hobbesian 'realist' perspective only military power matters, combined with a determination to defend one's territory, people and values with political, economic and military means, through sanctions against the aggressor and use or transfer of intelligence, targeting of information, and the supply of heavy weapons to the victim of an aggression.

The Russian war against Ukraine has been a classic case of an aggression with the goal of defeating a former member country of the USSR and unilaterally revising borders in violation of international law and the purpose of the United Nations (of which both Russia and Ukraine were founding members in 1945). This aggression was not a consequence of a failed security dilemma; it was, rather, a consequence of the perception of Ukraine and of NATO members that considered such a major breach of trust unthinkable in the post-Cold War era.

2.5.2 Downgrading the Effects of Global Environmental Change

Regarding global environmental change, we as human beings and humankind as a species have become the threat in terms of the societal outcome of the physical effects we have caused as a result of the exponential increase in greenhouse gas (GHG) emissions. Hobbesian logic and military weapons are unable to contain the average temperature increase, precipitation variability, sea-level rise, and extreme weather events. The people and infrastructure in the countries with the highest number of nuclear weapons and huge military arsenals (USA and Russia, 2011, 2012, 2021, 2022) could not be protected from repeated intensive forest fires. Extensive droughts, harvest failures, food insecurity and hunger are other challenges triggered by anthropogenic climate change. The consequences of the trends that have intensified with the Anthropocene require international cooperation in addressing and coping with these natural and societal consequences (Fig. 2.3).

As a result of Russia's war against Ukraine since 2022, food prices (for corn such as wheat and maize, and sunflower oil, etc.) experienced large increases and food has become unaffordable for poor people in countries that have been heavily affected by extreme precipitation shortages, drought and very low food production which has already caused food shortages, undernourishment and hunger.

Global environmental change and climate change require cooperation and cooperative solutions for these pressing environmental challenges. Since the war broke out, climate talks have been on hold and no major progress has been achieved by the member states of the Paris climate treaty of 2015 since the last COP in Glasgow in 2021. No substantial talks occurred in 2022, while in 2023 some progress was

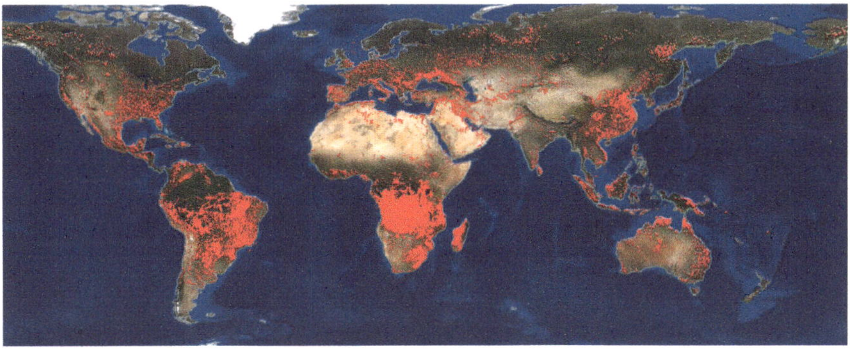

Fig. 2.3 NASA: *Fire Information Resource Management System* (FIRMS) based on data 7 July–5 August 2022; at: https://firms.modaps.eosdis.nasa.gov/map/#t:adv;d:2022-07-06.2022-08-05,2022-07-06;@31.6,6.2,2z (5 August 2022). The figure is in the public domain see at: https://firms.modaps.eosdis.nasa.gov/map/#d:24hrs;@0.0,0.0,3.0z

achieved with regard to unilateral commitments of states supporting loss and damages in the framework of the G7 and G20. Due to the increase in tension between China and the US, the bilateral climate talks between the two major emitters were also interrupted on 5 August 2022, but were resumed in July 2023.[31]

2.6 Towards an Integrated Perspective of the Peace and Security Narrative with the Ecological Narrative

During the Anthropocene the "peace and security narrative" has been dominant both in Political Science and in International Relations. Between 1945 and 1950 "the environment" was not yet socially constructed, and ecological concerns had no impact on the development of post-World War II planning in the US Department of State (Notter 1949), on the international negotiations on the United Nations Charter or on the establishment of the Bretton Woods institutions and policymaking in these institutions that became the legal framework for the liberal international order (see above).

The environmental agenda, issues of sustainable development and challenges posed by global environmental and climate change gradually evolved as new policy areas from the 1960s onwards and were added to the international agenda. They thus

[31] A. de la Garza, 2023: "Climate Geopolitics: These Two Charts Show Why U.S.-China Cooperation on Climate is so Challenging", *Time Magazine* (20 July); at: https://time.com/6295941/us-china-climate-cooperation-challenge/ (7 October 2023); T. Blaine; C. Freeman; H, Tugendhat, 2030: "What Does John Kerry's Visit Mean for U.S.-China Climate Cooperation? While the trip didn't yield any breakthroughs, it did reestablish climate as a shared issue amid broader diplomatic tensions", in: US Institute of Peace (20 July); at: https://www.usip.org/publications/2023/07/what-does-john-kerrys-visit-mean-us-china-climate-cooperation.

became objects of national and international governance (*politics*) and of international negotiations and have resulted in treaties and new international institutions and regimes (*polity*) on these topics since:

- the 1960s on national environmental issues (waste, pollution of air, water, soil etc.);
- the 1970s on international environment policies (after the UN conference in Stockholm in June 1972) with the establishment of UNEP in Nairobi;
- the 1980s on sustainable development; and on the ozone layer depletion (The Montreal Protocol on Substances that Deplete the Ozone Layer that signed in 1987 and entered into force in 1989);
- the 1990s with the creation of selected international environmental regimes on climate change (UNFCCC), on biodiversity (CBD) during the Earth summit in Rio de Janeiro (in June 1992), and on desertification and drought (UNCCD 1994);
- Since February 2000 this agenda may also be discussed in the framework of the Anthropocene Epoch of Earth history.

While greenhouse gases, and in particular carbon dioxide, have exponentially increased since the start of the Cold War (1945–1950), the mandate for negotiations on a United Nations Framework Convention on Climate Change (UNFCCC) was only adopted in December 1990 by the UN General Assembly, approved during the UN Earth Summit in Rio de Janeiro in June 1992 and entered into force on 21 March 1994. Between then and April 2023 CO_2 emissions further intensified, and the measures introduced so far have so far failed to freeze or reduce the global GHG emissions. The summary for Policy Makers of the Synthesis Report of the 6th Assessment Report in March 2023 reported this average temperature increase:

> Global surface temperature was 1.09 [0.95 to 1.20] °C higher in 2011–2020 than 1850–1900, with larger increases over land (1.59 [1.34 to 1.83] °C) than over the ocean (0.88 [0.68 to 1.01] °C). Global surface temperature in the first two decades of the 21st century (2001–2020) was 0.99 [0.84 to 1.10] °C higher than 1850–1900. Global surface temperature has increased faster since 1970 than in any other 50-year period over at least the last 2000 years (high confidence).[32]

Both the peace and security narrative and the environmental narrative and ecological concerns have coexisted since the post-Cold War liberal order, both in national and international politics and in the social sciences that have analysed it in two different research programmes. In the Anthropocene, an integrated analysis should be aimed at from 1945, when this new epoch in Earth history started. This should be the goal of an integrated peace and ecology or peace ecology approach (Brauch 2020).

[32] IPCC, 2023: "Summary for Policymakers", in: Climate Change 2023: Synthesis Report. Contribution of Working Groups I, II and III to the Sixth Assessment Report of the Intergovernmental Panel on Climate Change [Core Writing Team, H. Lee and J. Romero (eds.)] (Geneva: IPCC, Switzerland): 1–34 (4).

2.7 Fragmentation and Uncertainty During the Third Phase of the Anthropocene

There exists a consensus among most social and political scientists that the first 'bipolar' phase of the post-WW II international order ended symbolically with the fall of the Berlin Wall on 9 November 1989, and the second phase started with the Paris Summit of the Conference on Security and Cooperation in Europe on 19–21 November 1990, with the signing of the Charter of Paris for a New Europe. However, disagreement exists about when this second 'unipolar' phase ended and the third 'multipolar' phase began.

From a US security perspective, this brief transition phase ended with the attack on the US on 11 September 2001 and the US invasion of Afghanistan in autumn of 2001 and the war against Iraq in 2003. Some analysts claim that the liberal trade order of Bretton Woods was unilaterally eroded by the Trump administration (2017–2021). Others argue that the US bipartisan foreign policy consensus in place since 1941 ended during the second Clinton administration, when the Republican party opposed the ratification of both the Comprehensive Test Ban Treaty and the Kyoto Protocol in 1999 and the US sent its Air Force to stop the internal conflict in Bosnia Herzegovina in 1995 and acted in 1999 against Serbia to support the independence of Kosovo.

From a European peace and security perspective, this phase ended with the Russian annexation of the Crimea in February 2014 and with Russia's war of aggression against Ukraine on 24 February 2022 and with the annexation of some oblasts, including Donetsk, in eastern Ukraine on 30 September 2022. For many European observers a *Zeitenwende* (turning point) occurred on 24 February 2022 with Russia's aggression against Ukraine that ended the cooperative security order in Europe and froze several of its key institutions (especially the Organization for Security and Co-operation in Europe [OSCE] and the Council of Europe).

With the US-led military interventions in the Balkans in the post-Yugoslav space (1995–1999), the US occupation of Afghanistan in 2001, the war against Iraq in 2003 based on falsified intelligence claims, and the US-led NATO bombing against Libya (all without a mandate by the UN Security Council) and the Russian annexation of Crimea in 2014 and its war against Ukraine, both former superpowers have unilaterally violated international law and the UN Charter.

However, from a wider peace and ecology perspective, the US-led military interventions in the Balkans, 9/11 and the wars in Afghanistan (2001), Iraq (2003) and the US-led NATO bombing of Libya (2011) did not challenge the 'polity' of the European post-Cold War order of peace and security although the US sanctions against China (since 2017) and Russia (since 2022) and the struggle between the US defending its sole hegemony and China challenging this hegemony in the economic and political area has fostered the formation of an enlarged group of BRICS countries (5 + 6) which may help in establishing a multipolar world order.

The trends of global environmental change (climate change and biodiversity loss) that have been triggered by the "great acceleration" since 1950 and increased exponentially between 1990 and 2022 as a result of the widening of economic globalisation have continued unabated. This is despite the establishment of environmental regimes (climate change, biodiversity, desertification et al.) in 1992, and the consequences of the increase in extreme weather events—both in number and intensity—in hothouse Earth (see Fig. 2.4 based on Steffen et al. 2018). As a result of the present wars, cooperative efforts on climate change issues between the US, China and Russia have been frozen since 2022 and progress in climate policies at the COPs of the Paris Climate Change Treaty have been slowed down and many unilateral commitments have not been met.

In the first chapter this author referred to two features of the emerging disorder during the third phase of the Anthropocene: the fragmentation of a coalition of states and the increasing uncertainty and the reduced probability of achieving cooperative policy agreements in a hostile world. This will most likely result in the postponement of needed decisions in the area of climate change which the IPCC has called for in the Synthesis Report of its 6th Assessment Report (20 March 2023). This lack of trust in achieving joint policy agreements may make it less likely to achieve and preserve a 1.5 °C world and will probably further increase the number and intensity of extreme weather events in those parts of the world with a low societal and economic vulnerability (Bohle 2009).

This section will address: (a) the dominance of the Hobbesian security discourse over a cooperative discourse in achieving a "sustainability transition", (b) the fragmentation of international order in the third phase of the Anthropocene, (c) new risks

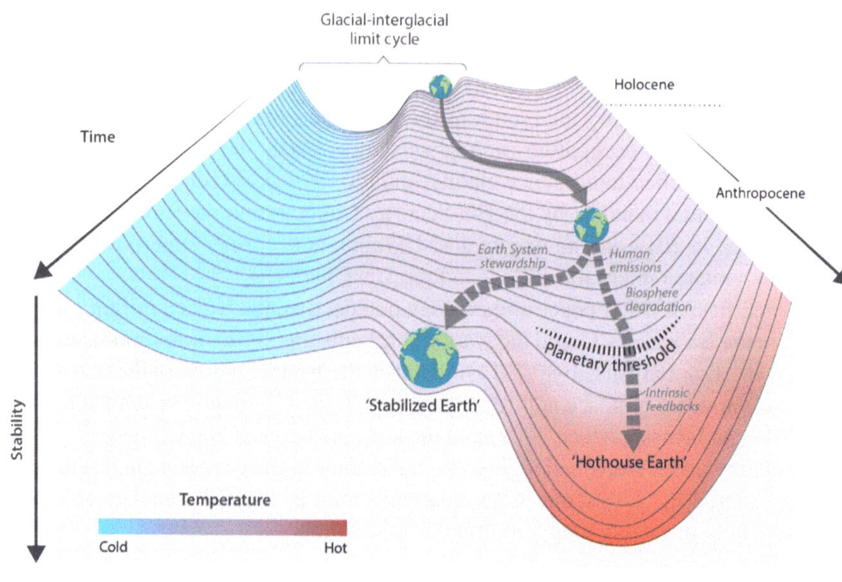

Fig. 2.4 Hothouse Earth. *Source* Steffen et al. (2018). The figure is in the public domain

and increased uncertainty as expressed in the concerns in the Risk Reports of the World Economic Forum (WEF 2022, 2023), (d) the probable costs of Hobbesian geopolitics due to the impacts on the Anthropocene discourse and the postponement and slowdown of strategies of economic transition and (e) the consequences of lacking an integrated holistic perspective from the vantage point of a peace ecology and ecological peace policy.

2.7.1 Dominance of the Hobbesian Security Discourse Over a Discussion on Cooperation on Sustainability Transition

Since the Russian aggression against Ukraine, the 'idealist' liberal peace and security discourse has been criticised—for example in Germany—for the 'naivety' of their analyses and interpretations of the cooperative phase of political relations with the Soviet Union and the Russian Federation in the framework of détente during the Cold War and after its end.[33] Others have called for an adjustment of the focus and an assessment of the new circumstances after "the turn" (or *Zeitenwende*).

Across the political spectrum there exists a wide consensus that the cooperative liberal security order in Europe has collapsed, and that at present there is no trust in a revival of the past order as long as the aggressor is in power and its policies are implemented by military means and war crimes. Given these circumstances, a return to policies expecting changes from economic trade seems to be illusory. As long as no ceasefire or stable and just peace agreement accepted by both parties exists, any discussion of or preparation for a new "cooperative order" will be premature.

However, in contrast with the detailed US post-war planning during World War II (Notter 1949), no such planning existed in 1990, either in the US or in the European Union and its member states, for example for dealing with the violent conflicts in the post-Yugoslav space. Similarly, there is at present no planning with regard to the future security framework and security guarantees for Ukraine, beyond the consensus that Ukraine has already become a candidate for EU membership, while no similar decision has been made about whether it should also become a full NATO member—which would bind all NATO members to directly support Ukraine if it should be attacked for a third time after 2014 and 2022.

In summer 2022 Kipping (2022), head of climate policy office in the German Development Ministry, discussed whether climate change has already become a "collateral damage" of the Russian aggression against Ukraine (Kipping 2022). He pointed to the dual impact of Russia's war on the rule-based multilateral order on which international climate policy is based and to the increased geopolitical power rivalry between the US and China, which may increase the likelihood of tipping points

---

[33] I limit myself to the German debate in a few selective journals and magazines, e.g. Mölling (2022: 71–76). Similar positions can be found in policy journals and reports by think tanks in the United States, the UK, France, Italy and in Brussels, related to the EU and NATO.

in the climate system. The war may also speed up the move towards renewables due to the high costs of imported fossil fuels. But Russia is indispensable for international climate diplomacy, as is China, which contributes at present one third of global CO_2 emissions.

In many countries, especially among the conflict partners with democratic and authoritarian systems of rule, geopolitical strategic and military goals have replaced the concerns regarding global climate change and the readiness to enter into major and costly climate initiatives. However, Russia's war against Ukraine has already become a driver to reduce Western Europe's energy dependence on Russian gas, oil and coal and has been a major push towards a large increase in renewables and longer-term imports of green hydrogen from countries in Australia, Africa and South America.

2.7.2 The Anthropocene and Uncertainty of International Order

During 2022 a move towards a fragmentation of international order became visible in the voting behaviour in the UN General Assembly on Russia's war of aggression against Ukraine[34] While two thirds of the UN members condemned the Russian aggression, only a much smaller majority supported the expulsion of Russia as a member of the UN Human Rights Council in 2022.[35]

In 2022, three (China, India, South Africa) and, after the election of president Lula four (Brazil) of the five BRICS countries abstained from condemning Russia, along with many larger countries of the global South that depend on Russian exports of food, fertilizers and weapons. Other countries too, for example in North Africa and South Asia, depend on food imports from Ukraine and some also on energy, food, fertilizer and arms exports from Russia.

The heavy US pressure on its allies to reduce their economic dependence on China may interrupt the free flow of goods the US has promoted since the 1990s (in the context of globalisation), which may result in a transfer of Western production capabilities from China to India and other countries in the Global South, especially towards Southeast Asia, South Asia, Latin America and Africa. The US has implemented partially discriminatory policies towards Chinese scholars and students at US universities which may further speed up the process of China's aiming at scientific excellence in science and technology and in medicine, trying to attract the best and brightest students and scholars from the Global South to get their education and obtain academic employment in China in areas where the US has, in the recent past, been highly dependent on immigration – especially in the natural sciences—of well-trained Chinese natural scientists and engineers.

---

[34] UNGA March 2022.

[35] UNGA 2022.

> China has made rapid advances in areas such as education, infrastructure, high-tech manufacturing, academic publishing, patents, and commercial applications and is now in some areas and by some measures a world leader. China is now increasingly targeting indigenous innovation and aims to reform remaining weaknesses. Per the Global Innovation Index in 2022, China was considered one of the most competitive in the world, ranking 11th in the world. …
>
> The 21st century has thus seen a series of central government initiatives designed to promote 'indigenous innovation' and technological development more generally in China. These include the National Medium- and Long-Term Program for Science and Technology Development (2006–20), the Strategic Emerging Industries initiative, the Internet Plus initiative, and the Made in China 2025 Program, among others. … China aims to increase the share of basic research to 15% of total research spending by 2020.[36]

Regarding Western plans for a sustainability transition in renewables, China has become an indispensable supplier and partly already a technology leader in solar photovoltaic energy panels and electrical car batteries. According to a "Research Report on China's Solar Cell Export Industry 2023–2032" released by Business Wire on 3 January 2023:

> China is the world's leading exporter of solar cells. … In 2021, China exported 3.201 billion solar cells, up 17.56% year-on-year, with an export value of US$ 28.460 billion, up 43.79% year-on-year. From January to October 2022, China exported 3.294 billion solar cells, up 24.06% year-on-year, with an export value of US$ 40.033 billion, up 75.58% year-on-year. In 2018–2022, the average price of China's solar cell exports showed an overall change trend of first decrease and then increase. In 2018–2020, the average price of China's solar cell exports fell continuously, from US$ 12.18 each in 2018 to US$ 7.27 each in 2020. … China is a major global producer and exporter of solar cells, and the publisher expects China's solar cell exports to continue to rise from 2023–2032 as the global new energy sector grows.[37]

According to *Taiyang News* of 2 March 2023: "Germany Imported €3.6 Billion Solar PV Systems in 2022"

> from across the globe. With an 87% share, China was its largest supplier shipping €3.1 billion worth PV products into the European country. … The German Federal Statistical Office (Destatis) counts the country to have imported €3.6 billion worth of solar PV cells, modules and the like in 2022, while it exported PV products for a total of €1.4 billion. China accounted for the bulk of German imports, €3.1 billion. China represented 87% market share of imported solar power products into Germany … Taiwan was next on the list importing 3% capacity representing €94 million. Remaining 2% came from Vietnam and rest 3% from other regions. …[38]

---

[36] See: "Science and technology in China"; at: https://en.wikipedia.org/wiki/Science_and_technology_in_China (5 April 2023).

[37] See: "Research Report on China's Solar Cell Export Industry 2023–2032 - ResearchAndMarkets.com"; at: https://www.businesswire.com/news/home/20230103005281/en/Research-Report-on-Chinas-Solar-Cell-Export-Industry-2023-2032---ResearchAndMarkets.com (5 April 2023).

[38] See: Taiyang News: "Germany Imported €3.6 Billion Solar PV Systems in 2022" (2 March 2023); at: https://taiyangnews.info/markets/germany-imported-e3-6-billion-solar-pv-systems-in-2022/ (5 April 2023).

The increased US strategic pressure on its allies, especially to reduce their economic, energy and economic dependence on Russia and China and to strengthen the US position as the sole hegemon in a 'unipolar world is likely to increase the process of fragmentation of the international order. However, progress in the area of global environmental change and climate change requires an international order that *includes* China, as the largest emitter of CO_2, and Russia, as the largest territorial country and one of the major fossil energy exporters, to cope with the projected consequences of global warming.

As long as the Hobbesian geopolitical mindset prevails, the Anthropocene narrative will be downgraded. As a direct consequence, the dangers of both the existential anthropogenic human challenges of a major nuclear war and of the outcome of "business as usual" in the climate realm may increase. The result of both evolving trends in Western policies is an increase in the sense of uncertainty and risk, to which the World Economic Forum 2023 (based on its review of the economic community) has referred.[39] At present no strategy seems to exist to address, face, cope with and solve these contradictions.

2.7.3 Costs of Hobbesian Geopolitics: Anthropocene Discourse and Postponement of Socio-economic Transition

In the autumn of 2023, Hobbesian Cold War-type mindsets[40] and worldviews[41] have returned and dominate not only political discussions, media attention, and public perceptions, but also scientific discourses on international politics and international relations. Since the Russian aggression against Ukraine, in many countries of the Global North issues of security policy, military issues, arms production, rearmament, and exports of heavy weapons to the victim of the aggression have moved to the top of the policy agenda, while themes of global environmental change and climate change have been downgraded on the list of policy priorities and in public attention.

[39] World Economic Forum, 2023.

[40] The concept of a 'mindset' has been discussed by Booth (2017) with regard to the mainstream Cold War thinking; see more generally on 'Mindset'; at: https://en.wikipedia.org/wiki/Mindset. A mindset is an "established set of attitudes, esp. regarded as typical of a particular group's social or cultural values; the outlook, philosophy, or values of a person (now also more generally) frame of mind, attitude, [and] disposition". It may also arise from a person's worldview or beliefs about the meaning of life.

[41] The concept of worldview was used by Kuhn (1962, 1965) in his study on scientific revolutions; see: "Worldview", at: https://en.wikipedia.org/wiki/Worldview. "A worldview or a world-view or *Weltanschauung* is the fundamental cognitive orientation of an individual or society encompassing the whole of the individual's or society's knowledge, culture, and point of view. A worldview can include natural philosophy; fundamental, existential, and normative postulates; or themes, values, emotions, and ethics."

This may also be reflected in related national budget priorities in the years to come, while policy issues related to "socio-ecological transition" and spending on the necessary transformation of society and the economy may be reduced or decline. The impact of such policy decisions and public spending may make the realisation of a "1.5 °C world" less likely and the prospect of an increase in the number and intensity of extreme weather events more probable.

2.7.4 Consequences of the Lack of an Integrated Holistic Perspective on a Peace Ecology and Ecological Peace Policy

In 2023 there is a lack of an integrated holistic perspective and policy discussion that both integrates the dominant peace and security narrative with the Anthropocene narrative, and addresses simultaneously the two existential anthropogenic threats humankind faces in the third phase of the Anthropocene since February 2022 with the (presently still very low) probability that the war of Russia in Ukraine may escalate into a limited nuclear war. Thus, humankind may be moving from a "Stabilized Earth" to a "Hothouse Earth" (Fig. 2.4).

However, it is more likely that the postponement of the necessary decisions to address the causes and consequence of global climate change may result in continued "business as usual" with an increased number and intensity of hydro-meteorological hazards.

In the social sciences, the scientific discourses addressing the peace and security problematique separately from environmental policy issues and ecological concerns continue. There are still only a few conceptual bridge-building efforts in both research and practice, such as:

- The Environmental Peacebuilding Association (EnPAX)[42];
- The trilateral Israeli, Palestinian and Jordanian NGO: Ecopeace Middle East.[43]

Dalby (2022), a contributor to this book, has offers several slightly different approaches for bridge-building in his book on *Rethinking Environmental Security*.[44] In his conclusions Dalby (2022: Chap. 1) suggests that: "rethinking security now requires confronting these implicit assumptions quite directly. Taking ecology at various scales seriously requires nothing less than rethinking assumptions of force controlling territory as the basis of security". He notes that "mainstream international

[42] See at: https://www.environmentalpeacebuilding.org/about/.

[43] See at: https://ecopeaceme.org/about/.

[44] Dalby (2022) addressed the following issue areas of 1. Realism, Firepower and Insecurity; 2. Sustainable Development/Environmental Insecurity; 3. Geostory: Deep Time and History; 4. The Geopolitics of Colonizing Nature; 5. Global Security/Environmental Conflict; 6. Catastrophic and Existential Risks; 7. Whole Earth Security: An Engineered World; and 8. Environmental Peacebuilding.

relations … has been mostly uninterested in the rapidly changing planetary context and the Anthropocene debate in particular (Simangan 2020)." Dalby (2022: Chap. 1) argues that

> discussions of peace and environment draw from both practitioner perspectives and disciplinary representatives from diverse sources. Development studies, geography, environmental studies and other social sciences focus on these themes. Perhaps this is an appropriate division of academic labour, but given the importance of conflict prevention as a theme in security studies, further developments of this field might reasonably be expected in the leading journals in the field in coming years. (Ide et al. 2021)

Dalby (2022: Chap. 8) concludes that

> security is now about the ability to adapt to changing circumstances and to simultaneously shape economic activity to focus on human needs within planetary boundaries (Raworth 2017), not about the ever larger appropriation of firepower to dominate external environments. In terms of the scholarly study of international relations, where earlier generations of realists worried about the firepower of nuclear arsenals, the current generation now has to worry about the consequences of the unrestricted use of firepower in its other sense, as lithic fire and extensive burning of fossil fuels. Environmental security now requires engaging with firepower in all its manifestations; human survival depends on coming to terms with the consequences of partially domesticating combustion, a key geophysical force in the planetary system, and dramatically reducing its future use in the novel circumstances it has recently wrought.

Jürgen Scheffran (see Chap. 8 below) calls in this book for a slightly different bridge building within the framework of an "integrated geography". Taking proposals on "peace ecology" into account (Kyrou 2007; Amster 2015[45]; Oswald Spring et al. 2014), Brauch (2021) has outlined possible features of a "peace ecology in the Anthropocene".

More recently, Hakala and Kononenko (2022) propose in *Re-securitizing Climate – From 'Climate Security' to 'Ecology of Peace'*[46]:

> One implication of the war in Ukraine is that collective solutions to climate security will be ever more difficult to forge. Earlier hopes for cooperation were focused on the 'securitization' of climate change—that is, presenting it as an issue for security policy. The expectation was that the security implications of climate change would be recognised as a shared threat and dealt with through multilateral cooperation under UN leadership. The securitisation of climate change took place at the same time as a decline in liberalism and multilateralism in international politics. The rise of conflictual and disintegrative tendencies in international relations creates a dangerous geopolitical context for climate change. Climate security is part of great-power politics. Recognising this requires a deep rethinking of how cooperation can be achieved if this issue cannot be decoupled from geopolitical gridlocks. Climate and environment issues should be 're-securitized', for example, by further embedding them into the few existing multilateral regimes, such as the Paris Agreement and the UN Convention on Biodiversity.

Hakala and Kononenko (2022) argue that at the initiative of the UK, Germany, and Sweden, climate change has been repeatedly addressed in the UN Security Council

[45] See at: https://www.peaceecology.com/.

[46] See at: https://www.fiia.fi/wp-content/uploads/2022/09/bp348_re-securitizing-climate_emma-hakala-vadim-kononenko.pdf.

while "an actual UNSC climate security resolution has yet to be adopted due to consistent opposition and vetoing by Russia and China". Both permanent members argued "that climate change is not a matter of international security and therefore should not be included in the UNSC mandate".

Hakala and Kononenko (2022) compare four different approaches towards climate security from hyper-conservative (Russia, China, India), conservative (US DoD), through moderate-orthodox (EU's global strategy) to emancipatory (OSCE ministerial decision on strengthening co-operation to address the challenges caused by climate change) and note that "a new agenda appears to be emerging primarily within the expert community", whose agenda may be described as an "ecology of peace" approach that could be integrated into the UNFCCC's annual COP meetings. They argue that "any attempt to 're-securitize' climate and environment, to decouple the issue from the geopolitical gridlock while sustaining cooperation, would require incremental steps".

> The war in Ukraine has accelerated the shift from a liberal to a conservative notion of climate security, but the evolution was already underway. This seems to feed into the kind of perpetual state of emergency that may follow from the blurring of lines between the realms of normal and exceptional politics. … However, securitisation as emergency may also prompt change. As climate catastrophes multiply and extend into their territory, even climate-security sceptics like China, India and Russia might feel under pressure to support collective responses in the UNSC. This may be framed as developing their national resilience and will be linked to their domestic interests and needs. Therefore, the focus should be on recognising moments of possible repositioning, and using it to expand the understanding of climate security – in other words to 're-securitize' by bringing these understandings closer to the vision of an ecology of peace.

Such an approach—as suggested by Hakala and Kononenko—may overcome the geopolitical gridlock between the USA, Russia and China and may, under optimistic circumstances, contribute to an environmental peace-making or contribute to an ecology of peace. At the very least it is a recent effort of bridge-building between the two narratives that have influenced the discourses in International Relations.

Research on a peace and security narrative may have to go hand in with the Anthropocene narrative, thus addressing the two existential threats of a nuclear war and the possible physical and societal effects of global climate change and possible tipping points triggered by human intervention in the Earth system. Such a holistic perspective may require policy makers to think together the security dilemma and the survival dilemma they may face in the twenty-first century.

İşeri (2020) has taken up this conceptual idea in a text published by the Heinrich Böll Foundation in Istanbul, titled *From the security dilemma to the human-centered survival dilemma in the Eastern Mediterranean: The environmental security approach.*[47] In applying Brauch's concept to the Eastern Mediterranean (Greece, Turkey, Cyprus), İşeri concludes

[47] See Emre İşeri; at: https://tr.boell.org/en/2020/11/30/security-dilemma-human-centered-survival-dilemma-eastern-mediterranean-environmental (12 November 2023).

> that the parties of the conflict, while trying to guarantee their energy security, are stuck in a 'security dilemma'—just as the state-centered security approach would predict. If the current situation should continue, the military, economic and environmental insecurities of both sides will increase, let alone ensuring security. A continuous, mutual increase in military power as well as military cooperation will further augment insecurity. Additional military expenditures will also deepen the economic insecurity in both Turkey and Greece. … Furthermore, the ongoing tension will prevent the optimum development of joint projects for making use of the rich, potential renewable energy resources, particularly in Cyprus.
>
> The environmental insecurity of the parties will escalate in two ways: while developing resources and constructing transmission lines will destruct the marine ecosystem, the struggle against climate change will be put on ice. Beyond these, one point is obvious: The plan for feasible production of securitized natural gas reserves, their transmission through the 1900 km long East Med pipeline, which will cost 6 billion Euros to build, and finally, their export to the European Union that is aiming at moving to a carbon-free economy and whose energy market is already shrinking, contains a lot of incalculability.
>
> As one of the regions that will be affected most by climate change, the East Mediterranean is face to face with various existential risks, water shortage being in the first place. Decision-makers of the involved states should be aware of the fact that the main security threat for the common fate of the people living in the region is sustainability. In other words, they should act with responsibility to address the 'survival dilemma' for the well-being of the whole region. For this, it is required that decision-makers move beyond narrow security perspectives and that they—parallel to the enlarged security approach—securitize climate change, so that they can immediately work on joint, renewable energy projects for the construction of a sustainable East Mediterranean region. This paradigmatic change suggested for the region can have a solid base with the support of great powers. Thus, if the EU, in coordination with the USA under the presidency of Biden, increases its support for sustainability projects … that could create the real 'positive agenda' in the region. Eventually, this would enhance prospects for building sustainable Eastern Mediterranean region.

This is an example of moving from a consideration of peace and ecology to an ecological peace policy that could create a framework for cooperative sustainable development policies in areas of conflict. Similar ideas have been presented by Brauch (2006, 2007) for the Gulf of Aqaba between Jordan, Israel and Egypt, and by Ecopeace Middle East for a *Green Blue Deal for the Middle East* (Bromberg et al. 2020). Ecopeace Middle East has proposed:

> a regional water-energy-nexus between Jordan, Israel and Palestine—a nexus that would create healthy and sustainable regional interdependencies based on critical natural resources, with Israel and Palestine producing desalinated water and selling it to Jordan, and Jordan selling Palestine and Israel renewable energy, thereby enabling each partner to harness their comparative advantage in the production of renewable energy and water.
>
> While the EcoPeace-led UAE/Israel/Jordan water-energy agreement brokered in November 2021 was a significant breakthrough, the potential impact of the WEN goes beyond Jordan and Israel, and must include Palestine. It could also expand to include additional countries such as Lebanon (adding coastline), Egypt/Sinai (adding both coastline and desert areas), Saudi Arabia, and parts of Syria and Iraq. In fact, the WEN model for optimizing natural resource-management via collaboration between hinterland and coastal areas, is relevant worldwide.
>
> Creating an integrated water and energy economy, for example between Israel and Jordan, could help meet the growing energy needs of both countries—linked to their needs for more fresh water through desalination—and counter the effects of climate change and its potential implications for security and stability in the region. …

> The long term relationship suggested is one of interdependence with Israeli, Palestinian and potentially Lebanese Mediterranean shores providing the much needed additional water to the Levant as a whole, in the short term fuelled by natural gas reserves, but in the longer term to be fueled by extensive investment in solar energy production in Jordan. While the Memorandum of Understanding signed in December 2013 between Jordanian, Palestinian and Israeli Water Ministers, witnessed a new era for potential water exchange, the inclusion of energy into the mix makes for a far more sustainable interdependence of geopolitical interests in the Levant, based on the water and renewable energy nexus.
>
> Advancing such regional cooperation, whether on limited cross border solar-based desalination or broader water-energy needs in general, requires a multi-disciplinary multi-country effort of further studies to identify strategies to help overcome political obstacles and generate political will, based on studies related to environmental, economic and technical feasibility.

The outburst of violence during a surprise attack by Hamas against Israel on 7 October 2023 and Israel's massive military reprisals against Gaza have both undermined the trust that would be needed for a peace settlement with justice in the Middle East, which is required for the specific proposals that have been developed by Ecopeace Middle East since its foundation in 1994. The proposals of Ecopeace Middle East[48] were supported by a wide coalition of sponsors, including USAID[49] and the US Institute for Peace,[50] the Konrad Adenauer Foundation,[51] and the United Arab Emirates.[52] The realisation of the proposals would require a wider political framework where the USA, the EU, the Arab League and BRICS countries from the Middle East (Egypt, Saudi Arabia, the United Arab Emirates), and possibly also China, would have to be involved.

2.8 Conclusions: From a Security to a Survival Dilemma in the Anthropocene

Both the outcome of Russia's war against Ukraine, and the effects of the future Russian government on the new post-war order in Europe are highly uncertain and cannot be foreseen or predicted in autumn of 2023, nor can the outcome of the new war in the Middle East be projected or the prospects for an ecological peace

[48] See the website of Ecopeace Middle East; at: https://ecopeaceme.org/ and on publications: https://ecopeaceme.org/publications/.

[49] Ecopeace Middle East/USAID; at: https://ecopeaceme.org/2022/03/30/wonderful-news-for-eco peace/.

[50] Ecopeace Middle East/US Institute for Peace; at: https://www.usip.org/publications/2022/12/water-can-be-rare-win-win-israelis-palestinians-and-region.

[51] Ecopeace Middle East/Konrad Adenauer Foundation, 2020: *The Climate Crisis and the Changing Security Environment of the Middle East*; at: https://ecopeaceme.org/wp-content/uploads/2022/03/the-climate-crisis-and-the-changing-security-environment-of-the-middle-east.pdf.

[52] Ahav Harkov, 2021: "Israel, Jordan to sign UAE-mediated energy and water agreement - The agreement is meant to help Israel and Jordan deal with the challenges of climate change", in: *The Jerusalem Post*, 17 November; at: https://www.jpost.com/middle-east/israel-jordan-to-sign-uae-mediated-energy-and-water-agreement-685230.

that includes functional cooperation on the water and energy nexus. Most likely, the three competitive narratives—the Hobbesian security and Grotian peace narratives on the one hand, and the Anthropocene ecological narrative on the other—may continue to coexist. The "security dilemma" may remain for many years a rationale for the legitimisation of higher military spending and for a major new build-up of armaments. This may result in a new arms race with Russia and China in the Global North that may increasingly be justified in Europe by Chinese military activities. From the dominant Hobbesian security perspective, a "new security dilemma" may be stressed, while a "survival dilemma" within a human security framework may be ignored.

2.8.1 *Towards Rethinking Politics, Policy and Polity in the Anthropocene: Responding to Crutzen's Scientific Challenges*

The mainstream thinking in Political Science and International Relations has not yet recognised, understood or analysed the triple impact of Crutzen's claim of 2000 that we are now living in the Anthropocene (Crutzen/Stoermer 2000). His thesis (Crutzen 2002) is that humankind has directly interfered in the Earth system by its burning of fossil fuels (coal, oil, gas) and the changes in production, consumption, transportation, housing, and agriculture, and that the scientific interpretation is that the Anthropocene started with the first test of a nuclear bomb on 16 July 1945 (AWG) and with the processes of the "Great Acceleration" (Steffen/McNeill 2007; Steffen et al. 2011). The start of the Anthropocene between 1945 and 1950 triggered a unique change in Earth history, from the Holocene to the Anthropocene, which coincided in human and in political history with a change in the US's global role after 1941 and a change in the international order between 1945 and 1950.

2.8.2 *Thinking 'Geological' and 'Geographic' Time Together*

Jürgen Renn's proposal (2020, 2022) that "geological time" and Braudel's (1966, 1968, 1972) "geographic time" of the *longue durée* should be thought together implies that the approaches in Earth systems science since the 1980s, and the knowledge developed in the natural sciences on global environmental change and on global climate change, have fundamentally changed the context for scientific analysis, triggering calls for a new "Copernican Revolution" in thinking (Schellnhuber 1999) that requires a fundamental change in the "scientific worldview" (Kuhn 1962, 1965) and in the "political mindset" (Booth 1990).

2.8.3 *Combining the Peace and Security Narrative with the Ecological and Anthropocene Narrative*

The process of specialisation and fragmentation of research in the social and political sciences has contributed to a lack of political and societal relevance of the professionalised scientific output in sophisticated analyses and scientistic journals in a self-referential system where citations collated in multiple indexes (e.g. in the Social Citation Index) have become instrumental for the scientific careers of many scholars. The price has been a decline in public reception of controversial scientific debates that in the past inspired an 'enlightenment' of the "attentive audience".

Therefore, a new enlightenment is needed, and a wider framework for the reconceptualisation of scientific concepts in the second '*Sattelzeit*' (Koselleck 1979, 1998, 2006) since 1940/1945, that coincided with the American century (Luce 1941, 1999) and the emerging US world leadership role since 1941 with the planning for and establishment of the new US-designed international political, security and economic world order (1945–1950). This US global leadership position has been eroded in the US Congress with its refusal to ratify the CTBT (1996) and with the non-ratification of the Kyoto Protocol (1998). The erosion of the rule-based global liberal order started in the "third phase of international order" in the Anthropocene, with the weakening of the arms control regime since the turn of the century by both the US and Russia, and by the neo-isolationist and protectionist policy of US President Donald Trump (2017–2021) with its call to "make America great again".

The cruel and violent policies of Russian President Vladimir Putin (since 2000 and especially since 2014 with the annexation of the Crimea), the start of a Russian-supported war in the Donetsk region of Ukraine (since 2014) and the start of an aggression and war against Ukraine (since 2022) to revise the outcome of the implosion of the Soviet Union have been driven by Putin's nationalist visions. They have been instrumental in the collapse of the rule-based security order in Europe that had gradually developed since the bilateral US-Soviet limited détente (1963), the détente in Europe (since 1975) and the reunification of Germany and Europe (since 1990) that opened a short "window of opportunity" for the emergence of cooperative arms control and global environmental change regimes (UNFCCC 1992; Kyoto Protocol 1997; Paris Agreement 2015; CBD 1992; UNCCD 1994).

2.8.4 *Integrating Earth Systems Science and History of Events*

While the knowledge produced by the Earth system sciences could well outline and address the geological context of humankind through human-intervention in the Earth system by leaving 'politics' out of the analysis, it could not explain the "climate paradox" (Brauch 2012), For these reasons Brauch et al. (2011) and Brauch

(2016) suggested a "Political Geoecology for the Anthropocene" to integrate scientific approaches in the natural sciences, including physical geography, and a "Peace Ecology in the Anthropocene" (Brauch 2020).

Since President Putin's aggression against Ukraine on 24 February 2022 and the collapse of the European and global cooperative peace and security order, the future is uncertain; a new additional dual focus is needed that combines an analysis of the policy decisions of the three presidents of the USA, Russia and China. If President Donald Trump should again win the US presidential election in November 2024, the disagreement between the yearning for international peace and the realisation of cooperative solutions on issues of security, trade and the environment, including global climate change and biodiversity, will further intensify.

Thus, an analysis of political actions and decisions (political events) and of longer-term changes in the formation of new international alliances (e.g. between China and Russia) and groupings (BRICS + 6) may help to address the question of why the regression in global environmental and climate change policies may not be revised in the near future, and why the economic and humanitarian price for non-action or postponement of decisions on climate change may also increase (for the world at large, but particularly also for the USA, Russia and China), with an increase in the number and intensity of extreme weather events, such as forest fires (Dalby 2023), against which the increase in military spending and military arms will offer no solution.

A holistic peace ecology in the Anthropocene (Brauch 2020) may also inspire conceptual designs for turning from a chaotic and unpredictable Hobbesian world towards strategies of ecological peace policies in the Anthropocene whenever a new "window of opportunity"—as during the decade from 1990 to 2000—should open again during the twenty-first century.

The social and political sciences, International Relations, and especially peace, security, environment and development studies can contribute towards widening the scope during a second phase of the enlightenment in the Anthropocene for which the Club of Rome (2017), von Weizsäcker (2019) and Pelluchon (2021) have called.

2.8.5 Developing Concepts Further: Towards a "Second Sattelzeit*" with a "New Enlightenment"*

Historians of science have proposed a "Rethinking of Science for the Anthropocene" (Renn 2020, 2022), and, inspired by R. Koselleck, some specialists of concept history have called for (Geulen 2010; Décultot/Fulda 2016), or commented critically on, analyses on a second '*Sattelzeit*' (Nolte 2010)—however, without any reference to the Anthropocene.[53]

[53] Brauch et al. (2008, 2009, 2011) have coedited three volumes with 270 peer reviewed English chapters in the *Global Environmental and Human Security Handbook for the Anthropocene* (GEHSHA) and a *Handbook on Sustainability Transition and Sustainable Peace* (Brauch et al.

The social and political sciences, International Relations, and especially peace, security, environment and development studies can contribute towards widening the scope during a second phase of the enlightenment in the Anthropocene, for which the Club of Rome (2017, 2018), von Weizsäcker (2019) and Pelluchon (2021) have called. The Anthropocene epoch of Earth history, starting between 16 July 1945 at Alamogordo and 1 November 1952 on atoll Eniwetok, could be a temporal context for a rethinking of "politics, polity and policy" and for a reconceptualisation of key concepts in the social, cultural and political realms in a "second *Sattelzeit*" from 1940 to the present.

2.8.6 *Earth Systems Science, Geoecology, Integrative Geography: New Multidisciplinary Approaches in the Anthropocene*

Since the 1980s at least four integrative concepts have been proposed for the natural sciences and Geography: (a) *Earth Systems Science* (ESS) or *Analyses* (ESA) in the framework of the natural sciences; (b) *Geo-ecology* (Huggett 1995); (c) Anthropocene Geopolitics (Dalby 2020) and (d) *Integrative Geography* (Scheffran in this volume). All four still lack a systematic political dimension. Geography combines knowledge from both the natural sciences (physical geography) and social and political science (geopolitics, environmental geopolitics and Anthropocene Geopolitics). Among these approaches only two have partly addressed a political dimension: "Integrative Geography" (Scheffran) and "Anthropocene Geopolitics" (Dalby).

2.8.6.1 Multidisciplinary Approaches for Earth Systems Science

Since the 1980s the concept and approach of an "Earth Systems Analysis" (Schellnhuber and Wenzel 1998; Schellnhuber et al. 2004) has been developed and promoted at the Potsdam Institute for Climate Research (PIK) by its former director H. J. Schellnhuber and colleagues. A few social and political scientists attended the Dahlem Workshop on Earth Systems Analysis for Sustainability; among them were W. C. Clark (Harvard University), R. B. Mitchell (University of Oregon), P. Romero Lankao (Autonomous University of Mexico at Xochimilco), and O. R. Young (University of California, St. Barbara). A key conceptual thinker behind the Earth

---

2016). However, at least in the Global North, since the Russian aggression against Ukraine a return to a narrow realist or Hobbesian security concept can be observed, where the environmental dimension does not matter anymore.

Systems Science (ESS) and the Anthropocene approach has been W. Steffen (Steffen et al. 2020). The Proceedings of the US National Academy of Science (PNAS 2018) published an article by top Earth Systems Scientists[54] that was edited by a Harvard political scientist (Clark 2018).

In 2022, the German Academy of Science Leopoldina[55] published a Report on *Tomorrow's Science: Earth System Science: Discovery, Diagnosis, and Solutions in Times of Global Change*.[56] In his Foreword, the President of Leopoldina, Gerald Haug, outlined the goals of the Report,[57] which is structured in six chapters dealing with: (1) Science for our Planet in Times of Change; (2) The Earth as a System;

---

[54] See: "Trajectories of the Earth System in the Anthropocene" by W. Steffen, J. Rockström, K. Richardson, T. M. L. F. Donges, I. Fetzer, S. J. Lade, M. Schefferl, R. Winkelmann, and H. J, Schellnhuber, Ed. by W. C. Clark, Harvard Univ., Cambridge, MA.

[55] See on the two major objectives of Leopoldina: "representing the German scientific community internationally, and providing policymakers and the public with *science-based advice*. The Leopoldina … promotes a *scientifically enlightened society* and the responsible *application of scientific insight for the benefit of humankind and the natural world*. In its *interdisciplinary discourse*, the Academy transcends thematic, political and cultural boundaries. … It is the role of the Leopoldina, to identify and analyse scientific issues of social importance. The Leopoldina presents its policy recommendations in a scientifically qualified, independent, transparent and prospective manner, ever mindful of the standards and consequences of science" (emphasis added in italics by hgb).

[56] See: German National Academy of Sciences Leopoldina (2022): *Report on Tomorrow's Science. Earth System Science – Discovery, Diagnosis, and Solutions in Times of Global Change*. Halle (Saale); at: https://www.leopoldina.org/fileadmin/redaktion/Publikationen/Zukunftsreport/2022_Zukunftsreport_Erdsystemwissenschaft_EN_web.pdf (27 September 2023).

[57] See Prof. Gerald Haug, President of Leopoldina, Foreword to the abovementioned *Report*: "Planet Earth is a complex system. The interaction and dynamics of its components shape our planet and have thus a significant influence on its habitability for humans. However, our understanding of the Earth system is incomplete. Many big questions in the Geosciences yet remain unanswered. Some of those are: Which conditions defined the origin and evolution of life on Earth? Are earthquakes reliably predictable? How long will oceans continue to buffer climate change? How do biodiversity and climate interact? Are the planet's resources sufficient to ensure the survival of the Earth's population?

Since humans massively intervened in the structure of the Earth and became a decisive force whose influence is drastically seen in all parts of the planet, it has become particularly clear how fragile the dynamics in the Earth system are. Nobel laureate Paul Crutzen ML coined the term Anthropocene for this new age of Earth history. Humans have initiated global processes threatening the environment and living conditions of a large part of the world's population, making entire regions of the Earth uninhabitable, impairing global health and depleting natural resources to such an extent that the development of future generations is compromised. Developing the scientific basis for an understanding of these processes and for risk assessments and solutions are the central challenges of the geosciences today. This requires an approach characterized by systemic thinking – the Earth System Science. However, the study of the Earth system also faces challenges within academia. The complexity of this field of research encounters fragmented academic training and research of a wide variety of disciplines. This Report on Tomorrow's Science is a plea to develop Earth System Science as a conceptual and structural framework to give new impetus to basic research and to address societal challenges. It calls for questioning traditional patterns of thinking in teaching and research and building sustainable structures that meet the major challenges. This is the only way to gain the necessary understanding of the changes in our living environment and to develop appropriate science based options for action." Prof. (ETHZ) Dr. Gerald H. Haug, President of the German National Academy of Sciences Leopoldina.

(3) The Grand Challenges of Earth System Science; (4) The Data Challenge in Earth System Science; (5) Academic Education for Earth System Research and (6) Recommendations for Decision-Makers and for the Geoscience Community. The Leopoldina's Report noted:

> ... The recent phase of unprecedented, rapid Earth System transitions, called the Great Acceleration, marks the beginning of the Anthropocene. In this new geological epoch, *humankind constitutes the largest driver of change on Earth.* Humans are making an imprint on the Earth System that is significant, measurable, and may cross thresholds that put Earth System stability at risk at a pace that is faster than human civilisations can adapt [to]. Yet it remains *essential to gain a deeper understanding of the societal challenges of the Anthropocene* and *develop potential mitigation and adaptation solutions and actions.* Thus, we need a more transdisciplinary perspective and novel approaches to identify physical, ecological, and societal thresholds and avoid crossing them to shape a future in which the next generations can thrive. ...
>
> The *Anthropocene constitutes a manifestation of new pressures (forcing) exerted by humans on the Earth System.* The climate system has not yet crossed a hard threshold that would propel the Earth System into a new state. However, there are indications that Earth is following a trajectory that will shift its climate away from relatively stable interglacial Holocene like conditions. ...

The Leopoldina's *Report* points to the following needs of future research, calling for multidisciplinary research, noting especially that "a holistic understanding and a sound diagnosis of the *social environmental aspects of the Earth System*" is needed and "three main courses of action" should be considered:

(1) To properly assess *biogeophysical risks*, we need to determine the tipping points of subsystems and the danger of crossing them to gain a deeper understanding of the dynamics of the integrated Earth System. This requires improved and novel model architectures, increased computing power, better parameterization of critical processes, and better model calibration with abundant data collected from a large array of new near real time observation systems. These data will be used to test and improve the fully scale bridging next generation Earth System Models, ... enabling us to recognize feedbacks, thresholds, and tipping points in the Earth System. ...

(2) Two approaches to mitigate global warming are currently envisaged: radiation management and negative CO_2 emission technologies, sometimes called '*geoengineering*'. ... In contrast, according to the IPCC projections, humanity is unlikely to limit global warming to 1.5 °C unless substantial amounts of CO_2 (about 10 Gt/yr) are withdrawn from the atmosphere. Thus, research on *technical negative emission technologies* (e.g., direct air capture linked to geologic carbon capture; biologically enhanced carbon capture and storage) and enhancing natural sinks (e.g., biochar formation and dispersal, enhancing soil organic carbon storage, enhancing weathering by dispersal of basalt powder onto croplands, and replanting seagrass) *should be pursued with high priority.* ...

(3) *Biogeophysical risks and tipping* points that are mediated by the biosphere are highly interwoven with *societal tipping points*. There is an urgent need [to] ...

> *include the human domain as a major planetary system in holistic studies of the Earth System.* Because our knowledge of the present state and future trajectory of the anthroposphere is still in its infancy, we must explore *when and which positive societal tipping elements are likely to achieve rapid transformative change.* Such change may emerge in different sectors and lead humans to pursue sustainable trajectories, preventing the crossing of biogeophysical tipping points. Importantly, major undesirable societal tipping elements (e.g., related to collapse of societal complexity, political polarization, conflicts) that lead towards the deterioration of other Earth tipping elements must be determined. Identifying examples for these types of societal thresholds can be a first step towards keeping essential social dynamics in balance, such as population and urban development, and the financial system, with special reference to global inequality and injustice, norms and values of human environmental behaviour including global governance, and environmental education and information feedbacks on best practice solutions. *Because biogeophysical and socio economic risks and tipping points are highly interlinked, the development of next generation Earth System Models must include the full dynamics of social ecological networks* [*emphasis* added by hgb].

2.8.6.2 Geoecology

Influenced by Troll's (1939, 1968) studies on landscape ecology, 56 years later the British physical geographer Huggett (1995) published a book on 'Geoecology' that

> investigates the structure and function of geoecosystems, their components and their environment. The author develops a simple dynamic systems model, the 'brash' equation, to form the conceptual framework for the book suggesting an 'ecological' and 'evolutionary' approach. Exploring [the] internal [or] 'ecological' interactions between geoecosystems and their near-surface environments—the atmosphere, hydrosphere, toposhere, and lithosphere—and external influences, both geological and cosmic. *Geoecology* presents geoecosystems as dynamic entities constantly responding to changes within themselves and their surroundings. An 'evolutionary' view emerges of geoecological systems, and the animals, plants, and soils comprising them, providing a new way of thinking for the whole environmental complex and the rich web of interdependencies contained therein.[58]

In several earlier texts Brauch (2011, 2016) has summarised the conceptual context and the research goal of Geoecology that has been influential in Geography departments, primarily in German speaking countries, that offer degree courses on Geoecology, while in the US, Canada, Australia and other countries degrees in Earth Systems Science have widely proliferated (Table 2.3).

---

[58] See R. J. Huggett (1995): *Geoecology* (London: Routledge, Taylor & Francis). See the blurb on Amazon's website at. https://www.amazon.de/Geoecology-Evolutionary-Richard-John-Huggett/dp/0415086892.

Table 2.3 Scope and focus of new multidisciplinary concepts on Earth-human systems linkages

Concepts	Emergence since	Scope: disciplines methods	Authors	Journals	Geographic spread	Degree programmes
ESA/ESS (rapidly emerging transdisciplinary endeavour aimed at understanding the structure and functioning of the Earth as a complex, adaptive system)	1980s	Physics Chemistry Geology Hydrology Ecology Meteorology Oceanography modelling	Schellnhuber P. J. Crutzen W. Steffen K. Richardon J. Rockström	*J. of Earth Systems Science, India* *Earth System Science Data, ESSD* *Natural Hazards and ESS* *Hydrology and ESS, EGU* *Earth Systems Dynamics* *Earth Science, Systems, Society* *Journal of Advances in Modeling Earth Systems* *Earth System Science Frontiers* *Nature* *Nature Climate Change*	USA Canada Australia India Europe Germany Denmark Sweden	Stanford University; Univ. of California, Irvine; Univ. of North Dakota; GMU (USA); McGill Univ.; Queens Univ. (Canada); San Jose Univ.; Univ. of Alaska; Univ. of Massachusetts; Cornell Univ.; Univ. of Zürich; Univ. Hohenheim; Hamburg Univ.
Geoecology (interdisciplinary/multidisciplinary science, integrates geosciences with life sciences, focusing on influencesof geological processes on historical and contemporary patterns of biogeography, causes and consequences of geoegraphics on biota)	Troll (1968) Huggett (1995)	Physical Geography	Geoecology Huggett (1995) Political Geoecology Brauch (2003) Brauch et al. (2011) Dalby (2020)		Germany Austria Switzerland	Karlsruhe Univ.; Univ. Tübingen; Univ. Potsdam; Univ. Halle; Univ. Wien; Oxford Univ.; Basel Univ.; Free Univ. of Berlin; et al.

(continued)

Table 2.3 (continued)

Concepts	Emergence since	Scope: disciplines methods	Authors	Journals	Geographic spread	Degree programmes
Integrative geography (Integrated Geography (see: Integrative Geography, Environmental Geography or Human–environment Geography) is the branch of Geography that describes and explains the spatial aspects of interactions between human individuals or societies and their natural environment, called coupled human–environment systems)	Since 2014	Geography • Systems theory • Complexity science • Modelling	B. M. W. Ratter J. Scheffran		Germany	Hamburg Univ., Geography Dep.
Environment security to Anthropocene geopolitics	2020	Political Geography (Sociology, Political Science)	Dalby (1999, 2020, 2022) Oswald Spring et al. (2009) Ide (2023)	Environment and security	UK Ireland The Netherlands USA Australia	University of Liverpool University College Dublin University of Leeds Johns Hopkins University Univ. of Chicago Bonn and UNU-EHS Monash University Murdoch University

Source The author

2.8.6.3 Integrative Geography

Since 2014, there have been two professorships in "Integrative Geography" at the Institute of Geography of Hamburg University that are

> [l]inking the areas of physical and human geography as well as re-integrating both areas into research and teaching practice. ... We focus on the interactions between humans and nature and the interdependency between the environment and society. As a hub between disciplines, we deal with several different topics. Some of these are resource conflicts and resource management, sustainable development of regions, livelihood and vulnerability concepts, security risks and conflict potential of climate change and energy systems, natural hazards and the conservation of nature, risk perception and management, power and dominance relationships in the context of nature occupation processes, as well as discourses on the perceptions, problem constellations, and solutions of ecological crises. Among other topics, our research focuses on empirical field studies, especially in coastal areas and on islands, along rivers and in mountain regions, in rural peripheral regions and the southern hemisphere. On one hand, the theoretical background is drawn from general system theory, specifically from complexity theory. At its core is the pursuit of the analysis and modelling of non-linear dynamic systems, such as integrated land, water, and energy use, or the causes, effects, and strategies regarding climate change. On the other hand, political and socio-ecological approaches are an important part of the topics mentioned above. Post-colonial and post-structuralist perspectives complete this field of work.[59]

Until the retirement of J. Scheffran in October 2023, the Department of Integrative Geography at Hamburg University had two full professorships. In 2023, the website listed Prof. Dr. Beate M. W. Ratter as a Professor for Integrative Geography and Prof. Dr. Benno Fladvad as Junior Professor for Natural Science Peace Research with focus on Climate and Security.[60]

2.8.6.4 Limited Integration in Bridge Building Approaches

Earth Systems Science has become a multidisciplinary approach to the study of Earth systems and of Earth–human linkages that is being taught in many universities in the USA and Canada, and also in Europe. In contrast Geoecology was originally influenced by Troll's (1968) landscape ecology and has become the theme of many multi- and interdisciplinary approaches in Physical Geography, primarily in German-speaking universities in Germany, Austria and Switzerland.

Up till now, Integrative Geography has been taught only at Hamburg's Geography Institute by two professorships, and has been much influenced by systems analysis and complexity studies. So far, conceptual suggestions to offer degrees in Political Geoecology (Brauch 2003; Brauch et al. 2011), Ecological Geopolitics (Dalby 2014) or Anthropocene Geopolitics (Dalby 2020) have not been taken up by mainstream Geography departments.

---

[59] See the description of the mission of the Hamburg University's Institute of Geography; at: https://www.geo.uni-hamburg.de/en/geographie/abteilungen/integrative-geographie.html.

[60] See at: https://www.geo.uni-hamburg.de/en/geographie/mitarbeiterverzeichnis.html.

Simon Dalby (1998, 2000, 2002, 2002a, 2008, 2009, 2020, 2022) has been a pioneering thinker in bridge building between security and environment research on the Anthropocene in Human Geography and has also widely published on Political and Social Geography in Canada. He has contributed to discourses in other social science disciplines and has become one of the globally most-cited authors on the Anthropocene and on climate change and conflicts (Brauch 2021). In *Anthropocene Geopolitics* (Dalby 2020) and in *Rethinking Environmental Security* (2022) he has integrated and systematised much of his work.

For Dalby the "political dimension" has been crucial in his thinking and writing. However, none of the four bridge-building efforts have specifically focused on a need to rethink 'politics', 'policy' and 'polity' due to the transition from the Holocene to the Anthropocene epoch of Earth history. As of 2023, mainstream authors in Political Science and International Relations have so far not discussed these multi- and interdisciplinary approaches in the natural sciences, or in Geography, nor have they noted the integration efforts in Political Science and International Relations. One of very few younger researchers, Hardt (2021), discussed in an edited book on the Anthropocene the "Encounters between Security and Earth Systems Sciences: Planetary Boundaries and Hothouse Earth".[61]

2.8.7 Addressing the Two Existential Threats in the Anthropocene in Political Science and International Relations

In 1945 a dual turning point occurred both in Earth history from the Holocene to the Anthropocene and in Braudel's *longue duree* that was analysed above as a multiple turn in international order. The first test of nuclear weapons on 16 July 1945 as the starting point of the Anthropocene Epoch. This historical event marks the emergence of two anthropogenic threats of human-built nuclear weapons and as a consequence of humankind becoming a 'geological actor' impacting on the physical effects of global climate change.

Both existential threats may, for the first time, directly threaten the survival of humankind. They differ fundamentally in how they were dealt with politically and analysed in IR in the peace and security narrative. In the peace and security narrative, "the other" poses the existential threat while in the Anthropocene narrative *we* (as part of humankind) pose a fundamentally different existential threat.

[61] See Hardt (2021: 41). She has structured her chapter in three sections: "The first section describes the evolution, basic assumptions, and central concepts of Earth System Science. The second section analyses Earth Systems Science security logics through the examination of the concepts and the interrelations that emerge in recent research reports. The third section highlights that the until now separate research communities, that both engage 'hard' (political and social) sciences, cannot continue to avoid engaging in common research and outlines some future challenges and pathways that arise at the nexus between ESS and IR approaches to security."

In International Relations, authors primarily addressed existential threats in the framework of the peace and security narrative. This also applies to Wæver's (1995, 1998, 2008, 2023) "securitisation theory", from which he has also addressed the impact of anthropogenic climate change on conflicts as an issue of "utmost importance" that requires "extraordinary [political and military] measures" (see his lecture in Helsinki in May 2023).[62]

However, if *we* pose this existential threat in the Anthropocene narrative, the consequences should be addressed in the framework of both narratives simultaneously. Waever has so far neither addressed the Anthropocene narrative nor has he tried to think the geological time and the political time of the *longue durée* together, as Renn (2022) has suggested. Neither has he dealt with the relevance of Earth Systems Science for IR, as has been proposed by the *Report* of Leopoldina (2022) that suggests focusing on two components of a new type of existential threat due to *biogeophysical risks*, where "biogeophysical risks and tipping points that are mediated by the biosphere are highly interwoven with *societal tipping points*."

The threats posed by the societal effects of extreme weather events are outside the logic of military threats and means. Intensive storms, floods, droughts, and forest fires cannot be contained with classical military means and sophisticated armaments. Sears (2020)[63] has suggested the concept of "existential security" in a "framework for the survival of humanity":

> Humankind faces a growing spectrum of anthropogenic existential threats to human civilization and survival. [His] article therefore aims to develop a new framework for security policy—'existential security'—that puts the survival of humanity at its core. It begins with a discussion of the definition and spectrum of 'anthropogenic existential threats', or those threats that have their origins in human agency and could cause, minimally, civilizational collapse, or maximally, human extinction. It argues that anthropogenic existential threats should be conceptualized as a matter of 'security', which follows a logic of protection from threats to the survival of some referent object. However, the existing frameworks for security policy—'human security' and 'national security'—have serious limitations for addressing anthropogenic existential threats; application of the 'national security' frame could even exacerbate existential threats to humanity. Thus, the existential security frame is developed as an alternative for security policy, which takes 'humankind' as its referent object against anthropogenic existential threats to human civilization and survival.

---

[62] See: Wæver, O. (2023): Public lecture: climate change as a security issue. Helsinki Collegium for Advanced Studies. May 12; at: https://www.youtube.com/watch?v=QE52mGzIIsU8 (28 September 2023).

[63] See: Sears, N. A. (2020): "Existential security: towards a security framework for the survival of humanity", in: Global Policy, 11, 2: 255–266; at: https://doi.org/10.1111/1758-5899.12800; Sears, N. A., 2022: *Great power rivalry and macrosecuritization failure* (Toronto, ON: University of Toronto); at: https://www.researchgate.net/publication/366020757_Sears_Nathan_2022_Great_Power_Rivalry_and_Macrosecuritization_Failure-_PhD_Thesispdf. I am grateful to S. Dalby who on 27 September 2023 brought the article by S. Bernstein in August 2023's *Global Policy* to my attention. In which he reviewed Sears' innovative conceptual thinking in: "Existential security and the governance challenge: Confronting the antinomies of securitisation"; at: https://doi.org/10.1111/1758-5899.13289.

Sears (2020: 255) points to these policy implications of his argumentation:

- *Anthropogenic existential threats* refer to a class of threats that have their origins in human agency and could bring about, at the minimum, the *collapse of civilization*, or, at the maximum, *human extinction*. The spectrum of anthropogenic existential threats is broad and growing.
- *Anthropogenic existential threats* should be conceptualized as a matter of 'security'; that is, through a *logic of protection from threats to the survival of some referent object*.
- *Existing security frames are inappropriate for security policy* towards anthropogenic existential threats: a 'human security' frame would lead to policy incoherence, while a 'national security' frame could even exacerbate the threats to humanity. Policy makers should resist making political decisions on the basis of a national security frame in response to anthropogenic existential threats.
- 'Existential security' represents a new framework for security policy; *security is about the survival of humanity. Its referent object is humanity, the dangers are anthropogenic existential threats*, the *modes of protection are mutual restraint and resilience*, the security actors are nation states and global political institutions, and the timeframe is intergenerational (*emphasis* added by hgb).

In his research on existential threats to humankind in the context of IR theory, Sears (2000: 266) explores "the intersection between international security and existential threats to humanity". He (2000: 260) argues that in dealing with "anthropogenic existential threats … military balancing [is] irrelevant for combatting climate change or superintelligence, but a relative gains/losses perspective of security would most likely exacerbate the risks." For instance, if climate change comes to threaten the national survival of some states more than—or before—others, then they may perceive 'geoengineering' as their only option for national security (Corry 2017). Geoengineering presents a potentially serious "security dilemma" (Jervis 1978), since one state's attempts to unilaterally alter the climate may inadvertently threaten others. It is even conceivable that the geopolitical struggle for power and security in the Anthropocene (Crutzen 2002) could extend to control over Earth's climate system. Such a world would leave all states less secure than one in which global cooperation succeeded in curbing the security consequences of climate change for humanity. Both Waever (1995, 1998, 2023) with his "theory of securitization" and Sears (2000) with his "existential security" concept remain within the mainstream of IR and have not problematised Crutzen's (2002) thesis in his brief comment on a "geology of mankind". There have been repeated efforts to problematise multiple linkages between peace research, security studies, development research and the environment (Brauch 2008).

The Brundtland Report (1987) introduced into the global discourse in the social sciences and in discussions within and publications by international organisations the concept of "sustainable development" and in 1999 the UNGA adopted the Millennium Development Goals and in 2015 fifteen Sustainable Development Goals (SDGs) to be realised by 2030.

Brauch (2021: 111) proposes building research linkages between peace research and security studies on the one hand and environmental studies and ecological concerns on the other in the framework of a proposed "peace ecology in the Anthropocene" that should establish thematic and institutional bridges between a peace and security narrative and an Anthropocene narrative, addressing:

- the early conceptual debate on the linkages between peace and the environment;
- the study of the impacts of weapons and wars on the environment;
- the conceptual debate on environmental and ecological security;
- the impact of environmental degradation and stress on environmental conflict;
- environmental peace-making and the role of the environment in post-conflict peacebuilding;
- the emerging discourse and policy debate on climate change and conflicts; and
- the early approaches to peace ecology and their shortcomings.

2.9 Towards an Ecological Peace Policy in the Anthropocene

"Peace ecology" and "ecological peace policy" are two sides of the same coin. The tasks of an ecological peace policy are to project the probable ecological consequences of the physical effects of global environmental change; to launch early preventive actions to avoid environmental crises, conflicts or wars; and to solve these environmental consequences peacefully without any outbreaks of violence (Brauch 2002). Since the Russian aggression against Ukraine on 24 February 2022, the European security regime has broken down and uncertainty prevails. In this chapter a contemporary diagnosis and a preliminary policy assessment has been offered.

The political attention and the financial resource allocation in the Global North (USA, Canada, EU and NATO, Japan, South Korea, Australia) has shifted from a "climate change priority" to issues of increased military spending in EU and NATO countries, and military assistance for exports of military weapons and ammunition to Ukraine. The peace and security narrative—or its military dimension—has acquired top priority, while the Anthropocene narrative has been downgraded and the urgency of its policy agenda (e.g. of the EU's European Green Deal, and of national climate policies and expenditures) has declined in public opinion polls, despite the call for urgency in the IPPC's Sixth Assessment Report (2022, 2023).

At the same time, extreme weather events, such as tropical storms, floods, heatwaves, droughts, and forest fires have increased. In several reports, MunichRe has documented that previous predictions of the IPCC of an increase in the frequency and intensity of extreme weather events have materialised between the years 1900 and 1950, 1950 and 2000, 2000 and 2020, and to 2023.[64]

[64] See: Em-Dat – The International Disaster Database (Louvain la Neuve: Centre for Research on the Epidemioloy of Disasters); at: https://www.emdat.be/; MunichRe NatCat Service: "Data

In December 2023, 78 years have passed since the first nuclear test in Alamogordo, the first eight decades of the Anthropocene may be studied, and modelling trends for the next eight decades—i.e. until 2100—have been projected and assessed. This date is the horizon of most climate models that were assessed by the IPCC in six assessment reports between 1990 and 2023. The outcome for the planet—and humankind—depends on the political strategies and programmes that are launched now, or later, or not at all.

It is impossible to be sure how countries which have previously only partly implemented or failed to implement their obligations under climate change agreements will behave in the future. However, deductions can be made about societal and political outcomes by using climate models which correlate projections of future population levels with the production and consumption levels associated with different lifestyles, including possible tipping pints and their possible geopolitical outcomes.

An ecological peace policy should aim to provide a preventive strategy so that (1) the projected physical effects of anthropogenic climate change can be contained, resolved, and structurally prevented through multilateral diplomacy and cooperation, and (2) domestic and international violent crises, conflicts, and wars can be avoided (Dunn et al. 2024). Jointly and proactively addressing resource scarcity, resource pollution and stress (of water, soil, and air) will make it possible to enhance the health and security of the affected people.

The Russian war in Europe and the US's interest in remaining the sole hegemon by trying to contain and counter China further complicate future climate trends, especially whether announcements and commitments under the Paris Climate Change Treaty will be implemented by the respective parties.

The second existential threat for humankind in the Anthropocene, due to the impacts of human actions (lifestyle, production, consumption etc.) and interference in the Earth system, cannot be addressed with the military means that are likely to increase further in the years to come. Climate change breakthroughs in mitigation, adaptation and resilience measures require a cooperative framework in the relationship between the US, China and Russia and a resolution of the many ongoing conflicts on the globe. An integration of the results of social science research into the modelling done by natural scientists in the framework of Earth Systems Science is needed. This requires a cooperation between natural and social scientists, a wider knowledge integration, and a gradual transition from multidisciplinary and interdisciplinary towards transformative research (Oswald Spring et al. 2016).

on Disasters since 1980" (Munich: MunichRe); at. https://www.munichre.com/en/solutions/for-industry-clients/natcatservice.html.

2.10 From a Revival of Hobbesian Geopolitics Towards an Ecological Peace Policy for the Anthropocene

International politics and the study of international relations have reached a turning point with a collapse of a cooperative security and peace system in Europe and with a possible transition towards an authoritarian century (Menzel 2023) with a possible return to Cold-War type Hobbesian geopolitical rationales and mind-sets which may also reduce the willingness to move ahead with cooperative policies involving all major GHG emitting countries. The US strategy to persuade and force its partners to reduce their economic and political cooperation with China may result in changes in supply lines.

US domestic politics has been, is and will remain fragile and relatively unpredictable with regards to climate policies and its policies with China and Russia. Hobbesian geopolitical mind-sets could distort the global priority goals to be achieved in the realm of cooperative global climate policy.

In the worst case the prediction of the Stern Report (2006) that the costs for the violent impacts of the physical impacts on humankind may rise for the common generations. This implies to a need of research on planning of the outcome of the present war in Europe including diplomatic initiatives of nations and mediations among the war parties and their supporters.

References

Anders, G., 1956, [7]1987, [4]2018: *Die Antiquiertheit des Menschen.* Vol I: *Über die Seele im Zeitalter der zweiten industriellen Revolution* (München: C. H. Beck).

Anders, G., 1972: *Endzeit und Zeitenende. Gedanken über die atomare Situation* (München: C. H. Beck).

Anders, G., 1980: *Die Antiquiertheit des Menschen.* Vol. II: *Über die Zerstörung des Lebens im Zeitalter der dritten industriellen Revolution* (München: C. H. Beck).

Anders, G., 1982: *Hiroshima ist überall* (München: C. H. Beck).

Antweiler, Christoph, 2022: *Anthropologie im Anthropozän. Theoriebausteine für das 21. Jahrhundert* (Darmstadt: WBG Academic).

Antweiler, C., 2023: "Apocalypse", in: Wallenhorst, N.; Wulf, C. (Eds.), 2023: *Handbook of the Anthropocene - Humans between Heritage and Future* (Cham: Springer Nature): 1473–1476.

Antweiler, C., 2023a: "Society", in: Wallenhorst, N.; Wulf, C. (Eds.), 2023: *Handbook of the Anthropocene - Humans between Heritage and Future* (Cham: Springer Nature): 1305–1308.

Axelrod, R., 1984: *The Evolution of Cooperation* (New York: Basic Books).

Badash, L., 1995: *Scientists and the Development of Nuclear Weapons: From Fission to the Limited Test Ban Treaty 1939–1963* (Amherst, N.Y.: Humanities Press International Inc.).

Benner, S.; Lax, G.; Crutzen, P. J.; Pöschl, U.; Lelieveld, J.; Brauch, H. G. (Eds.), 2021: *Paul J. Crutzen and the Anthropocene: A New Epoch in Earth's History* (Cham: Springer Nature Switzerland).

Bernstein, S., 2023: "Existential security and the governance challenge: Confronting the antinomies of securitisation", in: *Global Policy*: at: https://doi.org/10.1111/1758-5899.13289.

Bohle, H.-G., 2009: "Sustainable Livelihood Security. Evolution and Application", in: Brauch, H. G.; Oswald Spring, U.; Grin, J.; Mesjasz, C.; Krummenacher, H.; Chadha Behera, N.; Chourou,

B.; Kameri-Mbote, P. (Eds.): *Facing Global Environmental Change: Environmental, Human, Energy, Food, Health and Water Security Concepts* (Berlin – Heidelberg - New York: Springer-Verlag).

Booth, K., 1990: "The Concept of Strategic Culture Affirmed", in: Jacobsen, C.G. (Eds.): *Strategic Power: USA/USSR* (London: Palgrave Macmillan); at: https://doi.org/10.1007/978-1-349-205 74-5_8.

Booth, K., 2017: *Theory of World Security* (Cambridge: Cambridge University Press).

Brauch, H. G., 1976: *Struktureller Wandel und Rüstungspolitik der USA (1940–1950): Zur Weltführungsrolle und ihren innenpolitischen Bedingungen* (Heidelberg: Ruprecht-Karl Universität, PhD thesis).

Brauch, H. G., 1977: *Struktureller Wandel und Rüstungspolitik der USA (1940–1950): Zur Weltführungsrolle und ihren innenpolitischen Bedingungen* (Ann Arbor, Michigan: Xerox University Microfilms).

Brauch, H. G., 1996: *Klimapolitik – Naturwissenschaftliche Grundlagen, internationale Regimebildung und Konflikte, ökonomische Analysen sowie nationale Problemerkennung* (Berlin – Heidelberg – New York).

Brauch, H. G., 2000: "Partnership Building Measures to Deal with Long-term Non-military Challenges Affecting North-South Security Relations", in: Brauch, H.G.; Marquina, A.; Biad, A. (Eds.): *Euro-Mediterranean Partnership for the 21st Century* (London: Macmillan – New York: St. Martin's Press, 2000): 281–318.

Brauch, H. G., 2002: "Climate Change, Environmental Stress and Conflict: AFES-PRESS Report for the Federal Ministry for the Environment, Nature Conservation and Nuclear Safety", in: Federal Ministry for the Environment, Nature Conservation and Nuclear Safety (Ed.): *Climate Change and Conflict: Can Climate Change Impacts Increase Conflict Potentials? What is the Relevance of this Issue for the International Process on Climate Change?* (Berlin: Federal Ministry for the Environment, Nature Conservation and Nuclear Safety, 2002): 9–112; at: http://www.afes-press.de/pdf/Brauch_ClimateChange_BMU.pdf.

Brauch, H. G., 2003: "Worldviews and Mindsets: American vs. European Perspectives on Mediterranean (Environmental) Security Policy", in: Brauch, H. G.; Liotta, P. H.; Marquina, A.; Rogers, P.; Selim, M. El-Sayed (Eds.): *Security and Environment in the Mediterranean: Conceptualising Security and Environmental Conflicts* (Berlin – Heidelberg: Springer): 237–266.

Brauch, H. G., 2004: "From a Hobbesian Security to a Grotian Survival Dilemma", 40th Anniversary Conference of IPRA, Peace and Conflict in a Time of Globalisation, Sopron, Hungary, 5–9 July 2004.

Brauch, H. G., 2005: *Threats, Challenges, Vulnerabilities and Risks in Environmental and Human Security*, Source, 1/2005 (Bonn: UNU-EHS).

Brauch, H. G., 2006: "Environmental threats, challenges, vulnerabilities and risks", in: *Encyclopedia of Life Support Systems* (Paris: UNESCO; Oxford: Eolss Publishers, 2006).

Brauch, H.G., 2006: "Potential of solar thermal desalination to defuse water as a conflict issue in the Middle East - Proposal for functional cooperation in the Gulf of Aqaba": in: Linkov, I.; Morel, B. (Eds.): *Role of risk assessment in environmental security and emergency Preparedness in the Mediterranean Region* (Dordrecht: Springer): 25–48.

Brauch, H. G., 2007: "Impact of Global Warming and Non-Conventional Water Resources: Potential of solar thermal desalination to defuse water as a conflict issue - A conceptual contribution to conflict resolution and long-term conflict avoidance", in: Shuval, H.; Dweik, H. (Eds.): *Israel-Palestinian Water Issues – from Conflict to Cooperation* (Berlin – Heidelberg – New York – Hong Kong – London – Milan – Paris – Tokyo: Springer-Verlag): 357–378.

Brauch, H. G., 2008: "Introduction: Globalization and Environmental Challenges: Reconceptualizing Security in the 21st Century", in: Brauch, H. G.; Oswald Spring, Ú.; Mesjasz, C.; Grin, J.; Dunay, P; Behera, N. B.; Chourou, B.; Kameri-Mbote, P.; Liotta, P. H. (Eds.): *Globalization and Environmental Challenges - Reconceptualizing Security in the 21st Century* (Berlin: Springer-Verlag): 27–43.

Brauch, H. G., 2008a: From a Security to a Survival Dilemma", in: Brauch, H.G.; Oswald Spring, U.; Mesjasz, C.; Grin, J.; Dunay, P.; Chadha Behera, N.; Chourou, B.; Kameri-Mbote, P.; Liotta, P. H. (Eds.): *Globalization and Environmental Challenges: Reconceptualizing Security in the 21st Century* (Berlin – Heidelberg – New York: Springer-Verlag): 535–550.
Brauch, H. G., 2008b: "Conceptual Quartet: Security and its Linkages with Peace, Development, and Environment", in: Brauch, H.G.; Oswald Spring, U.; Mesjasz, C.; Grin, J.; Dunay, P.; Chadha Behera, N.; Chourou, B.; Kameri-Mbote, P.; Liotta, P. H. (Eds.): *Globalization and Environmental Challenges: Reconceptualizing Security in the 21st Century* (Berlin – Heidelberg – New York: Springer-Verlag): 65–98.
Brauch, H. G., 2009: "Introduction: Facing Global Environmental Change and Sectorialization of Security", in: Brauch, H.G.; Oswald Spring, U.; Grin, J.; Mesjasz, C. Kameri-Mbote, P.; Chadha Behera, N.; Chourou, B.; Krummenacher, H. (Eds.): *Facing Global Environmental Change: Environmental, Human, Energy, Food, Health and Water Security Concepts* (Berlin – Heidelberg – New York: Springer-Verlag, 2009): 21–42.
Brauch, H. G., 2009a: "Securitizing Global Environmental Change", in: Brauch, H. G.; Oswald Spring, U.; Grin, J.; Mesjasz, C.; Kameri-Mbote, P.; Behera, N. C.; Chourou, B.; Krummenacher, H. (Eds.), 2009: *Facing Global Environmental Change: Environmental, Human, Energy, Food, Health and Water Security Concepts* (Berlin – Heidelberg – New York: Springer-Verlag): 65–102.
Brauch, H. G., 2011: "Political Geoecology for the Anthropocene", 16 March, ISA Convention in Montreal.
Brauch, H. G., 2012: "Climate paradox of the G-8: legal obligations, policy declarations and implementation gap", in: *Rev. Bras. Polít. Int.*, 55 (special edition): 30–52.
Brauch, H. G., 2016: "Sustainable Peace in the Anthropocene: Towards Political Geoecology and Peace Ecology", in: Brauch, H. G.; Oswald Spring, Ú.; Grin, J.; Scheffran, J. (Eds.), 2016: *Handbook on Sustainability Transition and Sustainable Peace* (Cham: Springer International Publishing): 187–236.
Brauch, H. G., 2020: "Peace Ecology in the Anthropocene", in: Oswald Spring, U.; Brauch, H.G. (Eds.): *Decolonizing Conflicts, Security, Peace, Gender, Environment and Development in the Anthropocene* (Cham: Springer): 51–185.
Brauch, H. G., 2021: "The Anthropocene concept in the natural and social sciences, the humanities and law – a bibliometric analysis and a qualitative interpretation (2000–2020)", in: Benner, S.; Lax, G.; Crutzen, P. J.; Pöschl, U.; Lelieveld, J.; Brauch, H. G. (Eds.): *Paul J. Crutzen and the Anthropocene: A New Epoch in Earth's History* (Cham: Springer Nature Switzerland): 289–438.
Brauch, H. G.; Oswald Spring, Ú.; Mesjasz, C.; Grin, J.; Behera, N. B.; Chourou, B.; Kameri-Mbote, P.; Liotta, P. H. (Eds.), 2008: *Globalization and Environmental Challenges: Reconceptualizing Security in the 21st Century* (Berlin – Heidelberg – New York: Springer-Verlag, 2008).
Brauch, H. G.; Oswald Spring, U.; Grin, J.; Mesjasz, C.; Kameri-Mbote, P.; Behera, N. C.; Chourou, B.; Krummenacher, H. (Eds.), 2009: *Facing Global Environmental Change: Environmental, Human, Energy, Food, Health and Water Security Concepts* (Berlin – Heidelberg – New York: Springer-Verlag).
Brauch, H. G.; Dalby, S.; Oswald Spring, Ú., 2011: "Political Geoecology for the Anthropocene", in: Brauch, H. G.; Oswald Spring, Ú.; Mesjasz, C.; Grin, J.; Kameri-Mbote, P.; Chourou, B.; Dunay, P.; Birkmann, J. (Eds.): *Coping with Global Environmental Change, Disasters and Security – Threats, Challenges, Vulnerabilities and Risks* (Berlin – Heidelberg – New York: Springer-Verlag, 2011): 1.453–1.486.
Brauch, H. G.; Oswald Spring, Ú.; Mesjasz, C.; Grin, J.; Kameri-Mbote, P.; Chourou, B.; Dunay, P.; Birkmann, J. (Eds.), 2011: *Coping with Global Environmental Change, Disasters and Security – Threats, Challenges, Vulnerabilities and Risks* (Berlin – Heidelberg – New York: Springer-Verlag, 2011).
Brauch, H. G.; Oswald Spring, Ú.; Grin, J.; Scheffran, J. (Eds.), 2016: *Handbook on Sustainability Transition and Sustainable Peace* (Cham: Springer International Publishing).

Braudel, Fernand, 1966: *La Méditerranée et le Monde Méditerranéen à l'époque de Philippe II* (Paris: Armand Colin).
Braudel, Fernand, 1969: "Histoire et Science Sociales: La Longue Durée", in: *Écrits Sur l'Histoire* (Paris: Flammarion): 41–84.
Braudel, Fernand, 1972: *The Mediterranean and the Mediterranean World in the Age of Philip II*, 2 volumes (New York: Harper & Row).
Breitmeier, H.; Young, O. R.; Zürn, M., 2006: *Analyzing International Environmental Regimes: From Case Study to Database* (Cambridge, MA – London: MIT Press).
Bromberg, G.; Majdalani, N.; Abu Taleb, Y., 2020: A Green Blue Deal for the Middle East (Tel Aviv, Ramallah, Amann: Ecopeace Middle East, December); at: https://ecopeaceme.org/wp-content/uploads/2021/03/A-Green-Blue-Deal-for-the-Middle-East-EcoPeace.pdf.
Brown, L., 1977: "Redefining Security", *WorldWatch Paper 14* (Washington, D.C.: WorldWatch Institute).
Butterfield, H., 1951: *History and Human Relations* (London: Collins).
Buzan, B.; Wæver, O.: de Wilde, J., 1998: *Security: A New Framework for Analysis* (Boulder: Lynne Rienner).
Caldwell, L. K., [3]1996: *International environmental policy: from the twentieth to the twenty-first century* (Durham: Duke University Press).
Cardwell, Curt, 2011: *NSC 68 and the Political Economy of the Early Cold War* (Cambridge: Cambridge University Press).
Carson, R., 1962: *Silent Spring* (Boston: Houghton Mifflin).
Casey, Steven, 2005: "Selling NSC-68: The Truman Administration, Public Opinion, and the Politics of Mobilization, 1950–51", in: *Diplomatic History*, 29, 4: 655–690.
CBD, 1992: "The Convention on Biological Diversity: Text of the Convention" (Montreal: CBD); at: https://www.cbd.int/convention/text/.
Chandler, D.; Müller, F.; Rothe, D., 2021: *International Relations and the Anthropocene. New Agendas, New Approaches* (Cham: Palgrave Macmillan).
Chrowder Chamberlin, T., 1899: "An Attempt to Frame a Working Hypothesis of the Cause of Glacial Periods on an Atmospheric Basis", in: *Journal of Geology*, 7, 8: 751–787. For a review see: "History of climate change science", at: https://en.wikipedia.org/wiki/History_of_climate_change_science.
Churchill, R. P. 1983: "Nuclear Arms as a Philosophical and Moral Issue", in: *The Annals of the American Academy of Political and Social Science*, vol. 469, Nuclear Armament and Disarmament (September): 46–57.
Clark, W. C.; Crutzen, P. J.; Schellnhuber, H. J., 2004, 2005: "Science and Global Sustainability: Toward a New Paradigm", in: Schellnhuber, H. J.; Crutzen, P. J.; Clark, W. C.; Claussen, M.; Held, H. (Eds.): *Earth System Analysis for Sustainability* (Cambridge, MA; London: MIT Press): 1–28.
Clark, W. C. (Ed.) [Steffen, W.; Rockström, J.; Richardson, K.; Lenton, T. M.; Folke, C.; Liverman, D.; Summerhayes, C. P.; Barnosky A. D.; Cornell, S. E.; Crucifix, M.; Donges, J. F.; Fetzer, C., I.; Lade, S. J.; Scheffer, M.; Winkelmann, R.; Schellnhuber, H. J.], 2018: "Trajectories of the Earth System in the Anthropocene", in: *Proceedings of the National Academy of Sciences*, USA, 115, 33: 8252–8259.
Clausewitz, C. v., 1832, 1984 [Howard, M.; Paret, P. (Eds.)]. *On War* [Vom Krieg] (New Jersey: Princeton University Press).
Conca, K., 1994: "In the Name of Sustainability: Peace Studies and Environmental Discourse", in: Käkonen, J. (Ed.): *Green Security or Militarized Environment* (Dartmouth: Aldershot): 7–24.
Conca, K., 2002: "The Case for Environmental Peacemaking", in: Conca, K.; Dabelko, G. (Eds.): *Environmental Peacemaking* (Baltimore: Johns Hopkins University Press): 1–22.
Conca, K.; Carius, A.; Dabelko, G. D., 2005: "Building Peace through Environmental Cooperation", in: The Worldwatch Institute (Ed.): *State of the World 2005: Redefining Global Security* (New York – London: W.W. Norton): 144–157.

Corry, O. 2017: "The International Politics of Geoengineering: The Feasibility of Plan B for Tackling Climate Change", in: *Security Dialogue*, 48, 4; 297–315.
Crutzen, P. J., 2002: "Geology of mankind – The Anthropocene", in: *Nature*, 415: 23.
Crutzen, P. J.; Birks, J.W., 1982: "The Atmosphere After a Nuclear War: Twilight at Noon", in: *Ambio*, 11 (2/3): 114–125.
Crutzen, P. J.; Birks, J.W., 1983: "The Atmosphere after a Nuclear War: Twilight at Noon", in: Peterson, J. (Ed.): *The Aftermath: The Human and Ecological Consequences of Nuclear War* (New York: Pantheon): 73–96.
Crutzen, P.J.; Brauch, H. G. (Eds.), 2016: *Paul J. Crutzen: A Pioneer on Atmospheric Chemistry and Climate Change in the Anthropocene* (Cham – Heidelberg – New York – Dordrecht – London: Springer International Publishing).
Crutzen, P. J.; Stoermer, E. F., 2000: "The 'Anthropocene.'", in: *Global Change Newsletter*, 1: 17–18; at: http://tinyurl.com/crtzen-igbp.
Czempiel, E.-O., 1966: *Das amerikanische Sicherheitssystem 1945–1949: Studie zur Außenpolitik der bürgerlichen Gesellschaft* (Berlin: de Gruyter).
Dalby, S., 1998: "Ecological Metaphors of Security: World Politics in the Biosphere", in: *Alternatives*, 23, 3: 291–319.
Dalby, S., 2000: "Geopolitics and Ecology: Rethinking the Contexts of Environmental Security", in: Lowi, M. R.; Shaw, B. R. (Eds.): *Environment and Security. Discourses and Practices* (Basingstoke – London: Macmillan; New York: St. Martin's Press): 84–100.
Dalby, S., 2002: "Environmental Change and Human Security", in: *Canadian Journal of Policy Research*, 3, 2: 71–79.
Dalby, S., 2002a: "Security and Ecology in the Age of Globalization", in: Woodrow Wilson International Center for Scholars (Ed.): *Environmental Change & Security Project Report*, Issue No. 8 (Summer): 95–108.
Dalby, S., 2002b: *Environmental Security* (Minneapolis: University of Minnesota Press).
Dalby, S., 2008: "Security and Environment Linkages Revisited", in: Brauch, H. G.; Oswald Spring, Ú.; Mesjasz, C.; Grin, J.; Dunay, P.; Behera Chadha, N.; Chourou, B.; Kameri-Mbote, P.; Liotta, P. H. (Eds.): *Globalization and Environmental Challenges: Reconceptualizing Security in the 21st Century* (Berlin – Heidelberg: Springer-Verlag): 165–172.
Dalby, S., 2009: *Security and Environmental Change* (Cambridge: Polity).
Dalby, S., 2013: "Peace in the Anthropocene" in: *Peace Review*, 25, 4 (October).
Dalby, S., 2014: "Environmental Geopolitics in the Twenty-first Century", in: *Alternatives: Global, Local, Political*, 39, 1 (February): 3–16.
Dalby, S., 2020: *Anthropocene Geopolitics: Globalization, Security, Sustainability* (Ottawa: University of Ottawa Press).
Dalby, S., 2022: *Rethinking Environmental Security* (Cheltenham: Edward Elgar).
Dalby, S., 2023: *Pyromania: Fire and Geopolitics in a Climate-Disrupted World* (Agenda, November).
Dalby, S.; Brauch, H. G.; Oswald Spring, Ú., 2009: "Environmental Security Concepts Revisited During the First Three Phases (1983–2006)", in: Brauch, H. G.; Oswald Spring, Ú.; Grin, J.; Mesjasz, C.; Kameri-Mbote, P.; Behera, N. C.; Chourou, B.; Krummenacher, H. (Eds.), 2009: *Facing Global Environmental Change: Environmental, Human, Energy, Food, Health and Water Security Concepts* (Berlin – Heidelberg – New York: Springer-Verlag): 781–790.
Décultot, E.; Fulda, D. (Eds.), 2016: *Die Vielfalt der Sattelzeit. Strukturen und Tendenzen des Historischen Erzählens um 1800 im Deutsch-Französischen Vergleich Veranstaltung 2013 Halle (Saale)* (Berlin-Boston: Walter de Gruyter).
Dixson-Declève, S.; Gaffney, O.; Ghosh, J.; Randers, J.; Rockström, J.; Stoknes, P.E., 2022: *Earth for All – A Survival Guide for Humanity – A Report to the Club of Rome* (Gabriola Island, BC, Canada: New Society Publishers): 99–111; at: https://doi.org/10.1007/978-3-540-75977-5_4.
Dunn, D. J.; Brauch, H. G.; Burton, P. (Eds.), 2024: *John W. Burton: A Pioneer in Conflict Analysis and Resolution – Key Contributions on International Relations, Peace Theory, World Society, and Human Needs* (Cham: Springer Nature Switzerland).

Ecopeace Middle East, 2021: *Water energy nexus*, at: https://ecopeaceme.org/water-energy-nexus/.
EUR-Lex, 2015: *The Paris Agreement* [on Climate Change]; at: https://eur-lex.europa.eu/legal-content/EN/TXT/?uri=CELEX:22016A1019(01).
Gaddis, J. L., 1980: "NSC 68 and the Problem of Ends and Means", in: *International Security*, 4, 4 (Spring 1980): 164–170.
Geulen, C., 2010: "Plädoyer für eine Geschichte der Grundbegriffe des 20. Jahrhunderts", in: *Zeithistorische Forschungen/Studies in Contemporary History*, 7 (Göttingen: Vandenhoeck & Ruprecht): 79–97.
Grotius, H., 1622: *De jure belli ac pacis libri tres* [On the Law of War and Peace] (Paris: 1625; 2nd ed. Amsterdam 1631).
Grotius, H., 2012: *Hugo Grotius: On the Law of War and Peace*. Student edn. Neff, S.C. (Ed.), 2012 (Cambridge: Cambridge University Press, 2012).
Guerrier, Steven Warren, 1988: "NSC-68 and the Truman Rearmament, 1950–1953" (PhD dissertation, U. of Michigan).
Häckel, E., 1866: *Generelle Morphologie der Organismen*, 2 vol., vol. 1: *Allgemeine Anatomie der Organismen*; vol. 2: *Allgemeine Entwicklungsgeschichte der Organismen* (Georg Reimer).
Hahn, O.; Straßmann, F., 1939: "Nachweis der Entstehung aktiver Bariumisotope aus Uran und Thorium durch Neutronenbestrahlung; Nachweis weiterer aktiver Bruchstücke bei der Uranspaltung", in: *Naturwissenschaft*, 27: 89–95 (6 January).
Hakala, R.; Kononenko, V., 2022: "Re-Securitizing Climate from 'Climate Security' to 'Ecology of Peace'", in: *FIIA Briefing Paper 348* (Helsinki: Finnish Institute for International Affairs, September).
Hammond, P. Y., 1962: "NSC-68: Prologue to Rearmament", in: Schilling, Warner R.; Hammond, P.Y.; Snyder, G. H. (Eds.): *Strategy, Politics, and Defense Budgets* (New York: Columbia University Press): 267–378.
Hardt, J. N., 2021: "Encounters between Security and Earth System Sciences: Planetary Boundaries end Hothouse Earth", in: Chandler, D.; Müller, F.; Rothe, D., 2021: *International Relations and the Anthropocene. New Agendas, New Approaches* (Cham: Palgrave Macmillan): 39–57.
Hardt, J.N.; Harrington, C.; von Lucke, F.; Estève, A.; Simpson, N.P., 2023: *Climate Security in the Anthropocene - Exploring the Approaches of United Nations Security Council Member-States* (Cham: Springer Nature Switzerland).
Herz, J., 1951: *Political Realism and Political Idealism: A Study in Theories and Realities* (Chicago: University of Chicago Press, 1951).
Herz, J., 1961: *International Politics in the Atomic Age* (New York: Columbia University Press, 1961); 241.
Hillmann, K.-H., 1998: *Überlebensgesellschaft. Von der Endzeitgefahr zur Zukunftssicherung* (Würzburg: Carolus).
Hogan, M. J., 1998: *A Cross of Iron: Harry S. Truman and the Origins of the National Security State, 1945–1954* (Cambridge: Cambridge University Press): 12.
Huggett. R. J., 1995: *Geoecology* (London: Routledge, Taylor & Francis).
Ide, T., 2023: *Catastrophes, Confrontations, and Constraints: How Disasters Shape the Dynamics of Armed Conflict* (Cambridge, Mass.: MIT Press).
Ide, T.; Bruch, C.; Carius, A. et al., 2021: "The past and future(s) of environmental peacebuilding", in: *International Affairs*, 97, 1: 1–16.
IPCC, 1990: *Climate Change: The IPCC Impacts Assessment* (Geneva: WMO; UNEP; IPCC).
IPCC, 1996: *Climate Change 1995: The Science of Climate Change: Contributions of Working Group I to the Second Assessment Report of the Intergovernmental Panel on Climate Change* (Cambridge: Cambridge University Press).
IPCC, 2001: *Climate Change 2001: The Scientific Basis* (Cambridge – New York: Cambridge University Press).
IPCC, 2007: *Climate Change 2007: The Physical Science Basis: Working Group I Contribution to the Fourth Assessment Report of the IPCC* (Cambridge – New York: Cambridge University Press).

IPCC, 2011: *Special Report: Renewable Energy Sources and Climate Change Mitigation* (Cambridge: Cambridge University Press, April).
IPCC, 2012: *Managing the Risks of Extreme Events and Disasters to Advance Climate Change Adaptation* (Cambridge: Cambridge University Press, April).
IPCC, 2013: *Climate Change 2013: The Physical Science Basis: Working Group I Contribution to the Fifth Assessment Report of the Intergovernmental Panel on Climate Change* (Cambridge – New York: Cambridge University Press); at: https://www.ipcc.ch/report/ar5/wg1/.
IPCC, 2014: *Climate Change 2014: Synthesis Report: Summary for Policymakers* (Geneva: IPCC).
IPCC, 2014: IPCC, *Climate change 2014: Impacts, adaptation, and vulnerability* (Geneva: IPCC).
IPCC, 2018: *Global Warming of 1.5°C* (Cambridge: Cambridge University Press, October).
IPCC, 2019: *Climate Change and Land* (Cambridge: Cambridge University Press, August).
IPCC, 2021: *Climate Change 2021: The Physical Science Basis. Contribution of Working Group I to the Sixth Assessment Report of the Intergovernmental Panel on Climate Change* (Cambridge: Cambridge University Press).
IPCC, 2022: *Climate Change 2022: Impacts, Adaptation and Vulnerability*. 6th Assessment Report, WG II (Geneva: Intergovernmental Panel on Climate Change).
IPCC, 2022a: *Summary for Policymakers Headline Statements*. 6th Assessment Report, WG II (Geneva: Intergovernmental Panel on Climate Change).
IPCC, 2023: *AR6 Synthesis Report: Climate Change 2023* (Geneva: IPCC, 20 March).
IPCC, 2023: *Climate Change 2023. AR6 Synthesis Report* (Geneva: Intergovernmental Panel on Climate Change).
IPCC (Intergovernmental Panel on Climate Change), 2018: *Global Warming of 1.5°C* (Geneva: World Meteorological Organization).
IPCC (Intergovernmental Panel on Climate Change), 2022: *Climate Change 2022: Mitigation of Climate Change*. Working Group III Report (Geneva, 2022, April); at: https://www.ipcc.ch/report/ar6/wg3/.
Jervis, R., 1976: *Perception and Misperception in International Politics* (Princeton, NJ: Princeton University Press).
Jervis, R., 1978: "Cooperation under the Security Dilemma," in: *World Politics*, 30, 2 (January): 167–214.
Jervis, R., 1982: "Security Regimes", in: *International Organization*, 36, 2 (Spring): 36.
Jervis, R., 1999: "Realism, Neoliberalism, and Cooperation: Understanding the Debate", in: *International Security*, 24: 49.
Jungk, R., 1956: *Heller als tausend Sonnen - Das Schicksal der Atomforscher* (Reinbek: Rowohlt).
Jungk, R., 1958: *Brighter than a Thousand Suns: A Personal History of the Atomic Scientists* (New York: Harcourt Brace).
Kant, I., 1795: *Zum ewigen Frieden. Ein philosophischer Entwurf* (Königsberg: Friedrich Nicolovius).
Kipping, M., 2022: "Klimaschutz als Kollateralschaden?", in: *Internationale Politik*, 77, 4 (July/August): 90–94.
Kords, U., 1996: "Tätigkeit und Handlungsempfehlungen der beiden Klima-Enquête-Kommissionen des Deutschen Bundestages (1987–1994)", in: Brauch, H. G. (Ed.): *Klimapolitik – Naturwissenschaftliche Grundlagen, internationale Regimebildung und Konflikte, ökonomische Analysen sowie nationale Problemerkennung* (Berlin – Heidelberg – New York): 203–214.
Koselleck, R. (Ed.): 1979: *Historische Semantik und Begriffsgeschichte* (Stuttgart: Klett-Cotta).
Koselleck, R., 1998: "Social History and Begriffsgeschichte," in: Hampsher-Monk, I.; Tilmans, K.; Vree, F. van (Eds.): *History of Concepts: Comparative Perspectives* (Amsterdam: Amsterdam University Press, 1998): 23–35.
Koselleck, R., 2006: *Begriffsgeschichten: Studien zur Semantik und Pragmatik der politischen und sozialen Sprache (Frankfurt: Suhrkamp).*
Kuhn, T., 1962, 1965: *The Structure of Scientific Revolutions* (Chicago: Chicago University Press).

Lanchberry, J.; Victor, D., 1995: "The role of Science in the global Climate Negotiations", in: Bergesen, H. O.; Parman, G. (Eds.): *Green Globe Yearbook of International cooperation on Environment and Development 1995* (Oxford: Oxford University Press).

Letter by Albert Einstein to President Franklin Delano Roosevelt, 2 August 1939; at: https://en.wikipedia.org/wiki/Einstein%E2%80%93Szil%C3%A1rd_letter.

Luce, H.R., 1941: "The American Century", in: *Life* (17 February).

Luce, H.T. 1999: "The American Century", in: *Diplomatic History*, 23, 2 (Spring): 159–171.

Machiavelli, N., 1532: *Il Principe*.

Mathews, J. T., 1989: "Redefining Security", in: *Foreign Affairs*, 68, 2 (Spring): 162–177.

May, E. R. (Ed.), 1950, 1993: *American Cold War Strategy: Interpreting NSC 68* (Palgrave Macmillan, 15 March).

McNeill, J. R., Engelke, P., 2014, 2016: *The Great Acceleration: An Environmental History of the Anthropocene since 1945*. Cambridge MA: Harvard University Press.

Megginson: 'Lessons from Europe for American Business', in: *Southwestern Social Science Quarterly*, 1963, 44, 1: 3–13: 4.

Meitner, L.; Fritsch, O., 1939: "Disintegration of uranium by neutrons: a new type of nuclear reaction", *Nature*, 143, 3615 (11 February).

Menzel, U., 2023: *Wendepunkte: Am Übergang zum autoritären Jahrhundert* (Berlin: Suhrkamp).

Mische, P., 1990: "Ecological Security and the Need to Reconceptualize Sovereignty", in: *Alternatives*, 14, 4: 389–427.

Møller, B., 2003: "National, Societal and Human Security: Discussion – Case Study of the Israeli-Palestinian Conflict", in: Brauch, H. G.; Liotta, P. H.; Marquina, A.; Rogers, P.; Selim, M. E.-S. (Eds.): *Security and Environment in the Mediterranean: Conceptualising Security and Environmental Conflicts* (Berlin – Heidelberg: Springer): 277–288.

Mölling, C., 2022: "Keine Gewalt ist keine Lösung – Frieden nur zu wollen, vermeidet keinen Krieg. Deutschlands Friedenspolitik muss sich neu aufstellen", in: *Internationale Politik*, 77, 4 (July/August): 71–76.

Morgenthau, H. 1948: *Politics Among Nations: The Struggle for Power and Peace* (New York: A.A. Knopf).

Myers, N., 1986: "The Environmental Dimension to Security Issues", in: *The Environmentalist*, 6: 251–257.

Nichols, N., 1972: "Climate: Sawyer predicted rate of warming in 1972", in: *Nature*, 448, 7157 (30 August): 992.

Nitze, P. H., 1980: "The Development of NSC 68", in: *International Security*, 4, 4 (Spring): 170–176.

Nitze, P. H., 1989: *From Hiroshima to Glasnost: At the Center of Decision* (New York: Grove Weidenfeld).

Nolte, N., 2010: "Vom Fortschreiben und Umschreiben der Begriffe. Kommentar zu Christian Geulen", in: *Zeithistorische Forschungen/Studies in Contemporary History*, Online-Ausgabe, 7, 1: 98–103; at: https://zeithistorische-forschungen.de/1-2010/4616 and https://doi.org/10.14765/zzf.dok-1789 (24 September 2023).

Notter, H., 1949: *Postwar Foreign Policy Preparation, 1939–1945* (Washington, D.C.: U.S. Government Printing Office).

Oberthür, S., 1993: *Politik im Treibhaus – Die Entstehung des internationalen Klimaschutz-regimes* (Berlin: edition sigma): 25.

Oswald Spring, Ú., 1991: *Estrategias de supervivencia en la Ciudad de México* (Cuernavaca, CRIM-UNAM).

Oswald Spring, U.; Brauch, H. G. (Eds.), 2021: *Decolonizing Conflicts, Security, Peace, Gender, Environment and Development in the Anthropocene* (Cham: Springer).

Oswald Spring, U.; Brauch, H. G.; Scheffran, J. 2016: "Sustainability Transition with Sustainable Peace", in: Brauch, H. G.; Oswald Spring, Ú.; Grin, J.; Scheffran, J. (Eds.), 2016: *Handbook on Sustainability Transition and Sustainable Peace* (Cham: Springer International Publishing): 887–928.

Pelluchon, C., 2021: *Das Zeitalter des Lebendigen – Eine neue Philosophie der Aufklärung* (Darmstadt: wbg Academic).
PNAS, Steffen, W.; Rockström, J.; Richardson, K.; Lenton, T. M.; Folke, C.; Liverman, D.; Summerhayes, C. P.; Barnosky A. D.; Cornell, S. E.; Crucifix, M.; Donges, J. F.; Fetzer, C., I.; Lade, S. J.; Scheffer, M.; Winkelmann, R.; Schellnhuber, H. J. [ed. by Clark, W. C.], 2018: "Trajectories of the Earth System in the Anthropocene", in: *Proceedings of the National Academy of Sciences*, USA, 115, 33: 8252–8259.
Porter, G.; Brown, J. W., 1991: *Global Environmental Politics* (Boulder: Westview Press).
PSI, 2016: *Planetary Security: Peace and Cooperation in Times of Climate Change and Global Environmental Challenges - Conference Report* (The Hague: Netherlands Ministry of Foreign Affairs, 2–3 November, Peace Palace); at: https://climateand-security.files.wordpress.com/2012/04/planetary-security-2015.pdf.
Rademaker, M.; Jans, K.; Rõõs, H.; Borum, A.; van Schaik, L.; Slingerland, S.; Frattina della Frattina, C., 2016: *The Economics of Planetary Security: Climate Change as an Economic Conflict Factor* (The Hague: The Hague Centre for Strategic Studies, 13 December).
Raworth, K., 2017: *Doughnut Economics: Seven Ways to Think Like a 21st Century Economist* (London: Random House).
Renn, J., 2020: *Evolution of Knowledge: Rethinking Science for the Anthropocene* (Princeton - Oxford: Princeton University Press).
Renn, J., 2022: "From the History of Science to Geoanthropology", in: *Isis*, 113, 2 (June): 377–385.
Rosenberg, D. Alan, 1983: "The Origins of Overkill. Nuclear Weapons and American Strategy, 1945–1960", in: *International Security*, 7, 4 (Spring): 3–72.
Russell, R. B., 1958: *United Nations: A History of the United Nations Charter* (Washington D.C., The Brookings Institution).
Sawyer, J., 1972: "Man-made Carbon Dioxide and the 'Greenhouse' Effect", in: *Nature*, 239, 5366(1 September): 23–26.
Schellnhuber, H. J., 1999: "'Earth System' Analysis and the Second Copernican Revolution", in: *Nature*: 402: C19–C23 (2 December); https://doi.org/10.1038/35011515.
Sears, N. A. (2020): "Existential security: towards a security framework for the survival of humanity", in: *Global Policy*, 11, 2: 255–266; at: https://doi.org/10.1111/1758-5899.12800.
Sears, N. A., 2022: *Great power rivalry and macrosecuritization failure* (Toronto, ON: University of Toronto); at: https://www.researchgate.net/publication/366020757_and_Sears_Nathan_2022_Great_Power_Rivalry_and_Macrosecuritization_Failure-_PhD_Thesispdf.
Sellers, W. D., 1969: "A Global Climate Mod4el Based on the Energy Balance of the earth-Atmospheric System": in: *Journal of Applied Meteorology*, 8, 3: 392–400.
Shahbaz, M.; Khan, S.; Ali, A.; Mita Bhattacharya. M., 2015: *The Impact of Globalization on CO_2 Emissions in China* (Munich: Munich Personal RePEc Archive); at: https://mpra.ub.uni-muenchen.de/64450/1/MPRA_paper_64450.pdf.
Simangan, D., 2020: "Where is the Anthropocene? IR in a new geological epoch", in: *International Affairs*, 96, 1: 211–224.
Simma, B. (Ed.), 1994: *Charter of the United Nations: A Commentary* (Oxford – New York: Oxford University Press).
Steffen, W.; Crutzen, P.J.; McNeill, J.R., 2007: "The Anthropocene: Are Humans Now Overwhelming the Great Forces of Nature?", in: *Ambio*, 36: 614–621.
Steffen, W.; Grinevald, J.; Crutzen, P.; McNeill, J., 2011: "The Anthropocene: Conceptual and historical perspectives", in: *Philosophical Transactions of the Royal Society A: Mathematical, Physical and Engineering Sciences*, 369, 1938; 842–867.
Steffen, W.; Rockström, J.; Richardson, K.; Lenton, T. M.; Folke, C.; Liverman, D.; Summerhayes, C. P.; Barnosky A. D.; Cornell, S. E.; Crucifix, M.; Donges, J. F.; Fetzer, C., I.; Lade, S. J.; Scheffer, M.; Winkelmann, R.; Schellnhuber, H. J. [ed. by Clark, W. C.], 2018: "Trajectories of the Earth system in the Anthropocene", in: *Proceedings of the National Academy of Sciences*, USA, 115, 33: 8252–8259; at: https://doi.org/10.1073/pnas.1810141115.
Stern, N., 2006, 2007: *The Economics of Climate Change* (Cambridge: Cambridge University Press).

Stevenson, H., 2017: *Global Environmental Politics Problems, Policy and Practice* (Cambridge: Cambridge University Press).
Tang, S., 2009: "Security Dilemma: A conceptual Analysis", in: *Security Studies*, 18: 587–623.
The Committee for the Compilation of Materials and Damage Caused by the Atomic Bomb in Hiroshima and Nagasaki, 1981: *Hiroshima and Nagasaki: The Physical, Medical, and Social Effects of the Atomic Bombings* (New York: Basic Books).
Tickell, C., 2003: "Risks of Conflict: Population and Resource Pressures", in: Brauch, H. G.; Liotta, P. H.; Marquina, A.; Rogers, P.; Selim, M. El-S. (Eds.): *Security and Environment in the Mediterranean: Conceptualising Security and Environmental Conflicts* (Berlin – Heidelberg: Springer): 13–18.
Troll, C., 1968: "Landschaftsökologie", in: Tuxen, R. (Ed.): *Pflanzensoziologie und Landschaftsökologie* (Essen: IVF).
Ullman, R. H., 1983. "Redefining Security", in: *International Security*, 8, 1 (Summer 1983): 129–153.
UNCCD, 1994: *United Nations Convention to Combat Desertification* (New York: UN).
UNDP (United Nations Development Programme), 1994: *Human Development Report 1994* (New York: Oxford University Press).
UNFCCC, 1992: *UN General Assembly, United Nations Framework Convention on Climate Change: resolution/adopted by the General Assembly*, 20 January 1994, A/RES/48/189; at: https://www.refworld.org/docid/3b00f2770.html.
UNFCCC, 1997: "Kyoto Protocol to the United Nations Framework Convention on Climate Change" (Bonn: UNFCCC); at: https://unfccc.int/resource/docs/convkp/kpeng.pdf.
US EPA, 1962: "Health implications of fallout from nuclear weapons testing through 1961", Report of the Federal Radiation Council (May 1962); at: https://www.epa.gov/sites/default/files/2015-05/documents/frc_rpt3.pdf.
US EPA: "Radioactive Fallout From Nuclear Weapons Testing", at: https://www.epa.gov/radtown/radioactive-fallout-nuclear-weapons-testing.
Vennemo, H.; Aunan, K.; He, J.; Hu, T.; Li, S., Rypdal, K., 2005: *Environmental implications of China's WTO accession*, Working Paper 2005-003, Abstract (Oslo: ECON Analysis); at: https://inis.iaea.org/collection/NCLCollectionStore/_Public/38/013/38013647.pdf.
Volger, H., 2002: *A Concise Encyclopedia of the United Nations* (The Hague London – New York: Kluwer Law International).
Von Weizsäcker, E.U., [4]1994: *Erdpolitik. Ökologische Realpolitik an der Schwelle zum Jahrhundert der Umwelt* (Darmstadt: Wissenschaftliche Buchgesellschaft).
Wæver, O., 1995: "Securitization and Desecuritization", in: Lipschutz, R. G. (Ed.) *On Security* (New York: Columbia University Press.).
Wæver, O., 2008: "Peace and Security: Two Evolving Concepts and Their Changing Relationship", in: Brauch, H. G.; Oswald Spring, Ú.; Mesjasz, C.; Grin, J.; Dunay, P.; Behera, N. C.; Chourou, B.; Kameri-Mbote, P.; Liotta, P.H. (Eds.): *Globalization and Environmental Challenges: Reconceptualizing Security in the 21st Century* (Berlin – Heidelberg: Springer-Verlag).
Wæver, O., 2023: Public lecture: climate change as a security issue. Helsinki Collegium for Advanced Studies. May 12; at: https://www.youtube.com/watch?v=QE52mGzHsU8 (28 September 2023).
WBGU, 2007: *Climate Change as a Security Risk* (London: Earthscan).
Weizsäcker, E. U. von, 1993: *Erdpolitik* (Darmstadt: WBG).
Weizsäcker, E. U. von, 2019: "Eine neue Aufklärung für eine volle Welt. Klimaschutz, Politik und Technologie im Erdzeitalter des Anthropozäns", Hannover, Leibniz Campus Lecture, 21 April; at: https://www.uni-hannover.de/de/universitaet/aktuelles/veranstaltungen/leibniz-campus-lecture/weizsaecker.
Weizsäcker, E. U. v.; Wijkman, A., 2017: *Wir sind dran. Club of Rome: Der große Bericht: Was wir ändern müssen, wenn wir bleiben wollen. Eine neue Aufklärung für eine volle Welt* (Gütersloh: Gütersloher Verlagshaus).

Weizsäcker, E. U.; Lovins, A. B.; Lovins, L. H., 1995: *Faktor 4. Doppelter Wohlstand – Halbierter Naturverbrauch. Der neue Bericht an den Club of Rome* (München: Droemer Knaur): 249.
Wells, Jr., Samuel F., 1979: "Sounding the Tocsin: NSC 68 and the Soviet Threat", in: *International Security*, 4, 2 (Autumn): 116–158.
Westing, A. H., 1986: "An Expanded Concept of International Security", in Westing, A. H. (Ed.): *Global Resources and International Conflict* (Oxford: Oxford University Press).
White House, 1965: *Restoring the Quality of our Environment* (Washington, D.C.: White House).
Wittner, L. S., 1993: *One World of None: A History of the World Nuclear Disarmament Movement Through 1953* (Stanford: Stanford University Press).
Wittner, L. S., 1997: *Resisting the Bomb: A History of the World Nuclear Disarmament Movement, 1954–1970* (Stanford: Stanford University Press).
Wittner, L. S., 2003: *Toward Nuclear Abolition: A History of the World Nuclear Disarmament Movement, 1971–Present* (Stanford: Stanford University Press).
Wittner, L. S., 2009: *Confronting the Bomb: A Short History of the World Nuclear Disarmament Movement* (Stanford: Stanford University Press).
Young, O. R., 2011: "Effectiveness of international environmental regimes: Existing knowledge, cutting-edge themes, and research strategies", in: *PNAS*, 108, 50 (5 December): 19853–19860.
Young, K., 2013: "Revisiting NSC 68", in: *Journal of Cold War Studies*, 15, 1.
Zheng, J.; Mi, Z.; Coffman, D'M.; Shan, Y.; Guan, D.; Wang, S., 2019: "The Slowdown in China's Carbon Emissions Growth in the New Phase of Economic Development", in: *One Earth*: 240–253; at: https://www.cell.com/one-earth/pdf/S2590-3322(19)30088-0.pdf.

Part II
Politics, Policy and Polity in the Anthropocene

Chapter 3
The Anthropocene and Complexity: A Survey of Ideas

Czeslaw Mesjasz

Abstract Complexity-related ideas were directly referred to in the discussion on the Anthropocene a few years after its origins. Later, after 2010, the number of publications on the Anthropocene and complexity rapidly increased. They cover a great many ideas, beginning from mathematical models and simulation models through various applications of complexity-related analogies and metaphors to qualitative concepts reflecting the complexity of interacting natural and human systems. The aim of this chapter is to identify, systematise and evaluate preliminarily the applications of broadly defined complexity-related ideas in the development of the concept of the Anthropocene and in the Anthropocene-related areas of environmental studies. The chapter can be treated as a preliminary guide enumerating selected issues, providing directions for more profound studies of the links between complexity-related research and the Anthropocene.

Keywords Autopoiesis · Anthropocene · Chthulucene · Complexity · Cybernetics · Complex Adaptive Systems · Earth System · Earth System Science · Earth System Analysis · Gaia · Global Human System · Mathematical modelling · Metaphors · Systems thinking · Viable System Model

Czeslaw Mesjasz, Assoc. Prof. Dr. Habil., Cracow Economic University (ret.), Email: mesjaszc@uek.krakow.pl.

C. Mesjasz (✉)
Cracow Economic University, Kraków, Poland
e-mail: mesjaszc@uek.krakow.pl

H. G. Brauch (ed.), *Towards Rethinking Politics, Policy and Polity in the Anthropocene*, The Anthropocene: Politik—Economics—Society—Science 35,
https://doi.org/10.1007/978-3-031-71807-6_3

3.1 Introduction

3.1.1 Background

The current state of knowledge about the Anthropocene, the concept proposed in 2000 by Nobel Prize Laureate in Chemistry, Paul Crutzen and Eugene Stoermer (Crutzen/Stoermer 2000; Crutzen 2002), shows that the vast and growing number of studies of this topic embodies a broad collection of ideas deriving from almost all domains of science, from astrophysics, geology, biology, atmospheric science, hydrology, and hydrobiology, through broadly defined social studies to anthropology, cultural studies, psychology, and philosophy. The Anthropocene has also become an important theme in arts, law, and theology.

Seen from a cognitive point of view, it is natural that 'system' and 'complex system', with various interpretations, have become an indispensable part of the language of the studies of Earth, its physical and biological components, its global society, and its astronomical environment. From the systems studies perspective, the main question is how the practitioners and theoreticians understand the utterances: system, complexity and other akin terms.

A significant new phenomenon typical in the modern social sciences can be observed in studying the relations between the Anthropocene and complexity-related ideas. The notions – the Anthropocene and complexity – have a broad scope of meaning. Additionally, while the Anthropocene can be defined in a more precise way, complex, complexity and chaos are typical polysemous terms with multiple interpretations (Vicente/Falkum 2017: 1–3). In consequence, the studies of relations between the Anthropocene and complexity and complexity-related terms are becoming ever more intricate.[1]

[1] Author's remarks concerning the dispute among geologists on the recognition of the Anthropocene as a geological epoch in Earth history. The discussion in March 2024 concerning the rejection of the Anthropocene as a geological epoch (see the discussion in the introduction) impacts on the applications of the ideas taken from complexity studies on the theory and policy of the Anthropocene. These impacts will be assessed from two perspectives: validity of complexity-based mathematical models, analogies, and metaphors. An official acceptance of the Anthropocene as a geological epoch would have enhanced the epistemological and methodological impact of complexity-based mathematical modelling in environmental studies. Some models of complex systems could have been strengthened with conceptual and methodological ideas taken from geology. However, even without this approval, complexity-related mathematical models of environmental phenomena are one of the fundamental arguments in showing the scale of humanity-induced changes on ecosystems at all scales. As is emphasised in this chapter, the complexity-related analogies and metaphors are also a significant epistemological contribution to a better understanding of the Anthropocene in the initial meaning proposed by Crutzen and Stoermer and in the later discussions. Similarly, as in the case of mathematical modelling, official geological acceptance of the Anthropocene could have enhanced some complexity-related analogies and metaphors. However, the core complex ontological and epistemological problems of the Anthropocene remain unchanged: the need for improved identification of interactions between natural and human systems and assessment of the possibility of reducing and reversing the damage caused by human activities on ecosystems at all scales, including the global one. Consequently, no matter how the Anthropocene is treated in

3.1.2 Problem

The Anthropocene, defined as the geological epoch in which humans and societies have become a global geophysical force, was developed without any direct impact of complexity-related ideas already known at that time. These ideas began to be applied in the studies of the Anthropocene with a relatively short delay, e.g. by Ehlers and Krafft (2006). Later, the number of such publications rapidly increased. They cover a large number of ideas, beginning from mathematical models and simulation model sources through various applications of complexity-related analogies and metaphors, and ending with qualitative concepts reflecting the complexity of biological and human systems – *autopoiesis* and *sympoiesis* (Haraway 2015, 2016, 2016a). Due to the number and variety of interpretations of complexity and the breadth of the meaning of the Anthropocene, the applications of complexity-related ideas in the origins and development of the Anthropocene have not yet been systematised and assessed.

The appraisal of the applications of broadly defined complexity-related ideas in the discussions concerning the Anthropocene has an important normative aspect, necessary in environmental policy. One of the arguments supporting the validity of applications of complexity-related ideas in environmental studies is the epistemological and methodological quality of those ideas. The theory and empirical research of environmental studies are to a large extent built on complexity-related ideas. In the majority of the environmental research, the complexity-related concepts are used in the correct way. However, similarly to in other areas of social studies, in some instances they are abused, misused, and simplified. The reason for such a situation is because such terms as, for example, complexity, chaos, the edge of chaos, etc. have a strong emotional appeal – the world is complex, uncertain and chaotic.

All the above advantages and disadvantages of applications of complexity-related ideas are relatively well-known in management (Checkland 1999; McKelvey 2012; Boulton et al. 2015; Jackson 2019) and are treated in this chapter as a reference in studying the use of those ideas in the discussions on the Anthropocene and selected environmental problems.

A proper understanding of applications of complexity-related ideas in the environmental discussions has significant meaning for politics, policy and polity. In this chapter, the Anthropocene is applied as a representative example of the above phenomenon affecting environmental studies. A preliminary systematisation and critical assessment of applications of complexity-related ideas in the discussions relating to the Anthropocene can positively influence its value as a theoretical and policy-oriented idea.

There is an additional specific opportunity for filling, at least preliminarily, this research gap. The time perspective of the third decade of the twenty-first century allows for a more profound and balanced assessment concerning the evolution and the status quo of both complexity studies and the Anthropocene.

---

geological discussions, the above problems must be studied with a conceptual background rooted in complexity-related research.

3.1.3 Research Questions

Taking into account the above considerations, the following research questions can be formulated:

(1) What are the systemic characteristics of the Anthropocene?
(2) What was the impact of complexity-related ideas on the theoretical and policy-oriented discussions concerning the Anthropocene?
(3) How is this impact exerted in description, methodology, normative approach, and action-oriented recommendations?
(4) How can the epistemological and methodological attributes of complexity studies enhance the validity of politics, policy and polity connected with the Anthropocene, and in a broader sense, with the practice of environment policy in all areas of social life?
(5) What could be the directions of the future applications of broadly defined complexity-related ideas in the research on the Anthropocene?

3.1.4 Aim of the Study

The aim of the study is to identify, systematise and evaluate the cognitive validity of applications of broadly defined complexity-related ideas in the development of the Anthropocene and in selected Anthropocene-related areas of social research. The results of the study can be helpful in enhancing the role of the Anthropocene in shaping environment-oriented politics, policy and polity in all areas of social life and at all levels of the societal hierarchy. Special attention is given to the systemic core of the Anthropocene-associated phenomena, i.e., the coupling of natural and social systems.

From an epistemological and methodological perspective, this chapter can be treated as an introductory guide enumerating selected issues, and providing directions for more profound studies of the links between the Anthropocene and complexity-related research.

This study is addressed to a broad audience of theoreticians (scholars, students and practitioners) as well as policymakers, managers, and activists. Due to its character – a survey and assessment – it is based on short introductory presentations of the ideas associated with complexity and the Anthropocene. Consequently, it can help the readers in developing a systematic search for a more profound knowledge of complexity-related ideas and their links with the Anthropocene, and in a broader sense, in other environmental studies. As a result, the readers can learn how well-grounded research on the role of complexity in the Anthropocene can contribute to enhanced validity of the latter in the theory and practice of politics, policy and polity in all areas of the social reality.

3.1.5 Epistemological and Methodological Assumptions

The identification and appraisal of the application of complexity-related ideas in the Anthropocene are built on a rank of assumptions. The broad scope of the meaning of multiple notions discussed in the chapter makes difficult, if not impossible, any attempts to provide unequivocal definitions. With some exceptions, when definitions can be found, the term 'interpretations' seems more relevant.

Epistemology and methodology of research on complexity demand a precise description of the applied terms due to the large number and variety of (sometimes even contradictory) interpretations. Therefore, as a fundamental epistemological and methodological approach to survey the applications of complexity-related ideas in the studies of the Anthropocene and the related conceptions, several proposals for ordering the area of the studies of complexity are borne in mind (Bar-Yam 1997; Checkland 1999; Warfield 2006; Meyers 2009; McKelvey 2012; Wells 2013; Jackson 2019; Castellani/Gerrits 2012; Snowden/Guh 2021; Mesjasz et al. 2022).

The methodology of this study embodies two components. First, a classical survey of the literature, or as it is now called, a narrative literature survey, as opposed to a more formalised *Systematic Literature Survey* (SLR), has been conducted. As usual, such an introductory survey must be selective, and can be viewed as arbitrary. However, it can be regarded as representative and sufficient for a preliminary study of applications of complexity-related ideas in theoretical and policy-oriented discussions about the Anthropocene. Obviously, this overview can also be treated as an opening to future, more comprehensive and advanced methods of literature studies, e.g. the already mentioned SLR.

Subsequently, a conceptual framework for studying those applications mentioned above is employed for an initial ordering and assessment of the role of complexity studies in supporting a deeper understanding of the Anthropocene. The ordering is done on a classical foundation: description, methodology, normative approach and action-oriented approach. Due to the normative character of the Anthropocene, special attention is given to normative approaches and action-oriented approaches.

The survey of applications of complexity-related ideas in the studies of the Anthropocene is limited to the large-scale systems embodying natural and social systems. In consequence, the applications of mathematical models of complexity in biological, geological, geophysical, and atmospheric phenomena still need to be explored, but they remain beyond the scope of this study.

3.1.6 Content of the Chapter

The remainder of the chapter is organised into three parts. In the first part, a short survey of definitions of the Anthropocene is presented. It provides a necessary minimum understanding of the Anthropocene, helpful in the analysis of its relations with complexity-related research. In order to avoid terminological and conceptual inconsistencies, a typology of interpretations of complexity and complexity-related ideas is presented in the second part of the chapter. Additionally, a typology of research domains directly and indirectly associated with complexity is depicted. In the third part, the relations between the Anthropocene and complexity-related studies are surveyed. Firstly, the Anthropocene is treated as a complex phenomenon, and the Earth System studies are described. Second, the impact of complexity-related ideas on normative and action-oriented ideas associated with the Anthropocene is illustrated with several examples.

3.2 Origins and Interpretations of the Anthropocene

Origins and more or less precise definitions and interpretations of the concept of the Anthropocene have been widely studied since its emergence in 2000. The surveys of the state-of-the-art knowledge about the Anthropocene can be found in multiple sources – books, journal articles, and websites. With necessary caution, it is also worthwhile to recall the very detailed literature accompanying the entry 'Anthropocene' in Wikipedia (2023).

For the purpose of identification and assessment of the role of complex systems studies in the theory of the Anthropocene, the following issues will be given attention: origins of the concept of the Anthropocene, interpretations of the Anthropocene, and the Anthropocene and related holistic ideas.

3.2.1 Origins of the Concept of the Anthropocene

The concept of the Anthropocene was preceded by several ideas exposing the growing impact of human activities on the natural environment. The predecessors of the Anthropocene are described in multiple works quoted by Brauch (2021, 2021a). The notion the *Anthrocene*, with a meaning similar to the Anthropocene, was, for the first time, used by the American journalist Revkin (1992). The term *Anthropocene* was also proposed by Eugene Stoermer in the 1980s, although it became popular after Crutzen contacted Stoermer (Revkin 2011). Anthropocene, in the meaning dominating the present theoretical discourse and practice, was first coined by Paul Crutzen on 23 February 2000 during a conference of the *International Geophysical and Biological Programme* (IGBP) in Cuernavaca in Mexico, where he claimed that the Anthropocene had replaced the Holocene as a "New Epoch of Earth History" (Brauch 2021: 53). Crutzen's proposal, disseminated in the early works (Crutzen/Stoermer 2000; Crutzen 2002), initiated both a global scientific discourse across many scientific disciplines and raised awareness of emerging societal and political challenges.

3.2.2 Interpretations of the Anthropocene

There have been multiple attempts to depict, define and develop various interpretations of the Anthropocene and to place it in geological history and in the history of humanity. For the purposes of this study, it is only necessary to recall the basic definition of the Anthropocene, the discussion about its time framework, the official opinion of geologists about the status of the Anthropocene as a geological epoch, the inter- and multi-disciplinary character of the Anthropocene, and the critics of the Anthropocene.

The proposed definition of the Anthropocene of Crutzen and Stoermer (2000) has been followed by multiple interpretations merging purely geological ideas with a large variety of inter- and multi-disciplinary approaches. The most influential in terms of geological extension of Crutzen's proposal is the definition adopted by the *Anthropocene Working Group* (AWG) of a *Subcommission on Quaternary Stratigraphy of the International Commission on Stratigraphy* (ICS) of *the International Union of Geologic Sciences* (IUGS) (AWG 2019): "The 'Anthropocene' is a term widely used … to denote the present geological time interval, in which many conditions and processes on Earth are profoundly altered by human impact. This impact has intensified significantly since the onset of industrialisation, taking us out of the Earth System state typical of the Holocene Epoch that post-dates the last glaciation …".

In their pioneering publication, Crutzen and Stoermer (2000: 17–18) proposed that the Anthropocene began with the Industrial Revolution, especially with Watt's invention of the steam engine in 1782. This initial description stirred multiple streams of discussion concerning the beginnings and interpretations of the Anthropocene. In addition to the initial proposal of Crutzen and Stoermer (2000), at least four different times have been identified as the beginning of the Anthropocene (Brauch 2021: 293): (a) the Neolithic or Agricultural Revolution (12,000 BP); (b) the Impact of the Columbian Exchange (1492–1600); (c) the beginning of the Nuclear Era and the fallout from nuclear testing (1945) that coincided with the end of World War II and the gradual evolution of a new political, economic, security and ideological order; and (d) the period of the Great Acceleration (since 1950).

The geological scale of changes caused by humanity was exposed in multiple works by one of the leading geologists studying and developing the idea of the Anthropocene (Zalasiewicz et al. 2011). On 21 May 2019, 29 of the 34 members of the AWG voted that an official proposal on the Anthropocene, as a new geological epoch with a starting point in the mid-twentieth century, should be made to the ICS of the IUGS. This proposal has not yet been accepted by relevant international scientific bodies. The Anthropocene, referring to geological timing, is still in the process of acceptance as a new geological epoch (Anthropocene Working Group [AWG] 2019).

The authors of the idea of the Anthropocene did not treat it as a polysemous term with multiple interpretations. However, in the further discussions after its introduction, the Anthropocene obtained several interpretations collected in the above-mentioned reference monographs (Brauch 2021, 2021a). They can facilitate identification of the systemic characteristics of the Anthropocene.

A new interpretation exposing the physicality of the Anthropocene was proposed by Shoshitaishvili (2022: 14):

> The Anthropocene as an epoch relies on the conception of humankind as a collective, cumulative, and out-of-control geophysical force. This framing is based on the physics concept of 'force,' which emphasises impact while hinting at the possibility of equal and opposite geophysical reactions. But this framing reveals little about the rationale for treating humankind as a collective entity (*anthropos*) coherent enough to be said to exert force in the first place. Couldn't a person avoid Anthropocene conclusions and debates by continuing to see Humanity as a completely unstructured multitude of billions of individuals and trillions of motivations and forces, none of which qualifies as being 'geophysical' in scale?

The Anthropocene as a social and cultural phenomenon was also confirmed by Crutzen: "The long-held barriers between nature and culture are breaking down. It's no longer us against 'Nature'. Instead, it's we who decide what nature is and what it will be" (Crutzen/Schwägerl 2011).

An interpretation of the Anthropocene and its duration proposed by Lewis and Maslin (2015) has been recalled by Harari (2017: 71): "Officially, we live in the Holocene epoch. Yet it may be better to call the last 70,000 years the Anthropocene epoch: the epoch of Humanity. For during these millennia, Homo sapiens became the single most important agent of change in the global ecology".

The discussion on the Anthropocene has been extended to philosophy and sociology (Latour 2014) and different aspects of theology (Sideris 2015; Bergmann, 2020). These applications also remain under the influence of complexity-related ideas, (e.g. Latour 2014), but this topic has to be treated separately.

An inter- and multi-disciplinary character is visible in the framework typology of disciplines in which the works on the Anthropocene have been published. The literature surveys of Brauch (2021, 2021a) and Simangan (in this volume) show the scope of discussions on the Anthropocene – from mathematical modelling to theology.

It is natural that critical ideas concerning the Anthropocene have also been developed. Some of them are only mentioned in this descriptive part, and further attention is only given to those cases where the criticism is associated with the uses of broadly defined complexity-related ideas. As an example of a criticism of the Anthropocene, the concepts of the Capitalocene and the Chthulucene are presented later in of this chapter.

A critical linguistic analysis and indirect impact of complexity-related research on the Anthropocene has been proposed by Crist (2016). Critical interpretation of the Anthropocene, treated as discourse, is extracted after the deconstruction of its meaning. The main negative aspect of the Anthropocene derives from the fact that this term reflects an absolute dominance of human civilisation over Nature.

> [W]e now live on a domesticated planet, with wilderness gone for good; we might put ecological doom-and-gloom to rest and embrace a more positive attitude about our prospects on a humanised planet; technology, including risky, centralised, and industrial-scale systems, should be embraced as our destiny and even our salvation; major technological fixes will

likely be needed, including engineering climate and life; the human impact is 'natural' (Crist 2016: 15).

The second example concerns the systemic meaning of the Anthropocene. When seen from a cultural and social perspective, the following weaknesses of the Anthropocene can be enumerated (Hartley 2016):

(1) The use of the ahistorical, abstract term 'humanity' that makes difficult, if not impossible, all more advanced analyses of causality.
(2) Technological determinism expressed in the isolation of the development of technology from social, economic, and political impact.
(3) The linear approach to human development, assuming that the events from the past unequivocally determine the present status quo – without considering the multitude of potential paths of development.
(4) A Whig view of history as one endless story of human progress and enlightenment.
(5) Apolitical technical and managerial (technocratic) solutions for the problems created by the Anthropocene which are presented under the guise of politically-neutral solutions of governance systems and geoengineering.

3.2.3 The Anthropocene and Related Holistic Ideas

As a part of the discussion following the birth and development of the idea of the Anthropocene, several alternative broad proposals have been suggested by various authors, predominantly in the social sciences. A preliminary survey of these concepts is presented to show a broad perspective of the studies associated with the Anthropocene. It gives additional arguments and inspiration for applying complexity-related ideas in the studies of the Anthropocene.

According to Chwałczyk (2020), there are at least 80–90 proposed alternatives to the term "the Anthropocene". The collection of ideas gathered here is developed on the basis of the first list proposed by Brauch (2021a: 303–304) and direct study of other sources. Each of them embodies other determinants of the impact of human activities on Earth and the natural environment. Some of them were proposed before 2000: the *Anthrocene* (Revkin 1992), and the *Homogenocene* (Samways 1999). The majority of the comprehensive ideas were developed after the first publications of Crutzen and his co-authors. They were not designed only as simple extensions of the Anthropocene, but rather as critical, if not competitive, ideas. It is also worthwhile to underscore their variety. They are quoted here in alphabetic order. The *Anthroposphere* (Baccini/Brunner 2012), the *AnthropoScene* and the *Anthropo-obScene* (Swyngedouw/Ernstson 2018), the *Capitalocene* (Haraway 2015, 2016, 2016a; Moore, 2016), the *Chthulucene* (Haraway 2015, 2016), the *Econocene* (Norgaard 2013), the *Gynocene* (Pirici/Voinea 2015; Demos 2017), the *Manthropocene* and the *Northropocene* (Raworth 2014) the *Manthropocene* (another concept) (Di Chiro 2017), the *Misanthropocene* (Patel 2013), the *Myxocene* (Pauly

2010), the *Necrocene* (McBrien 2016), the *Neganthropocene* (Stiegler 2018), the *Planthroposcene* (Myers 2017), the *Novacene* (Lovelock 2020), the *Plantationocene* (Haraway 2015; Davis et al. 2019), the *Plasticene* (Stager 2011), the *Plutocene* (Glikson 2017), the *Technocene* (Hornborg 2015).

A set of new terms has been added to this collection by Bonneuil and Fressoz (2016): the *Thermocene: A Political History of* CO_2; the *Thanatocene: Power and Ecocide*; the *Phagocene: Consuming the Planet*; the *Phronocene: Grammars of Environmental Reflexivity*; the *Agnotocene: Externalizing Nature, Economizing the World*; the *Capitalocene: A Combined History of Earth System and World-Systems* (extended interpretation); *the Polemocene: Resisting the Deterioration of the Earth since 1750.*

Following this long list of names with the suffix *-cene*, it must be added that in the studies of complexity and the Anthropocene, a necessary depth of linguistic analysis is necessary. It has to be remembered that the suffix *-cene* has a more sophisticated meaning (Wiktionary 2022). It is noted in a reflection by Haraway, the author of the term *Chthulucene* (Haraway 2015: 163): "The suffix '*-cene*' proliferates! I risk this overabundance because I am in the thrall of the root meanings of *-cene/kainos*, namely, the temporality of the thick, fibrous, and lumpy 'now', which is ancient and not".

3.3 Interpretations of Complexity of Natural and Social Systems

3.3.1 Typology of Interpretations of Complexity

The studies of the relations between complexity-related ideas and the Anthropocene call for a brief explanation of the meaning of the complexity of large-scale natural systems and broadly defined social systems. This minimal 'encyclopedic' introduction is necessary due to the large number of interpretations of complexity and complexity-related ideas and the specificity of their applications in the studies of environmental issues.

In environmental studies, three types of complex systems are scrutinised: natural systems, social systems, and coupled natural and social systems. The characteristics of natural systems are always operationalised and applied in various types of complexity-related mathematical models. Of course, in such cases everything depends on the availability and quality of the data, relevant mathematical models, and access to sufficiently powerful computers. The epistemological value of such models can also be critically assessed, but this topic remains beyond the discussion in this study.

In this chapter attention is paid to the influence of the phenomena associated with the Anthropocene on the complex societal systems at various scales. In such systems, both natural and social phenomena are coupled, but it is not always necessary or possible to develop sufficiently adequate mathematical models.

Creating interpretations of the complexity of various types of social systems is a more challenging task. Since the 1960s, the social sciences have been developed under a strong influence of various areas of knowledge in which interpretations of complexity and complexity-related concepts originated – cybernetics, systems thinking, and broadly defined complexity studies. These have been used as mathematical models, analogies, and metaphors. The mathematical models can be applied for description, explanation, prediction, and in normative (prescriptive) approaches. It must be remembered that analogies and metaphors can also be applied for similar purposes, although in different ways.

It needs be clearly explained that cybernetics and systems thinking were not directly declared as designed for studying various kinds of complex systems. Complexity was sometimes indirectly referred to in these domains of knowledge.

First and foremost, we must realise that there are no commonly-accepted definitions of complexity. In the complexity research community, the 45 interpretations of complexity by Lloyd (2001) are often recalled. Other multiple attempts were made in comprehensive syntheses (e.g. Gleick 1987, 2011; Bar-Yam 1997; Biggiero 2001; Warfield 2006; Meyers 2009; Jackson 2019; Castellani/Gerrits 2012; Snowden/Guh 2021; Mesjasz et al. 2022). In addition to these and other fundamental publications about complexity, in any study concerning the Anthropocene, attention must be given to the book of Wells (2013), exposing the links between complexity-related research and environmental theory and policy.

The typology of complexity-related ideas inspired by the above works is to some extent universal. It is adapted here for studying relations between the Anthropocene and variously defined complexity.

The proposed typology includes the following interpretations of complexity:

- intuitive complexity,
- complexity defined directly and indirectly in cybernetics, system thinking, and associated areas, e. g., systems dynamics,
- complexity science embodying quantitative and qualitative interpretations of complexity,
- qualitative indigenous concepts of complexity – the *autopoiesis* of Maturana and Varela (1980), applied in the studies of complexity of social systems directly or indirectly (e.g. Luhmann 1995, 2012, 2013), and *sympoiesis* (Dempster 1998).[2]

All the above types of complexity can be divided into quantitative 'hard' interpretations based on mathematical models, 'soft' qualitative interpretations, and mixed ones. Here they are presented only in general terms, but they are discussed in detail in the works quoted above.

[2] The complexity of social systems proposed by Luhmann (1995, 2012, 2013) is not often applied in the discussions relating to the Anthropocene and the environment, but it is not completely neglected there (e.g. Leach 2023).

The intuitive complexity embodies such characteristics, e.g. number of elements, differentiation of elements and their states, number and characteristics of internal interactions, internal hierarchy of the system, number and characteristics of interactions between the system and the environment, dynamics of the system, change, growth, decay, dissipation.

In cybernetics, references to complexity can be found in such cybernetical concepts as feedback, information, communication, learning, nonlinearity, self-organisation, and self-reproducing machines (Wiener 1948, 1961). Another pioneer of cybernetics, Ashby (1957), understood complexity in general terms as a property of the brain, computing machines, human organisms, and society and declared that his research was aiming at developing methods of studying them as complex systems. He proposed a range of complexity-related concepts such as adaptation, Black Box, feedback, equilibrium, homeostasis, stability, regulation, self-organisation, the law of requisite variety, and ultrastability.

The 'soft' interpretations of complexity derive from the 'second-order cybernetics' in which the role of the observer is taken into account (von Foerster 1982). In 'soft' complexity it is assumed that the object is not independent of the observer, which is reflected in the famous observation, 'complexity as beauty is in the eyes of the beholder'.

British scholar and practitioner, Stafford Beer, inspired by the ideas of cybernetics and neurology, developed a synthesis of managerial cybernetics called the Viable System Model (VSM), which can be treated as an instrument of effective management and governance of variety and complexity (Beer 1970, 1972, 1979, 1981). The VSM has already been applied in studying sustainability at the corporate level and in a universal sense (Espinosa/Walker 2017).

At the same time as cybernetics, the concept called systems thinking, systems theory or systems approach was developed in the 1950s by von Bertalanffy (1968) and Boulding (1956). The concept of soft systems methodology, developed as an opposition to hard systems engineering (Checkland 1999), applied in management theory and practice, can be viewed as an extension of classical systems thinking. *Soft Systems Methodology* (SSM) is an approach for tackling problematical, complex, messy situations of all kinds. SSM was successfully applied in environmental studies, including the Anthropocene (e.g. Ison 2016; Ison et al. 2018; Ison/Straw 2020).

System Dynamics is an example of linking cybernetics and systems thinking with mathematical models, including feedback and nonlinearity as the characteristics of complexity. The core element of System Dynamics is simulation models. The interpretation of complexity used in Systems Dynamics can be assigned to intuitive quantitative complexity, to quantitative complexity developed in systems thinking (feedback), and indirectly to complexity science (non-linear models). System Dynamics was used in the first models of global phenomena (Forrester 1971).

Due to its special role in the discussions on the Anthropocene and other areas of environmental theory and practice, more attention has to be given to the group of ideas called complexity science and/or complexity theory. There are two terms used for describing this collection of mathematical models – complexity science and complexity theory – without clearly delineated distinctions. Following the already mentioned studies of the meaning of complexity, it is proposed to apply the term complexity science.

In consequence, complexity science embodies two areas. First, hard complexity science, including mathematical models of complexity and complexity-related ideas, and second, soft complexity, including analogies and metaphors built with the terms in which they are treated as source domains. This interpretation of complexity science is included in multiple works; here only some of them are quoted: Prigogine and Stengers (1984), Kauffman (1995), Gell-Mann (1994), Holland (1995, 1998), Bar-Yam (1997), and a popular book by Gleick (1987, 2011).

Complex systems are usually recognised by the following features: (1) Nonlinearity of interactions among the elements, (2) Simple behaviour of elements contributing to the intricacy of the system, (3) Simple models describing and explaining the complex behaviour of systems.

The vocabulary of complexity science includes a large variety of terms intuitive, taken from other complexity-related areas – with the most popular ones being: adaptation, attractor, bifurcations, butterfly effect, catastrophe theory, chaos theory, coevolution, Complex Adaptive Systems, edge of chaos, emerging properties, far-from-equilibrium states, fractals, learning systems, nonlinearity, open system, path dependence, scale-free networks, self-organisation, self-reflexivity, self-similarity, strange attractor, synergy, synergetics, turbulence.

The above list is an introduction to further studies depicted in the multitude of works devoted to applications of complexity science in the natural and social sciences. These terms have also become popular in the vocabulary of modern social theory and policy making, because they include a strong appeal to intuition, sometimes even normative and emotional, e.g., chaos, the edge of chaos, emergence, self-organisation, synergy, etc. They must be treated with very high caution and without the shallow fascination visible in such expressions, as for example, the environment is turbulent and chaotic, we are in the situation at the edge of chaos, this is a strange attractor, etc.

The multiple sources of these and similar approaches easily found in social sciences terminology are purposively not quoted. Instead, the following reflections originating from the studies of complexity-related works in social theory and practice are presented.

(1) They refer to social systems, which in the ultimate resort, are social constructs.
(2) First and foremost, it must be remembered that these ideas are applied in social studies both as mathematical models, and as analogies and metaphors.
(3) In sciences dealing with environmental phenomena, complexity-related ideas are applied predominantly as mathematical models and simulation models.
(4) In the studies of the coupled natural and social systems, the complexity-related ideas are applied as mathematical models, simulation models, and analogies and metaphors.
(5) It should then be understood how those metaphors are created, and what their source domains and target domains are.

(6) Some names of the mathematical models used as the source domains for analogies and metaphors were coined incidentally, e.g., the chaos theory (Li/Yorke 1975), the chaos theory and the Butterfly Effect (Lorenz 1972, 1995) or the edge of chaos (Langton 1992; Waldrop 1992). It is then erroneous to apply them directly in social studies with an exact meaning as instruments of studying the 'chaotic world'.
(7) Some ideas of complexity studies have various interpretations, e.g. Complex Adaptive Systems. Their initial different mathematical interpretations, of Gell-Mann (1994) and Holland (1995, 1998), have turned into multiple quantitative, qualitative, and mixed interpretations.

3.3.2 Environmental Studies and Complexity: Supplementary Explanations

When seen from the current perspective, the typology of interpretations of complexity shows that the broadly defined studies of complex systems constitute a fuzzily-defined and sometimes even controversial area of research. However, after an initial fascination with the applications of complexity science in all areas of social life, and the impact of catchy metaphors such as the chaos theory, the edge of chaos, etc., a more critical and balanced picture is emerging. The following topics need additional explanation to facilitate a better understanding of the links between environmental studies, including the research on the Anthropocene, and complexity-related research:

- complexity-related research, environmental problems, and inter- and multidisciplinarity,
- terminological explanation,
- complexity and analogies and metaphors,
- complexity and nonlinearity.

3.3.2.1 Complexity-Related Research, Environmental Problems, and Inter- and Multi-disciplinarity

One of the most comprehensive epistemological and methodological studies of the role of complexity in environmental research, focused predominantly on sustainability, but also referring to the Anthropocene, was elaborated by Wells (2013).

Responding to the terminological chaos, in order to study the links between complexity and environmental studies, Wells has proposed the following terminological convention. In a study referring to almost all areas of complexity-related research, the term 'complexity theories' is presented as the most universal name, referring to all areas of complexity-related knowledge dealing with the environmental issues – from physics, geology, and biology, through the social sciences to ethics.

From a large number of examples of the use of complexity-related ideas in environmental studies, the approach elaborated by Jennifer Wells can be used as a representative example. The older, well-known ideas and the more recent ones compiled by Wells can be presented thus: a new definitional approach to complexity of natural and social systems in environmental studies, inter-, multi- and trans-disciplinarity of complexity-related environmental studies and the role of complexity-related research in studying selected aspects of the Anthropocene.

Conforming to Wells (2013: 93), the research on complexity is also stimulated by 'complexification', an idea proposed by Rescher (1998). This complexification, treated as a mixture of economic and natural phenomena, can be also found in Tainter's (1988) *Collapse of Complex Societies*. It may be then concluded that environmental problems derive from the increasing complexity (complexification) of natural and social phenomena and their interactions/interdependence. As a consequence of all these processes, the main advantage of the complexity theories is their inter-, multi- and trans-disciplinary character, making them the key ontological, epistemological, methodological, and even eschatological (ethical and environmental) background of environmental studies.

One of the examples considered by Wells (2013) is the study of the role of complexity in the analysis of sustainability in the Anthropocene. In a simplified description, the example reflects several systemic characteristics that can be viewed as the specific features of sustainability in the Anthropocene – stability, instability, equilibrium, punctuated disequilibrium, and network causality. From the point of view of the potential human impact on the Anthropocene, the distinction between irreversibility and non-renewability is of special importance.

3.3.2.2 Terminological Explanation

The multitude of interpretations of complexity and related terms leads to significant terminological ambiguities, which undermine the effectiveness of communication. The proposed typology of interpretations of complexity facilitates at least a preliminary ordering of terminology that is applied in this study and can be used for other purposes in complexity-related research.

The typology of interpretations of complexity shows that it is not possible to use the terms complexity science and complexity theory. Additionally, such areas as cybernetics, systems theory, etc., can be called complexity-related studies and complexity-related research. To remain coherent with the proposed typology, the terms complexity-related ideas or complexity-related concepts are used, e.g., chaos, edge of chaos, emergence, feedback, fractals, self-organisation, etc.

3.3.2.3 Complexity and Analogies, and Metaphors

In the most general sense, complexity-related concepts are used in all studies of nature as mathematical models and simulation models. In the studies of society, complexity-related ideas are applied as mathematical models, analogies, and various types of figures of speech, predominantly metaphors. The use of analogies and metaphors from physics, chemistry, biology, and engineering in studying social systems creates a specific demand. In all cases, the authors using those metaphors should be aware of the transfer of the meaning from the source domain to the target domain.

The role of metaphors was already identified in the early period of the development of complexity science in the title of one of the first books summarising the results of the initial conferences held at the Santa Fe Institute (Cowan et al. 1994). In the applications of complexity-related ideas in all areas of science, the existence of the following types of metaphors should be borne in mind: conceptual metaphors, live metaphors and dead metaphors (Lakoff/Johnson 1980/1995), and deliberate metaphors (Steen 2017). The complexity-related metaphors may also have persuasive functions (Sopory/Dillard 2002) and they can be applied as performative utterances (acts of speech) (Austin 1962; Searle 1995). Multiple metaphors taken from complexity-related research are treated as indispensable characteristics of the environment, e.g. edge of chaos, emerging properties, self-organisation.

3.3.2.4 Nonlinearity

In numerous studies of the Anthropocene and of environmental phenomena, the nonlinearity of natural systems, social systems, and their coupled systems is underscored. In many cases in the application of complexity-related ideas in the studies of natural and social systems, nonlinearity is presented as a kind of extraordinary property. It is necessary to recall Stan Ulam, who commented on the extraordinary attention given to the nonlinearity of mathematical systems and natural phenomena. "Stan once remarked that this was 'like defining the bulk of zoology by calling it the study of non-elephant animals'" (Campbell 1987: 218).

There is a significant simplification concerning reference to nonlinearity in the application of complexity-related ideas in the natural sciences and in the social sciences. Nonlinearity as such plays an important role in natural and social phenomena, but in relation to the so-called chaos theory (this description is not incidental). In a universal sense, the mathematical models called "chaos theory" refer to dynamical systems reflecting the time dependence of the position of a point in a geometrical space. This phenomenon was described in the above-mentioned works of Lorenz (1972, 1995) and Li and Yorke (1975). The genesis of the chaos theory and its links with non-linearity are only signalled here, but they are worth remembering to avoid any potential misunderstandings concerning their applications.

3.4 The Anthropocene as a Complex Phenomenon

3.4.1 Assumptions of the Study

Due to the multitude and scope of definitions of such terms as complex, complexity, complexity studies, complexity-related ideas, and similarly, the Anthropocene, Gaia, Capitalocene, Novacene, etc., additional assumptions of the study must be clarified.

The simple intuitive and self-evident reflection that the Anthropocene is a complex phenomenon brings about important theoretical questions and demands for necessary policy-making. To fulfil the aim of this study, and bearing in mind the overwhelming number of ideas associated with the utterances system and complexity, the following approach to ordering the discussion is applied. It is rooted in the philosophical foundations of science – ontological, epistemological, and methodological. Obviously, these issues are studied only in an introductory way, although sufficient for this study.

In the modern interpretations of all kinds of systems and their complexity, it is assumed that they are the constructs of the observers. Obviously, social systems are different constructs to natural systems, but the fundamental idea that "complexity as beauty is in the eyes of the beholder" is universally actual. In the literature on systems thinking, this problem has been broadly discussed, but as a representative argument, the following reflection of Murray Gell-Mann, one of the founding fathers of complexity studies, can be quoted: "There are various quantities labelled 'complexity'. In each case, the complexity of a thing is context dependent – in other words, dependent not only on the thing being described but also on who or what is doing the describing" (Gell-Mann 1994: 33). Obviously, these remarks have a different meaning in the cases of natural systems, social systems and in the discussions in quantum mechanics.

This quotation allows us to avoid long considerations concerning the fact that social systems are socially constructed and that broadly defined natural systems, although treated as independent from the observer, at least partly result from subjective, or more precisely, intersubjective discourse. It must be also added that in this study the analysis of relations between the ontological and epistemological character of complexity and complex systems is not scrutinised.

In the following parts of the text, the research questions formulated in 1.1. will be answered. Of course, the length of the chapter imposes several limitations. For instance, a thorough appraisal would demand the study of a large number of publications. Here only the representative ones are considered, and that is why only partial introductory analysis is applied. The choice of publications has been made with a full awareness that it can be perceived as arbitrary. However, the number of publications on the links between complexity-related ideas and environmental issues, including the Anthropocene, is very large. It makes the choice somewhat justifiable.

The questions are referred to the applications of complexity-related ideas in the following areas associated directly and indirectly with the Anthropocene. First, comprehensive methodological approaches: 'Earth System Science' or 'Earth System Analysis'. Second, normative ideas inspired by the Anthropocene:

Econocene and Chthulucene. Third, action-oriented ideas: governance of the Anthropocene and the use of the Viable System Model of Stafford Beer in the governance of sustainability.

The overall assessment of applications of complexity-related ideas in the studies of the Anthropocene is of a limited character due to the introductory nature of the chapter.

3.4.2 Anthropocene as a Complex Phenomenon

In the broadest sense, the Anthropocene can be treated as a complex collective phenomenon reflecting the impact of humanity on Earth. A question thus arises of whether it is possible – or necessary – to develop a classical unique systemic interpretation of the Anthropocene embodying borders, elements, interactions, internal structure, history, patterns of dynamics of elements and the whole system.

Contrary to the Gaia hypothesis, that originated as a comprehensive picture of the Earth system built on the foundation of first order cybernetics and then absorbing a majority of complexity-related ideas (Lovelock 1972, 1979; Lovelock/Margulis 1974; Clarke 2020), the founders of the Anthropocene did not take into account the complexity of their concept. Neither 'complex' nor 'complexity' was mentioned in the pioneering works on the Anthropocene (Crutzen/Stoermer 2000; Crutzen 2002). In its initial version, the Anthropocene was just a description of the state of the changes on Earth due to the impact of human activities, with implicitly included systemic interpretations. This fact was treated as self-evident by the authors of the idea of the Anthropocene. In the later studies, the Anthropocene was described more precisely in systemic terms, and it has been commonly accepted that the complexity-related ideas constitute its ontological, epistemological, and methodological foundations (Verburg et al. 2016).

In the opinion of Ehlers and Krafft (2006a): "The scientific term 'Anthropocene' is an almost self-evident proof of this new complexity of interactions, actions and interrelationships between the natural and the social systems." It was recalled that the complexity of the environmental phenomena caused by human activities could be studied only with the use of new, holistic approaches and methods. A similar opinion was expressed in the works with the joint authorship of Steffen and Crutzen (Steffen et al. 2007, 2011).

A new systemic character of the Anthropocene was exposed by Homer-Dixon et al. (2015), who referred to the ideas borrowed from complexity-related studies that treated the natural and social systems as tightly coupled in non-linear relationships. In consequence, the impact of the complexity of these relationships led to the synchronous failure of both systems.

The Anthropocene as a structural coupling taken from autopoiesis (Maturana/Varela 1980), one of the most influential complexity-related ideas, was proposed by Ison and Straw (2020).

Following the above and other examples, it can be concluded that in the systemic interpretation deriving from broadly defined complexity-related studies, the Anthropocene can be depicted with a collection of characteristics that are included in mathematical models, analogies, and metaphors as well as indigenous qualitative complexity-related ideas, e.g., autopoiesis (Maturana/Varela 1980; Luhmann 1995, 2012, 2013; Haraway 2015, 2016, 2016a; Clarke 2020) and sympoiesis (Dempster 1998; Haraway 2015, 2016, 2016a).

The Anthropocene refers to two interacting (coupled) systems: the Earth System as defined in a general sense (NASA 2020, 2022), and the broadly defined Global Human System.

The Global Human System embodies all areas of human activity – tangible and intangible. The tangible human activities directly affect the Earth System. The intangible, cognitive human activities affect the Earth System only indirectly. They result in and, at the same time, they determine, the natural, material, biological and technological impact of Humanity on the Earth System.

(1) The Anthropocene reflects a specific stage of development of the Earth System – the Global Human System as the (dominant?) source of changes in the Earth System at the geological level of impact. In this sense, the Anthropocene is a specific complex dynamic system.
(2) In terms of complexity-related understanding, the complexity of the Anthropocene includes a specific synthesis of interactions of the complexity of the natural Earth System and the Human System.
(3) The interactions between natural systems and human systems constituting the core of the Anthropocene allows for the disclosure of physical and thermodynamic aspects of human activity. These were studied by Georgescu-Roegen (1971), but the complexity-related research led to a more profound understanding of them (Hermann-Pillath 2017).
(4) It can also be concluded that the complex interactions between the Human System and the natural Earth System have created a new, complex system that can be treated as a meta-system or supra-system for all phenomena linking broadly defined physical attributes of Earth as well as physical and cognitive activities of humanity. Some authors even use the term hyper-complexity (Wells 2013) or complexity of complexities. This term is not discussed in this study.

The ordered set of characteristics can be treated as a point of departure for studying the role of complexity-related ideas in the epistemology, axiology, and methodology of the research on the Anthropocene. Similarly, as in all areas of knowledge, it is always expected that the complexity-related concepts can enhance the cognitive value of classical functions of science: description, prediction, the normative approach, and the action-oriented (regulatory) approach. Because of the deepest sense of the Anthropocene as a specific warning about the negative status quo of the impact of humanity on the Earth System, attention is paid to the use of complexity-related ideas in normative interpretations and action-oriented proposals associated with the Anthropocene. Exposing complexity-related descriptions and normative and action-oriented approaches may seem arbitrary, but it results in this study's aim and scope.

Discussion about complexity-related ideas and prediction demands further, separate considerations. As an example of the studies of complexity and prediction, the discussion about probability distributions (normal distribution and the power law) and scales can be quoted (Neugebauer 2006).

3.4.3 Complexity, Earth System Studies and the Anthropocene

Several complexity-related, systemic, holistic, and partial interpretations of the Anthropocene have been elaborated since its origins. In this section, only the broad theoretical concepts predominantly focused on ontology, epistemology and methodology are depicted. How has the Anthropocene affected the Earth System, and how has it been studied?

The studies of the coexistence of humans and the Earth's natural systems can be divided into three groups. In the first group, the Earth and its natural subsystems and environment constitute a system identified independently from human activities. The second group includes the approaches embodying both Earth and its ecosphere along with human systems, and in the third, the interactions between the Earth's natural system and society are studied. Similarly, as in other areas of environmental studies, these three types cannot be clearly separated; therefore, in our discussion the borderlines between the types are marked only preliminarily.

3.4.3.1 Earth System

A comprehensive concept of the complex Earth System was, for the first time, proposed by NASA in 1983. The Earth System (sum of its parts) includes the following subsystems (NASA 2022): atmosphere, hydrosphere, biosphere, cryosphere, and geosphere. Sometimes referred to as part of the exosphere (all space-related phenomena that affect Earth), energy from the Sun drives processes of the Earth system. The Earth System is also treated as a System of Systems (NASA 2020).

The most influential studies linking the Earth System and its complexity with the Anthropocene were described in multiple works by Joachim Schellnhuber and Will Steffen together with their co-authors (Schellnhuber/Wenzel 1998; Schellnhuber 1998, 1999; Schellnhuber et al. 2004, 2006; Steffen 2004; Steffen et al. 2004, 2017, 2018, 2020).[3]

According to one of their interpretations, the Earth System is usually defined as a single, planetary-level complex adaptive system with a multitude of interacting biotic and abiotic components. The Earth System is driven primarily by solar radiation.

[3] Schellnhuber and Steffen authored and co-authored multiple works. Sometimes they collaborated in the teams preparing the publications; therefore their collective works are quoted separately, since the teams of the co-authors were different each time.

It is influenced by other extrinsic factors, including changes in orbital parameters and occasional bolide strikes, as well as by its own internal dynamics in which the biosphere is a critical component (Steffen et al. 2016: 325).

The relations between the Earth System and the Anthropocene have been reflected in several studies. In the descriptive and explanatory interpretation of Hamilton and Grinevalt (2015), in the universal sense, the Anthropocene, as proposed by Crutzen in 2000, is based on the concept of the Earth System, a single complex system at the planetary level with its own emergent properties, states, and modes of functioning. The Anthropocene thus represented a state change in the Earth System (Waters et al. 2016) and is viewed as an interdependent social–ecological system, different from the Holocene.

3.4.3.2 Earth System Science and Earth System Analysis

The studies of the Earth System referring to global environmental issues are conducted under two names: Earth System Science and Earth System Analysis.

Earth System Science (ESS) has been described in various publications (Schellnhuber/Wenzel 1998; Schellnhuber 1999; Steffen et al. 2004, 2016). The publication by Steffen et al. (2020) can be recommended as the main source of synthetic information about ESS, its origins, and its evolution. The beginnings of ESS are traced to the already-mentioned ideas from classical studies of the global natural and social processes (Brauch 2021, 2021a). Additionally, the first international collaborative efforts aimed at studying the dynamics of the Earth phenomena, e.g., the *International Geophysical Year* (IGY) 1957–58 contributed to the origins of the ESS. In parallel, more rigorous environmental studies also originated. A detailed chronology points to the 1970s as the time of birth of the ESS (Steffen et al. 2020).

From the point of view of complexity-related research, as was mentioned earlier, one of the stimulants of the development of the first Report to the Club of Rome on *Limits to Growth* (Meadows et al. 1972) was based on the mathematical model of System Dynamics developed by Jay Forrester in *World Dynamics* (Forrester 1971). It is also necessary to take into account the systemic thinking about Earth and society in the Gaia hypothesis formulated in the 1970s (Lovelock 1972, 1979; Lovelock/Margulis 1974). The opinions about the impact of the Gaia hypothesis on the Anthropocene are sometimes contradictory (Clarke 2020).

Earth System Science was more precisely defined in the 1980s. The NASA Earth System Science Committee was established in 1983. Then other institutions and organisations became involved in this area of research in various reports, workshops, and conferences, as has been presented in detail in the paper by Steffen et al. (2020).

The development of various streams of Earth System Science has been inherently influenced by the ideas associated with broadly defined complexity-related studies. As early as 1986 the NASA Bretherton Committee Diagram, comprising a simplified Systems Dynamics model, was used to illustrate the interactions between the components of the Earth System (Steffen et al. 2020: 56).

From the wealth of areas of mutual interaction between Earth System Science and complexity-related studies, it is also necessary to underscore the acceptance of ESS's inter-, multi- and trans-disciplinary character.

Undoubtedly, mathematical models constituted the core element of ESS. As one of the pioneering models, the Gaia-inspired Daisyworld explains how homeostasis, built on the feedback processes between life and its environment, could lead to global-scale temperature regulation (Watson/Lovelock 1983). It is recalled by Steffen et al. (2020).

The following models, presented in a very brief manner, can be treated as the most representative of the ESS. *General Circulation Models* (GCMs) focus on the fundamental physics and chemistry of the climate system, without a more profound analysis of the human system (Dahan 2010). Human dynamics are the domain of *Integrated Assessment Models* (IAMs), which link economic models and climate models built on complexity-related principles (Dahan 2010; Heymann/Dahan Dalmedico 2019; Steffen et al. 2020). As the last exemplary complexity-based modelling, the Earth System Models of Intermediate Complexity (EMICs) is recalled (Claussen et al. 2002; Steffen et al. 2020). The EMICs models are built on complexity-related ideas, but their advantage is a specific simplification allowing for more universal applications of the models. Simulation of excessively-detailed models does not allow for time flexibility and could be prohibitively costly.

The development of Earth System Science was accompanied by the concept of Earth System Analysis, which was proposed by Schellnhuber (Schellnhuber/Wenzel 1998; Schellnhuber 1998, 1999) and in the book resulting from the Dahlem Workshop held 25–30 May 2003 in Berlin on the topic "Earth System Analysis for Sustainability" (Schellnhuber et al. 2004). While Earth System Science was treated as a more general concept, Earth System Analysis, especially after 2001, was related directly to the impact of the Anthropocene. It was underscored by Clark et al. (2004) in the Introduction to the book by Schellnhuber et al. (2004).

According to its principal author, Earth System Analysis was supposed to become the second 'Copernican revolution'. It is depicted with ideas taken from complexity-related studies (Schellnhuber 1999: C20): "it will enable us to look back on our planet to perceive one single, complex, dissipative, dynamic entity, far from thermodynamic equilibrium—the 'Earth system'".

The direct reference to the studies of the dissipative systems and the far-from-equilibrium systems developed by Ilya Prigogine (Prigogine/Stengers 1984) leads to another observation. The dissipative systems elaborated by Prigogine have been frequently used, not only in physics, chemistry, biology, and in studying the Earth System, but also in social sciences as mathematical models, analogies, and metaphors.

In purely methodological and terminological terms, Earth System Science and Earth System Analysis should be distinguishable. It must be borne in mind that the terms Systems Science and Systems Analysis are treated as distinct in broadly defined systems studies and, subsequently, in complex systems-related research (Checkland 1999; Wells 2013; Castellani/Gerrits 2021; Mesjasz et al. 2022). While the Systems Analysis originated in the 1950s by the RAND Corporation is treated as a rigorous methodological approach (Miser/Quade 1985), Systems Science has several interpretations assigned to both qualitative and quantitative methods (Meyers 2009).

The preliminary survey of the literature concerning Earth System Science and Earth System Analysis shows that no detailed study exposing their differences has yet been elaborated. Additionally, the same authors – Schellnhuber and Steffen – use both terms together in different multi-author books, e.g., Earth System Science (Steffen et al. 2016, 2020) and Earth System Analysis (Schellnhuber/Wenzel 1998; Schellnhuber 1998, 1999; Schellnhuber et al. 2004). It can be only concluded that at a certain stage of research Schellnhuber and participants of the Dahlem Workshop preferred Earth System Analysis. When seen from the perspective of studying systemic aspects of the global models and the Anthropocene, this distinction is of a purely terminological character and seems not to have any theoretical or policy-oriented impact.

In the assessment of the role of complexity-related theory and policy in the Anthropocene, attention is paid to the Anthropocene equation proposed by Gaffney and Steffen (2017). The reason for choosing this model is that in some sense it is approaching the borders between epistemologically well-grounded mathematical modelling and mathematics-like broadly defined analogies and metaphors. Such an approach demands special caution, because it could lead to multiple misunderstandings and criticism concerning the scientific value of the applications of complexity-related ideas in studies of the Anthropocene.

According to its authors, who follow the ideas of Schellnhuber (1998, 1999) and Steffen et al. (2016), the Earth System is a single, planetary-level complex system composed of the biosphere, (defined here as the sum of all biota living at any one time) and its interactions, including interactions and feedbacks with the geosphere (defined here as the atmosphere, hydrosphere, cryosphere and upper part of the lithosphere) (Steffen et al. 2016). The authors have made an attempt to use a specific general formalism showing the changes of the *Earth System* (E). The rate of changes of the Earth System can be described as follows:

$$\frac{dE}{dt} = f(A, G, I, H) \tag{3.1}$$

where A is astronomical forcing, and G is geophysical forcing; I is the internal dynamics of the Earth System, and H human activity.

A comment derived from the aim of this study is necessary. It is not assessed whether the highly abstract formula – as the authors declare – is relevant to reality and can be applied for receiving results expected from the uses of the mathematical models, i.e., predominantly prediction. It is used as the basis for developing more detailed models and some specific results concerning the Anthropocene that may appear disputable.

The equation was discussed and criticised after publication (Heijungs et al. 2017; Gaffney/Steffen 2017a). The authors are aware that the initial formula and the derivative formulae are simplifications. Leaving for a separate discussion the above criticism by Heijungs et al. (2017), and subsequent debate, it must be emphasised that the central question concerns the obstacles to the operationalisation of variables. Classical operationalisation, as in physics (Bridgman 1959), or operationisation as in the behavioural sciences, using the scales designed for social research (Stevens 1946), are difficult to apply to such broadly defined variables. That is why the Anthropocene equation, its variables, and associated formulae can be treated solely as specific analogies requiring a very cautious introductory mathematisation and quantification.

3.4.4 Complexity, the Anthropocene and Normative Approaches

The next question relevant to this study might be: What is the role of complexity-related ideas in normative considerations concerning the Anthropocene? As an introduction, the normative complexity-related ideas associated with the Anthropocene are preliminarily discussed. Then the role of complexity-related ideas in the normative interpretations of the Anthropocene is illustrated with the examples of the Capitalocene and the Chthulucene. Furthermore, two other complexity-related normative examples are briefly referred to: legal aspects of the Anthropocene and a normative sense of sustainability in the Anthropocene.

3.4.4.1 Complexity and Normative Character of the Anthropocene

In the normative sense of the Anthropocene, humanity is negatively influencing the Earth's natural systems (the Earth system) to such a degree that, as a result, the influence will become harmful to the world system, including the future of humanity itself. Crutzen and Stoermer treated the Anthropocene as a warning about the destruction caused by humanity (Crutzen/Stoermer 2000; Crutzen 2002). The consequences of the Anthropocene have been assessed from various normative perspectives – ethical, social, economic, political, and environmental. The theological discussions about the Anthropocene embodying the normative reflections (e.g. Sideris 2015; Bergmann 2020), need separate consideration.

A normative character of the Anthropocene also corresponds with the universal sense of sustainable development proposed in the report *Our Common Future*, prepared by the *World Commission on the Environment and Development* (WCED) under the leadership of Gro Harlem Brundtland, where sustainable development was defined as development that meets the needs of the present without compromising the ability of future generations to meet their own needs (WCED 1987: 37).

The normative character of the Anthropocene can be summarised in a series of fundamental questions. What are the postulated normative dynamics of the Earth system following the diagnosis concerning the negative consequences of the Anthropocene? Is it necessary for humanity to take any action to change the situation? Is such an intervention feasible? Can this intervention be treated as positive?

These questions are asked in the discussions about governance in the era of the Anthropocene.Dalby (2016) identified three different perspectives for the framing of the Anthropocene: "the good, the bad, and the ugly". The notion of a "good Anthropocene" was also recalled by Gleditsch (2003). The "bad Anthropocene" is associated with the negative consequences of 'dangerous' or 'catastrophic' climate change (Schellnhuber et al. 2006) and with "hothouse Earth" (Steffen et al. 2018). The impact of climate change on security, conflicts and wars was scrutinised by Scheffran et al. (2012). Normative aspects of the Anthropocene are associated with peace, security and gender issues (Oswald Spring 2017; Oswald Spring et al. 2009, 2017).

This normative character of the Anthropocene is particularly apparent in fundamental theoretical discussions and empirical studies of the Anthropocene which are directly concerned with complexity-related ideas (Schellnhuber 1998, 1999; Schellnhuber et al. 2004, 2006; Steffen 2004; Steffen et al. 2004, 2017, 2018, 2020).

Assuming that humanity may not passively observe the negative consequences of the Anthropocene, the following questions have to be asked. What actions must be taken to minimise or, perhaps, to terminate the negative impact of the Anthropocene? Is it possible to terminate, slow down or reverse the negative consequences of the Anthropocene? What about the reversibility and renewability of the consequences of the Anthropocene? If the Anthropocene is reversible/renewable, then to what degree is it possible to return to the previous, better (or at least 'not that negative') state of the past, from the era before the Anthropocene?

An explanation how the above fundamental questions have received new impulses deriving from complexity-related research can be provided only in a broad sense and with selected examples. A series of new universal ideas deriving from the complexity-related studies has enhanced normative approaches associated with Anthropocene. Paradoxically, some of them, e.g., the emerging properties, have improved normative approaches by showing new limitations of predictability, like the already-mentioned ideas of Lorenz (1972, 1995) (chaos theory), or emerging properties useful both in the mathematical models and as analogies and metaphors. The contribution of the complexity-related ideas to the normative aspects of the Anthropocene can be recapitulated in the following observations:

(1) The epistemology and methodology of research of the Anthropocene were enriched with new complexity-related mathematical models, new analogies, and metaphors created using those models. Additionally, the normative uses of qualitative ideas related to complexity, such as *autopoiesis* and *sympoiesis*, should be mentioned.
(2) These ideas have changed the interpretations of the norms in the social systems, with the emergence of norms in social systems resulting from feedback, learning, self-organisation, and homeostasis.
(3) There is limited predictability of spontaneous changes in natural and social systems: instability, non-equilibrium,

(4) There is limited predictability of the consequences of human intervention at all scales of the social systems, although the predictability at the local level seems more feasible ('think globally, act locally').
(5) There is awareness of the limitations of purposive human intervention deriving from limitations of predictability.

3.4.4.2 The Capitalocene

The Capitalocene can serve as an example of the impact of complexity-related ideas on the normative aspects of environmental research considered in a broader economic, political and ideological context. The authors of this concept are Jason Moore and Donna Haraway, although the notion itself was used for the first time by Andreas Malm (Haraway 2015, 2016; Moore 2016a).

The Capitalocene, in the interpretation presented by Moore (2016a, b), linking the development of capitalism with the changes on Earth, was not regarded as totally critical, but rather as an attempt to provide more profound views of the meaning of the Anthropocene:

> Like globalisation in the 1990s, the Anthropocene has become a buzzword that can mean all things to all people. Nevertheless, reinforced by earlier developments in environmental history, the Anthropocene as an argument has gradually crystallised: 'Human action' plus 'Nature' equals 'planetary crisis'. Green arithmetic, formulating history as the aggregation of human and natural relations, had triumphed (Moore 2016a: 3).

According to Moore (2016a), the main reason for developing the Capitalocene was the absence of identification of the causes of the idea of the Anthropocene. It is an alarming signal, but what are the causes of that alarm? The Anthropocene is too mechanistic and is missing the real causes of the growing impact of the development of capitalism as the main motive for exploitation of the resources of Earth in creating a new socio-economic order. The idea of "cheap Nature" that stimulated capitalism in its early development has led to excessive exploitation of all kinds of resources, and the present situation, depicted, among other ideas, by the Anthropocene, demands radical changes.

The second author promoting the Capitalocene, Haraway, openly declares, multiple times, the links between the Capitalocene and broadly defined complexity studies: "I think the issues about naming relevant to the *Anthropocene*, *Plantationocene*, or *Capitalocene* have to do with scale, rate/speed, synchronicity, and complexity" (Haraway 2015: 159). Additionally, when discussing the Capitalocene, Haraway refers directly and indirectly to the foundational ideas of complexity research – the far-from-equilibrium states developed by a Nobel Prize Laureate in Physics, Ilya Prigogine (Prigogine/Stengers 1984) and closely related to complexity and the environmental ideas of Stengers (2015).

The systemic interpretations of Haraway (2015, 2016, 2016a) in the discussion about the Capitalocene are also strongly influenced by the Gaia hypothesis (Lovelock 1972, 1979; Lovelock/Margulis 1974; Margulis 1998). They also include reference to the two fundamental qualitative complexity-related ideas of *autopoiesis* and *sympoiesis*. A normative, complexity-based interpretation of the Capitalocene has given ground to a more comprehensive normative character of humanity-induced changes on Earth that are studied and exposed in Haraway's concept of the Chthulucene.

3.4.4.3 The Chthulucene

The Chthulucene, one of the most thought-provoking and at the same time intricate ideas concerning the relations between humanity and Nature, and treated at the same time as an extension of and in opposition to the Anthropocene, was invented by Haraway (2015, 2016, 2016a). Its author discusses it in reference to the Anthropocene, the Capitalocene, and the Plantationocene, where the last name is associated with the development of plantation agriculture as the main factor of changes on Earth (Davis et al. 2019).

The name Chthulucene is rooted in the language of the Goshute people of Utah, and it derives from denizens of the depths, from the abyssal and elemental entities called chthonic (Haraway 2016: 35). The metaphor of the Chthulucene has its origins in reference to the tentacles of a spider. It constitutes a specific point of departure for the discussion on the Anthropocene and the Capitalocene.

It is not possible to encapsulate in this survey study a very broad and sophisticated definition of the Chthulucene and its systemic characteristics associated with complexity-related ideas. The quotation below serves solely as an invitation to more comprehensive studies and as an introduction to exposing the relations between various streams of complexity-related research and the Chthulucene (Haraway 2015: 160):

> 'My' Chthulucene, even burdened with its problematic Greekish tendrils, entangles myriad temporalities and spatialities and myriad intra-active entities – inassemblages – including the more-than-human, other-than-human, inhuman, and human-as-humus. … It matters which stories tell stories and which concepts think concepts. Mathematically, visually, and narratively, it matters which figures figure figures, which systems systematise systems.

While the Capitalocene is connected with complexity-related research in a somewhat fuzzy manner, the sources of inspiration of the Chthulucene can be found in the impact of the Gaia hypothesis and in the writings of Isabelle Stengers and Ilya Prigogine (Prigogine/Stengers 1984; Stengers 2015). The Chthulucene is to a large extent built upon two very specific concepts drawn from complexity studies: *autopoiesis* and *sympoiesis*. The latter is treated by Haraway as opposite to *autopoiesis* and as more relevant to the present environmental problems of Earth. Haraway treated these two concepts as the foundation and inspiration for studying the complexity of biological and social systems. They were directly applied in development of the Chthulucene, and indirectly in the Capitalocene.

In unfolding the Chthulucene, Haraway refers to the inclusion of autopoiesis (Maturana and Varela 1980) by Margulis in the studies of Gaia. For the first time, Margulis focused attention on autopoiesis in the second edition of the book

Symbiosis in Cell Evolution (Margulis 1981, 1993). Margulis extended the concept of *autopoiesis*, not only to the Gaia theory, but also to the entire biological system from the cell to the biosphere. It was later extended even further by Margulis (1998). As was exposed in the study of the ideas of Margulis conducted by Clarke (2020), this approach meant a neocybernetic reformulation as autopoietic Gaia, in relation to nonautopoietic systems.[4]

The criticism of the *autopoietic* Gaia and, indirectly, the Anthropocene, by Haraway led to the application of another complex systems-related idea: *sympoiesis*. Haraway (2016: 33; 2016a: 37) had used this term in a different context, but later discovered a comprehensively-developed idea of sympoiesis by Dempster (1998). In the rudimentary sense, sympoiesis is the opposite term to autopoiesis (Dempster 1998: 3):

> My primary thesis concerns the existence of *sympoietic* (collectively produced) systems. I define these as complex self-organising systems lacking spatial and temporal boundaries. They have cooperative synergistic characteristics and must be identified by the continuing interactions among components. In particular, I make a distinction between *sympoietic* and *autopoietic* (self-producing) systems. The key distinction – relating to the lack/presence of self-defined boundaries – gives rise to a variety of other characteristics significant for understanding both natural and social systems.

The interactions between all subsystems and elements in a *sympoietic* system are also described by the term 'entanglement', attributed by Haraway (2016, 2016a) to a comprehensive understanding of the contemporary complex world – Gaia and the Chthulucene. Haraway also argues that *sympoiesis* is more relevant to the ecological visions of Earth and society than systemic views of the Anthropocene. "The sciences of the Anthropocene are too much contained within systems theories of *autopoiesis* and within evolutionary theories called the Modern Synthesis, which for all their extraordinary importance have proven unable to think well about *sympoiesis*, symbiosis, symbiogenesis, development, webbed ecologies, and microbes" (Haraway 2016a: 53).

Another important factor is associated with sympoietic character of the Chthulucene. It inspired Haraway to abandon the post-humanist understanding of the Anthropocene and in a broader sense, the contemporary relations between humans and the environment. Before showing the relations between the Chthulucene and post-humanism, a few remarks on the latter and the Anthropocene should be made.

There are several streams of ideas that are declared as post-humanist, e.g., philosophical posthumanism, cultural posthumanism, transhumanism, consequences of the development of AI, etc. The key issues in this study are the relations between various trends of posthumanism and the Anthropocene. Two patterns of reasoning can be distinguished. In the first, the Anthropocene is treated as the foundation of a post-humanist approach. Humanity's capability of destroying Earth and itself led to the situation forcing a re-think of the basic tenets of human interactions with both

[4] Neocybernetic systems theory is the name used by Clarke (2020) for a range of second-order cybernetic concepts: recursion, reentry, closure, autonomy, cognition, self-reference, and self-referential systems.

human and non-human agents on a planetary scale (Braidoti 2013: 5–6; Ulmer 2017). In the second pattern of thinking, the Anthropocene is treated as strongly anthropocentric and denies post-humanist considerations. In this case, it can be added that the discussions about the Anthropocene stimulate criticism and the development of post-humanist ideas and their extensions.

Haraway, regarded by some authors as a leading figure in post-human discussions (Cielemęcka/Daigle 2019), in the complex *sympoietic* concept of the Chthulucene declares posthumanism as too narrow for understanding the modern world: "... human beings are not in a separate compost pile. We are humus, not Homo, not *anthropos*; we are compost, not posthuman" (Haraway 2016: 55). This short quotation reflects the essence of reasoning showing the universal and complex character of environmental problems, inspired by sympoiesis and going beyond the Anthropocene.

Without delving into further details of conceptually- and linguistically-advanced considerations, the most significant complexity-related normative aspect of the Chthulucene is the criticism of the *autopoietic* character of the Anthropocene and the use of mathematical models of complex systems. It is reflected in the following observations (Haraway 2016: 49):

(1) Despite its reliance on agile computer modelling and *autopoietic* systems theories, the Anthropocene relies too much on what should be an 'unthinkable' theory of relations, namely the old one of bounded individualism – pre-existing units in competition relations that take up all the air in the atmosphere.
(2) The sciences of the Anthropocene are too much contained within systems theories of *autopoiesis* and within evolutionary theories called the Modern Synthesis, which for all their extraordinary importance, have proven unable to think well about *sympoiesis*, symbiosis, symbiogenesis, development, webbed ecologies, and microbes.

Any profound assessment of the impact of complexity-related ideas on the Chthulucene would need another voluminous study. Some essential remarks concern only autopoiesis and sympoiesis. First and foremost, it has to be reiterated that the idea of autopoiesis was developed by Maturana and Varela for biological studies at the molecular level. From the beginning, the applications of autopoiesis outside that part of biology were contested. Maturana and Varela (1980: xxiv–xxv) were critical of the direct transfer of their idea from biology to the study of social systems. Their criticism was addressed predominantly at the complexity of social systems developed by Luhmann (1995, 2012, 2013). The universalisation of *autopoiesis* in the Gaia studies and in the Chthulucene coerces the readers to conduct more comprehensive research allowing them to understand the subtleties of its meaning. *Sympoiesis* has a broader meaning, but, unfortunately, it has not been studied more profoundly than in the initial work of Dempster (1998) and in the works of Haraway quoted above. Therefore, a deeper understanding of this term is also needed in the studies of the Chthulucene and its consequences.

The above list can be supplemented with other normative ideas which are only mentioned here: the role of complexity in the Anthropocene-related legal regulations

(Kotzé 2020; Leach 2021; Petersmann 2021) and relations between complexity, the Anthropocene and the normative character of sustainability. The key issue in these works is the consequences of applications of complex systems models for studying the environmental consequences of the Anthropocene. It is challenging to separate such models from the equally complex socio-political contexts in which they are made and those to which they are expected to be applied. This results from the fact that the Anthropocene is an interface of two complex systems. Developing environmental norms for two interacting complex systems or a Unitarian global system poses significant challenges in law-making.

3.4.5 The Anthropocene, Complexity and Action-Oriented Approaches

3.4.5.1 The Anthropocene as an Action-Oriented Idea

The normative sense of the Anthropocene provides ground for the development of action-oriented concepts indicating how to impede, terminate, and perhaps reverse the processes which led to the Anthropocene. Obviously, the detailed relations between specific normative approaches to the Anthropocene and action-oriented recommendations need be studied separately.

The action-oriented sense of the Anthropocene has already been expressed in the first text of Crutzen and Stoermer (2000: 17): "to develop a worldwide accepted strategy leading to sustainability of ecosystems against human induced stresses will be one of the great future tasks of mankind."

Similarly, complexity-related mathematical models, analogies, and metaphors have become indispensable, if not the core, elements of the language of studies of the Anthropocene, including action-oriented considerations. Multiple references in this chapter confirm this universal assertion (Schellnhuber et al. 2004; Schellnhuber 2006; Steffen 2004, 2010; Steffen et al. 2011, 2018).

The term 'governance' is used as a universal idea in action-oriented approaches to environmental problems, including the enhancement of sustainability capabilities and resilience. Adequate governance is also viewed as the main factor in dealing with the threats caused by the Anthropocene (Biermann 2014).

Governance has multiple meanings, and in all cases of action-oriented approaches to governance, the meaning of this term has to be explained separately. This English language archaism (and, at the same time, neologism in some modern contexts) has become a dominant term in multiple areas. In business management, corporate governance initially described the relations between shareholders and companies (shareholders company), and later, it was expanded as to describe the relations between all a company's stakeholders (stakeholders company). In politics, governance was interpreted as a new comprehensive pattern of political ruling, including more empowerment to all political actors (Kooiman 2003).

It is also evident that complexity-related research can be treated as a foundation for action-oriented and regulatory approaches to the Anthropocene. Three examples are presented: the Anthropocene as a 'wicked problem' and its governance, complexity-based governance in the era of the Anthropocene, and the Viable System Model of Stafford Beer as a basis for sustainability governance in the era of the Anthropocene. These ideas are described here in a synthetic manner and, as in other cases, the description can be used as an invitation to and encouragement for more profound studies.

3.4.5.2 The Anthropocene, Complexity and Action-Oriented Approaches

Wicked Problems and the Anthropocene

The term 'wicked problems' was described in detail for the first time in the paper on social policy (urban problems) published by Rittel and Webber (1973). It must be added that the term 'wicked' has no ethical connotation but reflects an emotionally-laden perception of difficulty. Rittel and Webber (1973: 160) concluded that: "… wicked problems stand in opposition with 'tame' or 'benign' problems which occur in mathematics – solving a mathematical problem or identifying a structure in organic chemistry. The solution is found sooner or later or it is waiting for a genius. The wicked problems are not solvable at all in this manner".

The wicked problems, constituting a challenge in all kinds of collective human activities, at all levels of societal hierarchy, can be seen as representative of difficulties associated with all interpretations of complexity. It can even be argued that at the time when they were invented (described) by Rittel and Webber almost 50 years ago, they already expressed the growing helplessness of social theory and practice facing the increasing complexity of social, economic, and sociotechnical systems. The state of the present knowledge of complex systems allows for concluding that 'wickedness' is a common feature of the majority of social systems at all scales of societal hierarchy. It means that wicked problems are just another description of complexity. This conclusion diminishes the cognitive and methodological value of this concept, making it just another set of characteristics of complexity or another interpretation of complexity. This observation means that the attempts to develop methods which allow coping with wickedness can be helpful, but their usefulness is rather limited.

From a systemic perspective, the phenomena associated with the Anthropocene are to a large extent wicked problems. The coupling of societal and natural systems in the Anthropocene contributes to multiple cases of this 'wickedness'. This observation is expressed in multiple writings on the Anthropocene (Smith-Nonini 2017; Snick 2020). Sometimes, climate change (Lehtonen et al. 2019) is even called a 'super wicked' problem. Extending this observation to the Anthropocene is self-evident.

This preliminary analysis is focused on the mutual relations between the perception of the Anthropocene as a wicked concept/system and applications of complexity-related ideas. The references to the wickedness of the Anthropocene can be preliminarily divided into two groups. The first group refers to wickedness in very general terms.

Snick (2020) recalls the visions of the collapsing complex world of Diamond (2004) and shows two characteristics of the wickedness of the Anthropocene. A broadly defined non-linearity and insufficient understanding of the Anthropocene by the modern techno-scientific system is not able to capture the systemic interdependence and complex dynamics of modern life. The complexity and wickedness of the Anthropocene also constitute an inspiration for artistic endeavours. Lonie (2022: 290) treats the wickedness of the Anthropocene as a premise for artistic exhibition and speaks of "our understanding of the Anthropocene as a wicked, complex network of present and emerging anthropogenic transformations of the ecosystems of the planet from which human values and decisions could not be separated". In this broad qualitative interpretation, wickedness is determined by entanglement, proposed by Haraway (2016, 2016a) as the determinant of overall interdependence determining the complexity of the Chthulucene.

In the second group of interpretations of the Anthropocene as the cause and effect of wicked problems, attention is paid equally to description and attempts to explain, and to showing the actions necessary to deal with, that wickedness (Ison 2016; Ison et al. 2018; Ison/Straw 2020; Lehtonen et al. 2019; Craig 2020).

Without quoting the ten characteristics of the wicked problems formulated by Rittel and Webber (1973), an attempt is made here to show some fundamental characteristics of the Anthropocene contributing to its 'wickedness'. Subsequently, some proposals on how to deal with the wickedness of the Anthropocene are presented. First and foremost, it must be underlined that all experiences deriving from applications of complexity-related ideas in human problem solving – and, yes, common sense – show that any attempt to find comprehensive or universal solutions to the wicked problems are unachievable.

A comprehensive approach to governing wicked problems associated with the Anthropocene and sustainability was proposed by Ison et al. (2018) and by Ison and Straw (2020). The authors describe a governance deficit deriving from the Anthropocene-related problems and propose how cyber-systemic thinking/practices can enhance the empowerment and creativity that underpins sustainability science.

The following definition of governance relevant to the challenges of the Anthropocene, its complexity, and its sustainability was applied by Ison et al. (2018: 1212):

> Governance is a significant expansion, broader than management, encompassing the totality of mechanisms and instruments available for directing and influencing society, including the entire cycles of adaptive planning, designing, regulating, legislating, budgeting and managing. Governance is not an abstraction; it is something that is done, enacted in theory-informed and context-specific ways that embed ideologies and power relations.

These interpretations are not the only ones, but they are sufficient for this chapter. In the most general sense, the governance capturing the relational dynamics between social and biophysical systems requiring systemic sensibility, combined with systems literacy, is described as a three-dimensional "governance diamond" (Ison 2016). The 'governance diamond' embodies the following elements: civil society, the techno-sphere, the biosphere, the state, the judiciary/law, the private sector, and the media, which are all contributing to the social purpose.

According to the results deriving from theoretical assumptions and case studies described by Ison et al. (2018) and Ison and Straw (2020), the key concept which allows dealing with the wicked character of interacting and coevolving social and biophysical systems in the Anthropocene is the structural coupling which occurs when two or more systems in recursive interactions defined by the properties of their components undergo congruent structural changes or mutual adaptation (Maturana/ Varela 1980). According to Ison et al. (2018; Ison/Straw 2020), structural coupling concepts are central to both Anthropocene governance and sustainability science and are the basis of placing 'the biosphere' at the centre of the Ison's three-dimensional governance diamond mentioned above.

Interpretation of Relations Between Governance, Complexity, Sustainability, and the Anthropocene

Young (2017) proposed an environmental governance rooted in complex systems studies closer to the systems thinking of von Bertalanffy and Karl Deutsch, and quantitative models of complexity science, as well as complexity science-based metaphors (e.g. Deutsch 1963; Janssen 2002; Mitchell 2009; Scheffer 2009). This interpretation of governance in the era of the Anthropocene has associated the ideas linking complexity with global system governance, sustainability, the Anthropocene, and, in a more general sense, with all environmental considerations (Schellnhuber 1998, 1999, 2006; Steffen 2004, 2010; Steffen et al. 2011, 2018; Biermann 2014; Erdelen/ Richardson 2019). The ideas of Young were developed in the long-term international research programme the "Earth System Governance Project" (Biermann 2014; Young 2017).

The approach developed by Young (2017) embodies two areas. First, the analysis of systems, and especially the coupled systems, that arise in interactions between humans and biophysical settings. Second, the study of the social institutions (both formal and informal) steering interactions among themselves and between themselves and the biophysical environment.

Referring to the various understandings of governance in a universal sense of global governance, Young has defined several ideas connected directly and indirectly with governance, institutions, and the environment. The definition of the most important concept, i.e., the governance system, was formulated as follows: "A governance system is an ensemble of elements performing the function of governance in a given setting. Institutional arrangements form the core of such a system, but the ensemble normally includes cognitive, cultural, and technological elements as well" (Young 2017, Kindle Edition, location 576). It must be added that in this definition the coupling of the institutional factors with other factors plays a decisive role. It allows for better identification of systemic properties.

The complexity of broadly defined social and natural systems is viewed by Young (2017) as the foundation of environmental governance, research on the Anthropocene, and sustainability. Interconnectedness between rank factors creates challenges to those seeking to design and operate governance arrangements capable of producing sustainable results on a large scale: (i) connectivity or tight coupling among system components, (ii) thresholds, triggers, and nonlinear patterns of change, (iii) dynamic and directional processes, and (iv) emergent properties and the frequency of surprises.

Young has used these factors in studying various combinations of such issues as inter-relations between natural systems and social systems, relations between institutions and individual actors' behaviour and cognition, the dynamics of the Anthropocene, achieving sustainability, and transition to sustainability.

Treating the above initial synthetic description of the ideas as an invitation and encouragement to further studies of Young's concept, the following comments can be made. The proposal of developing and studying the ideas associated with environmental studies, with the use of a wide variety of complexity-related ideas, can be seen as a continuation and extension of earlier studies of such authors. The concepts of Young relating to the environmental governance system are well-grounded, both in global environmental studies and in complexity-related research. Additionally, this approach has been enriched by the ideas concerning the protection of the natural environment elaborated by Ostrom (1997, 2009) in the economic and institutional theory framework. As a result, the author has created carefully-defined conceptual foundations for studying global governance, sustainability, and the Anthropocene with a more precisely-defined conceptual framework. Subsequently, this framework was used for building a comprehensive model of governance of environmental issues in the Anthropocene, including various actors – governmental and non-governmental. Elaborating such a framework without the ideas (language) deriving from complexity-related research would not be possible.

Complexity, Sustainability, and the Viable System Model of Stafford Beer

The last action-oriented example of the links between complexity-related research and sustainability (directly) and the Anthropocene (indirectly) was developed by Espinosa and Walker (2017). They have applied the concept of the *Viable System Model* (VSM) of Stafford Beer.

The authors have applied two areas of complexity-related ideas: broadly defined complexity studies referring to management and to the complexity of social systems, and the then-universal complexity management cybernetics of Beer (1970, 1972, 1979, 1981).

The essential points of Espinosa and Walker's book concern the sources of complexity-related inspiration in management and in the social sciences, the meaning of the governance of sustainability, and the applications of the VSM in studies of sustainability. The book reflects state-of-the-art knowledge concerning complexity studies, predominantly in management. It results from the professional background of both authors. A series of publications devoted to complexity and management and complexity and social studies are grouped into four categories: (a) introduction, (b) foundations, (c) applications, and (d) reflections. They are illustrated with relevant sources available on the date of publication of the book. As examples, the following ideas are quoted: non-equilibrium thermodynamics, networks science, Boolean networks, *Complex Adaptive Systems* (CAS) and the NK model, chaos theory, and catastrophe theory. Artificial life, cybernetics and the particular part developed by Beer – management cybernetics (organisational cybernetics) – is the second source of inspiration presented in the book (Espinosa/Walker 2017: 16–22). Reference is made to the pioneering group of early cyberneticians: Norbert Wiener, Arturo Rosenblueth, Warren McCulloch, W. Ross Ashby, John von Neumann, Heinz von Foerster, Margaret Mead, and Gregory Bateson. The special role of the pioneering ideas of Bateson (1973, 1980) concerning ecology is underlined.

The study of the role of complexity-related ideas in governance and sustainability constitutes one of the key novel elements of this book. Espinosa and Walker's extended interpretation of sustainability embodies the following elements presented here in a significantly abbreviated way: the meta-systemic level, "organism in its environment" as the basic unit of analysis, sustainability as a co-evolutionary process, the negotiated and dynamic meaning of sustainability, and viable recursive systems as the key to understanding entangled hierarchical complexity (Espinosa/Walker 2017: 41–43). In the case of recursiveness, taken from Beer's VSM (1979), the Gaia concept is treated as an additional source of inspiration (Lovelock 1979).

Following the above considerations, the authors develop new interpretations of governance. They treat the ideas of corporate governance as the point of departure. They also refer to more universal interpretations of governance (Kooiman 2003). However, the authors observe that, at the time of writing, the vision of sustainability governance is not yet clear, although it has to be taken into account that bottom-up self-managed forms of governance have maintained the health of land and sea and forests for thousands of years (Ostrom 1990).

This initial introductory description is treated as an invitation and encouragement for a study of the applications of Stafford Beer's concept of the *Viable System Model* (VSM) and other cybernetical and complexity-related ideas in studying sustainability. In the remaining part of the book, studies of sustainability at the corporate level, the sustainable university, and a review of various systemic approaches and their potential contribution to environmental management are presented.

Summarising the applications of the VSM proposed by Espinosa and Walker and other authors referring to the VSM in studying sustainability and other environmental problems, it must be added that the ideas of Stafford Beer (and not only the VSM) have to a large extent not lost their currency and deserve more attention from a broader audience.

3.5 Conclusions and Further Research

3.5.1 Conclusions

It is very demanding, if possible at all, to present a coherent set of conclusions after a broad survey of literature on the links between complexity-related research and the Anthropocene.

As has been underlined in Sect. 3.1, the chapter is focused on epistemological and methodological aspects of applications of complexity-related ideas in the discussions on the Anthropocene, and on their assessment. This evaluation must provide additional arguments supporting the validity of the Anthropocene in politics, policy and polity in all areas of broadly defined social life at all levels of the societal hierarchy. Additionally, the specificity of the Anthropocene, i.e., the relations between natural systems and societal systems, or, in other words, their coupling, has been given special attention. Therefore, the conclusion and recommendations do not refer directly to the specific physical and social aspects of the Anthropocene. However, improved methodology always contributes to practice – identification, explanation, and setting the norms and actions. The survey developed in the chapter allows us to formulate the following conclusions.

First, as was already reiterated in the study, the number and scope of the works linking complexity-related research and the Anthropocene makes any comprehensive studies impossible. The reason for choosing the topic of the chapter is to develop an introductory survey giving ground for further, more advanced, and ordered research.

Second, the studies of the Anthropocene are affected by a new phenomenon embodying content and linguistic aspects. It is obvious that the consequences of the Anthropocene affect and will affect all areas of human life. Therefore, it is understandable, and it is easy to observe, that all areas of human intellectual activity, not only the artistic and theological ones, can be engaged in the discussion of the Anthropocene.

The same phenomenon, although on a different scale, can be observed in the studies of the complexity of natural and social systems. As has been shown, the number and variety of interpretations of 'complexity' is also tremendous, although less overwhelming than in the case of 'the Anthropocene'.

Looking at this phenomenon from a linguistic perspective leads to the observation that we have to deal with two terms with a very wide scope of meaning. Additionally, complexity is polysemous. The consequences of such a phenomenon are not new. Bearing in mind the fact that social systems are social constructs, and natural systems are also constructed (at least partly) by the observers, and adding to this the overabundance of information, it can be concluded that in the studies of the Anthropocene and of its relations with complexity, a special linguistic and hermeneutic caution must be applied. This also concerns all studies of the Anthropocene. Of course, it is not possible to make attempts to define everything, but at the same time, when writing about ideas with such a broad scope of meaning, a more careful approach is necessary.

Third, it must be concluded that the impact of complexity-related ideas on the Anthropocene, and, in a wider sense, on the discussions about environmental problems, is affected by similar weaknesses to those in other areas – economics, sociology, security studies, International Relations, and business management. There is only one very significant factor distinguishing environmental studies, i.e., the role of mathematical modelling. In the studies of social systems alone, the impact of constructivism is more significant. In the case of the Anthropocene and in the research on the environment, mathematical modelling of natural phenomena is less exposed to the distortions caused by constructivism: the parameters describing natural systems are more objective. In this case, however, the weaknesses and limitations of operationalisation and operationism (mentioned and distinguished purposively in this chapter) must also be taken into account.

Fourth, as it was mentioned earlier, the narratives applying analogies and metaphors taken from the 'hard' complexity studies can have different functions. Focusing attention on metaphors, it must be remembered that the mathematical complexity-related models are treated as source domains of conceptual and deliberate metaphors. They may also have persuasive and performative functions. This phenomenon has critical cognitive, normative, and policy-oriented consequences.

Fifth, the narratives based on such complexity-related ideas, for example, *autopoiesis* and *sympoiesis*, play a similar role. Although their persuasive meaning is limited – they are more subjective, although it must be borne in mind that the ideas of *autopoiesis* can be operationalised.

Sixth, it can be observed and hypothesised that the development of a broadly-defined language of the studies of the environment, including the Anthropocene, remains strongly influenced by complexity-related ideas. Without repeating all the Earth System vocabulary, such terms as emergence, feedback, self-organisation, chaos, the edge of chaos, and many others can be mentioned; and without delving into deep philosophical deliberations, it must be added that in this study, mathematics is treated as another language, not the essence of being.

Seventh, it must be remembered that the 'complexity language' enhances the persuasive capability of the areas where it is applied. It has an additional 'scientific' appeal as well as an emotional message: "it's chaos, we know the mathematical models, and we can help in preventing its emergence and consequences". It must be added that this phenomenon is commonly known in other areas, e.g., security studies or management, where the complexity specialist may enhance their persuasive capability by using the ideas from broadly-defined complexity research. This conclusion is not critical, but reflects the consequences of applications of complexity-related terms in the language of social practice.

The last conclusion, corresponding with the aim of the study, concerns the impact of the applications of complexity-related ideas in the studies of the Anthropocene. Taking into account all the results of the survey and assessment, it can be concluded that epistemologically- and methodologically-correct applications of complexity-related ideas significantly enhance the validity of the Anthropocene as the foundation of environmental politics, policy and polity. However, it must be added that the increased significance of the Anthropocene in environment-oriented politics, policy and polity can be supported by complexity-related ideas only when those ideas are sufficiently profoundly known. The necessary depth of understanding of complexity-related ideas constitutes a significant challenge for theory, policy and politics.

3.5.2 Further Research

The directions for further research should embody the following issues.

First, it is necessary to prepare more comprehensive studies of literature on the relations, or even better, the mutual impact of complexity research and environmental studies. This chapter is designed as a kind of guide and an initial ordering of the research area. The Anthropocene has been used as an example, or as is fashionable, a case study. At present, it is possible to merge the following factors: (a) the current perspective on the discussion of the idea of complexity *per se*; (b) the knowledge about the environment; (c) increased availability of information, and (d) availability of meta-analyses, like, for example, Systematic Literature Review (SLR) and similar methods.

Second, research on the links between complexity-related mathematical models associated directly and indirectly with the Anthropocene should embody more profound data acquisition and assessment studies. The processes of operationalisation should be given additional attention, guaranteeing the quality and verifiability of data.

Third, the deepened studies on the links between complexity-related ideas and the Anthropocene should embody more profound hermeneutic considerations allowing for a better understanding of the processes of intersubjective construction of the ideas associated with the Anthropocene.

Fourth, the practical role of a deepened knowledge of the role of complexity-related ideas in the studies of the Anthropocene should be studied, with the use of more advanced methods of data gathering and analysis. This would lead to a significant strengthening of the impact of the Anthropocene in politics, policy and polity in all areas of social life.

Fifth, this chapter is focused on the Anthropocene. Similar studies on the interrelationships between complexity-related ideas and sustainability, transition to sustainability, and resilience appear to be a very promising and important area of research.

The chapter presents the results of the Project no. 049/ZZP/2023/POT financed from the subsidy granted to the Cracow University of Economics.

References

Anthropocene Working Group (AWG), 2019: *Subcommission on Quaternary Stratigraphy, Results of binding vote by AWG Released 21st May 2019*, at: http://quaternary.stratigraphy.org/working-groups/anthropocene/ (23 November 2022).

Ashby, W. R., 1957: *An Introduction to Cybernetics* (London: Chapman & Hall).

Austin, J. L., 1962: *How to do things with words* (Oxford: Clarendon).

Baccini, P.; Brunner, P. H., 2012: *Metabolism of the Anthroposphere: Analysis, Evaluation, Design*, 2nd Edition (Cambridge, MA: The MIT Press).

Bar-Yam, Y., 1997: *Dynamics of Complex Systems* (Reading, MA: Addison-Wesley).

Bateson, G., 1973: *Steps to an Ecology of Mind* (London: Jason Aronson Inc.).

Bateson, G., 1980: *Mind and Nature a Necessary Unity* (New York: Dutton).

Beer, S., 1970: "Managing Modern Complexity", in: *Futures*, 2, 3: 245–257.

Beer, S., 1972: *Brain of the firm* (London: Allen Lane: The Penguin Press).

Beer, S., 1979: *The Heart of the Enterprise* (London - New York: John Wiley & Sons).

Beer, S., 1981: *Brain of the Firm*, 2nd edition (extended) (London - New York: John Wiley & Sons).

Bergmann, S., 2020: "Theology in the Anthropocene - and Beyond?", in: Bergmann, S.; Vähäkangas, M. (Eds.): *Contextual Theology: Skills and Practices of Liberating Faith* (London: Routledge): 160–180.

Biermann, F., 2014: *Earth System Governance: World Politics in the Anthropocene* (Cambridge, MA: The MIT Press).

Biggiero, L., 2001: "Sources of Complexity", in: *Human Systems. Nonlinear Dynamics, Psychology and Life Sciences*, 5, 1: 3–19; at: https://doi.org/10.1023/A:1009515211632.

Bonneuil, C.; Fressoz J.-B., 2016: *The Shock of the Anthropocene* (London: Verso, Kindle Edition).

Boulton, J, G.; Allen, P. M; Bowman, C., 2015: *Embracing Complexity* (Oxford: Oxford University Press).

Braidoti, R., 2013: *The Posthuman* (Cambridge: Polity).

Brauch, H. G., 2009: "Securitizing Global Environmental Change", in: Brauch, H. G.; Oswald Spring, Ú.; Grin, J.; Mesjasz, C.; Kameri-Mbote, P.; Behera, N. C.; Chourou, B.; Krummenacher, H. (Eds.), 2009: *Facing Global Environmental Change: Environmental, Human, Energy, Food, Health and Water Security Concepts* (Berlin—Heidelberg—New York: Springer-Verlag): 65–102.

Brauch, H, G., 2021: "Peace Ecology in the Anthropocene", in: Oswald Spring, Ú.; Brauch, H. G. (Eds.): *Decolonising Conflicts, Security, Peace, Gender, Environment and Development in the Anthropocene* (Cham: Springer Nature Switzerland): 51–186.

Brauch, H. G., 2021a: "The Anthropocene Concept in the Natural and Social Sciences, the Humanities and Law – A Bibliometric Analysis and a Qualitative Interpretation (2000–2020)", in: Benner, S.; Lax, G.; Crutzen, P. J.; Pöschl, U.; Lelieveld, J.; Brauch H. G. (Eds.), *Paul J. Crutzen and the Anthropocene: A New Epoch in Earth's History* (Cham: Springer Nature Switzerland): 289–438.

Brauch, H. G.; Oswald Spring, Ú.; Grin, J.; Scheffran, J. (Eds.) (2016): *Handbook on Sustainability Transition and Sustainable Peace* (Cham: Springer International Publishing Switzerland).

Bridgman, P. W., 1959: "P. W. Bridgman's: 'The Logic of Modern Physics' after Thirty Years", in: *Daedalus*, 88, 3: 518–526; at: http://www.jstor.org/stable/20026519 (8 May 2021).

Campbell, D. K., 1987: "Nonlinear Science from Paradigms to Practicalities", in: *Los Alamos Science Special Issue*, 15: 218–262.

Castellani, B.; Gerrits, L., 2021: *Map of Complexity Sciences. Art & Science Factory*; at: https://www.art-sciencefactory.com/complexity-map_feb09.html (17 November 2021).

Checkland, P., 1999: *Systems Thinking, Systems Practice* (Chichester: John Wiley & Sons).

Chwałczyk, F., 2020: "Around the Anthropocene in Eighty Names—Considering the Urbanocene Proposition", in: *Sustainability*, 12, 11: 4458; at: https://doi.org/10.3390/su12114458.

Cielemęcka O., Daigle, C., 2019: "Posthuman Sustainability: An Ethos for our Anthropocenic Future", in: *Theory, Culture & Society*, 36, 7–8: 67–87.

Clark, W. C., Crutzen, P. J., Schellnhuber, H. J., 2004: "Science for Global Sustainability. Toward a New Paradigm", in: Schellnhuber, H. J.; Crutzen, P. J.; Clark, W. C.; Claussen, M.; Held, H. (Eds.): *Earth System Analysis for Sustainability. Dahlem Workshop Reports* (Cambridge, MA - Berlin: The MIT Press in cooperation with Dahlem University Press): 1–28.

Clarke, B., 2020: *Gaian Systems: Lynn Margulis, Neocybernetics, and the End of the Anthropocene*, Vol. 60 (Posthumanities) (Minneapolis: University of Minnesota Press, Kindle Edition).

Claussen, M. et al., 2002: "Earth System Models of Intermediate Complexity: Closing the Gap in the Spectrum of Climate System Models", in: *Climate Dynamics*, 18: 579–586.

Cowan, G.; Pines, D.; Meltzer, D. (Eds.), 1994: *Complexity: Metaphors, Models, and Reality. Santa Fe Institute. Studies in the Sciences of Complexity, Proceedings* (Reading, MA: Addison-Wesley).

Craig, R. K., 2020: Resilience Theory and Wicked Problems, in: *Vanderbilt Law Review*, 73, 6: 1733–1775, *University of Utah College of Law Research Paper No. 376*; at: https://doi.org/10.2139/ssrn.3610420.

Crist, E., 2016: "On the Poverty of Our Nomenclature", in: Moore, J. M. (Ed.): *Anthropocene or Capitalocene? Nature, History, and the Crisis of Capitalism* (Oakland, CA: PM Press): 14–33.

Crutzen, P. J., 2002: "Geology of mankind – The Anthropocene", in: *Nature*, 415: 23.

Crutzen, P. J., 2006: The "Anthropocene", in: Ehlers, E.; Krafft, T. (Eds.): *Earth System Science in the Anthropocene: Emerging Issues and Problems* (Heidelberg: Springer): 13–18.

Crutzen, P. J., 2010: "Anthropocene Man", in: *Nature*, 467: S10.

Crutzen, P. J.; Schwägerl, C., 2011: "Living in the Anthropocene: Toward a New Global Ethos", in: *Yale Environment* 360, 24 January; at: https://e360.yale.edu/features/living_in_the_anthropocene_toward_a_new_global_ethos (28 November 2022).

Crutzen, P. J.; Stoermer, E. F., 2000: "The Anthropocene", in: *IGBP Newsletter*, 41: 17–18.

Dahan A., 2010: "Putting the Earth System in a Numerical Box? The Evolution from Climate Modelling toward Global Change", in: *Studies in History and Philosophy of Science Part B: Studies in History and Philosophy of Modern Physics*, 41: 282–292.

Dalby, S., 2016: "Framing the Anthropocene: The Good, the Bad and the Ugly", in: *The Anthropocene Review*, 3, 1: 33–51.

Davis, J.; Moulton, A. A.; Van Sant, L.; Williams, B., 2019: "Anthropocene, Capitalocene, … Plantationocene?: A Manifesto for Ecological Justice in an Age of Global Crises", in: *Geography Compass*, 13, 5: e12438, https://doi.org/10.1111/gec3.12438.

Demos, T. J., 2017: *Against the Anthropocene Visual Culture and Environment Today* (Berlin: Sternberg Press).

Dempster, M. B., 1998: *A Self-Organizing Systems Perspective on Planning for Sustainability* (Waterloo: MSc thesis, University of Waterloo, Department of Environmental Studies); at: https://citeseerx.ist.psu.edu/document?repid=rep1&type=pdf&doi=8207c73dfb568eae136c3ab9d8d549a3379f402a (18 November 2024).

Deutsch, K. W., 1963: *The Nerves of Government: Models of Political Communication and Control* (New York: The Free Press).

Diamond, J., 2004: *Collapse. How Societies Choose to Fail or Succeed* (London: Allen Lane).

Di Chiro, G., 2017: "Welcome to the White (M)Anthropocene? A Feminist-Environmental Critique", in: MacGregor, S. (Ed.): *Routledge Handbook of Gender and Environment* (London: Routledge): 487–505.

Ehlers, E.; Krafft, T., (Eds.) 2006: *Earth System Science in the Anthropocene: Emerging Issues and Problems* (Berlin-Heidelberg: Springer, Kindle Edition).

Ehlers, E.; Krafft, T., 2006a: "Managing Global Change: Earth System Science in the Anthropocene," in: Ehlers, E.; Krafft, T., (Eds.): *Earth System Science in the Anthropocene: Emerging Issues and Problems* (Berlin-Heidelberg: Springer, Kindle Edition): 5–12.

Erdelen, W. R.; Richardson, J. G., 2019: *Managing Complexity: Earth Systems and Strategies for the Future* (Abingdon – New York: Routledge).

Espinosa, A.; Walker, J., 2017: *A Complexity Approach to Sustainability: Theory and Application* (London: World Scientific Press).

Forrester, J. W., 1971: *World Dynamics* (Cambridge, MA: Wright-Allen Press).
Gaffney, O.; Steffen, W., 2017: "The Anthropocene Equation", in: *The Anthropocene Review*, 4, 1: 53–61.
Gaffney, O.; Steffen, W., 2017a: "Response to Heijungs et al.", in: *The Anthropocene Review*, 4, 3: 264–265.
Gell-Mann, M., 1994: *The Quark and the Jaguar: Adventures in the Simple and the Complex* (New York: W.H. Freeman and Company).
Georgescu-Roegen, N., 1971: *The Entropy Law and the Economic Process* (Cambridge, MA: Harvard University Press).
Gleditsch, N. P., 2003: "Environmental Conflict: Neomalthusians vs. Cornucopians", in: Brauch, H. G.; Liotta, P. H.; Marquina, A.; Rogers, P.; Selim, M. El-Sayed (Eds.): *Security and Environment in the Mediterranean. Conceptualising Security and Environmental Conflicts* (Berlin-Heidelberg: Springer): 477–485.
Gleick, J., 1987: *Chaos: Making of a New Science* (New York: Viking Press).
Gleick, J., 2011: *Chaos: Making a New Science. Enhanced Edition* (Open Road. Integrated Media, Kindle Edition).
Glikson, A. Y., 2017: *Plutocene: Blueprints for a Post-Anthropocene Greenhouse Earth* (Cham: Springer International Publishing).
Hamilton, C.; Grinevalt, J., 2015: "Was the Anthropocene anticipated?" in: *The Anthropocene Review*. 2, 1: 59–72; at: https://doi.org/10.1177/2053019614567155.
Harari, Y. N., 2017: *Homo Deus (A Brief History)* (New York: HarperCollins, Kindle Edition).
Haraway, D., 2015: "Anthropocene, Capitalocene, Plantationocene, Chthulucene: Making Kin", in: *Environmental Humanities*, 6: 159–165.
Haraway, D. J., 2016: *Staying with the Trouble* (Experimental Futures) (Durham, NC: Duke University Press, Kindle Edition).
Haraway, D. J., 2016a: "Staying with Trouble: Anthropocene, Capitalocene, Chthulucene", in: Moore, J. M. (Ed.): *Anthropocene or Capitalocene? Nature, History, and the Crisis of Capitalism* (Oakland, CA: PM Press): 34–76.
Hartley, D., 2016: Anthropocene, Capitalocene, and the Problem of Culture, in: Moore, J. M. (Ed.): *Anthropocene or Capitalocene? Nature, History, and the Crisis of Capitalism* (Oakland, CA: PM Press): 154–165.
Heijungs, R.; Boersema, J. J.; Huppes, G., 2017: "Mathematics and the Anthropocene Equation: Comment on Gaffney, O. and Steffen, W., 2017: The Anthropocene Equation, The Anthropocene Review 4, 1: 53–61", in: *The Anthropocene Review 2017*, 4, 3: 259–263.
Hermann-Pillath, C., 2017: *Economics of the Anthropocene: An Exploratory Essay* (August 23); at: https://ssrn.com/abstract=3024511, https://doi.org/10.2139/ssrn.3024511.
Heymann, M.; Dahan Dalmedico, A., 2019: "Epistemology and Politics in Earth System Modeling: Historical Perspectives", in: *Journal of Advances in Modeling Earth Systems*, 11: 1139–1152; at: https://doi.org/10.1029/2018MS001526.
Holland, J. H., 1995: *Hidden Order. How Adaptation Builds Complexity* (New York: Basic Books).
Holland, J. H., 1998: *Emergence: From Chaos to Order* (Reading, MA: Addison-Wesley).
Homer-Dixon, T.; Walker, B.; Bigg, R.; Crépin, A.-S.; Folke, C.; Lambin, E. F.; Peterson, G. D.; Rockström, J., Scheffer, M.; Steffen, W.; Troell, M., 2015: "Synchronous Failure: The Emerging Causal Architecture of Global Crisis", in: *Ecology and Society*, 20, 3: 6; at: https://doi.org/10.5751/ES-07681-200306.
Hornborg, A., 2015: "The Political Ecology of the Technocene. Uncovering Ecologically Unequal Exchange the World-System", in: Hamilton, C.; Bonneuil, C.; Gemenne, F. (Eds.): *The Anthropocene and Global Environmental Crisis. Rethinking Modernity in a New Epoch* (London-New York: Routledge): 57–69.
Ison, R., 2016: "Governing in the Anthropocene: What Future Systems Thinking in Practice?", in: *Systems Research and Behavioral Science*, 33, 5: 595–613; at: https://doi.org/10.1002/sres.2421.
Ison, R.; Straw, E., 2020: *The Hidden Power of Systems Thinking* (Abington – New York: Routledge).

Ison, R.; Alexandra, J.; Wallis, P., 2018: "Governing in the Anthropocene: Are There Cyber-systemic Antidotes to the Malaise of Modern Governance?", in: *Sustainability Science*, 13: 1209–1223; at: https://doi.org/10.1007/s11625-018-0570-5.
Jackson, M. C., 2019: *Critical Systems Thinking and the Management of Complexity. Responsible Leadership for a Complex World* (Hoboken, NJ: John Wiley & Sons).
Janssen, M. A., (Ed.) 2002: *Complexity and Ecosystem Management: The Theory and Practice of Multi-Agent Systems* (Cheltenham, UK: Edward Elgar).
Kauffman, Stuart A., 1995: *At Home in the Universe. The Search for Laws of Self-Organization and Complexity* (New York – Oxford: Oxford University Press).
Kooiman, J., 2003: *Governing as Governance* (London: Sage Publications).
Kotzé, L. J., 2020: "Earth System Law for the Anthropocene: Rethinking Environmental Law Alongside the Earth System Metaphor", in: *Transnational Legal Theory*, 11, 1–2: 75–104.
Lakoff, G.; Johnson, M., 1980/1995: *Metaphors we live by* (Chicago: University of Chicago Press).
Langton, C. G., 1992: Life at the Edge of Chaos in: Langton, C. G.; Taylor, C.; Farmer, J. D.; Rasmussen, S. (Eds.): *Artificial Life II. Santa Fe Institute Studies in the Sciences of Complexity*, Proceedings, vol. 10 (Proceedings of second Artificial Life workshop, 1990, Redwood City, CA: Addison-Wesley): 41–91.
Latour, B., 2014: "Agency at the Time of the Anthropocene", in: *New Literary History*, 45: 1–18.
Leach, M. C., 2021: "Negotiating the Descriptive–Normative Frontier of Complexity Research in the Anthropocene", in: *Frontiers in Physics*, 9: Article 665727.
Leach, M. C., 2023: "Seeking 'Systems' in Earth System Law: Boundaries, Identity, and Purpose in an Emergent Field", in: *Earth System Governance*, 15: 100162; at: https://doi.org/10.1016/j.esg.2022.100162.
Lehtonen, A.; Salonen, A. O.; Cantell, H., 2019: "Climate Change Education: A New Approach for a World of Wicked Problems", in: Cook, J., *Sustainability, Human Well-Being, and the Future of Education* (Cham, Switzerland: Palgrave Macmillan): 339–374; at: https://doi.org/10.1007/978-3-319-78580-6.
Lewis, S. L.; Maslin, M. A., 2015: "Defining the Anthropocene", in: *Nature*, 519, 7542: 171–180.
Li, T.-Y.; Yorke, J. A., 1975: "Period Three Implies Chaos", in: *The American Mathematical Monthly*, 82, 10: 985–992.
Lloyd, S., 2001: "Measures of Complexity: A Nonexhaustive List", in: *IEEE Control Systems Magazine*, 21, 4: 7–8; at: https://doi.org/10.1109/MCS.2001.939938.
Lonie, B., 2022: "Art, Complexity and Sustainability in the Anthropocene. A Case Study", in: *Futures of Education, Culture & Nature – Learning to Become*, 1: 289–294; at: https://tidsskrift.dk/FECUN/article/view/130256/176024 (5 January 2023).
Lorenz, E. N., 1972: "Predictability: Does the Flap of a Butterfly's Wings in Brazil Set off a Tornado in Texas?", in: 139th annual meeting of the American Association for the Advancement of Science (29 December 1972), in: *Essence of Chaos* (1995), Appendix 1, 181.
Lorenz, E. N., 1995: *The Essence of Chaos* (Seattle: University of Washington Press).
Lovelock, J. E., 1972: "Gaia as Seen Through the Atmosphere", in: *Atmospheric Environment* (1967), 6, 8: 579–580.
Lovelock, J. E., 1979: *Gaia A New Look at Life on Earth* (Oxford: Oxford University Press).
Lovelock, J. E., 2020: *Novacene. The Coming Age of Hyperintelligence* (London: Penguin: Allen Lane, Kindle Edition).
Lovelock, J. E.; Margulis, L., 1974: Atmospheric Homeostasis by and for the Biosphere: the Gaia hypothesis, in: *Tellus*, 26: 1–2, 2–10; at: https://doi.org/10.3402/tellusa.v26i1-2.9731.
Luhmann, N., 1995: *Social Systems* (Stanford, CA: Stanford University Press).
Luhmann, N., 2012: *Theory of Society*, vol. I. (Stanford, CA: Stanford University Press).
Luhmann, N., 2013: *Introduction to Systems Theory* (Cambridge: Polity).
Margulis, L., 1981: *Symbiosis in Cell Evolution: Life and Its Environment on the Early Earth* (San Francisco: Freeman).
Margulis, L., 1993: *Symbiosis in Cell Evolution: Microbial Communities in the Archean and Proterozoic Eons*. 2nd edition (New York: Freeman).

Margulis, L., 1998: *Symbiotic Planet, a New Look at Evolution* (New York: Basic Books).

Maturana, H.; Varela, F., 1980: "Autopoiesis and Cognition. The Realisation of the Living", in: *Boston Studies in the Philosophy of Science*, # 42 (Boston, MA: Riedel Publishing).

McBrien, J., 2016: "Accumulating Extinction: Planetary Catastrophism in the Necrocene", in: Moore, J. M. (Ed.): *Anthropocene or Capitalocene? Nature, History, and the Crisis of Capitalism* (Oakland, CA: PM Press): 78–115.

McKelvey, B. (Ed.) 2012: *Complexity. Critical Concepts* (New York: Routledge).

Meadows, D. H.; Meadows, D. L.; Randers, J.; Behrens III, W. W., 1972: *The Limits to Growth* (New York: Universe Books).

Mesjasz, C.; Bartusik K.; Małkus, T.; Sołtysik, M., 2022: *Agile Project Management and Complexity. A Reappraisal* (London – New York: Routledge, Taylor & Francis Group).

Meyers, R. A. (Ed.) 2009: *Encyclopedia of Complexity and Systems Science*, 2009 (New York: Springer Science + Business Media, LLC); at: https://doi.org/10.1007/978-0-387-30440-3.

Miser, H. J.; Quade, E. S. (Eds.) 1985: *Handbook of Systems Analysis: Overview of Uses, Procedures, Applications, and Practice* (New York: North-Holland).

Mitchell, M., 2009: *Complexity: A Guided Tour* (Oxford: Oxford University Press).

Moore, J. M. (Ed.) 2016: *Anthropocene or Capitalocene? Nature, History, and the Crisis of Capitalism* (Oakland, CA: PM Press).

Moore, J. M., 2016a: "Introduction", in: Moore, J. M. (Ed.), *Anthropocene or Capitalocene? Nature, History, and the Crisis of Capitalism* (Oakland, CA: PM Press): 1–11.

Moore, J. M., 2016b: "The Rise of Cheap Nature", in: Moore, J. M. (Ed.), *Anthropocene or Capitalocene? Nature, History, and the Crisis of Capitalism* (Oakland, CA: PM Press): 78–115.

Myers, N.: 2017: "From the Anthropocene to the Planthroposcene: Designing Gardens for Plant/People Involution", in: *History and Anthropology* 28, 3: 1–5; at: https://doi.org/10.1080/02757206.2017.1289934.

NASA 2020: *Earth as System of Systems*; at: https://svs.gsfc.nasa.gov/31139 (7 December 2022).

NASA 2022: *About the Earth as a System: Background Information*; at: https://mynasadata.larc.nasa.gov/print/pdf/node/56 (7 December 2022).

Neugebauer H. I., 2006: "What about Complexity of Earth Systems?", in: Ehlers, E.; Krafft, T. (Eds.): *Earth System Science in the Anthropocene: Emerging Issues and Problems* (Berlin-Heidelberg: Springer, Kindle Edition): 27–38.

Norgaard, Richard B., 2013: "The Econocene and the Delta", in: *San Francisco Estuary and Watershed Science*, 11, 3: 1–5.

Ostrom, E., 1990: *Governing the Commons: The Evolution of Institutions for Collective Action*, Political Economy of Institutions and Decisions (Cambridge: Cambridge University Press).

Ostrom, E., 2007: "A Diagnostic Approach to Going Beyond Panaceas", in: *Proceedings of the National Academy of Sciences of the United States of America*; at: 104: 15181–15187.

Oswald Spring, Ú., 2017: "Development with Sustainable-Engendered Peace: A Challenge during the Anthropocene", Brauch, H. G.; Oswald Spring, Ú.; Grin, J.; Scheffran J. (Eds.): *Handbook on Sustainability Transition And Sustainable Peace* (Cham: Springer International Publishing): 161–185.

Oswald Spring, Ú.; Brauch, H. G.; Dalby, S., 2009: "Linking the Anthropocene, HUGE and HESP in a Fourth Phase of Human, Environmental and Gender Security and Peace Research", in: Brauch, H. G.; Oswald Spring, Ú.; Grin, J.; Mesjasz, C.; Kameri-Mbote, P.; Behera, N. C.; Chourou, B.; Krummenacher, H. (Eds.): *Facing Global Environmental Change: Environmental, Human, Energy, Food, Health and Water Security Concepts* (Berlin – Heidelberg – New York: Springer-Verlag): 1277–1294.

Oswald Spring, Ú.; Brauch, H. G.; Scheffran, J., 2017: "Sustainability Transition with Sustainable Peace: Key Messages and Scientific Outlook", in: Brauch, H. G.; Oswald Spring, Ú.; Grin, J.; Scheffran, J. (Eds.): *Handbook on Sustainability Transition and Sustainable Peace* (Cham – Heidelberg – New York – Dordrecht – London: Springer): 887–927.

Patel, R., 2013: "The Misanthropocene?", in: *Earth Island Journal*, 28, 1: 21–21.

Pauly, D., 2010: *5 Easy Pieces: The Impact of Fisheries on Marine Systems* (Washington, D.C.: Island Press).
Petersmann, M.-C., 2021: "Sympoietic Thinking and Earth System Law: The Earth, its Subjects and the Law", in: *Earth System Governance*, 9: 1–8 [100114]; at: https://doi.org/10.1016/j.esg.2021.100114.
Pirici, A.; Voinea R., 2015: *Manifesto For The Gynecene - Sketch For A New Geological Era*; at: https://ro.tranzit.org/en/lecture/0/2015-10-29/manifesto-for-the-gynecene---sketch-for-a-new-geological-era (11 November 2022).
Prigogine, I.; Stengers, I., 1984: *Order out of Chaos* (New York: Bantam).
Raworth, K., 2014: "Must the Anthropocene be a Manthropocene?", in: *The Guardian* (20 October).
Renn, J., 2020: *Evolution of Knowledge: Rethinking Science for the Anthropocene* (Princeton - Oxford: Princeton University Press).
Rescher, N., 1998: *Complexity: A philosophical overview* (New Brunswick, NJ: Transaction Publishers).
Revkin, A. C., 1992: *Global Warming. Understanding the Forecast* (New York: Abbeville Press).
Revkin, A. C., 2011: *Confronting the 'Anthropocene'*, in: *The New York Times*, May 11.
Rittel, H. W. J.; Webber, M. M., 1973: "Dilemmas in a general theory of planning", in: *Policy Sciences*, 4: 155–169; at: https://doi.org/10.1007/BF01405730.
Samways, M. J. 1999: "Translocating fauna to foreign lands: Here comes the Homogenocene", in: *Journal of Insect Conservation*, 3: 65–66.
Scheffer, M., 2009: *Critical Transitions in Nature and Society* (Princeton: Princeton University Press).
Scheffran, J.; Brzoska, M.; Brauch, H. G.; Link, P. M.; Schilling, J. (Eds.), 2012: *Climate Change, Human Security and Violent Conflict: Challenges for Societal Stability* (Berlin – Heidelberg - New York: Springer-Verlag).
Schellnhuber, H. J., 1998: "Earth System Analysis – The Scope of the Challenge", in: Schellnhuber, H. J.; Wenzel, V. (Eds.): *Earth System Analysis: Integrating Science for Sustainability* (Berlin/Heidelberg: Springer-Verlag): 3–195.
Schellnhuber, H. J. 1999: "'Earth System' Analysis and the Second Copernican Revolution", in: *Nature*, 402: C19–C23; at: https://doi.org/10.1038/35011515.
Schellnhuber, H. J.; Wenzel, V. (Eds.) 1998: *Earth System Analysis. Integrating Science for Sustainability* (Berlin – Heidelberg: Springer-Verlag).
Schellnhuber, H. J.; Crutzen, P. J.; Clark, W. C.; Claussen, M.; Held, H. (Eds.), 2004: *Earth Systems Analysis for Sustainability. Dahlem Workshop Reports* (Cambridge, MA: The MIT Press in cooperation with Berlin: Dahlem University Press).
Schellnhuber, H. J.; Cramer, W.; Nakícénovic, N.; Wigley, T.; Yohe, G. (Eds.), 2006: *Avoiding Dangerous Climate Change* (Cambridge: Cambridge University Press).
Searle, J. R., 1995: *The Construction of Social Reality* (New York: The Free Press).
Shoshitaishvili, B., 2022: "Is Our Planet Doubly Alive? Gaia, Globalization, and the Anthropocene's Planetary Superorganisms", in: *The Anthropocene Review*, Online First: 1–21; at: https://doi.org/10.1177/20530196221087789.
Sideris, L. H., 2015: "Science as Sacred Myth? Ecospirituality in the Anthropocene Age", in: *Journal for the Study of Religion, Nature and Culture*, 9, 2; 136–153.
Smith-Nonini, S., 2017: "Making Complexity Your Friend: Reframing Social Theory for the Anthropocene", in: *Weather, Climate, and Society*, 9, 4: 687–699.
Snick, A., 2020: Redefining Sustainability in the Anthropocene A Complexity-based Paradigm for Responsible Research and Regenerative Economics; at: https://www.researchgate.net/publication/338395824 (21 September 2022).
Snowden, D.; Goh, Z., 2021: *Cynefin - Weaving Sense-Making into the Fabric of Our World* (Singapore: Cognitive Edge Pte Ltd., Kindle edition).
Sopory, P.; Dillard, J. P., 2002: "The Persuasive Effects of Metaphor: A Meta-analysis", in: *Human Communication Research*, 28, 3: 382–419; at: https://doi.org/10.1111/j.1468-2958.2002.tb00813.x.

Stager, C., 2011: *Deep Future. The Next 100,000 Years of Life on Earth* (New York: Thomas Dunne Books).

Steen, G., 2017: "Deliberate metaphor theory: Basic assumptions, main tenets, urgent issues", in: *Intercultural Pragmatics*, 14, 1: 1–24.

Steffen W, 2004: "Group Report: Earth Systems Dynamics in the Anthropocene", in: Schellnhuber, H. J.; Crutzen, P. J; Clark, W. C.; et al. (Eds.): *Earth System Analysis for Sustainability* (Cambridge, MA - Berlin: The MIT Press – Freie Universität Berlin): 313–340.

Steffen, W.; Sanderson, R. A.; Tyson, P. D.; Jäger, J.; Matson, P. A.; Moore III, B.; Oldfield, F.; Richardson, K.; Schellnhuber, H. J.; Turner, B. L.; Wasson, R. J., 2004: *Global Change and the Earth System: A planet under pressure* (Berlin - Heidelberg: Springer-Verlag).

Steffen, W.; Crutzen, P. J.; McNeill, J. R., 2007: "The Anthropocene: Are Humans Now Overwhelming the Great Forces of Nature", in: *Ambio*, 36, 8: 614–621.

Steffen, W.; Grinevald, J.; Crutzen, P. J.; McNeill, J., 2011: "The Anthropocene: Conceptual and Historical Perspectives", in: *Philosophical Transactions of the Royal Society A: Mathematical, Physical and Engineering Sciences*, 369, 1938: 842–867.

Steffen, W.; Persson, Å.; Deutsch, L.; Zalasiewicz, J; Williams, M.; (…); Schellnhuber, H. J.; Svedin, U., 2011a: "The Anthropocene: From Global Change to Planetary Stewardship", in: *Ambio*, 40, 7: 739–761.

Steffen, W, et al., 2016: "Stratigraphic and Earth System Approaches to defining the Anthropocene", in: *Earth's Future*, 4: 324–345: https://doi.org/10.1002/2016EF000379.

Steffen, W.; Rockström, J.; Richardson, K.; Lenton, T. M.; Folke, C.; Liverman, D.; Summerhayes, C. P.; Barnosky, A. D.; Cornell, S. E.; Crucifix, M.; Donges, J. F.; Fetzer, I.; Lade, S. J.; Scheffer, M.; Winkelmann, R.; Schellnhuber, H. J., 2018: "Trajectories of the Earth System in the Anthropocene", in: *Proceedings of the National Academy of Sciences of the United States of America*, 115, 33: 8252–8259; at: https://www.pnas.org/doi/pdf/10.1073/pnas.1810141115 (8 January 2023).

Steffen, W.; Richardson, K.; Rockström, J., Schellnhuber, H. J.; Dube, O. P.; Dutreuil, S., Lenton, T.; Lubchenco, J., 2020: "The Emergence and Evolution of Earth System Science", in: *Nature Reviews Earth & Environment*, Nature, 1, 1: 54–63.

Stengers, I., 2015: *In Catastrophic Times: Resisting the Coming Barbarism* (Open Humanities Press in collaboration with meson press); at: https://openhumanitiespress.org/books/download/Stengers_2015_In-Catastrophic-Times.pdf (18 November 2024).

Stevens, S. S. 1946: "On the Theory of Scales of Measurement", in: *Science*, 103, 677–680.

Stiegler, B., 2018: *The Neganthropocene. Edition. Translation and introduction. Daniel Ross* (London: Open Humanities Press).

Swyngedouw, E.; Ernstson, H., 2018: "Interrupting the Anthropo-obScene: Immuno-biopolitics and Depoliticizing More-than-Human Ontologies in the Anthropocene", in: *Theory, Culture & Society*, 35, 6: 3; at: https://doi.org/10.1177/0263276418757314.

Tainter, J. A., 1988: *The Collapse of Complex Societies* (Cambridge: Cambridge University Press).

Ulmer, J. B., 2017: "Posthumanism as Research Methodology: Inquiry in the Anthropocene", in: *International Journal of Qualitative Studies in Education*, 30, 9: 832–848; at: https://doi.org/10.1080/09518398.2017.1336806.

Verburg, P. H.; Dearing, J. A.; Dyke, J. G.; van der Leeuw, S.; Seitzinger, S.; Steffen, W.; Syvitski, J., 2016: "Methods and Approaches to Modelling the Anthropocene", in: *Global Environmental Change*, 39: 328–340; at: https://doi.org/10.1016/j.gloenvcha.2015.08.007.

Vicente, A.; Falkum, I. L., 2017: "Polysemy", in: *Oxford Research Encyclopedia of Linguistics*; at: https://doi.org/10.1093/acrefore/9780199384655.013.325 (12 February 2023).

Von Bertalanffy, L., 1968: *General System Theory: Foundations, Development, Applications* (New York: George Braziller).

Von Foerster, H., 1982: *Observing Systems. A Collection of Papers by Heinz von Foerster* (Seaside, CA: Inter-Systems Publications).

Waldrop, M. M., 1992: *Complexity: The Emerging Science at the Edge of Order and Chaos* (New York: Simon & Schuster).

Warfield, J. N., 2006: *An Introduction To Systems Science* (Singapore: World Scientific Publishing Co. Pte. Ltd.).
Waters, C. N.; et al., 2016: "The Anthropocene is Functionally and Stratigraphically Distinct from the Holocene", in: *Science*, 351, 6269, aad2622; 1–10.
Watson, A. J.; Lovelock, J. E., 1983: "Biological Homeostasis of the Global Environment: The Parable of Daisyworld", in: *Tellus B: Chemical and Physical Meteorology*, 35, 4: 284–289; at: https://doi.org/10.3402/tellusb.v35i4.14616.
WCED (World Commission on Environment and Development), 1987: *Our Common Future, Report of the World Commission on Environment and Development (United Nations)*; at: http://www.un-documents.net/our-common-future.pdf (17 January 2017).
Wells, Jennifer, 2013: *Complexity and Sustainability (Routledge Studies in Ecological Economics)* (London – New York: Routledge/Taylor and Francis Group, Kindle Edition).
Wiener, N., 1948: Cybernetics: Or Control and Communication in the Animal and the Machine (Paris: Hermann & Cie – Cambridge, MA: The MIT Press).
Wiener, N., 1961: Cybernetics: Or Control and Communication in the Animal and the Machine. 2nd revised ed. (Cambridge, MA: The MIT Press).
Wikipedia (English), 2023: "Anthropocene"; at: https://en.wikipedia.org/wiki/Anthropocene (10 April 2023).
Wiktionary (English), 2022: "-cene"; at: https://en.wiktionary.org/wiki/-cene (10 April 2023).
Young, O. R., 2017: *Governing Complex Systems: Social Capital for the Anthropocene* (Cambridge, MA: The MIT Press, Kindle Edition).
Zalasiewicz, J.; Williams, M.; Haywood, A.; Ellis, M., 2011: "The Anthropocene: A New Epoch of Geological Time?", in: *Philosophical Transactions of the Royal Society A: Mathematical, Physical and Engineering Sciences, A*, 369: 835–841.

Chapter 4
Politics and Geology in Anthropocene Geopolitics

Simon Dalby

Abstract The Anthropocene concept requires an extension of thinking well beyond traditional environmental formulations to recontexualise humanity as a major earth system force that is transforming the biosphere. This new context requires moving beyond conventional social science assumptions that the planet is a stable stage for human rivalries and engaging with the world making implications of understanding humanity as a geological actor. If humanity is to shape the future so that life operates in conditions loosely analogous to the last dozen millennia then a 'planet politics' that focuses on production processes that transcends the systems of fossil fueled capitalism of the last couple of centuries is essential. The task for social scientists is to facilitate such transitions, and the Anthropocene concept, if it is used carefully is a key formulation to facilitate transitions to more sustainable futures.

Keywords Anthropocene · Continuationism · Critique · Earth system · Epistemology · Neo-Malthusian · Non-stationarity · Safe operating space · Planet politics · Terrestrocentism

> The fraying thread that connects our past to our future is not limited to the flux in the natural order. The ecological shake-up wrought by climate change is also shaking up our economic and political-order. In the financial realm, as in the natural realm, the past provides fewer and fewer clues to our future. Like the migration patterns of songbirds that no longer correlate to the hatching patterns of their insect prey, or the mountain snowpacks that no longer store water for the dry summer months, the economy is facing miscues born of the feedback loop between tumult in the atmosphere and tumult on the earth. Rapid changes in the weather and temperature are outpacing our traditional ideas for assessing risk, redefining the calculus for economic success, shaking up the geopolitical status quo. (Schapiro 2016: xi)

Simon Dalby, Prof. emeritus, Balsillie School of International Affairs, Wilfrid Laurier University: Email: sdalby@gmailcom.

S. Dalby (✉)
Balsillie School of International Affairs, Wilfrid Laurier University, Waterloo, Canada
e-mail: sdalby@gmail.com

H. G. Brauch (ed.), *Towards Rethinking Politics, Policy and Polity in the Anthropocene*, The Anthropocene: Politik—Economics—Society—Science 35,
https://doi.org/10.1007/978-3-031-71807-6_4

4.1 Welcome to the Anthropocene

In March of 2011 the *National Geographic* made the discussion of the Anthropocene its cover story. Two months later the *Economist* did likewise with the headline 'Welcome to the Anthropocene'. Although the term had been in circulation in the earth system science community since 2000, since its appearance on these magazine covers the term has proliferated widely in numerous genres. At least in its original formulation the term was an explicitly geological one. Frustrated that some of his peers were not grappling with the scale and speed of contemporary transformations Crutzen (2002) suggested that the Holocene conditions of the last ten millennia were no longer the appropriate contextualization for discussions of the planetary biosphere. The new context of rapid change, in contrast to the relative stability of the Holocene, required a term that both distinguished from the Holocene and emphasised the role of the novel forcing mechanism in the earth system, the actions of industrial humanity. The term Anthropocene captures both in an elegantly concise formulation.

While the precise details of how this new geological era, or should that be epoch, are designated, and marked with an appropriate 'golden spike' to denote its beginning in sediments around the world, it remains a matter of intense debate within the earth sciences community (Zalasiewicz et al. 2019). The term has proliferated and migrated into at least some contemporary social science discussions, with as this chapter emphasises, some unsettling implications. It has also become a matter of contention in the humanities and debates about de-colonising the assumptions in both academic studies and practical policies dealing with climate, environment, extinction, and development (Giuliani 2021). How to 'get the Anthropocene right' has become a focus for interdisciplinary discussions (Thomas 2021). In the process it has also highlighted the difficulties of importing concepts from one field of inquiry into others, but likewise been a fertile focus for debates about the current human condition and our place in a rapidly changing world.

The term Anthropocene captures the scale of current transformations in a single simple term, a geological term that focuses attention on the crucial fact that human actions are shaping the future configuration of planet earth in profound ways. The Anthropocene also focuses attention on the crucial point that the modern assumption of humanity separate for nature is untenable (Hamilton et al. 2015). This simultaneously points to the fact that humanity has already so transformed its circumstances that the key geopolitical questions of our time now are about how the future earth system is to be shaped and who will decide (Dalby 2022). But such questions are a long way from traditional geological concerns, and the term Anthropocene is one that clearly evokes earth system science as the framework within which the term emerged.

Where many social sciences assume relative stability in background material conditions or take non-human factors as a given context for their investigations, physical and biological changes are intruding forcefully, and upsetting assumptions that the future will be a continuation of recent trends. A condition of non-continuity now must be worked into analysis where 'continuationism' no longer holds as a valid starting assumption (Albert 2020). This is especially the case in economics where future projections of climate impacts have so frequently assumed incremental rather than disruptive changes are what is coming (Keen 2021).

In hydrological matters, and water is now a major issue in terms of climate security (Selby et al. 2022), conditions of 'non-stationarity' now apply (Milly et al. 2008). The range of rainfall and temperature from the past is no longer the appropriate range to expect in future and hence can no longer be the criteria for designing bridges, dams, and other engineering projects. The sheer scale of industrial and engineering activities as well as the transformation of the human habitat as we become an urban species suggests that this novel element in the earth system, the emergence of a huge technosphere (Zalasiewicz et al. 2017), needs to be addressed as key to numerous aspects of the contemporary human condition.

4.2 The Anthropocene as Critique

But how to grapple with this new, dramatically unsettling recontextualisation of humanity isn't so simple. As Chakrabarty (2017) makes clear this is about more than extending the social science preoccupations with critiques of the deleterious consequences of globalization, themes that have frequently been fed into discussions of the Anthropocene. Yes, it's obviously about capitalism and the extraordinary inequities in the existing system and the history of the displacement and subjugation of peoples that have been, and continue to be, in the way of European imperial ambitions and its American extension (Moore 2015). But it's about much more than this if one takes the geological discussion seriously on its own terms. While there has been a productive sub discipline of international relations over the last few decades focused on global environmental politics, with its eponymous scholarly journal which deals with some of these matters, mainstream international relations, where one might expect a substantial engagement, has been nearly silent on the topic (Burke et al. 2016; Simangan 2020). It is also worth noting that the reverse is sometimes the case too; Clark/Szerszynski's (2021) extensive engagement with social sciences, and the necessity of envisioning the appropriate context in terms of a dynamic planet, is silent on international relations.

Here lies a fundamental epistemological difficulty, and one that appears in the attempts to bring social science into the discussion without fully appreciating the ontological shift implied in the Anthropocene discussion. The earth system science requires us to take the biosphere as a totality, as an interconnected system that is much more than the sum of individual environmental matters (Hamilton 2017). It also requires us to see humanity as a planet shaping force long before the advent of capitalism, even if there are undoubted qualitative differences between the early human efforts to change environments on the local scale and the dramatic transformations of the last few centuries (Davies 2016).

But to grapple with this agenda requires writing in the genre of critique. This chapter isn't science strictly speaking. Neither is it conventional social science. Its focus on the implications of novel formulations requires thinking about the political implications of these terms and that is something that is frequently anathema to classically trained scientists, and ironically too, to social scientists focused on either human institutions or immediate policy matters rather than the bigger picture. Critique is about examining assumptions and ontological contextualisations. In Foucault's (1988: 154–5) terms:

> A critique is not a matter of saying that things are not right as they are. It is a matter of pointing out on what kinds of assumptions, what kinds of familiar, unchallenged, unconsidered modes of thought the practices that we accept rest. . . . . Criticism is a matter of flushing out that thought and trying to change it: to show that things are not as self-evident as one believed, to see that what is accepted as self-evident will no longer be accepted as such.

The Anthropocene is a complex formulation that generates intense controversy not least because it forces researchers across the disciplines to rethink some key assumptions in their endeavours. In Foucault's terms it is no longer self-evident that the categories of 'human' and 'nature' can be treated separately. Scientists are struggling to cope with the recognition that they no longer work in a sphere outside human actions; this is especially tricky for geologists who, given their primary concerns with the ancient past of the planet, have good reasons to resist tackling contemporary concerns. Social scientists are confronted with the reality that the material context that modernity has so frequently taken for granted can no longer be assumed as the stable backdrop of the human drama. Social systems are not separate from either their specific contexts, or the consequences of their activities for future generations. Thoughtful engagement here requires a recognition that normative and conceptual matters are unavoidably intermingled; in terms of critique, it must be about rethinking the implicit assumptions about the planet and what it means to live in a planetary system being remade very quickly by the actions, and the consequences thereof, of the rich and powerful parts of humanity.

4.3 Earth System Boundaries

The Anthropocene is thus both about the new circumstances emerging by that human remaking how the planetary system operates and about how it is that sciences and social sciences come to understand these processes. The subtitle to Lewis/Maslin's (2018) volume on *The Human Planet* poses this double operation pithily in terms of 'How we Created the Anthropocene'. The concept of the Anthropocene and the material transformations it designates are inextricably linked and tracing some of the key parts of this discussion allows an engagement with the assumptions brought to bear by various protagonists in the debates.

One of the key assumptions, which conflates the scientific aspects of the transformation of both the earthly context with matters of human security is the matter of how dangerous the current trajectories of the Anthropocene are to humanity in general, and the need to drastically curtail at least some of the industrial and economic disruptions to natural processes. The focus on the earth system boundaries, and the related matter of a safe operating space for humanity, taken to be conditions analogous to those of the Holocene which have clearly demonstrated that humanity can thrive in those circumstances, suggests that the fate of the earth is a collective human concern.

Critics argue that this is only a concern for the rich and powerful who might be endangering their own long-term survival, but indigenous peoples, and those subject to violent colonization by Europeans over the last few centuries, have long had to deal with existential threats to their societies and calamity from afar is a matter familiar to many peoples. Focusing on this in terms of geopolitics, in Grove's (2019) terms, a matter of the expansion and imposition of particular modes of life on territories, suggests that present circumstances are better understood in terms of the Eurocene, given that the causal agent is violent European expansion. Moore's (2015) focus on the rise of capitalism and the novel economic dynamics set in motion by its particular mode of accumulation suggests that the designation Capitalocene is more appropriate.

Nonetheless, whatever about the attribution of responsibility for present circumstances, and hence the appropriate designation of the contemporary era, the discussion of earth system boundaries continues among scientists (Rockström et al. 2009; Steffen et al. 2015). The assumption, clearly stated in this literature, is that humanity has thrived in the last ten thousand years in circumstances of remarkably stable climatic conditions, the era of the Holocene when, unlike the previous hundreds of thousands of years, global temperatures, sea levels and glaciation have been remarkably stable. Humanity's actions, dramatically accelerating in the last few hundred years, have now pushed some of the parameters of the biosphere out of the range that human history has known, with as yet, less than clear implications for the functioning of the system. While Johan Rockström has emphasised the importance of keeping the earth system in conditions analogous to the Holocene, already humanity is pushing past some of the boundaries of this system (Rockström/Malm 2015). Although in one case, that of ozone depletion in the stratosphere, it may be that we have pulled back within the safe operating space, that is to approximately the Holocene conditions.

The Anthropocene formulation emphasises that the rich and powerful in humanity have already changed enough facets of the earth system that the changes are of an order large enough to be considered analogous to previous major shifts in the earth system. Discussions of the sixth extinction, new geological formulations such as 'plastiglomerates', parallels with Eocene conditions, or runaway climate change being imminent, all emphasise the scale of the current changes. Frequently too these point to the sheer speed of transformation now underway, something that appears unprecedented at the global scale (Bonneuil/Fressoz 2016). Earth system thinking likewise emphasises that changes frequently do not come in linear and predictable transitions, but ruptures, tipping points and phase changes may be relatively sudden and interconnected in complex ways (Barnosky 2012). These are the new circumstances set in motion by carboniferous capitalism. Where the tipping points are for some processes isn't clear in advance, but clearly the Arctic is warming rapidly as the shrinking polar sea icecap shows.

While climate change gets much of the attention, and given how it causes numerous terrestrial changes, and with rising carbon dioxide levels causing worrisome ocean acidification, it deserves this emphasis, other things are in motion too caused by the extraordinary expansion of human activity (McNeill/Engelke 2016). In addition to ocean acidification and the ozone depletion problem, the safe operating space formulation points to the dramatic changes to natural cycling of nitrogen and phosphorous due to the widespread application of artificial fertilizers. Atmospheric aerosols are an additional problem generated by inefficient fuel combustion and industrial pollution. Freshwater is being used on a scale that endangers ecosystems in many places and large-scale contamination presents numerous health hazards where potable water is in short supply. Much of the terrestrial land surface that can be used for agriculture is already being utilised which has involved dramatic changes of species. Parts are also now urban spaces and industrial sites, so much so that many of the larger terrestrial land areas are effectively artificial ecosystems. Related to that is the dramatic loss of species and reduction of biodiversity as habitat is removed for human activity. Chemical pollution because of industrial activity is also widespread, and humanity has introduced so many artificial entities into the biosphere that these have been added as a class of their own in the updated (Steffen et al. 2015) earth system boundaries framework.

In summary the extent of human alteration of the biosphere now means that humanity is living in artificial circumstances. This suggests that the traditional notions of environment are no longer appropriate categories for thinking about the future and what kind of planet human actions are producing.

Environmentalist traditions have long called for a halt to human interference in ecology and the Earth system. In the Anthropocene, the anthropogenic biosphere is permanent, the legacy of our ancestors, and our actions as human systems a force of nature, making the call to avoid human interference with the biosphere irrelevant. The implication is clear; the current and future state of the terrestrial biosphere is up to us, and will be determined by human systems of one form or another, whether it is the momentum of our past or new pathways we are able to achieve in the future (Ellis 2011: 1027).

Thus, the context in which questions of geopolitics and security are discussed are now increasingly matters of artificial environments. This is the context for thinking about the future, and for both considerations of what is to be sustained by whom for what purpose, and how human conflicts and rivalries may turn violent, and what to do about preventing this. If, as the modern formulations of politics repeatedly assert, the first responsibility of states is to provide security, then the novel formulations of the Anthropocene, or Eurocene, or Capitalocene, profoundly trouble the taken for granted assumptions of how this is to be provided and by whom. More specifically this requires a rethinking or environmental security to deal with the earth system science formulations and simultaneously the critical arguments from scholars such as

Grove and Moore that demand a more careful evaluation of the implicit assumptions about humanity as a singular actor with a common fate. Likewise, the legacy of colonial formulations of 'environment' as something external to humanity, a source of resources and a repository for waste, also need to be tackled; this powerful dichotomy which structures so much modern political discourse is also rendered inadequate by earth system thinking.

4.4 Geological Politics: Whither Security?

While traditional environmental concerns with nature preservation are still relevant in terms of habitat protection to keep biological options open for future generations, it is now understood that conservation is increasingly about matters of re-wilding and the migration and making of new habitats in a rapidly changing biosphere (Lorimer 2015). Climate change needs to be both reframed as a matter of much more than traditional environmentalist thinking (O'Lear/Dalby 2016) and reimagined in numerous cultural and political registers to grapple with what is coming (Wapner/Elver 2016). The task for international studies is to think about how to do these things in ways that offer useful political insights into how to move innovation forward in a rapidly changing world, one for which traditional notions of territorial sovereignty, national security and geopolitical order are often less than helpful. Perhaps the threats to national survival will yet motivate effective action on climate change (Lieven 2020)? But frequently the multiplicity of states and their rivalries acts to stymie cooperation, at least in terms of addressing rapid climate change, so far this mode of governance has usually proved to be pathological rather than helpful (Harris 2021).

Focusing on climate change as a matter of global security rather than national security is a helpful step, as Goldstein (2016) suggests, but much more is needed. What exactly is to be secured by who is much less than clear. While aspirational statements help, and environmental studies has much to offer here, clearly too analysis of success stories in facilitating transitions to less disruptive modes of political economy are important. How they translate from national experience into international innovation is a key question for work on sustainability transition that now links environmental and international studies in the new context of the Anthropocene (Hamilton et al. 2015). The notion of the Anthropocene is most useful because it simultaneously incorporates the key points about global scale rapid change and the focus on how human production decisions are shaping the technosphere and hence the future configuration of the earth system. The United Nations Development Programme (2020) has already begun using the Anthropocene contextualisation, but the tensions and contradictions between the various Sustainable Development Goals remain to be effectively dealt with nonetheless (Dalby et al. 2019). Empirical research on cases where the Goals have been effectively coordinated would seem to be a key task for social scientists.

As the discussion of the Anthropocene in terms of the new geological circumstances for humanity proceeds, some key insights are emerging that matter to policy discussions and to how international studies tackle global questions (Mitchell 2014). Many insights suggest that transformations in thinking are needed even if how these might be translated into mainstream understandings that facilitate political action is as yet unclear. Traditional environmental formulations remain useful for many things but the Anthropocene suggests more profound shifts in thinking are needed if the future is to be remade in more peaceful ways (Hamilton 2017). Traditional environmentalist movements have not been able to grapple with either the scale of the human enterprise in changing the planet nor the historical legacy of colonial appropriations of resources and the related disruptions of aboriginal societies (Dauvergne 2016). A much more drastic recontextualisation is needed to underpin innovative political thinking; the formulation of the Anthropocene offers this possibility.

The geological phrasing suggests looking to what is being made, and how, rather than presupposing, as much environmentalism does, a given context that is being damaged by human actions. While the climate change discussion has started to make it clear that humanity is changing the context in which future geopolitics will play out, the shift to focusing on environmental production, rather than environmental protection, needs further elaboration and incorporation in thinking about transitions to more peaceful societies globally (Brauch et al. 2016). In practical terms this requires abandoning the culturally popular, but politically defeatist, tropes of imminent apocalypse in favour of discussions of shaping the future and building flexible social and ecological arrangements rather than trying to use force to preserve things in the face of changes already set in motion.

Not least the focus on production is important because geological formulations of our present circumstances emphasise the dynamism of the earth system, and the fact that it can rapidly change and fluctuate between various states. The scale of the alterations already undertaken by humanity mean that the future will be much less stable than the recent past, and as such sustainability discussions need to focus on securing the ability to adapt on the part of both human and non-human entities in the biosphere (Dalby 2020). This will require both flexibility and the ability to innovate governance arrangements in ways that are very different from the persistent competitive rivalries of traditional geopolitical thinking. It requires thinking about security in terms of connections, consequences, and the future rather than trying to perpetuate a series of supposedly stable geographical features.

Focusing on protecting the past, and perpetuating practices that are becoming dangerous, rather than shaping the future in ways that both buffer societies from the worst extremes of climate and other disruptions and facilitate innovations to cope with new circumstances, is to dangerously misrepresent currently urgent political tasks (Dalby 2022). The central dilemma for security scholars is that protecting the present fossil fuelled order of things simply makes the future harder to deal with because present trajectories are set to transgress what earth system scientists see as a crucial set of boundaries for the safe operating system for planet earth.

If the planetary system is to remain in the 'safe operating space', something approximating the conditions of the last ten thousand years in the era of the Holocene, then it's clear that substantial changes in direction for parts of the global economy are needed quickly. This is not the sort of thing normally considered under the rubric of global security but such things are unavoidable in coming decades. This is precisely what discussions of transformation and sustainable transitions are starting to engage, but time is short and the geopolitical premises in these discussions and the security dimensions need much further attention (Brauch et al. 2011).

The alternative is to ignore the earth system boundaries and push the global system into a future with numerous unknown consequences. Rapidly fluctuating temperatures and shifting weather patterns will likely overwhelm contemporary agricultural systems as well as the lengthy commodity chains and complicated infrastructures that keep urban life functioning, and there is no clear indication that human systems are resilient enough to cope with rapid and unpredictable climate fluctuations. While some political elites may assume that they can use wealth and force, quite literally their 'firepower' (Dalby 2022) to evade the consequences of such disruptions, this is a gamble that is unlikely to succeed given the inter-connectedness of human and ecological systems. Such a refusal to 'face the Anthropocene' (Angus 2016) and confront the legacy of fossil fuelled capitalism, may well abandon much of humanity to a violently dystopian future. Security research now must be about how to remove obstacles to rapid adoption of renewable energy, and to rethinking urban life with very much less dependence on car culture and excessive consumption.

4.5 New Contexts, Peace and Sustainability

Hence the necessity of thinking about geopolitics in new ways, ones that think across the very substantial gap between what earth systems science indicates is coming and the capabilities of existing institutions to cope (Galaz 2014). Crucially such thinking must also recognise that the twentieth century solution to one of the key causes of European warfare, that of fixing state boundaries, and making wars of territorial aggression unacceptable international conduct, is now presenting governance problems in so far as it makes migration, the most basic form of adaptation to environmental change, into a problem that apparently needs a security response. In addition, new geopolitical thinking must recognise that the supposedly stable conditions of the Holocene that were understood to be the backdrop to human conflict, are themselves now being changed by human conduct (Dalby 2020). Anthropocene geopolitics is about living in a rapidly changing world, not a matter of permanent geographical verities providing the stage for human rivalries.

Transitions to sustainability must engage with both the possibilities of rapid phase shifts in the earth system and the simple fact that sustainability has to be thought of in terms of the ability to continue adapting to novel circumstances, rather than assuming a new fairly stable state after transition awaits in the future (Brauch et al. 2016). It's not clear that the current geological phase shift is likely to be of short duration, nor that it will result in a stable situation conducive to human flourishing (Barnosky 2012). If the next phase of the Anthropocene turns out to be a much more dynamic climate regime, with rapidly fluctuating conditions, then human planning will have to rethink everything from agriculture to infrastructure and buffer both systems against extremes and adopt modes of conduct that recognise that change rather than stability is the new normal. What needs to be focused on is ecology and maintaining fecund ecosystems (McDonald 2021). The Anthropocene, and discussions of phase shifts in the biosphere suggests that rapid shifts are occurring and that the earth system may respond in unpredictable, and crucially in non-linear ways, as it reconfigures itself due to rising greenhouse gas levels simultaneously with ocean acidification, terrestrial land use changes and numerous other matters.

The Paris Agreement in late 2015 on climate change suggests that political leaders are beginning to recognise that climate needs attention, but rapid progress there to has repeatedly run in to opposition from fossil fuel interests in particular putting obstacles in the way of rapid energy system innovation. As Joshua Goldstein (2016) puts it, international institutions would be quick to react if a large meteor were heading towards earth; they have failed to address the equivalent dangers on earth because in part they are self-imposed, but likewise because the context of rapid biospheric change has yet to register as such. Hence the utility of the formulation of the Anthropocene in terms of geological scale transformation. Simply collapsing the Anthropocene back into traditional environmental managerialist tropes is a major conceptual and political mistake. Indeed, such thinking may undermine the more promising innovations in environmental peacebuilding (Ide 2020). The current transformation requires much more from contemporary scholars, but the category of the Anthropocene won't deliver on its promise if some obvious bad intellectual habits from the past persist.

4.6 The Dangers of Anthropocenology

Bonneuil/Fressoz (2016) emphasise the dangers of the scientific 'view from nowhere' frequently implicit in how the Anthropocene is formulated. Likewise the assumption in some of the discourse implies that the Anthropocene is a novel formulation that will save humanity from its prior ignorant folly. The assumption of detachment, of the earth as an external entity to be managed by those with the power to do so reinvents an imperial view of things, of a superior civilization determining the fates of lesser beings supposedly in the service of the common good. The fraught debates about geoengineering frequently invoke just this mode of thinking neglecting to engage the politics or the history of climate change (Baskin 2019). Thus, a cautionary tale must be inserted in the grand narratives of stages of human evolution, the supposed progress of the great acceleration and the rhetoric of sustainable development with its promise of progress for all (Luke 2015).

In short, the Anthropocene concept has a politics, and Anthropocenologists, to use Bonneuil/Fressoz's (2016) apt term, are often apparently oblivious to the key questions as to who decides on the configuration of the next stages of the Anthropocene. While the Anthropocene provides a powerful recontextualisation of the global security agenda, and adds urgency to thinking about transition strategies, the term itself doesn't suggest either political strategies for the transition or how the geopolitics of the present might be remade to prevent rapid change causing conflict, or even the return of major war should mutual suspicion and regional rivalries be dramatically accentuated. These difficulties need to be directly confronted in any attempt to think through sustainable transitions at the global scale. How the Anthropocene is invoked, by whom and with what political consequences matters. Five points in particular are germane to Anthropocene geopolitics.

First is the temptation to rework neo-Malthusianism and the more pernicious versions of limits to growth arguments in ways that fail to grapple with the dynamics in play (Selby/Hoffman 2015). In terms of scarcity of resources arguments the key point about the Anthropocene is that climate change and many of the other earth system boundaries are in danger of being violated by the production of too many things. This isn't a population problem, it's a production problem coupled to wasteful use of materials and fuels in feeding and fuelling the current urban system. Pollution is the problem, resulting from inappropriate production methods; not a shortage of resources. In the case of carbon fuels in particular the problem is the overabundance of fossil fuels, not their shortage. Keeping carbon in the ground is what has to be accomplished, while not adding to the hardships and instabilities of those least able to move on to other energy sources.

Second, is the matter of how security and climate is framed. Thinking, as much of the literature on national security currently does, about a climate change as a conflict multiplier or conflict catalyst for peripheral instabilities, maps onto the residual geographical formulations of the war on terror but often misconstrues the causalities and the consequences of changes in contemporary rural political economy (Klare 2019). The assumption of far-away places as inherently violent is a powerful trope that normalises warfare as a permanent condition and hence facilitates justifications for inappropriate military interventions (Welzer 2012). In so far as climate is a catalyst for conflict the causes are in metropolitan uses of fossil fuels that are indirectly causing difficulties and directly in so far as the global economy, including attempts to adapt to climate change, is causing fluctuations in land use prices, property speculation, land grabbing and related disruptions that make life and adaptation strategies of subsistence farmers more difficult. Climate change is playing out in landscapes already being reshaped by commercial agriculture, and that context needs to be kept clearly in mind (Parenti 2011).

Third, and related to this is the importance of remembering that the notion of adaptation to environmental change is not a new invention driven by concerns over climate; they have much longer intellectual genealogies in discussions of disasters and the literature on political ecology (Watts 2015). The politics of centralised technocratic attempts to remake rural economies and societies in line with urban preconceptions of appropriate modernisation strategies have often exacerbated rural conflict, and climate smart agriculture driven by corporate technological innovations rather than a more comprehensive understanding of agro-ecology may, in the name of climate policy, replicate the injustices and displacements that prior efforts at the commercialisation of agriculture have done. Climate adaptation premised on the engineering prerogatives of central governments rather than the ecological contexts of human vulnerabilities show alarming propensities to perpetuate violent development practices (Sovacool/Linner 2016). In so far as the resultant disruptions are then seen as a security threat, especially when they displaced populations feed insurgencies, climate adaptations may generate 'backdraft' phenomena that simply aggravate difficulties rather than facilitating innovation (Dabelko et al. 2013).

Fourth, while handling new security issues by innovating with transnational cooperation and new management systems, and in the process transforming states to deal with borderless threats, shows some promise, so far, these efforts have had very mixed success (Hameiri/Jones 2015). Frequently local political economies either stymie innovation or misdirect management to enhance economic advantages of those with most to gain by new practices. While some global disease issues are handled better by careful cooperation across frontiers, the Ebola case in particular suggests that health innovations are patchy at best, not least because poor and conflict-ridden societies frequently don't have the resources to follow the supposedly universal health protocols agreed a decade ago. The COVID-19 pandemic and its dramatic economic disruptions starkly revealed the limitations of global economic security in terms of its provision of safety for marginal populations and metropolitan ones that were caught unprepared (Tooze 2021). Neither is there a clear indication that the new security agenda has anything like the capabilities to deal with such complex borderless threats presented by climate change.

All these difficulties are compounded by, fifth, what Chaturvedi/Doyle (2015) call climate terror in the sense of an invocation of imminent apocalyptic events generated by a supposedly universal state of climate vulnerability. The temptations to both artificial climate engineering, and draconian measures regardless of the local consequences, are implicit in much of the climate security discourse, and likely to simply aggravate rather than ameliorate the difficulties of marginal peoples and those in harm's way. The global inequalities that have been accentuated by the very uneven growth in the world economy in the last few decades perpetuate vulnerabilities, as the case of Bangladesh, vulnerable both because it is in political economy terms in the global south, and in geographical terms a low-lying state in the path of tropical storms, emphasises. Policies that fail to comprehend the specific vulnerabilities and diverse modalities of insecurity in particular places by invoking a general sense of alarm using the language of the Anthropocene will be worse than useless in working on sustainability transitions.

4.7 Anthropocene Social Science

The crucial point about the Anthropocene discussion for social scientists is that it requires social scientists to rethink the implicit contextualisations that inform their analyses. The earth system is being transformed at a rapid rate; the key determinant of the future climate conditions of the planet is now humanity and the decisions the rich and powerful make in coming decades about such things as what carbon dioxide concentrations in the atmosphere will be. Climate on this planet is no longer a matter of orbital dynamics and Milankovich cycles (Rockström/Gaffney 2021); its increasingly being shaped indirectly by human actions and what the rich and powerful decide to produce.

The transformation underway is causing the sixth extinction event in planetary history (Kolbert 2016), and the movement of species by humans in the last few centuries has rearranged biological geographies in a way probably not seen since Pangaea split up hundreds of millions of years ago. Thinking about human affairs in that context is key; the implicit geopolitical assumptions of most of International Relations is that there is a given context for human rivalries. The Anthropocene discussion makes that an impossible formulation as the premise from which to begin investigations or think politically.

It also raises great difficulties with the formulation of 'saving the planet'. In Hayhoe's (2021) formulation, the current debate about climate change is much more about saving civilization, or 'saving us' than it is about saving a planet. Humanity has already dramatically altered the planetary system and has done so by making all sorts of new things, and moving and reassembling others, not least animal and plant species. In the last couple of centuries, but most noticeably in the last couple of generations, we have been making carbon dioxide in prodigious quantities. And cities, roads, the ecological deserts that are agricultural mono-cropping systems, as well as plastics and all sorts of novel entities in the earth system term, that are making the future context for humanity very different from the past.

In so far as what needs to be 'saved' can be clearly articulated, it's the key elements of biodiversity and the habitats that these species live in, as well as a fairly stable climate, something akin to the Holocene conditions that humanity has known through history. Despite periodic fluctuations in the middle of the Holocene, and the more recent 'little ice age', the last 12,000 years have had a very unusually stable climate. The prospect of rapid climate change introducing a fluctuating, rather than a fairly predictable, climate isn't something that an urban humanity of 9 or 10 billion people can reasonably expect to cope with given its needs for food which requires reliable production and an economy to move it around.

This isn't about saving a given context but a matter of thinking intelligently about how we make multiple socio-ecological systems that allow humanity to, in various diverse ways, live well in new contexts without burning large quantities of material (biomass or fossil fuels) to do so. It's also about thinking about ecologies that can absorb the huge amounts of carbon dioxide that already are in the atmosphere, not least to slow, and hopefully eventually reduce the problems of ocean acidification, that so threaten numerous aspects of Oceanic life. This point is often overlooked; being a land-based species social scientists are usually guilty of a profound 'terrestrocentrism' that ignores the fact that life is an aqueous phenomenon first and foremost. The future condition of the oceans matters greatly for terrestrial life.

This is all a matter of doing more than debating the significance of the Anthropocene for our various disciplines. It's about how we can usefully research transition strategies and tell stories outside the academy that facilitate ecological flourishing. To reiterate Grove's (2019) formulation of geopolitics in terms of modes of life and their imposition on territories; sustainable ecologies must be the modes of life that are encouraged, but these will have to be sensitive to specific contexts, rather than imposed through one size fits all applications. And they will have to do so in ways that transcend the geopolitical imagination of competing powers that still dominates so many retrograde political narratives (Toal 2024). This does suggest that we need to think well beyond the categories available in contemporary international relations scholarship and formulate matters in terms of planet politics or something analogous (Burke et al. 2016).

Planet politics which takes the geological conditions of the Anthropocene seriously must be about production, about who decides what gets made, where, and how. The rich and powerful have disproportionate influence on these matters through their investment strategies and economic decisions. It matters greatly if fossil fuels power our cities or if instead windmills and solar panels do. If we use fossil fuels we make yet more carbon dioxide; if we use non-combustive systems climate change will be slowed allowing easier adaptation. No economic system is innocent; mining raw materials will continue to be a dirty business, but how we do these things, and which ore we use, dug from where, and processed how, has consequences that need to be worked into production decisions. These are now far too important to be left to the vagaries of so called 'markets' (Göpel 2016). The 2022 disruptions to fossil fuel prices round the world, and Europe in particular, makes this point unavoidable.

Planet politics must be about trying to think about systems that deal with the worst human suffering caused by economic inequities and doing so while ensuring that what we make doesn't preclude living well for future generations. It's obviously about technical change, but it's also about social changes that operate on the clear understanding that we are earth, are a part of earth, not apart from earth, and our societies are dramatically altering both human and non-human habitats. Updating the discussion of sustainable development requires the recognition that industrial policies matter, and that assumptions about a given environmental context within which they operate is no longer tenable. In terms of security specifically, the topics that need urgent attention from social scientists relate to political strategies to remove the roadblocks to rapid innovation in energy systems, and a focus on successful transition strategies to post-carbon fuelled economic systems.

This all requires abandoning the ontological premise of clear distinctions between Nature and Society, and understanding that we are moving species around, literally making new ones by selective breeding and building artificial habitats in a series of new ecological arrangements that are just that, new! International relations and its security subfield in particular has been mostly concerned about how to dominate a divided world or how to deal with the consequences of efforts to do so. Planet politics now must be about figuring out how to share a crowded world, one constituted of many new 'worlds', in the sense of interconnected assemblages, that operate in ways that allow others to flourish too in a biosphere that humanity is rapidly remaking.

References

Albert, M., 2020: "Beyond continuationism: Climate change, economic growth and the future of world (dis)order", in: *Cambridge Review of International Affairs*; at: https://doi.org/10.1080/09557571.2020.1825334.

Angus, I., 2016: *Facing the Anthropocene: Fossil Capitalism and the Crisis of the Earth System* (New York: Monthly Review Press).

Barnosky, A. D., 2012: "Approaching a state shift in Earth's biosphere", in: *Nature*, 486: 52–58.

Baskin, J. 2019: *Geoengineering, the Anthropocene and the End of Nature* (Cham: Palgrave Macmillan).

Bonneuil, C.; Fressoz, J.-B., 2016: *The Shock of the Anthropocene* (New York: Verso).

Brauch, H.G.; Dalby, S.;, Oswald Spring, U., 2011: "Political geoecology for the Anthropocene", in; Brauch, H.G.; Oswald Spring, Ú.; Kameri-Mbote, P.; *et al.* (Eds.): *Coping with Global Environmental Change, Disasters and Security Threats, Challenges, Vulnerabilities and Risks* (Berlin–New York: Springer-Verlag): 1453–1485.

Brauch, H.G.; Oswald Spring, Ú.; Grin, J.; Scheffran, J. (Eds.), 2016: *Sustainability Transition and Sustainable Peace Handbook* (Heidelberg–New York–Dordrecht–London: Springer-Verlag).

Burke, A.; Fishel, S.; Mitchell, A.; Dalby, S.; Levine, D., 2016: "Planet Politics: A Manifesto from the end of IR", in: *Millennium* 44,3: 499-523.

Chakrabarty, D., 2017: "The Politics of Climate is More than the Politics of Capitalism", in: *Theory, Culture & Society*, 34,2-3: 25-37.

Chaturvedi, S.; Doyle, T., 2015: *Climate Terror: A Critical Geopolitics of Climate Change* (London: Palgrave Macmillan).

Clark, N.; Szerszynski. B., 2021: *Planetary Social Thought: The Anthropocene Challenge to the Social Sciences* (Cambridge: Polity).

Crutzen, P. 2002: "Geology of Mankind – The Anthropocene", in: *Nature* 415,6867: 23.

Dabelko, G.; Herzer, L.; Null, S.; Parker, M.; Sticklor, R. (Eds.), 2013: "Backdraft: The Conflict Potential of Climate Change Adaptation and Mitigation" in: *Report* (Washington DC: Woodrow Wilson Center, Environmental Change and Security Program), 14,2: 1–60.

Dalby, S., 2020: *Anthropocene Geopolitics: Globalization, Security, Sustainability* (Ottawa: University of Ottawa Press).

Dalby, S., 2022: *Rethinking Environmental Security* (Cheltenham: Edward Elgar).

Dalby, S.; Horton, S.; Mahon, R.; Thomaz, D. (Eds.), 2019: *Achieving the Sustainable Development Goals: Global Governance Challenges* (London: Routledge).

Dauvergne, P., 2016: *Environmentalism of the Rich* (Cambridge MA, MIT Press).

Davies, J., 2016: *The Birth of the Anthropocene* (Berkeley: University of California Press).

Ellis, E. C., 2011: "Anthropogenic transformation of the terrestrial biosphere", in: *Philosophical Transactions of the Royal Society A*, 369: 1010–1035.

Foucault, M., 1988: *Politics, Philosophy, Culture: Interviews and Other Writings, 1977-1984* (New York: Routledge).

Galaz, V., 2014: *Global Environmental Governance, Technology and Politics: The Anthropocene Gap* (Northampton, MA: Edward Elgar).

Giuliani, G., 2021: *Monsters, Catastrophes and The Anthropocene: A Post-Colonial Critique* (Abingdon: Routledge).

Goldstein, J. S. 2016: "Climate Change as a Global Security Issue", in: *Journal of Global Security Studies*, 1,1: 95-98.
Göpel, M., 2016: *The Great Mindshift: How a New Economic Paradigm and Sustainability Transformations Go Hand in Hand* (Berlin: Springer).
Hameiri, S.; Jones, L., 2015: *Governing Borderless Threats: Non-Traditional Security and the Politics of State Transformation* (Cambridge: Cambridge University Press).
Hamilton, C., 2017: *Defiant Earth* (Cambridge: Polity).
Hamilton, C.; Bonneuil, C.; Gemenne, F. (Eds.), 2015: *The Anthropocene and Global Environmental Crisis: Rethinking Modernity in a New Epoch* (Abingdon: Routledge).
Harris, P., 2021: *Pathologies of Climate Governance: International Relations, National Politics and Human Nature* (Cambridge: Cambridge University Press).
Hayhoe, K., 2021: *Saving Us: A Climate Scientist's Case for Hope and Healing in a Divided World* (New York: Simon and Schuster).
Ide, T., 2020: "The Dark Side of Environmental Peacebuilding", in: *World Development*, 127: 104777.
Keen, S., 2021: "The appallingly bad neoclassical economics of climate change", in: *Globalizations*, 18,7: 1149-1177.
Klare, M., 2019: *All Hell Breaking Loose: The Pentagon's Perspective on Climate Change* (New York: Metropolitan).
Lewis, S.; Maslin, M. 2018: *The Human Planet: How We Created the Anthropocene* (London: Pelican Books).
Lieven, A., 2020: *Climate Change and the Nation State: The Case for Nationalism in a Warming World* (New York: Oxford University Press).
Lorimer, J., 2015: *Wildlife in the Anthropocene: Conservation after Nature* (Minneapolis, MN: University of Minnesota Press).
Luke, T., 2015: "On the politics of the Anthropocene", in: *Telos*, 172: 139–162.
McDonald, M., 2021: *Ecological Security: Climate Change and the Construction of Security* (Cambridge: Cambridge University Press).
McNeill, J.R.; Engelke, P., 2016: *The Great Acceleration: An Environmental History of the Anthropocene since 1945* (Cambridge, MA: Harvard University Press).
Milly, P.C.D.; Betancourt, J.; Falkenmark, M.; Hirsch, R.M.; Kundzewicz, Z.W.; Lettenmaier, D. P.; Stouffer, R.J., 2008: "Stationarity Is Dead: Whither Water Management?", in: *Science*, 319,5863: 573–574.
Mitchell, A., 2014: "Only human? A worldly approach to security", in: *Security Dialogue*, 45,1: 5–21.
Moore, J., 2015: *Capitalism in the Web of Life: Ecology and the Accumulation of Capital* (London: Verso).
O'Lear, S.; Dalby, S. (Eds.), 2016: *Reframing Climate Change: Constructing Ecological Geopolitics* (London: Routledge).
Parenti, C., 2011: *Tropic of Chaos: Climate Change and the New Geography of Violence* (New York: Nation Books).
Rockström, J., Gaffney, O., 2021: *Breaking Boundaries: The Science of our Planet.* (New York: DK Publishing).
Rockström, J.; Kulm, M., 2015: *Big World Small Planet: Abundance within Planetary Boundaries* (Stockholm: Max Strom).
Rockström, J. *et. al.*, 2009: "Planetary boundaries: exploring the safe operating space for humanity", in: *Ecology and Society*, 14,2: 32.
Schapiro, M., 2016: *The End of Stationarity: Searching for the New Normal in the Age of Carbon Shock* (Vermont: White River Junction: Chelsea Green).
Selby, J.; Hoffman, C. (Eds.), 2015: *Rethinking Climate Change, Conflict and Security* (London: Routledge).
Selby, J.; Daoust, G.; Hoffman C., 2022: *Divided Environments: An International Political Ecology of Climate Change, Water and Security* (Cambridge: Cambridge University Press).

Simangan, D., 2020: "Where is the Anthropocene? IR in a new 'geological epoch'", in: *International Affairs*, 96,1: 211-224.
Sovacool, B.K.; Linner, B.O., 2016: *The Political Economy of Climate Change Adaptation* (London: Palgrave Macmillan).
Steffen, W.; *et al.*, 2015: "Planetary boundaries: Guiding human development on a changing planet", in: *Science* 347,6223: https://doi.org/10.1126/science.1259855.
Thomas, J.A. (Ed.), 2021: *Altered Earth: Getting the Anthropocene Right* (Cambridge: Cambridge University Press).
Toal, G., 2024: *Oceans Rise, Powers Fall* - Why Geopolitics Hastens Climate Catastrophe (Oxford: Oxford University Press).
Tooze, A., 2021: *Shutdown: How COVID Shook the World's Economy* (New York: Viking).
United Nations Development Programme, 2020: *The Next Frontier: Human Development and the Anthropocene* (New York: United Nations).
Wapner, P.; Elver, H., (Eds.), 2016: *Reimagining Climate Change* (New York: Routledge).
Watts, M., 2015: "Now and Then: The Origins of Political Ecology and the Rebirth of Adaptation as a mode of Thought", in: Perreault, T.; Bridge, G.; McCarthy, J. (Eds.): *Routledge Handbook of Political Ecology* (New York: Routledge): 19–50.
Welzer, H., 2012: *Climate Wars: Why People will be Killed in the Twenty First Century* (Cambridge: Polity).
Zalasiewicz, J.; Williams, M.; Waters, C.N.; et al., 2017: "Scale and diversity of the physical technosphere: A geological perspective", in: *The Anthropocene Review*, 4,1: 9-22.
Zalasiewicz, J.; Waters, C.N.; Williams, M.; Summerhayes, C.P. (Eds.), 2019: *The Anthropocene as a Geological Time Unit: A Guide to the Scientific Evidence and Current Debate* (Cambridge: Cambridge University Press).
Zografos, C.; Goulden, M.C.; Kallis, G., 2014: "Sources of human insecurity in the face of hydro-climatic change", in: *Global Environmental Change* 29: 327-336.

Chapter 5
Thinking Politically About the Anthropocene

Richard Ned Lebow

Abstract I examine possible interactions between the Anthropocene and the study of international relations. I ask how the concept of the Anthropocene encourages new and productive frameworks for realism, liberalism, and constructivism. I suggest that suitably revised paradigms can help us reduce and cope with the effects of climate change.

Keywords Anthropocene · Climate change · Forecasting · International relations theory · Non-linearity

In this chapter, I examine possible interactions between the concept of the Anthropocene and the study of international relations. They are linked by a common concern for climate change and its possibly catastrophic consequences. The Anthropocene is a formulation intended to emphasise the effects humans have on our planet, and by doing so promote a greater commitment to forestall climate change. International Relations theory aims to understand patterns of relations among states and other international actors, but also has a normative agenda. Its practitioners seek to reduce war and promote economic development. Many among them are increasingly aware of the ways in which climate change impedes both goals. So, important parallels exist and are worth exploring.

Several questions come to the fore. First and foremost, how do we think about and study the consequences of environmental change for international relations? Second, can the concept of the Anthropocene encourage new and productive frameworks for studying international relations? Thirdly, can these concepts and resulting research

---

Prof. Dr. Richard Ned Lebow, Professor Emeritus of International Political Theory in the Department of War Studies, King's College London (part-time) and James O. Freedman Presidential Professor Emeritus at Dartmouth College. He is also an Honorary Fellow of Pembroke College, University of Cambridge. Email: nedlebow@gmail.com.

R. N. Lebow (✉)
Department of War Studies, King's College London, London WC2R 2LS, UK
e-mail: Richard.Lebow@kcl.ac.uk

H. G. Brauch (ed.), *Towards Rethinking Politics, Policy and Polity in the Anthropocene*, The Anthropocene: Politik—Economics—Society—Science 35,
https://doi.org/10.1007/978-3-031-71807-6_5

have any effect in reducing or coping with the effects of climate change? I am going to pass on the first question, as there is already a literature on international relations and the environment, and I have nothing to add to it as I do not work in this field. Instead, I will direct my attention to the second and third questions.

The Anthropocene Working Group has defined the Anthropocene as:

> the present time interval, in which many geologically significant conditions and processes are profoundly altered by human activities. These include changes in: erosion and sediment transport associated with a variety of anthropogenic processes, including colonisation, agriculture, urbanisation and global warming. The chemical composition of the atmosphere, oceans and soils, with significant anthropogenic perturbations of the cycles of elements such as carbon, nitrogen, phosphorus and various metals. Environmental conditions generated by these perturbations; these include global warming, ocean acidification and spreading oceanic 'dead zones'. The biosphere both on land and in the sea, as a result of habitat loss, predation, species invasions and the physical and chemical changes noted above.

5.1 International Relations

The Anthropocene refers to a geological era, and its supporters are thinking of having it recognised as such by an official international body. It nicely marks off an era or age in which the interaction of our species with its planet has been pronounced in multiple dimensions. The first problem such a concept poses for social science, and for international relations scholars in particular, is the time scale in question. We have never thought in terms of geological ages as we focus on much more immediate problems. Even such questions as the rise of the state, or the emergence of modern warfare, encompass centuries at most, not millennia or millions of years. Most social scientists would resist adopting such a perspective, because they do not see it relevant to the kinds of problems they address and the questions they ask about them. The Anthropocene might find a better reception in global history, but even it begins relatively recently, with the first signs of human settlements and agriculture.

This does not mean that we face an impossible task. A parallel might be drawn with the efforts of some international relations scholars to introduce concepts from evolutionary psychology. It too adopts a long time line, at least as far back as the Pleistocene, where some of its proponents allege humans acquired key characteristics relevant to present-day politics. Other international relation scholars consider evolutionary psychology a false science, and some point to it irrelevance on the grounds that evolutionary change is slow and political change fast. The practice of politics and war, as well as the frequency of the latter, has 'evolved' considerably over the past several hundred years, while the human genome has not.[1]

At least some scholars in my field are willing to consider longer time perspectives if they can be made germane to present-day problems. I think a stronger case can be made for the Anthropocene than for evolutionary psychology. In the first instance, this

[1] For a discussion of this controversy, Lebow, R.N. (2017): "Evolution, Adaption, and Imitation", in: Thompson, W.P. (Ed.): *The Oxford Encyclopedia of Empirical International Relations Theory* (Oxford: Oxford University Press) Vol. I, pp. 799–819.

geological frame of reference captures and directs our attention to some very short-term changes. The rise in global temperature and its effects really began to increase dramatically with the industrial revolution and the human population explosion – that is, since the mid-nineteenth century. This long perspective nevertheless focuses our attention on recent events. It also is relevant to the kinds of problems that concern international relations scholars. With regard to substantive questions, our field is sub-divided into security and international political economy. Environmental change affects both in important ways.

Unlike evolutionary psychology, the Anthropocene also has something to tell us about the future. There is a band of uncertainty about the increase of carbon dioxide in the Earth's atmosphere and the global warming that will result. And also about the many environmental consequences that will follow upon the degree of warming that actually occurs. Even relying on linear measures and consequences, we can make some general predictions about what will happen. They in turn provide the input for scholars in security and international political economy to think about their respective consequences. Prediction is out of the question – as it is more generally in international relations – but forecasting is an important tool. We can think about how these environmental changes will have political and economic impacts and construct narratives about them.

Non-linear events are common in international relations, and are one of the reasons why prediction is a false goal. Confluences of relative independent chains of causation can combine to have consequences much greater than the sum of their expected consequences. I have argued that World War I was the product of such a confluence because in the summer of 1914 it made leaders in Berlin, Vienna and St. Petersburg far more risk prone than they had been six months earlier – or might have been two years hence.[2] The same argument can be made about the end of the Cold War.[3]

Climate scientists warn that global warming may have similar effects, including the possibility that the Gulf Stream could suddenly reverse its course and flow back from Canada to the Caribbean instead of crossing the North Atlantic to Europe. If so, Europe would freeze and the glaciers would cover again much of the continent in possibly as little time as a century. Climate science accordingly provides models and data that can form the basis of non-linear political and economic narratives. What would be the political and economic consequences of such a catastrophe? Would Europeans flee to more hospitable regions? Would 'boat people' from southern Europe seek refuge in countries like Libya and Syria?[4]

The Anthropocene is relevant to international relations because it is an eye-opening framework that dramatizes the extent to which humans have transformed the planet. The concept of the Anthropocene enhances our normative commitment to do something about climate change before it is too late. It provides an incentive to focus

[2] Lebow, R.N., 2010: *Forbidden Fruit: Counterfactuals and International Relations* (Princeton: Princeton University Press): Chap. 3.

[3] *Ibid.*: Chap. 4.

[4] For a short story that imagines this reversal, Lebow, R.N, 2022): "Rough Waters", in: Lebow, R.N., 2022: *Rough Water and Other Stories* (Cambridge: Ethics International Press).

more research on the ways in which international politics is a contributing cause of climate change, and also on its consequences of climate change for international relations. It should push us to develop new concepts appropriate to studying these interactions between international relations and the environment and the environment and international relations.

These concepts can be developed within most existing paradigms of international relations. I will examine three in this paper: realism, liberalism, and constructivism.

Realism assumes that power is critical to relations between and among states. Sophisticated realists recognise that power is constrained by norms in some regional societies and that states, while legally sovereign, are increasingly interdependent. Interdependence has security and economic components, at times mutually supportive and on other occasions not. It generally has the effect of restraining actors and making their interactions more rule based. Liberals stress the role of institutions in this regard. They are said to make cooperation more attractive and increase the costs of defection.

The environmental threat should provide just as great an incentive as security and wealth do for international cooperation and interdependence. To the extent that environmental degradation is seen as a serious threat to national security and material well-being, states will want to cooperate to address the threat. Perhaps even more than security and material well-being, it can only be addressed collaboratively. Individual states, especially large and economically powerful ones that continue to pollute can do extensive damage to other states even if they are in other parts of the globe. This is hardly news to either scholars or policymakers. States have acted, sometimes successfully (e.g., the Montreal Protocol) and sometimes not (e.g., the Paris Accords) to limit damage to the environment. One of the critical questions of international relations is explaining this differential level of success, and many scholars from around the world already address this question. Certainly, one reason is the failure, especially in the U.S., of denial of climate change and political opposition to doing anything to address by some very powerful actors. Adoption of the Anthropocene by scholars will, alas, have no impact on their thinking and political preferences.

Classical realism contends that great powers are their own worst enemies because they succumb to hubris. Successful states and their leaders convince themselves that they can manipulate others and their environment to advance their parochial ends. Their actions often have tragic and unintended consequences, and may lead to the decline, or even collapse of their power. Beginning with Thucydides, tragedy in international relations has focused on efforts by powerful actors to manipulate, influence, cajole, or coerce other actors.[5] Human manipulation and exploitation of the earth can also be constructed as a tragedy for much the same reason. Studies that draw on the Anthropocene to frame it this way have the potential to make us more self-conscious of the longer-term and often unpredictable consequences of our behaviour.

[5] Erskine, T.; Lebow, R.N. (Eds.), 2011: *Tragedy and International Relations* (London: Palgrave-Macmillan).

Classical realists are more interested in influence than in power. They accuse postwar realists of focusing on power because they think it can be objectively measured and used as the basis for a parsimonious, systems level theory of international politics. Classical realists like Hans Morgenthau emphasise influence because material capabilities are only one source of power and power is only one source of influence. Influence, moreover, is context dependent. State A may have influence over B and B over C, and C, in turn, over A. This is because influence rests on many considerations other than power (e.g., depends on the interests of the states involved, is path dependent, has an important rhetorical component, and requires skilful actors).

The Anthropocene might lead some realists to recast their paradigm. There will be much resistance to doing so, as there is already to recognition that in a robust regional society like Europe power can only effectively be wielded through institutions and in accord with their rules. It is not that power is no longer important, but that it is constrained and more dependent on political skill, good arguments, and coalition building. The next generation of realists may grasp this truth, and also the need for transnational cooperation to address the environment. If aliens were about to invade or destroy the earth – as every Hollywood producer recognises – most human beings would put aside their parochial differences to counter the threat. Realists would cite this cooperation – as they do alliances – as confirming evidence of their claim that security must be the first concern of states. The environment, and the threat climate change and its degradation pose, is no different.

States that act to reduce their carbon dioxide emissions, act to reduce other environmental threats and hazards, and provide incentives to business and citizens to act more responsibly with respect to the environment, can emerge as role models for others. They may acquire soft power in the sense of being respected and admired for what they are accomplishing, and this in turn can become a powerful source of influence.

Liberals believe that human beings are primarily motived by wealth. As noted above, believe that the so-called anarchy of the international environment can be reduced, if not overcome, by institutions. The thicker they are, and the greater incentives actors have to obey their rule packages, the more international society comes to resemble its domestic counterparts. The immediate question that springs to mind for liberals is the extent to which growing concern for the environment and environmental change – they are not the same thing – are likely to lead to more or less cooperation among states, and more or less honouring of the norms and rules of existing institutions. Once again, the concept of the Anthropocene is helpful in the sense that it can provide a more integrated understanding of environmental change and accordingly provide the basis for better forecasting and political analysis.

The Anthropocene might also encourage a more radical reformulation of liberalism. Liberals stipulate a number of conditions that are essential to investment and economic development. They include physical security, stable currency and financial institutions, enforcement of contracts, and free movement of goods and money. Environmental security should surely be added to this list, as it is essential in its own right but also increasingly a prerequisite of other conditions on the list. Liberals favour state intervention to buttress, even create, fair markets, robust and secure money supply, enforceable contracts, so too should they support whatever is necessary to maintain environmental stability. Just as they expect governments to go after free-riders and other cheats, they should expect green legislation and action against polluters. Many liberals – in the academy and business worlds do favour such action – but liberalism has failed to provide a solid theoretical justification. Adoption of the Anthropocene would encourage a broader foundational framework for liberalism that would explicitly connect environmental stability to other core assumptions.

Constructivism is an epistemology, and unlike realism and liberalism, makes no substantive claims about the world. Although, some constructivists who focus on identity, do assert – wrongly, in my view – that stable, consistent identities are necessary to human contentment and international security.[6] For most constructivists, what matters are the reasons people have for acting as they do, as these reasons must be discovered through a bottom up, not top down process. This must start with the recognition that actors are embedded in societies, and that these societies confer identities, help to shape actor goals and understandings of the contexts in which they operate. Influenced by the linguistic turn, constructivists further assert the importance of intersubjective understandings that impart meaning to all action.

It is difficult to wed substantive concerns, like that for the environment, with epistemological ones. But identity might provide a bridge, and the Anthropocene the catalyst. Constructivists understand that self-identifications are socially imposed but to some degree matters of choice. They recognise, and have extensively studied, the extent to which local and trans-national identities have emerged as important, especially in Europe. They have added or strengthened these layers of identity and have helped reframe the meaning of national identity to make it more consistent with them. Presumably, a global identity could supplement these other ones and make those who incorporate it more committed to protecting the environment. The invention and propagation of the Anthropocene could be described by constructivists as a political act intended to do precisely this. The more central it becomes to scholarship the more likely students are to incorporate some kind of global identity. Constructivists should have an interest in studying this process, the ways in which changes at one level of identity require changes at others, and how additions, or strengthening, of other levels of identity may require prior changes in the definitions of existing ones.

Social science differs from the physical sciences because its concepts are all reifications. Consider the balance of power and temperature. Temperature ultimately refers to the energy level of molecules, and for this reason all temperature scales are interchangeable. The balance of power, by contrast, is ideas all the way down, never connecting to anything physical that can be measured with any degree of precision. For this reason, realists disagree among themselves about whether the current world is unipolar, bipolar, or multipolar.

---

[6] Lebow, R.N., 2012: *The Politics and Ethics of Identity* (Cambridge: Cambridge University Press): esp. Chaps. 1 and 3.

All social science concepts are subjective in this way because they are ideas resting on other ideas. The Anthropocene, like many scientific concepts has a more substantive foundation. Dimensions of the environment that interest us are indeed concepts of our making, *artifacts* of the mind. However, useful ones connect to physical features of our planet that can be measured with some degree of precision. There will, of course, always be problems of access to good data and measurement error. As noted, problems also arise in trying to assess the effects of change for the planet, and more so when there are non-linear interaction effects. Many of the latter are almost, by definition, unknown until they occur.

Human actors, unlike subatomic particles, are also reflexive. They learn, or think they have learned, from experience, and for this reason are often unlikely to respond to the same challenge or problem as they, or others did the last time. The very categorisation of last time is, of course, also reflexive, and subjective, as are categorisation and assessments of the consequences of previous responses. The concepts we invent and the knowledge we claim they help produce influence behaviour once they are known. Social knowledge – or what passes for it – can undercut some patterns of behaviour and strengthen others. We still know little about these diverse affects, but with respect to climate change, are particularly interested in those likely to change behaviour. Still, climate and environmental science provide a 'real' foundational base lacking in ordinary social science. Philosophical realists would claim otherwise, insisting for example that markets are 'real'. However, this is a highly questionable move that is rejected by most social scientists.[7]

The Anthropocene accordingly provides an appealing foundation for social science. We must recognise, of course, that climate and environmental science – like geology more generally – are better at explanation than prediction. Too many variables are in play to make predictions, and outcomes will often depend on confluences and non-linear interactions. More uncertain still, are efforts – important as they are – to understand the consequences of environmental change for human physical and material wellbeing. Here, we are carrying out social science research and must introduce other assumptions, none of them based on anything real. Nevertheless, tacking back and forth between the physical and the social, between the real and the imagined, is an improvement on what we currently do.

5.2 Political Will

All the scientific knowledge in the world will not make a difference in the absence of political will to act on it. Here, let me make an analogy to the threat of human extinction posed by nuclear weapons. The problem was not a scholarly one, although to be sure, there were considerable differences within the academic community about how to cope and whether nuclear deterrence was a threat or a solution. There were

[7] For a discussion of these claims and resulting controversy, see Lebow, R.N., 2014: *Constructing Cause in International Relations* (Cambridge: Cambridge University Press), Chaps. 1 and 2.

many non-scientific obstacles to doing away with nuclear weapons or even instituting a builddown through arms control. SALT I and SALT II did not due the latter; they slowed the development of new weapons and delivery systems and made the process more transparent. It was not until the end of the Cold War, that any meaningful progress was made, and it turned out to be short-lived given the current arms buildups in the U.S., Russia, and China.

Meaningful responses to the nuclear threat were hindered by Cold War fears, a large military-industrial complex in both superpowers, and widespread fear of nuclear war. Paradoxically, the latter made people seek refuge in deterrence – as many do in prayer – despite evidence in the case of the former that it was more provocative than preventative in its effects.[8]

A similar problem exists in the U.S. today with respect to gun violence. Social science research has made it abundantly clear that one variable explains the level of gun violence around the world: gun ownership. The more guns that are owned the more violence, and the more mass shootings. No other explanation comes anything close to explaining the variance.[9] Knowledge of this kind has done little to change public opinion in the U.S. or to convince the Congress to tighten gun laws. Rather, the National Rifle Association has succeeded in convincing many more people that more shootings require more guns. Here too, the evidence is unambiguous: gun ownership, which carried enormous risks of its own, does nothing to prevent mass shootings.[10]

Media barrages, advertising, fake news, and special interest lobbying are far more effective than scientific knowledge in building public support and influencing policy. Those with an economic interest in the status quo routinely exaggerate the cost of changing their way of doing business, and some may commission 'bad science' to counter the growing evidence that what they are doing is harmful to the environment, or peoples' health in the case of smoking. Good science can expose the 'it costs too much' argument, but arguments alone are generally ineffective. Companies have to learn from experience that they have built a false barrier to change and may make money from going green. Nothing so effectively illustrates this truth as the experience of the hole in the ozone layer and subsequent Montreal Protocol to ban the production of CFCs.

At present, the corporate world is divided in their approach to climate change. Some – largely American companies– stonewall it. Others are concerned and acting in various ways to go green and make profit by doing so. This movement has gone furthest in Germany and Northern Europe. So the problem is not capitalism, but politics. In Germany and Northern Europe, political systems and public opinion are supportive of these moves, but not in the U.S. The Trump administration has cut

---

[8] See Lebow, R.N.; Gross Stein, J., 1994: *We All Lost the Cold War* (Princeton: Princeton University Press, 1994); Lebow, R.N.; 2017 *Avoiding War, Making Peace* (London: Palgrave-Macmillan).

[9] Fisher, M.; Keller, J., 2017: "What Explains U.S. Mass Shootings? International Comparisons Suggest an Answer", in: *New York Times*, 7 November; at: https://www.nytimes.com/2017/11/07/world/americas/mass-shootings-us-international.html?hp&action=click&pgtype=Homepage&clickSource=story-heading&module=b-lede-package-region&region=top-news&WT.nav=top-news (17 November 2017).

[10] *Ibid.*

all tax benefits for going green and is doing its best to reward the coal industry. The West coast of the US is democratic and the governors of Washington, Oregon, and California are cooperating to establish common environmental standards. As they constitute a giant market, many companies will be compelled to conform to them. This happened in the past with the automobile industry when California issues regulations concerning auto emissions.[11]

There are numerous existing efforts to influence policy around the world. In the U.S., these include scientific groups like the National Research Council and the Union of Concerned Scientists, wildlife organizations like the Sierra Club and World Wildlife Fund, and a slew of lobby groups specially focused on climate change. The National Research Council in the US is an official institution that publishes studies on critical issues and controversies in which science can help in making educated policy choices. The Union of Concerned Scientists is a non-profit organisation with the same ends. Presumably, most other developed countries have parallel organisations.

Given these efforts, professionally led by organisations with widespread credibility, there is no reason for those of us interested in the Anthropocene to attempt to duplicate their efforts. Rather, we should devote our efforts to reordering the problems our several disciplines currently prioritise, developing concepts germane to the study of the interaction of environment and politics, and above all, thinking about how we can bring the resulting research findings to the attention of relevant political actors and audiences.

References

Erskine, Toni; Lebow, Richard Ned (Eds.), 2011: *Tragedy and International Relations* (London: Palgrave-Macmillan).

Fisher, Max; Keller, Joseph, 2017: "What Explains U.S. Mass Shootings? International Comparisons Suggest an Answer", in: *New York Times*, 7 November, https://www.nytimes.com/2017/11/07/world/americas/mass-shootings-us-international.html?hp&action=click&pgtype=Homepage&clickSource=story-heading&module=b-lede-package-region&region=top-news&WT.nav=top-news (17 November 2017).

Gareau, Brian J., 2013: *From Precaution to Profit: Contemporary Challenges to Environmental Protection in the Montreal Protocol* (New Haven: Yale University Press).

Lebow, Richard Ned, 2010: *Forbidden Fruit: Counterfactuals and International Relations* (Princeton: Princeton University Press).

Lebow, Richard Ned, 2012: *The Politics and Ethics of Identity* (Cambridge: Cambridge University Press).

Lebow, Richard Ned, 2014: *Constructing Cause in International Relations* (Cambridge: Cambridge University Press).

Lebow, Richard Ned, 2017: *Avoiding War, Making Peace* (London: Palgrave-Macmillan).

Lebow, Richard Ned, 2022: "Rough Waters", in: Lebow, Richard Ned: *Rough Water and Other Stories* (Cambridge: Ethics International Press).

[11] Gareau, B.J. 2013: From Precaution to Profit: Contemporary Challenges to Environmental Protection in the Montreal Protocol (New Haven: Yale University Press).

Lebow, Richard Ned, 2018: "Evolution, Adaption, and Imitation", in: Thompson, William P. (Ed.): *The Oxford Encyclopedia of Empirical International Relations Theory* (Oxford: Oxford University Press).
Lebow, Richard Ned; Stein, Janice Gross, 1994: *We All Lost the Cold War* (Princeton: Princeton University Press).

Chapter 6
Peace and Security Discourses on the Anthropocene: An International Relations Perspective

Dahlia Simangan

Abstract The Anthropocene has become more than a philosophical byword; it has also entered the vernacular of international policy circles. What are the implications of the Anthropocene for our understanding and pursuits of peace and security, and how does the International Relations (IR) discipline engage with peace and security discourses in this new geological age? A combination of bibliometric and thematic analyses presented in this chapter reveals that prominent discourses on the Anthropocene are generally anchored in traditional notions of peace and security. However, the IR discipline provides a space for a critique of these notions, thereby expanding conceptions of agency, challenging linearity and modernity, and unpacking the political motivations and implications of conceptual expansion. This chapter closes with several recommendations for advancing interdisciplinary conversations among IR, Peace Studies, and Security Studies, as well as with other disciplines.

Keywords International relations · Anthropocene · Peace · Security · Bibliometric analysis · Environment · Climate change

6.1 Introduction

The Anthropocene is the new geological epoch proposed to replace the current Holocene to mark humanity's unprecedented and growing impact on the Earth's biosphere. Scientists have been quantifying the thresholds of the Earth system's resilience to the impact of social and economic development. For instance, the concept of planetary boundaries connotes the 'safe operating space for humanity with respect to the functioning of the Earth system' (Rockström et al. 2009: 2). In 2015,

Dahlia Simangan, Assoc. Professor, Hiroshima University, Hiroshima, Japan; Email: simangan@hiroshima-u.ac.jp.

D. Simangan (✉)
The IDEC Institute, Hiroshima University, Hiroshima, Japan
e-mail: simangan@hiroshima-u.ac.jp

H. G. Brauch (ed.), *Towards Rethinking Politics, Policy and Polity in the Anthropocene*, The Anthropocene: Politik—Economics—Society—Science 35,
https://doi.org/10.1007/978-3-031-71807-6_6

human activities crossed four of the nine planetary boundaries—climate change, biosphere integrity, biochemical flows of phosphorous and nitrogen, and land use (Steffen et al. 2015). More recently, we have also exceeded the planetary boundary on environmental pollutants, especially plastic (Persson et al. 2022). Stratigraphic evidence of human-driven climatic, biological, and geochemical changes signifies the advent of the Anthropocene (Waters et al. 2016). The devastating floods in Pakistan in June 2022 and the 'heat apocalypse' in Europe a month later are just some of the recent tell-tale signs of impending doom in the Anthropocene.

Discussions on the Anthropocene, however, are not all about doomsday scenarios. There are several proposed interventions to natural processes using knowledge and technology to avert looming catastrophes while maintaining the socio-economic status quo of development. According to ecomodernists or the proponents of a 'good Anthropocene', our geological reality now 'demands that humans use their growing social, economic, and technological powers to make life better for people, stabilise the climate, and protect the natural world' (Asafu-Adjaye et al. 2015: 7). These propositions have been met with criticism for re-asserting human domination over nature—the root cause of the Anthropocene. Unless the problematic view of human-nature dualism is overcome, the long shadow of colonialism, and present forms of marginalisation, such hubristic propositions can easily lead to further marginalisation and other irreversible moral hazards (Biermann/Möller 2019; Grunwald 2018). Without completely abandoning the prospects of a 'good Anthropocene, McPhearson et al. (2021) argue the importance of radically rethinking the concepts and practices behind growth, efficiency, state governance, resource governance, and justice when it comes to decisions and policies in the Anthropocene. These are opportunities for shifting away from the 'business and politics as usual' and path-dependent practices that have dominated international negotiations surrounding global environmental change.

The impact of human activities on the habitability of our planet also extends to the stability of our societies. Study upon study shows that even though climate change does not necessarily lead to conflict directly, its impacts and consequences can exacerbate existing (risks of) conflicts (Koubi 2019; Sharifi et al. 2021). In most cases, it is the failure of governance that links climate to conflict. For example, several analyses of the role of climate in the war in Syria see the government's political and economic policies as more significant in fuelling the conflict than droughts and drought-driven migration (Selby et al. 2017 contra Kelley et al. 2015; see also Ide 2018 for a review of climate-conflict studies on Syria).

The indirect relationship between climate and conflict also confounds the conditions for international peace and security. For instance, while competition over scarce transboundary resources does not, by itself, fuel inter-state conflict (e.g., Swain 2015), and with the evidence of cooperation over resource management being more common than violent conflicts (e.g., Bernauer/Böhmelt 2020), short-term political interests regarding collective and inter-generational responsibility may create a hotbed for geopolitical rivalries (Buhaug et al. 2022; Colgan et al. 2021).

The confluence of the abovementioned factors that threaten humanity's survival is metaphorically depicted in the Doomsday Clock developed by the *Bulletin of the Atomic Scientists*. In 2020, scientists set the clock to 100 seconds to midnight, with midnight symbolising the end, when humanity ultimately destroys the world. The clock serves as a warning of potential global catastrophes brought by nuclear threats, climate change, global pandemics, and widespread disinformation (Bulletin of the Atomic Scientists 2022). This is one of the manifestations of the growing recognition within policy circles that social systems and the Earth system do influence each other. For instance, also in 2020, the United Nations Development Programme (UNDP) incorporated the term "Anthropocene" into its annual Human Development Report for the first time since the inaugural publication in 1990. Its policy recommendations take into account how natural and social systems are linked, interdependent, and embedded in each other (UNDP 2020). The UNDP published a special report in 2022 about the human security threats from climate change, biodiversity loss, COVID-19, and digital technologies (UNDP 2022)—the same threats raised by the scientists behind the Doomsday Clock. The report suggests a rethinking of the concept of human security in the context of high levels of insecurity despite high levels of human development. Recognising that 'higher incomes, for instance, do not automatically bring about peace and that a society without violent conflict is not a sufficient condition for people to live in dignity' (UNDP 2022: 21), the report calls for greater solidarity across all levels—from local communities to inter-governmental organisations—to guarantee human security amidst complex and cascading challenges.

The challenges in the Anthropocene are its shifting environmental, social, and political landscapes, and this is where this chapter comes in. It asks the question of how IR scholars approach peace and security in the context of the Anthropocene. To answer this question, I employ a combination of bibliometric and thematic analyses of selected scholarly articles to identify dominant discourses and marginal perspectives within IR's scholarship on peace and security in the Anthropocene.

6.2 Dominant IR Discourses on the Anthropocene

How do IR scholars make sense of the Anthropocene? According to Lövbrand et al. (2020), three prominent discourses emerge in the discipline: the endangered world, the entangled world, and the extractivist world. In the endangered world, IR scholars take a problem-solving stance towards saving the planet and securing humanity. In the entangled world, IR scholars problematise the consequences of modernity and champion co-existence with nonhuman nature. And in the extractivist world, IR scholars call for a break from the status quo and demand a transformative politics that will disrupt processes that lead to environmental degradation and the injustices it brings. Lövbrand et al. (2020) recognise that these discourses can still change and evolve as debates continue within the IR discipline, but their review shows that the Anthropocene demands a paradigm shift away from politics-as-usual and a radical rethinking of values, practices, and institutions that contribute to the challenges in the Anthropocene. These are some of the scholastic debates and policy discussions where IR scholarship can provide added value to mitigating the harmful consequences of global environmental change.

There are also complementary reviews that provide an overview of the broader literature. The most comprehensive of these was undertaken by Brauch (2021), who not only provided a bibliometric/scientometric analysis of all types of publications, drawn from various databases, but also scoped the impact of the concept on a wide range of disciplines, including Political Science, Security Studies, and IR. Within IR, the Anthropocene has become a valuable theoretical landscape for rethinking the ontologies and epistemologies within the discipline and infusing a sense of urgency in dealing with traditional IR subjects on politics and security (Simangan 2020). However, the dominant discourses fail to represent the perspectives of the Global South in the Anthropocene debate (Marquardt 2019; Simangan 2021b). This calls for further examination of the Anthropocene debate, including discourses on peace and security.

In her review of Security Studies' engagement with the Anthropocene, Hardt (2019) argues that the appeal of the Anthropocene to Security Studies lies in the threat it carries, specifically the threat to human existence. However, she also raises concerns about the securitisation of these threats, mainly due to the lack of definition and debate on the Anthropocene within Security Studies, and emphasises the need for a more transdisciplinary approach to responding to these threats. Despite references to peace, conflict, and security within the broader Anthropocene discourse, Hardt also observed that such a discourse needs further interdisciplinary exchange with IR to enrich these bodies of knowledge. This chapter contributes to this knowledge gap by identifying existing Anthropocene discourses in IR, Peace, and Security Studies, and mapping potential pathways for deeper interdisciplinary engagement.

At this point, it is imperative to define peace and security. Among definitions of peace, Johan Galtung's (1969) conceptualisation of negative and positive peace has become a common definition in peace research. Briefly, negative peace is the absence of war and direct or organised violence, when violence ceases due to certain efforts to end the conflict. On the other hand, positive peace refers to the creation and maintenance of conditions that enable human flourishing by eliminating structural violence. In this chapter, I refer to both negative and positive peace to provide a holistic understanding of peace in the Anthropocene. The selected articles will be categorised according to negative or positive peace, or both, based on the types of conflict they deal with.

The concept of security, like peace, is also multidimensional. It is a product of negotiations and contestations along political and ethical lines beyond politico-military concerns (Buzan et al. 1997; McDonald 2008). The most prominent concepts of security are traditional security, which pertains to military threats to state sovereignty and territorial integrity, and human security, which encompasses various aspects of human well-being. The concept of human security rose to prominence after the Cold War, especially with the publication of the UNDP Human Development Report in 1994. The report renewed the UN's commitment to 'freedom from fear' (national security, state sovereignty, and territorial integrity) and 'freedom from want' (e.g., job security, income security, environmental security, and security from crime; UNDP 1994). The report reminded the international community that security is not only threatened by competing militarised nations, but also by the daily concerns of people. Human security, which 'lies in development, not in arms,' is the real and sustainable foundation of peace (UNDP 1994). Indeed, peace and security are intrinsically linked, and they are also embedded in research agendas and policy discussions on development and the environment (Brauch 2008).

For the purpose of this chapter, I differentiate environmental security from human security for a more granular analysis of the selected articles. The concept of environmental security steers away from the state-centrism of traditional security by focusing on environmental degradation as a threat to human security, a cause of conflict, and a means for a sound security policy. However, it has also been criticised for relegating the environment to a mere tool for human advancement that justifies military means toward environmental protection (e.g., Graeger 1996). In contrast, ecological security secures the entire ecosystem, for the benefit not only of human beings, but of all living beings (McDonald 2021). According to McDonald (2021), ecological security also focuses on the most vulnerable populations, including future generations. It aligns with the concept of planetary security, which broadens responses to include the future of humankind and the planet. McDonald's (2021) conceptualisation of ecological security is not only informed by climate-conflict perspectives and militaristic responses to these threats, but focuses more on ecosystem resilience and the rights and needs of the vulnerable. For these reasons, and although some of the articles deploy the term 'planetary security', the analysis for this chapter will include national or state security, human security, environmental security, and ecological security.

6.3 A Systematic Review of Literature

This section lays out the methods for conducting a systematic review of IR discourses on peace and security in the Anthropocene. I combined bibliometric and thematic analyses of scholarly articles to provide an overview of publication trends and identify dominant topics or issues, respectively.[1] While a narrative or traditional literature review is vital for any research, the transparent and reproducible review strategies in bibliometric analysis complement traditional reviews and are widely used across disciplines, including in Politics and International Relations (e.g. Glänzel et al. 1999; Palomo et al. 2017; Sillanpää/Koivula 2010; Simangan 2020). The selection process for scholarly articles was guided by a modified version of ProKnow-C, an instrument for a structured selection of bibliographic materials that ensures completeness and the

[1] The method used in this chapter for studying Anthropocene discourses is not the type of discourse analysis that looks into the production, preservation, and propagation of power and knowledge through language and communication.

relevance of the bibliographic portfolio to the review purpose (Ensslin et al. 2015). This instrument is valuable in organising a large quantity of information, especially when scoping the state of knowledge and identifying knowledge gaps (e.g., dos Muchangos/Vaughter 2018; Ensslin et al. 2015; Simangan, 2021b). The process consists of '(1) the selection of bibliographic portfolio; (2) bibliometric analysis of portfolio; (3) systemic analysis; and (4) definition of the research question and its objectives' (Ensslin et al. 2015: 785). These steps can be modified according to the focus and objectives of the review.

For the selection of the bibliographic portfolio, I used the *Web of Science* (WoS) database. The comprehensive, peer-reviewed, and organised collection of academic journal titles in WoS makes it an appropriate source for selecting the materials for analysis. The topic search selected articles with titles, abstracts, and keywords containing both the terms Anthropocene and peace. Peace, however, is related to other terms, such as security, conflict resolution, or justice, and their opposites, such as war, crimes, and disputes, for example. To ensure the inclusion of relevant articles in the portfolio, I drew on the search string used in Sharifi et al.'s 2021 study on climate and conflict, and modified it based on terms related to the Anthropocene, peace, and security.[2] I then applied the inclusion criteria of research articles and the English language. While there are books, commentaries, and other grey literature discussing peace and security in the Anthropocene, the application of a document-type criterion for research articles helped ensure uniformity of the selected materials in terms of peer review. Meanwhile, the exclusion of other document types, such as book reviews and editorial materials, avoided the potential double-counting of research publications. The language criterion, which is due to my inability to analyse articles written in languages other than English, is one of the limitations of this search. The topic search was done on 27 September 2022, and resulted in 939 articles. It is important to note at this point that this search string could have excluded related discussions from other fields (e.g., International Political Economy) and topics (e.g., disasters and energy geopolitics) that discuss similar issues but do not usually invoke the term Anthropocene.

For the second step, I selected the WoS categories (or disciplines) related to IR to narrow down the bibliographic portfolio and ensure the relevance of the selected materials to the focus of this chapter.[3] This step resulted in 277 articles (hereinafter referred to as IR-related articles) that were used for the bibliometric analysis. Of these articles, 73 are categorised under International Relations and Political Science (hereinafter referred to as IR articles). I included the Political Science category

---

[2] The search string used for the initial topic search is: TS=(("Anthropocene") AND ("peace*" OR "resolution" OR "justice" OR "negotiat*" OR "securit*" OR "insecurity" OR "mediat*" OR "diplomacy" OR "diplomatic" OR "conflict" OR "dispute" OR "tension" OR "rebel*" OR "arms" OR "armed" OR "assault" OR "militar*" OR "war" OR "warfare" OR "terror*" OR "refugee*" OR "weapon" OR "riot" OR "unrest" OR "protest" OR "strike" OR "repression" OR "crime" OR "police" OR "homicide" OR "murder" OR "robbery")).

[3] The included *Web of Science* categories are International Relations, Political Science, Geography, History, Law, Economics, Social Psychology, Philosophy, Interdisciplinary Social Sciences, Area Studies, Asian Studies and Development Studies.

because of its close complementarity with IR. For the third step, the abstracts of these articles were reviewed to verify that they satisfied all the inclusion criteria. Two articles were review essays and were excluded from the portfolio. The remaining 71 IR articles informed the thematic analysis to define dominant discourses and knowledge gaps in the literature. This step provided an in-depth qualitative review of the articles to complement the broader bibliometric analysis. An information extraction sheet was created in Microsoft Excel to organise the information from the 71 IR articles and categorise them according to the types of peace, types of conflict, security dimension, environmental issues, and scale of analysis, as explained earlier.[4] Based on this analysis, and for the final step, I will present themes and questions relevant to IR, peace, and security as potential pathways for further interdisciplinary engagement.

6.4 Bibliometric Analysis

The data pertaining to the publication year, authors, categories, journals, and authors' institutional affiliations of the 277 IR-related research articles are analysed in this section. During the selection process, there was no limitation on the year of publication, but the earliest article included in the bibliographic portfolio was published in 2006. Since then, we see a growing number of publications on peace and security in the Anthropocene within IR and its associated disciplines (Fig. 6.1). However, the numbers also show that the topic is relatively new, considering that Crutzen/Stoermer (2000) popularised the term as far back as 2000.[5]

International Relations, however, is not the discipline most engaged with the Anthropocene, based on the bibliographic portfolio. Figure 6.2 depicts the distribution of selected articles according to the relevant WoS categories, wherein Geography has the greatest number of articles, with 29% (n = 105). This is not surprising, given the field's focus on the relationship between human societies and their environments. Environmental Studies is the second on the list, with 15% (n = 54), followed by International Relations with 13% (n = 48) and Political Science with 12% (n = 45).

The disciplinary engagement with peace and security in the Anthropocene is also reflected in the top 10 authors and top 10 journal titles in the bibliographic portfolio. Geography professor Simon Dalby has published widely on the topic (Fig. 6.3), and *Geoforum*, an interdisciplinary journal on human geography, published the most research articles (Fig. 6.4). *Millennium*, the only IR-dedicated journal in the list, has

[4] The bibliographic citations files for the selected articles are available upon request to the author.

[5] My earlier review of IR's engagement with the Anthropocene shows that the term only appeared in scholarly articles in 2011 (Simangan 2020), instead of 2006 as indicated in this study. This can be explained by the inclusion of more *Web of Science* categories in this study.

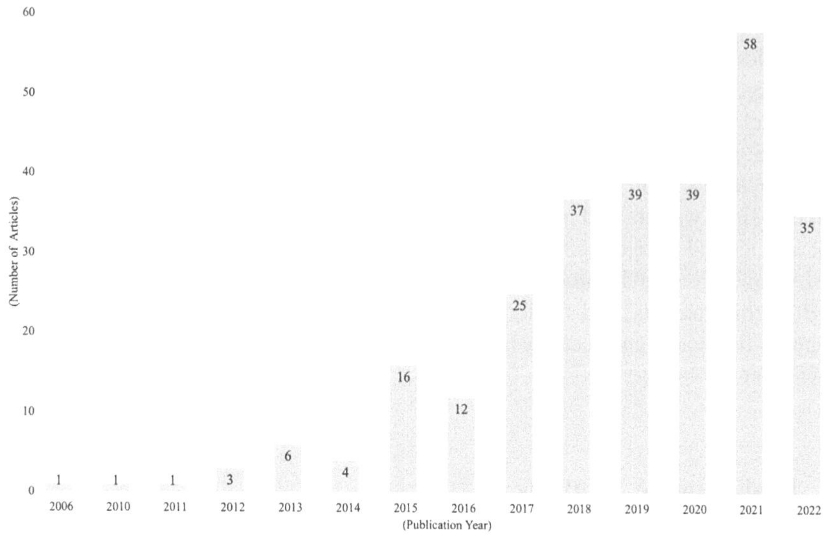

Fig. 6.1 IR-related research articles published from 2006 to 27 September 2022. *Source* compiled by the author based on bibliographic data from the *Web of Science* database

7 articles. The distribution of articles according to authors and journal titles indicates that the Anthropocene has become a useful concept and context in various disciplines, and the issues related to peace and security are relevant across the social sciences.

In terms of geographic distribution, it is not surprising that most of the articles have institutional affiliations in the Global North (Fig. 6.5). The United States leads the group with 77 articles, followed by the United Kingdom with 68, Australia with 44, and Canada with 31. European countries, specifically Germany (n = 28), Sweden (n = 19), the Netherlands (n = 13), and Spain (n = 9), also contribute to research publications. South Africa (n = 15) and Brazil (n = 7) are the only Global South countries in the top 10. Figure 6.5 also locates the top 10 institutions with the most publications on the topic, with the University of London (n = 14) publishing the greatest number. Each country in the top 10, except for Germany, the Netherlands, Spain, and Brazil, houses institutions with high numbers of publications, with United Kingdom and Australia each having three such universities. This geographic distribution reflects the broader disparity between Global North and Global South institutions in academia when it comes to research publications.[6]

To identify dominant research topics and illustrate the links of the IR-related articles in the bibliographic portfolio, I created a network of their keywords using *VOSviewer*, an open-source software tool for visualising bibliometric networks (van Eck/Waltman 2010). I reviewed duplicate keywords and ensured uniformity of the terms (e.g., climate-change to climate change; geographies to geography). Of the 1359 keywords from the 277 IR-related articles, 67 have at least 5 co-occurrences

[6] This is partly due to the English language criterion applied during the selection process.

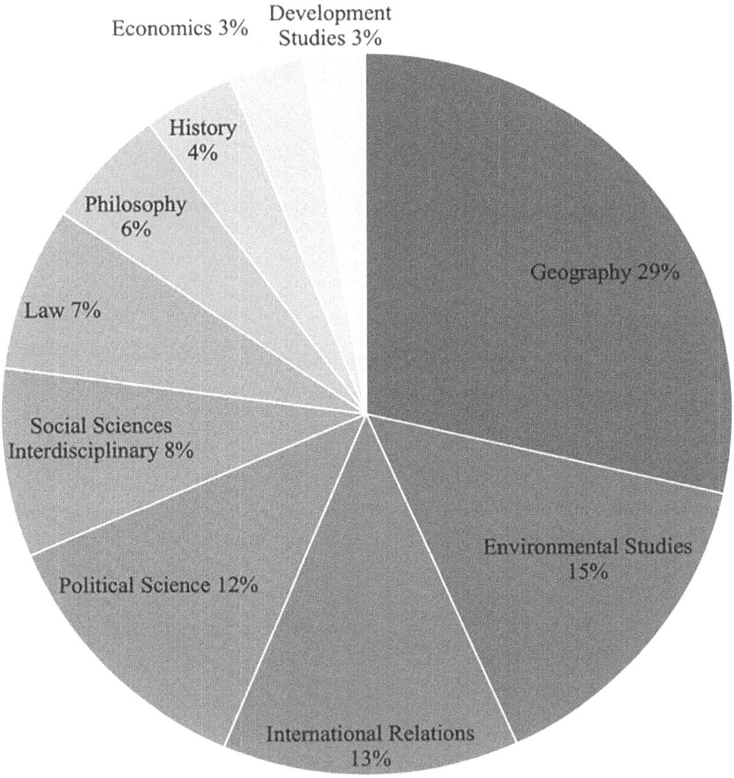

Fig. 6.2 Top 10 IR-related *Web of Science* categories with the most publications on peace and security in the Anthropocene. *Source* compiled by the author based on bibliographic data from the *Web of Science* database

(i.e., the number of documents in which the keywords appear together). The distance of the keywords from one another indicates their relatedness, the size of the rectangles represents the number of times the keywords have been used, and the variation in the thickness of the lines represents the number of co-occurrences or link strength (van Eck/Waltman 2010). The linked terms are grouped into six clusters with their respective colours (Fig. 6.6).

The red cluster consists of 18 keywords: biopolitics; conservation; ecosystem services; environment; environmental change; environmental security; extinction; food security; framework; land; life; management; resilience; resources; risk; vulnerability; and water. Most of these keywords relate to nonhuman nature, suggesting that the concept of the Anthropocene broadens our analysis beyond traditional and

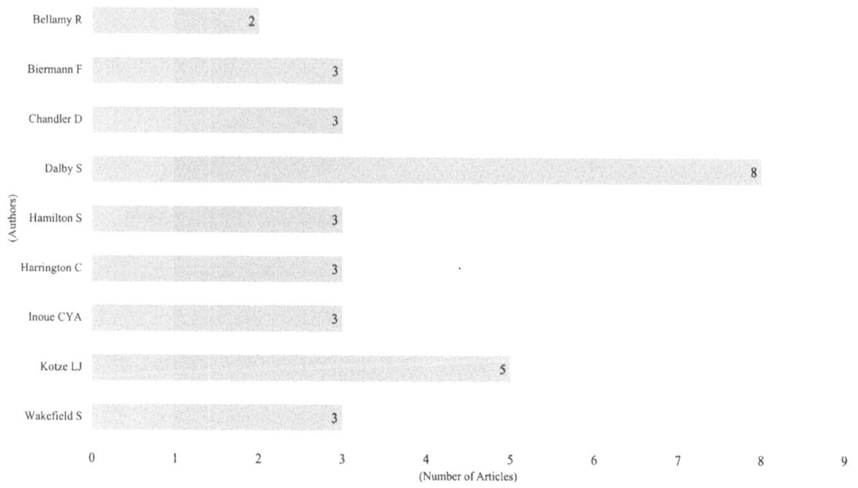

Fig. 6.3 Top 10 most-published IR-related authors on peace and security in the Anthropocene: *Source* compiled by the author based on bibliographic data from the *Web of Science* database

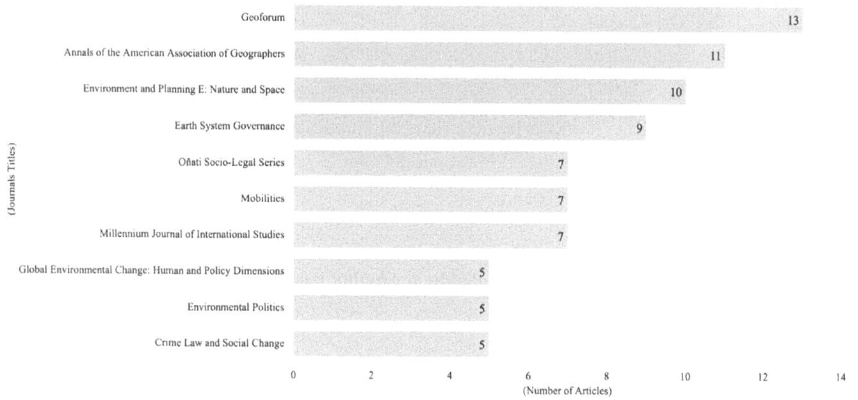

Fig. 6.4 Top 10 IR-related journal titles with the most research articles on peace and security in the Anthropocene. *Source* compiled by the author based on bibliographic data from the *Web of Science* database

human-centric subjects in IR-related disciplines. This broadened scope of subject, however, is still largely linked to issues concerning human societies, such as food security, risk, and vulnerability.

The green cluster contains 11 keywords: adaptation; capitalism; climate change; critique; environmental justice; migration; mobility; nature; politics; refugees; and settler colonialism. These keywords can be categorised as either causes or consequences of the Anthropocene, reflecting the critical and problem-solving components of theoretical frameworks used in IR and its related disciplines.

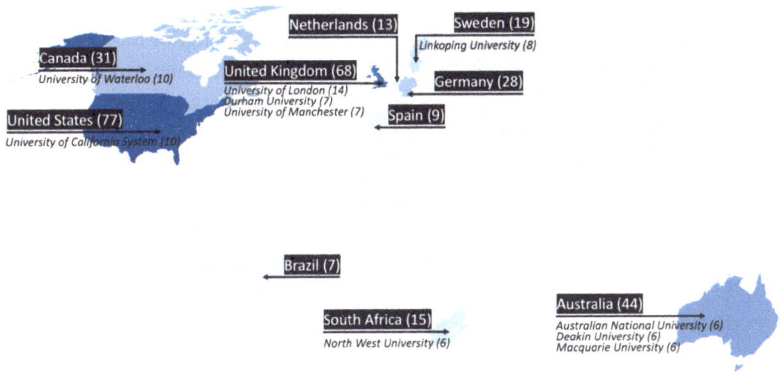

Fig. 6.5 Top 10 countries and top 10 institutions with the most IR-related research articles published on peace and security in the Anthropocene. *Source* compiled by the author based on bibliographic data from the *Web of Science* database

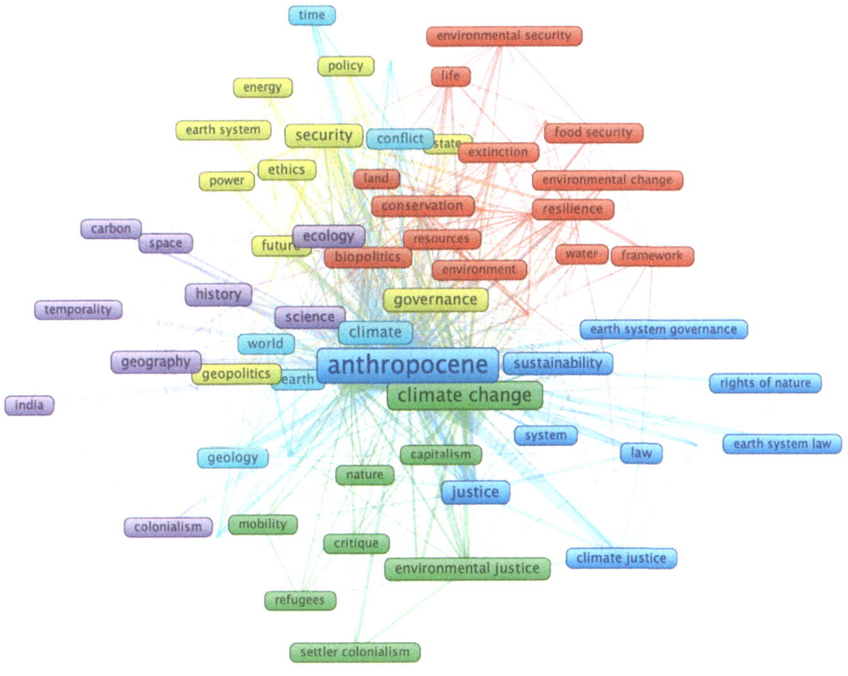

Fig. 6.6 Network visualisation of co-occurring keywords. *Source* compiled by the author based on bibliographic data from the *Web of Science* database and prepared using *VOSviewer.* Note that some of the keywords are not visible in the network map to avoid overlapping labels

The blue cluster has 10 keywords: Anthropocene; climate justice; Earth system governance; Earth system law; human rights; justice; law; rights of nature; sustainability; and system. These keywords indicate aspects of governance in the Anthropocene. Given that the articles are from IR-related disciplines, including political science and law, governance in the Anthropocene is not an unsurprising cluster to emerge in this network mapping.

The yellow cluster also includes 10 keywords: Earth system; energy; ethics; future; geopolitics; governance; policy; power; security; and state. I label this cluster as the politics of the Anthropocene, representing the unsurprising dominance of political aspects of any topic studied within IR-related disciplines. These keywords are also reminders that research on the politics of the Anthropocene gravitates not just toward conventional IR topics, such as geopolitics, power, security, and the state, but also nonconventional but Anthropocene-relevant topics, such as energy, ethics, and the future.

The cyan cluster has 9 keywords: Capitalocene; climate; conflict; Earth; geology; international relations; knowledge; time; and world. Most of these keywords characterise the globality of the Anthropocene, and how it transcends time and space and encompasses a wide variety of knowledge and relations vis-à-vis geological and atmospheric changes. This cluster is also closely linked to the purple cluster, with 9 keywords: carbon; colonialism; ecology; geography; history; India; science; space; and temporality. Most of these keywords can be considered as philosophical concepts or conceptual frameworks that help in *making sense of the Anthropocene.*

This network mapping of keywords shows commonly interconnected topics in peace and security research within IR and the prominent discourses scholars are engaging in about the Anthropocene.

6.5 Thematic Analysis

Having mapped the keywords, I then used the 71 IR articles as sources for a more focused assessment of peace and security concepts or approaches in the Anthropocene. I applied the concepts of negative peace and positive peace in categorising the articles (Galtung 1969). The majority do not have a clear link to either negative or positive peace ($n = 26$), but a significant number of them focus on positive peace ($n = 24$), followed by both types of peace ($n = 12$), with negative peace having the fewest articles ($n = 9$) (Fig. 6.7). The articles dealing with negative peace examine geopolitical tensions, war, nuclear threats, transboundary water conflicts, and migration caused by climate change. Studies that focus on war discuss its implications for human-nonhuman nature relations across levels. For instance, Leep's (2018) article on the killing of stray dogs during the Iraq war raises the question of human responsibility towards animals caught in war zones. Along the same lines, van Munster and Sylvest's (2021) reappraisal of Jonathan Schell's book, *The Fate of the Earth*, contributes to the rethinking of human-Earth relations in the nuclear age, and by extension, in the Anthropocene. While these articles focus on traditional IR subjects,

Fig. 6.7 Pathways between peace and security within IR's engagement with the Anthropocene. *Source* Author-generated Sankeys diagram using *MS Excel* based on bibliographic data from the *Web of Science* database

most of them challenge the centrality of these subjects in IR discourse by recentring their analysis away from human-centric institutions and toward human-nonhuman nature relations.

The articles relevant to positive peace deal with issues such as colonialism, marginalisation, class conflict, and various forms of injustice and structural violence. The scale of analysis of these articles also covers multiple levels, bringing to the fore the voices of indigenous groups and other marginalised groups, which are usually silenced when it comes to global discussions about the challenges of and responses to the Anthropocene (e.g. Amoureux/Reddy 2021; Johnson/Sigona 2022). These discussions, as we are reminded by some of the articles' authors, must be designed and implemented with recognition, inclusivity, empowerment, and the needs of future generations in mind (e.g. Biermann 2021; Christoff 2016; McDonald 2018; Pellizzoni 2021; van Dijk 2021). They also denounce the unsustainability of Western-based structures and institutions of production and consumption, calling for more inclusive and just modes of environmental or Earth system governance (Chandler 2020; Johnson/Sigona 2022). These articles contribute to the critique of the universalising tendencies of the Anthropocene discourse (see also Simangan 2019, 2021b). The articles that are relevant to both negative and positive peace are concerned with issues related to colonialism, racism, and other structural forms of injustice and indirect violence. Thus, I posit that IR discourses on peace and security in the Anthropocene are predominantly engaged with issues that undermine positive peace, while drawing on critical IR theories, such as postcolonialism, postmodernism, and feminism, in discussing the Anthropocene.

While few of the IR articles have an explicit link to either type of peace, their analyses speak to a wide range of security issues—state, human, environmental, and ecological (Fig. 6.7). Some articles are relevant to more than one security issue: for instance, an article might critique state security while advancing human security (in this case, it is assigned under human security). For the purposes of analysis, I categorised all 71 IR articles according to the most prominent security issue in each one. State security is assigned for articles dealing with military, political, and economic dimensions of (inter)national institutions; human security for human needs and freedoms; environmental security for environmental protection and climate action; and ecological security for articles analysing both human and nonhuman nature relations.

From this categorisation, state security's dominance in the literature can be seen (n = 25), followed by ecological security (n = 20), human security (n = 18), and environmental security (n = 8). Figure 6.7 illustrates this distribution; it also shows that negative peace is linked more with state security, and positive peace with human security. Articles relevant to both types of peace are linked more with human and ecological security, and articles without specific discussion on either type of peace have more discussions on state security. The difference between the articles, however, is not wide enough to suggest significant dominance of one security perspective over another. Nevertheless, it is clear that, with the multiple security perspectives considered in the articles, environmental security received the least attention. Meanwhile, state security, which is also conceptually narrower compared to ecological and human security, remains relatively dominant compared to the others. However, this engagement is not necessarily about advocating or preserving state security: many of the articles either critique the limitations of state security or discuss the implications of the Anthropocene for state security. In addition, the majority of the articles on state security do not focus on a specific type of conflict, although some of them examine potential or current conflicts arising from geopolitical tensions in the Anthropocene (e.g., Hansen-Magnusson 2022; Proedrou 2020; Rossi 2017), with some recommending a rethinking or revising of existing political or economic institutions of governance to make them more responsive to global environmental change (Biermann 2021; Burke et al. 2016; Conway 2020; Dalby 2020; Green 2022; Kotze/French 2018; Simangan 2021a). In sum, the IR discipline's engagement with the Anthropocene signifies a critical agenda that challenges the traditional conception and policy translations of peace and security.

6.6 Peace and Security in the Anthropocene

The Anthropocene has become an important contextual backdrop in understanding the causes of and solutions to a wide array of peace and security issues. The challenges in the Anthropocene are considered by many scholars as potential sources of insecurity that can undermine peaceful conditions or exacerbate existing conflicts. A review of IR's articulation of these peace and security issues in relation to the Anthropocene reveals three key observations. Firstly, discussions in IR about expanded notions of agency in the Anthropocene are also reflected in peace and security discourses. With the Anthropocene serving as a reminder of human-nature entanglement, some IR scholars challenge the institutions and systems built around human-nature/culture-geology binaries and adopt a broader understanding of who needs security in the Anthropocene (e.g., Burke 2022; Dalby 2014; Fagan 2017; Gellers 2021; Leep 2018; Lobo 2019; Simangan 2021a; van Munster/Sylvest 2021). This conceptual broadening, however, is accompanied by a disagreement on objectives, according to an earlier review by Lövbrand et al. (2020). Such a disagreement might have to do with the increasing recognition of the multiple experiences of human societies in relation to the environment/nature/the planet—from various objects of security to different ontologies of peace (e.g., Amoureux/Reddy 2021; Chandler 2020; Inoue 2018). The challenge for IR scholars is to connect these multiple experiences (and ontologies): not to assimilate one into the other, but to imagine just possibilities and welcome previously unexplored modes of relationality. As in Conway's vision of cosmopolitical planet politics, this would "[require] the invention of speculative intellectual and political practices that maintain the possibility of unlikely alliances, not only between nations but between planetary agents of all sorts" (Conway 2020: 158).

Secondly, IR's critique of binaries also extends to linearity and modernity, thereby expanding the political notions of time and space. The possibilities of extinction (Fagan 2017; Hamilton 2018; Harrington 2016), including the realities of the environmental crisis and political chaos that would precede it (Burke 2022; Houston 2013), prompt this expansion. Death (2022), for instance, argues that depoliticised and linear visions of the future may reproduce the colonial past and its present remnants, and suggests the welcoming of plural and radical temporalities in theorising climate futures. Another example is Wolff's article on the *Svalbard Global Seed Vault* (SGSV), the operation of which represents the 'politics of reversibility that can extend the present by reversing the outcomes of catastrophic events (Wolff 2021). Relatedly, but on spatial rather than temporal expansion, Megoran and Dalby (2018) suggest that spaces and places can inform peace constructions and peace formations. This expansion helps in reimagining a justice that is intergenerational, socio-ecological, and planetary, and therefore more able to confront the Anthropocene (e.g., Caney 2018; Death 2022; Gupta et al. 2021; Pope et al. 2021). These reimaginations overturn the state-centric and even anthropocentric constructions of temporality and spatiality in relation to security. IR scholarship has made advancements in the critique of neoliberal institutions and how they inform security narratives. For example, Ferguson (2019) argues that there remains a need for a practical clarification of how resilience discourses can actually benefit natural ecosystems and the most vulnerable, with the lack of clarity mainly due to low levels of institutionalisation and formal authority within existing forms of governance. These discourses are often met with a shortfall in implementation—from the inability of state-centric legal frameworks to protect future generations (van Dijk 2021) to the UN's slow progress towards responding to climate security (Hardt 2021b), the mismatch between the EU's policy initiatives on climate security and its traditional geopolitical mindset

(Proedrou 2020), and the continued complicity of capitalist institutions and practices in environmental degradation and disasters (O'Lear et al. 2022), among others.

Finally, this review also highlights ongoing discussions about the perils of conceptual expansion. The debate ensuing from the publication of Burke et al.'s (2016) 'Planet Politics: A Manifesto from the End of IR,' is an example of this. Sounding an alarm amidst environmental crises, the manifesto calls for IR's return to its interdisciplinary roots, but away from its anthropocentric focus and toward restructuring international politics. However, for Chandler et al. (2018), this kind of enthusiastic call is at risk of being another liberal, cosmopolitan and modernist solution to the very problem the liberal cosmopolitan approaches had produced. They counter-warn that 'instituting global governance in "firm and enforceable" ways, as if there were universal solutions that could be imposed from above, is a recipe for authoritarianism and new hierarchies and exclusions' (Chandler et al. 2018: 193).[7] Similar scepticism about reactionary rather than emancipatory approaches to climate change can also be found in warnings about unexpected consequences of impulsive responses to the climate crisis. Blumenfield (2022: 2), for example, argues that 'climate change awareness can just as well equally motivate heightened divisions of humanity into anti-egalitarian, xenophobic, class-differentiated zones of competitive survival.' This 'climate barbarism is a form of adaptation to climate change by prioritising the needs of the in-group and disregarding those of "others" through violence (Blumenfeld 2022). This debate highlights the importance of considering the political motivations for and the implications of dealing with the Anthropocene. As Meyer (2020: 408) warned, post-political interpretations and approaches to climate change and the Anthropocene, as manifested in 'technocratic management and objective scientific determination', limit political (re)imaginations and weaken resilience to uncertainties. On the whole, this particular discourse underscores the different strands of disagreement within the critical research agenda on the Anthropocene.

6.7 Conclusion

This review of peace and security discourses within IR's engagement with the Anthropocene demonstrates that the IR discipline provides a space for a critique of the dominance of traditional conceptions of peace and security. Most of the articles reviewed deal with issues of injustice and structural violence, thereby linking challenges in the Anthropocene to the need to uphold positive peace. Similarly, security in the Anthropocene is understood more broadly than state or human security, contributing to expanded notions of political agency, time, and space. Most importantly, this review reveals the presence of disagreements within IR's critical agenda on the Anthropocene. I argue that this critical agenda is important in theorising politics and power when reimagining the international system while ensuring a pluralistic

---

[7] Although not included during the selection process for this chapter, see also Fishel et al.'s (2018) response to Chandler et al. (2018).

and just living in this new geological epoch. These are potential intersections for deepening interdisciplinary conversations surrounding the Anthropocene.

Future IR research on the Anthropocene may wish to engage with underexplored components within these key discourses on peace and security. First, while the bibliographic portfolio in this review does contain some regional or national perspectives (e.g., Bezci 2021; Elliott 2021; Pereira/Viola 2022; Simangan 2019; Suliman 2016), more context-specific analysis from other regions or locales will enhance the principles of plurality and justice. This is important, given that some perspectives (especially those from the Global South, where the majority of the population is vulnerable to the impact of climate change) are marginal in the Anthropocene discourse (Marquardt 2019). Secondly, other environmental manifestations of the Anthropocene, such as biodiversity loss and pollution, did not receive as much attention as climate change (except for Dalby 2018; Wolff 2021). This is clearly understandable, given that climate change research and policy discussions have only recently gained traction. Nevertheless, other human-induced environmental changes could provide new insights into the peace and security landscape in the Anthropocene and prompt more collaborative research with other disciplines. Finally, this review suggests that there are insufficient peace articulations within IR's engagement with the Anthropocene. While the concept of (in)security is prominent in most of the articles, only five of the 71 IR articles have an explicit connection with peace, either in their title or abstract (i.e. Burke 2022; Hardt 2021a; Ide et al. 2021; Megoran/Dalby 2018; Simangan 2021a). This could be attributed to the dominance of research on conflict, rather than on existing peaceful conditions, both in peace and conflict studies and in climate-conflict research (Barnett 2019; Diehl 2016). While it is apparent in the articles that (in)security is being challenged or reimagined, particularly in the context of climate change, peace as a conceptual lens or even as a normative ideal seems to be inconsequential to these reimaginings (or at best implicit in the discourses identified in this review). What are the conditions that will support peace in the Anthropocene, and how can IR advance the conversations surrounding these conditions? In this 'planetary age', is peace conceived of and experienced differently compared to the previous century? How can the broadening of the politics of agency, time, and space present in critical IR intersect with notions of peace, and what are the implications of such intersections for different modes of peace formation and responses to challenges in the Anthropocene? Answers to these questions may come from a deeper convergence of IR with other disciplines in making sense of the Anthropocene.

References

Amoureux, J.; Reddy, V. 2021: "Multiple Anthropocenes: Pluralizing space-time as a response to 'the Anthropocene', in: *Globalizations*, 18,6: 929–946.

Asafu-Adjaye, J.; Blomqvist, L.; Brand, S.; Brook, B.; DeFries, R.; Ellis, E.; Foreman, C.; Keith, D.; Lewis, M.; Lynas, M.; Nordhaus, T.;, Pielke Jr., R.; Pritzker, R.; Roy, J.; Sagoff, M.; Shellenberger, M.; Stone, R.; Teague, P., 2015: "An Ecomodernist Manifesto"; at: https://www.ecomodernism.org/.

Barnett, J., 2019: "Global environmental change I: Climate resilient peace?", in: *Progress in Human Geography*, 43,5: 927–936.

Bernauer, T.; Böhmelt, T., 2020: "International conflict and cooperation over freshwater resources", in: *Nature Sustainability*, 3,5: 350–356.

Bezci, E., 2021: "Seeking shelter in the Anthropocene: Challenges and opportunities for Taiwan", in: *The Pacific Review* (online first); at: https://doi.org/10.1080/09512748.2021.1948912.

Biermann, F., 2021: "The future of 'environmental' policy in the Anthropocene: Time for a paradigm shift", in: *Environmental Politics*, 30,1–2: 61–80.

Biermann, F.; Möller, I., 2019: "Rich man's solution? Climate engineering discourses and the marginalization of the Global South", in: *International Environmental Agreements: Politics, Law and Economics*, 19,2: 151–167.

Blumenfeld, J., 2020: "Climate barbarism: Adapting to a wrong world. Constellations-An", in: *International Journal of Critical and Democratic Theory* (online first); at: https://doi.org/10.1111/1467-8675.12596.

Brauch, H.G., 2021: "The Anthropocene concept in the natural and social sciences, the humanities and law – a bibliometric analysis and a qualitative interpretation (2000-2020), in: Benner, S.; Lax, G.; Crutzen, P.J.; Pöschl, U.; Lelieveld, J.; Brauch, H.G. (Eds.): *Paul J. Crutzen and the Anthropocene: A New Epoch in Earth's History* (Cham: Springer Nature Switzerland): 289–438.

Brauch, H.G., 2008: "Conceptual quartet: security and its linkages with peace, development, and environment", in: Brauch, H.G.; Oswald Spring, U.; Mesjasz, C.; Grin, J.; Dunay, P.; Chadha Behera, N.; Chourou, B.; Kameri-Mbote, P.; Liotta, P.H. (Eds.): *Globalization and environmental challenges: reconceptualizing security in the 21st century* (Berlin-Heidelberg: Springer): 65–98.

Buhaug, H.; Benjaminsen, T.A.; Gilmore, E.A.; & Hendrix, C.S., 2022: "Climate-driven risks to peace over the 21st century", in: *Climate Risk Management*, 100471.

Bulletin of the Atomic Scientists, 2022: "At doom's doorstep: It is 100 seconds to midnight", at: https://thebulletin.org/doomsday-clock/current-time/.

Burke, A., 2022: "Interspecies cosmopolitanism: Non-human power and the grounds of world order in the Anthropocene", in: *Review of International Studies* (online first); at: https://doi.org/10.1017/S0260210522000171.

Burke, A.; Fishel, S.; Mitchell, A.; Dalby, S.; Levine, D.J., 2016: "Planet Politics: A Manifesto from the End of IR", in: *Millennium*, 44,3: 499–523.

Buzan, B.; *et al.*, 1997: *Security: A New Framework for Analysis* (Boulder, CO: Lynne Rienner).

Caney, S., 2018: "Justice and Future Generations", in: *Annual Review of Political Science*, 21: 475–493.

Chandler, D., 2020: "Security through societal resilience: Contemporary challenges in the Anthropocene", in: *Contemporary Security Policy*, 41,2: 195–214.

Chandler, D.; Cudworth, E.; Hobden, S., 2018: "Anthropocene, Capitalocene and Liberal Cosmopolitan IR: A Response to Burke et al.'s 'Planet Politics'", in: *Millennium*, 46,2: 190–208.

Christoff, P., 2016: "Renegotiating nature in the Anthropocene: Australia's environment movement in a time of crisis", in: *Environmental Politics*, 25,6:, 1034–1057.

Colgan, J.D.; Green, J.F.; Hale, T.N., 2021: "Asset Revaluation and the Existential Politics of Climate Change", in: *International Organization*, 75,2: 586–610.

Conway, P., 2020: "On the way to planet politics: From disciplinary demise to cosmopolitical coordination", in: *International Relations*, 34,2: 157–179.

Crutzen, P.J.; Stoermer, E.F.. 2000: "The Anthropocene", in: *International Geosphere-Biosphere Programme Newsletter*, 41: 17–18.
Dalby, S., 2014: "Environmental Geopolitics in the Twenty-first Century", in: *Alternatives*, 39.1: 3–16.
Dalby, S., 2018: "Firepower: Geopolitical Cultures in the Anthropocene", in: *Geopolitics*, 23,3: 718–742.
Dalby, S., 2020: "Bordering sustainability in the Anthropocene", in: *Territory, Politics, Governance*, 8,2: 144–160.
Death, C., 2022: "Climate Fiction, Climate Theory: Decolonising Imaginations of Global Futures", in: *Millennium*, 50,2: 430–455.
Diehl, P.F., 2016: "Exploring Peace: Looking Beyond War and Negative Peace", in: *International Studies Quarterly*, 60,1: 1–10.
dos Muchangos, L.S.; Vaughter, P., 2018: "Are gender perspectives included in education for sustainable consumption and waste education programs? A systematic literature review", in: *Detritus*, 4: 164–177.
Elliott, L., 2021: "Australian IR scholarship on the environment: The recent past and the possible future", in: *Australian Journal of International Affairs*, 75,6: 604–618.
Ensslin, L.; Dutra, A.; Ensslin, S.R.; Chaves, L.C.; Dezem, V., 2015: "Research Process for Selecting a Theoretical Framework and Bibliometric Analysis of a Theme: Illustration for the Management of Customer Service in a Bank", in: *Modern Economy*, 6, 6.
Fagan, M., 2017: "Security in the Anthropocene: Environment, ecology, escape", in: *European Journal of International Relations*, 23,2: 292–314.
Ferguson, P., 2019: "Discourses of Resilience in the Climate Security Debate", in: *Global Environmental Politics*, 19,2: 104–126.
Fishel, S.; Burke, A.; Mitchell, A.; Dalby, S.; Levine, D., 2018: "Defending planet politics", in: *Millennium*, 36,2: 209–219.
Galtung, J., 1969: "Violence, Peace, and Peace Research, in: *Journal of Peace Research*, 6,3: 167–191.
Gellers, J., 2021: "Earth system law and the legal status of non-humans in the Anthropocene", in: *Earth System Governance*, 7.
Glänzel, W.; Schubert, A.; Czerwon, H.-J., 1999: "A bibliometric analysis of international scientific cooperation of the European Union (1985–1995)", in: *Scientometrics*, 45,2: 185–202.
Graeger, N., 1996: "Environmental Security?", in: *Journal of Peace Research*, 33,1: 109–116.
Green, J., 2022: "Comparative capitalisms in the Anthropocene: A research agenda for green transition", in: *New Political Economy* (online first); at: https://doi.org/10.1080/13563467.2022.2109611.
Grunwald, A., 2018: "Diverging pathways to overcoming the environmental crisis: A critique of eco-modernism from a technology assessment perspective", in: *Journal of Cleaner Production*, 197,2: 1854–1862.
Gupta, J.; Liverman, D.; Bai, X.; Gordon, C.; Hurlbert, M.; Inoue, C.; Jacobson, L.; Kanie, N.; Lenton, T.; Obura, D.; Otto, I.; Okereke, C.; Pereira, L.; Prodani, K.; Rammelt, C.; Scholtens, J.; Tabara, J.; Verburg, P.; Gifford, L.; Ciobanu, D., 2021: "Reconciling safe planetary targets and planetary justice: Why should social scientists engage with planetary targets?", in: *Earth System Governance*, 10.
Hamilton, S., 2018: "Foucault's End of History: The Temporality of Governmentality and its End in the Anthropocene", in:. *Millennium*, 46, 3: 371–395.
Hansen-Magnusson, H., 2022: "Making Polar and Ocean Governance Future-Proof", in: *Politics and Governance*, 10,3: 60–69.
Hardt, J.N., 2021a: "Research Perspectives and Boundaries of Thought: Security, Peace, Conflict, and the Anthropocene", in: *Revista de Estudios en Seguridad Internacional*, 7,1: 11–28.
Hardt, J.N., 2021b: "The United Nations Security Council at the Forefront of (Climate) Change? Confusion, Stalemate, Ignorance", in: *Politics and Governance*, 9,4:, 5–15.

Hardt, J.N., 2019: "Security Studies and the discourse on the Anthropocene", in: Hickmann, T.; Partzsch, L.; Pattberg, P.; Weiland, S. (Eds.): *The Anthropocene Debate and Political Science.* (London: Routledge): 85–102.

Harrington, C., 2016: "The Ends of the World: International Relations and the Anthropocene", in:. *Millennium*, 44,3: 478–498.

Houston, D., 2013: "Crisis Is Where We Live: Environmental Justice for the Anthropocene", in: *Globalizations*, 10,3: 439–450.

Ide, T., 2018: "Climate War in the Middle East? Drought, the Syrian Civil War and the State of Climate-Conflict Research", in: *Current Climate Change Reports*, 4,4: 347–354.

Ide, T.; Ensor, M.; Le Masson, V.: Kozak, S., 2021: "ender in the Climate-Conflict Nexus: 'Forgotten' Variables, Alternative Securities, and Hidden Power Dimensions", in: *Politics and Governance*, 9,4: 43–52.

Inoue, C.Y.A., 2018: "Worlding the Study of Global Environmental Politics in the Anthropocene: Indigenous Voices from the Amazon", in: *Global Environmental Politics*, 18,4: 25–42.

Johnson, A.; Sigona, A., 2022: "Planetary justice and 'Healing' in the Anthropocene", in: *Earth System Governance*, 11.

Kelley, C.P.; Mohtadi, S.; Cane, M.A.; Seager, R; Kushnir, Y., 2015: "Climate change in the Fertile Crescent and implications of the recent Syrian drought", in: *PNAS*, 112,11: 3241–3246.

Kotze, L.; French, D., 2018: "A critique of the Global Pact for the environment: A still-born initiative or the foundation for Lex Anthropocenae?", in: *International Environmental Agreements-Politics Law and Economics*, 18,6: 811–838.

Koubi, V., 2019: "Climate Change and Conflict. Annual Review of Political Science" in:, 22,1: 343–360.

Leep, M., 2018: "Stray Dogs, Post-Humanism and Cosmopolitan Belongingness: Interspecies Hospitality in Times of War", in: *Millennium*, 47,1: 45–66.

Lobo, M., 2019: "ffective ecologies: Braiding urban worlds in Darwin, Australia", in: *Geoforum*, 106: 393–401.

Lovbrand, E.; Mobjork, M.; Soder, R., 2020: "he Anthropocene and the geo-political imagination: Re-writing Earth as political space", in: *Earth System Governance*, 4.

Marquardt, J., 2019: "Worlds apart?: The Global South and the Anthropocene", in: Hickmann, T.; Partzsch, L.; Pattberg, P.; Weiland, S. (Eds.): *The Anthropocene Debate and Political Science* (London: Routledge): 200–218.

McDonald, M., 2008: "Securitization and the Construction of Security", in: *European Journal of International Relations*, 14.4: 563–587.

McDonald, M., 2018: "Climate change and security: Towards ecological security?", in: *International Theory*, 10,2: 153–180.

McDonald, M., 2021: *Ecological Security: Climate Change and the Construction of Security* (Cambridge: Cambridge University Press).

McPhearson, T.M.; Raymond, C.; Gulsrud, N.; Albert, C.; Coles, N.; Fagerholm, N.; Nagatsu, M.; Olafsson, A.S.; Soininen, N.; Vierikko, K., 2021: "Radical changes are needed for transformations to a good Anthropocene", in: *npj Urban Sustainability*, 1,1, Article 5.

Megoran, N.; Dalby, S., 2018: "Geopolitics and Peace: A Century of Change in the Discipline of Geography", in: *Geopolitics*, 23,2: 251–276.

Meyer, J., 2020: "The politics of the "post-political" contesting the diagnosis", in: *Democratization*, 27,3: 408–425.

O'Lear, S.; Masse, F.; Dickinson, H.; Duffy, R., 2022: "Disaster Making in the Capitalocene", in: *Global Environmental Politics*, 22,3: 2–11.

Palomo, J.; Figueroa-Domecq, C.; Laguna, P., 2017: "Women, peace and security state-of-art: A bibliometric analysis in social sciences based on SCOPUS database", in: *Scientometrics*, 113,1: 123–148.

Pellizzoni, L., 2021: "Reconfiguring non-domination: Green politics from pre-emption to inoperosity", in: *Critical Review of International Social and Political Philosophy*, 24,5: 743–760.

Pereira, J.; Viola, E., 2022. "Brazilian climate policy (1992-2019): An exercise in strategic diplomatic failure", in: *Contemporary Politics*, 28,1: 55–78.
Persson, L.; Carney Almroth, B.M.; Collins, C.D.; Cornell, S.; de Wit, C.A.; Diamond, M.L.; Fantke, P.; Hassellöv, M.; MacLeod, M.; Ryberg, M.W.; Søgaard Jørgensen, P.; Villarrubia-Gómez, P.; Wang, Z.; Hauschild, M.Z., 2022: "Outside the Safe Operating Space of the Planetary Boundary for Novel Entities", in: *Environmental Science & Technology*, 56,3: 1510–1521.
Pope, K.; Bonatti, M.; Sieber, S., 2021: "The what, who and how of socio-ecological justice: Tailoring a new justice model for earth system law", in: *Earth System Governance*, 10.
Proedrou, F., 2020: "Anthropocene Geopolitics and Foreign Policy: Exploring the Link in the EU Case", in: *Alternatives*, 45,2: 83–101.
Rockström, J.; Steffen, W.; Noone, K.; Persson, Å.; Chapin, F.S.I.; Lambin, E.; Lenton, T.; Scheffer, M.; Folke, C.; Schellnhuber, H.J.; Nykvist, B.; de Wit, C.; Hughes, T.; van der Leeuw, S.; Rodhe, H.: Sörlin, S.; Snyder, P.; Costanza, R.; Svedin, U.; … Foley, J., 2009: "Planetary Boundaries: Exploring the Safe Operating Space for Humanity", in:. *Ecology and Society*, 14,2.
Rossi, C.R., 2017: "The Transboundary Dispute over the Waters of the Silala/Siloli: Legal Vandalism and Goffmanian Metaphor", in: *Stanford Journal of International Law*, 53,1: 55–88.
Selby, J.; Dahi, O.S.; Frohlich, C.; Hulme, M., 2017: "Climate change and the Syrian civil war revisited", in: *Political Geography*, 60: 232–244.
Sharifi, A.; Simangan, D.; Lee, C.Y.; Reyes, S.R.; Katramiz, T.; Josol, J.C.; Muchangos, L.D.; Virji, H.; Kaneko, S.; Tandog, T.K.; Tandog, L.; Islam, M., 2021: "Climate-induced stressors to peace: A review of recent literature", in:. *Environmental Research Letters*, 16,7: 073006.
Sillanpää, A.; Koivula, T., 2010: "Mapping Conflict Research: A Bibliometric Study of Contemporary Scientific Discourses1", in: *International Studies Perspectives*, 11,2: 148–171.
Simangan, D., 2019: "Situating the Asia Pacific in the age of the Anthropocene", in: *Australian Journal of International Affairs*, 73,(6: 564–584.
Simangan, D., 2020: "Where is the Anthropocene? IR in a new geological epoch", in: *International Affairs*, 96,1: 211–224.
Simangan, D., 2021a: "Reflexive Peacebuilding: Lessons from the Anthropocene Discourse", in: *Global Society*, 35,4: 479–500.
Simangan, D., 2021b: "Where is the Asia Pacific in mainstream international relations scholarship on the Anthropocene?", in: *The Pacific Review*, 34,5: 724–746.
Steffen, W.; Richardson, K.; Rockström, J.; Cornell, S. E.; Fetzer, I.; Bennett, E. M.; Biggs, R.; Carpenter, S.R.; de Vries, W.; de Wit, C.A.; Folke, C.; Gerten, D.; Heinke, J.; Mace, G.M.; Persson, L.M.; Ramanathan, V.; Reyers, B.; Sörlin, S., 2015: "Planetary boundaries: Guiding human development on a changing planet", in: *Science*, 347,6223: 1259855.
Suliman, S., 2016: "Rethinking about Civilizations: The Politics of Migration in a New Climate", in: *Globalizations*, 13,5:, 638–652.
Swain, A., 2015: "Water Wars", in: Wright, J.D. (Ed.): *International Encyclopedia of the Social & Behavioral Sciences*, 2nd ed., vol 25. (Oxford: Elsevier): 443–447.
UNDP, 1994: *Human Development Report 1994: New Dimensions of Human Security*. (Oxford: Oxford University Press).
UNDP, 2020: *Human Development Report 2020. The next frontier: Human Development in the Anthropocene* (New York: United Nations Development Programme).
UNDP, 2022: *New threats to human security in the Anthropocene: Demanding greater solidarity* (New York: United Nations Development Programme).
van Dijk, N., 2021: "From exacerbating the Anthropocene's problems to intergenerational justice: An analysis of the communication procedure of the human rights treaty system", in: *Earth System Governance*: 10.
van Eck, N.J.; Waltman, L., 2010: "Software survey: VOSviewer, a computer program for bibliometric mapping", in: *Scientometrics*, 84,2: 523–538.
van Munster, R.; Sylvest, C., 2021: "Nuclear weapons, extinction, and the Anthropocene: Reappraising Jonathan Schell", in: *Review of International Studies*, 47,3: 294–310.

Waters, C.N.; Zalasiewicz, J.; Summerhayes, C.; Barnosky, A.D.; Poirier, C.; Gałuszka, A.; Cearreta, A.; Edgeworth, M.; Ellis, E.C.; Ellis, M.; Jeandel, C.; Leinfelder, R.; McNeill, J.R.; Richter, D. deB, Steffen, W.; Syvitski, J.; Vidas, D.; Wagreich, M.; Williams, M.; … Wolfe, A.P., 2016: "The Anthropocene is functionally and stratigraphically distinct from the Holocene". In: *Science*, 351: 6269.

Wolff, L., 2021: "The past shall not begin: Frozen seeds, extended presents and the politics of reversibility", in: *Security Dialogue*, 52,1: 79–95.

Chapter 7
The Anthropocene and German Political Science: A Case Study

Georg Simonis

Abstract This article attempts to explain why political science in the German-speaking research area has so far only marginally dealt with the concept of the Anthropocene, although a drastic structural change between the Holocene and the Anthropocene is becoming apparent. At the centre of the explanatory approach is the hypothesis of an incongruence between political science and Earth science research. In the dominant political science discourse, environmental conditions are seen as constant and the focus is on political actors coping with the problems of the present. Earth science, on the other hand, ignores political actors and their struggle for political consensus as restrictive variables for planetary governance. The third part of the article presents a political science perspective that addresses the problems facing politics in the new Anthropocene era. Here, the thesis is developed that in the Anthropocene, the structural conditions of politics and political governance are undergoing lasting change.

Keywords Actors · Anthropocene · Democracy · Governance · Political science · Security · Scientification · Security

---

Prof. Dr. Georg Simonis, (ret.); Institut für Politikwissenschaft der FernUniversität in Hagen, Deutschland; Email: Georg.Simonis@fernuni-hagen.de. The text was translated from German into English by Elizabeth Norton Amor with the assistance of the free deepL translation software.

G. Simonis (✉)
Institut für Politikwissenschaft der FernUniversität in Hagen, Hagen, Germany
e-mail: Georg.Simonis@fernuni-hagen.de

H. G. Brauch (ed.), *Towards Rethinking Politics, Policy and Polity in the Anthropocene*, The Anthropocene: Politik—Economics—Society—Science 35,
https://doi.org/10.1007/978-3-031-71807-6_7

7.1 Introduction[1]

In political science in the German-speaking research area[2], the Anthropocene debate, which has been going on internationally in the field of natural sciences (*Earth sciences*) for a good twenty years and in the field of cultural and social sciences for at least ten years, is only slowly beginning to attract interest. In Autumn 2020, four younger scholars (Hickmann et al. 2020) wrote a commentary in the *Politische Vierteljahresschrift* (PVS), which is published by the *Deutsche Vereinigung für Politikwissenschaft* (DVPW) [German Association for Political Science], in which they call on political science to 'participate more strongly in the debate on the so-called human age and its far-reaching implications' (Hickmann et al. 2020). The four authors do not explain why German political science has been slow to participate in this international discussion. They limit themselves to outlining three areas in which political science could be expected to make relevant contributions to the debate, namely the concept of the Anthropocene, governance and political theory in the Anthropocene.

The commentary by Hickmann et al., which has so far not led to any recognisable reactions[3], refrains from giving a reason why political science here, apart from exceptions that will be presented later, does not deal with the Anthropocene findings and their interpretation. In the following, an attempt is made to find reasons for the weak resonance of the Anthropocene within political science in Germany; in doing so, the text restricts itself to plausibility considerations. Solid empirical evidence cannot be provided here. The thesis is developed that political science at German universities has only developed a weak interest in the topic because it is committed to a theoretical thinking that considers the environment, including natural conditions, to be of little relevance for the interpretation and explanation of political processes and therefore externalises them. Politics is analysed as an *autopoietic* functional system in which the possibility of self-destruction, whether through the potential for military violence or through unintended side effects of social action on the stability of planetary environmental systems, is ignored. Other reasons are added, such as the theoretical model of Earth sciences and the focus on empirical processes of the present. For the German

[1] I would like to express my sincere gratitude for the innovative suggestions and the time spent by the two anonymous reviewers and the editor in improving the text. The paper follows on from a lecture I gave at the Saxon Academy of Sciences and Humanities in Leipzig in Autumn 2017, a revised version of which appeared in the Academy's *Denkströme* (Simonis 2019). Some passages from the 2017 essay appear in this text.

[2] The text refers primarily to political science research results (publications) at universities within Germany; the German-speaking research area (Austria, Switzerland) is only included in part. However, the study also takes into account the research work of academics who have earned an academic degree in political science at a German university, but who teach and do research at foreign universities, especially European universities in neighbouring countries, during the study period (approx. 2010–2021/22).

[3] Recently, another article (Pauls et al. 2022) on the Anthropocene problem was published in the *PVS Politische Vierteljahresschrift* under the title *Das Planetarische Politisch(e) Denken* [Planetary Political Thought]. However, the Forum article by Hickmann et al. (2020) is not mentioned. Yet the debate article takes up an important aspect of the Anthropocene discourse (see below), namely the question of the actor status of non-human agents.

political public, there is no debate about the Anthropocene problem that could be a worthwhile object of study for political science as an empirical social science. The Anthropocene findings are still outside the topics politicised in the public sphere. Empirical political science research related to the present can treat the findings of the Earth sciences, like many other problems that are not present in the space of politicised topics, with *benign neglect.* However, this attention profile seems to be slowly changing. Starting from the interdisciplinary research margins of political science, such as peace studies (Brauch et al. 2011; Scheffran 2016) and international environmental governance (Bierman 2014; Pattberg/Zelli 2017), the political explosiveness of the 'geological' finding is being acknowledged by individual research groups (Rothe et al. 2022; Pauls et al. 2022) and universities in German-speaking academia (Sippl/Scheuch 2019; Hanusch et al. 2021; Leggewie et al. 2022).

Political science rightly sees itself as not responsible for confirming or refuting the disturbing empirical claims of geologists (Leinfelder 2017). Earth geoscience is oriented towards the determination of stratigraphic facts (Zalasiewicz et al. 2017; Zalasiewicz et al. 2017a, b; Will 2021; Zalasiewicz et al. 2021). *Homo sapiens* has only played a role in Earth's geological history in very recent times (Hamilton 2017; Irvine 2020). If the geological research community comes to the expected conclusion that the life forms of human civilisation are changing the evolutionary conditions of the planet, the Holocene geological era will give way to the Anthropocene.

Political science can also contribute little to clarifying two downstream questions: neither to answering the question of which anthropogenic forces have been modifying Holocene conditions on the planet and since when, nor to clarifying the question of what geo-ecological consequences anthropogenic intervention triggers for life on the planet. The social sciences only come into play when it comes to determining the range of response forms and possible actions of humanity to the changed living conditions on the planet. Alongside other social sciences, such as sociology, economics or social psychology, political science is called upon when answers are sought to the question of how the world, fragmented into states, reacts or could react to the dangerous changes in environmental and living conditions. In what way is coexistence on the planet changing? Does humanity have the capacity to have a corrective effect on disturbed planetary functional systems? Political science studies could be helpful in answering these and similar questions with societal relevance.

After this introduction, the chapter is divided into three sections. In the *first section*, the Anthropocene debate is briefly recapitulated. First, the Anthropocene findings of geology and the resulting discussion about the beginning of the Anthropocene and the crisis of the Holocene are summarised briefly. The section then deals with the beginning of the Anthropocene from a political science perspective, for which primarily two points in time, that choose the anthropogenically-generated self-destruction potential as indicators, come into consideration: the testing and first use of nuclear weapons or/and the growing evidence at the beginning of the 21st century of the anthropogenically-caused endangerment of several functional systems relevant for human well-being on the planet (ozone, climate, biodiversity, nitrate, etc.). Finally, reference is made to the collective pressure for action (governance problem) that has arisen with the possibility and increasing probability of anthropogenic self-destruction. In the *second section*, an attempt is made to explain why political science in the German-speaking academic world reacts only with weak resonance and with a delay to the Earth science anthropocide debate. Here, the innovation-theoretical hypothesis is confirmed that innovations tend to emanate from the periphery of an established discipline. From peripheral areas (peace research, young researchers, environmental research, researchers at foreign universities), the realisation that a new age is also emerging for political science is slowly penetrating the established discipline without irritating it to any significant degree. This will probably only be the case when the Anthropocene issue reaches the public consciousness. The debates about the civil society protest of the 'Last Generation' could mark a turning point. The *third section* deals with political science questions that arise from the Anthropocene findings and in research on which political scientists from German-speaking universities are participating. The aim is not to be exhaustive. Rather, the intention is to highlight areas of investigation and research that analyse new structural elements of the changing international system in the Anthropocene era. In this way, a panorama emerges from the disparate research works in the synopsis, which makes the contours of the discipline of international relations in the Anthropocene recognisable. In the *summary*, reference is made once again to the diffusion thesis; in addition, following the research presented, two research horizons are outlined whose perception in the core areas of established political science research would accelerate the diffusion process.

7.1.1 Anthropos as a Geological Force

The contribution to the discussion by the atmospheric chemist and Nobel laureate Crutzen (2002) and the marine biologist Eugene Stoermer (2000) with the thesis that humankind had left the stable environmental conditions of the Holocene that had prevailed for about 10–12,000 years and was in a new geological age, the Anthropocene, initially triggered a geological-stratigraphical debate. Of the related natural sciences, Earth and climate scientists took up the discussion triggered by Crutzen and Stoermer[4] (Steffen et al. 2005, 2007) and developed the hypothesis of

[4] In a bibliometric analysis of the Web of Science for the period 2000–2015, Brondizio et al. identified a considerable 1066 publications that used the term 'Anthropocene' in the title, abstract or text. Until 2005, only isolated publications were registered. The exponential increase in the number of publications as well as their citation (N = 8451) occurred from 2010 onwards. The research revealed that 64% of the publications could be attributed to the disciplines 'Earth and environmental sciences' and 24% to 'humanities and social sciences' (Brondizio et al. 2016: 320–321). Krämer presented similar data in a BA thesis. Furthermore, on the subject analysis of the publications, she came to the conclusion that political science is represented with a share of 2.2%. When determining the countries of origin of the authors, the USA leads with 42%, ahead of the UK with 18%, Australia with 13% and Germany with just under 11% (Krämer 2016: 76–77). In her bibliometric analysis of articles from the field of international relations on the Anthropocene topic in peer-reviewed journals conducted in the Web of Science in 2019, Dahlia Simangan identified

planetary boundaries (Rockström et al. 2009). Humanity, through its own actions, would endanger the Holocene conditions favourable to human life. There was a risk that the safe habitat, within which humanity could develop dynamically, would be abandoned. A dramatic deterioration of living conditions on the planet could only be avoided through quickly effective mitigation of anthropogenic environmental impacts (Steffen et al. 2015). The cultural and social sciences reacted to this finding of the Earth sciences with specific contributions to the debate (Brondizio et al. 2016; Krämer 2016; Nicholson/Jinnah 2016b; Zalasiewicz et al. 2021), which are condensed into independent discourse strands (narratives) (Dürbeck 2020). The destruction of local and regional ecosystems through the exploitation of natural resources and the use of the countryside as a dumping ground are traditional and widespread human behaviour patterns (Choucri 1995; Simonis 1996; Scheffran/Vogt 1998; Radkau 2000; Conca 2016). Also, the model-theoretical findings of the Club of Rome (Meadows et al. 1972; Meadows et al. 1992; Randers/Bus 2013; Weizsäcker/ Wijkman, 2017; Dixson-Declève et al. 2022) have been well known in science and by the public for decades. The 'great acceleration' of economic-technological (capitalist), and demographic growth has massive non-intended ecological consequences for human coexistence. Against this backdrop, the clear trends of quantitative measurement data series indicating that humanity has indeed embarked, with increasing certainty, in terms of empirical verifiability, on a development path that acutely and irreversibly endangers planetary environmental conditions conducive to human prosperity, constitute a new scientific and political horizon of knowledge (Bonneuil/Fressoz 2013, 2017: Chap. 2; Hamilton/Grinevald 2015; Hamilton 2016; Leinfelder 2017; IPCC 2021). Increasing degradation of Holocene conditions of the biosphere, atmosphere, hydrosphere and cryosphere favourable for human development must be expected (Zalasiewicz et al. 2019; UNEP 2019, 2020). Their restabilisation and governance becomes the major human task in the Anthropocene (Nakicenovic et al. 2016; Kotzé/French 2022).

While the attitude and action profile prevalent in the Holocene accepted the plundering of natural resources and the intensive use of the commons, in the Anthropocene the exploitation paradigm becomes increasingly fragile. The concept of sustainability (Commission on Environment and Development 1987) becomes an ambivalent guiding principle (Knaus/Renn 1998; Feiler 2003; Renn et al. 2007; Grunwald/ Kopfmüller 2022; Franz 2014) that could build a bridge between the epochs. European modernity's dream of being able to free itself from the constraints of nature is partially withdrawn in the demand for sustainable development. At the end of the Holocene is the realisation that social development at the expense of natural conditions is not possible in the long term. The stability of the planet's natural conditions should not be endangered by resource-consuming growth but secured through

134 articles with a broader IR reference and 36 with a narrower IR reference, 9 of which were on regional topics, from a population of 1286 articles with a social science reference for the period 2011 to 2019 (Simangan 2020: 213). The authors of the 134 articles are predominantly from Western industrialised countries: USA (51), UK (30), Canada (22), Australia (21), and France (17). The only developing country is Brazil, with 9 authors. The number of German authors is not mentioned (*ibid.* 215). For further information, see the introductory chapter in this volume and Brauch (2021a).

sustainable growth (UN 2015; Commission on Global Security, Justice & Governance 2015; Bückmann 2015; Hermanns/Li 2019); environmental destruction can be repaired. In the Anthropocene era, this growth and techno-optimism has given way to the realisation that humanity will have to live with planetary unstable environmental conditions (Loske 2015) and that controlling instability has become a permanent challenge for society and politics.

Important distinguishing features between the Holocene and the Anthropocene can be seen in Table 6.1. The overview draws attention to the major differences between the two epochs (Albert 2020). They can be used to justify the metaphor of historical rupture (Hamilton 2016), even though there are overlaps and challenging continuities. The age determined by the anthropos has a long and contested genealogy in explanation (Steffen et al. 2014; Moore 2016; Hamilton 2017; Chakrabarty 2021; Rothe/Benner 2021; Will 2021). In the following explanations, based on the finding that humanity has become the dominant geological force (Crutzen 2002), the anthropogenically-generated potential self-destructiveness with regard to beneficial living conditions for humans is chosen as the criterion for the beginning of the Anthropocene.

This is not the place to reconstruct the various strands of discourse in the international anthropocene debate (origins, manifestations, consequences, options for action). The accent of this study is on the hesitant participation of German-speaking political science (or political science that has emerged from German academia) in this field of research, which, however, seems to be generating growing interest in this country as well. In order to clarify the academic context against which the political science anthropocene debate is taking place, three fields of discussion, which are not discussed further here, should be pointed out beforehand (Table 7.1).

(1) The geological question to be clarified by international geological research, that of whether the hypothesis of the Anthropocene as a new man-made geological age can be considered empirically confirmed, has not yet been conclusively answered. No final consensus has yet been reached among geologists on the beginning of the new age and its stratigraphic features (Zalasiewicz et al., 2017a, b; Zalasiewicz et al., 2021). For the natural science-oriented subjects of the Earth sciences, the planetary environmental problems caused by anthropogenic interventions in the new age represent a challenge to the traditional research paradigm (Dürbeck 2020). Insofar as human activities have a formative influence on present and future geological and planetary conditions, the research paradigm of natural forces (laws of nature) acting independently of anthropogenic interventions no longer applies to scientific Earth research. In the future, anthropogenic forces and mechanisms of action must be taken into account as additional (new) geophysical variables in the Earth science explanatory models.

(2) The stratigraphic discussion of geology is struggling to reach a consensus on the beginning of the new Earth age (Zalasiewicz et al. 2017a, b). The intractable problem of defining a broadly accepted metric is understandable, as there are a number of possible leading indicators to choose from, such as post-1945 nuclear fallout, the disrupted carbon cycle or mass extinctions of faunal and floral

Table 7.1 Distinguishing features between the Holocene and the Anthropocene

Holocene (post-glacial period)	Anthropocene
Since approx. 11000 years ago	Since approx. 1850: Beginning of the great acceleration Since the mid-1940s (A-weapons)
Humanity has no great influence on the development of the planet Mankind has a safe operating space	Humanity became the strongest geological force (Crutzen/Störmer 2000) Humanity leaves the safe operating space (Rockström et al. 2009)
Separation between man and natural relations (Cartesianism)	Interdependence of nature and humans
Intrinsic development of humanity: 'liberation from nature'	Entangled development: humanity as part of the Earth system (Theory of Evolution)
Unlimited use (exploitation) of nature possible (technical progress)	Transformation of planetary nature relations through exploitation and non-intended emissions
Stability of planetary natural conditions	Instability of anthropogenically modified natural conditions with tipping points

Source Compilation by the author

species. From the perspective of political science, additional differences exist between the Holocene and Anthropocene epochs that are difficult for natural scientists to recognise, since in social science not only empirically observable (geophysical, geobiochemical) environmental changes are included in the analysis, but also potentials for action, such as nuclear arsenals that can lead to planetary destruction. The comparison of structural features of the threat constellation of a nuclear deterrent system with anthropogenic climate change may provide initial indications for an explanation of different thematisation and political response patterns of the community of states to the crisis of the Holocene (Sect. 6.1.2). The Anthropocene era may have already begun with the nuclear age. The social-historical and anthropological debate[5] on when the Anthropocene began to grow within the Holocene can only be referred to here (Ellis 2020; Chakrabarty 2021; Will 2021; Antweiler 2022).

(3) Largely independent of the formation of opinion in geosciences and cultural studies, the concept of the Anthropocene has met with considerable scientific resonance in international environmental science circles and has led to

[5] See also the recent bibliographical survey by Brauch (2021).

a research boom[6] that has been dynamically increasing in scope and depth since 2010. Fundamental questions are being raised in this debate, such as the understanding of nature and the environment in the modern age, the driving forces behind the growth dynamic, the transformation requirements and the existing and seemingly realisable technological potentials for the sustainable transformation of the fossil fuel industrial and growth society, as well as the change in values required for the great sustainability turnaround (Bonneuil/Fressoz 2017, 2013; Hamilton et al. 2015; Nakicenovic et al. 2016; Deudney/Mendenhall 2016; Latour 2017; Biermann/Lövbrand 2019). The political science research perspective is only marginally present in this debate, which focuses on the social diagnosis of the era and the changing circumstances of life (Laux/Henkel 2018; Adloff/Neckel, 2020). In cultural studies and sociology, an Earth science finding is taken up on a topic which they are researching anyway, such as (in particular) the environmentally-harmful consequences of unbridled industrial (capitalist) growth.

7.1.2 *Anthropocene I: Nuclear Self-destruction*

With the detonation of the atomic bombs of Hiroshima and Nagasaki in the Summer of 1945 and the subsequent nuclear arms race, it was immediately clear to the political public and the community of states that a new era of potential military force had begun (Acheson-Lilienthal Report 1946). The development of nuclear armament and the arms race of the nuclear powers (since 1949) was accompanied almost from the beginning by initiatives to establish controlling institutions (Baruch Plan 1946). In particular, disarmament and arms control were intended to prevent nuclear emergencies.

Until today, more than 75 years later, this attempt has been successful, despite highly dangerous moments of tension, such as during the Cuban Missile Crisis. Although the major nuclear powers are interested in maintaining the threat and destructive potential of nuclear weapons and are modernising their nuclear arsenals again after the détente phase (Kristensen/Norris 2022; SIPRI 2022), deterrence has not failed. The limiting conditions of the nuclear age are still being met. The ability to destroy human civilisation through the use of weapons of mass destruction has been established as a basic constant of the Anthropocene in the field of military security.[7]

[6] Four scientific journals specialising in the Anthropocene have already been published: *The Anthropocene Review* (Sage), *The Anthropocene* (Elsevier), *Elementa: Science of the Anthropocene* (open-access journal), *Earth's Future* (American Geophysical Union) and the internet magazine *Anthropocene: Innovation in the Human Age* (published by Future Earth). The book series *The Anthropocene: Politik—Economics—Society—Science* (https://www.springer.com/series/15232) by the editor of this volume (H. G. Brauch) should also be mentioned here (see also the Web of Science).

[7] Whether the threats of using nuclear and other weapons of mass destruction in the war against Ukraine on the part of the incumbent Russian President Vladimir Putin and former President Dmitry

A geologically-stratigraphically striking criterion for identifying the historical beginning of the nuclear age can be given by the radioactive fallout of extensive nuclear weapons testing (Zalasiewicz et al. 2017a, b; van Munster 2021). From a political science (action theory) understanding, the (potential) tension between empirically observable changes in nuclear arsenals and anthropogenic possibilities of action to use them forms a characteristic, structural feature of the new age. From this perspective, the Anthropocene era would have begun with the testing and subsequent dropping of the bombs on Hiroshima and Nagasaki. A condensed overview of the political forms of processing global military insecurity in the Anthropocene as a result of the permanent risk of the use of means of mass destruction is provided in Table 6.2 under the heading Anthropocene I.[8] The following sub-chapter deals with the ecological, geochemically-based threat potentials, which are founded on a completely different constellation of technologies and actions (Zalasiewicz et al. 2019). In the policy area of military (nuclear) security, when the development and deployment of nuclear weapons escalated dangerously and a nuclear catastrophe threatened, the reactions of policy-makers were completely different from those in the problem area of climate policy, where threatening border crossings are announced.

7.1.3 Anthropocene II: Planetary Boundary Violations

The global environmental risks diagnosed for the Anthropocene era mark a historical turning point, as does the destructive potential of nuclear weapons. Anthropogenically caused (and mostly unintentional) side effects are insidiously, yet existentially threatening and irreversible, changing the planetary environmental conditions of the Holocene that were conducive to human development. In the longer term, without societal transformation (mitigation of harmful emissions, new forms of land use, sustainable management of biodiversity), planetary environmental catastrophes must be expected (Steffen et al. 2007; Semal 2015; Northcott 2015; IPCC 2018). The key text on threats to Holocene conditions appeared under the leadership of Rockström in 2009. The *Nature* article on quantified planetary boundaries also involved Paul Crutzen, the originator of the Anthropocene debate, as a co-author. In this paper the following statement/hypothesis is formulated:

Medvedev will be followed by deeds is still highly unlikely, but must be taken very seriously for situations in which a clear military defeat of Russia becomes apparent or the Russian autocrats feel directly militarily attacked by the West. The failure of a deterrence system is unlikely, but still possible (Simonis 1976).

[8] The comparison of the security problem of nuclear deterrence with climate change emerged from the American debate about the securitisation of climate change (CNA 2007; Scheffran 2012; Diez et al. 2016).

> Since the Industrial Revolution, a new era has arisen, the Anthropocene, in which human actions have become the main driver of global environmental change. This could see human activities push the Earth system outside the stable environment state of the Holocene, with consequences that are detrimental or even catastrophic for large parts of the world. (Rockström et al. 2009: 472)

The geological-stratigraphical changes caused by human civilisation are thus only one aspect of the new geological age; the ecological consequences of humanity's actual impact on natural conditions are considered to be more serious, as catastrophic consequences for large parts of humanity are unavoidable without a change in institutionalised everyday routines. The publications of the working group around Johan Rockström and Will Steffen (Steffen et al. 2005, 2007, 2015; Rockström et al. 2009) on the emerging limits to the resilience of anthropogenically-beneficial environmental conditions (Steffen et al. 2014) have given real momentum to the geostratigraphic debate on the beginning of a new, anthropogenically-determined age (Zalasiewicz et al. 2017a, b; Zalasiewicz et al., 2019; Weizsäcker/Wijkman 2017; Brauch 2021).

In the 2009 *Nature,* nine environmental areas were identified, three of which had already left the zone of the „safe operating space"(biodiversity, climate change, nitrogen cycle), which is beneficial for humanity because it is risk-tolerant. In 2015, an international working group led by Will Steffen, also from the Stockholm Resilience Centre, presented an update of the 2009 research results in *Science* (Steffen et al. 2015). The thresholds and control variables for determining stable environmental conditions were further specified, critical uncertainty zones were defined, and updated values were assigned to the control variables. According to the 2015 study, four Earth system processes were now outside the boundaries of their safety zones, although these have not yet been conclusively measured in terms of their tolerance limits. The integrity of biodiversity, and the services of the phosphorus and nitrogen cycles in the reproduction of biodiversity, land use and the climate system are threatened. The accumulation of greenhouse gases in the atmosphere, in relation to CO_2, was still within the risk zone between 350 and 450 ppm CO_2, which can be controlled with some probability, with a value of about 400 ppm CO_2.[9]

The field of action addressed by the Earth system researchers around Rockström (Rockström et al. 2009), the boundary conditions of ecological functional systems of the planet, which are necessary for the stability of human welfare, differs in essential determinants from Anthropocene I, military security (see Table 7.2). The possible and increasingly probable environmental catastrophes are not the result of intended actions, but rather of unintended side effects of production and consumption processes, which creep up and have irreversible consequences on a global scale, with highly varying local effects (Zalasiewicz et al. 2017a, b). Individual consumption styles (Brand/Wissen 2017), but above all a multitude of organised social, economic and political actors, are involved in the emergence of the planetary threat situation

[9] In 2021, according to NOAA (www.noaa.gov), CO_2 levels averaged 414.7 ppm. (see: Atmospheric greenhouse gas concentrations | German Federal Environment Agency). In the first half of 2022, the value was almost 421 ppm (NOAA).

(Beck 2007). The collective action issues resulting from this problem constellation are well known in the social sciences (Hardin 1968; Beck 1986; Young 1997; Ostrom/Schöller 1999; Radkau 2000; Beck 2007).

In contrast to social science analyses of environmental hazards and impending environmental catastrophes, the Earth science research initiated by Paul Crutzen is primarily interested in disruptions caused by anthropogenic action to equilibrium states of planetary circulation processes that have remained stable over long historical periods, and only secondarily in the consequences of these planetary changes for the reproduction conditions of humankind. We are dealing with a natural science perspective that is concerned with the temporal development of the planet and the change of Earth ages. The increasing intensity of anthropogenic interventions during the Holocene epoch, which is short from an Earth-age perspective, modifies system states (environmental conditions) that threaten the well-being of humanity, but not life on the planet (Chakrabarty 2021, 2022).

From this perspective, CrutzenStoermer (2000) came to the conclusion that humanity, in the form of modern scientific and technological civilisation, has become a dominant factor influencing current geological and planetary conditions. With this observation, based on empirical research, they laid the foundation for a revision of the dual worldview of humans and nature that has been widely accepted in science and society in the age of modernity (Purdy 2015; Crist 2016; Arias-Maldonado 2017). The *anthropos*—the aggregate of humanity—endangers the natural (planetary) preconditions of its existence through its (industrial capitalist) mode of production, associated with increasing emissions. Currently, in 2022, the research networks coordinated by

Table 7.2 Comparison of military and ecological threat potentials in the Anthropocene

Comparative dimensions	Anthropocene I	Anthropocene II
Policy area	Global security	Global environment
Threat type	Direct and indirect mass destruction	Dysfunctions of planetary systems, geoecological uncertainty
Time course	Explosive destruction	Creeping change
Historical beginning	Trinity Test (16 July 1945); Hiroshima (6 August1945)	Age of great acceleration from 1950 onwards (Fordism)
Release mechanism	Crisis decision by governments	Non-intended side effects of social action
Political responsibility	Nuclear powers	Fossil fuels drive major powers, companies and consumers
Problem of action	Deterrence logic	Tragedy of the commons
Policy advice	Analysis of deterrence policies, identification of options for action	Analysis of environmental models, identification of options for action
Governance	Arms control/peace	Regime building/transformation

Source Compilation by the author

the Stockholm Resilience Centre for the Empirical Assessment of Anthropogenically Influenced Functional Systems, who have found that humanity is exceeding or threatening to leave safe operating space, conclude that this is likely to be the case in at least six environmental domains (see Fig. 7.1): Biosphere Integrity, Climate Change, Novel Entities, Biogeochemical Flows, Land-System Change, and Fresh Water (Steffen et al. 2015; Persson et al. 2022; Wang-Erlandsson et al. 2022).

This Earth-scientific finding differs from social-scientific environmental and risk analyses (Beck 1986, 2007; Bonß 1995; Renn et al. 2007) in terms of its (1) planetary scope and (2) anthropogenically generated threat potential not only for the present, but especially for future generations, as well as (3) its model-analytical and at the same time empirically verifiable, (4) prognostic (medium-term) and (5) area- and

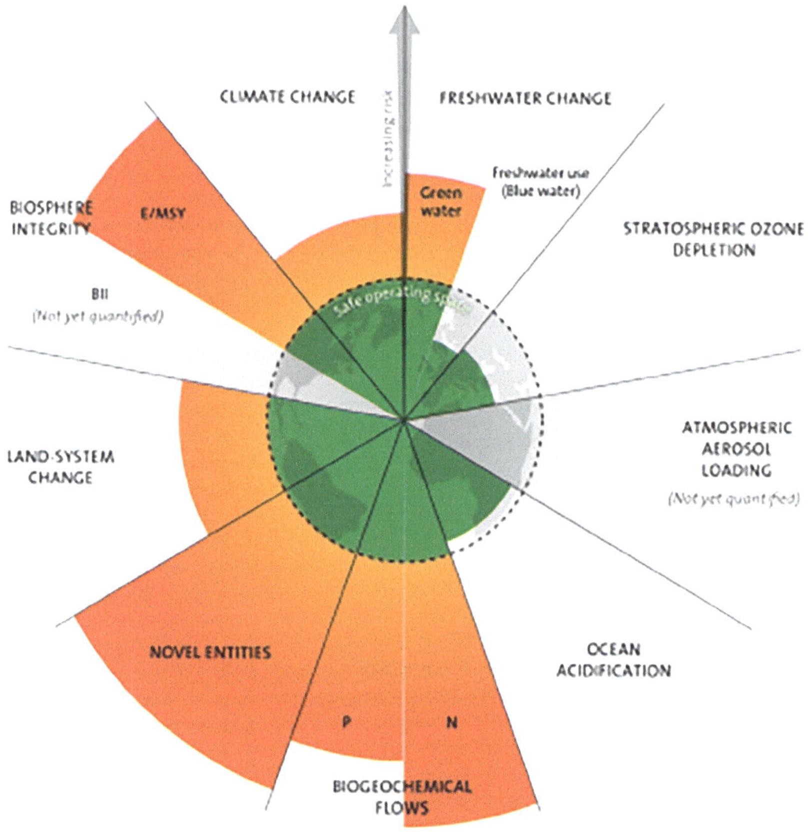

Fig. 7.1 Status of thresholds and control variables for seven out of nine planetary boundaries. *Source* Wang-Erlandsson et al., 2022 (reprinted with permission from Stockholm Resilience Centre, CC-BY-SA 4.0)

function-specific character. It is a status quo prognosis, which is based on observable action, and which applies solely on the assumption that the average action of the *anthropos* does not change. The prognosis has a normative character and is formulated with the intention of informing humanity and its collective decision-makers about what could occur with some (high) probability, provided they continue with their usual behaviour (lifestyle).

From the perspective of political science, of the questions and problems raised by Paul Crutzen's Anthropocene findings, it is primarily the gloomy and threatening prognosis of the planetary boundaries theorem that is of interest (Rockström et al. 2009; Steffen et al. 2015). To this day, politics and society are dominated by technological optimism that the environmental costs incurred in the (innovative) use of natural resources are not entirely unavoidable *ex ante,* but are technically and economically controllable *ex post* (Bonneuil/Fressoz 2013; 2017: Chap. 9). This basic trust in technical progress is further shaken by the Anthropocene hypothesis and its interpretation as a planetary transgression.[10]

In addition to the historical simultaneity, there are interactions between Anthropocene I and Anthropocene II, which can only be summarised and briefly referred to here: (1) The construction and maintenance of military, as well as civilian, production facilities for the extraction of nuclear fuel and bombs require an extensive infrastructure that changes the Earth's surface (mines, transport facilities, energy supply, enrichment facilities, production sites, reprocessing, disposal, final storage) and release climate-damaging CO_2 emissions along the production chain (Brunnengräber/Görg 2017; Brunnengräber 2021). (2) The fallout and destructive effects of nuclear bomb tests have disturbed the ecological balance in the long term where they were carried out (van Munster 2021). (3) Military equipment generates large amounts of CO_2 emissions through its production and maintenance, as well as through preparation for emergencies and its deployment—as seem currently in Ukraine as a result of the Russian attack. As a side effect, they also (!) pose a threat to climate and biodiversity (military training areas, port facilities, marine pollution). The five veto powers of the UNSC have 90% of the world's nuclear weapons (Kristensen/Norris 2022) and secure their military apparatuses through their possession; this means that in Anthropocene I the planetary destruction potential is highly unequally distributed (SIPRI 2022). The states that have the highest global military destruction potential are also the greatest threats to climate stability, not least because of the emissions from their military apparatuses (UNEP 2021).

---

[10] Within the Anthropocene debate, there are optimistic and pessimistic currents regarding the assessment of technological progress (Dalby 2016).

7.1.4 Interim Summary (1)

Political science research offers the anthropogenic capacity for self-destruction as a criterion for the beginning of the new Anthropocene era. The potential for self-destruction can be triggered by intentional or unintentional action with rapid (explosive) or slow consequences.[11] There are points of contact and differences between political science and natural science (geology or Earth science) approaches. As far as result-oriented (empirical), stratigraphic or geochemical/geoecological indicators are used to measure epochal change, common features (radiation exposure, CO_2 concentration) can be found. However, as soon as anthropogenic action potentials serve as a measurand, the canon of natural science methods is abandoned and the realm of actually existing options for action, each with intended consequences and a host of unintended side effects, is included in the diagnosis. The tension immanent in the anthropogenic technosphere between self-destructive potentials and their control through rational and institutionalised action becomes the central political and political science problem of the Anthropocene, an age in which humanity both has the technical potential to destroy its own livelihood and to develop the ability to control the destructive potential.

It is not yet clear how long the balancing act will succeed. Humanity—primarily the great powers and the emission-intensive corporations—are overstretching planetary boundaries in the areas of biodiversity and climate variability with regard to their anthropogenic tolerability; moreover, the stability of deterrence is being put to the test by the dominant nuclear powers through military technology innovations, new arms programmes and weapons tests in regime-change attempts (Iraq, Syria) and imperial wars of conquest (Ukraine). In the age of the Anthropocene, political science could—and should (?)—contribute to tackling the Anthropocene problem, with the help of situation analyses and the political enlightenment of the public, with the development of options for action or in the institutionalisation of democratic control of knowledge-based governance.

[11] The broad spectrum of planetary risks of different combinations of technologically-determined patterns of action with their temporal consequences cannot be dealt with here; nevertheless, it is a feature of the Anthropocene (MacAskill 2022). The present text confines itself to capturing the central structural feature.

7.2 Benign Neglect by (German) Political Science—Building Blocks of an Explanation

7.2.1 *Introduction*

With the end of the Cold War (Westad 2019), interest at German universities and demand from the politically-interested public regarding the problem area of nuclear security, which appeared guaranteed and well-researched through arms control and nuclear disarmament, steadily declined.[12] The nuclear deterrence system, which continued to exist at a lower level of deployed nuclear weapons (Kristensen/Norris 2022), employed only a few specialists in Germany. Despite the Balkan wars in the first half of the 1990s, interstate wars in Europe were now considered historically outdated and 'unthinkable'. Politics and political science had learned to live with the bomb. The debate on the forces and possibilities of anthropogenic self-destruction migrated from the policy area of military security to the field of environmental security (Brauch 2011). Since the beginning of the new millennium, *Earth System Research* (ESR) has drawn attention to the increasingly vulnerable living conditions on the planet, endangered by human action due to planetary environmental changes (ozone hole, climate change), and to the Anthropocene hypothesis it developed. Cultural studies took up the debate initiated by the ESR and contributed, among other things, the question of when humanity began to become the dominant geological force, i.e. when the Anthropocene should be considered to have begun (Horn/Bergthaller 2019; Ellis 2020; Antweiler 2022). The relationship between humans and nature has been widely discussed (Purdy 2015). In contrast, political science research and teaching in the Anglo-American tradition has found it difficult to take up the Anthropocene findings.[13] This also applies to the German-speaking academic world.

The following considerations refer exclusively to political science, with general arguments prevailing, but the German situation is also considered. For the German-speaking higher education area, it is instructive to trace the diffusion process of the new theorem in order to clarify the distribution of innovation potentials within the

---

[12] Research interest now turned to the analysis of the new wars that resulted from the dissolution of the Soviet Union (Yugoslavia, Chechnya, Georgia, Ukraine) and that have been increasingly fought in the Global South since 9/11 (keyword: war on terror) (Iraq, Afghanistan, Syria, sub-Saharan Africa).

[13] Bornemann et al. (2019) note that, in contrast to the reception in humanities and cultural studies, 'the Anthropocene has so far received rather little attention in political science' and refer to a bibliographic query on JSTOR, which in 2019 identified 432 publications with the search combination 'Anthropocene AND Political Science' (see also footnote 4). A repeat search on 17/12/2022 yielded 698 publications (2000–2010: 14; 2010–2015: 118; 2015–2020: 433; 2020–12/2022: 133). A random analysis of the literature documented at JSTOR in the subject area 'Anthropocene and Political Science' revealed that many publications are only marginally relevant to political science, even if the texts deal with political aspects. The comprehensive bibliographic analysis by Brauch (2021: 336) also comes to the conclusion that there is no focus on Anthropocene research in the field of political science.

discipline. It is not the established centre of the discipline that takes up the Earth science findings. The great challenge of the Anthropocene hypothesis for political science is first recognised and addressed by interdisciplinary frontier researchers, as well as by representatives of environmental politics with academic roots in Germany who conduct research in other European countries, and finally by young scholars.[14]

In German political science, peace and environmental security are widely recognised as basic values of political coexistence, alongside freedom, social justice and democracy. For the normal operation of research and teaching, peace and geopolitical environmental security represent background conditions that have not been addressed since the end of the Cold War. The daily routine is determined by institutional framework conditions (study programmes, teaching offerings, equipment) and interested demand for political science services, as well as by thematic cycles that promise innovativeness. The following section addresses the question of why Anthropocene research has not yet led to a broader thematisation within political science, provided that related discourses (climate change, biodiversity, societal transformation), which, however, do not refer to the Anthropocene hypothesis, are left aside. In order to explain the subject's far-reaching reticence, three mutually complementary and reinforcing causal contexts are considered: (1) political science's concentration on the present, (2) ESR's low connectivity with regard to political science, and (3) political science's self-referential orientation. In the mainstream of (German-language) political science, this constellation of disciplinary practices has given rise to a resonance profile that takes note of the research results of the Earth sciences with *benign neglect. There* is no debate or even rejection[15] of the Anthropocene hypothesis within the discipline in German-speaking academia.

7.2.2 The Present in the Focus of Political Science

The historical change in Holocene conditions and the forces causing it are at the centre of anthropological and socio-cultural Anthropocene research, followed by the question of how a new ecological equilibrium corresponding to human reproductive needs can be established in the Anthropocene through anthropogenic governance. For political science, which is primarily concerned with the forms of exercising authority and the continuous struggle for positions of power, this historical and at the same time Earth science research perspective is *prima vista* not very attractive. The research interest of political science focuses on the present, deals with present visions of the future and discounts the future. If a position of power—an electoral decision or a war—is lost today, it is highly uncertain whether it can be regained

[14] The authors of the volume edited by Hickmann et al. (2019) can serve as an indicator for this observation. In the volume edited by Chandler et al. (2021), only a few representatives from German-speaking higher education institutions are represented.

[15] The works by Altvater (2016), Mahnkopf (2022) and Bornemann (2019) deal critically with the Anthropocene hypothesis without rejecting it. Truly dismissive criticism tends to come from sociology (Manemann 2014; Matejovski 2016).

tomorrow. The political constellations of the present dominate political calculations and, accordingly, political science research. With the increasing political pressure in Germany for societal transformation and adaptation to climate change, however, future issues are gaining importance in the political public sphere, which is also likely to have an impact on political science theory formation (Bornemann 2022; Tremmel 2022).

In this country, academic political research still keeps its distance from future topics, since future political constellations (events, people, political conjuncture and power relations) are considered unpredictable (Flechtheim 1972; Son 2015; Urry 2016; Delanty 2020; Adloff Fladvad et al. 2020; Grunwald/Kopfmüller 2022; Zilles et al. 2022). In order not to endanger the trademark of the scientific approach, political phenomena that are factually observable and intersubjectively verifiable are at the centre of the researching gaze. There are well-established reservations regarding risky statements about future developments (forecasts, trend analyses, scenarios) that could turn out to be frivolous speculations. The focus on the present is only extended in situations of upheaval, when the future is being negotiated (formation of a government, peace treaties) or a reappraisal of historical events is pending. However, even in these exceptional situations, the dominance of the present is only partially corrected. Problems that only arise in the future are assessed under the institutional and power-political constellations of the present and dealt with accordingly (MacKenzie 2016; Smith 2021; Zilles et al. 2022).

With its strong focus on conditions, events and dynamics observable in the present, the empirical-analytical (primarily quantitative) theoretical approach of political science, which is very widespread at German-speaking universities, largely ignores the management of future problem situations. Institutionalist theories, such as historical institutionalism and the English School of International Relations, or explanatory approaches critical of domination (political economy, discourse theory) are also more concerned with the past than with the future. However, the focus on the present is being broken up by real political events in the current situation of upheaval.[16] Demands for a more sustainable orientation of state action programmes, no matter by which actors they are raised, irritate the power relations and find their way into the power calculations of the political-administrative elites (Zilles et al., 2022). Democratic relations with open institutions vis-à-vis civil society and science form a relevant supporting condition for the emergence of a state policy that, following the Brundtland Report (World Commission on Environment and Development [WCED], 1987; United Nations 2012), strives to reconcile the interests of the present with those of future generations (González-Ricoy/Gosseries 2016; Tremmel 2019; Smith 2021; Gumbert et al. 2022).

[16] Civil society movements, such as Fridays for Future, are a good example of this (Dietz/Garrelts 2013; Deudney/Mendenhall 2016; Fraundorfer 2022). The international climate movement triggered by Greta Thunberg in 2018/2019 could mark a turning point in global public thematisation of the risks of climate change (Rucht/Sommer 2019).

7.2.3 *Weak Resonance Between ESR and Political Science*

In this section, a complementary argument is presented: It could be, after all, that the weak resonance in German political science with regard to the important results of Earth science research on planetary boundary conditions is also partly caused by the theoretical-methodological approach of Earth science climate research. One reason for the lack of interest could be that while Earth system research addresses policy makers (politics) (Verburg et al. 2016b; Donges et al. 2017a), fundamental findings of political science on political action and decision-making processes remain unconsidered (including the need for consensus and legitimacy of political decisions), making interdisciplinary dialogue more difficult. The understanding between Earth science and political science is blocked by mutual resonance problems, which would explain the 'benign neglect' of political science.

The fact that political science in German-speaking universities only marginally takes up the results of Anthropocene research, and especially of *Earth Systems Science* (ESS), with a 'benign neglect' would therefore not only be a consequence of the theoretical approach prevailing in political science, but also a result of the research design of Earth sciences. From the perspective of the social sciences, specifically political science, the natural science modelling approach has serious weaknesses. Humanity, the *anthropos,* becomes a statistical aggregate in the ESS models. Its behaviour (e.g.CO_2 emissions in the Kaya identity), without taking the level of action into account, is greatly simplified and statistically represented with a few equations. A modelling of political formation and decision-making processes is not included in the analyses of the ESS. Thus, the deepening of scientific cooperation between political science and Earth science remains a desideratum.

7.2.3.1 Models with an Undifferentiated Concept of Politics

Earth science anthropocene research (Steffen et al. 2007, 2015; Nakicenovic et al., 2016) uses a systems theory explanatory approach to model global processes (carbon cycle, water cycle, energy balance, climate system, etc.). Integrated or coupled (Robinson et al. 2017) nature-human system models are usually used to simulate anthropogenically-influenced geo-ecological development options. Social dynamics are modelled on the basis of simple socio-economic functions and impact mechanisms (O'Neill et al. 2014, 2017). As far as the mapping of political processes is concerned, the models are still under-developed. Politics only takes place externally and indirectly. The models used do not explain how the political input into the respective model comes about. Depending on the "political" input of the modellers, the socio-economic development (demography, economic growth, use of fossil energy resources, energy efficiency, emission intensity) is modified politically. To improve this weakness, a further system element for modelling the policy dimension was added to the IPCC models for climate change research, which are composed of the sub-models for radiative forcing and socio-economic development (van Vuuren et al.

2014; Kriegler et al. 2014). With the chosen methodological approach, however, it is not possible to model quantitatively how the input of policy into the climate model is transformed into the output of political decision-making and will-forming processes. Policy forms something about which one can rejoice, as for example after the Paris Agreement (2015), or about which horror spreads, as after President Donald Trump's announcement of an end to the US membership of the Paris climate regime.

As Verburg et al. (2016a) have pointed out, the modelling of anthropogenically-induced dynamics of the climate system is carried out with a scientific-analytical interest on the one hand, but with the aim of policy advice on the other. In relation to this goal, four different tasks for policy-oriented model simulations of the Anthropocene are ideal-typically elaborated, derived from the analytical concept of the policy cycle (see Fig. 7.2). The scientifically-elaborated model simulations should help to identify environmental problems, formulate policy programmes, estimate effects of the programmes *ex ante* and then support the evaluation of the programmes *ex post*. The objective is honourable and correct, but at the same time structurally naïve, since it is connected to the claim to represent real political processes. The policy cycle is an analytical model without reference to real political decision-making processes. The addressees in politics and political science remain unclear. There is hardly any incentive for political science to get involved, since politics is only simulated in the abstract. In order for the political system and especially its political (administratively responsible) actors to feel compelled to engage with the information and advice offered by Anthropocene research, the assumptions and the results modelled on their basis should not have a hypothetical (artificial) character.

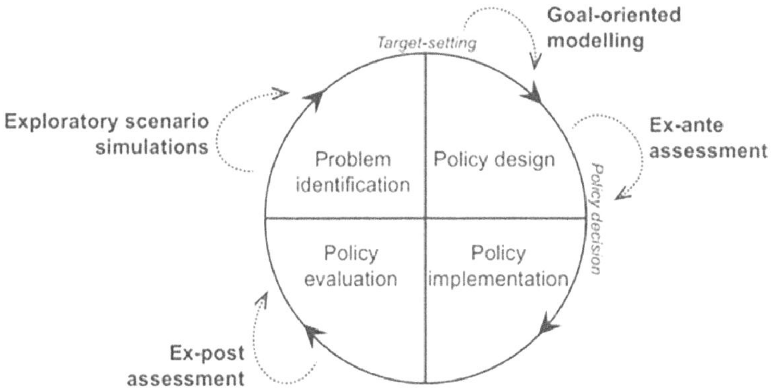

Fig. 7.2 Different roles of models in the policy cycle (Differential roles of models in policy and management designs and implementation). *Source* Verburg et al. 2016a. (Reprinted in accordance with Creative Commons CC-BY)

7.2.3.2 Functional Differentiation of Society

For modelling impending ecological boundary transgressions, Earth system research must take into account that modern societies are functionally differentiated and composed of complex institutions and organisations. Social reality is greatly simplified for theoretical (questioning) and practical (computer capacity) reasons (Mooney et al. 2013). The social and political differentiations, in particular the global differences in terms of anthropogenic influences and possibilities for action (developed North versus Global South; major powers vs. small states) are now being taken into account to some extent in more recent research (Rounsevell et al. 2014; Donges et al. 2017b). However, the orientation towards the individual as consumer, producer of infrastructures and forms of land use, and producer of side effects remains. Although more complex behavioural models are increasingly being taken into account in system simulations (Brown et al. 2016; Motesharrei et al. 2016; Gerten et al. 2018), a methodological individualism continues to dominate, which ignores political and social structures, and the complex differentiation of international society (Buzan/Schouenborg 2018).

7.2.3.3 Agency of the Anthropos

Although politics is the preferred target of the results of the ESR, the research results of political science with regard to state rule, consensus and will formation, international conflicts, coordination of action, security, power and legitimation are not even rudimentarily integrated into the model-theoretically conceived analyses of planetary functional systems. Thus, humanity (*anthropos*) in its entirety is understood as an actor that consumes energy, food and resources and produces ecological consequences in the process. Viewed from a geological or planetary ecological perspective, it is primarily the relatively probable, especially risky, anthropogenic effects that can be attributed to the aggregate of human society in toto that are of interest. From a research perspective that turns to potential for actions to process environmental risks, these general behavioural attributions are not sufficient. The options for action that can actually be realised by the actors with the power to act—states, urban metropolises, associations and corporations—to secure the planetary boundaries with their respective institutionally-limited and controlled options for action should be the focus of research.

For social science research, the aggregate of the *anthropos* makes little sense because the analytical construct of an apparent actor lacks real agency. As soon as possibilities are sought to avoid or at least reduce planetary risks, potentially capable actors (states, corporations, civil society) must be sought (Biermann 2014, 2016; Simonis 2017c; Pattberg/Zelli 2017a). In realistic scenarios, the options for action of different, politically-influential (powerful) actors would have to be modelled depending on their own ecological (climatic), institutional, societal and economic framework conditions, while simultaneously taking into account the likely response profiles of politics and society. Climate simulations could gain political attention by integrating model components on political power and decision-making processes.

7.2.3.4 Forgetting the Power of the ESR

Anthropocene research must ask itself how it can generate resonance among relevant political actors with simulation models and global risk scenarios that dispense with the modelling of concrete political-administrative decision-making processes. It must resolve the contradiction that, on the one hand, politics is seen as the saving anchor for coping with the risks of the Anthropocene, but that, on the other hand, politics is ignored as a social system that has to cope with the power and domination problems of societies organised as nation states. In the simulation models of *Earth Systems Science* (ESS) to date, only the policy output recorded by means of the agreed state action programmes for stabilising environmental security is registered and brought to bear as input for calculating environmental scenarios. Finally, the scenarios are held up to the political actors as the expected or possible outcome of their activities and as a mirror of their irresponsible actions.[17] However, a research programme that deals with the endangerment of global commons (Nakicenovic et al., 2016) should be expected to take into account and investigate the problems of domination, power and governance that stand in the way of simple solutions at the level of available scientific knowledge (Ostrom/Schöller 1999; Luhmann 2004; Beck 2009b; Radkau 2000; Moore 2016; Latour 2017; Vig/Kraft 2018).

7.2.4 Political Science Without Natural Variability

Although the geostratigraphic finding has resonated in many scientific disciplines (Zalasiewicz et al. 2021), German-language political science has only rarely participated in the global discussion, with a few exceptions, two productive representatives of which teach in the Netherlands (Biermann 2014, 2016; Biermann et al. 2016; Pattberg/Zelli 2017a). The exceptions include in particular Brauch et al. (2011); Brauch (2011); Brauch et al. (2016); Altvater (2016); Scheffran (2016); Pattberg and Zelli (2017a); and Hickmann et al. (2019). This reticence requires explanation. Basic norms of political science are touched on and questioned by the Anthropocene

[17] In this context, it is worth mentioning UNEP's Emissions Gap Reports (UNEP 2021), which hold up a mirror to the members of the Climate Convention without analysing why the agreed targets have not been met.

findings. The well-being of all humankind is recognised to depend on the stability of deterrence and the observance of anthropogenically beneficial planetary functional conditions.[18]

7.2.4.1 Political Science as an Autopoietic System

Contrary to the Anthropocene findings, political science at German-speaking universities has, with few exceptions (see above), made a clear distinction between environment and society, between active politics on the one hand, and natural ecological site conditions with specific natural resources that are more or less suitable for exploitation on the other. To explain the low interest from the core areas of political science in Earth science research results that challenge traditional assumptions of political science, plausible hypotheses can be formulated with reference to political science, arguing from a systems theory perspective. Like other sciences, politics can be understood with Luhmann (1971, 2013, 2004) as an autopoietic system that is relatively isolated from its environment, autonomously pursuing its social function according to an institutionalised decision-making mechanism (code of action). The analysis of politics as an *autopoietic* system with functional specialisation in the exercise of rule and the securing (acquisition) of power allows typical features of political science to be explained: the assumption of stable environmental conditions, the fixation on the present with the priority of dealing with acute problem situations, the path dependency of decisions and the underestimation/overestimation of one's own ability to act. Since these and other explanatory factors (including the media landscape) vary from country to country, only comparative studies could clarify which of the factors discussed here are particularly relevant for which countries.

7.2.4.2 Stability of Environmental Conditions and Externalisation

In accordance with this theory, political science is concerned with the analysis of the internal relations of political systems, with the power struggles for political positions (*politics*), the institutional (democratic or authoritarian) structures (*polity)* and the various arenas and institutionalised policy areas in which power struggles are fought out, consensus-building processes take place and certain political outcomes (*policies)* are produced. As a sub-discipline of political science, international politics examines the various forms of interaction between political systems of rule, in particular security, peace, conflict and integration, whereby the state's logic of action directed outwards is determined on the one hand by the structure of the international system (anarchy, polarity, hegemony) and on the other hand by the internal power

[18] See, among others, Jungk (1956); Senghaas (1969); Steffen et al. (2005); Schellnhuber (2015); Nakicenovic et al. (2016); IPBES (2019); and IPCC (2021).

relations.[19] In all variants of this analysis of politics based on systems theory, the decision criterion dominates: power or loss of power, whereby an instrumental relationship is assumed in relation to the 'environment'. The only thing of interest about the environment (raw materials, logistics, spatial factors) is its potential to increase or limit power.

This instrumental view of the environment is subject to increasing criticism in the age of the Anthropocene. The relationship of the national level of the political-administrative system to the local and to the global level of action is changing. The respective local ecological conditions are, in many of their manifestations, the result of the interaction of anthropogenic—social and political—activities with the natural resources and natural forces of the globe. Natural conditions untouched by human hands, whose services (resource supply and regulating functions) are freely appropriated and used, are increasingly disappearing (Millennium Ecosystems Assessment 2003; Primmer et al. 2015; Mehring et al. 2017). Environmental conditions, which are considered by policy-makers as nature's given endowment (soil fertility, rainfall, biodiversity, water quality, etc.), are themselves, to an increasing extent, the result of human activity (Nakicenovic et al. 2016; Purdy 2015).

To the extent that local environmental conditions are not only understood as increasingly threatening and risky, but also as anthropogenically co-caused, the tension between the (already problematic) actual state and the (further) deteriorating development conditions will trigger political conflicts of varying virulence.[20] The form and intensity of conflict will be shaped not least by the patterns of interpretation that can be established in the local space about the causative forces, which actors appear to be responsible for the hazards and which authorities (public administration, companies) are held liable for damage. The search for the polluters and the debate about which political actors have failed has only just begun. The big polluters in ecological hazard situations, such as the oil companies and states with highly-externalised emissions, have neither an increased interest in expensive requirements to avoid external costs nor in legal disputes (up to and including political conflicts). They react—at least very often—with a public communication of dethematisation that presents fears as unfounded, claims risks as exaggerated and classifies damages as inflated (Oreskes/Conway 2011; Bradley 2011; Dunlap/McCright 2015).

There are obstacles to the public and political discussion of Anthropocene natural relations that are difficult to overcome: The actors in local as well as national politics point to the responsibility of the voters, to gaps in knowledge and to higher political authorities; the multinational but locally-active companies justify their actions with world market constraints, their limited knowledge and consumer wishes. The blame game continues at the international level. The countries of the Global South see

---

[19] This statement is generally true. The dominant focus of the discipline at any given historical point in time is on current events, wherever they are on the globe.

[20] A distinction must be made between resource conflicts over raw materials that have only been slightly processed (transformed) and resource conflicts in which the naturally available resources are changed and upgraded by industrial processes.

the developed industrialised countries as responsible (cbdr principle)[21]; the latter accept the principle only as a general norm, buy their way out by making (small) payments into funds, and in turn make demands on the countries of the Global South to accelerate mitigation.

7.2.4.3 Stable Institutions (Path Dependency)

A third factor that complicates the thematisation and politicisation of the threat to global environmental limits is the path dependency of political action (Werle 2007). The institutions established in the Holocene era, whose programmes of action are supposed to simultaneously secure economic performance and protect the anthropogenic habitat, prove to be stable and only capable of learning to a limited extent. They are therefore less and less able to meet the triad of goals of sustainable ecological, social and economic development (Franz 2014; Bückmann 2015; Dauvergne 2016; Grunwald/Kopfmüller 2022). In order for the Anthropocene not to take the predicted catastrophic turn for humanity (Steffen et al. 2007), the global environmental limits would have to be recognised as a political guideline for a global (!) socio-ecological transformation (WBGU 2011). This normative desideratum is functionally evident, provided that the research results of Earth system research are politically accepted. However, since the institutions of the Holocene continue to steer action in a path-dependent manner, conflictual learning processes, which are further complicated by the strongly unequal distribution of the respective national—as well as above all international—potentials for action, are on the agenda. The collective result of these appropriation processes of the Anthropocene, consisting of countless individual achievements, is completely open (Bammé 2014a). In modern, highly-differentiated societies, the individual social systems adapt to the new conditions, and to new demands emanating from other social subsystems, according to their own traditional systems of rules. In order to keep the costs of adaptation as low as possible, and the internal order of rule as stable as possible, politics tends to adapt only slowly to changing environmental conditions.

7.2.4.4 Environmental Policy—One Policy Area Among Others

German political science has so far hardly registered the changes in political structures and processes induced by the challenges of the Anthropocene, and has tended to deal with consequential phenomena in which inner-societal conflict constellations dominate the political stage (energy policy, transport policy, glyphosate conflict). A possible cause for this diagnosis of deficits could also be the disciplinary parcelling of

---

[21] This refers to the fundamental principle of the United Nations Framework Convention on Climate Change (UNFCCC), which assumes that member states have '*common but differentiated responsibilities*'. Since its agreement in 1992 at the Rio Earth Summit, the cbdr principle has shaped all environmental negotiations between North and South.

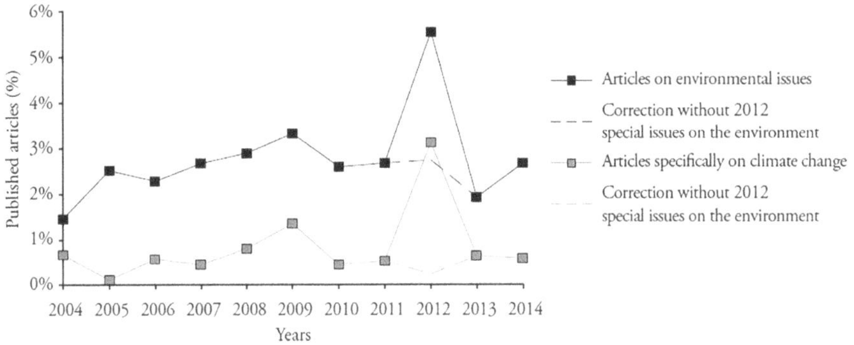

Fig. 7.3 Published articles directly related to environmental issues in selected journals. *Source* Pereira (2017: 11) (Open Source)

political science: The various sub-fields of political science conduct research largely isolated from one other, without mutually taking up research results. In a bibliographic study, Joana Castro Pereira (2017) determined the extent to which articles on environmental topics were published in the twenty most important English-language journals in the field of international relations (IR) in the years 2004–2014. With the exception of 2012, when two special issues on climate change were published, the environmental theme accounted for no more than three per cent of the total number of articles published (see Fig. 7.3). Scholarly attention in the IR field does not focus on environmental policy, which is given only marginal importance by the core of the discipline as *low politics*, with little influence on international security. The data from Pereira's study confirm the thesis that environmental governance did not play a major role in the field of IR research, at least until the mid-2010s. Pereira's analysis can serve as an example of the low preoccupation within international and global environmental governance with the political science sub-discipline of International Relations, and thus also of the *benign neglect of* the Anthropocene issue.

Another reason for the low degree of willingness to take up the Anthropocene hypothesis may be the low level of inter- and transdisciplinarity (Schneidewind/Singer-Brodowski 2013; Wissenschaftsrat 2015) in university foreign policy and IR research. Research cooperation with non-university research institutions is still a rarity in the policy area of political science IR research in Germany, especially since university capacities for establishing cooperative relationships are very limited.

7.2.4.5 Overestimation of Agency: Power Over Nature

Political science analyses focus on the political actors. Only when changing environmental conditions affect the actors does the environment come into the focus of politics. However, in political science policy advice the situation is different. There, environmental policy actors want to be advised on how they can assert their interests in the face of increasingly complex political and socio-ecological problem situations

(WBGU 2011). But even in policy advice, the capacities of governments to act are easily overestimated, but also underestimated. Economic policy advice may overestimate the leakage effect of national mitigation measures and block a national-centred climate policy with this assessment of the situation. However, omitted mitigation measures assume that their temporal shift is ecologically unproblematic. They are based on the assumption that, on the one hand, the absorption capacity of the planetary environmental systems for national emissions is arbitrarily large at a given point in time, and that, on the other hand, the state's capacity to act is sufficient to repair or at least compensate for damage. In relation to the damage caused, both environmental economics and environmental policy greatly overestimate the ability of states to remedy the damage. For political science, as for real politics, neither irreversibilities nor tipping points exist. The logic of politics and the basic understanding of stable planetary environmental conditions inscribed in political science do not correspond to the functional laws of the planet.

States seem well advised, at least superficially, to externalise the costs of global environmental problems and to accept abatement costs only under the condition of the existence of collective and legally enforceable regulations. In an economic environment in which an increase in production costs easily works to the advantage of competitors, the reluctance of companies and states to internalise environmental costs is rationally justified. Mainstream economic reason (Sinn 2009) advocates cautious tactics, until a reliable and sanctionable international regime is established, since assuming a pioneering role in global environmental and climate protection would relieve third countries via the price mechanism, trigger leakage effects, and serve only good conscience, but not environmental security (Nordhaus 2020). In this calculation, the potential for action by individual states is not sufficient to compensate for the (anticipated) negative consequences of a unilateral environmental protection strategy. However, this strategy of 'wait and see' clearly underestimates a country's own potential for action, as positive side effects for the environment, health, innovation potential and jobs, as shown above all by Elinor Ostrom (Ostrom/Schöller 1999; Ostrom 2009a, 2009b), are not taken into account.

However, as long as the positive side effects are ignored, it is hardly possible for the individual company, like the individual state, to escape the rationality trap and the drama of the common good (Hardin 1968). If actors do not assess their potential for action with comprehensive rationality, they will refrain from actions that appear to benefit their state competitors and, as a result of politico-legally enforced environmental protection requirements, possibly (*ceteris paribus*) lead to rising costs and reduced consumption on their part. With this argument, in preparation for the climate negotiations in Kyoto in 1997, the U.S. Senate passed a resolution (Byrd-Hagel) with 95:0 votes against 'one-sided' mitigation obligations favouring emerging countries (China, India) to the detriment of industrialised countries, above all the United States (Simonis, 2017a). The American Senate overestimated its ability to act here, and its threat potential to induce the emerging countries to turn away from the CBDR principle (*common but differentiated responsibilities*, Art. 3/1 of the Rio Convention of 1992). Together with the developing countries of the Global South, the USA adhered to this basic norm of the international climate regime, which therefore also found its way into the Paris Agreement (2015). Overestimating their capacity to act to enforce an international regime that avoids the likelihood of leakage through binding and verifiable mitigation commitments by emerging economies, the senators underestimated the ecological consequences of climate change, not only for the planet, but also for the US.

7.2.4.6 Interim Summary (2)

It is not a good sign for the innovative capacity of political science at German universities that for over ten years the findings of the Earth sciences regarding the threat to humanity's planetary *safe operating space*—apart from singular exceptions (Brauch, Scheffran, Roth) and the environmental political scientists based abroad—have not been taken note of by the core of the discipline. The benign neglect seems to be slowly coming to an end (see Sect. 7.3), but the new era is still marginal in research and teaching. At present, the 'turning point' (Federal Chancellor Olaf Scholz on 27 February2022 in the Bundestag) dominates attention. As after the implosion of the Soviet system of rule, the question arises as to why fundamental changes in the international environment, although long in the offing, are not recognised and brought to public attention by German political science.

The explanations in Sect. 7.2 can be condensed into four findings:

(1) Political science is fixed on the present with the temporary having priority. The temporalities between ESR and political science diverge.
(2) There is little fit between Earth system research and political science. The *anthropos* is a statistical aggregate without actor quality. Historical differences between the former colonial states and the Global South are ignored. The concept of politics used by the ESR has little to do with reality and overestimates the capacity and potential for action of states.
(3) The theoretical conception of politics as a relatively autonomous social subsystem, which makes binding decisions for the respective nation-state society in the struggle for power, focuses on the present, concentrates analyses on the management of crises, instrumentalises the environment, is not interested in the consequences of decisions for neighbours and future generations (externalisation), and overestimates its own power to act.
(4) With its theoretical approach focused on power relations, political science is able to explain which hurdles have to be overcome in the sphere of politics in order to cope politically with the challenges of the Anthropocene. In this way, political science can explore where power games provide options for action, for which political actors (parties, governments, city administrations) windows of opportunity are opened in a particular historical constellation, and how transformation processes can be helped along.

7.3 Contributions of Political Science from German-Speaking Academia

The endangerment of the living conditions on the planet, which is scientifically expressed by the Anthropocene hypothesis of the Earth sciences, is changing the environmental as well as the social framework of politics. While the globe's political actors began to respond to the new, scientifically diagnosed (Rockström et al. 2009), situation with the UN World Conference on Human Development in 1972 and the Stockholm Declaration and Action Plan for the Human Environment (Lövbrand/Mobjörk 2021; Töpfer 2013; Conca 2016; Guterres 2020), the window of opportunity to avoid catastrophic change already seems to be closing (Darby 2018; Bammé 2014a; IPCC 2021). Policymakers are too slow to react. Irreversible environmental changes, through loss of biodiversity and natural *greenhouse gas* (GHG) reservoirs, are beginning to limit the options for action in Earth governance. It is all the more surprising that political science in German-speaking academia is only just beginning to discover the challenges of epochal change (Brauch 2011; Hickmann et al. 2020).[22]

At the international level, a broad and intensive debate on the Anthropocene findings has already emerged in the cultural and social sciences, with several strands of discourse which Bonneuil (2015; Bonneuil/Fressoz 2013, 2017) refers to as 'narratives' (see also Dürbeck 2020). He distinguishes between a naturalistic, an eco-catastrophic, an eco-Marxist and a 'post-nature' narrative (see also Lövbrand et al. 2015; Brondizio et al. 2016; Steffen et al. 2016; Dürbeck 2020). Admittedly, there are also contributions to the debate that are critical of the Anthropocene concept (Castree et al. 2014; Manemann 2014; Matejovski 2016; Moore 2016; Mahnkopf 2022; and see also Dalby 2016; Wissenburg 2017). These social-theoretical, natural-philosophical and ideology-critical research findings cannot be addressed in the following overview, which focuses on political science contributions to the Anthropocene discussion.

Within political science, as generally within the Anthropocene debate (Dalby 2016; Dürbeck 2020; Brauch 2021; Zalasiewicz et al. 2021), several strands of discourse can be observed. Bornemann, Glaab and Partzsch (Bornemann 2019: 331f.) distinguish between (1) an 'instrumental' use of the Anthropocene concept, (2) a 'substantive', crisis-diagnostic and prescriptive variant, as well as (3) an interpretative, albeit heterogeneous, research group positioned critically towards the Anthropocene diagnosis, whose work cannot be qualified as political science interpretations in the narrow sense. For the following argument, the instrumental strand of discourse, which uses the Anthropocene concept to signal the context of a work to the audience,

[22] These include the 'International Politics in the Anthropocene' workshop of the Institute for Peace Research and Security Policy (Detlef Rothe) at the University of Hamburg (April 2016), the author conference organised by Hans Günter Brauch (May 2017) on "Politics' in and for the Anthropocene: Obstacles, Challenges, Opportunities and Tasks for the Social Sciences in the 21st Century' and the April 2017 Potsdam workshop 'Environmental Policy in the Anthropocene. Problem Areas, Challenges and Research Perspectives' of the DVPW's Working Group on Environmental Policy and Global Change (Hickmann et al. 2019).

is irrelevant. The interpretative discourse strand does not need to be of further interest either, since a new social and political subject area is analysed here using familiar methodology (discourse analysis), but does not affect the theoretical-methodological approach.[23] The following analysis is interested in the 'substantive' discourse strand, which does not exclude an interaction between the subject area and theory from the outset.

It can be assumed that with increasing social science environmental research (economics, sociology) on the 'new' age, and corresponding research activities in neighbouring disciplines (history, law, geography, cultural studies) (Bammé 2014b; Hamilton et al. 2015; Hamilton 2017; Nicholson/Jinnah 2016a; Renn/Scherer 2017; Will 2021), political scientists who teach at German-speaking universities or have completed their academic degrees there will also recognise the political relevance of the Anthropocene hypothesis. Interest in the new problem is likely to increase as soon as concerned political actors in the public sphere make the planetary challenges the subject of political debate. In a scientifically- and politically-changed context, political science, which is fixated on current conflict constellations and usually conducts research in a decontextualised and theory-based manner, will also take up the issue triggered by the Earth system sciences. The emerging planetary boundaries of the non-sustainably regulated post-modern risk society (Beck 2007) are modifying political conditions on such a scale that even research branches with no affinity to the environment will have to take the framework conditions of the new age into account.

However, in several thematic areas with a clear focus on the environment, there are contributions from political scientists who are employed at German-speaking universities and research institutions or who work at foreign universities but have German academic provenance. The research interest can be assigned to six thematic fields: (1) scientification of environmental governance; (2) age of planetary insecurity; (3) contributions to global governance; (4) contemporary diagnostics; (5) contributions to IR theory; (6) contributions to political education. The following overview does not aim at an exhaustive presentation of the individual contributions. Rather, the analysis is guided by the consideration that the thematic fields can also be understood as indications of the structural change in international politics in the age of the Anthropocene. Characteristic features of the emerging political structures of the Anthropocene can be identified by looking at the problems that political scientists in Germany are working on in the academic sphere.

Based on the ever-closer link between Earth science research and politics, the focus is on the expanded security problem in the Anthropocene, which arises from the increasing threat of environmental systems to state integrity and sovereignty (*state security*), to human reproduction (*human security*) and to ecological stability and biological diversity (*ecological security*). The growing insecurity is fuelling

[23] See also the call: *Herausforderung Nachhaltigkeitstransformation: Mehr Politikwissenschaft wagen!* by Thomas Hickmann et al. (2022), which points to a 'broad methodological, empirical and theoretical fundus' (p. 154) 'to face the social debates in the age of the Anthropocene'. The existing canon of theory and methods in political science seems to be ideally suited here to grasping the changing conditions in the new age and to interpreting sustainability norms (Hickmann et al. 2022:150ff.).

increasing political pressure to defuse the growing risks to the survival of humankind through political measures (governance). Three main strategies can be distinguished: Mitigation of environmental burdens, adaptation to the new socio-ecological conditions, and transformation of social and technological conditions. The mutual interdependence of the three options for action and their interaction with traditional conflict situations results in a great need for new analyses of the present and for processes of self-understanding. The research results of the Earth sciences, in particular the threat situations, risks and options for action they have identified, must be comprehensibly transferred to the political sphere, just as, conversely, political government programmes must stand up to scientific scrutiny in order to prove their practicability and secure their acceptance (legitimacy). Contemporary diagnostics is part of an institutionalised transnational political-ecological complex (Simonis 2018), which fulfils a mediation task between science, society and politics in the Anthropocene. The analysis of the new planetary conditions with their manifold consequences for international politics has led to the demand for a reorientation of IR theory and to the first attempts to implement this demand (Chandler et al. 2021). The traditional focus on national interests and national political systems is modified by a view of planetary politics and global governance as well as a re-evaluation of the treatment of other living beings. Finally, political education is receiving strong impulses from Anthropocene research in the Earth, social and cultural sciences. First concepts and educational programmes have been developed, primarily in Austria, and are now being tested and form an entry point into a transdisciplinary practice of democratic virtues in the Anthropocene.

7.3.1 Scientification of Environmental Governance

Two decades before the planetary boundaries hypothesis (2009), the *Intergovernmental Panel on Climate Change* (IPCC) had already been established by the United Nations General Assembly in December 1988, at the urging of climate scientists (Bolin 2008), and in 1990 had presented its first report (FAR/AR1) on the situation of the global climate. The IPCC report poses an old problem in a new form. Research results of the natural sciences (Earth sciences) have to be translated from the sphere of science into the political arena in order to be perceived and interpreted there as politically relevant facts. The example of climate research clearly shows the central interface between science and politics in the age of the Anthropocene (Beck 2009a, b; Edwards 2010; Hulme 2011). Without the research results of the Earth sciences on climate change[24] and its anthropogenic drivers, and without the political interventions of individual climate researchers, such as Bolin (2008), James Hansen (2006), Hans-Joachim Schellnhuber (2015), and Michael E. Mann (2012, 2021), the global public would know nothing about the climate problem and the international climate regime would not have come into being (Beck eck 2009a; Hulme 2011;

[24] See the IPCC Assessment Reports, AR1 (1990) to AR6 (2021/22).

Simonis 2017). Climate change is a theoretical construct the existence of which is recorded and proven with the help of certain theoretically-determined indicators and observational data aggregated over longer periods of time (30 years) (Edwards 2010).

The thesis of the end of the Holocene and the beginning of a new Earth age was formulated by geoscientists (Crutzen/Stoermer 2000) and taken up by climate researchers (Rockström et al. 2009), who base their analysis on the self-endangerment of the *Anthropos*. In contrast to the stratigraphic Anthropocene finding (Zalasiewicz et al. 2017a, b), the Rockström model has a normative message. To avoid catastrophic boundary transgressions, the *anthropos* must change its behaviour. With this insight, and the demand on national governments and all political actors to ensure the necessary behavioural changes, climate research actors inadvertently assumed a political function: to educate humanity (and especially governments) about the non-intended side effects of the fossil fuel social formation and to develop options for action for state policies. For this complicated translation process to succeed (Jasanoff 1994, 2004), the separate worlds of science and politics need to work together in trust (Hoppe et al. 2013; Arias-Maldonado 2020; Jasanoff 2021; Kohler 2022). In the German research space, Silke Beck has primarily dealt with the cooperation between Earth sciences and politics using the example of the IPCC—and later also the IPBES (Beck 2009a, 2016; Beck/Mahony 2017; Beck/Forsyth 2018; Beck/Oomen 2021).

There is a wide range of literature on the relations between science and politics, including international politics, as two functional systems with entirely different codes of action. From the perspective of constructivist institutionalism and Anthropocene research, both the linear model, which asserts the one-sided reception of (objective) scientific (especially natural science) research results by political actors, and the deficit model, which is convinced of the existence of a lack of scientific knowledge on the part of politicians, have little explanatory power. More convincing is the analytical model of co-production of knowledge (Jasanoff 2004; Hoppe et al. 2013; Beck/Forsyth 2018; Posner/Cvitanovic 2019; Harvey et al. 2019), which assumes a productive collaboration between science and policy. In this model, both functional areas have relative autonomy, but are in an exchange relationship (norms, rules, finances) and generate action-guiding knowledge in the common communication space, which on the one hand provides science with urgent questions and on the other hand recommends governance goals (governance objects[25]) and options for action to politics (Posner/Cvitanovic 2019).

In the age of the Anthropocene, the co-production of scientifically—model-theoretically and empirically—validated knowledge for action is gaining political significance. The planetary boundaries of *safe operating space* are governance goals determined by Earth science, which also have a social science component that allows statements about the environmental compatibility of living conditions. In order to

[25] For Lukas Hermwille (2019), the concept of 'global climate governance' forms a 'boundary object'. In Fig. 7.3, I use the term 'governance object' to characterise the results of cooperation between science and politics, which in turn influence or even control science and politics. Governance objects are a certain class of boundary objects, i.e. arrangements 'that allow different groups to work together without consensus' (Star 2010: 602).

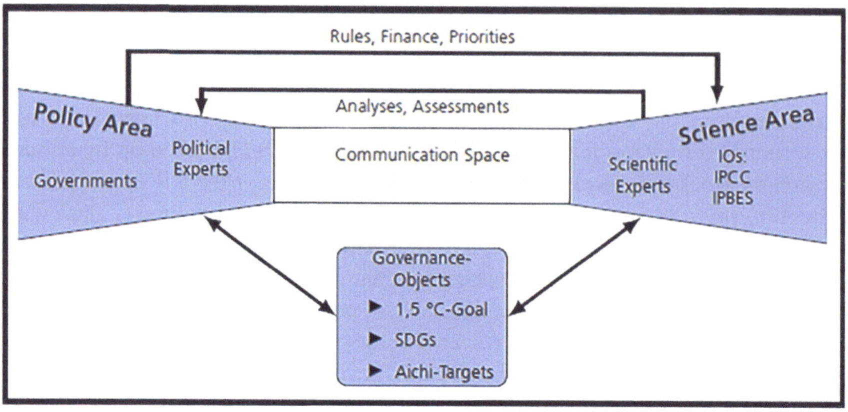

Fig. 7.4 Environmental knowledge: Co-production of policy and science. *Source* Figure by the author

create trust in international actors and their agreements on the part of national governments, which is necessary to deal with the collective problem of action, the binding collective political recognition of Earth-science boundary conditions as goals guiding action by the political representatives themselves is also necessary at the international level (Victor et al. 2022). If the assumption is correct that in the Anthropocene, out of pure self-interest on the part of states, measures of planetary governance, as soon as the stability of governments becomes dependent on them, acquire overriding political relevance, it can be assumed that the co-production of governance objects and their scientific control will also gain political weight. The climate regime can serve as an example (see Fig. 7.4).

In the Paris Agreement (2015), international politics assigned climate research an expanded range of tasks. In addition to the traditional, but ever-expanding, climate reporting and the assessments of the IPCC on the state of climate research, the consequences of climate change and possible options for action, the Earth sciences were assigned two further tasks:

(1) The voluntary commitments on mitigation and adaptation are to be submitted by the members of the regime at least every five years and, in accordance with the Treaty, should be increasingly ambitious, and must not only be reviewed, but their collective contribution to the achievement of the climate targets of Art. 2 and Art. 4/1 of the Paris Agreement must also be determined on the basis of the 'best available knowledge' (preamble, sentence 4). The results calculated by international climate research form the information basis for the subsequent negotiations on stricter mitigation and adaptation measures and agreements on their (voluntary) implementation by the member states.
(2) In the area of adaptation to climate change, climate research is given additional tasks, with the development of options for action, monitoring and evaluation of measures. Since the manifestations and impacts of climate change differ greatly in local areas, the regionalisation of climate change scenarios and the modelling

of the respective local conditions, as well as the elaboration of locally-optimised adaptation strategies, are gaining in importance.[26]

For policy formulation, implementation, monitoring and international reporting, institutionalised coordination procedures between politics and science are necessary, as analysed by Silke Beck. National governments need up-to-date analyses of the effectiveness and efficiency of aggregate, national and sectoral measures to prevent environmental damage and to adapt to changing environmental conditions. The reference to scientific studies is an important building block in the legitimisation of government action vis-à-vis other states and the public. International cooperation to secure planetary environmental boundaries within the framework of multilateral regimes (global governance) requires extensive scientific-technical cooperation between experts with their own specialist languages, which are not understood by third parties without appropriate training (Wyborn 2015; Haas 2016).

The existence and habitus of a transnational scientific-technocratic elite, supported by the institutionalisation of enhanced scientific-political cooperation at the international level through regime-building, complicate domestic consensus-building for global environmental governance. The climate sceptic movement has found a grateful victim with its criticism of internationalised Earth Systems Science (ESS) (Oreskes/Conway 2011; Dunlap/McCright 2015). Nevertheless, it must be stated: Compliance with the planetary boundaries hypothesis cannot succeed without close cooperation between science and politics. However, the dovetailing of politics and Earth sciences has the effect of blurring the traditional—and functionally necessary dividing lines between the two social systems (Jasanoff 1994, 2004; Miller/Edwards 2001; Edwards 2010; Beck 2009a; Beck/Mahony 2017; Beck/Oomen 2021).

The rapid increase in scientific knowledge on the topics and fields of action of climate change represents a major challenge for the climate policies of nation states. The number of scientific articles published annually can serve to illustrate the rapidly-growing scientific knowledge in the field of Earth sciences. Fig. 6.4 shows the dynamic increase in scientific papers on the topic of climate change, broken down by subject area (Haunschild et al. 2017). For the period 1980–2014 alone, 22 2060 publications were recorded. The quantity of annual scientific publications in a thematic field is not necessarily an indicator of the changing relationship between science and politics. However, insofar as the publications contain demands on policy or deal with political action itself, the flood of publications indicates that climate science is increasingly advising policy and making demands on it, just as, conversely, political actors expect services from science, such as risk analyses, options for action, problem diagnoses or the evaluation of policy programmes. The increase in publications with mutual references can serve as an indicator (a) of the growing interaction between the two spheres of politics and science and (b) of the increasing dependence of the political system on scientific inputs.

[26] See Art. 7 of the Paris Convention, in particular Art. 7/5, 7/7 and 7/9.

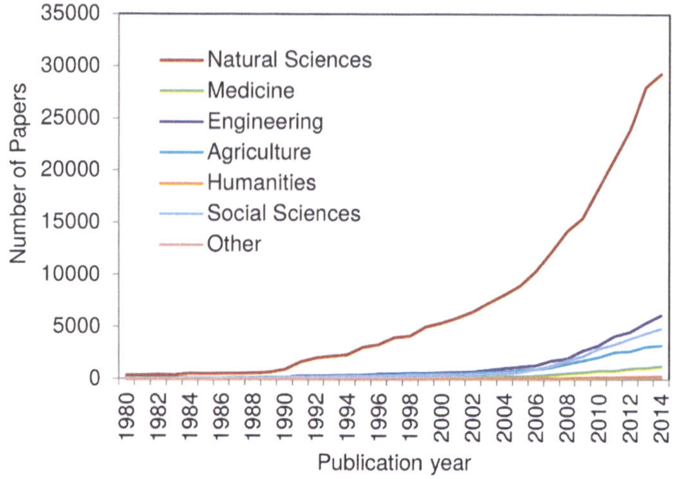

Fig. 7.5 Increase in scientific papers (articles and reviews) in the field of climate change. *Source* Haunschild et al. (2016) (Reprinted under Creative Commons Attribution Licence)

In the Anthropocene era, the close coupling of Earth sciences and politics is likely to form a typical (since structure-determining) feature of the global governance system (Zürn 2018a; Weiss/Wilkinson 2019; Simonis 2022). It implies an increasing dependence of politics on scientific services, which have a normative quality because of their reference to the survival or good life of the *anthropos*. However, this also increases the likelihood of the politicisation of the scientific system. Political actors might be inclined to exert influence on the normative statements of science. The work of Silke Beck deals with this constellation, typical for the Anthropocene, using the example of the IPCC and the *Intergovernmental Science-Policy Platform on Biodiversity and Ecosystem Services* (IPBES), in which critical social and political science is particularly in demand (Bornemann et al. 2019). Political options for action can be opened up through reflexive science, provided that enlightenment is provided on possible attempts to instrumentalise politics through politicised Earth sciences and, conversely, Earth sciences through politics (Beck et al. 2014; Beck 2016, 2019; Beck/Mahony 2017; Beck/Oomen 2021) (Fig. 7.5).[27]

7.3.2 *Anthropocene: Age of Planetary Insecurity*

When a team of internationally-renowned Earth scientists published alarming findings in the two globally-recognised scientific journals *Nature* (2009) and *Science*

[27] Bornemann et al. (2019b) see the task of interpretative policy research, among other things, as being to show 'the objectifying adoption of natural science diagnoses' and the crypto-normativity associated with 'crisis diagnoses' for the political sphere.

(2015) about the planetary environmental risks threatening humanity, generated by humanity itself, and in addition a political appeal for urgent action (the State of the Planet Declaration) (Brito/Stafford-Smith 2012), the warning bells should have rung in the political arena—but they did not! Instead, the hypothesis of human-induced changes to the planetary system states of the biosphere (e.g. biodiversity, gas concentrations in the atmosphere), which are necessary for beneficial living conditions for humans, diffused only slowly into the political public understanding. The social and political implications of the thesis of planetary boundaries, as dangerous for human well-being, have been classified in the sphere of politics more as alarmism than as a call to action (Matejovski 2016). Nevertheless, the findings of the ESR mark an epochal shift in which the Holocene age of relative environmental security is being replaced by the age of anthropogenically-induced environmental uncertainty (the Anthropocene). Humanity has become a dominant geological influence during the Holocene (Crutzen 2002). The production processes and technical tools it has developed threaten the planet's life support system (Rockström et al. 2009; Steffen et al. 2015). On a planetary scale, technologically-induced growth processes generate environmental and life risks with catastrophic potential, which, however, only ever have an effect that can be experienced locally (IPCC 2022).

A broad and rapid reception of the Rockström-Steffens hypothesis naturally took place in the Earth and climate sciences, as well as in the social and cultural sciences cooperating with them (Zalasiewicz et al. 2021; Brauch 2021; Luciano 2022). The connection of the geological-stratigraphical thesis of the beginning of a new Earth age determined by human activities, with the hypothesis of an anthropogenically-caused and threatening violation of the planetary boundary conditions beneficial to humankind has given rise to a debate in the scientific community concerned with environmental issues as to the extent to which Holocene environmental security will give way to increasing destabilisation, and what consequences are to be expected. In addition to the further elaboration, concretisation and verification of the Earth science core statement within the ESR (Steffen et al. 2015; Wang-Erlandsson et al. 2022; Persson et al. 2022), a polyphonic discourse developed on the consequences of the planet-wide threat to the environmental security of human existence, associated with considerable local, temporal and social differences (Scheffran et al. 2012; Lövbrand et al. 2020; Lövbrand/Mobjörk 2021; Trombetta 2021; Rothe/Benner 2021).

These debates were fertilised and stimulated by insights into the concept of political security that had developed in the social sciences in the decades after the beginning of the nuclear age (Senghaas 1969). A theoretical explanation of the security discourse was also provided by the constructivism of the Copenhagen IR school (Buzan et al. 1998; Buzan/Hansen 2011), which recognised the securitisation of political problems as a political strategy to increase legitimacy and justify extraordinary measures. Political science reacted with a differentiation of security theory: state security, human security, military security, nuclear security, economic security, environmental security. The results of Earth science research have expanded the security discourse and have produced further interpretations of the security concept: ecological security, global security, ontological security, Earth system security (Brauch et al. 2009; Oswald Spring et al. 2009; Brauch 2011; Scheffran et al. 2012a; Scheffran

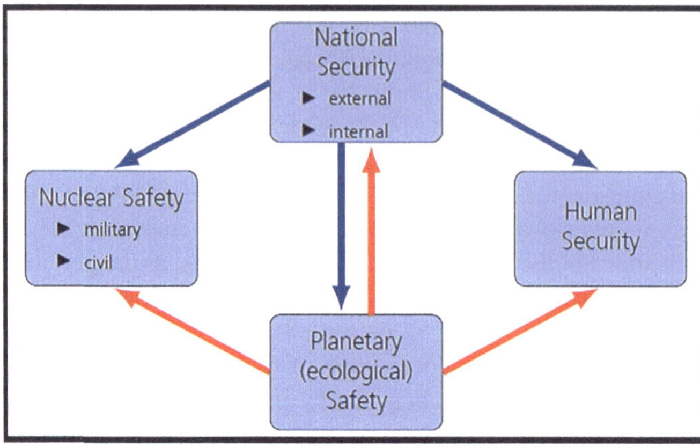

Fig. 7.6 Perspectives of political science security research. *Source* Compilation by the author

2016; Hardt 2019). Scholars from the German-speaking research area (e.g. Brauch, Hardt, Roth, Scheffran) also participate in the international discourse (Lövbrand et al. 2020) on the interpretation of the security problems of the Anthropocene.

Figure 7.6 shows the change in perspective associated with the Anthropocene paradigm in the field of security research. While in classical IR research the national interest and national security, which could be endangered by hostile actions of other states, form the normative basis of theory and empiricism, the new paradigm changes the perspective. The focus of research interest is now on securing 'safe operating space' by respecting the ecological boundary conditions of the planet. The threat to nuclear security and the threat to human security, like national security, are primarily viewed from the perspective of the anthropogenically-induced change in planetary environmental conditions. The power games and violent conflicts associated with the anarchy of the state system recede into the background and are—falsely—overshadowed by the ecological problem.[28] Ecologically-oriented Anthropocene research would be well advised, in the author's opinion, to keep an eye on the entire spectrum of destruction potentials and their consequences for human as well as societal (national) security. The security issue in the Anthropocene has developed into a highly complex inter- and multidisciplinary research subject.

7.3.2.1 Understanding Security in the Anthropocene

In her work, Judith Nora Hardt has elaborated the differences and similarities existing between the security conceptions of *Earth system research* (ESR), IR realism and liberal institutionalism, using the analytical questions of critical (constructivist)

[28] In the US security strategy revised in autumn 2022 (White House 2022), reference is made to both the changing military (Russia, China) and the worsening ecological security situation. However, the two problem areas are not linked, but analysed in parallel.

Copenhagen security research (Buzan et al. 1998; Hardt 2017, 2019, 2021). In doing so, she arrives at results that suggest a stronger cooperation between the various schools of security research, in particular classical realism, and Earth system research. ESR, like IR realism, is committed to a political 'threat-response logic' (Hardt 2019: 95) of security, in which rationally-behaving actors, which include states, take carefully-calculated countermeasures in the face of external threats. Even in the case of threatening planetary boundary transgressions, counter-strategies might fail or succeed. The Anthropocene discourse is historically open and cannot be dismissed as dystopian catastrophism. Judith Hardt (2019: 93–94) elaborates security policy weaknesses of the Earth system research-oriented environmental policy discourse: the implicit stability and status quo assumption with regard to political conditions, the fading out of power relations, the assumption of the controllability of social conditions, unbroken 'Holocene thinking' (progress, growth, feasibility), the recommendation of top-down control to governments and international organisations, the modelling of the Earth system as a 'natural' entity that humanity is confronted with. The author concludes that even Earth science research dealing with *environmental security* has not yet grasped the radical nature of the Anthropocene thesis on human-environment integration.

Director of the *Stockholm International Peace Research Institute* (SIPRI), Dan Smith, has made a thought-provoking contribution to the security policy debate (Smith 2021). The three security theories of realism, institutionalism and ESR could be pragmatically combined. He suggests introducing the concept of an *operating sphere* of security policy. Security, he argues, forms an independent policy sphere that is constructed by the actors involved, depending on the respective historical, social and political conditions. Following the Copenhagen School (Buzan et al. 1998), it is assumed that the actors make different problems the object of security policy, i.e. they make them secure; these can be conflicts of interest fought out with military-technological means, but also conflicts over scarce resources, political ideologies and claims or over how to deal with environmental risks. In the respective *operating sphere,* completely different problems may be curated and take political priority. With increasing environmental insecurity as a result of the violation of planetary boundary conditions, conflicts over military (classical) security would be supplemented, superimposed or replaced by disputes over the consequences of planetary insecurity in the *operating sphere.*

7.3.2.2 Human (In)security

Since the end of the Cold War, the publication of the Brundtland Report (Commission on Environment and Development, 1987: *Our Common Future*) and the first *Human Development Report* (HDR) by the *United Nations Development Programme*

(UNDP) (1994)[29], the concept of *human security* has been increasingly discussed at the international level and in research. At least since the first HDR, research has also been conducted into the areas and ways in which environmental changes affect human security.[30] It is not possible here to adequately address the connection between human security, conflict action and variance in environmental conditions, which depends on many contextual factors (Schlag et al. 2015). It should only be noted that the topic had attracted research interest long before the formulation of the Anthropocene thesis and planetary boundaries (MacFarlane/Khong 2006; Chandler 2012; Forsythe 2018a; Raymond/Neumeier 2018a; Penny 2020). Thus, the question arises whether the Anthropocene finding has in fact stimulated research in this thematic field and triggered research activities at German-speaking universities.

My impression is that neither the stratigraphic findings nor the research results of the ESR on the threatened planetary boundaries triggered a particular push in political science research in the German-speaking university area on threats to human security. However, already-existing research foci, such as, in particular, the Hamburg-based Research Group on Climate Change and Security (CLISEC in CLISAP[31]), saw their research activities confirmed and interdisciplinary cooperation strengthened, and intend to take up transdisciplinary research approaches (Scheffran et al. 2012b; Hardt/Scheffran 2019; Brzoska/Scheffran 2020; Scheffran et al. 2020). Only a few researchers dare to explore the new field of research that is fed by the Anthropocene thesis.

The research activities of Hans Günter Brauch (Brauch et al. 2009; Brauch 2011; Brauch et al. 2016) are particularly worthy of mention here. As early as 2011, in a paper on 'Political Geoecology for the Anthropocene, written together with Simon Dalby and Ursula Oswald Spring, he stated that the Anthropocene epoch requires a 'fundamental rethinking of security' (Brauch et al. 2011: 1457). In order to meet this demand, a research programme is presented in which the geoecological research approach is combined with a political science perspective that focuses on political processes (*politics*), fields of action (*policy*) and political institutions (*polity*). The new research approach should promote anticipatory thinking and action on the part of political actors. A reorientation of research and teaching is necessary with regard to the spatial, social and human consequences of the new planetary framework. The working group's considerations are plausible, but have not yet been taken up by the German research community (Hickmann et al. 2020).

Anthropogenically-generated, predicted as well as real, planetary boundary transgressions, which (can) lead to a permanent deterioration and endangerment of human living conditions, are to be classified as the normal state of affairs in the new Anthropocene era. In the Holocene, unstable and threatening environmental conditions, as well as environmental catastrophes, were consequences of natural constellations

[29] See also the UNDP report (UNDP 2020).

[30] See the Hexagon Series on Human and Environmental Security and Peace edited by Hans Günter Brauch (Berlin/Heidelberg: Springer).

[31] See the report of the research group Brzoska/Scheffran (2020) with the historical overview of the development of the research focus (Brzoska/Scheffran 2020).

of forces; now they are the result of an inextricable bundle of geo-ecological and anthropogenic factors and constantly (and not, as before, singularly) threaten human security. The thesis of climate change as a trend amplifier of extreme weather events is plausible (Scheffran 2020; Scheffran et al. 2012a; Brzoska/Scheffran 2020), but still falls short. It is not individual weather phenomena that are characteristic of the change, but rather the increasing potential threat to human life on the planet through the rising (!) and (as a result of possible tipping events) even increased risk of the destruction of anthropogenically-beneficial climate and biosphere conditions.

Ulrich Beck's concept of risk society (2007) refers to the technical risks for humans and the environment that are self-generated in the globalised world as side effects of capitalist modernity. The Anthropocene findings radicalise the thesis of the risk society because humanity, as an independent geological force, makes its mark on the planet and in the process modifies the planetary reproduction conditions for human existence, tends to endanger them or completely destroys them through the spasmic use[32] (*Spasm War*) of weapons of mass destruction (Kahn 1965). Human security is likely to become an increasingly scarce commodity for the few in the Anthropocene, in contrast to the Holocene, in which human security improved due to technological progress and rising prosperity, predominantly, but not only, in the industrialised countries. In the new age, structural insecurity prevails as a non-intended consequence of the technosphere, which was only weakly controlled during the Great Acceleration (Steffen et al. 2014) and whose unsustainable functioning now threatens to destroy the planetary framework of human existence (Steffen et al. 2007, 2015; Dalby 2020). The comprehensive and general improvement of human security in the Holocene has given way to a new threat situation with structural uncertainty of human existence. While different development paths can be followed, the potential for human existence to be threatened by anthropogenic action can no longer be eliminated.[33]

7.3.2.3 Ecological Safety

For Earth science Anthropocene research, the concept of ecological security is of central importance. Numerous research papers (including those of the *Potsdam Institute for Climate Impact Research* [PIK]) point to the growing threat to the stability of ecosystems. Climate change will irreversibly endanger functionally-relevant ecosystems of the planet, as *tipping points* are reached that lead to system changes that cannot be reversed (Lenton/Schellnhuber 2007; Lenton et al. 2008; Cai et al. 2016). Social science research on the Anthropocene has only just begun to address this hypothesis. The separation between humans and nature, which determines the Western worldview, is found to be no longer tenable. Humans have become an integral

[32] The highest of 44 levels of nuclear escalation devised by the nuclear age theorist to rationally control thermonuclear war (Kahn 1960, 1965).

[33] Schmelzle (2020) has drawn attention to the political pressure for action that arises in situations of existential endangerment with a critical view of the Anthropocene discourse.

element of ecosystems whose stability, in the sense of security, now depends on their actions (Bonneuil/Fressoz 2013, 2017; Horn/Bergthaller 2019). This hypothesis has opened up a post-human field of enquiry for the cultural and social sciences, in which factual and normative research is being conducted on the close bond between humans and their co-environment (Rothe 2017; Rothe/Benner 2021). This in turn raises the question of what significance the post-human research approach has played so far in political science, especially at German-speaking universities.

Political science research on the Anthropocene, when it is not working on the general problem of the human-nature relationship, focuses on the problem of resilience. With critical and practical intent, it asks about political strategies to improve the resilience of ecosystems—and thus their security (Oels 2012; Scheffran et al. 2012a, b; Chandler 2021). Most political science research dealing with ecosystem resilience refers to political-practical and administrative adaptation to climate change or to the protection of biodiversity (Fichter et al. 2010; Lalyer et al. 2020). They make no or only marginal reference to the Anthropocene findings.

Political science is interested in ecosystem security in the Anthropocene from either a discourse theory or governance perspective. The thematic field of social science analysis of systemic risks (Steffen et al. 2005; Cai et al. 2016; Renn 2020a, b), which arise as a consequence of the changing coupling of the Earth-human system in the Anthropocene, is rather left to sociology (Rothe/Benner 2021) and the Earth sciences. The de- and reconstruction of the Anthropocene discourse can draw attention to the norms used in this debate, to underlying power relations and to possible (alternative) campaigns of action (Folkers 2018, 2020; Bornemann et al. 2019a; Bornemann 2019). As a critical theory, discourse analysis is not suited to empirically exploring the new risk and security landscape and to addressing the question of which societal and geopolitical problems and danger situations—independent of their politicisation in earth science and political discourses—produce the destabilisation of planetary and regional ecosystems. Without the normatively guided and at the same time empirical (case-analytical) determination of the role of political 'mismanagement' in processes of ecosystem destabilisation, it cannot succeed in finding ways out of danger.

The reflections of the literary scholars Horn/Bergthaller (2019), who distinguish between 'incommensurable temporalities' (Will 2021, 202) in their analysis of ecosystem crises of the Anthropocene, oriented on the broad debate between Earth and cultural sciences about the time scales of the Anthropocene (Görg 2016; Ellis 2020; Chap. 3; Irvine 2020; Will 2021; Chap. 5.3), go a little way in this direction. A differentiation is made between a geological deep time, long-term historical developments (latency) and temporally condensed crisis phases when system states change because tipping points are exceeded (Horn 2021). Political science could contribute to the analysis of crisis dynamics on the one hand and to their prevention (resilience policy, safeguarding biospheres) and their management (renaturation) on the other, and does so—but without reference to the Anthropocene concept and its questioning. Until today, political science research has distinguished between locally/regionally polluted environments, those threatened by pollution, renaturable (!) and globally available 'natural' environments. The fact that the planetary biosphere in its current state is itself the result of anthropogenic influences is not yet addressed, or only marginally.

7.3.2.4 Nuclear Safety

Since the anthropocene discourse focuses on planetary environmental problems, primarily climate change, nuclear safety plays a secondary role. However, this narrowing of the epochal shift is by no means inevitable. The radiation released by nuclear energy, be it from weapons or from nuclear power plant accidents, can be used as a geological marker to define the beginning of the Anthropocene epoch. Nuclear weapons testing in particular has left a permanent geological as well as socio-ecological fingerprint (van Munster/Sylvest 2016; van Munster 2021). If the anthropogenically-generated destructive potential were used to determine the beginning of the Anthropocene epoch, August 1945 would be a plausible date. However, the current Anthropocene discourse is less interested in the damaging potential of weapons of mass destruction than in the observable, anthropogenically-caused changes in planetary functional systems such as the atmosphere, the biosphere, the hydrosphere or the cryosphere.

An exception is the working group around Achim Brunnengräber (Brunnengräber/Görg 2017; Themann/Brunnengräber 2019; Brunnengräber 2021), which deals with the final disposal of nuclear high-level radioactive waste resulting from the civilian use of nuclear energy in Germany. Brunnengräber concludes that the Anthropocene should also be understood as an 'Age of Waste' (Brunnengräber/Görg 2017: 98), which extends far into the future and is associated with high uncertainties. The research on final disposal not only draws attention to the legacy of the nuclear age, which has not yet been overcome, but also to the disposal problems of the Anthropocene in general. Because of the advancing exploitation of nature and its use as a dumping ground, which builds on unequal power relations, production relations and imperial lifestyles, according to Brunnengräber we should speak of the Capitalocene 'in critical distinction to the Anthropocene (Brunnengräber 2021: 57). With this proposal, he follows the authors of the volume edited by Jason Moore, 'Anthropocene or Capitalocene?' (Moore 2016), to which Elmar Altvater (2016) also made an important contribution. Because of the rational externalisation of all possible costs in capitalism, the Anthropocene is better described as the Capitalocene.[34]

[34] See also Mahnkopf (2022).

7.3.2.5 Interim Summary (3)

As a conclusion to Sect. 7.3.2, it can be stated that the Anthropocene hypothesis has an impact on the policy area of security, be it state, human or ecological security, and should be a reason to reconsider its ontological, analytical, empirical and normative design. The interface between science and politics discussed above (7.3.1) acquires a growing political significance through real (ontological) change in human-environment relations as well as through the securitisation of ecological security problems, which in turn can only be discerned by means of (Earth) scientific methods of analysis. Academic political science in Germany is only gradually beginning to address the largely neglected ecological security problem (Brauch 2011; Scheffran et al. 2012a; Hardt 2017). However, addressing Anthropocene security problems can only succeed in an interdisciplinary manner and with broad participation of the Earth and life sciences. The integration of different temporalities (Irvine 2020), in particular the linking of latency periods with crisis phases, in the analysis of geoecological security constellations is likely to gain importance for empirical and normative analyses of system resilience (Chandler 2012). In this context, it will be important to include the future dimension. The Anthropocene findings compel us, as the study of governance will also show, to consider the dimension of the future, especially in analyses of development processes with risk potentials. Whichever reference object—state, human, ecosystem—is the focus of path analyses and development projections, political science will have to deal with assessing the political consequences of destabilising environmental conditions (Brauch et al. 2011; Bai et al. 2016).

7.3.3 Contributions to Governance[35]

Even the internationally growing cultural and social science literature on the ecological stress limits of the planet (Laux/Henkel 2018; Horn/Bergthaller 2019; Chakrabarty 2020; Adloff/Neckel 2020; Ellis 2020; Zalasiewicz et al. 2021; Brauch 2021) has not triggered a broader resonance in academic political science in this country with regard to the problems of planetary governance.[36] This statement is especially true with regard to the Rockström-Steffen theorem of the imminent (and partly already present) endangerment of the planetary system services that are beneficial for human reproduction, and the governance problems that have arisen to mitigate this risk. There is broad research on institutionalised regimes, regime configurations, and regulatory systems for the prevention of global and local environmental disasters, but only in exceptional cases do these take into account the Anthropocene findings,

[35] For an introduction to the political science concept of governance, see Benz et al. (2007); Benz/Dose (2010).

[36] Exceptions and pioneers include Brauch (2011); Biermann (2014); Scheffran (2016); Kluge/Schramm (2016); and Biermann (2016a).

tipping points, irreversible processes of change in habitats and different (historical) responsibilities of national societies for mitigating the consequential costs of damage that has occurred and of preventing future damage.

The Earth science finding of anthropogenically-endangered functional conditions of planetary ecosystems that are still beneficial to humans implies the governance policy question of which measures and instruments can be used to avoid the catastrophe. Furthermore, the Anthropocene findings suggest reconstructing the evolution of nature-human relations with regard to political variables of influence. In addition, normative questions arise about the distribution of costs in the elimination of climate damage or precautionary measures in the Global South. The following chapter focuses solely on the problem of governance. In particular, the work of Frank Biermann (Biermann 2014, 2016; Biermann/Kim, 2020b), Thomas Hickmann (Hickmann 2017; Hickmann et al. 2021) Philipp Pattberg (Pattberg et al. 2017; Pattberg/Zelli 2017b; Pattberg/Davies-Venn 2020) and Fariborz Zelli (Zelli/Pattberg 2017) will be taken into account and questioned with regard to their contribution to the analysis and treatment of climate change.

Analytically, the concise review of research results is based on a simple scheme of global climate governance (Simonis 2017d). The governance system is conceived as a multi-level political system with three dominant levels of action (international, national, subnational), which in turn is embedded in and interacts with the functional systems: Science, Technology, Economy and Environment. The multi-level system is dominated by the community of states, which consists of competing sovereign nation states endowed with highly unequal potentials for action. Since there will be no world state in the foreseeable future, planetary governance cannot do without successful state cooperation, even in the age of the Anthropocene. Humanity itself has no capacity to act. It acts at the political level through sovereign nation states and their alliances, and increasingly through sub-state public and private actors. And only in a few exceptional cases have nation states, such as within the framework of the EU, renounced sovereign rights in favour of international organisations and regimes. The international climate regime (UNFCCC) agreed at the so-called Earth Summit in Rio (1992) is no exception as far as the renunciation of sovereign rights is concerned. The member states of the regime have to cope with the collective action problem that no effective coercive measures are available against free riders (Ostrom/Schöller 1999; Nordhaus 2013).

In modern, functionally-differentiated societies, the international political community, in addition to many other political actors working at different geographical-administrative levels (international, regional, national, sub-state, local) has to vouch for global environmental security. This general statement, which is silent about real conditions, is based on the omnipotence of politics, justified by the principle of sovereignty. At the international level, nation-states have the legal and political competence, but also the responsibility, to ensure environmental security, vested in them by the United Nations Charter and numerous conventions that supplement and clarify it. In the present context, it is significant that with the declarations of the environmental conferences of Stockholm (1972) and Rio (1992), the community of states organised in the UN system explicitly and codifiedly assumed responsibility for the protection of the global environment (Stockholm Declaration[37], Agenda 21[38]). Due to competence and political responsibility, there is an institutionalised receptivity in the organisations of international environmental policy (including UNEP, WMO, IPCC, IPBES) for politically-relevant news from science regarding the state of the ecologically functional systems of the globe.

It took until the Climate Change Conference in Paris (2015) for states to agree on a compromise on which regulations of the 1992 *Framework Convention on Climate Change* (UNFCCC) impose binding restrictions on state sovereignty and which regulations are of a voluntary, yet internationally agreed, nature (Bodansky 2016a, 2016b). The compromise enables the institutionalisation (from 2020) of a long-term regime structure (Aykut/Dahan 2015; Simonis 2017b; Klein et al. 2017). On the one hand, procedural norms and guidelines binding under international law were established for all states; on the other hand, the member states commit to voluntary measures (*nationally determined contributions*) to reduce the risks of climate change, to adapt to its risks and to finance transformation processes in the Global South. The regulations for the national implementation of the *Paris Agreement* (PA) are concretised by unanimous decisions of the annual COPs. Based on contributions by authors with a German academic background who participate in the Anthropocene debate, three governance mechanisms of the Paris climate regime are highlighted below, which are likely to point the way for the institutionalisation of new forms of global governance in the Anthropocene: governance through agreed targets, governance through institutionalised planning for options for action, and polycentric forms of governance. Previously, reference was made to the critique of environmental policy, whose assumptions still reflect the political framework of the Holocene and do not yet capture the governance constellation in the Anthropocene.

7.3.3.1 Criticism of Environmental Policy

Frank Biermann (2021) has critically examined the programme of 'environmental policy' and assigned its assumptions and implications to the Holocene era. Policy area analyses on environmental policy conducted on the basis of the concept of 'environmental policy' would turn out to be under-complex in the Anthropocene era and would be based on assumptions that convey a research and policy perspective that is no longer appropriate to the problem. In Table 7.3, typical differences between *environmental policy* and *the Earth system policy* appropriate for the Anthropocene have been compiled in a dichotomously-pointed manner, based on Biermann's considerations.

[37] See UNEP (1972); Corry/Stevenson (2018b); Lövbrand/Mobjörk (2021).

[38] See UN General Assembly (1992).

Table 7.3 Criticism of environmental policy

Comparison criteria	Holocene	Anthropocene
Theoretical perspective	Environmental policy (Policy area analysis)	Earth system policy (Governance analysis)
Human-nature relationship	Dichotomous	Integrative
Temporal focus	Current problems	Future problems
Spatial focus	Primarily local (national)	Primary planetary
Political integration	Isolated policy area	Multidisciplinary policy area
Political relevance	Low Politics	High Politics
Growth model	Sustainable development(Green growth)	Dematerialisation (Degrowth)
Social change	Status quo orientation	Transformation
Governance approach	Problem solving in the national space	Global targeting (multi-level governance)
Participation	National stakeholders	Transnational civil society
Normative basis	Human securitySustainable use	Ecological safetyPost-Humanism
Technical mission statement	Production-integrated environmental protection, internalisation	Circular economy (Zero Waste, Zero Emissons)

Source Compilation by the author, inspired by Frank Biermann (2021): The future of 'environmental' policy in the Anthropocene: time for a paradigm shift

The concept of *Earth system policy* focuses on the planetary system, while *environmental policy* looks at one specific policy area among many others, where the analysis is concentrated on the national container. The serious differences between the two analytical perspectives can be found in Table 7.3. It should be emphasised that the problem constellation typical of environmental policy in the Holocene does not disappear in the Anthropocene. In the Anthropocene, too, current, local and politically low-threshold problems must be dealt with by local (regional) participation in the nation-state framework. However, as long as the local and politicised environmental problems are recognisably connected to the planetary functional systems and there are interactions with the *safe operating space,* the Holocene and Anthropocene perspectives should be linked, which is often still a desideratum. However, it must be noted that there are conflicting interests between the long-term planetary and the short-term local perspectives of action that are difficult to bridge—and such bridgings are always temporary. Furthermore, it must be noted that in the Anthropocene a change in values is required that enables an integrative human-nature relationship. How to protect the geosphere and biosphere from 10 billion human predators, all claiming a good life, and keep it in a sustainable state of equilibrium, is an open question. That all anthropogenic potentials for action are required to cope with this task seems indisputable. The ESR and Anthropocene research call for science to participate with its resources in coping with this planetary task.

7.3.3.2 Governance Through Target Agreements

Global governance of the planet to comply with geophysical, geochemical and geo-biological boundary conditions for a population growing to 10 billion people requires the knowledge-based translation of anthropogenical beneficial boundary values recorded by Earth science methods into targets with metrics on which political agreements can be reached (Bai et al. 2016). Finding target values suitable for global governance is itself a political process (Bierman/Kanie 2017), as targets must be suitable from an Earth science perspective, politically plausible, technically easy to measure and administratively easy to control. The long-standing discussion about a quantitative upper limit of 2 °C of global warming that can be tolerated by the international community illustrates the complexity of the task of agreeing on consensual targets at the global level (Geden 2010). Earth science research has been instrumental in this discussion process. The IPCC's special report on the possible consequences of global warming of more than 1.5 °C (IPCC 2018) has led to the Paris Agreement's target definition being sharpened in the public debate.

Article 2 of the Paris Agreement aims to keep the 'increase in average global temperature well below 2 °C above pre-industrial levels'. By ratifying the Paris Agreement, the member states undertake to strive for this (and other) goals by means of the national commitments provided for in the agreement, while observing the procedural coordination procedures binding under international law, as well as agreed principles[39], standards and incentives. In this context, targeting is to be supported by a whole battery of closely interlinked coordination procedures, such as the Global Stocktake with Ratchet Mechanism (Klein et al. 2017; Friedrich 2017; Rajamani et al. 2022; Simonis 2022: chap. 12). The latter has so far been singular in character. Governance through target agreements (Young 2010; Keohane/Victor 2010; Biermann 2014; Kanie/Biermann 2017; Biermann/Kim 2020a), on the other hand, is also applied in other international regimes that secure (biodiversity regime, Aichi Targets) or promote (sustainable development regime, 2030 Agenda) global common goods.

Agreeing on concrete targets with indicators and measurement methods on the future state of global commons at the international level is obviously considerably easier than negotiating national commitments that each individual regime member must comply with in a verifiable and, as in the case of the Kyoto Protocol, punishable manner (Holtwisch 2006; Oberthür/Lefeber 2012). Since the community of states cannot agree on the distribution of national commitments to guarantee common goals

[39] Of outstanding importance is again the CBDR principle (see note 21 above).

in the fields of action of biodiversity (CBD), sustainable development (SDGs) and anthropogenic climate change (UNFCCC), it reduces the need for intergovernmental consensus and the level of ambition of hierarchical global governance by introducing decentralised procedures of polycentric governance.

Arguments against hierarchical governance institutions in the international system include the high degree of consensus required for agreements in the context of international negotiations, the weak democratic legitimacy via the delegation of responsibilities, the costly implementation and control of governance decisions, and the weak sanction potential that rewards deviant action (Bodansky 2016a; Stavins/Stowe 2016; Zürn 2018a). Decentralised governance via target agreements does not *per se* offer a (higher) guarantee of effective and efficient target achievement. Here, there is a risk that the goals agreed at the international level will be implemented at the regional, national and local levels in highly divergent and generally inadequate ways (Biermann et al. 2017; Bierman/Kanie, 2017; Breitmeier et al. 2021; Schwindenhammer et al. 2017). In terms of international law and tax policy, the chances of free-riding are not small, even with decentralised control. In the many years of negotiations that finally led to the Paris Agreement (COP 21), pragmatism won out in order to break the negotiating deadlock that occurred at the COP 15 in Copenhagen (2009).

Whether the mobilisation effect of the governance system agreed in Paris will be sufficient to achieve the agreed goals remains to be seen. Scepticism is warranted (Geden 2010; Victor 2014; Simonis 2017b; Nordhaus 2018, 2020). Six years (COP26, Glasgow) after the entry into force of the Convention, it is clear that the agreed goals will only be realised if there is an increased willingness in business, society and politics to assume transformation costs (OECD/IEA 2021; UNFCCC 2021b; Yanosek/Victor 2022). The latter will also depend on how future adaptation costs are perceived. Environmental and climate communication on the latest research results, development trends, and options for action in society, the public and politics will play a central role in estimating these costs. Even if international target agreements are oriented towards planetary boundary conditions and are designed to be evidence-based, they require accompanying national information and mobilisation strategies. Global target agreements can lay the institutional foundation for planetary governance. However, to increase their likelihood of success, they must be supported by other governance instruments (Biermann/Kim 2020a, 2020b; Weiland et al. 2021; Breitmeier et al. 2021).

7.3.3.3 Institutionalisation of Planning Governance

The member states of the environmental regimes are and remain responsible for implementing the goals agreed at the international level. In exceptional cases, a regional organisation, such as the EU, can take on coordinating tasks for its members and exercise rights of representation vis-à-vis the international regime (Tröltzsch 2017). The actors committed to implementing the agreed goals within the regime decide independently on the measures they consider appropriate to achieve the goals. The more long-term and the more ambitious the goals are, the greater the political task of designing, agreeing on politically-administratively and socially enforcing effective and cost-efficient bundles of measures that are legally and legitimately secured and that are accepted by the population and those affected. In the case of climate governance, not only is a long-term and politically-binding goal being pursued; achieving the goal is also highly demanding technologically, economically, socially and politically (WBGU 2011; Bulkeley 2013; Brunnengräber/Di Nucci 2014; Turnheim 2016; Bataille et al. 2016; OECD 2019; IPCC 2022). The transformation of the modern industrial and growth society required for this cannot be achieved through short-term interventions. What is required are long-term political-administrative measures in the economy and in society, that need to be politically coordinated with regard to the respective social situation, if a stable political consensus is to remain assured. A political-administrative control system with this task fulfils the function of a governance mode that can be described as planning governance (OECD 2021).

Planning governance can be understood as government action programmes that set binding milestones for the achievement of a long-term goal and provide for measures (instruments) with which the milestones are to be achieved. The bundle of measures intended to realise the nearest intermediate goal in terms of time is usually agreed upon in a legally-binding manner in the national or also in the regional space of the EU (Simonis 2018). The emergence of planning governance can be observed in the implementation of the Paris climate regime. Member states are guided by the formal (procedural) requirements of the Paris Agreement and start developing implementation programmes to mitigate GHG emissions or to adapt their societal structures to climate change (see UNFCCC 2021a).[40] Planning governance is carried out with the aim of providing planning certainty for the political and, above all, the social actors of the jurisdiction concerned. In addition, third countries and members of the climate regime are to be informed about planned measures and confidence in the regime's fidelity created. Whether planning governance can achieve these goals depends on many factors, not least on the fact that it does not allow itself to be diverted from its chosen course of action by societal disputes about the respective transformation policies.

Currently, the establishment of planning governance action programmes for the implementation of the Paris Agreement can be observed at the level of the European Union as well as in various member states (European Commission 2016; Eu Commission 2016; Delbeke 2017; EU 2021). The measures aim to meet the voluntary commitment made on behalf of the EU by the 27 EU states to reduce greenhouse gases by 55 per cent by 2030 compared to 1990 levels.[41] With this interim target, the EU wants to come closer to its long-term goal of striving for climate neutrality by 2050. If this goal is actually to be achieved, comprehensive planning governance would have to be institutionalised at the level of the Community and its members, involving several sectors besides the energy sector, including transport, construction (heating), industry (chemicals and primary industries) and agriculture. Within

[40] Here, the assumption is made that the voluntary pledges of the states are also followed by real implementation processes.

[41] See: http://data.europa.eu/eli/reg/2021/1119/oj.

the next 30 years, the Community's economic system will have to be largely decarbonised. However, between 1990 and 2020, only about 26% of GHG emissions were saved within the EU. With the currently planned mitigation measures, the EU would not even reach the interim reduction target of 40% compared to 1990 by 2030 (European Environment Agency 2020). If the 55% GHG emission mitigation by 2030 agreed in 2021/22 is to be realised in practice, planning governance at all political levels within the community with measures and target controls must be much stricter.

Taking the EU as an example, the hypothesis was developed that political actors use a governance mode that can be analysed with the concept of planning governance to implement agreed long-term goals for securing planetary boundaries, such as stabilising the CO_2 concentration in the atmosphere. Since it can be assumed that targets corresponding to the climate regime will also be agreed in the future to secure the anthropogenically-beneficial stability of geoecological planetary systems[42], the thesis seems plausible that planning governance is developing into a typical feature of global governance in the Anthropocene era. With the establishment of planning governance, politics partially changes its character and, in order to ensure survival on the planet, integrates an institutional apparatus that incorporates the dimension of the future into the political events of the present. This functionalist hypothesis, due to its optimistic approach, could be refuted historically. In fact, many reasons can be put forward against its tenability. But science and politics in the Anthropocene are trying to make low probabilities a little more probable.

7.3.3.4 Polycentric Governance

Another structural feature of global governance with planetary scope in the Anthropocene, which can also be observed in other environmental regimes, such as the regime for the protection of biodiversity (CBD), relates to the rapidly growing number of political and social actors who participate in the preparation and implementation of decisions of the regime bodies (Willetts 2011; Higham 2017; Schroeder/Wapner 2018; Scholte 2018). Within the framework of the climate regime, they take on voluntary commitments, oriented towards its agreed goals, primarily for the mitigation of greenhouse gases. The NAZCA platform[43] shows that 11355 cities, 270 regions, 12957 companies, 1529 investors, and 3349 civil society organisations and cooperative initiatives are publicly and verifiably committed to the goals of the climate regime.[44] For their part, the sub-state actors participating in the climate regime are organised in numerous national and transnational networks, each with its own objectives (Dietz/Garrelts 2013; Schaefer Caniglia et al. 2015; Widerberg/Pattberg 2017).

[42] In autumn 2022, for example, the so-called Montreal Targets (among other things, 30% of the planet's surface should become nature reserves) were agreed, to implement the biodiversity regime as a continuation of the Aichi Targets.

[43] The Global Climate Action (GCAP) Platform was adopted by COP20 in Lima in December 2014 under the indigenous name 'Nazca' and documents the activities of NGOs under the UNFCCC climate regime (UNFCCC-Home).

[44] The figures given refer to 7 July 2022.

In doing so, they do not only pursue climate policy goals with their networks, but also goals that go beyond this, such as securing and promoting jobs or improving air quality through the diffusion of renewable energy technologies. The concept of polycentric governance (Ostrom 2009a) makes use of diverging motivations and interests in environmental policy measures, i.e. their multifunctionality due to their embedding in different socio-ecological systems. As a positive side effect of interest-led measures, the climate and biodiversity can be promoted at the same time and acceptance can be increased.

To ensure the effectiveness of planetary governance regimes that are institutionalised according to procedural specifications, polycentric governance approaches with decentralised nation-state responsibility can be institutionalised, not only on a small-scale, but also on a global scale (Ostrom/Schöller 1999; Ostrom 2009a, 2009b). In order to increase regime effectiveness, other actors (including local political organisations, regions, cities, civil society, associations, companies, farmers, craftsmen), who benefit from the regime materially, ideally and in terms of power politics, are involved in its governance. This seems to be the direction in which the climate regime of the Paris Agreement is developing (Biermann et al. 2010; Stavins/Stowe 2016; Widerberg/Pattberg 2017; Klein et al. 2017). Within the framework of the annual Conferences of the Parties[45], societal actors are granted increasing opportunities for participation, such as the formulation of proposals for further shaping the regime, presentation of new technologies, participation in special networks for the reduction of greenhouse gases or the financing of adaptation projects.

The importance that non-state actors have already acquired in the governance of the climate regime can be seen in their contribution to regime stability after Donald Trump announced the withdrawal of U.S. participation in the Paris Agreement on 1 June 2017 (Trump 2017). Under the aegis of the UN Special Envoy for Cities and Climate Change, former New York Mayor Michael R. Bloomberg, and the then UN Special Envoy for Cities and Regions and Governor of California, Edmund G. Brown, the organisation 'America's Pledge' was formed, which 20 U.S. states and 110 U.S. cities joined in 2017 (Bloomberg Philanthropies 2017; Bloomberg/Pope 2017). In addition, the 'We are still in' network, founded by environmental organisations, was formed and joined by more than 200 cities, 1700 companies and numerous universities and colleges, as well as quite a few states and tribes.[46] In his capacity as UN Special Envoy, Bloomberg wrote to the UN Secretary-General, Antonio Guterres, and to the Executive Director of the UN Climate Change Secretariat, Patricia Espinosa,

[45] Non-state actors are also systematically involved in the initial review process of the commitments made by the PA member states ('stocktaking'). A newly established stocktaking governance process (Global Stocktake governance and facilitation | UNFCCC) is used.

[46] See the movement's homepage/organisation: https://www.wearestillin.com/about (19 February 2018).

to express the desire of 'U.S. subnational and non-state actors to affirm and demonstrate Americans' collective commitment to the Paris Agreement and to supporting climate action to meet the nationally determined contribution made by the United States under that accord.'[47]

Under international law, this letter had no significance. Politically, however, it demonstrated the will of the American opposition to adhere to the Paris Agreement, as well as the ability of sub-national political actors to secure compliance with the mitigation contributions promised by the central state by assuming voluntary commitments (Karapin 2016; Rabe 2018). After his election, and one day after taking office (20 January 2021), US President Joe Biden reversed the withdrawal from the Paris Agreement (21 January 2021). To support his policy, 'America's Pledge' and 'We are Still In' have merged into the new initiative 'America is All In'.[48]

The example of the American 'All In' initiative (America Is All In 2021) points to the growing potential for action by subnational political and economic, as well as civil, society actors. Together with the globalised Earth and social sciences, the sub-state political and social actors form a communication space in which options for action are developed and the consequences of social action are clarified (Simonis 2018). There should be no illusions about the political steering effect of the civil society communication space. The actors involved each have very different power potentials to enforce their investment decisions and political programmes. Nevertheless, civil society public discourses are of great political relevance. At the various international, national and sub-state political levels of action, possible consensus zones of cooperative action are formed through social discourses, into which the analyses of risks and opportunities of evidence-based options for action flow (see Sect. 7.3.1), which in this way indirectly contribute to overcoming the collective action problem of international cooperation. In the Anthropocene, it is assumed, civil society actors and discourses at the international and sub-state levels, in which the findings of Earth sciences and the options for action of international or local politics are negotiated and coordinated, gain in importance for governance policy.

7.3.3.5 Evolution of Democracy

The political governance of the Anthropocene will hardly be successful without an institutional evolution of democracy. Jörg Tremmel, for example, points out in several papers that the political management of the challenges of the Anthropocene can only succeed with a transformation of democracy (Tremmel 2019, 2022). So far, the democratic institutions of horizontal and vertical separation of powers do not allow to take in the political process of decision-making the concerns of future generations adequately and systematically into account. In order to limit the fixation of politics to the present, the parliamentary system would have to be expanded by

[47] See: https://3blmedia.com/News/Americas-Pledge-Letter-Michael-R-Bloomberg-United-Nations-Secretary-General-Antonio-Guterres, 13 June 2017 (accessed 21 Feb. 2018).

[48] See: Home—America Is All In (Hultman et al. 2021).

a third chamber in which the interests of future generations are already articulated in the present and recognised in political-legal terms. A political science discussion on the specific design of a 'third chamber' has begun in the context of the anthropocene debate and can look at a long tradition. Tremmel (2022) focuses on the institutionalisation of future councils, which could act as a source of impulses without having the right of veto to block legislation. Tremmel's ideas can be linked to Böhret's proposals on legislative impact assessment (Böhret/Konzendorf 2001) or to the extensive debate on technology assessment (Bröchler et al. 1999; Grunwald 2010; Simonis 2013), which had already led to the founding of the *Office of Technology Assessment at the German Bundestag* (TAB) in 1990 (Meyer 1999; Simonis 2001). The point of departure in the current reform discussion seems to me to be that the interests (needs) of future generations have come into the focus (Commission on Environment and Development 1987; Grunwald/Kopfmüller 2022). In the Anthropocene, not only the consequences of technologically induced innovation and diffusion processes should be considered in the democratic (parliamentary) decision-making process, but the entire range of anthropogenic planetary risks to the future should be taken into account (Galaz 2019; Irvine 2020). As plausible as this normative demand may be, there is no way around the political significance of the interests of the living generation. Therefore, it would already be a big—and surprising—step forward if new substantial action programmes were examined systematically and without exception with regard to their sustainability within the framework of the legislative process.

7.3.3.6 Interim Summary (4)

In the Anthropocene, the mode of global governance is changing. In the transition phase from the Holocene to the Anthropocene, a system of global governance has been developing since the end of the Cold War, in which regimes and networks of planetary governance (ozone regime, climate regime, biodiversity regime) are gaining political importance (Biermann 2014, 2016; Zürn 2018a; Biermann/Kim 2020b). The fact that older regimes, such as the trade and development regimes, are contested and challenged by countervailing power formation (multipolarity) cannot be further explored here (Cooley/Nexon 2020). At this point, we are concerned with diagnosing the beginnings of a long-term structural change in which the three modes of governance outlined form elements of a new governance system, with a planetary scope, that overlap with and modify the internationally-established governance system, which also tends to be multilateral and global (Simonis 2022: Chap. 13). According to the functionalist hypothesis, a new planetary governance system with multi-level governance, consisting of target governance, planning governance and polycentric governance, is likely to emerge in the Anthropocene epoch, as can already be observed at present. This prognostic assumption is not linked to a reliable statement as to whether this new global governance system with planetary scope will be able to avoid dystopian crisis phenomena of the planetary environmental systems.

7.3.4 Diagnosis of the Present

It cannot be the task of the social sciences to empirically substantiate the geological and/or earth-scientific Anthropocene findings (Leinfelder 2017). However, the social sciences, as well as cultural studies and historiography, are in demand when it comes to (1) reconstructing the history of its formation, (2) clarifying the socio-political conditions that have allowed the *anthropos* to become a planetary force, and (3) exploring possibilities for action to avoid anthropogenically-caused planetary catastrophes (Zalasiewicz 2017; Zalasiewicz et al. 2021). Providing the right diagnosis for the emergence of the Anthropocene from the Holocene is crucial for a promising therapy. There is no lack of factors discussed in the literature to explain the emergence of the Anthropocene; among others, the following are mentioned: the Western growth and consumption model, colonialism, the capitalist growth imperative, reliance on fossil fuels, the great acceleration, the exploitation of nature with the onset of sedentarisation, the scientific-technologically supported will to power, etc. All these factors have their specific share in the emergence of the Anthropocene. Depending on the normative orientation and the research question, one factor or a combination of factors can be highlighted to explain the destruction of the equilibrium conditions of the Holocene.

Researchers from the German-speaking university system have also made important contributions to this debate, which cannot be recapitulated here (Altvater 2016; Görg 2016; Brand/Wissen 2017; Pattberg/Davies-Venn 2020; Görg et al. 2020; Will 2021; Mahnkopf 2022). In the context of this text, it should only be pointed out that the Anthropocene findings are likely to be causally overdetermined. Several forces (factors) worked together to establish the *anthropos* as a geological force. There are overlaps between broad explanatory approaches, such as reliance on fossil fuels, colonialism, capitalism or the great acceleration since 1950. From the perspective of governance research, which is concerned with the prevention of anthropogenically-caused disasters, it is important to diagnose factors (forces) that can be influenced with some probability by collective action, despite their historical and present-day impact. Colonialism, racism and capitalism may have done much to threaten the ecological balance of the planet and enable the rise of the Anthropocene; but they do not constitute effective platforms for combating climate change or the biodiversity crisis. Without mitigation of pollutants that alter the energy balance of the atmosphere, climate change that is detrimental to humans cannot be limited. Without changing forms of land use, the biodiversity necessary for human life cannot be maintained. Therefore, it is important to develop ecologically-effective options that are at the same time politically feasible, since they enjoy broad (!) approval, and to implement them politically with Earth science monitoring.

The identification of options for action, like the implementation of action programmes, is both an Earth science and a political activity. Present-day diagnostics has precisely this function. In cooperation and close interaction with earth sciences, social sciences (economics, sociology and politics) and political actors, it must discover viable paths out of danger and critically accompany their implementation. Without an institutionally-secured diagnosis of the present in the public arena at all levels of politics (supranational, national, local), mitigation and adaptation to the new planetary conditions cannot succeed. Diagnosis of the present should therefore be an indispensable element of political governance in the Anthropocene.

In the transition phase from the Holocene to the Anthropocene, the influence of the politicised Earth sciences as well as the pressure of the environmental movements on governments, parties and associations is increasing. Without the diagnostic power of Earth science, political actors will not be able to recognise and react to anthropogenic global environmental threats. Arguing in the tradition of critical—observational and enlightening—political science, Bornemann et al. (2019) draw attention to the fact that the research results of ESR could be hastily recognised and adopted by political science, leading to objectification, crypto-normativity and performativity. Instead, interpretive political research (IPF) should and "could clarify what meaning the Anthropocene takes on and assumes in the social and political debate on the one hand and in the political science discussion on the other, which actors with which normative as well as factual convictions create these meanings under which conditions, and what implications and effects these meanings have". (Bornemann et al. 2019: 341)

The reconstruction of the anthropocene discourse makes an important contribution to the diagnosis of the present and its political conditions. In particular, IPF makes it possible to open up room for manoeuvre and facilitates analyses of changing political conditions. We agree with this assessment. The investigation of and reflection on (natural) scientifically-determined options for action for politics, as well as the empirically-observable (institutionalised) system of scientific diagnosis of the present, are indispensable tasks of critical social science research. However, it should be noted that without the existence of an institutionalised and socially-accepted system of transferring ESR research results to the political public, reflective criticism of the needs for action and options for action identified by the ESR community in summary reports for policy[49] would be pointless.

7.3.5 Contributions to IR Theory

The Anthropocene has consequences for the normative system of international relations and thus for theory formation (Harrington 2016, 2020). The national interest as a standard of orientation for foreign policy and the discipline of international relations loses normative power in relation to cosmopolitan values, such as respect for planetary limits that are beneficial for humanity and for ecological stability (Simangan 2020; Rothe et al. 2021). In line with this shift in norms, scholars' view of the existing

[49] The Summary for Policy Makers of the IPCC reports, which precede the respective synthesis reports, should be mentioned here, e.g. IPCC (2015).

system of international relations is changing, and is understood here as a politically- and functionally-fragmented system of global governance[50] (Zürn 2018a; Weiss/Wilkinson 2019; Simonis 2022).

Hans Günter Brauch was quick to recognise that the Anthropocene findings call into question the basic ontological assumptions and the normative foundation of the academic IR research that predominates in the Western world (Brauch et al. 2009; Brauch 2011; Brauch et al. 2011). Continuing his peace studies research, he has elaborated the normative challenges of the new planetary age determined by *anthropos*. In doing so, he emphasises the destabilising consequences of the anthropogenically-modified environment, both for political relations at all levels of institutionalised politics (local, national, international), and with regard to the human-nature relationship. Brauch developed the research approach of political geoecology and drafts the programme of a peace ecology as a framework concept for single-disciplinary research approaches in the Anthropocene (Brauch 2021). Geoecology is primarily focused on the local/regional space, in which adaptation policies with a planetary perspective should be institutionalised to avoid human, ecological and political hazards. The peace ecology programme, on the other hand, presents a bridging and framework concept that brings together research approaches and bodies of knowledge that could contribute to the transformation of unsustainable social and political structures. The Anthropocene finding is understood by Brauch as a contextual variable that allows for the integration of research branches from the circle of peace sciences that emerged after the Second World War from a peace ecology perspective. However, Hans Günter Brauch also states:

> ... these multiple research projects and studies have not yet resulted in a joint comprehensive peace ecology research programme within the framework of the Anthropocene Epoch in earth and human history. (Brauch 2021: 111)

The reformulation of the theoretical foundations of the discipline of International Relations, taking into account the Anthropocene findings, received new impetus only a few years ago (Burke et al. 2016; Harrington 2016; Corry/Stevenson 2018b; Müller 2019; Simangan 2020; Chandler et al. 2021). However, it is still unclear whether the suggestions from the margins also resonate in the core of the discipline at German-speaking universities. Remarkably, scholars from German academia are also participating in this debate, which is focused on the developed Western world. However, with the exception of the Hamburg network, the international policy established at German universities is holding back from contributing.[51]

On the other hand, a group of researchers who are well-networked at European universities, but whose academic roots lie in Germany, are engaged in research projects, books and articles in important international journals (this group includes Frank Biermann, Markus Fraundorfer, Thomas Hickmann, Philipp Pattberg, Sabine

[50] On the global governance discussion, see in particular Zürn (2018a, b), and Weiss and Wilkinson (2019), who, however, analyse the concept without reference to the Anthropocene findings.

[51] The Panel on Planetary Thinking launched by Claus Leggewie, who cannot be considered a typical IR researcher, at the Justus Liebig University in Giessen could become another crystallisation point for Anthropocene research in political science (Leggewie et al. 2022).

Weiland, and Fariborz Zelli). At this point, their contributions to the theoretical discussion cannot be acknowledged in detail. Here it must remain with the observation that the substantive focus of this research lies in the field of international environmental policy. New forms of global climate and environmental governance are dealt with (see above, Sect. 6.3.3) and questions are asked about their effectiveness and democratic legitimacy. With the focus on institutionally-controlled political action, the theoretical perspective questioning the conditions for the possibility of planetary (global) governance in the Anthropocene may not always have the weight it deserves, according to Basil Bornemann (2019; Bornemann et al. 2019a).

Meanwhile, the recently published anthology by David Chandler, Franziska Müller and Delf Rothe, *International Relations in the Anthropocene* (Chandler et al. 2021) contributes to the reorientation of international relations that has been deemed necessary. Several authors from Hamburg in the field of political science and peace research have contributed to the volume (including Judith Nora Hardt, Delf Rothe, Ann-Kathrin Benner, and Franziska Müller). In didactically well-prepared contributions, they demonstrate the explosive power of the Anthropocene concept for basic assumptions about the established spectrum of theories of international relations (realism, institutionalism, the English school, international political economy, constructivism). The crux of the theoretical challenge for the discipline of International Relations is to mediate the new, now anthropogenically-reshaped, risky environmental conditions of the planet with the traditional theoretical approaches to the legalities of inter-societal and inter-state relations (Corry/Stevenson 2018a). The planetary (ecosystemic) reference point has implications for the analysis of the temporal and spatial dimensions of the international system, its institutions and power relations, as well as for the normative frame of reference for assessing international processes and national options for action.

Finally, three problems for political science theorising in the field of international relations after the end of the Holocene should be pointed out (Müller 2019; Marquardt 2019; Pereira 2021):

(1) Since neither the externalisation of planetary environmental relations practised by IR realism nor IR institutionalism can any longer be justified theoretically, empirically or normatively, the reintegration of planetary environmental relations, services and functional conditions into the respective theories becomes a challenge (Hardt 2019). The change in the international state system, mediated by the rational self-interest of states in keeping the anthropogenically-generated potentials for destruction under control, must be taken into account in the realist as well as in the institutionalist theory building, insofar as these paradigms still claim scientific validity and do not only wish to grant an apparently scientific justification to populist arguments in political-ideological disputes. The reintegration of anthropogenically-modified natural relations into the theoretical edifice of international relations leads to new normative and analytical questions with explosive force, such as how *anthropos* deal with their non-human

co-world (Gupta 2016; Burke et al. 2016; Hamilton 2017; Burke/Fitzi 2019; Leggewie et al. 2022).

(2) Above all, the continuity assumption must be revised (Meadowcroft 2019; Albert 2020; O'Brien 2021). The international system is in flux, and traditional patterns of behaviour will not be modified overnight and institutions transformed; however, new world views, political constellations and forms of governance are emerging in the Anthropocene. The fragility of the international system is reinforced by the multipolarisation of the state system and expected hegemonic conflicts (Cooley/Nexon 2020). But at its core is a new line of conflict, the control of anthropogenically-generated risks with catastrophic potential (Harrington 2020; Hardt 2021). International relations theory is not set up to deal with planetary discontinuities, such as tipping points of planetary environmental systems, pandemics, terrorist violence or famines. The highly unlikely exceptional case could become the normal case if alternatives to Holocene regulatory patterns are not tested and institutionalised. So far, IR theory has only responded to this upheaval in exceptional cases (Biermann 2014; Dalby 2020).

(3) International state society is integrated into numerous functional systems, such as trade, finance, security and health (Albert 2016; Buzan/Schouenborg 2018; Simonis 2022). The planetary environmental system has become more politically and socially influential in the Anthropocene because of the risky transgressions of the *safe operating space. The* consequence of this development, as already explained above in section 7.3.3.1, is that the policy area of planetary environmental systems can no longer be analysed autopoeitically (Biermann 2021). Because of the increasing interaction with more or less all other international functional systems (policy areas), their forms, mechanisms and consequences and the resulting networks of complex regime formation should not be excluded from policy analyses. In the Anthropocene, not only the vertical (global, national, local), but also the horizontal policy interdependence between different functional systems, such as between climate, health and security, will have to be addressed in theory and policy debate.

7.3.6 Contributions to Political Education

The scientific findings on the new Anthropocene era and its governance problems are now also beginning to radiate onto the established concepts of civic education. Initially, the relevance of the Anthropocene findings for the field of education, as well as for civic education in general, was recognised by representatives of the earth sciences, cultural studies and science studies (Leinfelder 2017; Roso 2021). It is only with some delay that the established discipline of academic subject didactics for civic education is beginning to address the topic. The concepts of *Education for Sustainable Development* (ESD) still dominate the discourse space. Therefore, the question arises as to how the conceptual differences between ESD and 'civic education in the context of the Anthropocene and the climate crisis' (Seider 2021) can be grasped (Friedrich 2021a; b).

Even if the discussion in the field of political science about a possible change to the spectrum of tasks of political education in the Anthropocene is still in its infancy, two differences compared to the ESD programme stand out:

(1) The Earth science (geological) concept of the Anthropocene defines an Earth age or epoch. The Anthropocene is thus a temporally, spatially (planetary) and socially defined subject area with humanity as the geological force; ESD refers to social development processes that are not sustainable, but which can be corrected by politics if they aligns their actions with the canon of norms of sustainability. ESD is able to contribute to this directional correction without calling development, understood as increasing goods production per capita, into question.

(2) The structural characteristic of the Anthropocene is humanity's ability, acquired during the Holocene epoch, to endanger the anthropogenically-required living conditions on the planet through its own actions, and even in extreme cases to destroy them completely. In the Anthropocene era, the negative consequences and non-intended side effects of the forces released in the Holocene must be recaptured under penalty of self-destruction. Political education in the Anthropocene is therefore faced with the task of, on the one hand, restraining the built-up potentials (weapons of mass destruction, resource consumption) and forces of acceleration and, on the other hand—and at the same time—finding ways of peaceful coexistence between people and between people and their anthropogenically-modified "natural" environment. While ESD sets out to make development ecologically, socially, economically and also politically (!) sustainable (Biermann 2006), political education in the Anthropocene must also banish the powers of the past (destruction and externalisation) in order to give humanity, which will soon (in 2100) number ten billion, the chance of a future. ESD, rooted in the Holocene, has an optimistic undertone, whereas that of political education in the Anthropocene is pessimistic (Sippl/Scheuch 2019; Leinfelder 2020; Seider 2021; Friedrichs 2021b).

The discussion about the new conception of civic education in the Anthropocene era, in which humans, and not least politics, must assume obligations in all fields of action for the observance of planetary stress limits, has yielded its initial results in this country (Friedrichs 2021b). The volume *Learning and Teaching the Anthropocene* edited by Sippl et al. (2020) gives hints on how political education in schools could be reformed. The way in which the Anthropocene findings and the research based on them will have consequences for the political science courses offered at universities can only be guessed at present. It can be assumed that with increasing instability of environmental conditions caused by human activity, and with a corresponding increase in mitigation and adaptation measures by governments and other political actors, the teaching of political science will also change. It is only under the pressure of the changing framework conditions that the range of courses is likely to be oriented towards the new planetary conditions and inter-, multi- and transdisciplinary courses in the field of political science are likely to be offered to a greater extent. The orientation of political science teaching and research towards the (democratic) nation-state should remain, but will be supplemented and reshaped by the increasing consideration of supranational, regional, global (communication, economy, finance, migration, tourism) and planetary functional systems (including the phosphorus cycle, climate, water, biodiversity, and health) (Simonis 2022: 13). In the extremely unstable global, and at the same time planetary, governance system (Chandler et al. 2021: 17), destructive forces can easily gain the upper hand and determine day-to-day business, as the Russian war of aggression against Ukraine shows.

7.4 Conclusion

Political science in the German-speaking academic world and with academic roots in Germany is only slowly beginning to address the challenges of the Anthropocene. With a few exceptions (peace research, international environmental policy, climate governance), the new Earth Age of the Anthropocene is only just being discovered as a subject for political science. German research is still dominated by a few innovative pioneers who are well-networked internationally. Their publications draw attention to the turning point in time with the anthropogenically-changed nature-human relationship and the resulting problems for all areas of politics, from political theory (democratic theory) to nation-state organisation, international politics and political ethics. However, the political science establishment at German universities has not yet perceived this turning of the times.

Using an institutionalist explanatory approach, it can be said that science develops in a path-dependent manner. Anthropocene research is taking shape in contrast to the established fields of research, although there are points of contact in terms of content and even personnel. An example of this is the thematic issue of the journal *Politics and Governance* (vol. 9/1, 2021) edited by Thomas Hickmann, Markus Lederer, Jens Marquardt, Sandra Schwindenhammer, and Sabine Weiland: *The 2030 Agenda for Sustainable Development: Transformative Change through Sustainable Development Goals*. Although some of the editors and authors of the issue advocate for greater consideration of the results of Anthropocene research in political science, they publish research results from a topic area that would have suggested the inclusion of results from the Anthropocene discussion.

Mono-, inter- or even transdisciplinary detailed studies from the field of political science on the political challenges of the new age are still exceptions. Besides self-imposed barriers, there are institutional obstacles. At German universities, the political science research capacities required for long-term and methodologically-elaborate cooperation projects are underdeveloped. Thus, the conclusion of the above considerations is hardly surprising: For lack of mass, a critically-reflective assessment of the way in which German political science has perceived and processed the epochal break between modernity (Holocene) and the present time (Anthropocene) cannot be put up for discussion. Political science in the German-speaking world has not yet taken note of the new age and its challenges.

The extent to which the forms of political reaction to the threat of climate change (outlined in Chap. 7.3.3) constitute a generalisable example for political processing on the part of the (currently six) global transgressions diagnosed by Earth system research cannot be further clarified here. However, the example does draw attention to a dilemma for policy advice in the transition to the Anthropocene. If policy is understood as path-dependent in the models, scenarios and forecasts of climate and Anthropocene research, as is usually the case in all baseline scenarios (*business as usual*), policy responses in the sense of learning processes and regime breaks are systematically excluded. The modelled scenarios are therefore systematically under-complex. If, however, an attempt is made to model politics, the scenarios quickly become over-complex, as the number of actors and decision points exceeds computer and interpretation capacities. This is not to say that simple models cannot also produce important results; but as soon as international (global) actor constellations and multi-level decision-making processes are modelled, the complexity increases rapidly.

However, reducing politics to a social aggregate of Pavlovian individuals is such a drastic reduction of complexity that it makes no sense for social scientists. The reverse assumption of a problem-appropriate learning capacity of the international community and national, social and political actors makes no sense either, of course. Although it is called for by Earth system research, it cannot be observed empirically. From the perspective of a social and political scientist, the only course of action that remains is the one proposed by Max Weber (1919) for politics: 'the drilling of hard boards' (Weber 1919). In relation to the case discussed here, this could mean, with a normative conclusion: On the one hand, it would be desirable for Earth science Anthropocene research to open up to political science and begin to recognise and systematically include in its models and scenarios the *inherent logic of* fragmented politics as an enabling and as a limiting factor, even if political will-forming and decision-making processes would have to be greatly simplified. On the other hand, political science would be expected to engage with the research results of Earth system research and increasingly ask how politics actually deals with impending environmental catastrophes and how, from a counterfactual perspective, problem-solving potentials of political-administrative, civil society and economic actors could be mobilised to avert the catastrophic scenario after all.

References

Adloff, F.; Fladvad, B.; Neckel, S.; Hasenfrath, M. (Eds.), 2020: *Imaginationen von Nachhaltigkeit: Katastrophe, Krise, Normalisierung, Zukünfte der Nachhaltigkeit, Vol. 2* (Frankfurt, New York: Campus).

Adloff, F.; Neckel, S. (Eds.), 2020: *Gesellschaftstheorie im Anthropozän* (Frankfurt am Main: Campus).

Albert, M., 2016: *A theory of world politics. Cambridge studies in international relations: Vol. 141* (Cambridge, UK: Cambridge University Press).

Albert, M. J., 2020: "Beyond continuationism: Climate change, economic growth, and the future of world (dis)order", in: *Cambridge Review of International Affairs*; at: https://doi.org/10.1080/09557571.2020.1825334.

Altvater, E., 2016: "The Capitalocene, or, Geoengineering against Capitalism's Planetary Boundaries", in: Moore. J. W. (Ed.): *Anthropocene or Capitalocene? Nature, History, and the Crisis of Capitalism* (Oakland, CA: PM Press/Kairos): 138–152.

America Is All In, 2021: *Blueprint 2030: An All-In Climate Strategy For Faster, More Durable Emissions Reductions* (America Is All In).

Antweiler, C., 2022: *Anthropologie im Anthropozän: Theoriebausteine für das 21. Jahrhundert* (Darmstadt: WBG Academic).

Arias-Maldonado, M., 2020: "Bedrock or social construction? What Anthropocene science means for political theory", in: *The Anthropocene Review*, *7*, 2: 97–112.

Arias-Maldonado, M., 2017: "Nature and the Anthropocene: The sense of an ending?", in: Pattberg, P.; Zelli, F. (Eds.): *Environmental Politics and Governance in the Anthropocene. Institutions and legitimacy in a complex world* (London/New York: Routledge): 31–46.

Aykut, S. C.; Dahan, A., 2015: *Gouverner le climat?: 20 ans de négociations internationales. Références. Développement durable* (Paris: Presses de Sciences Po).

Bai, X.; van der Leeuw, S.; Brondizio, E.S.; Duraiappah, A.; Revkin, A.; Biermann, F.; *et al.*, 2016: "Plausible and desirable futures in the Anthropocene: A new research agenda", in: *Global Environmental Change, 39:* 351–362.

Bammé, A. (Ed.), 2014b: *Schöpfer der zweiten Natur: Der Mensch im Anthropozän* (Marburg: Metropolis).

Bammé, A., 2014a: "Fünf Gründe, warum die Menschheit den Herausforderungen des 21. Jahrhunderts nicht gewachsen ist", in: Bammé, A. (Ed.): *Schöpfer der zweiten Natur. Der Mensch im Anthropozän* (Marburg: Metropolis): 49–62.

Bataille, C.; Waisman, H.; Colombier, M.; Segafredo, L.; Williams, J.; Jotzo, F., 2016: "The need for national deep decarbonization pathways for effective climate policy", in: *Climate Policy, 16*, (supl): S7-S26.

Beck, U., 1986: *Risikogesellschaft: Auf dem Weg in eine andere Moderne. edition suhrkamp Neue Folge, vol. 365 (Frankfurt am Main: Suhrkamp).*

Beck, U., 2007: *Weltrisikogesellschaft: Auf der Suche nach der verlorenen Sicherheit, Edition Zweite Moderne* (Frankfurt am Main: Suhrkamp).

Beck, S.; Mahony, M., 2017: "The IPCC and the politics of anticipation", in: *Nature Climate Change, 7*, 311-313.

Beck, S.; Oomen, J., 2021: "Imaging the corridor of climate mitigation - What is at stake in IPCC's politics of anticipation?", in: *Environmental Science & Policy*, 123: 169–178.

Beck, S.; *et al.*, 2014: "Towards a reflexive turn in global environmental expertise - the cases of the IPCC and IPBES", in: *GAIA - Ecological Perspectives for Science and Society, 23*,2: 80–87.

Beck, S.; Forsyth, T., 2018: "Environmental science and International Relations", in: Corry, O.; Stevenson, H. (Eds.): *Routledge research in global environmental governance. Traditions and trends in global environmental politics. International relations and the earth* (London, New York: Routledge Taylor & Francis Group): 81–99.

Beck, S., 2009: "*The Climate Experiment and the IPCC: Interfaces between Science and Policy in International Relations*", in: *Ecology and Economic Research, vol. 84* (Marburg: Metropolis).

Beck, S., 2009a: *Das Klimaexperiment und der IPCC: Schnittstellen zwischen Wissenschaft und Politik in den internationalen Beziehungen. Ökologie und Wirtschaftsforschung, Vol. 84* (Marburg: Metropolis).

Beck, U., 2009b: *Macht und Gegenmacht im globalen Zeitalter: Neue weltpolitische Ökonomie, Edition Zweite Moderne, Vol. 4099* (Frankfurt am Main: Suhrkamp).

Beck, S., 2016: "Hybride Organisationen als Schnittstellen - Der Fall IPCC", in: Besio, C.; Romano, G. (Eds.): *Umweltsoziologie: Band 1. Zum gesellschaftlichen Umgang mit dem Klimawandel. Kooperationen und Kollisionen* (Baden-Baden: Nomos): 151–185.

Beck, S., 2019: "Coproducing Knowledge and Politics of the Anthropocene: The Case of the Future Earth Program", in: Biermann, F.; Lövbrand, E. (Eds.): *Earth system governance. Anthropocene Encounters. New Directions in Green Political Thinking* (Cambridge: Cambridge University Press): 191–211.

Benz, A.; Lütz, S.; Schimank, U.; Simonis, G. (Eds.), 2007: *Handbuch Governance: Theoretische Grundlagen und empirische Anwendungsfelder (*Wiesbaden: VS Verlag für Sozialwissenschaften).

Benz, A.; Dose, N., 2010: *Governance - Regieren in komplexen Regelsystemen: Eine Einführung, Lehrbuch: Vol. 1., 2nd ed.* (Wiesbaden: VS Verlag für Sozialwissenschaften).

Biermann, B., 2006: *Nachhaltige Nahrungs-Netzwerke: Eine Analyse der prozeduralen und der materiellen Dimension lokaler Nachhaltigkeitspolitik, Dissertation (*Hagen: Fernuniversität in Hagen).

Biermann, F., 2014: *Earth system governance: World politics in the anthropocene. Earth system governance (*Cambridge, Massachusetts: The MIT Press).

Biermann, F., 2021: "The future of 'environmental' policy in the Anthropocene: time for a paradigm shift", in: *Environmental Politics, 30,*1-2: 61–80.

Biermann, F.; Zelli, F.; Pattberg,P. H. (Eds.), 2010: *Global climate governance beyond 2012: Architecture, agency and adaptation* (Cambridge: Cambridge University Press).

Biermann, F.; Bai, X.; Bondre, N.; *et al.*, 2016: "Down to Earth: Contextualizing the Anthropocene", in: *Global Environmental Change, 39*: 341–350.

Biermann, F.; Kanie, N.; Kim, R.E., 2017: "Global governance by goal-setting: The novel approach of the UN Sustainable Development Goals", in: *Current Opinion in Environmental Sustainability, 26-27:* 26–31.

Biermann, F.; Kanie, N., 2017: "Conclusion: Key Challenges for Global Governance through Goals", in: Kanie, N.; Biermann, F. (Eds.): *Earth system governance. Governing through goals. Sustainable development goals as governance innovation* (Cambridge, MA: The MIT Press): 295–309.

Biermann, F.; Kanie, N., 2017: "Conclusion: Key Challenges for Global Governance through Goals", in: Kanie, N.; Biermann, F. (Eds.): *Earth system governance. Governing through goals. Sustainable development goals as governance innovation* (Cambridge, Massachusetts: The MIT Press): 295–303.

Biermann, F.; Kim, R.E., 2020a: "The Boundaries of the Planetary Boundary Framework: A Critical Appraisal of Approaches to Define a 'Safe Operating Space' for Humanity", in: *Annual Review of Environment and Resources, 45,*1: 497–521.

Biermann, F.; Kim, R.E., 2020b: "The Boundaries of the Planetary Boundary Framework: A Critical Appraisal of Approaches to Define a 'Safe Operating Space' for Humanity", in: *Annual Review of Environment and Resources, 45,*1: 497–521.

Biermann, F.; Lövbrand, E. (Eds.), 2019: *Anthropocene encounters: New directions in green political thinking. Earth System Governance series* (Cambridge: Cambridge University Press).

Biermann, F.; Kim, R. E. (Eds.), 2020a: *Earth system governance. Architectures of Earth system governance: Institutional complexity and structural transformation (Cambridge: Cambridge University Press).*

Biermann, F.; Kim, R.E. (Eds.), 2020a: *Earth system governance. Architectures of earth system governance: Institutional complexity and structural transformation (Cambridge: Cambridge University Press).*

Biermann, F., 2016a: "'Erdsystem-Governance'. Plädoyer für ein neues Paradigma der globalen Umweltpolitik", in: Leitschuh, H., *et al.* (Eds.): *Jahrbuch Ökologie 2016. Gesucht: Weltumweltpolitik. Herausforderungen im Anthropozän* (Stuttgart: Hirzel): 107–115.

Biermann, F., 2016b: "Politics for a New Earth: Governing in the 'Anthropocene'", in: Nicholson; S.; Jinnah, S. (Eds.): *Earth system governance. New earth politics. Essays from the Anthropocene* (Cambridge, Mass.: The MIT Press): 405–420.

Bloomberg Philanthropies (Ed.), 2017: *America's Pledge: States, Cities, and Businesses in the United States Are Stepping Up on Climate Action.* Phase 1 Report; at: https://www.bbhub.io/dotorg/sites/28/2017/11/AmericasPledgePhaseOneReportWeb.pdf.

Bloomberg, M.; Pope, C., 2017: *Climate of hope: How cities, businesses, and citizens can save the planet* (New York: St. Martin's Press).

Bodansky, D., 2016a: "The Legal Character of the Paris Agreement", in: *Review of European, Comparative & International Environmental Law, 25,*2: 142–150.

Bodansky, D., 2016b: "The Paris Climate Change Agreement: A New Hope?", in: *American Journal of International Law, 110,*2: 288–319.

Böhret, C.; Konzendorf, G. (Eds.), 2001: *Handbuch Gesetzesfolgenabschätzung (GFA): Gesetze, Verordnungen, Verwaltungsvorschriften* (1st ed.), (Baden-Baden: Nomos).

Bolin, B., 2008: *A history of the science and politics of climate change: The role of the Intergovernmental Panel on Climate Change* (Cambridge: Cambridge University Press).

Bonneuil, C., Fressoz, J.-B., 2017: *The shock of the anthropocene: The earth, history, and us* (London, New York: Verso).

Bonneuil, C., 2015: "The Geological Turn: Narratives of the Anthropocene", in: Hamilton, C.; Bonneuil, F.; Gemenne, F. (Eds.): *The anthropocene and the global environmental crisis. Rethinking modernity in a new epoch* (Abingdon, Oxon: Routledge): 17–31.

Bonß, W., 1995: *Vom Risiko: Unsicherheit und Ungewißheit in der Moderne* (Hamburg: Hamburger Edition).

Bornemann, B.; Glaab, K.; Partzsch, L., 2019: "Politikwissenschaftliches Interpretieren des Anthropozäns – Zur Bedeutung interpretativer Politikforschung im Umgang mit Objektivität, Normativität und Performativität von Meta-Konzepten", in: *Zeitschrift für Politikwissenschaft, 29,* 2: 325–344; at: https://link.springer.com/article/10.1007/s41358-019-00184-5.

Bornemann, B., 2019: "The Anthropocene and governance: Critical reflections on conceptual relations", in: Hickmann, T.; Partzsch, L.; Pattberg, P. H.; Weiland, S. (Eds.): *The anthropocene debate and political science, Routledge research in global environmental governance* (London, New York: Routledge, Taylor & Francis Group): 48–66.

Bornemann, B., 2022: "Nachhaltigkeitstransformation in Zeiten des Populismus: Ansatzpunkte einer integrierten Betrachtung", in: Gumbert, T.; Bohn, C.; Fuchs, D.; Lennartz, B.; Müller, C.J. (Eds.), *Demokratie und Nachhaltigkeit. Aktuelle Perspektiven auf ein komplexes Spannungsverhältnis* (Baden-Baden: Nomos): 119–136.

Bradley, R.S., 2011: *Global warming and political intimidation: How politicians cracked down on scientists as the earth heated up* (Amherst: University of Massachusetts Press).

Brand, U.; Wissen, M., 2017: *Imperiale Lebensweise: Zur Ausbeutung von Mensch und Natur in Zeiten des globalen Kapitalismus* (München: oekom Verlag).

Brauch, H. G.; Oswald Spring, Ú.; Grin, J.; *et al.* (Eds.): 2009: *Facing Global Environmental Change: Environmental, Human, Energy, Food, Health and Water Security Concepts* (Berlin, Heidelberg: Springer Science & Business Media).

Brauch, H. G., Oswald Spring, Ú., Grin, J.; Scheffran, J. (Eds.), 2016: *Handbook on Sustainability Transition and Sustainable Peace* (*Cham:* Springer International Publishing).

Brauch, H. G.; Dalby, S.; Oswald Spring, Ú.; *et. al.* 2011: "Political geoecology for the Anthropocene", in: Brauch, H. G. (Ed.): *Coping with global environmental change, disasters and security. Threats, challenges, vulnerabilities and risks* (Berlin, New York: Springer): 1453–1485.

Brauch, H. G., 2021: "Peace Ecology in the Anthropocene", in: Oswald Spring, Ú.; Brauch, H. G. (Eds.): *Decolonising Conflicts, Security, Gender, Environment and Development in the Anthropocene* (Cham: Springer Nature): 51–185.

Brauch, H. G., 2021a: "The Anthropocene Concept in the Natural and Social Sciences, the Humanities and Law—A Bibliometric Analysis and a Qualitativ Interpretation (2000-2020)", in: Benner,

S.; Lax, G.; Crutzen, P. J.; Pöschl, U.; Lelieveld, J.; Brauch, H. G. (Eds.): *Paul J. Crutzen and the Anthropocene: A New Epoch in Earth's History* (Cham: Springer Nature): 289–438.

Brauch, H. G. (Ed.) et al., 2011: *Coping with global environmental change, disasters and security: Threats, challenges, vulnerabilities and risks* (Berlin, New York: Springer).

Breitmeier, H.; Schwindenhammer, S.; Checa, A.; Manderbach, J.; Tanzer, M., 2021: "Aligned Sustainability Understandings? Global Inter-Institutional Arrangements and the Implementation of SDG 2", in: *Politics and Governance, 9*,1: 141–151; at. https://www.cogitatiopress.com/politicsandgovernance/article/view/3591.

Breitmeier, H.; Schwindenhammer, S.; Checa, A.; Manderbach, J.; Tanzer, M., 2021: "Politicized Sustainability and Agricultural Policy: Comparing Norm Understandings of International Organizations", in: *Journal of Comparative Policy Analysis: Research and Practice, 23*, 5–6: 625–643; at: https://doi.org/10.1080/13876988.2020.1769480.

Brito, L.; Stafford-Smith, M., 2012: *State of the Planet Declaration.: New Knowledge: Towards Solutions.* Planet Under Pressure Conference (London: ICSU).

Bröchler, S.; Simonis, G.; Sundermann, K. (Eds.), 1999: *Handbuch Technikfolgenabschätzung* (Bonn: sigma).

Brondizio, E.S.; O'Brien, K.; Bai, X.; *et al.*, 2016: "Re-Conceptualizing the Anthropocene: A call for collaboration", in: *Global Environmental Change, 39:* 318–327.

Brown, C.; Brown, K.; Rounsevell, M., 2016: "A philosophical case for process-based modelling of land use change", in: *Modeling Earth Systems and Environment, 2*, 2: 50; at: https://doi.org/10.1007/s40808-016-0102-1.

Brunnengräber, A.; Di Nucci, M. R. (Eds.), 2014: *Research. Im Hürdenlauf zur Energiewende: Von Transformationen, Reformen und Innovationen; zum 70. Geburtstag von Lutz Mez.* (Wiesbaden: Springer VS).

Brunnengräber, A.; Di Nucci, M. R. (Eds.), 2014: *Research. In the hurdle race to the energy turnaround: Of transformations, reforms and innovations; on the 70th birthday of Lutz Mez* (Wiesbaden: Springer VS).

Brunnengräber, A.; Görg, C., 2017: "Nuclear Waste in the Anthropocene: Uncertainties and Unforeseeable Timescales in the Disposal of Nuclear Waste", in: *GAIA, 26,2*: 96–99.

Brunnengräber, A., 2021: "Das Kapitalozän ist das eigentliche Problem: Atommüll im Menschenzeitalter", in: oekom e.V. - Verein für ökologische Kommunikation (Ed.): *Menschengemacht. Streifzüge durch das Anthropozän* (München: oekom verlag): 51–57.

Brzoska, M.; Scheffran, J., 2020b: "Introduction: Research on climate change and security in Hamburg", in: *Brzoska, M.; Scheffran, J. (Eds.), Climate Change, Security Risks, and Violent Conflicts: Essays from Integrated Climate Research in Hamburg (Hamburg: Ham-burg University Press): 1–18.*

Brzoska, M.; Scheffran, J. (Eds.), 2020: *Climate Change, Security Risks, and Violent Conflicts: Essays from Integrated Climate Research in Hamburg (*Hamburg: Hamburg University Press).

Bückmann, W., 2015: *Die Vision der UNO für die Zukunft der Welt: die 2030-Agenda für nachhaltige Entwicklung. FAGUS-Schriften, Vol. 17* (Berlin: Universitätsverlag der TU Berlin).

Bulkeley, H. (Ed.), 2013: *Cities and low carbon transitions. Routledge studies of human geography, vol. 35* (London: Routledge).

Burke, A.; Fitzi, G., 2019: "Power, World Politics, and Thing-Systems in the Anthropocene", in: Biermann, F.; Lövbrand, E. (Eds.): *Earth system governance. Anthropocene Encounters. New Directions in Green Political Thinking* (Cambridge: Cambridge University Press): 87–108.

Burke, A.; Fishel, S.; Mitchell, A.; Dalby, S.; Levine, D. J., 2016: "Planet Politics: A Manifesto from the End of IR", in: *Millennium: Journal of International Studies, 44*, 3: 499–523.

Buzan, B.; Waever, O.; Wilde, J. de, 1998: *Security: A New Framework for Analysis* (Boulder, Col.: Lynne Rienner Publishers).

Buzan, B.; Hansen, L., 2011: *The evolution of international security studies* (Cambridge: Cambridge Univ. Press).

Buzan, B.; Schouenborg, L., 2018: *Global international society: A new framework for analysis (*Cambridge: Cambridge University Press).

Cai, Y.; Lenton, T. M.; Lontzek, T.S., 2016: "Risk of multiple interacting tipping points should encourage rapid CO2 emission reduction", in: *Nature Climate Change*, *6*, 5: 520–525; at: https://www.nature.com/articles/nclimate2964.

CAN, 2007: *National Security and the Threat of Climate Change* (Alexandria, VA: Center for Naval Analysis).

Castree, N.; Adams, W.M.; Barry, J.; Brockington, D.; Büscher, B.; Corbera, E.; *et al.*, 2014: "Changing the intellectual climate", in: *Nature Climate Change, 4*,9: 763–768.

Chakrabarty, D., 2021: *Climate of History in a Planetary Age (Chicago:* University of Chicago Press).

Chakrabarty, D., 2022: "The Planetary and the Political", in: *Politische Vierteljahresschrift*, *63*, 4: 708–711.

Chakrabarty, D., 2020: "Der Planet als neue humanistische Kategorie", in: Adloff, F.; Neckel, S. (Eds.): *Gesellschaftstheorie im Anthropozän* (Frankfurt am Main: Campus): 23–53.

Chandler, D., 2012: "Resilience and Human Security: The Post-interventionist Paradigm", in: *Security Dialogue, 43*,3: 213–229.

Chandler, D.; Müller, F.; Rothe, D. (Eds.), 2021: *International relations in the anthropocene: New agendas, new agencies and new approaches (*Cham: Palgrave Macmillan).

Chandler, D., 2021: "Security Through Resilience: Contemporary Challenges in the Anthropocene", in: Chandler, D.; Müller, F.; Roth, D. (Eds.): *International relations in the anthropocene. New agendas, new agencies and new approaches* (Cham: Palgrave Macmillan): 173–190.

Choucri, N. (Ed.), 1995: *Global accord: Environmental challenges and international responses. Global environmental accords series* (Cambridge, Mass: MIT Press).

Commission on Environment and Development, 1987: *Our Common Future,* Document A/42/427 (New York: UN, GA).

Commission on Global Security, Justice & Governance, 2015: *Confronting the Crisis of Global Governance (*Washington, D.C.: Stimson Center): at: https://www.stimson.org/sites/default/files/Commission-on-Global-Security-Justice-Governance-A4.pdf (16 December 2019).

Conca, R., 2016: "The Changing Shape of Global Environmental Politics", in: Nicholson, S.; Jinnah, S. (Eds.): *Earth system governance. New earth politics. Essays from the Anthropocene* (Cambridge, Massachusetts: The MIT Press): 21–42.

Cooley, A.; Nexon, D.H., 2020: *Exit from hegemony: The unraveling of the American global order (Oxford: Oxford University Press).*

Corry, O.; Stevenson, H., 2018a: "International Relations as if the Earth matters", in: Corry, O.; Stevenson, H., (Eds.): *Traditions and trends in global environmental politics. International relations and the earth. Routledge research in global environmental governance* (London, New York: Routledge Taylor & Francis Group): 187–196.

Corry, O.; Stevenson, H. (Eds.), 2018b: *Traditions and trends in global environmental politics: International relations and the earth. Routledge research in global environmental governance (London, New York: Routledge, Taylor & Francis Group).*

Crist, E., 2016: "On the Poverty of Our Nomenclature", in: Moore, J.W. (Ed.): *Anthropocene or Capitalocene? Nature, History, and the Crisis of Capitalism* (Oakland, CA: PM Press/Kairos): 14–33.

Crutzen, P.J.; Stoermer, E.F., 2000: "The 'Anthropocene'", in: *IGBP Newsletter*, 41: 17–18.

Crutzen, P. J., 2002: "Geology of mankind", in: *Nature, 415:* 23.

Dalby, S., 2016: "Framing the Anthropocene: The good, the bad and the ugly", in: *The Anthropocene Review, 3*,1: 33–51.

Dalby, S., 2020: *Anthropocene Geopolitics: Globalization, Security, Sustainability (*Ottawa: University of Ottawa Press).

Darby, M., 2018: *"11 takeaways from the draft UN report and a 1,5 °C global warming limit", in: Climate Home News*; at: http://www.climatechangenews.com/2018/02/13/11-takeaways-draft-un-report-1-5c-global-warming-limit/ (25 February 2018).

Dauvergne, P., 2016: "The Sustainability Story: Exposing Truths, Half-Truths, and Illusions", in: Nicholson, S.; Jinnah, S. (Eds.): *Earth system governance. New earth politics. Essays from the Anthropocene* (Cambridge, Massachusetts: The MIT Press): 387–404.

Delanty, G., 2020: "Wann beginnt die Zukunft? Überlegungen zu Temporalität, Nachhaltigkeit und Zukunftsszenarien", in: Adloff, F.; Fladvad, B.; Hasenfrath, M.; Neckel, S. (Eds.): *Imaginationen von Nachhaltigkeit. Katastrophe, Krise, Normalisierung. Zukünfte der Nachhaltigkeit, vol. 2* (Frankfurt, New York: Campus): 49–70.

Delbeke, J., 2017: *EU's Climate Policy: 2017-2018* (Brussels: CEPS).

Deudney, D.; Mendenhall, E., 2016: "Green Earth: The Emergence of Planetary Civilization", in: Nicholson, S.; Jinnah, S. (Eds.): *Earth system governance. New earth politics. Essays from the Anthropocene* (Cambridge, Massachusetts: The MIT Press): 43–72.

Dietz, M.; Garrelts, H. (Eds.), 2013: *Ein Handbuch, Bürgergesellschaft und Demokratie: Die internationale Klimabewegung, vol. 39* (Dordrecht: Springer).

Diez, T.; Lucke, F. von; Wellmann, Z., 2016: *The Securitisation of Climate Change: Actors, processes and consequences* (London, New York: Routledge, Taylor & Francis).

Dixson-Declève, S.; Gaffney, O.; Ghosh, J.; Randers, J.; Rockström, J.; Stoknes, P. E. 2022: *Earth for All: Ein Survivalguide für unseren Planeten. Der neue Bericht an den Club of Rome 50 Jahre nach 'Die Grenzen des Wachstums'* (München: oekom).

Donges, J.F.; Lucht, W.; Müller-Hansen, F.; Steffen, W., 2017a: "The technosphere in Earth System analysis: A coevolutionary perspective", in: *The Anthropocene Review, 4*,1: 23–33.

Donges, J. F.; Winkelmann, R.; Lucht, W.; Cornell, S. E.; Dyke, J. G.; Rockström, J.; *et al.*, 2017b: "Closing the loop: Reconnecting human dynamics to Earth System science", in: *The Anthropocene Review, 4*,2: 151–157.

Dunlap, R. E.; McCright, A. M., 2015: "Challenging Climate Change: The Denial Countermovement", in: Dunlap, R. E.; Brulle, R.J. (Eds.): *Climate change and society. Sociological perspectives* (New York, NY: Oxford University Press): 300–332.

Dürbeck, G., 2020: "Narratives of the Anthropocene: From the perspective of postcolonial ecocriticism and environmental humanities", in: Albrecht, M. (Ed.): *Postcolonialism Cross-Examined. Multidirectional Perspectives on Imperial and Colonial Pasts and the Neocolonial Present* (London/New York: Routledge): 271–288.

Edwards, P. N., 2010: *A vast machine: Computer models, climate data, and the politics of global warming* (Cambridge, Mass.: MIT Press).

Ellis, E. C., 2020: *Anthropozän: Das Zeitalter des Menschen - eine Einführung* (München: oekom).

European Environment Agency, 2020: "*Trends and projections in Europe 2020: Tracking progress towards Europe's climate and energy targets*", *in: EEA Report No. 13/2020* (Copenhagen: EEA), at: https://op.europa.eu/en/publication-detail/-/publication/590a9c36-383d-11eb-b27b-01aa75ed71a1/language-en.

EU Commission, 2016: *Implementing the Paris Agreement: Progress of the EU towards the at least -40% target* (Commission Staff Working Document No. SWD (2016) 349 final). (Brussels: EU Commission).

EU, 2021: "European Climate Law", Document 32021/R111; at: https://climate.ec.europa.eu/eu-action/european-green-deal/european-climate-law_en.

Europäische Kommission, 2016: Vorschlag für einen Beschluss des Rates: über den Abschluss des im Rahmen des Rahmenübereinkommens der Vereinten Nationen über Klimaänderungen geschlossenen Übereinkommens von Paris im Namen der Europäischen Union (No. COM(2016) 395 final) (Brussels: EU Commission).

Feiler, K. (Ed.), 2003: *Nachhaltigkeit schafft neuen Wohlstand: Bericht an den Club of Rome.* (Frankfurt am Main: Peter Lang).

Fichter, K.; von Gleich, A.; Pfriem, R.; Siebenhüner, B., 2010: *Theoretische Grundlagen für erfolgreiche Klimaanpassungsstrategien (Bremen:* Universität Bremen).

Flechtheim, O.K., 1972: *Futurologie: Der Kampf um die Zukunft* (Frankfurt/M.: S. Fischer).

Folkers, A., 2018: "Resilienz als Nomos der Erde", in: Laux, H.; Henkel, A. (Eds.): *Die Erde, der Mensch und das Soziale. Zur Transformation gesellschaftlicher Naturverhältnisse im Anthropozän* (Bielefeld: transcript): 137–160.

Folkers, A., 2020: "Was ist das Anthropozän und was wird es gewesen sein? Ein kritischer Überblick über neue Literatur zum kontemporänen Erdzeitalter", in: *NTM-Zeitschrift für Geschichte der Wissenschaft, Technik und Medizin, 2:* 589–604.

Forsythe, D.P., 2018a: "Human rights", in: Weiss, T. G.; Wilkinson, R. (Eds.): *International Organization and Global Governance,* 2nd ed. (Milton: Taylor and Francis): 511–522.

Franz, J.H., 2014: *Nachhaltigkeit, Menschlichkeit, Scheinheiligkeit: Philosophische Reflexionen über nachhaltige Entwicklung* (München: oekom).

Fraundorfer, M., 2022: *Global Governance in the Age of the Anthropocene* (Cham: Palgrave Macmillan).

Friedrich, J., 2017: "Globale Stocktake", Article 14, in: Klein, D.; Carazo, M. P.; Doelle, M.; Bulmer, J.; Higham, A. (Eds.): *The Paris agreement on climate change. Analysis and commentary* (Oxford, New York, NY: Oxford University Press): 319–337.

Friedrichs, W., 2021a: "Terrestrisch-Werden - Politische Bildung im Anthropozän", in: *Polis, 25,* 2: 11–13.

Friedrichs, W., 2021b: "Zur Neuvermessung der politischen Bildung im Anthropozän", in: Stainer-Hämmerle, K. (Ed.): *Glaube - Klima - Hoffnung. Religion und Klimawandel als Herausforderung für die politische Bildung* (Frankfurt/M.: Wochenschau Verlag): 45–59.

Galaz, V., 2019: "Time and Politics in the Anthropocene: Too Fast, Too Slow?", in: Biermann, F.; Lövbrand, E. (Eds.): *Earth System Governance series. Anthropocene encounters. New directions in green political thinking* (Cambridge: Cambridge University Press): 109–127.

Geden, O., 2010: *Abkehr vom 2-Grad-Ziel: Skizze einer klimapolitischen Akzentverschiebung,* Arbeitspapier der FG 1 (Berlin: SWP).

Gerten, D.; Schönfeld, M.; Schauberger, B., 2018: "On deeper human dimensions in Earth system analysis and modelling", in: *Earth System Dynamics, 9,2: 849-863.*

Geyer, C., 2013: *Niklas Luhmann: Die Knappheit der Zeit und die Vordringlichkeit des Befristeten* (Berlin: Kulturverl. Kadmos).

González-Ricoy, I.; Gosseries, A. (Eds.), 2016: *Institutions for future generations, 1st* ed. (Oxford, New York: Oxford University Press).

Görg, C., 2016: "Zwischen Tagesgeschäft und Erdgeschichte: Die unterschiedlichen Zeitskalen in der Debatte um das Anthropozän", in: *GAIA, 25,*1: 9–13.

Görg, C.; Plank, C.; Wiedenhofer, D.; Mayer, A.; Pichler, M.; Schaffartzik, A.; Krausmann, F., 2020: "Scrutinizing the Great Acceleration: The Anthropocene and its analytic challenges for social-ecological transformations", in: *The Anthropocene Review,* 7,1 (December 2019); at: https://doi.org/10.1177/2053019619895034.

Grunwald, A.; Kopfmüller, J., 2022: *Nachhaltigkeit, 3nd updated ed., Campus Studium* (Frankfurt, New York: Campus).

Grunwald, A., 2010: *Technikfolgenabschätzung: Eine Einführung* (2nd extended ed.). *Gesellschaft, Technik, Umwelt*: Neue Folge, 1 (Berlin: Edition Sigma).

Gumbert, T.; Bohn, C.; Fuchs, D.; Lennartz, B.; Müller, C.J. (Eds.), 2022: *Demokratie und Nachhaltigkeit: Aktuelle Perspektiven auf ein komplexes Spannungsverhältnis* (Baden-Baden: Nomos).

Gumbert, T.; Bohn, C.; Fuchs, D.; Lennartz, B., 2022: "Demoktatische Nachhaltigkeitspolitik im Anthropozän", in: Gumbert, T.; Bohn, C.; Fuchs, D.; Lennartz, B.; Müller, C.J. (Eds.): *Demokratie und Nachhaltigkeit. Aktuelle Perspektiven auf ein komplexes Spannungsverhältnis* (Baden-Baden: Nomos): 9–32.

Gupta, J., 2016: "Toward Sharing Our Ecospace", in: Nicholson, S.; Jinnah, S. (Eds.): *Earth system governance. New earth politics. Essays from the Anthropocene* (Cambridge, Massachusetts: The MIT Press): 271–291.

Guterres, A., 2020: *The State of the Planet.* Secretary-General's address at Columbia University (New York: UN, 2 December).

Haas, P. M., 2016: "Introduction: reconstructing epistemic communities", in: Haas, P. M. (Ed.): *Epistemic communities, constructivism, and international environmental politics* (London, New York: Routledge): 1–24.

Hamilton, C., 2016: "The Anthropocene as rupture", in: *The Anthropocene Review, 3*,2: 93–106.

Hamilton, C., 2017: *Defiant earth: The fate of humans in the Anthropocene*, 1st ed. (Cambridge, UK, Malden, MA: Polity).

Hamilton, C.; Bonneuil, C.; Gemenne, F. (Eds.), 2015: *The anthropocene and the global environmental crisis: Rethinking modernity in a new epoch* (Abingdon, Oxon: Routledge).

Hamilton, C.; Grinevald, J., 2015: "Was the Anthropocene anticipated?", in: *The Anthropocene Review*: 1–14.

Hansen, J., 2006: "The Threat to the Planet", in: *The New York Review of Books, 53*,12: 12–16.

Hanusch, F.; Leggewie, C.; Meyer, E. 2021: *Planetar denken: Ein Einstieg* (Bielefeld: transcript Verlag).

Hardin, G., 1968: "The Tragedy of the Commons", in: *Science, 162:* 1243–1248.

Hardt, J. N., 2017: *Environmental Security in the Anthropocene: Assessing Theory and Practice* (Abingdon: Routledge).

Hardt, J.N.; Scheffran, J., 2019: *Environmental Peacebuilding and Climate Change: Peace and Conflict Studies at the Edge of Transformation*, Policy Brief No. 68 (Tokyo: Toda Peace Institute).

Hardt, J.N., 2019: "Security studies and the discourse on the Anthropocene: shortcomings, challenges and opportunities", in: Hickmann T.; Partzsch, L.; Pattberg, P.H.; Weiland, S. (Eds.): *The anthropocene debate and political science, Routledge research in global environmental governance* (London, New York: Routledge): 85–102.

Hardt, J. N., 2021: "Encounters between Security and Earth System Sciences: Planetary Boundaries and Hothouse Earth", in: Chandler, D.; Müller, F.; Rothe, D. (Eds.): *International relations in the anthropocene. New agendas, new agencies and new approaches* (Cham: Palgrave Macmillan): 39–57.

Harrington, C., 2016: "The Ends of the World: International Relations and the Anthropocene", in: *Millennium: Journal of International Studies, 44*,3: 478–498.

Harrington, C., 2020: "A Quantum Anthropocene? International Relations Between Rupture and Entanglement", in: Pereira, C.; Saramgo, A. (Eds.): *Non-Human Nature in World Politics. Theory and Practice* (Cham, Switzerland: Springer Nature): 53–72.

Harvey, B.; Cochrane, L.; van Epp, M., 2019: "Charting knowledge co-production pathways in climate and development", in: *Environmental Policy and Governance, 29*,2: 107–117.

Haunschild, R.; Bornmann, L.; Marx, W., 2016: "Climate Change Research in View of Bibliometrics", in: *PloS one.*

Hermanns, T.; Li, Q., 2019: "Sustainable impact as sessment of land use changes in the Anthropocene", in: Hickmann, T.; Partzsch, L.; Pattberg, P.H.; Weiland, S. (Eds.): *The anthropocene debate and political science. Routledge research in global environmental governance* (London, New York: Routledge Taylor & Francis Group): 167–181.

Hermwille, L., 2019: "Global climate governance as a boundary object: Making the meaning of the Anthropocene", in: Hickmann, T.; Partzsch, L.; Pattberg, P.H.; Weiland, S. (Eds.): *The anthropocene debate and political science. Routledge research in global environmental governance* (London, New York: Routledge Taylor & Francis Group): 103–123.

Hickmann, T., 2017: "The Reconfiguration of Authority in Global Climate Governance", in: *International Studies Review, 19*,3: 430–451.

Hickmann, T.; Partzsch, L.; Pattberg, P.; Weiland, S., 2020: "Mehr Engagement der Politikwissenschaft in der Anthropozän-Debatte", in: *Politische Vierteljahresschrift, 61*,4: 659–670.

Hickmann, T.; Widerberg, O.; Lederer, M.; Pattberg, P., 2021: "The United Nations Framework Convention on Climate Change Secretariat as an orchestrator in global climate policymaking", in: *International Review of Administrative Sciences, 87*,1: 21–38.

Hickmann, T.; Partzsch, L.; Pattberg, P. H.; Weiland, S. (Eds.), 2019: *The anthropocene deba-te and political science, Routledge research in global environmental governance* (London, New York: Routledge).

Hickmann, T.; Lederer, M.; Marquardt, J.; Radtke, J.; Schwindenhammer, S.; Weiland, S., 2022: "Herausforderung Nachhaltigkeitstransformation: Mehr Politikwissenschaft wagen!", in: Bergem, W.; Schöne, H. (Eds.): *Wie relevant ist die Politikwissenschaft?* (Wiesbaden: Springer Nature): 143–158.

Higham, A., 2017: "Pre-2020 Climate Action and the Emergent Role of Non-party Stakeholders", in: Klein, D.; Carazo, M.P.; Doelle, M.; Bulmer, J.; Higham. A. (Eds.): *The Paris agreement on climate change. Analysis and commentary* (Oxford, New York, NY: Oxford University Press): 43–49.

Holtwisch, C., 2006: *Das Nichteinhaltungsverfahren des Kyoto-Protokolls: Entstehung - Gestalt - Wirkung* (Berlin: Duncker & Humblot).

Hoppe, R.; Wesselink, A.; Cairns, R., 2013: "Lost in the problem: The role of boundary organisations in the governance of climate change". *Wiley Interdisciplinary Reviews: Climate Change, 4:* 283–300; at: https://wires.onlinelibrary.wiley.com/doi/10.1002/wcc.225.

Horn, E.; Bergthaller, H., 2019: *Anthropozän zur Einführung* (Hamburg: Junius Verlag).

Horn, E., 2021: "Tipping Points: The Anthropocene and Covid-19", in: Delanty, G. (Ed.): *Pandemics, Politics, and Society. Critical Perspectives on the Covid-19 Crisis* (Berlin: De Gruyter): 123–138.

Hulme, M., 2011: *Why we disagree about climate change: Understanding controversy, inaction and opportunity* (Cambridge: Cambridge Univ. Press).

Hultman, N.; Kennedy, K.; Cyrs, T.; *et al.*, 2021: *An All-In National Climate Strategy to Delver Ambitious, Robust, And Credible U.S. Action (Regents Drive College Park:* University of Maryland: Center for Global Sustainability); at: https://spp.umd.edu/research-impact/publications/all-national-climate-strategy-deliver-ambitious-robust-and-credible-us (8 July 2022).

IPBES, 2019: *Global assessment report on biodiversity and ecosystem services of the Intergovernmental Science-Policy Platform on Biodiversity and Ecosystem Services.* (Bonn: IPBES Secretariat).

IPCC, 2015: *Summary for Policy Makers: AR5 Synthesis Report: Climate Change 2014.* (Geneva: IPCC).

IPCC, 2018: *Global warming of 1.5°C: An IPCC Special Report on the impacts of global warming of 1,5°C above pre-industrial levels. Special Report* (Geneva: IPCC).

IPCC, 2021: *Climate Change 2021: The Physical Science Basis*, Contribution of Working Group I to the Sixth Assessment Report of the Intergovernmental Panel on Climate Change (Cambridge: Cambridge Univ. Press).

IPCC, 2022: *Climate Change 2022: Mitigation of Climate Change*, Working Group III Contribution to the IPCC Sixth Assessment Report (Cambridge, UK: Cambridge Univ. Press).

Irvine, R. D. G., 2020: *An Anthropology of Deep Time: Geological Temporality and Social Life* (Cambridge: Cambridge University Press).

Jasanoff, S., 1994: *The fifth branch: Science advisers as policymakers* (Cambridge Mass.: Harvard Univ. Press).

Jasanoff, S., 2004: *States of Knowledge* (Abingdon, UK: Taylor & Francis).

Jasanoff, S., 2021: "Humility in the Anthropocene", in: *Globalizations, 18*,6: 839–853.

Jungk, R., 1956: *Heller als tausend Sonnen: Das Schicksal der Atomforscher* (Reinbek: Rowohlt).

Kahn, H., 1960: *On Thermonuclear War* (Princeton, N.J: Princeton University Press).

Kahn, H., 1965: *On Escalation: Metaphors and Scenarios* (New York: Praeger).

Kanie, N.; Biermann, F. (Eds.), 2017: *Earth system governance. Governing through goals: Sustainable development goals as governance innovation* (Cambridge, Massachusetts: The MIT Press).

Karapin, R., 2016: *Political opportunities for climate policy: California, New York, and the federal government* (New York: Cambridge University Press).

Keohane, R.O.; Victor, D.G., 2010: *The Regime Complex for Climate Change*, Discussion Paper No. 10-33: Harvard: The Harvard Project on International Climate Agreements.

Klein, D.; Carazo, M.P.; Doelle, M.; Bulmer, J.; Higham, A. (Eds.), 2017: *The Paris agreement on climate change: Analysis and commentary* (Oxford, New York, NY: Oxford University Press).

Knaus, A., Renn, O. (Eds.), 1998: *Den Gipfel vor Augen: Unterwegs in eine nachhaltige Zukunft. Ökologie und Wirtschaftsforschung, vol. 29* (Marburg: Metropolis-Verl).
Kohler, P. M., 2022: "*Science-Policy Interfaces: From Warnings to Solutions*", *in: Policy Brief 30* (IISD).
Kotzé, L. J.; French, D., 2022: "Staying within the Planet's 'safe operating space'? Law and the planetary boundaries", in: French, D.; Kotzé, L. J. (Eds.): *Research Handbook on Law, Governance and Planetary Boundaries. Research handbooks in environmental law (Cheltenham, UK:* Edward Elgar Publishing): 1–19.
Krämer, A. K., 2016: "Das "Anthropozän" als Wendepunkt zu einem neuen wissenschaftlichen Bewusstsein? Eine Untersuchung aus ethnologischer Perspektive zur Bedeutung und Verwendung des Konzeptes", Bachelorarbeit Universität zu Köln, in: *Kölner Ethnologische Beiträge*, vol. 45 (Cologne: University of Cologne).
Kriegler, E.; Edmonds, J.; Hallegatte, S.; Ebi, K.L.; Kram, T.; Riahi, K.; *et al.*, 2014: "A new scenario framework for climate change research: The concept of shared climate policy assumptions", in: *Climatic Change, 122*,3: 401–414.
Kristensen, H.M.; Norris, R.S., 2022: *Nuclear Notebook.* Federation of American Scientists.
Lalyer, C.; Gleich, A. V.; Giese, B., 2020: "Vulnerability Analysis of Ecological Systems", in: Gleich, A. V.; Schröder, W. (Eds.): *Gene Drives at Tipping Points* (Cham: Springer International Publishing): 57–78.
Latour, B., 2017: *Kampf um Gaia: Acht Vorträge über das neue Klimaregime* (Berlin: Suhrkamp).
Laux, H.; Henkel, A. (Eds.), 2018: *Die Erde, der Mensch und das Soziale: Zur Transformation gesellschaftlicher Naturverhältnisse im Anthropozän* (Bielefeld: transcript).
Leggewie, C.; Hanusch, F.; Bauer, L.; Hartl, C.; Finkelstein, C., 2022: "Der Wald als Mitakteur? Das Fallbeispiel einer planetaren Politik", in: *Politische Vierteljahresschrift (PVS), 63*,4: 718–723.
Leinfelder, R., 2017: "Das Zeitalter des Anthropozäns und die Notwendigkeit der großen Transformation: Welche Rollen spielen Umweltpolitik und Umweltrecht?", in: *Zeitschrift für Umweltrecht (ZUR), 28*,5: 259–266.
Leinfelder, R., 2020: "Das Anthropozän—mit offenem Blick in die Zukunft der Bildung", in: Sippl, C.; Rauscher, E.; Scheuch, M. (Eds.): *Das Anthropozän lernen und lehren. Pädagogik für Niederösterreich, vol. 9* (Innsbruck/Wien: Studien Verlag): 17–68.
Lenton, T. M.; Held, H.; Kriegler, E.; *et al.*, 2008: "Tipping elements in the Earth's climate system", in: *PNAS, 105,6:* 1786–1793.
Lenton, T. M.; Schellnhuber, H. J., 2007: "Tipping the scales", in: *Nature Climate Change, 1*: 97–98; at: https://www.nature.com/articles/climate.2007.65.
Loske, R., 2015: *Politik der Zukunftsfähigkeit: Konturen einer Nachhaltigkeitswende. Fischer Taschenbuch Forum für Verantwortung Vol. 03221 (*Frankfurt am Main: Fischer).
Lövbrand, E.; Beck, S.; Chilvers, J.; Forsyth, T.; Hedrén, J.; Hulme, M.; *et al.,* 2015: "Who speaks for the future of Earth? How critical social science can extend the conversation on the Anthropocene", in: *Global Environmental Change, 32:* 211–218.
Lövbrand, E.; Mobjörk, M.; Söder, R., 2020: "The Anthropocene and geo-political imagination: Re-writing Earth as political space", in: *Earth System Governance*, 4: article id 1000051.
Lövbrand, E.; Mobjörk, M. (Eds.), 2021: *Anthropocene (In)securities: Reflections on Collective Survival 50 Years After the Stockholm Conference. SIPRI Research Report, vol. 26* (Oxford: Oxford University Press).
Luciano, E., 2022: "The History of the Anthropocene: Meanings and Research Trajectories, in: *Global Environment, 15*,3: 573–601.
Luhmann, N., 1971, 2013: *Die Knappheit der Zeit und die Vordringlichkeit des Befristeten (Berlin: Kulturverlag Kadmos).*
Luhmann, N., [4]2004: *Ökologische Kommunikation: Kann die moderne Gesellschaft sich auf ökologische Gefährdungen einstellen?* (Wiesbaden: VS Verlag für Sozialwissenschaften).
MacAskill, W., 2022: "The Beginning of History: Surviving the Era of Catastrophic Risk", in; *Foreign Affairs*: 101,5.

MacFarlane, S.N.; Khong, Y.F., 2006: *The UN and Human Secuity: A Critical History (*Indianapolis: Indiana University Press).

MacKenzie, M.K., 2016: "Institutional Design and Sources of Short-Terminism", in: González-Ricoy, I.; Gosseries, A., (Eds.): *Institutions for future generations* (Oxford, New York: Oxford University Press): 24–47.

Mahnkopf, B., 2022: "Der große (Selbst-)Betrug: ‚Klimaneutralität' durch 'grünes Wachstum'"; in: *Aus Politik und Zeitgeschichte (APuZ), 72,*3-4: 42–46.

Manemann, J., 2014: *Kritik des Anthropozäns: Plädoyer für eine neue Humanökologie* (Bielefeld: transcript).

Mann, M.E., 2012: *The hockey stick and the climate wars: Dispatches from the front lines (*New York: Columbia University Press).

Mann, M.E., 2021: *The New Climate War: the fight to take back our planet (*London: Scribe Publications).

Marquardt, J., 2019: "Worlds apart? The Global South and the Anthropocene", in: Hickmann, T.; Partzsch, L.; Pattberg, P.H.; Weiland, S., (Eds.): *The anthropocene debate and political science. Routledge research in global environmental governance* (London, New York: Routledge Taylor & Francis Group): 200-218

Matejovski, D., 2016: *Anthropozän und Apokalypse: Zum Verhältnis von Ökologie, Theoriedesign und kommunikativer Hegemonie*, CSSA Discussion Paper No. 3 (Wiesbaden); at: http://www.cssa-wiesbaden.de/fileadmin/Bilder/B%C3%BCcher_Brosch%C3%BCren/Papers-cssa/cssa-paper_Anthropoz%C3%A4n_3_2016.pdf.

Meadowcroft, J., 2019: "Epilogue: Continuity and Change in the Anthropocene", in: Biermann, F.; Lövbrand, E. (Eds.): *Anthropocene Encounters. New Directions in Green Political Thinking. Earth system governance* (Cambridge: Cambridge University Press): 228–242.

Meadows, D.; Meadows, D.; Zahn, E.; Milling, P., 1972: *Die Grenzen des Wachstums: Bericht des Club of Rome zur Lage der Menschheit (*Stuttgart: DVA).

Meadows, D. H.; Meadows, D. L.; Randers, J., 1992: *Die neuen Grenzen des Wachstums: Die Lage der Menschheit: Bedrohung und Zukunftschancen (*Stuttgart: DVA).

Mehring, M.; Bernard, B.; Hummel, D.; Liehr, S.; Lux, A., 2017: "Halting biodiversity loss: How social–ecological biodiversity research makes a difference", in: *International Journal of Biodiversity Science, Ecosystem Services & Management, 13,*1: 172–180.

Meyer, R., 1999: "Das Büro für Technikfolgen-Abschätzung beim Deutschen Bundestag (TAB)", in: Bröchler, S.; Simonis, G.; Sundermann, K. (Eds.): *Handbuch Technikfolgenabschätzung* (Bonn: Edition Sigma): 459–467.

Millennium Ecosystems Assessment, 2003: *Ecosystems and human well-being: A report*. (Washington, D.C.; London: Island Press).

Miller, C.; Edwards, P.N. (Eds.), 2001: *Changing the Atmosphere: Expert Knowledge and Environmental Governance (*Cambridge, Massachussetts: MIT Press).

Mooney, H. A.; Duraiappah, A.; Larigauderie, A., 2013: "Evolution of natural and social science interactions in global change research programs", in: *Proceedings of the National Academy of Sciences of the United States of America, 110 Suppl. 1: 3665–3672.*

Moore, J.W. (Ed.), 2016: *Anthropocene or Capitalocene? Nature, History, and the Crisis of Capitalism (*Oakland, CA: PM Press/Kairos).

Motesharrei, S.; Rivas, J.; Kalnay, E.; Asrar, G. R.; Busalacchi, A. J.; Cahalan, R. F.; *et al.*, 2016: "Modeling Sustainability: Population, Inequality, Consumption, and Bidirectional Coupling of the Earth and Human Systems", in: *National Science Review*, 3,4: 470-494.

Müller, F., 2019: "International Theory in the Anthropocene: Moving Beyond Species, State and Governance", in: Hickmann, T.; Partzsch, L.; Pattberg, P. H.; Weiland, S. (Eds.): *The anthropocene debate and political science, Routledge research in global environmental governance* (London, New York: Routledge): 67–82.

Nakicenovic, N.; Rockström, J.; Gaffney, O.; Zimm, C., 2016: *Global Commons in the Anthropocene: World Development on a Stable and Resilient Planet,* IIASA Working Paper No. WP-16-019 (Laxenburg: IIASA).

Nicholson, S.; Jinnah, S., 2016a: "Living on a New Earth", in: Nicholson, S.; Jinnah, S. (Eds.): *Earth system governance. New earth politics. Essays from the Anthropocene* (Cambridge, Massachusetts: The MIT Press): 1–16.

Nicholson, S.; Jinnah, S. (Eds.), 2016b: *Earth system governance. New earth politics: Essays from the Anthropocene* (Cambridge, Massachusetts: The MIT Press).

Nordhaus, W. D., 2020: "The Climate Club: How to Fix a Failing Global Effort", in: *Foreign Affairs, 99*,3: 10–17.

Nordhaus, W. D., 2013: *The Climate casino: Risk, uncertainty, and economics for a warming world* (New Haven, Conn: Yale University Press).

Nordhaus, T., 2018: "The Two-Degree Delusion: The Dangers of an Unrealistic Climate Change Target", in: *Foreign Affairs (8 February).*

Northcott, M., 2015: "Eschatology in the Anthropocene: From the chronos of deep time to the kairos of the age of humans", in: Hamilton, C.; Bonneuil, C.; Gemenne F. (Eds.): *The anthropocene and the global environmental crisis. Rethinking modernity in a new epoch* (Abingdon, Oxon: Routledge): 100–111.

O'Brien, K., 2021: "Reflecting on the Anthropocene: The Call for Deeper Transformations", in: *Ambio, 50*, 1793–1797.

O'Neill, B. C.; Kriegler, E.; Riahi, K.; Ebi, K.L.; Hallegatte, S.; Carter, T. R.; *et al.*, 2014: "A new scenario framework for climate change research: The concept of shared socioeconomic pathways", in: *Climatic Change, 122*,3: 387–400.

O'Neill, B. C.; Kriegler, E.; Ebi, K. L.; Kemp-Benedict, E.; Riahi, K.; Rothman, D. S.; *et al.*, 2017: "The roads ahead: Narratives for shared socioeconomic pathways describing world futures in the 21st century", in: *Global Environmental Change, 42:* 169–180.

Oberthür, S.; Lefeber, R., 2012: "Das System der Erfüllungskontrolle des Kyoto-Protokolls: Eine Bestandsaufnahme nach sechs Jahren Praxis", in: *Archiv des Völkerrechts, 50,4:* 420–449.

OECD, 2019: *Accelerating Climate Action: Refocusing Policies through a Well-being Lens; at:* https://www.oecd-ilibrary.org/environment/accelerating-climate-action_2f4c8c9a-en (20 August 2022).

OECD, 2021: *Transport Strategies for Net-Zero Systems by Design (*Paris: OECD).

OECD/IEA, 2021: *Understanding Countries' Net-Zero Emissions Targets*, OECD/IEA Climate Change Expert Group Papers No. COM/ENV/EPOC/IEA/SLT(2021)3 (Paris: OECD/IEA).

Oels, A., 2012: "From 'Securitization' of Climate Change to 'Climatization' of the Security Field: Comparing Three Theoretical Perspectives", in: Scheffran, J.; Brzoska, M.; Brauch, H. G.; Link, P. M.; Schilling, J. (Eds.): *Climate Change, Human Security and Violent Conflict. Challenges for Societal Stability* (Berlin, Heidelberg: Springer):186–205.

Oreskes, N.; Conway, E. M., 2011: *Merchants of doubt: How a handful of scientists obscured the truth on issues from tobacco smoke to global warming* (New York, NY: Bloomsbury Press).

Ostrom, E.; Schöller, E., 1999: *Die Verfassung der Allmende: Jenseits von Staat und Markt. Die Einheit der Gesellschaftswissenschaften, Vol. 104 (*Tübingen: Mohr Siebeck).

Ostrom, E., 2009a: *A Polycentric Approach for Coping with Climate Change,* Policy Research Working Paper No. 5095 (Washington, D.C.: The World Bank).

Ostrom, E., 2009b: *Beyond Markets and States: Polycentric Governance of Complex Economic Systems* (Stockholm: Nobelprize.org); at: https://www.nobelprize.org/nobel_prizes/economic-sciences/laureates/2009/ostrom-lecture.html.

Oswald Spring, Ú.; Brauch, H. G.; Dalby, S., 2009: "Linking Anthropocene, HUGE and HESP: Fourth Phase of Environmental Security Research", in: Brauch, H. G.; Oswald Spring, Ú.; Grin, J.: *et al.* (Eds.): *Facing Global Environmental Change. Environmental, Human, Energy, Food, Health and Water Security Concepts* (Berlin, Heidelberg: Springer Science & Business Media): 1277–1294.

Pattberg, P.; Davies-Venn, M., 2020: "Dating the Anthropocene: Why Deciding on a Start Date for the Most Recent Geological Epoch Matters", in: Dürbeck, G.; Hüpkes, P. (Eds.): *The Anthropocenic Turn. The Interplay between Disciplinary and Interdisciplinary Responses to a New Age* (New York: Routledge): 70–89.

Pattberg, P.; Zelli, F. (Eds.), 2017a: Environmental Politics and Governance in the Anthropocene: Institutions and legitimacy in a complex world (London/New York: Routledge).

Pattberg, P.; Zelli, F., 2017b: "Global environmental governance in the Anthropocene: An introduction", in: Pattberg, P.; Zelli, F. (Eds.): *Environmental Politics and Governance in the Anthropocene. Institutions and legitimacy in a complex world* (London/New York: Routledge): 1–12.

Pattberg, P.; Kristensen, K.; Widerberg, O., 2017: *Beyond the CBD: Exploring the institutional landscape of governing for biodiversity*, Report No. R-17/06 (Amsterdam: Institute for Environmental Studies (IVM)).

Pauls, C.; Oppelt, M.; Weber, N.K.; Huni Kui, N.I.; Leggewie, C.; Hanusch, F.; *et al.*, 2022: "Das Planetarische Politisch(e) Denken", in: *Politische Vierteljahresschrift (PVS), 63*,4: 703–728.

Penny, C. K., 2020: "Human Security", in: Weiss, T. G.; Daws, S. (Eds.): *The Oxford handbook on the United Nations* (Oxford: Oxford University Press): 635–651.

Pereira, J. C., 2017: "The limitations of IR theory regarding the environment: Lessons from the anthropocene", in: *Revista Brasileira de Política Internacional 60*(1): eO18err.

Pereira, J. C., 2021: "Towards a Politics for the Earth: Rethinking IR in the Anthropocene", in: Chandler, D.; Müller, F.; Rothe, D. (Eds.): *International relations in the anthropocene. New agendas, new agencies and new approaches* (Cham: Palgrave Macmillan): 21–37.

Persson, L.; Carney Almroth, B. M.; Collins, C. D.; Cornell, S.; Wit, C.A. de; Diamond, M. L.; *et al.*, 2022: "Outside the Safe Operating Space of the Planetary Boundary for Novel Entities", in: *Environmental Science & Technology, 56*,3: 1510–1521; at: https://doi.org/10.1021/acs.est.1c04158.

Posner, S. M.; Cvitanovic, C., 2019: "Evaluating the impacts of boundary-spanning activities at the interface of environmental science and policy: A review of progress and future research", in: *Environmental Science and Policy*, 92: 141–151.

Primmer, E.; Jokinen, P.; Blicharska, M.; Barton, D. N.; Bugter, R.; Potschin, M., 2015: "Governance of Ecosystem Services: A framework for empirical analysis", in: *Ecosystem Services, 16*: 158–166.

Purdy, J., 2015: *After Nature: A Politics for the Anthropocene* (Cambridge, Mass.: Harvard University Press).

Rabe, B. G., [10]2018: "Racing to the Top, the Bottom, or the Middle of the Pack? The Evolving State Government Role in Environmental Protection", in: Vig, N.J.; Kraft, M.E. (Eds.): *Environmental policy. New directions for the twenty-first century* (Thousand Oaks: SAGE): 37–65.

Radkau, J., 2000: *Natur und Macht: Eine Weltgeschichte der Umwelt* (München: Beck).

Rajamani, L.; Oberthür, S.; Guilanpour, K., 2022: *Designing a Meaningful Global Stocktake*, C2ES Center for Climate and Energy Solutions, at: https://www.c2es.org/document/designing-a-meaningful-global-stocktake/.

Randers, J.; Bus, A., [2]2013: *2052: Der neue Bericht an den Club of Rome; eine globale Prognose für die nächsten 40 Jahre* (München: Oekom).

Raymond, M.; Neumeier, S., [2]2018a: "Human Security as a Global Public Good", in: Weiss, T.G.; Wilkinson, R. (Eds.): *International Organization and Global Governance* (Milton: Taylor and Francis): 574–586.

Renn, O.; Deuschle, J.; Jäger, A.; Weimer-Jehle, W., 2007: *Leitbild Nachhaltigkeit: Eine normativ-funktionale Konzeption und ihre Umsetzung* (Wiesbaden: VS Verlag für Sozialwissenschaften).

Renn, J.; Scherer, B. (Eds.), [2]2017: *Das Anthropozän: Zum Stand der Dinge* (Berlin: Matthes & Seitz).

Renn, O.; Schweizer, P.-J.; Dreyer, M.; Klinke, A., 2007: *Risiko: Über den gesellschaftlichen Umgang mit Unsicherheit* (München: oekom).

Renn, J., 2020a: *Surviving the Anthropocene*. Max Planck Research, No. 2 (Munich: Max Planck Society).

Renn, J., 2020b: *Überleben im Anthropozän*, Max Planck Forschung, No. 2 (München: Max-Planck-Gesellschaft).

Robinson, D.; Di Vittorio, A.; Alexander, P.; *et al.*, 2017: "Modelling feedbacks between human and natural processes in the land system", in: *Earth System Dynamics; at:* https://doi.org/10.5194/esd-2017-68.

Rockström, J.; Steffen, W.; Noone, K.; Persson, A.; Chapin, F. S.; Lambin, E. F.; *et al.*, 2009: "A safe operating space for humanity", in: *Nature, 461:* 472–475.

Rosol, C., 2021: "Finding common ground: The global Anthropocene Curriculum experiment", in: *The Anthropocene Review, 8*,3: 221–229.

Rothe, D.; Benner, A.-K., 2021: "Genealogies of the Anthropocene and How to Study Them", in: Chandler, D.; Müller, F.; Rothe, D. (Eds.): *International relations in the anthropocene. New agendas, new agencies and new approaches* (Cham: Palgrave Macmillan): 113–131.

Rothe, D.; Müller, F.; Chandler, D., 2021: "Introduction: International Relations in the Anthropocene", in: *Chandler, D.; Müller, F.; et al.: International Relations in the Anthropocene* (Cham: Palgrave): 1–16.

Rothe, D., 2017: "Global Security in a Posthuman Age? IR and the Anthropocene Challenge", in: Eroukhmanoff, C.; Harker, M. (Eds.): *Reflections on the Posthuman in International Relations. The Anthropocene, Security and Ecology* (Bristol, England: E-International Relations): 87–101.

Rounsevell, M. D. A.; Arneth, A.; Alexander, P.; Brown, D. G.; Noblet-Ducoudré, N. de; Ellis, E.; *et al.*, 2014: "Towards decision-based global land use models for improved understanding of the Earth system", in: *Earth System Dynamics, 5*, 1: 117–137.

Rucht, D.; Sommer, M., 2019: "Fridays for Future: Vom Phänomen Greta Thunberg, medialer Verkürzung und geschickter Mobilisierung: Zwischenbilanz eines Höhenflugs", in: *Internationale Politik, 74*,4: 121–125.

Schaefer Caniglia, B.; Brulle, R. J.; Szasz, A., 2015: "Civil Society, Social Movements, and Climate Change", in: Dunlap, R. E.; Brulle, R. J. (Eds.): *Climate change and society. Sociological perspectives* (New York, NY: Oxford University Press): 235–268.

Scheffran, J.; Vogt, W.R. (Eds.), 1998: *Kampf um die Natur: Umweltzerstörung und die Lösung ökologischer Konflikte* (Darmstadt: Wissenschaftliche Buchgesellschaft).

Scheffran, J.; Brzoska, M.; Brauch, H. G.; Link, P. M.; Schilling, J. (Eds.), 2012a: *Climate Change, Human Security and Violent Conflict: Challenges for Societal Stability (*Berlin, Heidelberg: Springer).

Scheffran, J.; Brzoska, M.; Kominek, J.; Link, P.M.; Schilling, J., 2012b: "Climate Change and Violent Conflict", in: *Science*, 336: 869–871.

Scheffran, J.; Balbo, A. L.; Rothe, D., 2020: "The Anthropocene: an opportunity for transdisciplinary and inclusive science?", in: Brzoska, M.; Scheffran, J. (Eds.): *Climate Change, Security Risks, and Violent Conflicts. Essays from Integrated Climate Research in Hamburg* (Hamburg: Hamburg University Press): 287-299.

Scheffran, J., 2016: "Governing the *Anthropocene: Complex Crises and Transitions to Sustainable Peace", in: Sustainable Security*; at: https://sustainablesecurity.org/2016/10/11/governing-the-anthropocene-complex-crises-and-transitions-to-sustainable-peace/ (23 February 2018).

Scheffran, J., 2020: "Climate Change and Weather Extremes as Risk Multipliers: Tipping points, cascading events, and societal instability", in: Brzoska, M.; Scheffran. J. (Eds.): *Climate Change, Security Risks, and Violent Conflicts. Essays from Integrated Climate Research in Hamburg* (Hamburg: Hamburg University Press): 19–48.

Scheffran, J. *et al.*, 2012: *Climate Change, Nuclear Risks and Nuclear Disarmament: From Security Threats to Sustainable Peace* (London: World Future Council).

Schellnhuber, H.-J., ^{2}2015: *Selbstverbrennung: Die fatale Dreiecksbeziehung zwischen Klima, Mensch und Kohlenstoff* (München: C. Bertelsmann).

Schlag, G.; Junk, J.; Daase, C. (Eds.), 2015: *Transformations of Security Studies: Dialogues, Diversity and Discipline (*London: Routledge).

Schmelzle, C., 2020: "Menschheit als Argument: Politisches Handeln angesichts existenzieller Risiken", in: *Zeitschrift für Politische Theorie*, 1: 45–58.

Schneidewind, U.; Singer-Brodowski, M., 2013: *Transformative Wissenschaft: Klimawandel im deutschen Wissenschafts- und Hochschulsystem (*Marburg: Metropolis-Verlag).

Scholte, J. A., 2018: "Civil Society and NGOs", in: *Weiss & Wilkinson (Eds.): International Organization and Global Governance (London/NewYork: Routledge):* 351–364.

Schroeder, M.; Wapner, P., ^{2}2018: "Non-Governmental Organizations", in: Weiss, T.G.; Daws, S. (Eds.): *The Oxford Handbook on the United Nations* (Oxford: Oxford Univ. Press): 346–357.

Schwindenhammer, S.; Breitmeier, H.; Kirf, B., 2017: "Die Norm der Nachhaltigkeit im globalen Regimekomplex für Ernährung – anerkannt und doch umstritten", in: *Zeitschrift für Außen- und Sicherheitspolitik, 10*,3: 353–371; at: https://link.springer.com/article/10.1007/s12399-017-0655-8.

Seider, T., 2021: "Politische Bildung im Zeichen von Anthropozän und Klimakrise", in: Deichmann, C.; Partetzke, M. (Eds.): *Demokratie im Stresstest: Reaktionen von Politikdidaktik und politischer Bildung* (Wiesbaden: Springer Fachmedien): 31–52.

Semal, L., 2015: "Anthropocene, Catastrophism and Green Political Theory", in: Hamilton, C.; Bonneuil, C.; Gemenne, F. (Eds.): *The anthropocene and the global environmental crisis. Rethinking modernity in a new epoch* (Abingdon, Oxon: Routledge): 87–99.

Senghaas, D., 1969: *Abschreckung und Frieden: Studien zur Kritik organisierter Friedlosigkeit* (Frankfurt/M.: Europäische Verlagsanstalt (EVA)).

Simangan, D., 2020: "Where is the Anthropocene? IR in a new geological epoch", in: *International Affairs, 96*,1: 211–224.

Simonis, U.E., 1996: *Globale Umweltpolitik: Ansätze und Perspektiven. Meyers Forum, vol. 40 (*Mannheim: B.I.-Taschenbuchverlag).

Simonis, G., 2001: "Die TA-Landschaft in Deutschland - Potenziale reflexiver Techniksteuerung", in: Simonis, G.; Martinsen, R.; Saretzki; T. (Eds.): *Politische Vierteljahresschrift: Sonderheft 31/2000. Politik und Technik. Analysen zum Verhältnis von technologischem, politischen und staatlichen Wandel am Anfang des 21. Jahrhunderts // PVS; Zeitschrift der Deutschen Vereinigung für Politische Wissenschaft* (Wiesbaden: VS Verl. für Sozialwiss.): 425–456.

Simonis, G. (Ed.), 2013: *Lehrbuch. Konzepte und Verfahren der Technikfolgenabschätzung*, 1st ed. (Wiesbaden: Springer VS).

Simonis, G., 2017a: "Die USA", in: Simonis, G. (Ed.), *Handbuch globale Klimapolitik. UTB L (Large-Format): vol. 8672,* Paderborn: Ferdinand Schöningh): 334–376.

Simonis, G., 2017b: "Komplexe Governance—Governanceversagen?", in Simonis, G. (Ed.): *Handbuch globale Klimapolitik* (Paderborn: Ferdinand Schöningh): 497–526.

Simonis, G., 2017c: "Theoretischer Rahmen: Governance", in: Simonis, G. (Ed.): *Handbuch globale Klimapolitik* (Paderborn: Ferdinand Schöningh): 138–175.

Simonis, G., 2017d: "Verflochtene Handlungsebenen", in: Simonis, G. (Ed.): *Handbuch globale Klimapolitik* (Paderborn: Ferdinand Schöningh): 211–259.

Simonis, G., 2018: "Klimaprognose und politische Macht: Annäherung an ein komplexes Verhältnis", in: *Peripherie, 38*,152: 416–449.

Simonis, G., 2019: "Politikwissenschaft im Anthropozän", in: *Denkströme*, 21: 9–43.

Simonis, G., 2022: *Global Governance - Begriff, Institutionen, Analyse (*Wiesbaden: Springer VS.)

Sinn, H.-W., ^{2}2009: *Das grüne Paradoxon: Plädoyer für eine illusionsfreie Klimapolitik* (Berlin: Econ).

Sippl, C.; Scheuch, M., 2019a: "The Anthropocene as a Framework for Thinking about Educational Processes - A Project Outline", in: *Journal of Agricultural and Environmental Education Research, 1*, 107-119.

Sippl, C.; Scheuch, M., 2019b: "Das Anthropozän als Denkrahmen für Bildungsprozesse - eine Projektskizze", in: *Zeitschrift für agrar- und umweltpädagogische Forschung, 1*: 107–119.

Sippl, C.; Rauscher, E.; Scheuch, M. (Eds.), 2020: *Das Anthropozän lernen und lehren. Pädagogik für Niederösterreich, vol. 9* (Innsbruck/Wien: Studien Verlag).

SIPRI, 2022: *SIPRI Yearbook 2022: Armaments, Disarmament and International Security.* (Oxford: Oxford University Press).

Smith, G., 2021: *Can democracy safeguard the future? (*Cambridge: Polity Press).

Smith, D., 2021: "The security space in the Anthropocene epoch", in: Lövbrand, E.; Mobjörk, M. (Eds.): *Anthropocene (In)securitirs. Reflections on Collective Survival 50 Years After the*

Stockholm Conference. SIPRI Research Report, vol. 26 (Oxford: Oxford University Press): 67–82.

Son, H., 2015: "The history of Western futures studies: An exploration of the intellectual traditions and three-phase periodization", in: *Futures, 66*: 120–137; at: https://www.sciencedirect.com/science/article/pii/S0016328714002079.

Star, S.L., 2010: "This is not a boundary object: Reflections on the origin of a concept", in: Stavins, R. N.; Stowe, R. C. (Eds.), 2016: *The Paris Agreement and Beyond: International Climate Change Policy Post-2020* (*Cambridge, Mass:* Harvard University Press).

Steffen, W.; Broadgate, W.; Deutsch, L.; Gaffney, O.; Ludwig, C., 2014: "The trajectory of the Anthropocene: The Great Acceleration", in: *The Anthropocene Review, 2*,1: 81–98.

Steffen, W.; Leinfelder, R.; Zalasiewicz, J.; Waters, C. N.; Williams, M.; Summerhayes, C.; *et al.*, 2016: "Stratigraphic and Earth System approaches to defining the Anthropocene", in: *Earth's Future, 4*,8: 324–345.

Steffen, W.; Crutzen, P. J.; McNeill, J. R., 2007: "The Anthropocene: Are Humans Now Overwhelming the Great Forces of Nature", in: *AMBIO: A Journal of the Human Environment, 36*,8: 614–621.

Steffen, W.; Richardson, K.; Rockström, J.; Cornell, S.E.; Fetzer, I.; Bennett, E. M.; *et al.*, 2015: "Planetary boundaries: Guiding human development on a changing planet", in: *Science, 347*,6223, 1259855: 1–10.

Steffen, W.; *et al.* (Eds.), ^{2}2005: *Global Change and the Earth System: A Planet Under Pressure* (Berlin: Springer-Verlag).

Themann, D.; Brunnengräber, A., 2019: "The nuclear legacy in the Anthropocene: interrelations between nature, technology and society", in: Hickmann, T.; Partzsch, L.; Pattberg, P.H.; Weiland, S. (Eds.): *The anthropocene debate and political science. Routledge research in global environmental governance* (London - New York: Routledge): 182–199.

Töpfer, K., 2013: "Nachhaltigkeit im Anthropozän", in; *Nova Acta Leopoldina, NF, 117*,398: 31–40.

Tremmel, J., 2019: "The Anthropocene concept as a wake-up call for reforming democracy", in: Hickmann, T.; Partzsch, L.; Pattberg, P.H.; Weiland, S. (Eds.): *The anthropocene debate and political science. Routledge research in global environmental governance.* (London, New York: Routledge Taylor & Francis Group): 219–236.

Tremmel, J., 2022: "Der Übergang ins Anthropozän erfordert eine Weiterentwicklung der Demokratie", in: Gumbert, T.; Bohn, C.; Fuchs, D.; Lennartz, B.; Müller, C.J., (Eds.): *Demokratie und Nachhaltigkeit. Aktuelle Perspektiven auf ein komplexes Spannungsverhältnis* (Baden-Baden: Nomos Verlag): 193–215.

Trombetta, M. J., 2021: "Security in the Anthropocene", in: Chandler, D.; Müller, F.; Rothe, D. (Eds.): *International relations in the anthropocene. New agendas, new agencies and new approaches* (Cham: Palgrave Macmillan): 155–172.

Trump, D., 2017: "*Statement by President Trump on the Paris Climate Accord" (Washington, D.C.:* White House).

Turnheim, B.; Berkhout, F., 2016: *Report of comparative analysis of transition Pathways, dynamics and governance,* Seventh Framework Programme (Pathways project N° 603942).

UN General Assembly, 1992: *Rio Declaration on Environment and Development,* A/CONF.151/26, Vol. 1 (New York: United Nations).

UN, 2015: *The Millennium Development Goals Report 2015,* No. 15-04513 (New York: UN); at: https://www.un.org/millenniumgoals/2015_MDG_Report/pdf/MDG%202015%20rev%20(July%201).pdf (12 August 2021).

UNDP, 2020: *Human Development Report 2020: The Next Frontier: Human Development and the Anthropocene (*New York: UN Development Programme).

UNEP, 1972: *Stockholm Declaration: Declaration on the Human Environment (*New York: United Nations Conference on the Human Environment).

UNEP, 2019: *Global Environment Outlook 6 (*Nairobi: UNEP).

UNEP, 2021: *Emissions Gap Report 2021: The Heat Is On - A world of climate promises not yet delivered*, Executive Summary (Nairobi: UNEP).

UNFCCC, 2021a: *Nationally determined contributions under the Paris Agreement: Synthesis report by the Secretariat* (Bonn: UNFCCC).

UNFCCC, 2021b: *Yearbook of Global Climate Action 2021: Marrakech Partnerships For Global Climate Action*. (Bonn: UNFCCC Secretariat); at: file:///C:/Users/User/Downloads/Yearbook_GCA_2021%20(1).pdf (13 December 2021).

United Nations, 2012: *The Future We Want: Outcome document of the United Nations Conference on Sustainable Development* (New York: United Nations); at: https://sustainabledevelopment.un.org/content/documents/733FutureWeWant.pdf (3 January 2020).

Urry, J., 2016: *What is the Future?* (Cambridge: Polity).

van Munster, R., 2021: "The Nuclear Origins of the Anthropocene", in: Chandler, D.; Müller, F.; Rothe, D. (Eds.): *International relations in the anthropocene. New agendas, new agencies and new approaches* (Cham: Palgrave Macmillan): 59–77.

van Munster, R.; Sylvest, C., 2016: *Nuclear Realism: Global Political Thought during the Thermonulear Revolution* (Abingdon: Routledge).

van Vuuren, D.P.; Kriegler, E.; O'Neill, B.C.; Ebi, K.L.; Riahi, K.; Carter, T.R.; *et al.*, 2014: "A new scenario framework for Climate Change Research: Scenario matrix architecture", in: *Climatic Change, 122*,3: 373–386.

Verburg, P. H.; Dearing, J. A.; Dyke, J. G.; van der Leeuw, S.; Seitzinger, S.; Steffen, W.; Syvitski, J., 2016a: "Methods and approaches to modelling the Anthropocene", in: *Global Environmental Change, 39:* 328–340.

Verburg, R.; Selnes, T.; Verweij, P., 2016b: "Governing ecosystem services: National and local lessons from policy appraisal and implementation", in: *Ecosystem Services, 18:* 186–197.

Victor, D. G., 2014: "Climate policy: Ditch the 2°C warming goal", in: *Nature*, 514: 30–31.

Victor, D.G.; Lumkowsky, M.; Dannenberg, A., 2022: "Determining the credibility of commitments in international climate policy", in: *Nature Climate Change, 12*,9: 793–800.

Vig, N. J.; Kraft, M. E. (Eds.), [10]2018: *Environmental policy: New directions for the twenty-first century* (Thousand Oaks: SAGE).

Wang-Erlandsson, L.; Tobian, A.; van der Ent, R.J.; Fetzer, I.; te Wierik, S.; Porkka, M., *et al.*, 2022: "A planetary boundary for green water", in: *Nature Reviews Earth & Environment, 3*(6): 380–392; at: https://www.nature.com/articles/s43017-022-00287-8.

WBGU, 2011: *Welt im Wandel: Gesellschaftsvertrag für eine Große Transformation* (Berlin: WBGU).

Weber, M., 1919: *Politik als Beruf* (München/Leipzig: Duncker&Humblot).

Weiland, S.; Hickmann, T.; Lederer, M.; Jens Marquardt, J.; Schwindenhammer, S., 2021: "The 2030 Agenda for Sustainable Development: Transformative Change through the Sustainable Development Goals?", in: *Politics and Governance, 9*,1: 90–95.

Weiss, T.G.; Wilkinson, R. (Eds.), 2018: *International Organization and Global Governance* (London: Routledge).

Weiss, T.G.; Wilkinson, R., 2019: *Rethinking Global Governance* (Cambridge: Polity).

Weizsäcker, E.U. von; Wijkman, A., 2017: Wir sind dran. Club of Rome: Der große Bericht: Was wir ändern müssen, wenn wir bleiben wollen. Eine neue Aufklärung für eine volle Welt (Gütersloh: Gütersloher Verlagshaus).

Werle, R., 2007: "Pfadabhängigkeit", in: Benz, A.; Lütz, S.; Schimank, U.; Simonis, G. (Eds.): *Handbuch Governance. Theoretische Grundlagen und empirische Anwendungsfelder* (Wiesbaden: VS Verlag für Sozialwissenschaften): 119–131.

Westad, O. A., 2019: *Der Kalte Krieg: Eine Weltgeschichte* (Stuttgart: Klett-Cotta).

White House, 2022: *National Security Strategy*. (Washington DC: The White House); at: https://www.whitehouse.gov/wp-content/uploads/2022/11/8-November-Combined-PDF-for-Upload.pdf, 24 Nov. 2022.

Widerberg, O.; Pattberg, P., 2017: "Accountability Challenges in the Transnational Regime Complex for Climate Change", in: *Review of Policy Research, 34*,1: 68–87.

Will, F., 2021: *Evidenz für das Anthropozän: Wissensbildung und Aushandlungsprozesse an der Schnittstelle von Natur-, Geistes- und Sozialwissenschaften. Umwelt und Gesellschaft vol. 24* (Göttingen: Vandenhoeck & Ruprecht).

Willetts, P., 2011: *Non-Governmental Organizations in World Politics: The Construction of Global Governance* (Abingdon, UK: Routledge).

Wissenburg, M., 2017: "The Anthropocene and the body ecologic", in: Pattberg P.; Zelli, F. (Eds.): *Environmental Politics and Governance in the Anthropocene. Institutions and legitimacy in a complex world* (London, New York: Routledge): 15–30.

Wissenschaftsrat, 2015: *Zum wissenschaftspolitischen Diskurs über große Herausforderungen: Positionspapier* (Köln: Wissenschaftsrat).

World Commission on Environment and Development (WCED), 1987: *Our Common Future: Brundtland-Report* (Oxford: Oxford University Press).

Wyborn, C.A., 2015: "Connecting knowledge with action through coproductive capacities: Adaptive governance and connectivity conservation", in: *Ecology and Society*, 20,1: 11; at: https://ecologyandsociety.org/vol20/iss1/art11/.

Yanosek, K.; Victor, D. G., 2022: "How Big Business is Taking the Lead on Climate Change: Harnessing Private Capital for the Public Good", in: *Foreign Affairs* (3 February).

Young, O. R. (Ed.), 1997: *Global governance: Drawing insights from the environmental experience. Global environmental accords series* (Cambridge, Mass.: MIT Press).

Young, O. R., 2010: *Institutional Dynamics: Emergent Patterns in International Environmental Governance* (Cambridge, Mass: MIT Press).

Zalasiewicz, J.; Waters, C. N.; Williams, M.; Summerhayes, C. (Eds.), 2019: *The Anthropocene as a Geological Time Unit: A Guide to the Scientific Evidence and Current Debate* (Cambridge: Cambridge Univ. Press).

Zalasiewicz, J.; Steffen, W.; Leinfelder, R.; Williams, M.; Waters, C., 2017a: "Petrifying Earth Process: The Stratigraphic Imprint of Key Earth System Parameters in the Anthropocene", in: *Theory, Culture & Society, 34*,2-3: 83–104.

Zalasiewicz, J.; Waters, Colin N.; *et al.*, 2017b: "The Working Group on the Anthropocene: Summary of evidence and interim recommendations", in: *Anthropocene, 19:* 55–60.

Zalasiewicz, J.; Waters, C.N.; Ellis, E.C.; Head, M.J.; Vidas, D.; Steffen, W.; *et al.*, 2021: "The Anthropocene: Comparing Its Meaning in Geology (Chronostratigraphy) with Conceptual Approaches Arising in Other Disciplines: Review Article", in: *Earth's Future*, 9: 1–25.

Zalasiewicz, J., [2]2017: "Die Einstiegsfrage: Wann hat das Anthropozän begonnen?", in: Renn, J; Scherer, B. (Eds.): *Das Anthropozän. Zum Stand der Dinge* (Berlin: Matthes & Seitz): 160–180.

Zelli, F.; Pattberg, P., 2017: "Conclusions: Complexity, responsibility and urgency in the Anthropocene", in: Pattberg, P.; Zelli, F. (Eds.): *Environmental Politics and Governance in the Anthropocene. Institutions and legitimacy in a complex world* (London, New York: Routledge): 231–242.

Zilles, J.; Drewing, E.; Janik, J. (Eds.), 2022: "*Umkämpfte Zukunft: Zum Verhältnis von Nachhaltigkeit, Demokratie und Konflikt", in: Soziale Bewegung und Protest: vol. 7,* 1st ed. (Bielefeld: transcript).

Zürn, M., 2018a: "Contested Global Governance", in: *Global Policy*, ***9***,1:138–145.

Zürn, M., 2018b: A theory of global governance: Authority, legitimacy, and contestation (Oxford: Oxford University Press).

Part III
Integrative Geography, Governance, and Gender

Chapter 8
Planetary Boundaries, Polycrisis and Politics in the Anthropocene: Climate Pathways, Tipping Cascades and Transition to Sustainable Peace in Integrative Geography

Jürgen Scheffran

Abstract Building on conceptions of human-nature relationships and the discourse on the limits to growth since Malthus and Marx, this chapter investigates the impacts of the "great acceleration" in human development on planetary boundaries and the polycrisis in the Anthropocene. An Integrative Geography perspective connects geographical concepts of landscape and interspace with multiple limits and pathways in the complex interaction between global expansion, multiple crises and geopolitical conflicts challenging the existing world order, leading to interconnected instabilities, compounding risks and tipping cascades. Major areas of risk and conflict are highlighted, including environmental and resource conflicts, the climate-migration-conflict nexus, disasters and extreme events, the Corona crisis, the Russia-Ukraine war, impacts of violent conflict on biodiversity and ecocides. Future trends challenging the world order include the transformation of fossil capitalism, power shifts in the North-South relationship, the influence of social networks and civil society on democracy. To address the multiple challenges, humanity can enforce a transformation to sustainable peace, merging solution concepts and synergies to stabilize expansive human development within the available environmental space in the common house of planet earth. To balance human needs and available natural resources, an efficient, sufficient and fair use and distribution is required. A key question is whether a transition can be achieved mainly by technical innovations within the existing capitalist economy or requires societal innovations and a fundamental system change of fossil capitalism towards a new socio-economic and political

Jürgen Scheffran, Professor in Integrative Geography (retired), Chair of the Research Group Climate Change and Security (CLISEC), Institute of Geography, University of Hamburg, Germany; Email: juergen.scheffran@uni-hamburg.de.

J. Scheffran (✉)
Research Group Climate Change and Security (CLISEC), Institute of Geography, University of Hamburg, Hamburg, Germany
e-mail: juergen.scheffran@uni-hamburg.de

H. G. Brauch (ed.), *Towards Rethinking Politics, Policy and Polity in the Anthropocene*, The Anthropocene: Politik—Economics—Society—Science 35,
https://doi.org/10.1007/978-3-031-71807-6_8

world order. Possible futures are shaped by critical thresholds between pathways of disruption and construction, conflict and cooperation, war logic and peace logic, risk and resilience, exclusion and coexistence, identity and diversity, tradeoffs and synergies. Adaptive and anticipative governance integrates social-ecological transformation and conflict transformation, sustainable and environmental peacebuilding, climate mitigation and adaptation, conflict-sensitive and resilient climate policies, climate justice and climate matching in North-South climate cooperation and the global energy transition.

Keywords Climate-conflict-migration nexus · Ecocide · Human-nature relations · Integrative geography · Limits to anthropocene · Planetary boundaries · Polycrisis · Sustainable peace · Transformation · World order

8.1 Introduction

Since the agricultural and industrial revolutions, humanity has increasingly shaped the face of the Earth. Crutzen/Stoermer (2000) coined the term 'Anthropocene' for the current human-made geological age following the Holocene, leading to a permanent change of the geological record resulting from actions of human beings (Brauch 2021b; Balbo et al. 2020). This adds a temporal dimension to the spatial dimension of 'Anthroposphere', i.e. the geographical space where the Earth's resources are used for the service to humanity while associated consequences, crises and conflicts of this endeavour have long played only a minor role. With the Great Acceleration (McNeill/Engelke 2016) of human development reaching limits to growth and planetary boundaries they turn into disruptive forces that can undermine the existential living conditions of humanity, as we are currently experiencing in the multicrises world, challenging the current Western-dominated world order which started with colonial expansion from Europe and the industrial revolution in Great Britain creating a global empire (Scheffran 2023b).

As the world wages ever new defensive battles against crises and disasters, from climate change to the Corona crisis, what room for manoeuvre remains to avoid or manage future crises? A broad spectrum of possible futures is conceivable, ranging from a collapse of human civilization to a profound transformation of the economy and society and a reshaping of human-nature relationships. As an alternative, humanity can enforce a transformation to sustainable development which is an attempt to steer expansive human development into orderly and stabilising channels. The available environmental space is to be used in a planned and sensible way and at the same time an acceptable life is to be guaranteed for all inhabitants in the common house of the earth in the long run. In addition to an efficient and fair use and distribution of resources, it is also a matter of adapting human needs to the available possibilities and of reorganising human coexistence in harmony with nature.

To address the challenges a key question here is whether the transition into a new world order can be achieved mainly by technical innovations and adjustments within the existing capitalist economic system and the liberal political world order, or whether it requires a fundamental system change in the underlying mechanisms of fossil capitalism towards a new socio-economic and political world order? Which pathway is still possible depends on whether humanity has already reached or passed critical thresholds beyond which destabilising tendencies undermine or disrupt the conditions of human existence. Can we expect more negative effects at the friction points of natural and societal dynamics, such as environmental degradation and waste pollution, disease and death, hunger and poverty, violence and conflict? Or will humanity find new pathways out of the vicious circles of problems and crises towards virtuous circles of solutions and synergies for resilience, sustainable development and peace? While these questions are in different ways facing normative challenges for policies, politics and polities, finding innovative solutions requires integrative and preventive governance approaches beyond crisis management.

It would be counterproductive if crisis management and associated conflicts continue to undermine the foundations for sustainable problem-solving in the context of a socio-ecological transformation, a neglected issue since three decades. The implementation of the *Sustainable Development Goals* (SDGs) can generates synergies but also trade-offs, on the one hand due to conflicting goals, and on the other hand due to tensions between the actors involved. Each policy measure has not only the intended outcomes but also side-effects and consequences, which may appear as positive or negative from different perspectives. This indicates the growing ambivalence of our increasingly complex world where nothing seems to be certain anymore (Scheffran 2018a; Dörre 2022). This raises the question how tensions between different policy fields can be reduced or constructively managed, in particular by making a social-ecological transformation and a conflict transformation compatible and mutually enhancing.

Addressing ambivalence and complexity of multiple crises needs novel contributions by interdisciplinary research and transdisciplinary science, in particular geographical science which traditionally studies human-nature interactions through physical and human geography. Increasingly the field of 'integrative geography' is emerging which explores the spatial and temporal interactions between the natural environment and human society. Terms like space and landscape, limit and boundary, connection and pathway, transition and transformation, Anthroposphere and Anthropocene are key concepts of an integrative geography that have in common that they investigate the main objects, subjects, systems, terms and statement as well as pathways, linkages and interactions between them. Going further is to study what happens in the 'interspace' between two different terms, which both connects and separates them, and thus is framed by both terms. Such connecting conceptions will serve as an overarching umbrella in the following, including human-nature relationships and scientific ambivalence, geographical concepts and debates on limits to growth, the origins of political economy, the Club of Rome's limits to growth and planetary boundaries. Current risks, crises and geopolitical challenges focus on interactions between climate change and conflict, including issues of biodiversity loss, ecocides and environmental migration. Integrative concepts and options in related interspaces provide opportunities for selecting appropriate policies, political governance measures and institutional polities to avoid negative tipping points and utilise positive tipping mechanisms for climate policy, cooperative governance, conflict transformation and sustainable peace.[1]

8.2 Conceptions of Human-Nature Relationships

The relationship between natural systems and human societies is affected by various experiences, perceptions, understandings, attitudes, practices and interactions which are represented by the following conceptions of nature (Scheffran 2000b).

8.2.1 Nature as a Habitat of Human Evolution, Experience and Knowledge

In the history of philosophy, the idea of a given nature is connected with the idea of an origin not influenced by humans. Human evolution emerged from and adapted to natural conditions, within the constraints and opportunities of natural laws, physical space and time, and the availability of natural resources. Like for other biological organisms, the human body is endowed with natural abilities such as sensory organs to feel, hear and observe the environment, hand and feet to physically interact with it, as well as a brain to connect perceptions, experiences and recognitions. Knowledge, rules and relationships drawn from this interaction form an internal model about the world and its intersection with the body, guiding human behaviour and communication with the natural and social environment expressed in an adequate language. States and changes in the environment are classified and evaluated positively or negatively, as desirable and undesirable, if not ignored as irrelevant to reduce complexity. The environment contributes natural resources vital for human life, survival and basic needs like material substances, water, energy and information gathered and extracted from nature, including ecosystems with their plants and animals. In contrast, other conditions in nature are evaluated as dangerous, from certain wild plants and animals to natural disasters, poisonous substances and lack of essential resources.

[1] This chapter integrates and synergises four decades of research by the author on the linkages between conflict, climate and complexity, to provide insights of synergistic knowledge integration, merging historic developments and recent events, demonstrating how future-oriented research expected todays polycris long ago. Part of the work was written in German language which is made here accessible to the English speaking community.

8.2.2 Nature as an Object of Human Construction, Action and Exploitation

The embedding of humans into nature and its resources provides possibilities for actions to change the environment and construct it according to human imagination and capabilities based on science and technology, to better meet human desires and purposes. Increasingly the world is created by human actions, where needs and desires are based on social conditions and human livelihoods, economic production and consumption as well as human waste and pollution degrading the natural environment. With more successful learning and practices in human-nature interaction, opportunities grow for systematic influence and purposeful intervention to use natural processes enshrined in scientific knowledge, technical tools and experience-based innovations to control, appropriate and render natural forces and services in order to enhance desired effects and reduce harmful ones. New capabilities improve the conditions of existence and find action pathways with a more effective and efficient relationship connecting ends and means, allowing for expansion even against resistance (Schäfer 1995).

Descartes imagined nature as a deterministic mechanical system contrary to the free will of human subject, and Kant distinguished nature from the freedom of human beings, following theoretical and practical reasoning. Exploring nature as an object of measurable experience, the experimental natural sciences systematically developed instruments for technical-industrial production (Francis Bacon) to increase health, wealth and material well-being. Science and technology designed a machinery exploiting natural resources to convert them into economic capital, political power, military force and social domination, driving economic production and consumption to maximise utility. Human interactions with the natural environment impose a wide range of physical experiences and sensory impressions on human bodies, feeding the human brain with an internal representation of how 'nature works'. Mental representations of nature as if-then relationships replicate what is possible, serving as a basis for human behaviour.

8.2.3 Nature in Need of Conservation and Its Sustainable Use

Contrary to the idea of unlimited scientific progress and technical omnipotence, natural resources and environmental boundaries constrain human expansion, living space and ecological footprint on earth. Taking resource consumption, physiological metabolism and ecosystem functions into account, human-nature relationship has transformed over time. Nature is no longer regarded merely as an object of exploration and an unlimited reservoir for resources and waste. Once considered inexhaustible and self-reproducing without limits, it is now experienced as vulnerable, as a precious commodity that must be preserved through guardrails. The biosphere is seen as a living network and a community of organisms, including *Homo Sapiens*, whose interventions must be considered in terms of their consequences. Depending on nature, human societies, infrastructures and lifecycles are also vulnerable and need to be sustainably embedded in the natural cycles (Johnke et al. 2004).

8.2.4 Co-evolution in Human-Nature Interaction

Human and natural spheres are separated by boundaries and interconnected through multiple pathways of exchange. Human-nature interactions range from natural constraints of human action to human control over the natural environment and its resources needed for human life and evolution. One challenge is to handle the complexity of this interaction and find a sustainable balance between natural and human systems. While assessments of human-nature interaction differ between disciplinary natural science and social science perspectives, the focus is here on interdisciplinary approaches exploring the middle-ground between disciplines. As humans are also grown up in a social environment, they can learn from others through observation, education and narratives to adapt and follow beliefs and rules guiding human action, in particular towards the natural environment. Human perceptions, imaginations and desires as well as future expectations and assessments of consequences are driving the process of human creation, innovation and intervention.

Human ability to work, extended and multiplied by technology expanding the action space of the human body, continuously shapes human-nature relationships. Individual and collective actions transform physical spaces, from envisioned to constructed landscapes that meet human needs and frame natural landscapes as 'nature by design' which however is an Oxymoron as nature disappears in the process of construction and becomes a cultural artifact. Apparently neither a natural nor a social science perspective alone can explain the complexity of human-nature interaction. Reductionist approaches, such as geo-determinism or human framings of nature, cannot explain the complex dynamics of one system mutually affecting the other. The challenge is to develop co-evolutionary conceptions of mutual and reciprocal transformations between both systems towards a sustainable balance of nature and society, taking into consideration the different time scales as social evolution usually can occur more rapidly than natural evolution. This calls for an integrative science of human-nature interactions to understand the stabilizing and destabilizing conditions of this relationship.

8.2.5 Mixed Methods in Human-Nature Interactions

To investigate complex human-nature interactions, mixed methods offer a toolbox of integrative approaches. Systems theory and complexity science analyse non-linear dynamics, complex adaptive and self-organising systems, game theory, agent-based modelling and social network analysis to study complex social interactions

(Scheffran 2006). Regarding the presentation and analysis of data widely used are remote sensing and geographic information systems (GIS). Integrative research methods such as interviews und focus groups, qualitative, participatory and agent-based modelling provide insights about stakeholder perceptions, assessments, actions and interactions (Bendor/Scheffran 2019). With mixed methods it is possible to frame geographic conceptions such as landscapes, interfaces, borders and diffusion processes from a natural and social science perspective. Models and simulations can help to understand how individual components become an integrated whole through self-organisation and synergies, although they represent only projections and not the full complexity of human-nature interactions (Scheffran 2016). The question is whether discrepancies, simplifications and uncertainties between the model and real world can lead to misleading projections about the future.

8.3 The Framework of Integrative Geography

8.3.1 What Is Integrative Geography?

Integrative geography is not a new discipline but an integrative approach within the discipline of geography which is inherently interdisciplinary through building bridges between physical and human geography and connections to other disciplines, such as energy geography, water geography, climate geography, economic geography, political geography, social geography, historical geography, to mention a few. This offers the opportunity to connect to other disciplines and research fields and merge knowledge to jointly address key challenges, raising the question what the core of geography is. In a widely held understanding geography is about the spatial dimensions of the world in general and science in particular, as indicated by terms like space, place, location, distance, contagion, etc. While integrative geography is not consistently defined,[2] it can be further developed. Many key terms and challenges

[2] Some indications can be found in the description of the research area of Integrative Geography at Universität Hamburg: "Linking the areas of physical and human geography as well as re-integrating both areas into research and teaching practice, as has often been discussed in the past, are the goals of this section. We focus on the interactions between humans and nature and the interdependency between the environment and society. As a hub between disciplines, we deal with several different topics. Some of these are resource conflicts and resource management, sustainable development of regions, livelihood and vulnerability concepts, security risks and conflict potential of climate change and energy systems, natural hazards and the conservation of nature, risk perception and management, power and dominance relationships in the context of nature occupation processes, as well as discourses on the perceptions, problem constellations, and solutions of ecological crises. Among other topics, our research focuses on empirical field studies, especially in coastal areas and on islands, along rivers and in mountain regions, in rural peripheral regions and the southern hemisphere. On one hand, the theoretical background is drawn from general system theory, specifically from complexity theory. At its core is the pursuit of the analysis and modelling of non-linear dynamic systems, such as integrated land, water, and energy use, or the causes, effects, and strategies regarding climate change. On the other hand, political and socio-ecological approaches are

in an ever complex world can be explained and embedded within an open framing of 'Integrative Geography'.

While some argue that geography is per se integrated, going back to the integrated perspectives of Alexander von Humboldt, others complain about a divide within geography between the natural science approaches of physical geography and the social science perspectives in human geography, which can be seen from the fact that geography belongs to different faculties at universities. Over the past decade integrative geography has been not so much a unifying concept but was occasionally considered as an opportunity respecting the diversity within geography and with other disciplines. While large multi-volume encyclopaedias have no entry on integrative geography (besides few mentions in the text), at least the term has found its way into *Wikipedia*, where it is also called integrated geography, environmental geography, human–environment geography (Rupke 2008). It is described as a third branch of geography "where the branches of human geography and physical geography overlap to describe and explain the spatial aspects of interactions between human individuals or societies and their natural environment" (Castree 2009; Garcia 2010; Moseley et al. 2014).

Integrative geography plays a role in exploring the interactions in coupled human–environment systems that form complex spatial structures in geographical landscapes by the flows of energy, matter and information and analysing human communications, interventions and feedbacks within and between these spaces. A common attribute of integrative geography is that it often studies intermediate phenomena, as well as transitions and transformations, between established concepts, such as physical and human geography, natural environment and human society, spatial and temporal processes, environment and conflict, conflict and cooperation, war and peace, environmental destruction and sustainable development, Anthroposphere and Anthropocene. This implies that not only these concepts and their specific aspects matter, but also the intermediate zone between them which separates and connects them. In a geographic terminology it can be called 'interspace' (in German *Zwischenraum*) which is the space or transition zone between structures, whether they are physical, biological, economic, social, political or explicitly geographical. Studying the dividing and interconnecting zones between these structures explicitly allows understanding the processes driving and blocking exchange and transition between them, including the discourses and dialogues of communication.

The concept of interspace may also draw upon and potentially enrich the philosophical foundations of logic. Philosophy has for a long time focused on dyads, pairwise relations between opposing terms, such as true-false, subject-object, large-small, finite-infinite, dry-humid, we-they, centre-periphery, or Ying–Yang. There have been many attempts to go beyond binary logic, from dialectic logic (thesis – anti-thesis – synthesis) to overcome contradictions and develop synergies between

an important part of the topics mentioned above. Post-colonial and post-structuralist perspectives complete this field of work." See: https://www.geo.uni-hamburg.de/en/geographie/abteilungen/integrative-geographie.html.

two opposing pairs, to fuzzy logic bridging the 0–1 dichotomy with a spectrum of numbers in between.

A new way of trialectical thinking was proposed by Henri Lefebvre (1981), assuming that there is something between two terms that is relevant. After his death in 1991, this concept was further developed and applied to space by Edward Soja (1996: 70) who argues that "the third term never stands alone, totally separate from its precedents or given absolute precedence on its own." Moving from binary to dialectic and trialectic thinking, it might be worthwhile to also consider 'interlectic' thinking to explore the interspace between multiple things and thus better understand the complexity of their relationships, which is becoming common in the 'nexus' approach studying interactions between multiple phenomena such as climate-conflict-migration or water-energy-food. This might also allow to consider the pathways and transitions between systems characterised by qualitative change in their attributes. Drawing on this philosophical embedding, conceptions of integrative geography are discussed in the following.

8.3.2 Space, Time and Landscapes

Cutting short the extensive literature on space in philosophy and science, space is considered here as a multi-dimensional frame where different things are located and moving across distances, changing their relative positions to each other at varying places. Changing relations in space occur as sequences in time, thus leading to shifting relational constellations of closeness and distance, location and place, spatiality and temporality. Within and against this framing different conceptions of space are highlighted: three-dimensional physical space in Newtonian mechanics based on a fixed grid of coordinates without relation to anything external; absolute space and time as an *a priori* form (Kant); relative space of location of things and network of places where motion is ascribed to a body (Leibniz); four-dimensional space-time continuum (Einstein); terrain where moving takes effort against friction and resistance, finding locations, places and pathways through mapping and navigation (geography); hot spots of climate and ecological change (environment); land allocated to property and ownership, production and consumption (economics); landscapes of colonialism, slavery and globalisation (history); territory for power projections and conflict at regional, national and international levels (politics); imaginary and behavioural spaces (psychology, sociology) (Thrift 2017).

Lefebvre (1991) describes space as a social product while overlapping social processes produce space in which the production of commodities and accumulation of capital takes place (Lukasz 2011). Lefebvre distinguishes three types of spaces within the trialectics of spatiality: objective space of spatial practices and actions in the real world; conceived representations of space formed in mind and not bounded by objective space; and lived space of representations resulting from objective and conceived spaces combined (imagination meets reality). Soja (1996) addresses how humans understand space, as material/physical or represented/imagined, and introduces the concept of "third space" corresponding to Lefebvre's "lived space" to account for the complex ways in which humans inhabit, experience and understand the world where the spatial dimension plays a role. This is also the case in many other conceptions of space in the capitalist world, for instance highlighting that changes in the modes of production and consumption of capital interact with developments in transportation and technology, affecting the perception and compression of time, space and distance (Harvey 1990).

A related concept is 'landscape' representing geographic spaces shaped and grown over long historic times by interactions of natural and human forces, processes and practices acting upon earth. Landscapes are a living synthesis of people and place, land and landforms. These include physical elements like mountains, hills, water bodies such as rivers, lakes, ponds and the sea, living elements of vegetation, human patterns including different forms of land use, buildings, and structures, and transitory elements such as light and weather conditions.

8.3.3 Interspaces, Boundaries and Pathways

Geographic spaces and landscapes are characterised by structuring features and patterns, separated by interspaces which include and exclude, connect and disconnect, integrate and divide, separate and join human and natural spheres. The German term 'Grenze' represents a wide range of related terms in English language, each with specific meanings in different contexts: border, barrier, bound/boundary, front/frontier, limit, contact, division, edge, interface, periphery/perimeter, guardrail, restriction, constraint, demarcation, confinement, threshold. Some of these terms are more exclusive (divisive), others more inclusive (connecting) (for selected definitions see Table 8.1).

Geographical boundaries and borders divide geographical spaces into historically grown units (such as nations) along which interactions, transformed by physical and anthropogenic drivers that determine the spatial forms and patterns. All systems have limits in which they operate and are viable. Value limits are defined by what is desirable and acceptable. Action limits are given by feasibility and capability in reaching targets or by inacceptable risk associated with pursuing certain actions. Steering a system needs to respect boundaries of stability and viability beyond which tipping to qualitatively different state can occur. There are static and dynamic interspaces which is a matter of degree. For instance a mountain may appear as static although

Table 8.1 Definitions of key terms related to interspace

Term	Definition
Barricade barrier	*Barricade*: to prevent someone or something from going somewhere *Barrier*: prevents people from doing something (trade barrier, language barrier, education barrier) • Fence that keeps people or things separate, or prevents people from entering a place • Physical object separating two areas, groups of people, etc. (mountain barrier between countries)
Border	Official line separating two countries, states, or areas; • Zone along each side of the boundary between states (synonymous with frontier)
Bound boundary	*Bound*: → constrained to do something (legal or moral duty or law to provide safety, cannot do what you want) → Constrained by something such as a wall, river, road, etc. (valley bounded by high mountains) → Legal or social limit or rule (e.g. bounds to imagination or law) *Boundary*: line marking the edge of a surface, space or land inside a country, e.g. river as boundary → Actual line of demarcation between two countries → Limit of acceptance or possibility (e.g. human knowledge) *Boundless*: without any limits or end
Edge	*Edge*: part of something that is farthest from the centre (e.g. the edge of a table, bed, street) → Edge over: advantage in competition, game, or fight (edge over/on competitors) → On edge: feel nervous over something bad expected: → On the edge of something: point at which something different will happen (e.g. collapse) → Cutting edge: highly developed and advanced → Curb: edge of the sidewalk, area beside a steep slope (e.g. cliff)
Frontier, front	*Frontier*: "narrow zone that 'fronts' or faces a neighbouring country" → Thinly populated pioneer zone in North American history on the margins of settled lands in the westward expansion of population and settlement (also applied to Siberia, Australian 'outback', harsh and largely unsettled environments) → Area beyond the places that people know well → Limit of what is known about something (frontiers of science) *Military front*: line of contact of two opposing forces; direction of the enemy or military unit; home front: places affected by fighting
Interface	• Point or junctions where two things link, connect or interact. • Device or system that unrelated entities use to interact (e.g. remote control of TV) or translate (e.g. language translator or protocol of behaviour)
Interspace	• *Space* or *interval* between two things https://en.wiktionary.org/wiki/interspace

(continued)

Table 8.1 (continued)

Term	Definition
Limit limitation	*Limit*: greatest or least amount, number, etc. that is allowed or is possible (speed limit, limited time, financial limit) → Furthest point or edge of a place that must not be passed (city limits) → Off limits: beyond the area allowed to go → Acceptable within limits: within time, level, amount, etc. → Stop an amount or number from increasing beyond a particular point (e.g. class size) → Allow a particular amount of something (e.g. alcohol, damage, emissions) *Limitation*: act or process of controlling or reducing something (e.g. treaty limiting nuclear weapons, risky technology)
Other terms	*Perimeter*: border around an area of land, area or shape (fence around school perimeter) *Rim*: outside edge of something (e.g. rim of the cup) *Margin*: empty space at the side of a printed page *Guardrail*: crash barrier, safety fence *Crisis*: Decisive or critical moment; turning-point for better or worse

Sources Witherick et al. (2001), Longman (2009) and Merriam (2007)

it has been (trans-)formed over geologic times of millions of years (sometimes with rapid eruptions), while rivers and ocean coasts are more fluid and can change their form on a daily basis.

Complementary to boundaries are pathways which are routes and channels that allow to connect, move into selected directions, enable and govern metabolic or other regulatory processes downstream, through exchange of energy, matter and information along or across contact surfaces. Boundaries can block pathways and slow down expansion due to resistance, cost or risk (e.g. when limits to growth are reached), which in turn can induce actions to overcome boundaries or find new pathways through innovations. Which one succeeds depends on the forces, capabilities and powers applied by the systems in the respective interaction to overcome the boundary as well as the resistance to 'defend' it. Ultimately changing a boundary is a question which force, power or capital dominates and the efficiency of their use. A few examples are highlighted here, where the concept of geographical interspace is relevant across disciplines:

1. Rivers may separate countries but also connect different geographical, cultural and political regions which can lead to interdependencies favouring conflict or cooperation. As an example, the Nile river connects the tropical areas of Central Africa with the arid zones of North Africa. The water courses of the Himalaya glaciers are a source of water for vast areas of South Asia, East Asia and Central Asia and connect high mountains with coastal regions.
2. Coasts connect and separate land and sea, thus are being formed by dynamical processes and features of both elements. This is relevant for coastal communities which have adapted to the specific livelihood conditions along and across

land-sea interfaces, including weather, climate, food, labour, transportation and culture.

3. The atmosphere is the interspace between Earth's surface and outer space along a gradient of temperature, pressure, humidity and other variables important for weather and climate patterns. Its composition and dynamics are driven by physical exchange processes and anthropogenic activities using this medium for life support, transportation and deposition of polluting substances that affect local and global climate which in turn affect earth system dynamics and living conditions for all organisms specific to different climatic zones.
4. Peri-urban spaces have features of both rural and urban areas, separating and connecting them, e.g. through movement of people, resources, goods and information.
5. Multiple levels from micro to macro levels are connected through the meso level in between; similarly, local and global levels interact with the national level.
6. Migration is a change in place over an extended period and across borders, driven by motivation and capability along migratory pathways which are the interspaces between origin and destination.
7. Conflict and cooperation are connected through decision-making and negotiation processes, selecting pathways of action according to individual or collective motivations and capabilities. Between conflict and cooperation mixed cases are possible, e.g. cooperation in conflict or conflict resolution in cooperative situations.
8. In the climate-conflict-migration nexus multiple pathways of interactions are possible, depending on intensity of change in key variables and their sensitivity to change, leading to escalation for risk multipliers or de-escalation through synergies.
9. The nexus of water, food and energy can aggravate a mutual decline when losses of one resource trigger declines of others or do the opposite through win-win synergies.
10. The ambivalence of science and the dual-use of technology demonstrate that often there are no clear dividing lines between positive and negative applications while mixed cases make social control of the consequences difficult.
11. Connecting spatial and temporal dimensions, the continued expansion of the Anthroposphere transforms the Earth system into the Anthropocene, running into conflict with planetary boundaries which manifest as limits to growth.
12. National borders allow for the inclusion of the 'own' by exclusion of the 'other', used to justify claims of influence. Western expansion before and after the Cold War has triggered opposition and resistance, contributing to geopolitical conflict.
13. Increasing human footprints put pressure on the interspace with the biosphere, leading to losses in ecosystems and biodiversity and related diseases and pandemics like Covid-19.
14. Destabilizing system changes include collapse of ecological and social systems, as well as transition and transformation processes towards sustainability,

conflict resolution and sustainable peacebuilding, driven by complex mechanisms of compounding risk, tipping points and cascading domino effects.
15. In large spatial and temporal perspectives relevant is the position of planet Earth within the space environment, in particular the manifold interactions between global interspace and outer space, with spaceflight connecting both and the potential exploitation of resources in space.
16. More broadly, the concept of interspace allows to study various phenomena where the term 'inter' plays a role, such as international, interdisciplinary, intercommunal etc.

Some of these examples and their implications will be discussed in the following within the framework of integrative geography, in particular those relevant to political dimensions. Starting from historical debates on colonial expansion and the beginning of the Industrial Revolution, the following section will highlight limits and crises between nature and society leading to tensions as well as cooperation, with a view on Anthropocene boundaries, environmental conflicts, and conditions for a cooperative transition to sustainable development, linking natural science system levels and social science actor levels in an interdisciplinary way.

8.4 Limits to Human Expansion in the Anthropocene

8.4.1 Colonial Expansion and the Division of the World

The colonial expansion from Europe has been an inseparable part of the Western success story which became a driver of human growth in the Anthropocene. Following Columbus the 'New World' was conquered through armed forces, technological advances and Christian missionaries. Around the world, millions of people and huge amounts of resources were subjugated to European exploitation (Reinhard 2016). The oppressed were often exposed to terrible atrocities, their labour and resources exploited. A chain of genocides exterminated millions of indigenous peoples and their natural living conditions, by violent force, diseases, invasive species (Zimmerer 2020), facilitating European expansion on a largely depopulated continent. Slavery brought more than twelve million people from Africa alone to America to tap the wealth of the colonies.

The colonisation and exploitation of other continents contributed significantly to the development of Europe. By appropriating foreign wealth, the colonial powers were able to mobilise enormous resources (from gold to crops) and achieve comparative advantages that continue to have an impact until today. From colonial profits, European powers were able to create economic surplus and foster industrialisation, feed a growing population and finance wars that partially absorbed the surplus and reduced population (only to mention the vast population losses by the Thirty Years

War; Menzel 2015). Poverty, hunger, persecution, and war in Europe pushed millions of people in several waves of emigration to leave the overcrowded old continent, externalising the failed politics of European powers (Scheffran 2021).

With the Industrial Revolution and the use of domestic coal, Great Britain was able to extend its lead and establish a world empire. Its dominant role was partly taken over by the USA as a hegemonic power in the 20th century. Germany found itself disadvantaged in the acquisition of colonies and tried in vain to shift the power game to its favour in both World Wars. Following the Russian colonial expansion since Tsar Peter the Great over vast parts of the Eurasian continent, with the October Revolution of 1917 these territories were integrated into the Soviet Union, providing a powerful counterweight to the West. After World War II, the United States was able to expand its power in the East-West conflict and the ideological battle between capitalism and communism which led to the nuclear arms race and near extinction of humanity through nuclear war. Even after that, violence remained a means of politics for former colonial powers, preserving neo-colonial structures until today (Scheffran 2021).

These support the prosperity of the rich countries at the expense of poorer regions, which are suffering from deprivation, debt, resource exploitation, low-wages, precarious living conditions, climate change and environmental degradation. They are caught in a poverty spiral of low income and productivity. This shapes perceptions in the Global South today, as does the collective memory of colonial times. Although the countries there never had the development opportunities of the colonial powers and were victims of foreign invasions, they were blamed for their weaknesses (hunger, poverty, conflicts, lack of democracy, corrupt elites, low efficiency, etc.). This explains the differences between North and South but does not justify failed politics by regimes in the Global South. Starting from colonialism and driven by the growth dynamics of the industrial revolution, the Western development model, with its accumulation of growth, power and violence, has unfolded an unprecedented expansion encompassing a globe that is full and reaching limits to its further expansion. Going beyond general conceptions of the Anthropocene, some actors and governments are more responsible for the deterioration of the Earth than others.

8.4.2 Roots of the Growth Debate: Smith, Malthus, Marx

We can learn from roots of the debate about the limits to growth dating back to the 18th century, when the Industrial Revolution began in England, fuelled by the exploitation of cheap coal resources driving long-distance transportation and mechanised mass production which secured England an enormous lead over other European countries. Until then, besides little wind and hydropower, energy supply was based primarily on biomass, which reached limits due to heavy deforestation and agricultural land constraints for a growing population. While there were about one billion people on the planet at the turn to the 19th century, exponential population and economic growth became apparent to political economists who thought about the consequences.

At first possible natural limits only played a marginal role in emerging capitalism. Adam Smith (1723–1790) assumed that resource scarcity would be regulated by markets and trades. In his famous treatise *The Wealth of Nations* he assumed that non-renewable raw materials could be exploited without being limited by geographic location or the yield of the own mines due to international transport possibilities (Smith 1905 [1776]: 325f.): "The quantity of the precious metals which is to be found in any country, is not limited by anything in its local situation, such as the fertility or barrenness of its own mines." Preempting globalisation, he assumed "the easy and cheap transportation of those metals, of their small bulk and great value." In contrast, Smith recognised the scarcity of renewable resources such as fish when fishery increased which revealed insights about the relationship between nature and economics relevant until today. While fish production is uncertain and limited by geographic conditions and the fertility of the fish population, demand and price in the fish markets grow with human population and economic production: "The fish must generally be sought for at a greater distance, larger vessels must be employed, and more expensive machinery of every kind made use of." Thus, as scarcity increases, it becomes more difficult to meet the demand, requiring more effort (labour) and increasing fishing costs (cf. *ibid.*), although technical tools can increase catch efficiency. The process continues until further resource output is no longer worth the input and additional efforts cannot be justified when the resource base is severely destroyed, requiring a transition of a new one.

Smith also grappled with the biological and economic bases of population growth. He believed that the scarcity of food would place barriers to further human increase, while abundant rewards for labour would provide for more children and expand the barriers: "The demand for men, like that for any other commodity, necessarily regulates the production of men, quickens it when it goes on too slowly, and stops it when it advances too fast" (Smith 1776: 111). Thus, the increase of a population would be naturally proportionate to the means of its subsistence.

The relationship between population and resources was further pursued by Thomas Robert Malthus (1766–1834), British pastor and political economist. In his work *An Essay on the Principle of Population* (Malthus 1798), he expected population growth to exceed resource availability in the foreseeable future. He postulated that the human population would increase in exponential steps (geometrically) (1, 2, 4, 8, 16 …), while food production would increase only linearly (arithmetically) (1, 2, 3, 4, 5 …). The result is a disproportion between the number of people and the amount of food, and "unless an emigration takes place, the deaths will shortly exceed the births", leading to a deterioration of living conditions of the poorer masses:

> The vices of mankind are active and able ministers of depopulation. They are the precursors in the great army of destruction; and often finish the dreadful work themselves. But should they fail in this war of extermination, sickly seasons, epidemics, pestilence, and plague, advance in terrific array, and sweep off their thousands and ten thousands. Should success be still incomplete, gigantic inevitable famine stalks in the rear, and with one mighty blow levels the population with the food of the world." (Malthus 1798: Ch. VII, p. 61 [1])

Facing these challenges, Malthus assumed that all social utopias ultimately must fail because of the insurmountable law of population. Even if the wealth of a nation increased, this would not benefit the masses, who lived in poverty despite hard work and struggled daily to survive. In his religious assumptions, he rejected contraception and abortion and advocated abstinence and continence. The plight of the working class could not be improved in the long run by poor government laws because the amount of food was limited and led to competition. Prosperity for all would allow the population to grow until hardship put an end to the ideal state. Accordingly, Malthus was critical not only of Adam Smith's promise of wealth, but also of England's intellectuals influenced by Enlightenment, whom he saw as threatening England's social order. In turn they rejected his assumptions and accused him of pessimism or lack of science, while some of the classical economists (such as David Ricardo or John Stuart Mill) largely agreed with Malthus's population law.

Charles Darwin also referred to Malthus' struggle for existence in his theory of evolution of natural selection. Intertwining social and natural science theories (see Remoff 2016), Darwin recognised the potential of geometric (exponential) growth for all organisms and noted that growing populations colonise new spaces where the best adapted survive, which opened up the possibility of satisfying humanity's growing demand for resources compatible with the living world, an early account of the concept of sustainability.

Much of the work of Karl Marx (1818–1883) was critical of Malthus. He considered him an apologist for capitalism and condemned the theory of overpopulation as reactionary, since it justified the misery of the working class and the ethics of abstinence primarily affected the poorest. What mattered for Marx was not the number of people, but the productive forces and the socio-economic conditions of the working class, which could very well be improved. In *Theories of Surplus Value*, Marx, however, conceded: "As compared to the wretched bourgeois economists who preach harmony, Malthus's only merit lies in his pointed emphasis on the disharmonies" (Marx 1919: 110). In his critique of Malthus, Marx also referred to James Anderson of Hermiston (1739–1808), who questioned the population theory and said explicitly that the land is capable of increasing and perennial improvement, because he assumed that the "soil can be continuously improved by chemical influences and cultivation" and "under a judicious system of management, that productiveness may be made to augment, from year to year, for a succession of time to which no limits can be assigned" (Anderson 1801: 35–36). Moreover, Marx referred to Darwin's 'geometrical' growth of life, which opened up the possibility of organically satisfying humanity's growing resource needs. Friedrich Engels, however, countered a supposed boundlessness with a warning in his *Dialectics of Nature*:

> Let us not flatter ourselves too much with our human victories over nature. For every such victory she takes her revenge on us. […] Thus at every step we are reminded that we by no means rule over nature like a conqueror over a foreign people, like someone standing outside nature—but that we, with flesh, blood and brain, belong to nature, and exist in its midst, and that all our mastery of it consists in the fact that we have the advantage over all other creatures of being able to learn its laws and apply them correctly […]. (Engels 1975 [1876]: 452)

8.4.3 The Club of Rome and the Limits to Growth

The relationship between population and resources remained contested until today. Despite an eightfold increase in the world's population since 1800, food production has largely been able to keep pace. Not foreseeable to Malthus were the increased production capacities of industrialised agriculture as a result of artificial fertilizers, seed breeding and machinery in the wake of the 'Green Revolution'. Contrary to his thesis, with growing prosperity in the industrialised countries, the birth rate declined as a result of the Demographic Transition. Thus, the limits to growth were pushed further and further into the future. Environmental problems were more a matter of national resource availability and access, making scarcity a driver of trade and conflict. Despite the actual improvements expected by Anderson and Marx, also disasters and disharmonies took place, such as famines, epidemics and major wars such as the two world wars, associated with violent struggles over natural resources such as land, coal and oil which cost the lives of millions of people. However, these events were not able to decisively reduce global population growth as expected by Malthus.

Malthus' theses continue to polarise until today, reflected in heated debates (Ide/Scheffran 2012). While some neo-Malthusians agree with his ideas, others find his political conclusions hard to accept such as the abstinence of the poor (e.g. Magnus 2009). Fundamentalist ecology, which saw overpopulation as an enemy of nature, or development policies aimed at population limitation, are criticised by leftists as fighting the poor more than poverty.

The scientific debate reached a new level with the Club of Rome's report *The Limits to Growth* (Meadows et al. 1972), which used a computer-based world model, represented by linear dynamic equations and available data, to simulate the future of the planet from 1900 to 2100. The interplay of five basic growing factors (world population, food production, natural resources, industrial production and pollution) was mapped in 12 different scenarios. If the growth trends continued unchanged, the planetary limits to growth would be reached within the next 100 years, likely resulting in a sudden and uncontrollable decline in both population and industrial capacity in the wake of natural resource depletion and environmental pollution. Such a collapse could be avoided by altering the growth trends and establishing a global equilibrium designed through resource efficiency, environmental protection and limits to growth that is ecologically and economically stable and sustainable far into the future, so that the basic material needs of each person on Earth are satisfied and each person has an equal opportunity to realise a certain human development potential. Which of these scenarios is realised is defined by decisions of the world's people to strive for the second outcome rather than the first, the sooner they begin working to attain it, the greater will be their chances of success.

Notwithstanding the simple model structure, which ignored environmental changes like global warming and socio-political processes like conflict and cooperation, the Club of Rome report, reinforced by the oil crisis and other crises in the 1970s, unleashed a wave of interest and prompted the U.S. government under

Jimmy Carter to produce the *Global 2000 report* published in 1980 which in great detail described the destabilising effects of environmental degradation and population growth (CEQ 1980). As a result, political efforts increased to reduce oil dependence through alternative energy sources, environmental agreements such as the Montreal Protocol, and the 1987 Brundtland Report on Sustainable Development.

After the end of the Cold War, the situation initially relaxed: the former adversaries disarmed[3], and many hoped for a peace dividend that would also benefit environmental protection and security conceptions to include ecological dimensions. The door was opened for the UN Rio Conference of 1992, which put climate and biodiversity on the agenda and laid down guidelines for sustainable development in Agenda 21, which in 2016 led to the Sustainable Development Goals (SDGs) under the 2030 Agenda for Sustainable Development. War and peace did not play a role at the Rio Summit despite attempts to suggest conversion of military budgets, infrastructures and resources for civilian purposes (UN 1991; Scheffran 1992).

While the Club of Rome report was criticized for its simplistic assumptions, it did not predict one future but project a wide range of possible futures. Despite hopes that the worst scenarios of could be avoided, globalisation accelerated the pace of worldwide trade, development and CO_2 emissions. By the end of this century, up to 12 billion people are expected to live on earth, many in poverty. The largest increase will be in countries of the global South (most particularly in Africa), where more than 80% of all people live. Without sustainable development this population growth will increasingly conflict with the earth's resources and endanger the natural foundations of life, particularly evident in pressures such as climate change and biodiversity loss, which were not yet a factor in the original Club of Rome report. The latest update expects a decline in growth rates of the world population and of wealth in the coming decades (Dixson-Declève et al. 2022). A transition to degrowth would require a radical change in the financial, economic, political and social system and a reorganization of the flow and usage of resources as well as the indicators of wellbeing, prosperity and innovation (Hickel et al. 2022).

8.4.4 Resource Limits and Planetary Boundaries

Since the availability and use of natural resources is an essential precondition of human existence, their depletion and scarcity is a threat to human livelihood and a potential source of conflict, making the protection of natural resources necessary for a sustainable human-nature relationship. The distribution and scarcity of resources is intertwined with socio-economic and geopolitical dynamics that exacerbate inequality and conflict as much as environmental degradation and climate

[3] Between 1990–1996, the USA and the successor states of the Soviet Union reduced their military expenditure until 1996, partly disarmed and sored the old equipment or increased their arms exports. For a few years, the USA reduced its military procurement but simultaneously increased its expenditures for Military Research and Develoment.

change. From the perspective of political ecology and integrative geography, there are significant differences for different resource types and their material and energy flows. Non-renewable resources (minerals, metals, rocks, sand, and fossil fuels) were created over long geological periods but are depleted over short geological periods. In contrast, inanimate (water, air, wind, sun, soils) and animate renewable resources (biodiversity, genetic resources, bioenergy, cereals, grasses, forests, fish, meat, wool, etc.) allow steady flows within limits.

Climate and environmental change affect the nexus of water, food and energy security which is essential for human livelihoods. While the exploitation of non-renewable resources such as minerals and fossil fuels is polluting the environment and reaching planetary boundaries, the challenge is to integrate resource flows of energy, matter and information into natural cycles through recycling and the shift to renewable resources. The limited range of material transport favours regional production of material goods while electricity-generating forms of energy through power grids allow for transcontinental networking and cooperation (for example between Europe and North Africa across the Mediterranean). Diffusion processes expand the border-crossing propagation and exchange of resources in space and time, which is shaping complex geographic resource landscapes. The total area of fertile arable land is declining, as well as the extent of forests and the number of species living in them. Without fundamental change, natural resources would be depleted, ecosystems overstressed and the climate significantly altered.

With its growing impact on nature, humanity is increasingly competing with other species and endangering their regenerative capacity by consumption and transformation of natural resources (input side), pollution by release of waste materials and waste heat into nature (output side); and the modification of the functioning of ecological systems (systemic destabilisation). Since organisms, ecosystems and societies have viable tolerance ranges within which they are capable of adapting and keeping damage in 'acceptable limits' and outside of which qualitatively different state occurs. Whether for better or worse, who is the winner or loser, also depends on the value criteria chosen. The growing human footprint leads to irreversible destruction of species, ecosystems and habitats, affects natural livelihoods and the coexistence of nature and society, degrades water, forests, soils and arable land, provokes climate change and species loss, overfishing of the oceans and the overexploitation of strategic raw materials. Natural resources become scarce through declining quality and quantity, caused by exploitation, waste, modifications of regenerative capacity or uneven distribution. Local and short-term changes usually have a more direct impact than global and long-term phenomena.

Planetary boundaries specify ecological stress limits of the Earth that endanger ecosystem stability and human livelihoods, and define a "safe operating space" that must be guaranteed in order to maintain security, resilience and sustainability (Rockström et al. 2009; Steffen et al. 2018). Certain thresholds and tipping elements must not be exceeded or undercut, triggering abrupt and irreversible changes that endanger global stability (Kim/Kotze 2021). Due to uncertainties in complex systems, however, these thresholds cannot be precisely determined, so that safety margins may be considered allowing time for action before a planetary boundary is exceeded. When a 'dangerous zone' is reached, the risk increases as the boundary is crossed. Nine *earth system boundaries* (ESBs) have been proposed in the dimensions of climate change, stratospheric ozone depletion, atmospheric aerosol loading, biogeochemical cycles, land-use change, biosphere integrity, novel substances, freshwater consumption and ocean acidification (ibid.), with limits not yet exceeded in the last two dimensions. A quantitative assessment finds that seven of eight globally quantified safe and just ESBs and at least two regional safe and just ESBs in over half of global land area are already exceeded (Rockström et al. 2023).

8.4.5 Socio-economic and Political Boundaries and Crises

Limits to growth do not only apply to environmental planetary boundaries but also to economic, social and political limits and boundaries which prevent further expansion on established pathways which are becoming too costly and too risky, triggering multiple crises (Scheffran 2020a, 2021, 2023a, b). The question is what limits of the previous growth-oriented and expansionist development model may have been reached? Or can the principle of expansion across borders be continued? These questions will be discussed for various dimensions of limits, contradictions and crises which are subject to uncertainty.

Attitudes towards nature also have a significant influence on its economic use, serving as a source of resources and a sink for waste. For a long time, nature played a subordinate role in economic theory; natural resources seemed to be available without constraints, ignoring that they are finite and essential for the functioning of the economy (Murphy 2022). The capitalist economy continues to rely on growth and aims to conquer global markets with new products. It contains growth- and power-promoting feedbacks: Consumers with higher incomes have more leverage to secure their income advantage; companies and transnational corporations with high profits have more resources to invest in new means of production. The accumulation of capital, which is part of the market economy, corresponds to the principles of exponential growth and the concentration of power, which is based on the ownership of wealth and means of production and influences political decisions. Accordingly, processes of concentration occur in economic competition, in which the strongest have the best chances. The globalisation of the Western growth model thus contributes to the accumulation of wealth and power. This reinforces the unjust gap between rich and poor as well as the unsustainable exploitation of nature. Falling costs and wages and the technical substitution of labour exclude large parts of the world's population from prosperity and drive entire regions into marginalisation. Lack of capital, debt and competitive pressures are blocking development in many countries of the Global South. In the globalised and interconnected world, the financial system is becoming increasingly dominant, rendering political control and governance mechanisms ineffective (Sachs 2020). Huge investment flows in digital worlds find no counterpart in material production or population needs. This became evident in the financial and banking crisis of 2008, with its manifold after-effects such as the Greek crisis or the price fluctuations that contributed to the Arabic Spring.

Never before has so much wealth been produced; the social market economy as well as globalisation have brought prosperity to many people. However, a large part of humanity cannot share the fruits of wealth, is unemployed and excluded or falls through the 'social net'. The gap between rich and poor, winners and losers, North and South is growing. A vicious circle of personal problems and social decline develops. Especially in poor countries, many people suffer from a bundle of interwoven problems: Hunger, poverty, lack of resources and environmental destruction, diseases and epidemics, repression and violence. Precarious living conditions and social problems promote uprooting and displacement of large populations. More and more people from poor rural regions are migrating to expanding megacities, where some find work and others contribute to worsening economic, environmental and social problems. Tens of millions of people are displaced each year by violent conflict and natural disasters. In the 2015 refugee crisis, hundreds of thousands migrated to the centre of Europe.

Industrialised countries are also affected by social disintegration and loss of control, including crime, violence, terrorism, drugs, diseases, social and political fragmentation. Increasing 'social disorder' endangers internal peace and the social structures that secure individual existence and strengthen social cohesion. Social cuts shift the limits of growth to the periphery of society, and the strongest have the best chances to survive. Violence is both a cause and a consequence of the other problem areas and acts like a disaster amplifier. Thus, ecological and social structures that are supposed to protect individuals and strengthen social cohesion are disintegrating. Expansive globalisation is reaching limits associated with a fragmentation and disintegration of society, represented by demarcations and a dismantling of social structures, which shifts problems and conflicts to social peripheries. The challenge is whether social structures are strong enough to mitigate and preserve peace against the tipping dynamics of growth, power and violence.

In November 1990, the "Charter of Paris for a New Europe" – supported by thirty European states, the USA and Canada – sealed the end of the bloc confrontation between East and West after four decades of the Cold War and was supposed to open opportunities for a pan-European peace order and a cooperative security system. However, hopes for a peace dividend have not been fulfilled and the opportunities were missed. Immediately after the end of the Cold War, international stability was undermined by new conflicts, including the Gulf and Balkan wars in the 1990s, the Afghanistan and Iraq wars in the wake of September 11, 2001, as well as the interconnected conflict landscapes of the 2010s, notably the 'Arabic Spring' and the wars in Syria and Ukraine. The refugee crisis of 2015 and terrorist attacks, the disintegration and isolationist tendencies of the EU and the West, autocratic governments and nationalist populist movements all had knock-on effects in interconnected crisis dynamics and chain reactions that reinforce each other and drive the world into chaos (Scheffran 2008, 2016). Globalisation out of control puts states under pressure, creates violence and conflict, and provokes resistance in civil society and governments.

In addition, climate and environmental changes are connecting with other crises. In 2010/11, for example, natural disasters in the wheat belts of China and Russia contributed to price increases in international food markets and triggered further chains of responses. At the same time, countervailing powers such as China, India, and Russia gained influence, as did environmental and climate movements and other civil society activities that challenged fossil capitalism's economic and lifestyle practices based on resource exploitation and natural stress. Critics come into conflict with political power relations and take high personal risks, as the examples of whistle-blowers like Edward Snowden or Julian Assange show. Thus, the liberal world order is coming under pressure from all sides, showing destabilising tendencies and conflicts both internally and externally. The current situation seems more dangerous than ever. The question is what security systems can hold in the face of this situation and prove capable of sustaining peace?

8.4.6 Ambivalence of Science and Technology Between Force Multiplier and Growth Limits

The degree to which humans can control and manipulate nature is strongly affected by science and technology which multiply abilities to act upon and control natural processes. Technical innovations can increase adverse human impacts on nature, even lead to technical disasters such as reactor accidents, dam bursts, aircraft crashes or chemical disasters, but also diminish risks through safety mechanisms or resource needs through sustainable and low-carbon energy systems. Landscapes are shaped and transformed by technology, such as built environments in urban areas, transportation networks and energy landscapes. As large distances can be overcome in short time by transportation and communication systems, geographical spaces are shrinking, accelerating the exchange across boundaries and transforming the human-nature interspace. Technology transfer enables the flow of technological capacity across borders and accelerates technology development in regions without access, overcoming large regional differences.

Science and technology play an ambivalent role, creating opportunities for problem solving as well as problem exacerbation (Scheffran 2018a). They have a share in the growth dynamic, as science strives to explore unknown territories in the world of knowledge, while technical innovations provide the means to 'conquer' the real world, changing it constructively or destructively. Thus, they can ease the hardships of human existence and allow more people to have a decent life and avoiding limits to growth and scarcity. At the same time, technical means have increased human impact on the natural world which threatens the living conditions. Thus, the power of scientific knowledge itself has limits and contributes to them (Lüthje et al. 2011).

Since science and technology expand the functions of the human body interacting with the outside world, they provide enhanced senses to perceive and understand things outside the everyday horizon of experience, and create instruments to cross borders and to intervene in more distant worlds. As more and more becomes 'comprehensible', science and technology develop into factors of power, for securing domination but also for overcoming it when new actors take advantage of it. Powerful technical instruments can serve as means of production, but also as destructive means, becoming involved in technology conflicts (Scheffran 2015a). In addition to more intensive exploitation of resources, acceleration of economic growth and increase in the means of violence, science and technology can be used to overcome problems.

The more scientific-technical civilization becomes a problem, the greater are the temptations for technical intervention, which in turn brings new problems. This is shown by critical discourses on genetic engineering, geoengineering or artificial intelligence. In the post-factual age of *fake news*, scientific knowledge also comes under pressure by nurturing doubts about its truth and validity. The complexity of the constructed world increases immeasurably, making social control and political governance more difficult. Complex systems tend to surprise, react sensitively to fluctuations, make the future uncertain and open up many options, which sets limits to knowledge and makes decisions more difficult (Scheffran 2008).

Experts are often called to the front when politics is stuck and needs them, mostly in crises, wars and catastrophes, less when it is a matter of avoiding the driving forces and power structures that lead to them. Provided it is in line with the interests of the population, in the long run science and technology can contribute to problem avoidance and improvement in order to secure life on earth and to change society in a sustainable way. Scientific responsibility also includes replacing the destructive exploitation of nature with constructive co-evolution of the biosphere and sociosphere that makes their respective inherent dynamics compatible. This includes the development and implementation of practical proposals for sustainable development in politics and economics, and the search for 'alternative' conceptions of nature and forms of traditional knowledge and alternative knowledge production. In addition to the science of inanimate nature, which like classical physics provides regularities for the development of technical innovations and mastery of nature, the sciences of animate nature (biology, ecology, social sciences) focus on complex interrelationships and social innovations that learn from nature and strive for integration with nature. This offers the chance to explore the scientific interspace and overcome the old dualism of subject and object, of body and soul, of nature and society

Science and technology have been a driving force in the climate debate since its inception. Scientific measurement data and theories have been used to create computer models to simulate future scenarios, depending on plausibility assumptions about economic and societal processes and policy responses. The state of climate research is summarised in reports of the *Intergovernmental Panel on Climate Change*

(IPCC), deriving proposals for mitigating climate change and feeding them into negotiations, such as the temperature target in the Paris Agreement. In parallel, technical instruments for emission reduction were developed for the energy, transportation and agricultural transitions, etc., but have so far been insufficiently implemented.

8.5 Complexity of Multiple Crises, Compound Risks and Tipping Cascades

8.5.1 Complexity, Security and Stability

During the 1980s, complexity emerged as a new paradigm in science and politics, expressed in the 'complexity turn' of international relations (Urry 2005; Mesjasz 2010). A spectacular example was the fall of the Berlin Wall in 1989, which served as a tipping point for the domino-like collapse of the Soviet world system in a matter of weeks, German unification, the breakup of the Eastern Bloc and chaotic breakdown of the Cold War, becoming a tipping point to an era of ever-growing complexity[4] (Jathe et al. 1995; Scheffran 2008). In the new world (dis)order, cascading chains of events emerged, including complex social interactions and self-reinforcing collective dynamics such as stock market crashes, social turmoil, mass migration and violent conflict that increasingly challenged international security and stability (Kominek/Scheffran 2012). A particular form of social instability is conflict, based in incompatible values, priorities, and actions of agents who undermine each other's values and provoke responses, leading to the waste of resources and escalating interactions if conflicts are not resolved.

A crucial issue is whether growing complexity breeds instability, a question that has been extensively discussed for ecosystems. Artificially constructed complex systems are dysfunctional if the components do not fit together, while unstable modes tend to disappear in evolving complex systems and those with better fitness or control mechanisms have higher chance of survival. Systems are often robust and adaptive against the most likely disturbances in the core region of stability (Held/Schellnhuber 2004); close to critical thresholds between regions of stability and instability small and rapid variation can lead to a systemic break down. This is symbolised by the famous butterfly effect in chaos theory, which may occur when a system is already 'on the edge', driven by other processes. A key term is the sensitivity of couplings between variables, which determines how changes spread through the network of interconnections (Scheffran et al. 2012c). Beyond a given sensitivity, threshold changes may trigger instabilities, tipping points and cascading sequences.

[4] Complexity expresses the difficulty to describe, understand, or explain something. A critical issue is to find a description of a system that contains the essential features and components (Scheffran 2015b). Systems that initially appear complex can become simple once they are understood or controlled.

To maintain stability, it is essential to understand the conditions under which micro-level events lead to qualitative changes at macro level that propagate in space and time. New and unforeseen 'disturbances' may endanger system stability and force it to adapt to changing conditions.

Methods of non-linear dynamics describe transition phenomena, such as self-organisation or micro-macro phase transitions which often occur beyond thresholds and tipping points when the dynamics accelerates and there is a qualitative switch of behaviour. A system is stable if essential system characteristics are preserved despite disturbances. Naturally evolved systems are often adapted to environmental change, which reduces their susceptibility to disturbance. At critical thresholds of instability, small changes can trigger qualitative system change. If the system tilts, upheavals and phase transitions occur, from collapse to transformation. Examples are transitions between war and peace or the change from exploitation to sustainable use of resources.

As described above, limits to growth and planetary boundaries drive the existing world order into multiple and complex crises. The world is at the edge of chaos where small causes have big impacts and spread across spatial and temporal scales. When everything is interconnected, changes in one part of the world propagate through systemic networks like a domino effect or chain reaction (Scheffran 2016a), accumulating in high risk hotspots.

8.5.2 *Polycrisis in an Interconnected World*

The more the planetary boundaries and limits to growth are reached or exceeded, the more the associated risks, crises and conflicts become relevant in different sectors. In the last decade the world has experienced multiple crises which combined in interconnected and compounding cascades, including hunger and poverty, refugees and pandemics, vulnerable supply systems and critical infrastructures, the gap between rich and poor, financial and economic crises, climate change and environmental degradation, destabilisation of social and political structures, national populism and terrorism, security risks and violent conflicts. While the primary consequences are often initially confined to affected areas or subsystems, they can spread in the globally interconnected world via long-distance effects (*teleconnections*) and global connectors, including globalisation and financial markets, infrastructures and supply chains, communication and transport systems, social media and networks, infections and diseases, climatic changes and resource flows, mobility and migration, markets and trade flows (Fig. 8.1).

Crises chains intensify in the network structures to complex multiple crises (Scheffran 1996, 2015c, 2016a, b), also called 'polycrisis' (Morin/Kern 1999; Tooze 2022; Homer-Dixon et al. 2022), driving the world into permanent crisis mode. Amplification triggers domino effects and risk cascades endangering stability of the global system and drifting the world into a self-reinforcing vicious circle of interconnected crises difficult to control. In network structures, there are coupling, spreading and

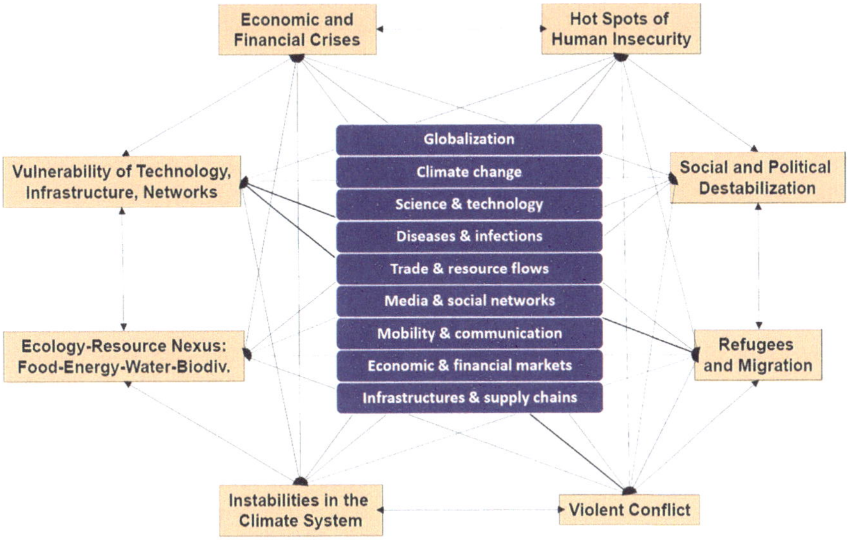

Fig. 8.1 Global connectors and compound risks in the global polycrisis. *Source* Scheffran (2017b), adapted from presentation at ISA conference in Atlanta, 17 March 2016

accumulating flows of information, capital and power that can multiply or contain problems. In the globalised world, domino effects, risk cascades and chain reactions endanger the stability of the global system and lead to chaotic dynamics in which small causes have a large effect and spread across spatial and temporal scales (micro to macro).

Instead of recognising own responsibilities for the causes of destabilisation, the consequences are blamed and the symptoms are fought – through isolation and exclusion, nationalism and interventionism, resulting in geopolitical tensions. Violence and war can be both cause and consequence of the other problem areas; they inhibit development and cooperative management of environmental problems. Globally interconnected processes such as resource flows or global production, consumption and supply chains can accelerate this development and undermine the conditions for peace. Conversely, violence and war stand in the way of sustainable development. Since the economic crisis of 2008, crisis events include the Arab Spring, the Syrian war, the refugee crisis, terrorist attacks, Brexit, the Trump election, weather extremes and, most recently, the Corona crisis and the Russia-Ukraine war, the Gaza war and the Trump reelection. Failure to contain crises appears as a loss of control for the prevailing neoliberal world order; the world seems to be falling apart.

8.5.3 Tipping Points, Compound Risks and Cascading Events

To ensure the functionality and viability of a system, exceeding critical tolerance limits must be avoided by practical control measures (Scheffran 2016a, b). A resilient system is able to restore itself after an external shock or to bring about a stabilising change. In the guardrail concept (*tolerable-windows approach*), climate consequences and chains of events that threaten the existence of the system are recognised at an early stage and avoided through appropriate actions (Petschel-Held et al. 1999). In this context, it is important to understand the conditions under which tipping elements, chain reactions and risk cascades are triggered that lead to complex transitions in the interspace between qualitatively different system states, both positive and negative (Scheffran 2020d). Environmental risks can be amplified by the combination of multiple stressors and hazards, the co-occurrence of which becomes more likely and contributes to societal and/or ecological risks (Zscheischler et al. 2018). Examples include the interaction of different weather and climate phenomena, such as extreme precipitation, storms and coastal flooding damaging infrastructures, or drought and heat leading to tree mortality and fires. The latter can cause air pollution, affect crops, and harm human health, as in the summer of 2010 in Russia (Reichstein et al. 2021). In hurricanes such as Katrina in 2005, Sandy in 2012, and Harvey in 2017, the coincidence of heavy rainfall and storm surge had a devastating effect, causing massive damage and loss of life in urban centres. Certain compound effects are represented in the nexus approach, such as the water-food-energy nexus or climate-conflict-migration nexus.

Bifurcation points facilitate switching between multiple equilibria (Scheffer 2009). Beyond tipping points a self-enforcing dynamics leads to qualitatively different states from which often there is no easy return. According to Milkoreit et al. (2018: 9), a tipping point is a "point or threshold at which small quantitative changes in the system trigger a non-linear change process that is driven by system-internal feedback mechanisms and inevitably leads to a qualitatively different state of the system, which is often irreversible". Thus, tipping points are characterised by four features of criticality: multiple stable states, non-linear change, driving feedbacks and limited reversibility. Tipping points often involve the three notions "that events and phenomena are contagious, that little causes can have big effects, and that changes can happen in a non-linear way but dramatically at a moment when the system switches" (Urry 2002: 8; Scheffran 2008: 14).

Beyond critical thresholds and tipping points, complex dynamics include phase transitions, risk cascades, and chain reactions (Scheffran 2015b; AghaKouchak et al. 2018). An example is the exponential chain reaction of nuclear fission, which is uncontrolled in the atomic bomb and held at the threshold of criticality in the nuclear reactor by control rods to extract energy. If an accident distracts reactor control, global cascades of consequences can be set in motion, as demonstrated by the nuclear disasters at Chernobyl in 1986 and Fukushima in 2011 (Scheffran 2016a). Sometimes the switching results from triggering events, such as natural disasters, mass migrations or social movements, leading to self-reinforcing cascading sequences, e.g. when an action taken by one actor provokes actions by other actors. Exponential cascades are demonstrated by the Corona pandemic, in which all humans are part of a virus chain reaction.

The climate system can also become unstable due to tipping elements (Lenton et al. 2008, 2023). These include self-reinforcing melting of the Greenland and West Antarctic ice sheets, release of frozen greenhouse gases such as methane, weakening of the North Atlantic Current, or changes in the Asian monsoon (IPCC 2019). Above a critical temperature threshold, amplification effects and event chains could lead to fundamental Earth system changes (Steffen et al. 2018), with profound consequences for global security and international stability. Certain thresholds and tipping points must not be violated, as they would trigger abrupt and irreversible changes that endanger global and regional stability (e.g. Franzke et al. 2022). Even in the absence of rapid and severe climate change, global warming can tip ecological and social systems (Rodriguez-Lopez et al. 2019; Otto et al. 2020). Whether these are 'negative' or 'positive' tipping points depends on an assessment of their advantages and disadvantages. Climate change can connect to cascades in social networks, in protest movements, elections, stock market crashes, revolutions, mass exodus, or violent conflicts (Kominek/Scheffran 2012). Such effects beyond tipping points are difficult to contain or control geographically. If the choices and actions of others influence our own decisions, tipping points in social interaction may become more likely that undermine the stability of the whole system. It is important to avoid exceeding adaptation limits with systemic risks and develop synergies triggering positive tipping points (Juhola et al. 2022).

8.5.4 Managing Complexity and Simplicity

Growing complexity provokes opposing trends of over-simplification, populism, nationalism, religious fundamentalism, illiberalism, anti-globalisation and anti-science attitudes that fail to address the underlying mechanisms. Without adequate management or reduction of complexity, the world may continue on a slippery slope of destabilisation. Instead, stability may be achieved by adapting the complexity of policies to the complexity of the systemic processes that they regulate. The challenge is whether humanity can anticipate and avoid hazardous pathways by counteracting forces that slow down and change course within the planetary boundaries of the Anthropocene. Throughout history, *Homo Sapiens* was able to overcome constraints on resources in crises and to expand into new spaces by applying problem-solving capabilities and developing technical and social innovations that address the limits of growth, allowing more wealth to be generated on a shrinking base of natural resources. The question is whether humanity will succeed facing today's complex world or whether disasters destroy any chances of success. Thus, stabilising human-nature interactions becomes a major challenge in international relations and global governance but there is little experience with integrative approaches to understand, manage and diminish high level of complexity.

A better scientific understanding of the underlying complex interactions is a prerequisite to stabilise the Earth system at a manageable and low-risk level and to enable adaptive and anticipative policies. Operating in a multi-risk environment needs a whole-system approach to analysing and measuring compounding risks (Zscheischler et al. 2018). It makes sense to identify criteria for stability, determine thresholds when transitions to instability occur and find mechanisms that facilitate the transition across thresholds. Regarding climate-conflict linkages, the question is how stable a certain level of conflict or cooperation is against climatic change and associated escalation. This is not a single-stage process but a continued interaction between climate stressors and governance mechanisms driving the dynamics through a sequence of decisions and tipping points between conflict and cooperation. Whether climate stress fuels a cycle of violence or climate governance facilitates a cycle of cooperation and sustainable peace depends on the effectiveness of human and societal responses (Scheffran et al. 2014).

8.5.5 Integrative Framework of Climate-Society Interaction

An integrative framework presents the complex interplay of systems, conditions, and actors in the Earth system, assessing multiple pathways connecting changes and impacts in the *climate system, natural resources, human security and societal stability* (CNHS; Scheffran et al. 2012a, c; Daoudy 2021). The couplings in this network of interactions can be characterised by sensitivities that represent the impact of a change in one variable on another variable, in particular, climate sensitivity or conflict sensitivity (Fig. 8.2). Risks are influenced by the vulnerability of subsystems as well as adaptive capacities (IPCC 2022a, b). Changes in the climate system (such as variations in temperature and precipitation) affect the functioning of ecological systems and natural resources (e.g. soil, water forests, biodiversity). Depending on vulnerability, environmental changes stress basic human needs and security (such as the availability of water, food, energy, health and wealth). Human responses to environmental change can affect the stability of societal structures, driving tensions and social disruptions in regional climate hot spots through instability events, such as migration, riots, insurgencies or violent conflict that spread and cascade in an interconnected world. In contrast, cooperative and sustainable countermeasures can reduce the causes and consequences of climate change, through mitigation of greenhouse gas emissions or adaptation to climate change. To be considered is the relationship with other indicators of planetary boundaries (biodiversity, land use, nitrogen, …), their prior pressures, interactions, and cumulative effects. For example, a relatively small climate variation could trigger a system change in the case of a large loss of biodiversity.

The challenge is to develop strategies to address complexity, avoid dangerous instabilities of climate change, and maintain stability despite system changes. Complexity plays a role in multi-level crises constellations linked through the global connectors. Due to non-linear effects, an increase in global temperature above a

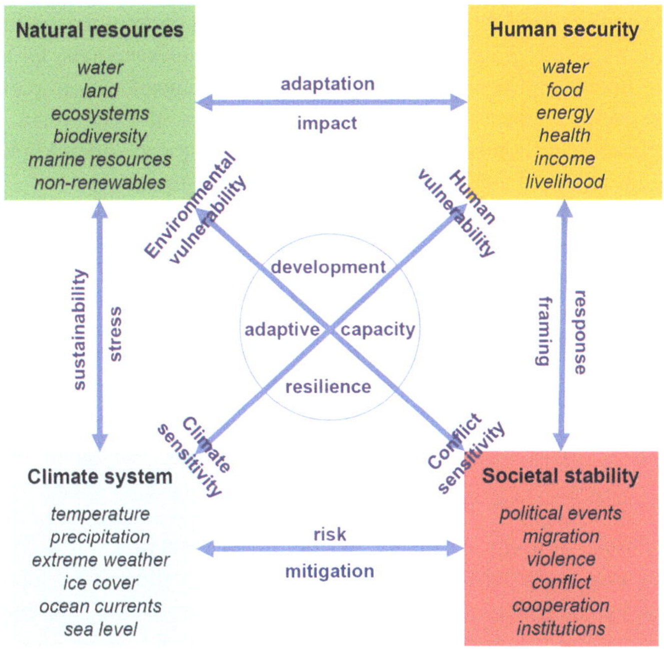

Fig. 8.2 Framework of complex interactions between climate, natural resources, human security and societal stability. *Source* Adapted from Scheffran et al. (2012a)

certain threshold may trigger instabilities, tipping points and cascading sequences that could exceed the adaptive capacity and resilience of natural and social systems. Operating in multi-risk environments requires a whole-system approach to analysing compound risks, such as experienced in the Corona or climate crisis. Complex crises amplify consequences through impact chains that affect the functioning of critical infrastructures and supply networks; intensify the nexus of water, energy and food; trigger economic crises across regions. The sign of sensitivities provides information to classify qualitative patterns or 'syndromes' of global change (Petschel-Held et al. 1999).

Beyond critical thresholds, amplifying effects from tipping elements, compound risks and cascading events can trigger vicious circles (Scheffran 2020d; Lenton et al. 2023). The destabilisation of society could also affect wealthy people who have better capabilities to survive but feel threatened if societies around them fall apart. The Corona crisis is a vivid example how even developed societies could reach their limits. Exceeding capacities of societies to respond effectively justifies additional efforts to build institutions and governance mechanisms to break the vicious circle. Using spatial vulnerability indicators provides a geographical representation of combined vulnerability to climate change and violent conflict which exceed adaptive and coping capacities in hot spots. If the primary consequences cannot be confined to the affected areas, they may proliferate and multiply through cross-regional effects and teleconnections such as flows of resources, refugees and arms, increasing prices, humanitarian aid and military interventions. Of particular interest is to understand potential pathways and path dependence, in particular of intermediate events connecting climate and conflict risks, and vice versa. New and better models and decision-support methods are necessary to enrich knowledge of the uncertainties, how to manage risks and how to avoid them.

8.6 Geopolitical Conflicts at the Frontiers of World Order

8.6.1 Geopolitical Origins and the World Wars

In contrast to political geography, studying the effect of geography on politics, geopolitics projects political actions in and through geographical spaces. Geopolitics evolved at the beginning of the 20th century in the wake of power politics in the colonial and imperial tradition emanating from Europe where it was used to justify imperial claims over distant territories. National borders allowed for the inclusion of the 'own' by exclusion of the 'other', aiming to control both. Starting from a biologistic view and inspired by Malthus, the zoologist and geographer Friedrich Ratzel (1897) established the "Basic Laws of the Spatial Growth of States" in his work *Political Geography* which was suitable to justify nationalist expansion (Bassin 1987; Stogiannos 2018). The Anglo-Saxon tradition of geopolitics was shaped by Alfred Thayer Mahan, Rear Admiral of the U.S. Navy, who saw the superiority of sea power over land power as the key to world power. In contrast, for the British geographer Halford Mackinder in 1904, the resource-rich 'heartland' of the Eurasian continent was the potential centre of power and gravity of the world's political power geography, from which world domination was possible.

This thinking has been dominant in the nationalist attitudes leading to World War I which was long expected by those who understood and criticised the logic of imperial powers competing for dominance. In the second half of the 19th century some recognized that Europe was heading for a devastating war shaped by the enormously increased technical and industrial capabilities which became apparent in the Crimean War of 1853–1856 and the American Civil War of 1861–1865 with more than 600,000 casualties. As early as 1888, Friedrich Engels predicted a future world war with eight to ten million soldiers killed,[5] and in 1893, the German Social Democrat August Bebel warned in the Reichstag of mass slaughter in the coming

[5] "And, finally, the only war left for Prussia-Germany to wage will be a world war … of an extent the violence hitherto unimagined. Eight to ten million soldiers will be at each other's throats …. The depredations of the Thirty Years' War compressed into three to four years and extended over the entire continent; famine, disease, the universal lapse into barbarism, both of the armies and the people, in the wake of acute misery irretrievable dislocation of our artificial system of trade, industry and credit, ending in universal bankruptcy collapse of the old states". In: Friedrich Engels, 1888: Einleitung zu Sigismund Borkheims Broschüre. Translation in: https://www.marxists.org/archive/marx/works/1887/12/15.htm.

war. Around the same time the Polish-Russian railroad industrialist Jan Bloch (also known as Ivan Bloch, Jean de Bloch or Johann von Bloch) painted the 'future war' in great detail on several thousand pages in his six-volume work (Bloch 1899). With scientific scrutiny he showed the consequences of strategic thinking combined with the industrial-technological dynamics for a future war, arguing that modern technology and its enormous destructiveness would give the defender such advantages that victory was not to be expected (Bloch 1899, Vol. 1: xv). Major wars between industrialised nations could now only be fought at the price of suicide (Scheffran 2014).

Because of the inability to simply decide wars by military-technical means, war became predominantly a matter of economic power. Industrial societies could mobilise mass armies with millions of people, involving whole economies and societies until they collapse politically, with the risk for ruling circles that a foreseeable defeat in the war could lead to violent revolutionary upheavals to create new political forms or a new social order. According to Bloch, the enormous devastation and economic shocks from the war would have grave consequences for the period after the war, including financial and economic disruption, famine, disease and epidemics. Bloch recognised that the war momentum could become a self-fulfilling prophecy as a collective expectation of catastrophe (Münkler 2014). He attempted to show the ruling forces the consequences of their destructive logic of power, which was damaging to themselves. In closing their eyes to the realities they acted like 'sleepwalkers' (Clark 2013), who followed a direction of movement colliding with opposing forces.

While Bloch's views were rejected by military strategists of his time, they found fertile ground in the peace movement. Russian Tsar Nicholas II took an interest in his work and supported the Hague Peace Conference in 1899, which did not bring much progress in disarmament. Being a friend of Bertha von Suttner, Bloch was nominated for the first Nobel Peace Prize in 1901, which after his early death was then awarded to Henri Dunant (Scheffran 2014).

This escalating logic of war was also recognised by British meteorologist Lewis Frye Richardson who used a model to study how the arms race before the war was mutually building up military arsenals and threat perceptions which then erupted in violence (Scheffran 2020f). Another study found how alliance formation between friendly and hostile states before World War I explained the cataclysmic diffusion of conflict in social networks which contributed to miscalculations by political and military leaders, slipping out of control. Rather than simple contiguity, territorial embeddedness and network density are conceived as components of tipping in interstate rivalries (Flint et al. 2009; Vasquez et al. 2011; Chi et al. 2014).

Just as the road to World War I was expectable, so was its catastrophic end. Albert Einstein was frightened by the patriotic mood of almost all his fellow scientists. Three weeks after the war started, together with two other colleagues, he signed an Appeal stating: "The struggle that rages today is unlikely to produce a victor; it will probably leave only the defeated." They expected that all European relations would fall into an unstable state. This became apparent when the daily routines of war increased mass unemployment, inflation, food prices and poverty. These cascading consequences laid the foundations for another war, through economic crashes and political radicalisation in the 1920s, leading to fascism and a new arms race in the 1930s, which Richardson again analysed with his model to warn of World War II (see Scheffran 2020f).

In this war, geopolitical thinking played a justifying role in the works of the geographer Karl Haushofer, and fell on fertile ground in Nazi Germany's 'Lebensraum' (living space) ideology to justify brutal expansion to the East (Herwig 1999, 2016). After World War II, geopolitical ambitions in Germany were discredited by the lost war and its monstrous crimes. It took until the end of the Cold War and German unification that geopolitical thinking gained new ground. Geopolitics was increasingly envisioned for Europe in the last decade, merging with unbroken geopolitical traditions in the Anglo-Saxon world, from Mackinder to Brzezinski, Huntington and Kaplan (Huntington 1996; Kaplan 2009), who saw the results of the two world wars and the Cold War from a winner's perspective. Geopolitical conflicts and strategies to justify violent expansion are continued, in the spirit of former U.S. national security adviser Zbigniew Brzezinski (1997), who declared the Eurasian region to be a chessboard for Western power projections.

One response to the current world in disorder is the revival of geopolitics to pursue partial interests and regain control in a world of limits and crises (Ioannides 2022). Shifting the coordinates of world politics towards confrontation resembles 20th century territorial conceptions of geopolitics, like territorial claims, control of national borders and resources, artillery and tank warfare. The world is in a transition phase (or interspace) between the old and a new world order, reminiscent of the destabilisation of the colonial world order a hundred years ago, when World War I erupted in a spiral of violence, followed by other crises such as the Spanish flu, the Great Depression and fascism, which led to the catastrophe of World War II (Scheffran 2009, 2014). Today it is about the crises of the fossil and neoliberal variants of globalized capitalism, which drive humanity down a path if it does not make a fundamental turnaround.

8.6.2 Power Shifts and Crisis in the Liberal World Order

The idea of a 'victory' of the liberal democracies in the Cold War shaped thinking and action in parts of the West after 1990. This corresponded with the assumption that Western values, models and claims for power could be brought to a worldwide breakthrough. Thus, the West continued its century-long history of expansion, driven by a combination of economic growth, political power and military force (Scheffran 1996), reinforced by scientific and technological innovation, prosperity, and Western values. This was fuelled by the expectation that the West could solve the world's problems in its own way: Enforcement of human rights, overthrowing dictatorships, democratic regime change and nation building, coping with climate change and other global problems. Following the 1992 Rio Summit, the issue of war and peace was neglected, although in a crowded, interconnected and multipolar world, armament and war can multiply insecurity and instability with other crisis drivers, such as globalisation and climate change.

Instead of ushering into a peaceful world order and using the peace dividend to tackle global problems, the U.S. and its allies used money and diplomacy as well as weapons procurement and military interventions to secure the liberal world order and further expand their lead. The justification by universal values was mixed with the assertion of own interests, which were usually prioritised. Since the terrorist attacks of September 11, 2001, and the economic crisis of 2008, the chain of crises also affected the West and contributed to its destabilisation. Thus, the world today is more concerned with crisis management than with shaping the future, which would be more urgent than ever in view of interconnected problems, described earlier.

There are many reasons why the crisis of the liberal world order (Rupnik 2015) is now following the epochal upheavals three decades ago. Among them are inherent contradictions and limits of Western expansionism. Globalised capitalism produces not only winners but also losers, creates suffering and inequality, fractures and turbulences, neglect of social needs and identities that work against a stable order (Klein 2007). While Fukuyama (1989) translated the Western supremacy into the end of history, Huntington (1996) opposed the democratic-capitalist alignment with a 'clash of civilizations', with asymmetric conflicts and authoritarian politics. The spiral of violence continued, with hundreds of thousands of casualties and trillions of dollars for military budgets. NATO remained a military alliance rather than converting into a system of collective and common security. Military interventions showed the limits of high-tech warfare (Neuneck/Scheffran 2000b), leaving many problems unsolved, most clearly in the Afghanistan mission (2001–2021), which failed to achieve its objectives but resulted in enormous losses.

Although the nuclear arsenals are lower than in the Cold War, nuclear deterrence and the nuclear arms race continues. With nearly 13,000 nuclear weapons, there is a risk of their use in crises, wars and terrorist attacks. This is complicated by the 'Revolution in Military Affairs', which involves all areas of high technology, including missile defence, space armament and cyberwar (Neuneck/Scheffran 2000a). After the U.S. abandoned or did not ratify arms control agreements (ABM, INF, Open Skies, CTBT), the nuclear arms race became less regulated. Opportunities for comprehensive nuclear disarmament and supporting the 2017 Treaty on the Prohibition of Nuclear Weapons (TPNW) were missed. Militarisation extends to missiles and missile defense systems, outer space and cyberspace, drone and hybrid warfare, vulnerability of military and civilian infrastructures, *fake news* and *hate speech* on social media, where the lines between war and peace are blurred (Scheffran 2018a, 2019b).

Global military spending rose to more than two trillion USD for the first time in 2021 (SIPRI 2022), and even more in 2022. The U.S. alone invested about $800 billion, as much as the following ten countries combined, about 2.7 as much as China ($293 billion), more than ten times as much as India ($76.6 billion), and twelve times as much as Russia ($65.9 billion). In addition came the military spending of other NATO countries, including the United Kingdom at $68.4 billion, and the sharp increase in German military spending (about $56 billion) which still missed NATO's two percent target (SIPRI 2022). Together with growing arms exports, this creates a direct link between economic and military growth. Global military spending reached record levels when the Russian-Ukraine war triggered a drastic increase in 2022, reaching USD 2240 billion (SIPRI 2023). Other crisis indicators also increased substantially in recent years, such as violent conflict and forced displacement.

The more the expansive growth model encounters limits, the more evident are marginal costs and risks in a world facing intertwined multiple crises, conflicts and catastrophes that appear as wicked problems in complex crisis landscapes (Rittel/Webber 1973; Scheffran 2008). The compounding problems reinforce the erosion of the rule-based international order and loss of control by the Western hegemony (Brzoska et al. 2019; Taylor et al. 2020). The world of 1990 has given way to a confusing situation in complex crisis landscapes, fractures of globalisation and systemic turbulence. One explanation is that we are experiencing a world in transition, an interim period in which the old order is challenged by existential problems that can no longer be solved within the existing framework before a new order is found (Schröder 2022).[6] Possible futures include global power shifts and geopolitical conflicts, especially between China and the United States, as well as multipolar power constellations with multiple competing orders where the liberal order could persist with limited scope (Flockhart 2016). Whether the limits to the Anthropocene are largely limits to the expansion of the Western world order or limits to humanity as a whole, depends on the policies pursued and their impacts on power structures. New solution pathways are required addressing the complexity of the challenges The chain of crises is reminiscent of the situation a hundred years ago, with World War I, the Spanish flu, the world economic crisis and fascism, which led to World War II. In today's world political, security, environmental and social crises converge (Spangenberg/Kurz 2023), densifying and intensifying long-term trends and short-term events to an epochal turn (*Zeitenwende*) (Scheffran 2021a, b).

8.6.3 Geopolitics and Zeitenwende *Between National Borders and Planetary Boundaries*

At the turn of the millennium, the rift between the West and its former partner Russia became entrenched, creating the basis for today's situation, reminiscent of the Cold War (Scheffran 2000a, b). Russia attempted to regain a voice in the international arena after a decade of weakness and ignorance of Russian security interests by the West in NATO expansion, Kosovo war, missile defence and arms control. Against

[6] In his famous statement in the Prison Notebooks nearly hundred years ago, Antonio Gramsci spoke of an interregnum: "The crisis consists precisely in the fact that the old is dying and the new cannot be born; in this interregnum a great variety of morbid symptoms appear" (English translation cited in: Hoare and Nowell-Smith 1971). See also Babic (2020).

this background, Russia increasingly pursued a more confrontational policy toward the West, through nuclear arms build-ups, hybrid wars and military interventions. Apparently the large Western dominance over Russia could neither guarantee security nor prevent Putin's attack on Ukraine, perhaps even incited it by raising Russian threat perceptions. Accordingly, Russia is pushing great power claims, not only in areas of the former Soviet Union, but also in the Middle East, Africa and the Arctic. Attempts to include Ukraine into the Western sphere of influence and expand it to the Russian frontier led to violent reactions and a hardening of authoritarian structures in Russia, from the annexation of Crimea in 2014 to the illegal war against Ukraine since February 24, 2022. The West adopted massive sanctions against Russia and provided arms exports to Ukraine. On 27 February 2022 Chancellor Olaf Scholz switched into crisis mode and in his speech to the German Parliament declared a response to the supposed *Zeitenwende* (epochal turn) induced by Russia's attack, providing 100 billion Euros in special funds for the German Armed Forces (Scholz 2023). It is questionable whether such a military-oriented turn can meet planetary security needs or rather feeds the arms race.

In Russia's attack on Ukraine old and new geopolitics interfere in complex ways (Scheffran 2023a, b). The revival of old geopolitical strategies of confrontation, militarisation, arms race, violence, hot and cold war activate multipolar power shifts and rivalries, between the European Union and Russia, China and the United States, Global North and South. Europe is facing geopolitical challenges in all geographic directions: in the East by Russian threats and Putin's power games, in the West by US nationalism and hegemony, in the South by the destabilisation of the Mediterranean region, and in the North by climate change, resource competition and rivalries in the Arctic. China is trying to reshape the international order and expand its global political influence. With the 'New Silk Road' on land and sea, a network of infrastructures is being created from East Asia to Europe and Africa. China challenges the West in key technologies and is preparing for military confrontation, especially in the South China Sea. The U.S., under President Joe Biden, is struggling to maintain its hegemony and forging alliances in the Indo-Pacific which is becoming increasingly armed (Brands/Gaddis 2021).

While the Ukraine war is a burning glass of old geopolitics, it also serves as a new geopolitical framing of emerging security issues, such as cyber and hybrid war, drone and space warfare, anti-globalisation and energy transition, environmental and climate change. The mindset of geopolitics is increasingly spreading in technically constructed spaces. New terrain is claimed not only in distant regions of the world or in outer space, but also in cyberspace of information war and in the nanoworld connecting physical, chemical and biological innovations at smallest scales (Ruhl et al. 2020). While distances are compressed in the micro-world of genomes and brains, the macro view of satellites and geographic information systems connects all areas globally. In the networks of social media internal and external worlds are bridged by interfaces potentially connecting and controlling humans (Zuboff 2019; Scheffran 2023b).

Corresponding to the declining relevance of national borders and shrinking spatial distances in a highly connected world society, the geo-economics of markets and geopolitical fault lines have gained relevance. Facing the limits of the Anthropocene the exploitation of natural resources, global warming and renewable energy are becoming new areas of geopolitics, giving territoriality a new meaning (Burles 2021; Scheffran 2023a, b). Climate change is drawing geographical maps of regional hot spots with high vulnerability and risk, while the quest for a green economy is selecting suitable places with high solar radiation and biomass productivity, strong flows of water, wind, geothermal, tidal, and ocean currents as well as related strategic materials. Driven by the social-ecological transformation new energy landscapes integrate natural and societal interactions in complex geopolitical frames of control, resistance and conflict, connecting local and global levels (Link et al. 2018). When natural limits are reached, efforts to adjust to the risk of these limits also grow, such as geoengineering to keep the climate system within boundaries which can create new side effects (Link et al. 2013; Maas/Scheffran 2012).

Critical approaches to geopolitics doubt that human behaviour is determined by geographical factors and criticise that borders, boundaries and barriers are used for political discrimination and exclusion, for example against poor, female, ethnic, religious or migrant groups (Scheffran 2023b). From a feminist and racial perspective, geopolitics serves to enforce patriarchal structures and white supremacy (Tilley 2014). In local contexts, participatory approaches and resistance can create spaces free of established power structures and help evolving alternative forms of *Anthropocene Geopolitics* (Dalby 2020). Integrative geography opens the possibility of developing interfaces with other disciplines and their interdisciplinary fusion. A key question is whether nation-states are able to manage today's global challenges and interconnected crises. Nationalism provokes geopolitical conflicts which are major obstacles to effective and coordinated global strategies and solutions against climate change and other crises (Conversi/Posocco 2022). The propagated ideological 'battle between democracy and autocracy' is dividing the world into 'good' and 'bad' according to Western standards which found a geopolitical representation in the Russian-Ukrainian border as a frontier between the transatlantic Western world and the Eastern Eurasian land mass. Such a rift is constructing and justifying a new Cold War narrative and its proxy wars (Scheffran 2000a, b, 2023a, b).

The Ukraine war follows a similar logic as other wars which mobilise the means of violence of collective agents which are polarised into friend and foe, good and evil, mutually justify increasing threats that trigger a spiral of violence, move up the escalation ladder until the limits are reached in total war or further acts of war do not seem reasonable because of the costs and risks involved (Scheffran 2023c). Different from moral justifications, realist thinking explains the mechanisms of escalation and how to prevent them (Mearsheimer 2014; Kissinger 2022). The consequences are obvious: massive destruction of cities and vital infrastructure, numerous dead and wounded people, millions of refugees and existential hardships, accusations of genocide and ecocide, attacks on nuclear facilities, spirals of escalation on the threshold of world or even nuclear war. While Russia and Ukraine deployed massive military force, the West supplied heavy weapons and imposed tough sanctions and other means of economic warfare, which increased inflation and supply chain disruptions, volatility of food and energy prices as well as increased military spending, hitting a global economy battered by the Corona pandemic, with the most vulnerable suffering most. The war distracts attention and resources away from planetary security and constrains solutions to the environmental and social crises, as well as the willingness to cooperate, further fuelling the crisis (Scheffran 2022a, d).

This war ignores historical experiences of previous wars, revives geopolitical power struggles, rearmament and other violent conflicts. The continuing logic of war destroys the conditions for a sustainable peace order, consumes enormous funds and resources that are lacking to solve urgent problems, and pollutes climate and the environment. Instead of overcoming crises, it multiplies them into a permanent crisis and a perpetual war ('sustained war'), which former U.S. President George W. Bush propagated two decades ago in Iraq and Afghanistan. Instead of getting used to crisis mode or fighting the consequences, it is more reasonable to prevent the origins and drivers. Without a self-critical analysis, the war could deepen the crisis of the Western world order (Neumann 2022; Koschorke 2023). The fossil war and the climate crisis confirm the conflict-prone nature of the fossil-nuclear energy system, re-enforcing the urgency to develop an alternative energy system that is less conflictive (Scheffran 2022a).

8.7 Violent Nature? Environment and Conflict

8.7.1 Conflicts in Human-Environment Interaction

The negative nexus of environmental destruction, underdevelopment and violence could drive the world into a self-reinforcing vicious circle of interconnected crises which undermine the conditions for peace and sustainable development (Scheffran et al. 2014; Buhaug/von Uexkull 2021). Climate and environmental change impact human needs, critical resources and infrastructures that are essential for the provision of human security and societal stability. On the other hand, climate change could set the course for a dynamic and globally coordinated climate policy, strengthened by risk reduction and conflict management. Whether conflict or cooperation prevails depends on the social and political conditions as well as human responses and societal interventions. While reactive measures to avert disaster risk can increase conflict risk, cooperative and preventive strategies that strengthen institution building and collaboration in environmental peacebuilding tend to diminish conflict risk.

Conflict generally refers to social or political incompatibilities of actors who fail to reduce their differences and tensions to tolerable levels and undertake actions to enforce or articulate their interests, including violence, protests, riots and other forms of resistance (Ide et al. 2016). If different interests and tensions cannot be overcome or mediated, conflicting actions may undercut each other's values and provoke escalating and extreme responses (Scheffran et al. 2014). Conflict risks combine the intensity and probability of harmful consequences, ranging from social disruption and small-scale communal violence to civil and interstate wars. The types of conflict vary with respect to the number of actors, casualties and the degree of violence. Collective violence consumes considerable resources and contains the risk of escalation.

Climate change is potentially associated with a wide range of conflictive issues, from disputes over scientific predictions, impacts und uncertainties of climate change to violent conflicts fuelled by the security risks of climate change. Most studies on climate-conflict linkages discuss the effects of various climate phenomena (e.g. change of temperature and precipitation, resource availability, weather extremes, sea-level change) on different phases of conflict (onset, initiation, prevention, escalation, prolongation, termination). In addition, measures to prevent and address climate risks can also become issues of conflict, in particular tensions over mitigation and adaptation, disaster management and damage limitation, climate geoengineering or the fair distribution of costs, risks, and benefits of climate change. Conflict parties can be nations, individuals, parties, companies, trade unions, activist groups and generations, among others. Most studies refer to armed conflict, in which actors use military force to achieve aims.

The concept of security has been expanded to include ecological dimensions and the availability of natural resources (Brauch et al. 2008, 2009, 2011). Environmental conflicts concern the use and degradation of exhaustible as well as renewable resources, regenerated in metabolic cycles, depending on the functioning and stability of ecosystems, which in turn are affected by conflicts. The unresolved problems of sufficient, equitable and future-oriented resource use can lead to significant conflicts. If it is not possible to achieve a lasting balance between human demands and usable resources, different resource conflict types are considered.

8.7.2 Conceptions of Resource Conflicts

Conflict-relevant factors are the quantity of available resources, their benefits and risks, as well as their distribution and needs formulated by the actors involved. Depending on these factors, different types of conflict can be distinguished. (Scheffran 1998, 2020a).

1. *Conflicts over resource scarcity and abundance*: Since the 1990s, there has been an extensive scientific debate on how the scarcity of natural resources such as minerals, water, energy, fish, and land affects violence and armed conflict (Homer-Dixon 1994; Bächler 1998). While many case studies suggest that environmental degradation and resource scarcity undermine human well-being, the effect on violent conflict "appears to be contingent on a set of intervening economic and political factors that determine adaptation capacity" (Bernauer et al. 2012: 1). Population growth, increased demand, and unequal distribution

can affect the availability of natural resources, as can overexploitation or pollution of resources and degradation of ecosystem services. Resource scarcity can weaken social and economic systems or undermine political agency and authority of governments. Associated restrictions on wealth, quality of life and basic needs, loss of identity and livelihood, economic stagnation and social disadvantage can increase the potential for conflict. Manifest conflict arises when people use means of conflict including violence to address scarcity. Often, it is not the scarcity of natural resources that drives conflict, but their abundance, which can result in a 'resource curse', for example when revenues from the extraction of valuable raw materials (diamonds, coltan, rare earths) are unequally distributed or are used to procure means of conflict (weapons, soldiers, equipment) by rebel groups or private security services.

2. *Resource availability, distribution and equity conflicts*: The distribution, access to and financing of resources are contested power factors on a national, regional or global scale. Conflicts are fought to increase or defend one's share of resources, or because of injustice in resource distribution – especially in the North-South context, where one fifth of humanity claims four fifths of the wealth and thus of natural resources. Ecological marginalisation and unequal resource distribution contribute to underdevelopment and impoverishment of populations, and the loss of vital resources (for example, agricultural land) leads to economic decline, weakens institutions, and provokes conflict tensions. While wealthy residents of industrialised nations currently live quite well at the expense of people who are spatially and temporally distant, people in developing countries currently fare poorly not so much because of natural limitations as because of unequal distribution. Who gets which share of the resource pie depends on power structures and institutional distribution and negotiation processes. Examples include the allocation of usage quotas, for example in fisheries or emissions trading. The extent to which environmental and resource problems contribute to violent conflict has been subject of scholarly debate for three decades. Studies examine the impact of climate change on conflict, from international tensions to intra-societal disputes.
3. *Conflicts arising from the risk of resource use*: Anthropogenic disruption of natural systems and risks arising from resource use can lead to security risks and political tensions, become conflict multipliers, e.g. security risks from dams, proliferation of nuclear weapons, nuclear accidents, resource depletion (oil, uranium, strategic metals), pollution of air, water and soil, or global warming, contributing to the global risk society (Beck 1992). The worst consequences can induce a collapse of resource production and damage to people and society, for example through overfishing, destruction of rainforests and biodiversity loss, in particular when carrying capacity limits are exceeded. Disruptions of natural systems have triggered political disputes, e.g. on pollution and waste disposal, construction of dams, chemical and nuclear plants, but have also triggered cooperative solutions and relief efforts. Environmental disasters can affect aid and resolution capacities, but they can also trigger cooperative solutions and relief efforts.

4. *Conflicting policy goals, strategies and actions:* Various conflicts arise over conflicting values and goals of actors in environmental policy, and the decision on actions taken to influence natural or social systems. These include differences between human resource claims and ecological regeneration, the enforcement of guiding principles (well-being, peace, sustainable development) or protests against measures in environmental, energy or climate policy.

In real conflict situations, different conflict types can influence each other. The extent to which environmental risks lead to conflict depends on societal conditions, in particular conflict history, group identities, income, institutions, organisation and equipment of conflict parties, and the importance of resources for group interests. How conflicts are fought and resolved is based on the willingness to use conflict resources: Conflicts tend to escalate when actors mutually harm each other, triggering violence and exhaustion. Although environmental crises rarely lead to violent conflict between countries, they can have indirect effects on the international community, especially when they affect vulnerable countries with a lack of water, forests and fertile land. Where institutionalised conflict-regulation mechanisms are underdeveloped, environmental conflicts carry the risk of violent conflict. Often, environmental conflicts appear as ethnic conflicts, centre-periphery conflicts, or long-distance conflicts due to transboundary risk factors such as climate change, natural disasters, epidemics, radioactive pollutants, or rising resource prices. When large numbers of people migrate to ecologically fragile and conflict-affected regions, this can amplify conflict, especially in the absence of institutionalised conflict regulation. In some cases, however, environmental problems also lead to greater cooperation, for example in agreements on water use (Ide 2023).

The complex influences of environmental change are evident in conflicts over strategic raw materials, e.g. natural gas and oil exploitation in the Arctic and Gulf regions, conflict diamonds in Sierra Leone and Angola, coltan and tin mines in Congo, Rwanda and Uganda, uranium mining in indigenous peoples' territories, rare earth mining in China or lithium in Bolivia, Chile and Argentina. Conflict potentials around renewable resources concern water scarcity in the Middle East and Central Asia, land use in Africa, bioenergy in North and South America, dams in South and East Asia; flooding and soil erosion in South Asia, deforestation in Haiti and Brazil, drug cultivation in Afghanistan and Colombia or fisheries in the North Atlantic.

8.8 The Climate-Conflict Nexus

8.8.1 Climate Vulnerability and Risk

Without strong reductions in CO_2 emissions and temperature limitations of 1.5–2 °C by the end of the 21st century (one of the planetary boundaries), anthropogenic climate change continues and has serious consequences around the world. Growing energy demand leads to shortages, environmental pollution and global warming through the greenhouse effect. Severe impacts on ecological and social systems are expected by climate change which undermines economic development and human security, especially for poor countries and populations. Risks include sea-level rise, storms, floods, droughts, wildfires, and other weather extremes which affects the lives of millions of people. Global warming is exacerbating environmental problems such as water scarcity, weather extremes and species extinction, threatening agricultural production and degrading the livelihoods of people worldwide. It can increase tensions and, together with other causes, become a driver of forced displacement.

Natural and social systems are exposed to climate-related hazards "that may cause loss of life, injury or other health impacts, as well as damage and loss to property, infrastructure, livelihoods, service provision, ecosystems and environmental resources" (definitions here are from the glossary of IPCC 2022a, b). The effect depends on their vulnerability which is the "propensity or predisposition to be adversely affected. Vulnerability encompasses a variety of concepts and elements including sensitivity or susceptibility to harm and lack of capacity to cope and adapt". While impacts are the consequences of interactions between hazards, exposure and vulnerability, risk is the "potential for adverse consequences for human or ecological systems, recognising the diversity of values and objectives associated with such systems" (IPCC 2022a, b). Adaptation is "the process of adjustment to actual or expected climate and its effects, in order to moderate harm or exploit beneficial opportunities" (IPCC 2022a, b). While hazard exposure can be considered less conflict related, vulnerability, impact and risk cannot, suggesting that the climate-conflict link is not a one-way road. Systems vulnerable to climate change experience risks to human life, income, well-being, health, migration and security and on societal infrastructures, assets, networks, stability and conflict. Vulnerability connects environmental risks with social conflict.

Serving as arisk amplifier or threat multiplier, climate change accelerates and combines different factors of social vulnerability (IPCC 2014), in particular for countries with primarily agrarian societies and a high economic dependence on natural resource markets. If tipping points are crossed, cascading effects can destabilise the Earth's climate (Steffen et al. 2018) and lead to global catastrophe (Kemp et al. 2022). Conflict-relevant mechanisms include lack of water and food, natural disasters, or environmentally induced migration. The combined effect of environmental change and social conflict can exceed adaptive and coping capacities of the affected systems which aim to maintain identity and resilience of affected units. Different geographical entities have diverse identities representing different nations, cultures and religions, levels of development and urbanisation which are interconnected. Interactions can be conflicting, cooperative and mutually adaptive, affecting societal stability in ambiguous ways.

8.8.2 Empirical Findings on Climate Change and Violent Conflict

While long-term historical studies suggest a coincidence between climate variability and armed conflict, empirical findings are less conclusive for recent periods. Empirical findings indicate that climate-conflict linkages vary significantly between the world's regions. Countries with low human development are particularly vulnerable to the double exposure of natural disasters and armed conflict (Ide 2023). Economic, political, and social factors on local, regional and global levels are interlinked with broader effects of climate change. The likelihood of violent conflict is moderated by human development, effective institutions and governance.

Frequently used data sources are the *Uppsala Conflict Data Program* (UCDP), the *Armed Conflict Location and Event Data Project* (ACLED) at the University of Sussex, the *Social Conflict Analysis Database* (SCAD) at the University of Texas and the *Conflict Monitor* of the *Heidelberg Institute for International Conflict Research* (HIIK). Most frequently analysed is communal conflict and armed conflict where the latter is defined by UCDP 2020 as the "incidence of the use of armed force by an organised actor against another organised actor, or against civilians, resulting in at least 1 direct death". After a temporary decline in the 1990s, the number of armed conflicts have increased again in recent years, to more than 55, more than in 1990. According to UCDP, 76 conflicts were classified as non-state violence, 54 state-based violence and 40 one-side violence in 2021. In comparison, the Conflict Barometer 2022 (HIIK 2023) observed a total of 363 conflicts worldwide, of which about 60% (216) were fought violently, 21 were full-scale wars and 21 were limited wars in 2022 (both one more than in 2021). Quantitative empirical studies increasingly use geocoded data for small regions on different forms of climate and conflict events, predominantly for intergroup conflict, less for conflicts against the state (Adams et al. 2018; Buhaug 2017; Ide 2017; von Uexkull/Buhaug 2021).

Over the past decade, considerable research has addressed the climate-conflict nexus, with mixed empirical evidence. Quantitative studies over all historical periods, world regions, forms of violence, and causal mechanisms remain controversial, partly due to diverse knowledge and considerable uncertainties and complexities. Significant links were found for the 'little ice age' between the 15th and 19th centuries. Since World War II, the rise in global mean temperature has not been clearly correlated with the number of armed conflicts which rose until the end of the Cold War, then fell and have been rising again. Studies on recent periods found significant, ambivalent or no clear links between climate and conflict which depends on contextual conditions, types of climate phenomena (e.g. temperature and precipitation change, resource availability, weather extremes, sea-level change) and different phases of conflict (onset, initiation, prevention, escalation, prolongation, termination). Whether climate change leads to violent conflict or war depends on political and socioeconomic contexts. Social inequality, weak government and other socioeconomic drivers have a stronger influence on violent conflicts than climate change. Many studies show that violent conflict is primarily exacerbated by climate change in places where there is a large population increase, low level of development, low economic growth, medium level of democracy, ethno-political differences as well as political instability and violence (see Scheffran et al. 2012b; Buhaug 2015; Ide et al. 2016; Detges 2017; Adams et al. 2018; IPCC 2022a, b).

A controversy emerged between those who attributed social instability and violence in past and present times to variations in temperature and precipitation (Hsiang et al. 2013) and those who remained skeptical about simple causalities (Buhaug et al. 2014). In the meantime, an expert elicitation documented the state of agreement and found that the risk of armed conflict would increase significantly with future warming (with large uncertainties), but that it could be contained through adaptation (Mach et al. 2019). Other drivers were judged more influential for conflict overall, where the specific mechanisms are uncertain. Future directions of research seek more consolidation and agreement on how the scope and severity of climate impacts matter for conflict risk (Mach et al. 2020). There is agreement that additional climatic changes, including climate extremes and their impacts on societies, will increase the risk of civil conflict with growing uncertainties, but there are divergent findings and controversies about the direct and identifiable causal mechanisms and their significance relative to other drivers of conflict which depends on the extent of moderation by political and socioeconomic conditions that could change these impacts over time (Theisen et al. 2013; Koubi 2019). The links play out differently across scales, from individual-level violence and civil war to interstate conflict and diverse forms of social contestation (e.g., Adger et al. 2014).

When climate change impacts hit critical resources, infrastructures and supply-chains that are essential for the provision of human needs, this undermines societal stability. Climate change interweaves resource issues (e.g. the water–food–energy nexus) with locally specific conditions of disaster, migration and violent conflict to increase the fragility of certain regions (Schäfer et al. 2015). Research increasingly focuses on the multiple connections between climate change and conflict (Brauch 2002; WBGU 2007; Scheffran/Battaglini 2011). Economic, political, and social factors on local, regional and global levels are interlinked with climate effects. The likelihood of violent conflict is moderated by human development, institutions and governance. Increasingly machine learning is used for climate-conflict pathway analysis (Ge et al. 2022; Xie et al. 2022).

When global warming alters human livelihoods and limits the availability of resources, such as water, food, biodiversity and land, prices and competition for these resources increase, dampen economic growth and weaken civil institutions, creating occasions for violent conflict. These developments can aggravate famine, economic crises, distributional conflicts, displacement, resource exploitation and environmental degradation, making conflict resolution and climate policy more difficult. A comparison of the number of fatalities from natural disasters and armed conflicts shows that countries with low levels of development are particularly vulnerable to this double threat, which can intensify in a spiral of escalation and spread to other regions (Scheffran et al. 2014; von Uexkull et al. 2016). Climate change is a connector that can trigger tipping elements and cascades through global networks of human-environment interaction. Armed conflict increases vulnerability to climate change, for example, by destroying water and energy infrastructure, causing skilled labour to migrate, and resulting in a lack of investment in green technologies, as demonstrated in recent violent conflicts in Syria and Ukraine (Scheffran 2022b).

Climate change could also strengthen international cooperation, risk reduction and conflict management in a coordinated climate policy. Some studies argue that cooperation prevails over conflict in crises (Slettebak 2012) which depends on certain conditions and how societies respond. Governments can support policies and institutional structures for social learning and innovations to prevent and anticipate risks. To cope with complex crises, humanitarian aid and disaster management, regulation of markets and prices, resilient and sustainable peacebuilding are needed, supported by a socio-ecological transformation and positive tipping points (Scheffran 2016b; Thonicke et al. 2020; Grêt-Regamey et al. 2019; Scheffran 2023c).

8.8.3 Hotspots of Climate Change

Climate change is a complex pattern, linking overlapping stressors and instabilities, tipping elements and cascading risks from global to local levels which combine in regional hot spots. Research in climate hot spots is often using qualitative case studies specific to circumstances, structural conditions, and social dynamics (Brzoska 2018; Scheffran/Battaglini 2011). Most affected are vulnerable countries that depend on agriculture, are exposed to droughts or floods, and cannot easily help themselves. Climate-related injustices concern the asymmetric distribution of costs and risks between those who pollute and those who suffer from climate impacts. In areas that are particularly vulnerable to climate stressors and have low adaptive capacities, human livelihoods are threatened by extreme events and the gradual destruction of natural resources that are essential to meeting needs, such as water, food, forests, or biodiversity. In vulnerable hot spots climate change can exacerbate existing problems, including violence, terrorism, civil war, and forced displacement, particularly in fragile regions where poverty, violence, injustice, and social insecurity are already prevalent (Detges 2017; WBGU 2007; Werrell/Femia 2015).

Many threats to human security are not caused solely or primarily by climate change. Rather, complex constellations of problems in the affected areas are responsible: the destruction of ecosystems, poverty, political instability, land-use change and the overexploitation of soils, or the lack of early warning and disaster prevention. Various destabilisation processes can intensify in hot spots and spill over into neighbouring regions. Coastal zones and river basins, hot and dry areas, and regions whose economies depend on climate-sensitive resources and agriculture are particularly vulnerable (Scheffran/Battaglini 2011; WBGU 2007). In addition to primary regional impacts, indirect effects in distant areas can trigger changes nearby, as the example of food prices shows. Some responses can exacerbate the situation, when people in need force the overexploitation of resources, migrate to other risk zones, or use violence to ensure survival. Examples include forcing the use of fossil energy (such as in the Arctic), cutting down rainforests, or air conditioning in hot regions, which would accelerate climate change worldwide. Because physical, economic and geopolitical risks are interlinked, climate-related events in a globally interconnected world can undermine social and political stability elsewhere (Scheffran 2023a, b).

In fragile states that cannot guarantee core government functions, climate change overwhelms problem-solving and adaptive capacities. Marginalisation, environmental degradation, dwindling and unequally distributed resources can exacerbate displacement and security risks (Molo 2015). If staple foods become scarce and more expensive, this can threaten the livelihoods of poor classes and trigger social upheaval and conflict. In fragile areas of the Mediterranean and Sahel of Africa, Latin America, South Asia, and the Pacific, climate impacts can contribute to complex conflict constellations. When, for example, river deltas in southern Asia are flooded or cultivated areas dry up, the livelihoods of many farmers are destroyed. Tensions occur over territorial claims and natural resources in the Arctic region and the Mediterranean. Melting polar ice and thawing permafrost affect the strategic interests of Europe, Russia and North America.

8.8.4 Future Climate-Conflict Risk

Analysing future conflict risk is inherently difficult, whether in the context of early warning or long-term scenario-based projections (Hegre et al. 2016; Witmer et al. 2017). While future climate change may significantly increase conflict risk over large spatio-temporal scales and various causal chains, considerable uncertainties remain which are hard to eliminate or reduce. Some system interactions are deeply complex, including the lack of historical precedents, low-probability outcomes, the extent and effectiveness of adaptation. Large-scale conflict risk and geopolitical interventions are hard to anticipate and face major issues of out-of-sample prediction (Bowlsby et al. 2019; Van Weezel 2019). Due to context-specific circumstances it is challenging to generalise results and theories (Mach et al. 2019, 2020; Scartozzi 2020). Existing models are better at long-term projections of conflict vulnerability for states or regions than at predicting violence and escalation at a given place or time. Relationships between climatic variability and instability may hold over a given window of time or context, but not beyond that (Abel et al. 2019). Forecasting future conflict risk is facing intrinsically unpredictable strategic adjustments by government and members of society, posing a challenge for proactive intervention, peacekeeping missions, development aid or food assistance.

To inform decision-making about possible futures and help to select goals and priorities requires exploring local-to-global interactions relevant to both climate change and conflict risk. Exchange between analysts and practitioners can help to advance modelling capacity as a basis for context-specific risk communication and policy analyses robust to the uncertainties (Mach et al. 2020). A range of analytical tools are relevant for stakeholders, including scenario simulations, preferences elicitation, agent-based modeling, dynamic adaptive pathways, or theories of tipping points, risk cascades and stability (Hegre et al. 2019; Kwakkel et al. 2016; Witmer et al. 2017). Conflict could arise in times and places that are distant from the actual trigger. For instance, a combination of weather shocks and export bans in major grain-producing countries led global food prices to spike in 2010, with destabilising effects for importers in the Middle East and North Africa (Costello et al. 2015), affecting agricultural markets or the flows of displaced people

8.8.5 Intermediate Factors and Pathways

Going beyond the current state of empirical research on the link between climate change and violent conflict, selected intermediate pathways and regional differences are considered in how climate change and violent conflict affect human security and how human development and institutional processes multiply or reduce potential conflicts. A reason for the different findings in the empirical literature on climate-conflict risk is the weak theoretical basis, which does not sufficiently consider the complexities of the issue. The multiple pathways between climate change and violent conflict can be analysed in the integrative framework of the linkages between *climate change, natural resources, human security and societal stability* (CNHS) (Scheffran et al. 2012a, b). Understanding the different pathways provides a foundation for the analysis of empirical results. More comprehensive approaches are needed to disentangle the complex climate-conflict nexus, discussing key intervening variables and causal pathways between precipitation changes, freshwater scarcity, and food insecurity as well as weather extremes and environmental migration. Pathways may differ in their relevance for different types of violence and regional contexts. Theoretical insights suggest that low-level violence is more likely to be linked to the effects of climate change than full scale wars.

Climate change has multiple relevant effects on the environment, economics, society, and politics, such as the frequency and intensity of extreme weather events, the variability and long-term changes of weather parameters such as temperature and precipitation. These variables may have direct impacts on social systems or indirect implications through other pathways affecting natural resources and human security, which together can have ambiguous results. For each pathway, the consequences of climate change depend on how vulnerable affected natural and social systems are and how sensitive they respond to the stress. At each level, human intervention can influence the systems, e.g. through mitigation and adaptation strategies to reduce risks, strengthen resilience and improve sustainability. The main focus here is on disasters and extreme events as pathways between climate and conflict.

8.8.6 Disasters and Extreme Events

When natural and adaptation limits are exceeded, the risks of environmental hazards and disasters tend to grow, on the one hand by a growing frequency and intensity of extreme events, on the other hand by people living in high-risk zones, increasingly turning 'natural' disasters into human-made events. Different weather-related hazards (e.g. droughts and wildfires, heat and cold waves, flash floods, landslides, mudslides, storms and storm tides) have diverse spatial and temporal consequences, with potential massive disruptions of affected infrastructures through large impacts that threaten their existence and exceed thresholds of adaptability, stability and resilience. Fragile hot spots are particularly vulnerable to and unable to cope with the consequences of disasters, leading to qualitative and irreversible systemic changes.

Climate change is projected to affect the frequency and intensity of weather extremes with rising economic and social costs in many areas of the globe (IPCC 2018, 2021), which kill and displace people, and contribute to social instability and conflict, especially in regions already affected by societal cleavages and political tensions. More likely and intense weather extremes may result in high economic damages of hundreds of billions of euros and threaten the health and lives of more than a hundred million people each year. In total, more than 475,000 people died from 11,000 extreme weather events in 2000–2019 (Germanwatch 2021). While absolute financial losses were higher in rich countries, deaths, misery and existential threats from extreme weather were more common in low-income countries due to higher vulnerability to damage and lower coping capacities. During the 2000–2019 period, Puerto Rico, Myanmar, and Haiti experienced the highest damages and deaths. In 2019, Mozambique, Zimbabwe, and the Bahamas suffered particularly from weather extremes (Germanwatch 2021). Examples from the last decade include the Indus flood in Pakistan in 2022, the drought in China in 2010/2011, and Typhoon Haiyan in the Philippines in 2013. Industrialised countries are also affected, e.g., by the European heat waves of 2003 and 2018, the Elbe floods in Germany in 2002 and 2013, a record number of storms in the North Atlantic in 2020, devastating fires in Russia, California, Brazil, and Australia (Howe et al. 2017, MunichRe 2021). The summer of 2021 showed devastating floods in the Eifel and other parts of Central Europe, and heat waves and large-scale fires in the Mediterranean, which can cause air pollution, crop loss and health problems. In the US, hurricanes such as Katrina in 2005, Sandy in 2012 and Harvey in 2017 resulted in large amounts of rain, extreme winds and severe flooding, causing massive damage and loss of life in urban centres.

By disrupting social livelihoods, climate extremes are a potential driver for instability and violent conflict, including harmful consequences such as economic crises, famines, displacement, resource exploitation, and environmental degradation, which in turn can increase climate vulnerability (Scheffran 2020a, b). There is no simple, static, and direct causal relationship, but a complex, dynamic, and mutually reinforcing interaction, depending on the type of climate or conflict events and the local context in which they occur. The impact of climate extremes on violent conflict is contested, depending on methodology, assumptions, and interpretations of causal pathways and cases. The question is: how sensitive to climate extremes are the different conflict types (e.g., armed conflict, military interventions, and other collective violence; riots, protests, disputes, and other social conflicts) and actors (e.g., rebel groups, protesters, militia, governments, military, police, United Nations, and others).

In a review of large-N studies by Ide et al. (2016; Ide 2023), most found climate-related disasters to significantly increase violent conflict risk while in slightly more studies precipitation extremes did not increase this risk. Empirical research allows for general conclusions (Brzoska 2018; Scheffran 2020b; Ide 2023):

1. Single weather events do not cause collective violence by themselves, but amplify conflict dynamics and contribute to its initiation, (de-)intensification, prolongation, and termination.
2. While large-N research comes to conflicting results, a growing number of studies found that extreme events and disasters weakly increase the likelihood of collective violence; some studies found a statistically significant relationship, others fail to do so (Sakaguchi et al. 2017).
3. Results differ with regard to the type and magnitude of climate extremes as well as the different fields, types, and stages of conflict. Larger disasters with long-term consequences are more likely a source of contention and collective violence than smaller events with short-term, yet corrosive effects on society and low-level violence (Boubacar et al. 2017).
4. Droughts are among the primary drivers of conflict initiation and aggravation, in particular in ethnically divided societies, adding to grievances and pushing conflict into violent and repressive directions (Schleussner et al. 2016). Without ethnic conflict, results are less clear.
5. Consequences vary in different phases of conflict, being more significant for conflict escalation and prolongation than for initiation and termination (Eastin 2016). On the other hand, extreme events can be opportunities to lower intensity of and terminate collective violence, for example, through local trust building or assistance. Qualitative studies are more concerned with the dynamics of low-level violence in specific regions (Mobjörk et al. 2016).

Regardless of historical data, the impact of future climate extremes goes beyond previous experiences, leaving space for scenarios, plausibility, and speculations. Even if societies were able to adapt to moderate climate change in the past, they may be overwhelmed in the future by extreme and rapid climatic changes exceeding adaptive capacities. Future studies consider both conflict and cooperation in extreme events and the role of institutions (Walch 2018; Ide 2023).

Disasters and weather extremes often exceed the coping ability of societies, particularly where socio-economic living conditions have already been affected by environmental degradation, poverty, hunger and conflict. In fragile and weak states with social fragmentation and inadequate governance capacities, the core functions of government cannot be guaranteed, such as law and public order, state monopoly on the use of force, welfare, participation and public services (infrastructure, health and education) (WBGU 2007). A breakdown of state functions and social systems after hurricanes can trigger looting and criminal activity, increase tensions, conflicts and refugee movements.

Disasters set relief measures and disaster management in motion to mitigate the consequences and prevent escalation. Sometimes the opposite can occur, such as in violent disputes over the distribution of relief supplies (Scheffran 2020b). Climate-related extremes can act as risk amplifiers in hotspots, where they undermine social and political stability through physical, economic and geopolitical risks in a globally interconnected world and set in motion global chains of events that can destabilise across borders and trigger or increase migration. In addition to storms and floods affecting food supplies temporarily and locally, droughts can affect agricultural production and water resources on a large spatial and temporal scale with persistent impacts on global food markets. Shortages of water and food and associated price increases can threaten the livelihoods of poor social classes. Bread protests and hunger revolts responded to global supply crises such as 2008 and 2011.

8.9 Human Migration Between Climate and Conflict

8.9.1 The State of Mobility and Migration

Similar to the diffusion of goods, technical innovations and human behavior in a globalized world, mobility within and between geographical places and spaces is a key phenomenon for all life-forms of plants, animals and humans, connecting the interspace between origin and destination through pathways crossing geographical borders and boundaries. Migration, displacement and flight are different forms of mobility which fully represent the geographical 'triality' concept introduced earlier. More recent studies characterise migration as a combination of the ability to move and the motivation to move, both of which are required to migrate. Climate impacts can affect both motivation and capability to migrate, both in re-enforcing and contradicting ways. For instance, if capability to move is non-existent or impaired by climate change, it becomes difficult for vulnerable parts of the population to migrate, whatever the motivation. Thus, migration is a complex phenomenon, where environmental drivers are often difficult to separate from economic, social, political and conflict drivers. Environmental degradation, resource scarcity and climate change are part of a complex nexus of forced migration and displacement that includes underdevelopment and impoverishment, oppression and persecution, war and violent conflict driving people away from their homes.

People migrate for many reasons, from worsening livelihood conditions (push factors such as exploitation, oppression, poverty, diseases, tensions and conflicts) to the search for new opportunities (pull factors like better income and living conditions, prosperity and innovation, profit and privileges, power and control), which can be voluntary and involuntary, short- and long distance, temporary and permanent. People can also migrate, relocate and resettle in response to climate, environmental or land use change (BMZ 2021). When people leave their home and cross borders in search for new opportunities, along the migration pathway they live temporarily in the interspace between old and new places where they are exposed to distress and social change (income, jobs, training, …). This can challenge their identity and well-being, in the process of integration and adaptation to the new environment, which can lead to conflicts among migrants and with local populations, but also offers opportunities of cultural enrichment. As communities remain connected between origin and destination, social networks contribute to exchange through the flow of funds, information and knowledge, as well as the spread of diseases and harmful substances. In this process migrants also change themselves (Scheffran 2017a).

In 2022 the number of forcibly displaced people worldwide for the first time exceeded 100 million, more than double the number a decade ago (see Fig. 8.3). Between 2010 and mid-2022, the number of refugees under the mandate of the *UN High Commissioner for Refugees* (UNHCR) or recognised by other countries as refugees more than tripled, from 10.5 million to 32.4 million people (UNHCR 2023). The total number of *internally displaced persons* (IDPs) increased from 22.4 million in 2011 to 71.1 million as of end of 2022. The number of new displaced persons per year reached 60.9 million in 2022 among them 32.6 million by disasters and 28.3 million by conflict and violence (IDMC 2023). By far most continue to live in developing countries and tend to stay close to their original homes. Displacement from natural disasters occurred mostly in Asia, and from conflict mostly in Africa.

In 2022 conflict, disasters, and the continued effects of the Covid-19 pandemic combined to aggravate people's pre-existing vulnerabilities and inequalities. Since Ukraine and Russia are among the world's largest producers of fertiliser and grain, their conflict fuelled cascading effects on global supply chains and food prices, heightening food insecurity for the poorest and most vulnerable, including IDPs, and undermined progress in reducing global hunger and malnutrition. The La Niña weather phenomenon was leading to record levels of flood displacement in countries including Pakistan, Nigeria and Brazil and aggravated the worst drought on record in Somalia, Ethiopia and Kenya (IDMC 2023). Conflict and violence left 62.5 million people displaced across 65 countries and territories at the end of 2022. Over the

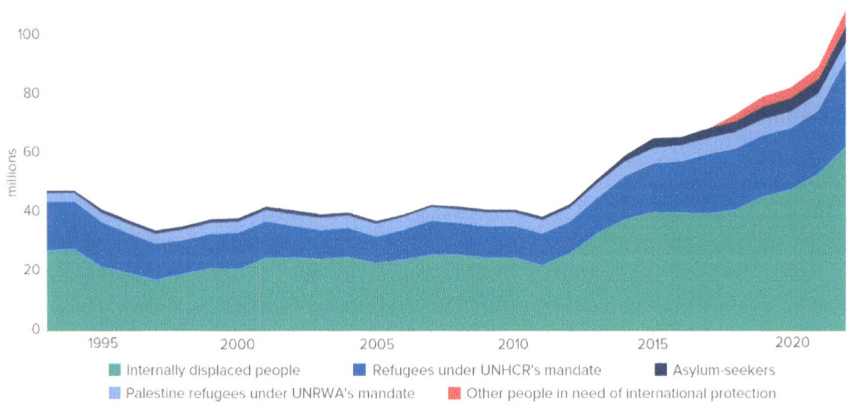

Fig. 8.3 People forced to flee 1993–2022. *Source* UNHCR (2023)

past 10 years, sub-Saharan Africa, the Middle East and North Africa had the highest numbers of IDPs. In countries such as the Philippines, Madagascar and South Sudan, consecutive disasters forced people to flee repeatedly. Beyond immediate humanitarian assistance, investments are needed for anticipatory action and risk reduction measures that strengthen displaced communities' resilience (IDMC 2023).

8.9.2 Migration and Violent Conflict

Alongside political persecution, armed conflicts have been a main driver of forced migration in history. Wars and civil wars have triggered large movements of refugees, including the two world wars, the end of the Cold War and armed conflicts in former Yugoslavia, Darfur, Afghanistan, Iraq, Libya, Syria and Ukraine. In 2022, about 28.3 million people were displaced by violence and conflict worldwide (IDMC 2023), with war-related proportions highest in Africa, the Middle East and Ukraine. The impact of violent conflict on refugee movements has long been studied more than the reverse links, when people fleeing can themselves become a conflict factor. For example, migrants experience conflict in their home countries and violence along migration routes, or overcrowded refugee camps can become hotbeds of conflict and violence. Relevant conflict dynamics also arise from problems and differences between native and immigrant populations, especially due to competition for scarce resources (water, arable land, housing, jobs and basic social services) or changes in the ethnic composition of the population.

In the past decade, the conflicts in the Arab world and the resulting increase in refugee movements have been associated with various crisis phenomena, such as terrorism, right-wing populism, nationalism and the political division of the European Union (EU) which was surprised by the 'refugee crisis' of 2015. Differences evolved within the EU due to the scale and speed of immigration, exacerbated by protests and resistance, including violent attacks on refugee shelters. Many of the refugees came from war and crisis zones, particularly the Syrian civil war, from which some 6.8 million people fled to neighbouring countries (UNHCR 2023), mainly to Lebanon, Iraq, Jordan and Turkey. A few hundred thousands reached central Europe through the Balkan route, connecting political problems from source to destination. In the face of media-amplified images of 'refugee flows', political reflexes and cracks opened up in Europe. The battle over the refugee issue led to perceptions of threat and heated sentiments against 'foreigners' in the mass media. The debate has been hyped up as a cultural and religious clash, conflated with terrorist attacks by the *Islamic State* (IS) in European cities which unleashed a political momentum (Bank et al. 2017) to establish borders blocking migration pathways. The EU's refugee deal with Turkey strengthened the autocratic Erdogan regime and reduced Europe's willingness to seek sustainable solutions in the region. The British Brexit played out immigrants against the EU, and U.S. President Donald Trump mobilised sentiments against immigrants, terrorists and Islam and for a border fortification between the U.S. and Mexico. If involuntary displacement and resettlement intensifies, it becomes urgent to evaluate approaches for absorbing displaced persons into communities to minimise frictions (Maystadt et al. 2019).

8.9.3 Environmental Migration and Disasters

In history humanity has responded to environmental degradation and climate variability in two ways, forced displacement and conflict or innovations and cooperation. More research has been focused on degrading living conditions forcing people to leave areas that have become inhospitable and migrate to other areas, in particular when resources and ecosystems become unsuitable to support human life, including severe physical, chemical, biological and environmental stresses such as weather extremes, disasters and threats to ecosystem services, which combined with social instability and violent conflicts are supposed to force people to move away from harm (Scheffran 1994; Richter 1998; in historical retrospect McLeman 2013).

There are no accepted definitions nor legal regulations of environmental migration which according to the *International Organization for Migration* (IOM) involves "persons or groups of persons who leave their place of residence, either temporarily or permanently, due to sudden or progressive changes in the environment that adversely affect their lives or living conditions, and who move within or outside their country" (IOM 2007: 33). The definition includes recognition of environmental migration due to extreme weather events and permanent environmental degradation. In addition, spatial and temporal facets of environmental migration are distinguished, as well as natural and anthropogenic causes (Scheffran 2017a). The UN *Inter-Agency Standing Committee* (IASC) identified four possible drivers of environmental migration: hydro-meteorological disasters; environmental degradation and/or long-lasting disasters; land loss due to rising sea levels; conflicts caused by resource scarcity (IASC 2008). One of the first comprehensive studies of environmental migration (EACH-FOR 2009) distinguished between people who (1) voluntarily leave in response to environmental change (environmental migrants); (2) leave their homes to avoid environmental risks (environmentally induced dislocation); or (3) are relocated in connection with a planned transformation (development-induced dislocation). Several specific case studies of the impact of environmental change on migration movements have been analysed in the annual State of Environmental Migration reports (Zickgraf et al. 2021).

In 2022 disasters displaced 32.6 million people, most in Asia (China, India, Philippines, Nepal, Myanmar), but also in parts of Africa, Australia, and North and Latin America (IDMC 2023). Many were short-term weather extremes that affect low- to middle-income countries. Droughts, species loss, or sea-level rise that build up and compress over longer periods of time can also bring catastrophic consequences. People who are negatively affected by environmental stress can stay and adapt or migrate when living conditions become unbearable and territorial integrity of states is threatened (e.g. small islands due to rising sea levels). Climate-related stress increases the pressure to migrate in fragile hotspots, but most do not flee but try to solve problems locally, or stay close to their origin to return as soon as possible. Migration planned in response to gradual environmental degradation is different from a sudden exodus after an extreme weather event. Empirical findings identified many linkages about environmental migration as a precursor, transmission mechanism or intermediary variable (Hoffmann et al. 2020). Affected regions are under pressure if they have to handle sudden, unexpected and large-scale migration that overwhelms their adaptive and management capacities.

Environmental change can not only promote migration, but also make it more difficult by increasing poverty among rural populations and limiting opportunities to migrate, leading to *trapped populations* (Hoffmann et al. 2020). Environmental pressures and vulnerabilities may increase as people migrate to exposed regions – such as coastal cities affected by storms or sea-level rise. At their destination, in-migration can exacerbate environmental problems and competition for scarce resources such as arable and pasture land, housing, water, jobs, and social services (Ionescu et al. 2017). Rural migration to metropolitan areas and their slums exacerbates environmental and social consequences to the brink of collapse.

8.9.4 Debates on Climate Migration

How climate change affects human mobility has been studied for three decades and was part of the discourse on climate change as early as 1990, when the first report of the IPCC warned that changes in precipitation and temperatures could lead to major migration movements, for example due to crop failures and sea level rise, or indirectly due to economic problems and conflicts. Some projected future climate change as one of the most important reasons for forced migration (Myers 2002) and climate migration as one of the key political challenges of the 21st century. The *German Advisory Council on Global Change* (WBGU) saw climate-induced migration as one of the future areas of conflict in international politics, especially in regional hotspots (WBGU 2007). A position paper of the European Commission anticipated a "substantially increased migration pressure" due to climate change (EU 2008). The UN High Commissioner for Refugees (UNHCR) recommended to avoid the terms 'climate refugees' and 'environmental refugees', which were perceived as inaccurate and misleading (UNHCR 2011).

It is much debated to which degree progressing climate change will increase the number of refugees and displaced persons. Estimates and forecasts range from 50 million to one billion affected people (Myers 2002; Black et al. 2011; Jakobeit/Methmann 2012). As long as robust estimates and scenarios are not available, it remains difficult to determine the future magnitude of climate-induced migration. A compilation in the IPCC's fifth Assessment Report of 2014 found evidence of increased migration in 18 case studies, a reduction in migration in six cases, and a reduction in migration in seven cases (IPCC 2014: 770). Critical studies challenge projections with huge numbers as speculative and question 'apocalyptic narratives' that justify extreme responses, such as anti-immigration discourses and military or police action for border control. Exaggerated assumptions can be explained by uncertainties and complex relationships that make it difficult to determine the environmental influence on migration. There is no clear demarcation between climatic, environmental and other reasons for migration.

Since 2015 the scientific discourse on climate-induced migration has grown considerably, becoming more multi-layered, differentiated, and controversial. Critics doubt that migration is primarily determined by climatic changes (Boas et al. 2019; Fröhlich/Klepp 2020). The alarmist tone of climate migration studies is criticised as neo-Malthusian. Narratives of refugees as security threats are rejected for victims who struggle for life and are instrumentalised in political power games. A defence-oriented European policy that sees migration as a trigger for security problems, political instabilities, and conflicts is unlikely to prevent the underlying causes of forced displacement, but is more likely to exacerbate them (Scheffran 2017).

8.9.5 Pathways and Hotspots of Climate Migration

In the future, the climate crisis is expected to be part of a complex set of drivers of displacement that undermine the livelihoods of people (Burrows et al. 2016; Abel et al. 2019). While storms, floods and landslides can lead directly and locally to displacement over short periods of time, drought, desertification and water shortages tend to have a regional impact and occur over longer periods of time. The pressure for displacement increases when rising temperatures, shifts in precipitation, and weather extremes affect the infrastructures of water, food, and energy essential for human life (McLeman 2013). Multiple factors can have compounding effects and exceed tipping points triggering self-reinforcing drivers of movements.

Impoverished and marginalised populations are most vulnerable, "because they are exposed to the greatest dangers or because they are destitute, do not have the means to flee, even though they want to" (Ionescu et al. 2017: 34). The lack of mobility increases the proportion of trapped populations caught in a downward spiral that threatens their very existence. Despite precarious environmental conditions, many affected people do not migrate long distances but stay close to home, contributing little to international south-north migration. People are not protected when they migrate to vulnerable regions, such as coastal cities affected by storms or sea-level rise (Scheffran 2017a). Impacts can outstrip governance and adaptive capacities of affected societies, especially in fragile states, where climate hazards act as risk amplifiers. To compensate, institutional mechanisms and legal regimes need to address climate-induced displacement, including prevention, protection and international cooperation among states and civil societies.

Prone to forced migration are fragile regions and hotspots where poverty, violence, injustice, social inequality and insecurity prevail, with low adaptive and coping capacity, economies dependent on climate-sensitive resources (Ionescu et al. 2017;

Hoffmann et al. 2020). Climate-sensitive coastal and river basins, hot and dry areas, and regions dependent on agriculture and resources are particularly vulnerable. Although most migratory movements tend to remain in the Global South, there is increasing migratory pressure on the Global North, in Europe mainly from sub-Saharan Africa and the Arab world, and in North America from the Caribbean, Central and South America. In the most sensitive and fragile hotspots, climate-related stress can exacerbate existing problems and increase migration pressure which can transfer human misery and tensions elsewhere. People in the Mediterranean region are affected by drought and water shortages, in the Sahel and Central Asia by droughts, in Central America and South and East Asia, and on small islands such as the Maldives by storms and floods, or in populous coastal zones that are already struggling with water and food problems. Due to climate change and population growth, overuse and land consumption, also by industrialised countries, less drinking water and fishing, arable and pasture land can be expected in large parts of Africa. In the Pacific and the Arctic environmental and climate change combine with social problems, geopolitical tensions, and the consequences of globalisation in complex constellations of conflict. In many cases, global and regional climate changes are intermingled with local environmental problems, such as deforestation, overuse of soils or coastal degradation, and political and socioeconomic conditions.

Migration pressure was intensified by 2011 upheavals and conflicts in the Arab world, which were exacerbated by global food crises and weather extremes. In addition, Syria suffered a devastating drought (2007–2010) shortly before that uprooted many people in rural areas, which increased discontent with the Assad regime. In the bloody civil war, the country became a pawn of regional and global powers, driving many refugees as far away as Europe. The climate role in civil war and flight in Syria is controversial (Kelley et al. 2015; Selby et al. 2017). In the Sahel, climate change, population growth, overexploitation, and local environmental degradation are causing prolonged droughts and desertification, with tensions between nomads and sedentary arable farmers over fertile land. The complex interplay of migration factors is evident in the Sudanese region of Darfur as well as at Lake Chad where water and land use conflicts intermingle with civil wars between governments and rebels, exacerbated by state failure, corruption and social exclusion (Scheffran et al. 2019; Froese et al. 2019; Kamta et al. 2021).

While climate-induced migration is expected mainly in the Global South, there are concerns in the Global North that the migration pressure on or within their own borders will increase. An example was Hurricane Katrina 2005 which cost the lives of more than 1800 people and forced hundreds of thousands to flee New Orleans, some of whom never returned (Palinkas 2020). This made clear that industrialised countries are not immune against the climate crisis, and even rich nations can be overwhelmed by natural disaster. The risk of flooding can make low-lying coastal regions uninhabitable in the long run and lead to migration, for example because real estates lose value in anticipation of future risks. While this still receives little attention, migration from crisis areas is leading to controversial debates, intensified by right-wing populist movements. This was evident in the 2015 'refugee crisis', when domino effects put pressure on Germany and the EU. Many of the immigrants come from the MENA region and the Sahel, which are directly affected by climate change and a resulting migration pressure (Scheffran 2017a).

8.10 Environmental Impact of Conflict

8.10.1 Military Impact on Climate and the Environment

Military, armament and war are not only dangerous for human society, but also for the natural environment which often enough in history fell victim to scorched earth warfare and repeatedly has been manipulated and used for military purposes (Scheffran 2019a). Agricultural fields and forests have been burned, waters poisoned, and land rendered unusable in order to deprive an adversary of its livelihood. Modern wars can wreak even greater destruction on ecosystems, affecting human life and health. Armed conflict consumes and strains natural resources (air, water, soil, land, forests, and oceans), damages public infrastructure and services (e.g., energy, health, sanitation, waste collection), and negatively impacts wildlife and ecosystem conservation. Large areas of land were destroyed and polluted by trench and poison gas warfare in World War I, while massive devastations followed from World War II, with area bombings, shipwrecks and mines, and the increase in arms production contaminated marine and terrestrial ecosystems. Multiple environmental stresses resulted during the Cold War, from large-scale use of herbicides such as Agent Orange in the Vietnam War to radioactive contamination of the nuclear arms race. The East-West conflict ended with dangerous legacies of the arms race. Hopes were dashed that conversion of the military infrastructure would free up funds for environmental protection. On the contrary, still high military spending, new arms races and warfare became an obstacle of environmental protection, sustainable resource use, biodiversity preservation, and pollution reduction. War and genocide led to massive refugee movements and camps that stressed surrounding ecosystems, forests, and national parks (Scheffran 2018b). The 1991 Gulf War saw severe environmental damage from oil fires in Kuwait and oil releases in the Persian Gulf.

In a crowded and resource-constrained world, high resource intensities and environmental impacts of armaments are an essential problem, even more with a resource competition between renewable energy, digitalisation, armament and other uses. War and the military are among the largest consumers of energy and other resources and release harmful environmental pollutants. Military land use affects the quality of life, health, and nutrition of society. Fuels and toxic chemicals can remain in the environment for a long time. Heavy military vehicles damage soils and infrastructure (Scheffran 2019a). Due to high dependence on fossil fuels, military activities and infrastructure also release considerable GHG emissions that fuel global warming. "Armed forces have a massive carbon footprint that is absent from global accounting."

(Rajaeifar et al. 2022). The carbon footprint of the EU military has been estimated 2019 as nearly 25 million tons CO_2e (Parkinson/Cottrell 2022), and the U.S. Department of Defense is the "world's single largest greenhouse gas emitter" (Crawford 2022: xvi), with more than 50 million tons CO_2e in 2020, nearly three-quarters of U.S. government emissions (van Schaik et al. 2022). If the world military were a country, it would be ranked fourth in aggregated annual emissions (Parkinson/Cottrell 2022).

Because of oil dependence, military forces and defence contractors support armed interventions or deployments to secure oil and energy supplies and defend tanker routes or pipelines. Conflicts and military spending are rising, which is diminishing the capability to address environmental challenges, such as sustainable resource use, biodiversity preservation, and pollution reduction. Armed conflicts are often financed through the exploitation of valuable resources, from forests to raw materials. Facing the security risks of climate change military actors call for actions which focus less on mitigating the threat (e.g. by reducing emissions) than on countering them, e.g. by securing borders and protecting trade routes, influencing resource conflicts, instability from weather extremes, fending off refugees, and containing social unrest.

Relying on military force to 'fight' the climate crisis would be counterproductive, given the environmental risks involved, and possibly reinforce a vicious circle. While military GHGs have been excluded from the Paris climate agreement, and there is no consolidated public reporting for national militaries and no overarching reduction goals, experts have called for decarbonising the military (Rajaeifar et al. 2022). To minimise fossil-fuel-related vulnerabilities, several initiatives to move to low-carbon energy, reduce dependency on Russia, and combat climate change (van Schaik et al. 2022) are underway. As military spending is already at an all-time high (SIPRI 2023), the rapid growth in military spending will draw funding away from ambitious climate protection and increase military emissions, thereby constraining deep decarbonisation. The consequences may lead to complex conflicts and provoke reactive technical measures, such as the expansion of nuclear energy and geoengineering (technical interventions in the climate system) or military responses like missile defence and nuclear weapons, which further exacerbate the problems (Scheffran 2019c). Thus crises are mutually enforcing: climate change creates new hotspots of conflict while military force exacerbates conflicts and pollutes the environment.

8.10.2 Environmental and Climate Damage of Nuclear Weapons

With the use of atomic bombs by the USA against the Japanese cities of Hiroshima and Nagasaki in August 1945, the destruction of humans and nature reached a new level. The combined effects of heat, pressure and radiation not only killed hundreds of thousands of people, but also local flora and fauna, contaminating large areas of land, water and the atmosphere with radioactivity for years. Of the more than 60,000 nuclear weapons worldwide at the end of the Cold War, nearly 13,000 still exist

30 years later, plus nuclear material from disarmed weapons, more than enough to wipe out life on Earth. With a half-life of 24,110 years, the impact of Plutonium-239 lasts far longer than the history of human civilization, including cancer and genetic damage (Scheffran 2020e).

The nuclear complex (civilian and military) has created multiple environmental burdens throughout the Earth system, radioactively contaminating thousands of square kilometres of soils, rivers, lakes, and oceans, as well as the atmosphere and biosphere. Nuclear fission products enter the bodies of all organisms through air, water, and food and continue to radiate in the body's cells. All components of the nuclear fuel spiral, from uranium mines to uranium enrichment, reactors and nuclear weapons to (final) storage, pollute the environment and health. Particularly serious is the mining of uranium ore, which leaves behind radioactive and toxic sludge and tailings. Radioactive substances enter the food chain through air, water and soil. Depleted uranium is used by the U.S. military in munitions and has been linked to the Gulf War Syndrome. Many uranium mines are located in areas of indigenous peoples (Scheffran 2020e).

More than 2000 nuclear weapons were tested in the nuclear arms race of the Cold War. The plutonium and other radioactive substances released during these tests circulate around the world and accumulate in the ecological food chain, resulting in estimated hundreds of thousands of deaths. Test sites have become uninhabitable for indigenous peoples (such as Bikini Atoll). Some nuclear weapons facilities affect drinking water and food supplies. Mortality is elevated near test and production sites. Finally, the released radiation causes genetic diseases and birth defects. Leaks and accidents often have an impact far beyond the immediate environment. The dangerous legacies of the nuclear arms race include global stocks of highly enriched uranium and plutonium, enough for more than 200,000 simple nuclear weapons. It is uncertain if and when a responsible solution will be found for the long-term disposal of nuclear weapons materials, a problem for many future generations. Cleaning up nuclear facilities costs hundreds of billions of USD until today (Scheffran 2020e). About one-seventh of Russia's land area has become an ecological emergency area, and seven million people live on radioactively contaminated soil.

The nuclear arms race is in full swing again and consumes enormous resources that are missing for environmental and climate policies and sustainable energy supply. Among the world's increased military spending, nuclear armament claims significant resources, in the U.S. alone, some USD 8.7 trillion (adjusted for inflation) have been spent on nuclear weapons since 1940, another trillion is planned in the next three decades, leading to more CO_2 emissions. Delivery systems, such as nuclear submarines and ships, have a high CO_2 footprint (Scheffran 2020e)

Thousands of nuclear weapons pose high risks of intentional or accidental nuclear war and an existential threat to the world's climate. Scientific studies show that even a 'limited' regional nuclear war could destabilise the global climate. Nuclear explosions in cities would ignite large fires that carry smoke, soot, and dust into higher atmospheric layers and block sunlight. As a result, the earth's atmosphere could cool down by several degrees (nuclear winter), and significantly reduce crop and food productivity leading to catastrophic famines, threatening the lives of more than a billion people. The climate-nuclear nexus is an essential threat to international stability, global security and peace (Scheffran et al. 2016). Climate-driven weather extremes threaten nuclear safety, e.g. of nuclear facilities in high-risk zones. In the 2018 heatwave summer, wildfires in California threatened military nuclear and missile test facilities. Some regions face a double threat by the combination of nuclear weapons and climate change. The Marshall Islands were devastated by nuclear weapons testing and are now under pressure from sea-level rise and storms, seeking defence against both dangers (Scheffran et al. 2016).

8.10.3 Environmental Impacts in the Russian-Ukraine War

The war in Ukraine poses severe risks for the environment, with long-term consequences, claiming energy, raw materials, and GHGs, through military operations and rearmament, and the destruction of infrastructure that must be rebuilt. Impacts are chemical and radioactive pollution, the flooding of areas, growing military spending, production, deployment, operations and destructions of infrastructure that have to be rebuilt (Tollefson 2022; Pereira et al. 2022a; Shumilova et al. 2023). The destruction of the Kakhovka dam on 6 June 2023 caused a major humanitarian and environmental disaster and affected the irrigation system important for agriculture in southern Ukraine. Like in other violent conflicts, large quantities of fossil fuels and GHGs are being released, through military operation, the arms buildup before and after the war, and the destruction of infrastructures that have to be rebuilt.

Determining systematically the direct and indirect emissions of the war, de Klerk et al. (2023) calculated that Russia's war of aggression in Ukraine caused about as many climate-damaging emissions in the first year alone as a country the size of Belgium in the same period. Accordingly, the Ukraine war emitted 120 million tons of CO_2 emission equivalents in the first year. The calculation estimates that the largest share of emissions – about 50 million tons – is caused by post-war reconstruction, when power plants, industry and buildings have to be rebuilt. The remaining share of emissions are caused by direct combat operations (19%), largely from fuel consumption, mainly by Russian but also Ukrainian troops, by fires that often break out near the front lines (15%), leaks from the Nord Stream pipelines (15 %) and the emissions caused by airplanes taking long detours across Asia due to sanctions. Others suggest to rebuild Ukraine in a sustainable way with low-emission. The war can both block and initiate cooperative efforts for environmental and climate policies and undermine progress in the *sustainable development goals* (SDGs) (Pereira et al. 2022b).

8.10.4 Biodiversity and Conflict

Violence and war affect living conditions, threaten biodiversity and its protection, diminish wildlife abundance and species richness. Environmental impacts of armed conflict also affect *ecosystem services* (ESS) and *Nature's Contributions to People* (NCP) in manifold ways, poorly addressed in the emerging field of "warfare ecology" (Hanson 2011). There are direct linkages of biodiversity and natural resources, with conflict in multiple forms, from local resistance and protest to civil and inter-state war. Both are indirectly connected through human development and *Good Quality of Life* (GQL) (economic production and consumption; resource infrastructure and distribution of goods; societal participation and political decision-making). In this triangular nexus, manifold cascading and contradictory interactions are possible (Scheffran 2018b).

Armed conflicts often occur in hot spots of high biodiversity (Schröter et al. 2017) where they weaken biodiversity and diminish wildlife abundance and species richness (Hanson et al. 2009), with negative effects on habitat preservation, through "herbicide use, wildlife mortality from landmines, hunting to nourish soldiers, opportunistic poaching, direct targeting of preservation experts by equipped groups, pest outbreaks" (Solomon et al. 2018). Examples are tribal regions in New Guinea and the Amazon, in Eastern Cambodia, the Western Area Peninsula Forest Reserve in Sierra Leone, and the Okapi Faunal Reserve in the *Democratic Republic of Congo* (DRC). In parts of Africa, war is a major factor in the decline of wildlife populations in protected areas. Military transportation systems contribute to the spread of exotic species that threaten local species. Noise pollution from military activities disturbs animals and plants (Scheffran 2018b).

Indirect effects of conflict include military expenditures (on arms race, wars and military intervention) at the cost of reduced funding for environmental protection, natural resource management, biodiversity preserves and emission reduction. War and genocide (as in Rwanda, Somalia and Syria) created massive migration flows and refugee camps (e.g. in Tanzania, DRC and Jordan) which put pressure on surrounding ecosystems. Conflict has an effect on forest cover with implication for biodiversity, e.g. on the Atlantic Coast of Nicaragua; in north-eastern India; in Colombia (where armed groups reduced forest cover in gold-rich areas and lands for cattle grazing); use of high-explosives and chemical agents (herbicides) during the Vietnam war (Westing 2011). In some cases, despite armed conflict forest cover increased and local institutions could still do biodiversity management. Although armed conflicts often result in massive damage to biodiversity and NCPs, they can also act as 'developmental barriers' to natural resource exploitation. Destruction of people, economic production and human development can lower adverse effects on biodiversity. Military installations and training ground in areas with low population density or agricultural, urban or other uses may become refuges for protected species. Disarmed or abandoned military areas could be converted into protected areas, e.g. along the former border area between East and West Germany (Green Belt).

Conflicting issues can be the degradation of biodiversity (Hummel et al. 2002; Young et al. 2010) and human-wildlife management (Marshall et al. 2007; Biru et al. 2017). Biodiversity loss and scarcity aggravate the gap between growing demand and diminishing supply, driving resource competition, economic stagnation and social deprivation. When power relations exclude people or social groups from access to natural resources or their (ecosystem) services, this is often charged with conflicts over rights, distribution, equity and justice, in particular between Global North and South. Further, inadequate resource usage may pose risks to wealth, health, life and the environment (from natural disasters to climate change) which may become a conflict driver. as well as resources themselves, e.g. selling resource revenues to buy arms (e.g. ivory or timber). On the other hand, environmental protection could provoke protests against biodiversity reserves or when impoverished and disempowered inhabitants are exposed to conservationist efforts. While biodiversity losses rarely trigger violent conflict they contribute to other conflict factors.

Whether tensions arise and conflicts are contained is a question of governance and institutional settings, designing participatory governance and conflict resolution strategies (Scheffran/Stoll-Kleemann 2003; Rauschmayer 2013). In a conflict situation the challenge is to address environmental concerns and negotiate cooperative agreements. The ecological dimension in conflict resolution and reintegration processes needs to be addressed from the outset to stabilise the solutions found. Most effective is a mix of policies: integration of biodiversity conservation into military, reconstruction, and humanitarian programmes in conflict zones (Hanson et al. 2009); coordination of national policies and international fragmented environmental agreements and institutions; biodiversity management through stakeholder involvement (NGOs, researchers, public authorities, companies) (Boiral/Heras-Saizarbitoria 2017). Indigenous people represent 5% of the world's population and care about 80% of biodiversity.

8.10.5 Vulnerability, Security and Conflict in the Corona Crisis

In the Corona pandemic (Sars-CoV-2) a virus spread through an unprepared globally networked world killing millions of people and subordinating all humans to the dictates of an exponentially growing pandemic once the limit of criticality was exceeded, similar to a nuclear chain reaction, cancer, or population growth (Scheffran 2016a). An alliance of science and politics reacted in disaster mode, multiplied through the interaction between media and the public. In a very short time, far reaching decisions were made under high uncertainty, to the point of a partial shutdown of the economy and society worldwide, which was hardly imaginable before. The crisis connected and confounded everything from private lives to the global economy, demonstrating the fragility of today's complex world and undermining international stability. It also confirms the earlier identified multiple limits and contradictions in the Anthropocene in ecology, economy, societal politics, science and technology (Scheffran 2020a)

These boundaries generated and intensified conflict potentials from the individual to the global levels. They concern the human-nature relationship (diseases in overlapping habitats), intra-societal tensions over conflicting goals between different interest groups (sick and weak, hospital staff, entrepreneurs, employees and consumers in closed sectors), the interplay between epidemics and violent conflicts (from the Spanish flu after World War I to today's pandemic spread in war zones in Africa or the Middle East), infection of refugee camps (for example, in the border area between Greece and Turkey), differences between national and multilateral responses (unilateral border closures), a global struggle for scarce protective resources in health care (protective masks, respirators, medicines), geopolitical power struggles (between the U.S. and China, possibility of biological warfare), and conflicts between North and South (lower resilience of poorer countries). Accordingly, the pandemic has been suggested as an issue of peace and security policy (Brzoska et al. 2021; Mondragón et al. 2022)

The Corona crisis demonstrated and reinforced the vulnerabilities of today's networked society to a spreading disease, caused by various factors, including the number and density of people and interconnections, the acceleration of exchange processes, the exploitative human-nature relationship, shortages of system-relevant resources and infrastructures due to globalisation and privatisation. To take pressure off the system, some of these developments were reversed, decelerated, decoupled, and de-globalised (Scheffran 2020a). Many expressed the view that nothing would be the same after this disruptive event and become a permanent crisis as a 'new normal'. At the same time, the system returned to the old way of life, or triggers a transformation into a more sustainable world. The attention to health issues has generated support for a solidarity transformation towards sustainability (WBGU 2023).

8.10.6 Ecocide, Climate Litigation and Rights of Nature

The debate on ecocides was triggered by the observation that human activities exploit natural resources and cause massive environmental degradation, of a magnitude similar to genocide (Zimmerer 2014). Human livelihoods are damaged by environmental degradation and climate change as well as by armament and war. Both can reinforce each other in a way that environmental degradation threatens peace and war threatens the environment. Nuclear war and nuclear winter would be extreme forms of ecocide, putting the existence of humanity and biosphere at risk. As planetary boundaries are reached, there is an increasing need to make environmental and climate protection mandatory international law. In the context of the debate on climate action and the rights of nature, new perspectives are opening up for using legal means to bring environmental protection and peacekeeping together with a socio-ecological transformation towards sustainable peace.

The term ecocide was used at the 1970 Conference on War and National Responsibility in Washington, D.C., where an international agreement to ban ecocides was proposed (Zierler 2011). A starting point was the use of Agent Orange by the United States in the Vietnam War to defoliate large areas of forests, with massive damage to flora, fauna, and humans. Besides 400,000 deaths and disabilities and genetic defects in 500,000 children, plant life was unable to regenerate for decades, and the number of animal species declined.

The 1972 UN Stockholm Conference on the Human Environment, with Olof Palme, Indira Gandhi and others, advocated ecocide as an international crime. In 1973, Richard A. Falk proposed an International Convention on the Crime of Ecocide, a petition and a Protocol on Environmental Warfare (Falk 1973: 26). One consequence was the 1977 ENMOD Convention on the Prohibition of Military or Any Other Hostile Use of Environmental Modification Techniques which banned intentional environmental destruction in war with "widespread, long-lasting or severe effects as the means of destruction, damage or injury", with 78 member states, including those with nuclear weapons, except France and Israel. That same year, the Additional Protocol to the Geneva Conventions reaffirmed the protection of the environment from combat operations that cause severe civilian casualties. While this regulates environmental change in wartime, there is as yet no international law regulating ecocide in peacetime (Scheffran 2022c).

In 1978, discussions on international crimes against peace and security started within the *International Law Commission* (ILC). Some states advocated to include the protection and preservation of the natural environment. Whitaker (1985) proposed that ecocide be prohibited as genocide or crime against humanity. Examples of harmful, often irreparable, changes to the environment include nuclear explosions, chemical weapons, severe pollution, acid rain, or rainforest destruction that threaten the existence of entire populations, intentionally or through criminal negligence. In 1991, the ILC identified deliberate and serious harm to the environment as one of the 12 crimes against peace and security of humanity (Article 26). The proposal was discussed during the negotiation of the Rome Statute establishing the *International Criminal Court* (ICC) which in its final form did not explicitly mention ecocide, except in the context of war crimes. In 2016 the ICC paid greater attention to crimes of environmental degradation.

With growing protests against environmental degradation, species extinction and resource exploitation, the call to condemn ecocides has been growing by civil society. Environmental lawyer Polly Higgins suggested to expand the Rome Statute to include the crime of ecocide as a fifth international crime against peace (along with genocide, crimes against humanity, war crimes, and crimes of aggression), defined as "the extensive loss or damage or destruction of ecosystem(s) of a given territory, whether by human agency or by other causes, to such an extent that peaceful enjoyment by the inhabitants of that territory has been or will be severely diminished" (Higgins et al. 2013: 257), with natural causes also considered. A legislative proposal was supported in 2019 at the 18th session on the Rome Statute by Vanuatu and the Maldives, who felt the time was ripe to criminalise ecocide. Initiatives for legislation from civil society such as the Stop Ecocide Foundation led to responses and diplomatic initiatives by the governments of France and Belgium, as well as in the European Parliament, to recognise ecocide as an international crime in the Rome Statute. In response to the significant environmental damage of the war in Ukraine its President Zelensky spoke of ecocide.

While ecocide is under discussion in international criminal law to strengthen the rights of nature, related crimes are already relevant under criminal law in a dozen states (are successor states to the Soviet Union, Ecuador and Vietnam). The debate on ecocides is gaining ground due to the climate crisis, which is the background for climate litigation initiatives (Scheffran 2022c). For example, in 2015 the environmental organization Urgenda successfully sued the Dutch government in the European Court on Human Rights to reduce CO_2 emissions by at least 25% from 1990 levels till the end of 2020. Other climate lawsuits are ongoing, including one filed by Peruvian farmer Saúl Luciano Lliuya in 2015 against the German energy company RWE because its CO_2 emissions threaten his livelihood through global warming. With scientific advances, causal attribution of CO_2 emissions to climate damage events is becoming possible (Stuart-Smith 2021). A new level was reached by the ruling of Germany's *Federal Constitutional Court* (Bundesverfassungsgericht) of March 24, 2021, judging the Federal Government's Climate Protection Act of 2019 as incompatible with the German Basic Law.

In view of the planetary risks of climate change, it is proposed to elevate climate protection to the highest level of international law (*ius cogens*), a fundamental principle that is accepted by the international community of states as a norm from which no derogation is permitted (Scheffran 2022c). In addition, it is suggested to establish new legal mechanisms and institutions for the protection of the environment. The limits of the international climate change regime and the international legal and governance systems to effectively address the impacts of climate change leaves States with little recourse to judicially resolve climate-related disputes (Davis et al. 2019). This weakens the ability of the international system to prevent climate-induced armed conflict. The *International Court of Justice* (ICJ) and the *UN Security Council* (UNSC) have limited applicability to climate-related norms, including the *United Nations Framework Convention on Climate Change* (UNFCCC) and subsequent decisions of its *Conference of the Parties* (COP). The ICJ does not have jurisdiction to resolve climate-related disputes, nor has the UNSC exercised its powers to recognise climate change as a threat to international peace and security. Thus the international climate change regime is lacking effective dispute resolution, and does not extend to the security risks of climate change. How can the gap be bridged between climate-related conflicts on the one hand and dispute resolution on the other hand?

In the past decade the efforts increased in the UNSC to acknowledging climate change as a threat to international peace and security (Hardt et al. 2023). The UN's institutional responses are to be strengthened to prevent the occurrence of climate induced armed conflict. The establishment of an *International Court for the Environment* (ICE) could provide an avenue for peaceful resolution of climate-related disputes (Davies et al. 2020).

8.11 Future Trends, Global Security and New World Order

8.11.1 Global Future Megatrends

Facing multiple limits, policy failures and crises, the world of 1990 has given way to a multipolar power constellation that challenges the liberal world order. In addition to the developments described, three major trends are shaping a future: the transformation of fossil capitalism, power shifts in the North-South relationship, and the influence of civil society and social networks between democracy and autocracy (Scheffran 2023a, b). All trends are reinforced by scientific-technological developments and challenged by geopolitical tensions.

1. *Transformation of fossil capitalism*: While the industrial revolution was fuelled by fossil fuels, the world is reaching the limits of fossil capitalism. Cheap oil and natural gas have been largely depleted, secondary sources (offshore, fracking, oil sands) are being developed at greater cost and risk. When the atmosphere is no longer available for carbon disposal, risks and costs are increasing, making other energy sources viable alternatives. The vast majority of countries accepted to address these challenges 2015 in the Paris Climate Agreement. It is open yet whether the socio-ecological transformation will be realised predominantly by technical means (renewable energy, efficiency, low-carbon technologies in agriculture, transportation and other sectors) within existing capitalism or whether it requires a major system change.
2. *Power shifts in the North-South relationship*: Of the more than ten billion people expected by end of this century, the largest increase will be in countries of the Global South, assembling more than 80% of the world population, compared to 10–15% in Western countries. While the Western relative share of the world population falls, this could exacerbate North-South conflicts, for example over trade, migration and proliferation of weapons of mass destruction. Part of the Global South is claiming more influence and objecting to further exploitation by the Global North, while also coming into conflict with natural boundaries if Western lifestyles are replicated. Whether this dilemma can be solved depends on the economic influence of the developing countries, their integration into cooperative structures and the chances of democratic participation of the population. Emerging economies such as the most populous countries of China and India are advancing from cheap production sites and sales markets to competitors for Western capitalism and are acting as strategic players in the North-South relationship without following the West. If the privileged status of the Global North declines, it becomes more difficult to satisfy its population via cheap production

in the South and mass consumption, which harbours social instability. Exacerbated by multiple crises and conflicts, there are signs of global power shifts in a multi-polar world that can be better managed with multilateral cooperation and coexistence, responding to complexity better than confrontation.

3. *Influence of social networks and civil society on democracy*: Connected via social media and networks, civil society is speaking out across all scales, from local to global, claiming participation, influencing public debates and supporting political decisions. Resentment is breaking out against failing regimes, poverty and unemployment, environmental destruction, war, discrimination and exploitation, for and against nationalism. The year 2019 was marked by mass protests worldwide, ending a decade of numerous protest movements, from the anti-regime protests in the Arab world in 2011 to transnational protests such as the *Occupy movement* or *Fridays for Future*. In 2019, there were 65 instances of mass mobilisation involving at least 50,000 people in 45 countries, including 48 protests in democracies (Friedensgutachten 2020). Protests continued in 2020, culminating in the *Black Lives Matter* mass movement against racism and discrimination in the United States. Civil society includes whistle-blowers such as Edward Snowden and Julian Assange, who, despite high personal risks, offended the most powerful country with little regard for human rights and freedoms. Not to be ignored are counter-movements, from right-wing populism and anti-Corona protests that are also connected through social media, to state responses that restrict the action space of civil society, notably against the Last Generation, Extinction Rebellion, etc. (Froese et al. 2023). The disputes cut across Western and non-Western societies, linking internal and external conflicts. The 2022 Ukraine war mobilised millions of protesters worldwide and is creating socio-economic disruptions and a climate of militarism that affect the future of civil society. In pursuit of collective interests, transnational governance coordinates, regulates, manages and controls interdependent multi-level relations among actors, including coalitions, social networks and organisations of governmental, intergovernmental and non-governmental actors (Engels et al. 2023). Cooperative governance is essential for providing a "bottom-up structure of local participation, which is essential to complement the top-down imposition of a set of global regulations" (Piazza 2021: 10).

8.11.2 Competition and Cooperation: Managing Multiple Crises

The aforementioned trends offer opportunities to contain the multi-crisis world, but also risk its aggravation if policies fail and social media together with mass media distract the truth and promote reactionary tendencies. There is the potential for a major shift in history, similar to the French Revolution at the beginning of the 19th century or with the First World War at the beginning of the 20th century. We are witnessing a world in upheaval, deciding whether the old system of fossil capitalism will plunge the world into catastrophe or whether smarter alternatives will prevail. Besides collapse, the global system could also tip in a positive direction through amplification and transformation processes, strengthening resistance against crises and enabling a sustainable transformation (Scheffran/Froese 2016), reinforcing the future trends.

The nested crises cannot be solved by perpetuating the underlying drivers. Instead of wasting more resources in fighting crises once they occur, from climate change to pandemics to violent conflicts, their preventive avoidance makes more sense in terms of cost, risk and efficiency. Hazard prevention is usually cheaper and more efficient than hazard management. To prevent multiple crises from becoming 'normal disasters', experience with complex systems can be used to avoid exponential growth, chain reactions and tipping points, slow down processes and decouple them from risk amplifiers. It is possible to learn from one crisis for another?

For instance, the experience of the Covid-19 pandemic is also relevant for the climate crisis, which is changing the planet over a longer time horizon. There are many connections between the epidemics and climatic conditions, or between the means to combat viruses and climate-relevant emissions. Whereas in the climate crisis effective measures were refused, in the case of Covid-19, politicians displayed proactive crisis management and cooperative governance on a large scale (Lin et al. 2021). In both cases concerns were raised about inter-generational solidarity, of the young with the old generation in Covid-19, and vice versa in climate change (Vinke et al. 2020). Climate change as well as biodiversity loss are breeding grounds for viral diseases in a new era of pandemics under business as usual (IPBES 2019). The pandemic has also learning and lessons that can be used for future events (Anjankar Ashish et al. 2020).

Rather than fighting crises and perpetuating their causes more reasonable is their preventive avoidance. The wrong measures and responses can make the problems worse. Using weapons against war is like fighting global warming with air condition systems to protect the own home against the heat. When the West pushes for military intervention against global crises, this undermines the foundations for peace, freedom, democracy, prosperity, development, climate and environmental protection, putting values at risk and limiting the scope for shaping the future, driving megatrends into a negative direction. In turn, the climate crisis and other planetary limits may also constrain military expansionism by limiting resources and capabilities in the military sector. Think tanks tend to focus on a more offensive, risk-taking foreign policy, calling for a greater European self-reliance with a stronger military orientation to compete in the global geopolitical power games, investing in multi-billion dollar projects without sense for security and peace, ignoring that military strength cannot solve the world's problems and conflicts which need to address the underlying root causes in the own system. The effectiveness of including the UN is seriously challenged in today's multiple crises world (Posocco/Watson 2023).

8.11.3 Coexistence and Cooperation in a Multipolar World

To face the common global challenges and enormous upheavals in a cooperative way, it is important to put hegemonic aspirations and geopolitical conflict aside. Like in Cold War times, peaceful coexistence can be established in a world with multiple orders, even between rivals like EU and Russia or U.S. and China. A fresh and self-critical look at previous policy failures over the last decades is needed to prevent them in the future and avoid unnecessary confrontations. It is important to understand why the current world order is facing so many crises and threats today, why it is repeatedly surprised, and what part its own policies play here. While diplomacy, development, peacebuilding, institutions and cooperation are under severe pressure, they are more needed than ever. A central question is how peaceful coexistence in a world with multiple orders can be shaped and how the polycrisis can be managed cooperatively (Scheffran 2023b).

While the world is becoming more multi-polar involving the Global South and civil society, increasingly diverse geographical dimensions play a role, such as urban-land-coastal interactions, the energy-water-food nexus or the connecting impact of climate change in the Arctic and Mediterranean sea. At the same time large-scale infrastructure projects are established, such as *China's Belt and Road Initiative* (BRI) or New Silkroad, Africa's Green Wall in the Sahel, the G-7 Partnership for Global Infrastructure and Investment (PGII) or the resource networks for hydrogen and critical materials as part of the renewable energy transition. China and Western countries are investing heavily in a transformation away from fossil energy sources which is more difficult for Russia because of its dependence on oil and gas and the consequences of the Ukraine war. While China, as a G-77 member, has so far been anchored in the Global South, its future influence will also depend on whether its relationship with its neighbours is conflictual or cooperative. A civil society that takes Western values seriously is at odds with authoritarian tendencies in the West as well as in Russia and China (Scheffran 2023b).

The interplay of competition and cooperation is illustrated by the U.S.-China relationship. Cooperation is urgently needed to overcome common global challenges such as pandemics and climate change in the multipolar world. But if the U.S. wants to eliminate China as a threat to its hegemony, there is the risk of an escalation to arms races and war. Taking responsibility for the future also means recognising one's own responsibility for the past. Is the model of Western expansion, dominance and values universalisable for the whole planet? Does the global spread of resource-intensive lifestyles, consumption and production irreversibly overstep planetary boundaries, or can new approaches offer a sustainable alternative? Can the value-oriented normative project of the West be further developed towards a sustainable and peaceful world?

Contrary to geopolitical conflict, cooperative governance can help to mitigate, adapt to and manage complex crisis landscapes which requires coordination among multiple actors on goals, efforts and actions. The limited success in reaching global cooperation corresponds to a lack of sanctioning mechanisms against those who violate the rules. Avoiding free rider and collective action dilemmas requires mechanisms that strengthen the ability to enforce decisions through rules and regulations, practices and guidelines, formal and informal institutions (Sun et al. 2021). Global institutions are complemented by a bottom-up, self-organisation approach of local institutions to sanction non-cooperation in a polycentric approach involving multiple institutions (Vasconcelos et al. 2013). Agents of transformation can play a moderating role and together develop building blocks for a viable world. Transformative governance can help to tip the global system into a constructive direction by reinforcing and synergising solution concepts (Scheffran 2016). The 2015 Paris Climate Agreement, the 2022 Montreal-Kunming Global Biodiversity Framework (the biodiversity pendant to the Paris agreement) and the 2017 Treaty on the Prohibition of Nuclear Weapons, are global building blocks of a world without nuclear weapons or CO_2 emissions, characterised by cooperation, institutions and law. They serve a transformation to sustainable peace, supported by an alliance of states and civil society, with active contributions from the Global South (Scheffran 2023a).

8.11.4 Cooperative Security Between Identity and Diversity

Security concerns the expectation to maintain desired identity-forming values against potential future threats, i.e. a difference between opportunity and risk. In a dynamically changing global risk society, new threats to security are emerging, leading to an expansion of the concept of security to include economic or ecological dimensions. The question arises who has to bear security risks and who benefits from opportunities, who causes and who suffers from events, and who classifies a risk as acceptable for whom. Of particular importance are the criteria that define the limits of identity to be secured and asserted. Responses can vary widely depending on whether identification refers to the individual, a social group, the nation, or the entire world. While the global crisis dissolves social bonds and identities, experiencing and provoking violence, peace and sustainable development aim at securing existing identities and creating new ones. Diversity of identities is a prerequisite to prevent the dominance of a 'mono-culture' (e.g. the Western one) and to avoid emerging enemy images among diverse cultures.

With rising political attention to global warming, its security implications have been addressed by political institutions, as well as in political discourses, with mixed results. When ecosystems and vital resources such as water and oceans, farmland and soil, forests and biodiversity are lost worldwide, or when weather extremes increase, ecological and human security are at stake. Different security concepts are important which could serve as building blocks of climate security discourses (McDonald 2013; see also Brzoska 2009). Different meanings of 'security' identify the relevant referent objects affected by climate-related security threats; the agents who suggest or take particular actions (i.e. responses to these threats based on values, capabilities and risk perceptions); and the issues that are at risk, such as valuable resources and goods (Brauch et al. 2007, 2009, 2011; Schäfer et al. 2015). Most studies on the securitisation of climate change refer to the levels of referent objects and responding agents, distinguishing between between 'national' (or 'state') security and 'human' (or 'livelihood') security, others 'international' and 'ecological' security (Brauch 2009:4).

A security order is "an order of rules, institutions [...] and instruments by which states, peoples and people protect themselves against the danger of violence, especially war" (translated from Lutz 1999: 1). Other definitions emphasise the internal maintenance of independence and identity against external threats. In international law, the term security refers to the state of a government in which it enjoys effective protection against external threats to its existence, independence and territorial integrity (see Deiseroth 2000: 110). The understanding of independence can also be understood more broadly than the preservation of territorial integrity. For Buzan (1991: 18f.), security is "the ability of states and societies to maintain their independent identity and functional integrity".

Less attention has been paid to the concepts of common security and cooperative security within global governance frameworks developing global or planetary security. In contrast to collective security (such as the system of United Nations) and collective self-defence (such as NATO), cooperative security does not include obligations to provide military assistance or to participate in coercive measures, nor does it necessarily exist within an alliance. Cooperative security can include membership in international organisations (for example, the OSCE) as well as concrete multi- and bilateral agreements, such as arms control. While security through deterrence and military force aims at military capability to repel or fight threats and armed aggression (negative security), cooperative security aims at reducing threats by changing the environment in which threats can develop and the creation of cooperative political, social and economic conditions to build lasting peace (positive security). Cooperative security aims to reduce threats, while comprehensive security aims to increase capabilities within various dimensions (military, but also political, social, societal, economic, environmental, and cultural) to counter threats (securitization) (for further discussion see OPIC 2022; Werkner/Fischer 2019).

8.12 Sustainable Development and Social-ecological Transformation

8.12.1 From Adaptation to Transformation

Facing climate and environmental change, humanity has the choice between risk cascades of environmental destruction, poverty and violent conflict, or a transformation of the socio-economic system towards a sustainable and peaceful world (Scheffran 2015b, c). System collapse is not a necessary outcome of the polycrisis, but can be avoided by future strategies for system change within a deep social-ecological transformation while respecting natural boundaries and social guardrails that avoid catastrophes at an early stage. To meet the challenges of the complex world at an early stage and to solve the problems, humanity must work more closely together.

To contain environmental risks, protective measures against catastrophic environmental and climate impacts, preventive steps to reduce vulnerability and risk, and improvement of adaptive capacity and resilience are needed. Global risk amplifiers must be avoided and decoupled, dynamics slowed down, and critical infrastructures for energy, water, food, health, etc. must be preserved and strengthened. Resilience is the persistence, resistence and renewal of system function and enables the preservation of systemic boundaries against external shocks. Adequate human responses and societal interventions can reduce risks, improve adaptation and resilience of human-nature interactions, and contribute to co-evolution, sustainability transformation and diffusion of social innovations. Rather than aiming for a control of nature in the Anthropocene, it is important to embed human development into the natural boundaries. Natural and social systems have the ability to cope with changing environmental conditions by adapting their pathways to new circumstances. Some of these pathways may aggravate problems (maladaptation), others could help to diminish harm and develop new opportunities. Natural and human systems have limits of adaptability and stability and can respond to disturbances to ensure its viability and survival against environmental change. Going beyond the limits contains the risk of system destabilization and break-down of key system functions. A question is to what degree of climate change societies can adapt and how effective and creative they are in developing strategies that are sufficiently fast and complex to cope with the speed and intensity of global change (Scheffran 2015a).

A social-ecological transformation relies on support from society, mobilising social networks and collective behaviour (Scheffran 2016b). In the transformation multi-agent settings and social networks represent the responses and interactions of multiple regions, countries, businesses or citizens affected by environmental change. At the global level of decision-making, main actors are governments of nation states or groupings among them, at local levels, citizens and consumers. The multi-level process between local and global decision-making is connected via several layers of aggregation, with each layer having its decision procedures for setting targets and implementing actions. To address environmental stress and social dilemmas, new rules, norms and innovations need to evolve, as well as social networks that are viable, resilient, cooperative, peaceful and sustainable (Ostrom 2000; Scheffran 2015b).

To cooperatively address environmental problems within the planetary boundaries, it is indispensable to control and reduce the means of violence through conflict prevention, arms control and disarmament, first and foremost the atomic bomb, which represents unrestrained growth in the nuclear chain reaction, utmost violence and omnipotent power perceptions. There is no place for nuclear weapons in a peaceful and sustainable world, demanding their abolition. Demilitarization and disarmament are also important to reduce the ecological footprint and environmental impact of the military. Military personnel, equipment, vehicles and technologies may be used in disaster management, e.g. to rescue people in danger. However, such activities should not justify an arms race or armed conflict, and could be carried out by civilian forces. To avoid geopolitical clashes, a 'culture of peace' satisfies the basic needs of all people and the material basis of peace. This includes an intact environment and sustainable use of resources. Sustainable peace aims at solidarity, cooperation and conflict resolution, justice, dialogue and participation, requiring institutions, negotiation processes and governance structures.

8.12.2 Integration of Sustainable Development Strategies

The concept of sustainable development aims to relax the fundamental conflict between nature and society and to establish a balance between human needs and natural limitations. Liveable conditions ensure that people can develop, unfold and shape their needs within limited environmental spaces without massively curtailing these spaces or getting into serious conflicts. The challenge is to permanently embed the socio-sphere within the limited framework of the ecosphere. The definition of the Brundtland Report formulates the goal of "meeting the needs of present generations without compromising the needs of future generations" (Brundtland/Hauf 1987). Human needs become the benchmark for sustainable development, based on physiological requirements (e.g. need for water and food) and value-oriented demands (e.g. property and social relations). The desired equality between the needs of present and future generations necessitates the conservation of environmental space (Spangenberg 2002). In order to establish a sustainable balance between human needs and natural possibilities (desire and reality), preventive and anticipative approaches combine several key strategies (Scheffran 2020a) (Fig. 8.4):

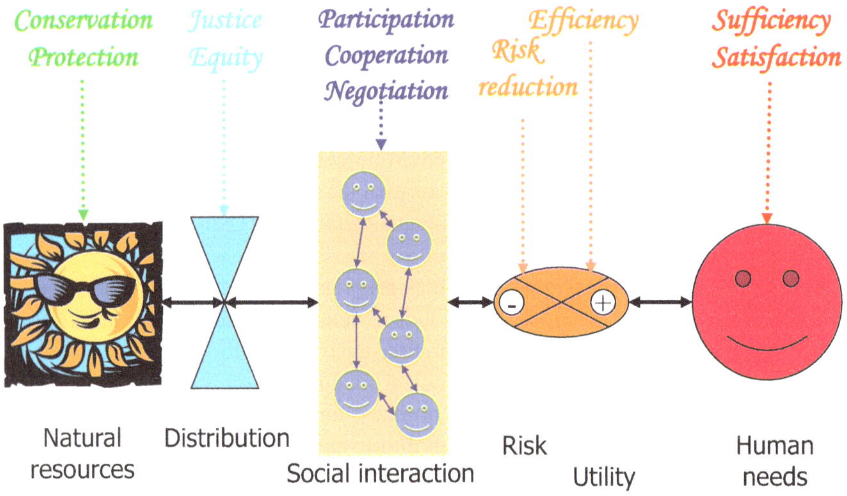

Fig. 8.4 Balancing natural resources and human needs with intermediate factors and sustainability strategies. *Source* Adapted from Scheffran (1998)

1. *Eco-consistent conservation of natural resources*: In order to make resource consumption compatible with natural material and energy flows, the availability of resources must be preserved through environmental protection and conservation of the regenerative capacity of renewable resources. Nature-based strategies include the creation of nature reserves, restrictions on the rates of use and harvesting, strengthening regenerative capacity and preserving endangered species. At the same time, waste quantities must be reduced and nature's capacity to absorb waste improved.
2. *More efficient use of resources*: How effectively a resource satisfies needs depends on the efficiency of its extraction, transformation, use and regeneration. Resource use becomes more efficient through conservation, structural changes, technical means, innovation and ingenuity, so that needs require fewer resources (decoupling of resource use and wealth/quality of life).
3. *Risk prevention*: If resource consumption is associated with hazards and damages to one's own needs or those of others (e.g. through released environmental pollutants and radioactivity), either the cause of the damage must be avoided (e.g. through agreements) and eliminated, or its consequences must be limited to a tolerable level through disaster prevention and cooperation.
4. *Justice in resource distribution*: Democracy, law and social systems allow for a more equitable and fair distribution of natural resources. However, overcoming unjust power and distribution relations requires disputes as well as the concretization of principles of justice and solidarity that enable a balance between rich and poor, North and South, present and future generations.
5. *Sufficiency of needs*: Apart from basic needs for which freedom of choice is not possible, needs can be adapted to natural conditions. Available resources may be considered adequate or sufficient, so that no further effort is required, or alternative demands with fewer or different resources may be pursued. The field of human needs is influenced by external limits (income, resources, time, social framework) and internal limits (self-limitation, inertia, lack of knowledge).
6. *Participatory decision-making and negotiation*: The shift from unsustainable to sustainable action pathways, behaviours and lifestyles, with fewer resources and less impact on nature requires the participation of large parts of the population. Participation in alternative consumption and production patterns, democratic decision-making and social negotiations are key in the sustainability transition (Brauch et al. 2016). This requires a social context in which mutual understanding, the ability to learn, conflict resolution and cooperation are highly valued.

Sustainable development policies revolve around these six strategies, which can be mixed in different combination. While scientific and technical measures focus on efficiency, eco-conservation, risk avoidance and post-disaster relief, justice, sufficiency and participation strategies aim at social sustainability. A challenge is to integrate all strategies to ease the conflict between nature and society. Participatory approaches in mediation, conflict resolution and cooperation contribute to implementing the other strategies. In this context, the conflict potential of a 'great transformation' from fossil capitalism to a sustainable society must also be taken into account. The goals of risk reduction, participation and justice directly affect the interests and power structures of society. Resistance is expected from those who have so far profited from the exploitation of natural resources and unsustainable production and consumption, and who might use their power to maintain existing structures, thus reinforcing path dependencies. On the other hand, potential beneficiaries of a sustainability transition are trying to increase the pressure, for example by forming new alliances of interests to implement the agreements of the Rio process,

Strategies for sustainable and adaptive governance to stabilise human-nature interaction range from climate mitigation, adaptation, and the building of social networks to new capabilities of disaster management, crisis prevention, conflict resolution, and environmental peacebuilding. Various measures can support the adaptive capacity of ecosystems and their human values, including the establishment of nature reserves, sustainable land use, preservation of endangered species, and the protection of terrestrial carbon stocks. Within limits, ecosystems can adapt to climate change. Consumption of natural resources should not exceed their carrying capacity, given by their limited regeneration and absorption abilities. Integrative and interdisciplinary knowledge helps to avoid dangerous pathways, influence critical decision points and develop collective adaptive strategies to a more sustainable, peaceful, and viable world (Scheffran et al. 2016).

To ensure the habitability in Earth's common house, the 17 Sustainable Development Goals aim to use the available environmental space for sustainably living of all inhabitants. In addition to an efficient and equitable use of resources, human needs are to be in balance with nature. Acting with nature is more sustainable than acting against it. This will require large-scale redistribution of wealth and consumption – provoking conflict with those who are unsustainable, in particular the rich and wealthy people with their extreme environmental footprint. To moderate conflict, six key characteristics need to be considered in sustainable development: limits to growth, equity, inclusion, reflexivity, participation, and international solidarity (Bake/Quinn 2022). While the SDGs can only be realised within the planetary boundaries, a global transformation must be specific to local conditions and the diversity of communities which requires dialogue to develop synergies of different designs of social nature relations (e.g. conceptions like like 'Buen Vivir' (Oswald Spring 2023), 'Ubuntu' (Kiyala 2023) and 'Degrowth'[7] (Hickel et al. 2022), focusing on life, common goods and needs, cooperation and redistribution, self-determination and recognition of diversity and the precautionary principle.

[7] See for an overview of the extensive scientific, political and activist debates on the conceptual issues on 'degrowth' at: https://en.wikipedia.org/wiki/Degrowth (16 August 2023); see Hickel et al. (2022).

8.13 From Climate Conflict to Cooperation

8.13.1 Synergies in the Climate-Conflict-Migration Nexus

The problem nexus of climate change, migration and conflict has attracted more research than the positive nexus of integrative synergies in climate, peace and migration policy which avert security threats, refugee vulnerability, and climate-related risks but take constructive solutions into account. Instead of securitising climate migration with surveillance and defence measures against refugees, it is more sensible to address the drivers and problems preventively through international cooperation. Preventing forced migration includes measures of curbing wars, civil wars and other acts of violence, as well as persecution, impoverishment and marginalisation of population groups, environmental degradation and climate change. An effective climate policy reducing GHG emissions is essential to mitigate climate change and related consequences such as violent conflict, forced displacement and irregular migration.

Appropriate institutional mechanisms and legal regimes need to address synergies in the climate-conflict-migration nexus (see the compilation in Scheffran 2017a). An integrative approach addresses the entire causal chain from the causes to the consequences of climate migration. As far as the causes are concerned, emissions can be reduced by means of an effective climate policy, through renewable energy systems and better efficiency as part of the energy transition and sustainability transformation between North and South under the Paris Treaty. The consequences can be mitigated through adaptation measures, i.e., disaster preparedness and development cooperation, strengthening the affected countries' own capacities, and improving the living conditions of people at risk of forced migration. Individual problem-solving can be supported and extended by international institutional mechanisms and cooperation between the countries of origin and destination that are based on joint development in the context of a major transformation (Scheffran 2017a). Conflict prevention combines peace and security policy, migration, refugee and asylum policy.

The interconnections between climate and conflict stretch from global-scale cooperation on GHG emission reductions to community-scale interventions to manage the impacts. Effective interventions can potentially break the links between climate change, armed conflict, and instability. The question is which policies against climate-related economic shocks, reduced agricultural production, or forced displacement reduce engagement in violence. Some measures are better understood than others such as transboundary institutions for managing conflict risks in resource-scarce regions water management within-country (Mach et al. 2020). Policy-focused analyses could evaluate specific interventions related to adaptation, insurance and risk sharing, humanitarian assistance, or development policy, considering implications for conflict risk.

Coproduction of research among researchers, stakeholders, practitioners, or organisations can potentially contribute to proactively address potential linkages between climate and conflict, provide early warning and contain climate conflict risk, encouraging implementation of adaptation in fragile states or conflict situations (Abrahams 2020). Climate-conflict research is relevant for practitioners such as peacekeeping groups, development banks, and disaster relief organisations who are interested in the return of their investments. A strategic and forward-looking approach is necessary for reducing forced displacement due to climate change. This takes into account the estimated impacts of climate change, future uninhabitability of regions, targeted adaptation in regions with development prospects, and foreseeable migrations to urban centres or more fertile land, where the infrastructure and social services can be oriented towards immigration. In addition to improved forecasting capacity, important is strengthening and networking of national and local institutions to implement disaster reduction in a targeted manner, if sufficient financial resources are available. It must be ensured that poor countries receive subsidies for their protection according to their economic performance to avoid debt.

The local population should be included in the planning and implementation of risk reduction measures. It can contribute by sharing the knowledge on which regional factors threaten their livelihoods or which measures are necessary to support them in the event of weather extremes or the long-term consequences of climate change. The diaspora can help increase the resilience in countries of origin to the climate impacts by importing knowledge, innovations, technologies and money. Many of the most affected countries cannot overcome these challenges by themselves and need help and funding from industrialised countries, e.g. with knowledge and technology transfer, by adapted early warning systems, dike building and planning processes, in strengthening and networking the responsible institutions and non-state actors (for example Red Cross and Red Crescent Societies). This can and should be further enhanced, strengthen autonomy of the partner governments and establish compensation funds for disaster loss and damage suffered and to start reconstruction as soon as possible.

Helping affected regions in adapting to climate change can at least mitigate the worst consequences, but not prevent them if the drivers continue. As long as this is the case, legal migration options can support adaptation. It is not necessary to wait for the global deal since pioneers can set positive examples for others and trigger partial strategies in an integrative framework. Simultaneous actions across levels involve civil society, governments and companies that cooperate and do not block each other. According to the *International Law Commission* (ILC) legal conditions are needed for climate-induced migration, such as improving civil protection against weather extremes and long-term climate-related events. These include co-development between developed and developing countries, cross-regional networking among populations, and integration of immigrants as part of a development strategy (Gioli et al. 2016).

Possible solutions were offered by various initiatives (such as the Nansen Initiative or the Sendai Framework for Disaster Risk Reduction), opening up various options for humanitarian interventions and responses for protection, risk prevention and resilience providing humanitarian assistance for displaced people in natural disasters. For years there have been efforts to make climate-induced migration a subject in climate policy and to take concrete steps in the UNFCCC (Nash 2019). Since the 2010

World Climate Summit in Cancún, migration was explicitly addressed, and in the 2015 Paris climate negotiations, a task force on climate-induced forced migration has been launched, in coincidence with the 2015 'refugee crisis' in Europe which brought the issue of climate-related migration into focus of politics, media and academia. Public representatives used the threat of climate-induced refugees to justify a climate agreement, blaming Europe for the 'migration flows' and calling for cooperation with Africa.

The widespread notions of climate migrants as threats or victims are too simple. Agents adapt their living conditions to environmental change within the boundaries of their livelihoods and networks. Focused adaptation measures can safeguard development prospects and also project future migration patterns while planning the infrastructure and basic services in urban centres. Rich countries could ensure increasing financial commitments for adaptation measures. These can contribute to stable structures and networks between regions, in which migrants can participate as catalysts and all sides benefit with their potentials and skills. In this way, the problem nexus of climate change, conflict and migration could be transformed into a nexus of the respective policy fields towards building synergies. One aim is to strengthen the capacities of migration networks between countries of origin and destination.

To ensure that people are not forced to move by climate impacts, adaptive capacity and social resilience of affected communities can be strengthened, to cope with risks and emergencies (Black et al. 2011). New economic and technical frameworks can strengthen societies' adaptive capacities to climate change, and those with more resources have better opportunities to do so, while poorer people have less coping capacities, and also fewer opportunities to migrate or become trapped. Where communities are threatened by ongoing climate change despite local adaptation and protection measures, migration can be seen as a legitimate adaptation measure to avoid climate impacts, strengthen the agency of displaced people, and protect them from risks, with additional financial, political, and legal support from the international community (Black et al. 2011), potentially opening up opportunities and resources elsewhere. Regulated forms of migration, for example circular migration of labour, could counteract a crisis-like mass exodus. The effectiveness and acceptance of such proposals is controversial and depends on the political and legal frameworks, including their embedding in development strategies (Vinke et al. 2022).

When people are affected by climate variability, they usually try to cope with problems locally. Technical and social innovations as well as cooperation and institutions have an edge over reactive and violent behaviours that focus only on short-term and individual gains. Migration networks can help create stable structures between countries of origin and destination (Scheffran et al. 2012d). Projects initiated and run by migrant organisations can be successful if they are supported by institutional frameworks and development cooperation, involving governmental and non-governmental organisations. Migrant networks can help build social capital to increase the resilience in the affected communities and spur cross-regional innovation through the transfer of knowledge, technology, money, and other resources. These could increase community flexibility, diversity, and creativity in coping with climate impacts and open up new pathways (Scheffran et al. 2012d) and their return to areas of origin.

The skills and transfer of resources in migrant networks, including knowledge, money and people, can lead to technical and institutional innovations in home communities that support climate adaptation, sustainable development and peacebuilding. Also can be the income from migrant remittances, which have increased globally and directly impact resource endowments, economic well-being, and resilience in the community of origin. Young, professionally trained returnees can support successful environmental initiatives. They are familiar with local conditions in home countries and are in a favourable position to support their communities. The use and transfer of technologies promotes improved resource efficiency, new forms of natural resource use and crop varieties, sustainable energy supply and disaster management.

8.13.2 Climate Mitigation and the Paris Treaty

International climate governance includes the 1992 UNFCCC, the 1997 Kyoto Protocol, and the 2015 Paris Treaty which defines boundaries of global mean temperature and national commitments of emission reductions as well as instruments for financial and technology transfer to developing countries. The Paris Climate Agreement is a central building block of climate policy and thus also an essential contribution against climate migration. Implementation of the Paris Treaty is critical to realize deep cuts in CO_2 emissions. Investing in emission reduction and climate adaptation would significantly diminish the consequences, being more efficient than post-emergency refugee and disaster management. 188 states and the EU are contractual partners of the Paris Convention of 2015, with the objective of limiting the rise in global temperature to well below two degrees, ideally to 1.5° as compared to pre-industrial levels. In order to achieve this goal, they must economise on a climate-neutral basis in the second half of the century, releasing less GHGs into the atmosphere than natural reservoirs on the planet can extract from the atmosphere, or compensating for unavoidable emissions through other means.

The commitments by the *Nationally Determined Contributions* (NDCs) of all countries are only partially met and may not be sufficient to prevent dangerous climate change. Opportunities were not seized, and valuable time was lost (IPCC 2022b, 2023). While climate stabilisation and complying with climate safety limits are not yet in sight, the limits and risks of climate engineering (Lawrence et al. 2018) leave no viable alternative to cutting down anthropogenic GHG emissions by half until 2030 and achieve climate neutrality by 2050, implementing new technologies and behaviours in energy, transport and agricultural sectors (Engels et al. 2023). A committed climate policy and a turnaround to a sustainable energy supply can help to mitigate and adapt to climate change and its consequences, including new economic and lifestyle patterns.

While the EU is committed to reduce GHG emissions by at least 55% by 2030 compared to 1990, the transformation to a low-carbon society needs to be accelerated to become climate neutral by 2050 and develop innovations that can create new economic sectors and jobs. It is a question of justice to enable poorer countries to undergo this modernisation, which is already underway in the wealthy countries. To achieve this, the main polluters must make and increase their respective NDC contributions for emission reduction. In addition to participation in the NDC partnership, the European Green Deal provides a framework to decisively reduce GHG emissions and accelerate the transformation to a climate-neutral and resilient society, promoting energy efficiency and recycling (European Commission 2021: 417). This requires to increasingly engage the private sector and civil society and invest in programmes to protect natural resources, forests and other ecosystems as well as biodiversity. The investments that have been made worldwide so far were not sufficient to achieve the goal of climate neutrality within three decades.

8.13.3 Climate Justice and Climate Matching: North-South Cooperation in Climate Policy

Climate change reveals an international equity problem, because industrialised countries and emerging economies have historically emitted most GHGs and are mainly responsible for human-made global warming and its prevention (OWD 2023). Their prosperity is significantly based on fossil fuel emissions. In some cases, emissions in industrialised countries are five to ten times higher per capita than in developing countries. While the main causes of the climate problem lie in the Global North, most affected are people in hot spots of climate change in the Global South, which reinforces existing injustice in the world. Therefore, industrialised countries are largely responsible to address both the climate problem and its consequences, such as climate-induced migration. They are commited to climate-neutral economic development and to significantly reduce their ecological footprint and GHG emission through their NDCs. On the other hand, the most vulnerable and poorest countries, for whom fossil fuels have been an important economic factor so far, are hit hardest by negative effects. They need support for restructuring their economies, adapt to changing climatic conditions, reduce damages and losses, and hedge against future climate risks. To achieve their climate objectives, they need access to the necessary technologies as well as financial support for investments that can create new industries and jobs.

This requires North-South cooperation for a climate-friendly and sustainable transformation that reduces economic disparities and distributive injustice in the world. The rich countries are supposed to help the most vulnerable countries to adapt to changing climatic conditions, manage losses and damages, and protect against future climate risks, advancing climate-neutral and environment-friendly development in the global South, and harnessing synergies of renewable energies. The challenge is to find a fair global deal between developed and developing countries that balances gains, costs and risks. Developed countries set a goal at the 2009 Copenhagen and 2015 Paris climate conferences to mobilise $100 billion annually from public and private sources for climate mitigation and adaptation in developing countries, taking into account their needs and priorities. The Climate Change Investment Platform, unveiled in 2019, aims to advance the use of clean energy in line with the Paris Agreement. Sub-regional investment fora help improve access to finance for projects. However, the current level of financing is not sufficient to implement the envisaged transformation. For this to happen, global financial flows in all economic sectors and financial markets must be aligned with these goals, including private investment and foreign direct investment (Santarius et al. 2012; Scheffran/Froese 2016).

Increased international cooperation is necessary to avert a climate catastrophe and achieve the global energy transition. Rapid decarbonisation will require unprecedented initiatives and investments that take into account different regional conditions. To reduce emissions, the implementation and further development of climate goals in the Global North at all levels must go hand in hand with countries in the Global South. To this end, it is necessary to support emerging and developing countries and mobilise international financial institutions and the private sector. Cooperation partnerships are supposed to support the NDC obligations of partner countries, taking the Global South into account for all climate change mitigation measures. Industrialised countries would support the climate and environment-friendly development and restructuring of the economies in the developing countries which can be supported by climate protection investments providing a significant share of funding for climate mitigation action.

Both can work together to trigger innovations for a global energy transition and effective climate adaptation to avoid driving people from their homes, including a mechanism of joint projects (*climate matching*) to promote climate-neutral development in the Global South together with the Global North (BMZ 2021: 105; Scheffran 2022d). To use the synergy effects of renewable energies includes different funding and financial instruments, ranging from full or partial cost absorption in bilateral projects to multilateral climate funds and debt relief for green investments. In joint projects where partners take financial shares, co-design, co-determination, and barrier removal must be ensured. In addition to the financial scope, other means are also decisive, such as the transfer of technology, infrastructure, knowledge and personnel. One focus here is on the use of renewable energies in the context of a global energy transition.

8.13.4 The Global Energy Transition

Access to energy is a prerequisite for education and healthcare, improving living conditions, creating jobs and reducing poverty. To reduce the number of people without access to electricity and clean cooking and increase the share of modern renewable energy, SDG-7 calls for "affordable, reliable, sustainable and modern energy for all" by 2030 which opens up other synergy effects besides climate protection, e.g. reduce climate conflict and displacement (Stöcker 2020: 424). Solar, wind and other renewable energy sources are abundant in most countries. For the promotion of climate-friendly development in the Global South, the expansion of renewable energies has a prominent role. Decentralised renewable energy solutions offer rural areas in Africa or South Asia, which have not yet been connected to electricity grids, new development opportunities to build local value chains and access to information and communication via Internet and mobile phones. In light of these synergy effects, governments continue and increase their efforts for a global energy revolution. In addition to the cooperation within the framework of the NDC partnership, it can also tie in with existing initiatives, such as the *International Renewable Energy Agency* (IRENA), the *Energising Development* (EnDev) programme, which is largely funded by the BMZ and involves several countries and donor states (BMZ 2022), or the *South East Asia Energy Transition Partnership* (ETP), which supports countries with ambitious climate targets in their energy revolution by the *International Climate Initiative* (ICI).

Compared to technological innovations that improve the availability, cost reduction, reliability and accessibility of renewable energy, less relevant have been social innovations which aim to improve livelihoods or services of end users (households, businesses, public institutions). They involve the active engagement of women, youth, and marginalised communities as agents of change. Global energy transition investments totalled \$755 billion in 2021, of which \$366 billion was in renewables. Asia-Pacific accounted for nearly half, with China investing nearly two-thirds, comparable to Europe, the Middle East, and Africa combined, compared to \$150 billion in North and Latin America. Associated with the strong growth in renewables has been a significant reduction in costs, making them more competitive with fossil fuels. Despite the growth, investment is nowhere near sufficient to achieve a secure, resilient, equitable, affordable, and climate-compatible energy supply (Scheffran 2022d). According to IRENA's *World Energy Transitions Outlook* 2022, investment needs are \$5.7 trillion per year through 2030, with a focus on decarbonisation, efficiency, electrification, green hydrogen, electro-mobility, bioenergy with carbon capture and storage, creating 85 million jobs worldwide.

The rapid rise of renewable energies is transforming the geopolitical map, in a race for technological innovation and dominance. Most countries have viable renewable energy potentials to become independent of fossil fuels, create energy security and improve their trade balance. Transformation offers strategic advantages for these countries, making them less vulnerable to supply shortages and price volatility, political instability, terrorist attacks and armed conflict. A fully renewable power supply based on different sources is technically feasible and the variability of power generation in the grid is mitigated by an energy mix (Breyer et al. 2022). A change in energy supply may be associated with a system change for cooperation and peacebuilding.

8.13.5 Resistance and Conflict-Sensitivity in Climate Policies

The question is how the transformation from conflict-prone fossil-nuclear to a more peaceful and sustainable energy supply can be achieved without opening up new environmental pressures and conflicts. Renewable energy sources and their infrastructures are not without conflict, including land use conflicts in the introduction of bioenergy, protests against wind turbines, hydropower dams and power grids, or resistance to strategic raw materials for photovoltaics or electric cars. They require important natural resources (land, water, food crops, minerals) whose competing uses create tensions and contestation of 'green colonialism'. Environmental impacts lead to local protests and resistance against power grids, dams, bioenergy, large wind and solar plants (Scheffran/Cannaday 2013; Scheffran 2020c). Solar panels, wind turbines, electric vehicles and energy storage require non-renewable minerals and metals for their production which are largely found in a few countries of the Global South and on the ocean floor. Strategies to control conflict minerals aim to improve transparency along global supply chains. Recycling and reusing critical minerals in a circular economy reduce dependence and counteract cartelisation. The spread of renewable energy stimulates electrification and electricity trade, promoting regional cooperation, power grids and balancing energy sources. Decentralised energy systems and intercontinental distribution networks can be combined, promoting cooperation and 'power to the people' between civil society, private actors and governments in North and South (Stephens 2019).

To be successful, climate policies need to be less conflictive and more conflict-sensitive. Accordingly, attention must be paid to social and ecological compatibility, political acceptance and conflict avoidance at an early stage, based on a holistic view. Thus, it is important to design mitigation, adaptation and protection measures, as well as climate engineering, in a conflict-sensitive manner (Nadiruzzaman et al. 2022) and respect adaptation limits that trigger systemic risks and tipping points (Juhola et al. 2022). Social acceptance of energy transition strategies and technologies is crucial for a successful implementation. Climate engineering to remove CO_2 from the atmosphere or to influence the radiation budget appear to be particularly conflictual, facing complex and contentious issues about feasibility and financing, environmental consequences, risks, and responsibilities between global, national, and local scales (Scheffran/Cannaday 2013). This also applies to differences over adaptation to climate change or equity conflicts over how to allocate costs or where the benefits and risks of actions lie for current and future generations. In contrast, there is a need for incentives and ways to solve problems cooperatively. A key question is who should be protected against impacts and who should bear the costs and risks.

8.14 Conflict Transformation, Environmental Peacebuilding and Sustainable Peace

Coping with the environmental and climate crisis can build on the long experiences about war and peace. To break the alliance of growth, power and violence in the fossil-nuclear age (Scheffran 1996), positive synergies of sustainability, development and peacebuilding are created, promoting a "climate for peace" (Rüttinger et al. 2015). Securitisation and militarisation of climate policy would be replaced by 'climatisation' of security policy (Oels 2012, Aykut/Maertend 2023). This could promote various constructive measures: disarmament, demilitarisation and redirection of military spending to productive areas of society (such as environment and health), cooperative governance and institution building, reduced vulnerability to climate change and strengthened adaptive capacity, resilience and distributive justice. Policy would aim for synergies between social-ecological transformation and conflict transformation (Pastoors et al. 2022), prevent negative and induce positive tipping points (Otto et al. 2020; Juhola et al. 2022; Eker et al. 2024) for sustainable and environmental peacebuilding.

8.14.1 Conflict Transformation and Management

In the social-ecological transformation a future challenge is to minimise its conflict potential, which is the tension between aspired and actually achieved human values and needs driving conflict action to bridge this difference. The conflict potential can be reduced by successful adaptation of invested resources to satisfy the needs, or by adaptation of aspiration to reality. While sustainable development aims at an equitable balance of needs and available resources among multiple actors, it can be an essential contribution to cooperative conflict avoidance and resolution. To manage related conflicts (e.g. between values, social groups, states and generations), key stakeholders are involved to negotiate strategies to balance their interests, investments and resources. An important question is whether established procedures of conflict management (e.g., mediation, peacekeeping and aid) are sufficient to reduce conflict potential or whether new interventions are needed to manage emerging security and conflict risks. By constructive conflict resolution, previously dormant intellectual and physical capabilities for solving problems would be released, using conflict as a "fertile and creative principle" (Dahrendorf 1963) and an opportunity for conflict resolution to address injustice and ecological problems simultaneously in socio-ecological transformations (Pastoors et al. 2022).

1. Since climate conflicts are often based on power inequalities, addressing and dismantling of environmental injustice is a central steps in conflict transformation (Holifield et al. 2018), which critically exposes and constructively changes power inequalities. It is also important to strengthen the position of people who are disproportionately affected but rarely heard.

2. Since sustainable development aims to balance the satisfaction of human needs with the continuation of life on Earth, this is closely related to concepts of peace that simultaneously unfold human life and social-ecological relations (Meyers 2019: 22) in which people make peace with themselves, with others, and with nature (Pastoors et al. 2022).

Collective efforts to address climate change are an opportunity not only for socio-ecological transformations of society, but also for *civil conflict transformation* (CCT). In the current discourse, most scientific and societal debates predominantly address the negative impacts of the climate crisis (Vinke et al. 2021). Underrepresented are perspectives that understand the climate crisis as an opportunity to address conflicts and power inequalities. As environmental change and socio-ecological transformations are associated with conflict, approaches to conflict resolution and transformation are gaining importance. With *civil conflict management* (CCM), appropriate tools and ideas are available to constructively address these conflicts (Pastoors et al. 2022).

CCT comprises different approaches to civil and nonviolent conflict transformation. Prominent examples are round tables and dialog forums, mediation and conciliation, e.g. between victims and offenders, but also other forms of violence prevention and diplomacy, peacebuilding and civil peacekeeping. A successful example in Germany is municipal conflict consulting which supports cities and municipalities in defusing conflict potential, shaping transformation processes constructively and promoting cooperation between different parties, also addressing climate. Examples abroad are provided by the *Civil Peace Service* (CPS), a constructive peacebuilding and violence prevention programme in crisis and conflict regions, using tools such as human rights monitoring, facilitation and dialogue, peacebuilding, conflict-sensitive journalism, remembrance work, and psychosocial support for victims of violence (Pastoors et al. 2022). In contrast to military intervention and technocratic management, the variety of outlined civilian conflict management instruments can make important contributions to sustainable peace, showing how important it is to focus attention on the design of peace processes and socio-ecological transformation processes. CCM can make important contributions to dealing with the climate crisis and the current crises of socio-natural relations, through constructively designed integration of social-ecological and conflict transformation, with linkages between climate policy and peace policy; sustainable peace and climate justice; peace, environment and development; violence prevention and diplomacy; peacebuilding and civil peacekeeping, (Pastoors et al. 2022).

8.14.2 Sustainable Peace and Environmental Peacebuilding

A critical overview and assessment of the literature on environmental conflict and peacebuilding traces the development of the field from its beginnings in the 1980s until today. It identifies several distinct stages, research gaps and shortcomings as well as problematic implications of the research to inform future Environmental Peace and Conflict Studies, making links to the field of Anthropocene Studies and the concepts of 'sustaining peace' and 'sustainable peace' (Hardt/Scheffran 2019). Environmental-conflict research has evolved in three waves, starting after 1989 with a focus on environmental resources and state security, followed since 1994 by a second wave on environment and socio-political dimensions of conflict and a third wave on climate change as a trigger for risks and conflicts (ongoing since 2007). Environmental-peace research emerged in parallel, with early studies after 1989 addressing the concept of sustainable peace and the first wave of environmental issues starting 2002 as constitutive for cooperation and peace, while the second wave since 2007 focused on cooperation through and against climate change, and the third wave started in 2015 to expand the concept of environmental peacebuilding. Since 2018 concurrent research in both research areas emerged (Hardt/Scheffran 2019).

In the discourse and practice of peacebuilding the blueprint of sustaining peace evolved which differently from the concept of sustainable peace makes almost no reference to environmental phenomena (UNSG 2015). Increasingly peacebuilding is aware of and addresses environmental challenges (Swain/Öjendal 2018). Environmental peace and peacebuilding can facilitate more general cooperation (Ide 2018), although the validity of the relationship between environment and peace in connection to climate change is questioned and not clearly defined (Ide 2018). The research field has received increasing visibility and acceptance (see Swain/Öjendal 2018), e.g. by the *Environmental Peacebuilding Association* and its first conference in 2019. Research is moving from theory to practice, focusing on specific tools and pathways to transform traditional peacebuilding projects and policies, implementing strategies against climate change and conflict (Matthew 2014), including climate mitigation and adaptation policies in peacebuilding processes (Besigye Bazirake 2013: 24). Obstacles include the difficulty of establishing an inventory of practical tools that would be easily accessible to practitioners of reconstruction and capacity-building (Matthew 2014). The peacebuilding sector should avoid "short-term urgency" and "long-term unsustainability" through "crude exploitation of natural resources, unsustainable environmental practices, and massive threats to resource-based local livelihoods" (Swain/Öjendal 2018: 10).

Directions of research include theoretical approaches, informing Peace and Conflict Studies with Political Ecology in human-nature relations (LeBillon/Duffy 2018: 247), conceptualising "scarcity, abundance and dependence temporally through historically-grounded analysis" and "within uneven power relations and resource entitlements" (LeBillon/Duffy 2018: 247). A broader and encompassing framework that bridges theory and practice seems to be on its way. Research also focuses on different world regions, dealing with climate–induced conflict and conflict-sensitive climate change policies. A critical analysis of rationales for environmental peacebuilding in the context of climate change highlights the pitfalls and blind spots of the literature (Ide 2020). Research gaps along the pathways of critical theory include a bias towards conflict, missing Anthropocene thinking and future pathways, de-politisation and alienation of political causes and actors, universalist versus pluralist visions of peace (Hardt/Scheffran 2019).

The literature lacks a holistic understanding and keeps humans and Earth as separate entities, addressing climate and environmental change as controllable. The field is dominated by problem-solving, technocratic approaches and managerial, de-politicising tendencies. Largely neglected are complex dynamic human-nature relations and connections to peace conceptions in relation to violence and conflict (or the absence thereof). A specific focus on critical approaches, as opposed to problem-solving approaches, would help to question power structures, discourse settings, political interests and biases and also challenge the mainstream universalism. A new integrative research wave would broaden the study field opening the debates on the linkages between different scientific fields and communities, including the emerging field of Anthropocene Studies, perceptions of the world and of humankind in and as part of the Earth, explicitly focus on bridging the isolated bodies of literature through broader transdisciplinary, integrative and in-depth approaches to analyse and assess the issues, focusing on interconnectedness and relations. This could lead to a new stage of Environmental Peace and Conflict Studies (Brauch 2020).

8.14.3 Enabling the Logic of Peace

Enabling cooperation and peace in climate-related shocks, conflicts and post-conflict situations is fundamental to interventions that work. These include the integration of displaced populations and support by social-safety nets to marginalised groups as well as natural resource management, environmental cooperation and environmental peacebuilding (Ide/Scheffran 2014; Schilling et al. 2017; Ide 2018; Ide et al. 2021), which depend on capabilities of people and government institutions (Mach et al. 2020). To escape the destructive logic of war and its escalation it is essential to create alternative directions for decision-making to enable the logic of peace, counteracting the conditions for geopolitical conflict, injust globalisation, or a destruction of livelihoods through climate change.

To think, enable and secure peaceful and sustainable coexistence in the common house of planet earth is a programme also for peace research, which replaces the logic of war with a logic of peace (Frey et al. 2014; Scheffran 2023c). This includes reduction and prevention of violence (instead of its propagation), minimisation of damage (instead of maximisation), de-escalation and disarmament (instead of escalation and rearmament), avoidance of polarisation (instead of creation of enemy images), common security with each other (instead of particular security against each other), cooperation (instead of confrontation), shaping of the positive (instead of the negative). In terms of power politics and society, it is about inclusion (instead of exclusion), universal values and norms (instead of the priority of one's own interests), broad distribution of power and participation (instead of concentration of power), flat hierarchy (instead of strong hierarchy), fair balance as well as symmetry and fairness (instead of dominance and injustice), (self-)critical reflection (instead of uncritical self-affirmation).

In order to create conditions for a peace logic in the context of the Ukraine war, it is important to mobilise civil society and the peace movement in all parties to the conflict, in addition to humanitarian aid for refugees and victims of violence. Through de-escalation, diplomacy, negotiation and mediation between the parties to the conflict, acts of war can eventually be stopped and weapons withdrawn. Principles of the peace logic can then be successively applied, for a European and global peace order that strives for common security for all actors and win-win solutions, strengthens international law, organises coexistence and joint conflict resolution. Instead of a 'Zeitenwende' for military force and geopolitical power struggles to defend the world order, the course of the war shows how necessary a *Zeitenwende* would be for a double transformation to peace and sustainable development. Peace research is needed, to understand links between environmental degradation and violence, and to proactively develop integrated concepts and synergies for sustainable peace (Brauch et al. 2016; Scheffran 2023c).

In general, peace encompasses a balanced social relationship in or between communities. While 'negative peace' is about preserving existence against physical violence, 'positive peace' is the absence of 'structural violence' against people and nature restricting development opportunities (Galtung 1995), which combines preservation and development (see contributions in Senghaas 1995). In Chinese culture, harmony of complementary forces (Ying and Yang) serves as a symbol of peace, combining the taming and nurturing activity of nature with the transforming work of humans. This duality is also found in Indo-Germanic peace in the sense of 'sparing' (letting live) and 'loving' (helping each other). Czempiel (1995: 170) characterises peace by "preservation of the existence of the individual due to decreasing violence" and "continuous unfolding of the existence of the individual due to equal distribution of opportunities for development." As in peace, conservation (of nature) and unfolding (of the individual) are also central categories of sustainable development (Scheffran 1998). In both cases, the balance of conflicting forces allows satisfaction through exchange, negotiation, and solidarity.

There are parallels, but also divergences between the concept of security which "refers to the quality of an existing or perceived state", and the concept of peace which "refers to the quality of a relationship between different actors or groups of actors" (Gießmann 2011: 543). Peace is not just a goal, it is a pathway and process supporting sustainable development and (conflict) transformation. The world is facing a double transformation to a sustainable and peaceful world which needs an integrated policy design (Sterner et al. 2019). As the socio-ecological transformation is also a peace project (Scheffran 2022a), conflict transformation and peace policy can promote the transition to a sustainable world (Pastoors et al. 2022). UN Secretary General António Guterres complains about humanity's "war against nature" and calls for "peace with nature" (Guterres 2021; UNEP 2021), with also needs "peace among people" (Scheffran/Jathe 1997).

8.15 Conclusion: Towards a Future-Oriented and Sustainable Peace Science and Policy

Based on Integrative Geography conceptions of landscapes, interspaces and pathways of complex human-nature interactions, this chapter investigates the impacts of humanity's global expansion on planetary boundaries and limits to growth, leading to multiple crises and geopolitical conflicts in the Anthropocene. When ecological and social systems are out of balance, the existing world order is under pressure by interconnected instabilities, compounding risks and tipping cascades. Drivers and consequences of exceeding thresholds are highlighted for several problem areas, including environmental and resource conflicts, disasters and extreme events, the climate-migration-conflict nexus, impacts of national rivalries and geopolitical competition on climate and environmental change. Future challenges and opportunities for multipolar world orders include the transformation of fossil capitalism, North-South power shifts, and the role of civil society in democracy. Established policies, politics and polities are not well prepared to preventively deal with the huge challenges of a highly complex and ambivalent polycrisis and need to develop new concepts to prepare for the necessary transformation.

Instead of continuing the fatal triumvirate of growth, power and violence, driving the Earth into a vicious circle between conflict, environmental degradation and underdevelopment, it is more constructive to establish a positive virtuous circle between sustainability, development and peace. Building blocks for a viable world include renewable energy for all, ecological footprint within ecological limits, clean prosperity for all, and cohabitation of nation-states within a world domestic policy (von Weizsäcker/Wijkman 2018; Scheffran et al. 2024). To ensure the Earth's habitability, 17 Sustainable Development Goals guide human development into orderly channels, using available environmental space wisely to endure life for all inhabitants in the Earth's common home. In addition to an efficient and equitable use of resources, human needs are to be adapted to an efficient, sufficient and fair use and distribution of resources.

The double transformation to a viable world requires a variety of concrete measures for a sustainable peace policy, including cooperation, disarmament and common security as well as a resilient and low-conflict and low-carbon energy supply. The coexistence and cohabitation of different world orders to address these problems is more promising than further geopolitical power struggles. Instead of simplistic friend-foe constructions, a democratic peace would rely on self-criticism and the ability to learn, adapt and anticipate, taking the future into account to enable a change towards preservation, development and transformation. This requires a mix of technical and social innovations within and beyond the existing capitalist economy, towards a fundamental system change and a new world order. For the time being the world will struggle between competing and ambivalent pathways and tipping points of disruption and construction, conflict and cooperation, war logic and peace logic, risk and resilience, exclusion and coexistence, identity and diversity, tradeoffs and synergies. Adaptive and anticipative governance approaches integrate different policy fields, seeking synergies between risk reduction and resilience, social-ecological transformation and conflict transformation, sustainable and environmental peacebuilding, climate mitigation and adaptation policies in North-South cooperation. An integrative geography framework could inform and contribute to sustainable peace policy and critical geopolitics expanding the action space within planetary boundaries.

References

Abel, G. J.; Brottrager, M.; Cuaresma, J. C.; Muttarak, R., 2019: "Climate, conflict and forced migration", in: *Global Environmental Change*, 54: 239–249; at: https://doi.org/10.1016/j.gloenvcha.2018.12.003.

Abrahams, D., 2020: "Conflict in abundance and peacebuilding in scarcity: Challenges and opportunities in addressing climate change and conflict", in: *World Development*, 132, 104998; at: https://doi.org/10.1016/j.worlddev.2020.104998.

Adams, C.; Ide, T.; Barnett, J.; Detges, A., 2018: "Sampling bias in climate-conflict research", in: *Nature Climate Change* 8,3: 200–203.

Adger, W. N.; Pulhin, J. M.; Barnett, J.; Dabelko, G. D.; Hovelsrud, G. K.; Levy, M.; Oswald Spring, Ú.; Vogel, C. H., 2014: "Human security", in: *Climate change 2014: Impacts, adaptation, and vulnerability. Part A: Global and sectoral aspects. Contribution of Working Group II to the Fifth Assessment Report of the IPCC* (Cambridge, UK: Cambridge University Press): 755–791.

AghaKouchak, A.; Huning, L.S.; Mazdiyasni, O.; Mallakpour, I.; Chiang, F.; Sadegh, M., *et al.* 2018: "How do natural hazards cascade to cause disasters?", in: *Nature*, 561, 7724): 458–460.

Anderson, J., 1801: *A Calm Investigation of Circumstances that have led to the Present Scarcity of Grain in Britain* (2nd ed., London).

Anjankar Ashish, P.; Anjankar Vaibhav, P.; Anjankar Anil, J; Lata, K., 2020: "Positive aspects of Covid 19 pandemic: A blessing in disguise", in: *International Journal of Research in Pharmaceutical Sciences, 11 (Special Issue 1): 187–191.*

Aykut, S.C.; Maertens, L., (Eds.), 2023: *The Climatization of Global Politics* (Cham: Palgrave Macmillan).

Babic, M., 2020: "Let's talk about the interregnum: Gramsci and the crisis of the liberal world order", in: *International Affairs*, 96,3: 767–786.

Bächler, G., 1998: "Why Environmental Transformation Causes Violence: A Synthesis", on: *Environmental Change and Security Project Report*, 4: 24–44.

Baker, S.; Quinn, M. J., 2022: "Populism, Austerity and Governance for Sustainable Development in Troubled Times. Introduction to Special Issue", in: *Sustainability*, 14,6: 14; at: https://doi.org/10.3390/su14063271.

Balbo, A.; Rothe, D.; Scheffran, J., 2020: "The Anthropocene – an opportunity for transdisciplinary and inclusive science", in: Brzoska, M., Scheffran, J. (Eds.): *Climate Change, Security Risks, and Violent Conflicts* (Hamburg: Hamburg University Press): 287–205.

Bank, A.; Fröhlich, C.; Schneiker, A., 2017: "The Political Dynamics of Human Mobility: Migration out of, as and into Violence", in: *Global Policy*, 8: 12–18.

Bassin, M., 1987: "Imperialism and the nation state in Friedrich Ratzel's political geography", in: *Progress in Human Geography*, 11,4: 473–495; at: https://doi.org/10.1177/030913258701100401.

Beck, U., 1992: *Risk Society: Towards a New Modernity*. (London: Sage Publications).

Bendor, T.; Scheffran, J., 2019: *Agent-based Modeling of Environmental Conflict and Cooperation* (Boca Raton, FL: CRC Press/Taylor & Francis).

Bernauer, T.; Böhmelt, T.; Koubi, V., 2012: „Environmental changes and violent conflict", in: *Environmental Research Letters,* 7, online; at: https://doi.org/10.1088/1748-9326/7/1/015601.

Besigye Bazirake, J., 2013: "Climate Change Discourse in Peacebuilding", in: *Peace Review: A Journal of Social Justice,* 25,4: 489-494.

Black, R.; Adger, W. N.; Arnell, N. W., *et al.*, 2011: "The effect of environmental change on human migration", in: *Global Environmental Change* 21(Suppl. 1), 3–11.

Bloch, J. 1899: *Der Krieg. Der zukünftige Krieg in seiner technischen, volkswirthschaftlichen und politischen Bedeutung.* 6 Vols. (Berlin: Puttkammer & Mühlbrecht; online at: archive.org/details/derkrieg05blocgoog.

Bloch, J. *The Future War in its Technical, Economic and Political Significance* [English translation in parts] (Boston: The World Peace Foundation, 1914).

BMZ, 2021: *Preventing Crises, Creating Prospects, Protecting People.* Report by the Commission on the Root Causes of Displacement (Berlin: Bundesministerium für wirtschaftliche Zusammenarbeit).

BMZ, 2022: *Cooperation in Action: Energising Development: Access to Energy for Millions of People* (Berlin: Bundesministerium für wirtschaftliche Zusammenarbeit).

Boas, I.; Farbotko, C.; Adams, H., *et al.*, 2019: "Climate migration myths», in: *Nature Climate Change,* 9, 898–903.

Boiral, O.; Heras-Saizarbitoria, I. 2017: "Managing Biodiversity Through Stakeholder Involvement: Why, Who, and for What Initiatives?", on: *Journal of Business Ethics* 140: 403–442. at:

Boubacar, S.; Pelling, M.; Barcena, A.; Mntandon, P., 2017: "The erosive effects of small disasters on household absorptive capacity in Niamey: a nested HEA approach in: Environ", in: *Urban,* 29,1: 33–50.

Bowlsby, D.; Chenoweth, E.; Hendrix, C. S.; Moyer, J. D., 2019: "The future is a moving target: Predicting political instability", in: *British Journal of Political Science,* 1– 13, at: https://doi.org/10.1017/S0007123418000443.

Brands, H.; Gaddis, J.L., 2021: "The New Cold War: America, China, and the Echoes of History", in: *Foreign Affairs,* 100,6: 10–21.

Brauch, H. G., 2021a: "Peace Ecology in the Anthropocene", in: Oswald Spring, Ú.; Brauch, H.G. (Eds.): *Decolonising Conflicts, Security, Peace, Gender, Environment and Development in the Anthropocene* (Cham: Springer Nature Switzerland): 51–186.

Brauch, H. G., 2021b: "The Anthropocene Concept in the Natural and Social Sciences, the Humanities and Law – A Bibliometric Analysis and a Qualitative Interpretation (2000–2020)", in: Benner, S.; Lax, G.; Crutzen, P. J.; Pöschl, U.; Lelieveld, J.; Brauch, H. G. (Eds.): *Paul J. Crutzen and the Anthropocene: A New Epoch in Earth's History* (Cham: Springer Nature): 289–438.

Brauch, H. G.; Oswald Spring, Ú.; Mesjasz, C.; Grin, J.; Dunay, P; Behera, N. B.; Chourou, B.; Kameri-Mbote, P.; Liotta, P. H. (Eds.), 2008: *Globalization and Environmental Challenges – Reconceptualizing Security in the 21st Century* (Berlin: Springer-Verlag).

Brauch, H. G.; Oswald Spring, U.; Grin, J.; Mesjasz, C.; Kameri-Mbote, P.; Behera, N. C.; Chourou, B.; Krummenacher, H. (Eds.), 2009: *Facing Global Environmental Change: Environmental, Human, Energy, Food, Health and Water Security Concepts* (Berlin – Heidelberg – New York: Springer-Verlag).

Brauch, H. G.; Oswald Spring, Ú.; Mesjasz, C.; Grin, J.; Kameri-Mbote, P.; Chourou, B.; Dunay, P.; Birkmann, J. (Eds.), 2011: *Coping with Global Environmental Change, Disasters and Security – Threats, Challenges, Vulnerabilities and* (Berlin – Heidelberg – New York: Springer-Verlag).

Brauch, H. G., Oswald Spring, Ú., Grin, J.; Scheffran, J. (Eds.), 2016: *Handbook on Sustainability Transition and Sustainable Peace (Cham:* Springer International Publishing).

Breyer, C., *et al.*, 2022: "On the History and Future of 100% Renewable Energy Systems Research", in: *IEEE Access,* vol. 10, 78176–78218; at: https://doi.org/10.1109/ACCESS.2022.3193402.

Brown, W., 2019: *In the Ruins of Neoliberalism: The Rise of Antidemocratic Politics in the West* (New York: Columbia University Press).
Brundtland, G. H.; Hauff, V. (Eds.), 1987: *Unsere gemeinsame Zukunft* (,Greven: Eggenkamp).
Brundtland, G. H, 1987: *Our Common Future* (New York: Oxford University Press).
Brzezinski, Z., 1997: *The Grand Chessboard: American Primacy and Its Geostrategic Imperatives (New York: Basic Books).*
Brzoska, M., 2009: "The securitization of climate change and the power of conceptions of security", in: *Security and Peace*, 27,3: 137–145.
Brzoska, M., 2018: "Weather extremes, disasters, and collective violence: conditions, mechanisms, and disaster-related policies in recent research", in: *Curr. Clim. Chang. Rep.*, 4, 320–329.
Brzoska, M.; Heller, R.; Kahl, M.; Kreikemeyer, A., 2019: "Das liberale Paradigma auf dem Prüfstand: Friedensforschung und Sicherheitspolitik an den Bruchstellen der Globalisierung"; IN. *Sicherheit und Frieden* [Security and Peace], 37,2: 53–61.
Brzoska, M.; Neuneck, G.; Scheffran, J., 2021: *Corona-Pandemie – Implikationen für die Sicherheitspolitik*, in: *IFSH Policy Brief 02/21* (Hamburg: Institut für Friedensforschung und Sicherheitspolitik an der Universität Hamburg); at: https://ifsh.de/news-detail/neuer-ifsh-policy-brief-corona-pandemie-implikationen-fuer-die-sicherheitspolitik.
Buhaug, H., *et al.*, 2014: "One effect to rule them all? A comment on climate and conflict", in: *Climatic Change*, 127: 391–397.
Buhaug, H., 2015: "Climate-conflict research: some reflections on the way forward", in: *WIREs Climate Change*, 6,3: 269–275.
Buhaug, H., 2017: "Climate change and conflict: taking stock", in: *Peace Econ. Peace Sci. Publ. Policy* 22,4: 331–338.
Buhaug, H.;Von Uexkull, N., 2021: "Vicious circles: violence, vulnerability, and climate change", in: *Annu. Rev. Environ. Res.* 46: 545–568.
Burles, R., 2021: "Another Geopolitics? International Relations and the Boundaries of World Order", in: *International Studies Review*, 23,4: 2108–2123; at: https://doi.org/10.1093/isr/viab038.
Burrows, K.; Kinney, P.L., 2016: "Exploring the Climate Change, Migration and Conflict Nexus", in: *International Journal of Environmental Research and Public Health*, 13: 443.
Buzan, B., 1991: "New Patterns of Global Security in the Twenty-First Century", in: *International Affairs*, 67:432.
Castree, N., *et al.*, 2009: *A Companion to Environmental Geography* (London: Wiley-Blackwell).
CEQ 1980: T*he global 2000 report to the President of the U.S. – Entering the 21st century. A report.* Council on Environmental Quality and the Department of State. Gerald O. Barney. (New York: Pergamon Press).
Chi, S.-H.; Flint, C.; Diehl, P.; Vasquez, J.; Scheffran, J.; Radil, A.M.; Rider, T.J., 2014: "The Spatial Diffusion of War: The Case of World War I", in: *Journal of the Korean Geographical Society*, 49,1: 57–76.
Clark, C., 2013: *Die Schlafwandler: Wie Europa in den Ersten Weltkrieg zog* (München: Deutsche Verlags-Anstalt).
Conversi, D.; Posocco, L., 2022: "Which Nationalism for the Anthropocene? A Comparative Study of Exemplary Green Nation-States", in: *Front. Polit. Sci.*, 27,4; at: https://doi.org/10.3389/fpos.2022.857597.
Costello, M.; Jenkins, J. C; Aly, H., 2015: "Bread, justice, or opportunity? The determinants of the Arab Awakening protests", in: *World Development*, 67: 90–100; at: https://doi.org/10.1016/j.worlddev.2014.10.002.
Crawford, Neta, 2022: *The Pentagon, Climate Change, and War: Charting the Rise and Fall of U.S. Military Emissions* (Cambridge, Massachusetts: The MIT Press).
Crutzen, P.J.; Stoermer, E.F., 2000: "The Anthropocene", in: *Global Change Newsletter*, 41, 17–18.
Czempiel, E.-O., 1995: "Der Friede – sein Begriff, seine Strategien", in: Senghaas, D. (Ed.), *Den Frieden denken* (Frankfurt: edition suhrkamp): 165–176.

Dahrendorf, R., 1963: "Konflikt und Kontrolle im internationalen System", in: Nerlich, U.; (Ed.): *Krieg und Frieden im industriellen Zeitalter*, Beiträge der Sozialwissenschaft, Vol. I (Gütersloh: Bertelsmann): 310–320 (Reprint from 1963).
Dalby, S., 2020: *Anthropocene Geopolitics: Globalization, Security, Sustainability* (Ottawa: University of Ottawa Press).
Daoudy, M., 2021: "Rethinking the Climate–Conflict Nexus: A Human–Environmental–Climate Security Approach", in: *Global Environmental Politics*, 21,3: 4–25.
Daoudy, M.; Schraven, B.; Kristiansen, M.; Swain, A.; Scheffran. J., 2022: "Migration and the 2030 Agenda", in: *One Earth*. 5 (19 Aug): 838–840; at: https://doi.org/10.1016/j.oneear.2022.08.001.
Davies, K.; Riddell, T.; Scheffran, J., 2020: "Preventing a Warming War – Protection of the Environment and Reducing Climate Conflict Risk as a Challenge of International Law", in: *Goettingen Journal of International Law*, 10,1: 307–343.
de Klerk, L.; Shmurak, A.; Gassan-Zade, O.; Shlapak, M.; Tomliak, K.; Korthuis, A., 2023: "Climate Damage Caused by Russia's War in Ukraine", in: *Initiative on GHG accounting of war, International Climate Initiative (IKI)*, 1 November 2022.
Deiseroth, D., 2000: "Die NATO – Ein System 'kollektiver Verteidigung' oder 'kollektiver Sicherheit'? Kritische Bemerkungen zur Rechtsprechung des Bundesverfassungsgerichts", in: *Die Friedens-Warte*, 75,1: 101–128.
Detges, A., 2017: "Droughts, state-citizen relations and support for political violence in Sub-Saharan Africa: A micro-level analysis", in: *Political Geography*, 61: 88–98.
Dixson-Declève, S.; Owen, G.; Ghosh, J.; Rockström, J.; Stoknes, P. E.; Randers, J., 2022: *Earth for All – A Survival Guide for Humanity* (Gabriola Island, BC: New Society Publishers).
Dörre, K., 2022: „Eine Epoche der Ambivalenz", in: Wissenschaft & Frieden, 2022/4: 14–18.
EACH-FOR, 2009: „Environmental Change and Environmental Migration Scenarios Synthesis Report".
Eastin, J., 2016: "Fuel to the fire: natural disasters and the duration of civil conflict", in: *International Interactions*, 42: 322–349.
Eker S.; Lenton, T.M.; Powell, T., Scheffran, J., Smith, S.R., Swamy D., Zimm, C., 2024: "Cross-system interactions for positive tipping cascades," *Earth System Dynamics*, 15. 789–800. at: https://doi.org/10.5194/esd-15-789-2024
Engels, F., 1886/1975: " Dialektik der Natur", in: *Werke*, Vol. 20 (Berlin: Dietz).
Engels, A.; Marotzke, J.; Gonçalves Gresse, E.; López-Rivera, A.; Pagnone, A.; Wilkens, J. (Eds.), 2023: *Hamburg Climate Futures Outlook 2023. The plausibility of a 1.5°C limit to global warming—Social drivers and physical processes* (Hamburg: Cluster of Excellence Climate, Climatic Change, and Society [CLICCS]).
EU, Rat der Europäischen Union, 2008: "Klimawandel und Internationale Sicherheit" (Brüssel: Rat der Europäischen Union); at: consilium.europa.eu und http://register.consilium.europa.eu/pdf/de/08/st07/st07249.de08.pdf.
European Commission, 2021: *A European Green Deal: Striving to Be the First Climate-Neutral Continent* (Brussels: European Commission).
Falk, R. A., 1973: "Environmental warfare and ecocide. Facts, appraisal, and proposals", in: *Bulletin of Peace Proposals*, vol. 1.
Flint, C.; Diehl, P.; Scheffran, J.; Vasquez, J.; Chi, S., 2009: "Conceptualizing ConflictSpace: Toward a Geography of Relational Power and Embeddedness in the Analysis of Interstate Conflict", in: *Annals of the Association of American Geographers* 99, 5,5: 827–835.
Flockhart, T., 2016: "The coming multi-order world", in: *Contemporary Security Policy*, 37,1: 3–30.
Franzke, C.L.E.; Ciullo, A.; Gilmore, E.A.; Matias, D.M.; Nagabhatla, N.; Orlov, A., *et al.* 2022:Perspectives on tipping points in integrated models of the natural and human Earth system: cascading effects and telecoupling. *Environmental Research Letters* 17 (1): 15004; at: https://doi.org/10.1088/1748-9326/ac42fd.
Frey, U.; Lammers, C.; Birckenbach, H.-M.; Jaberg, S.; Schweitzer, C.; Buro, A., 2014: . "Friedenslogik statt Sicherheitslogik: Theoretische Grundlagen und friedenspolitische Realisierung", Dossier 75, in: *Wissenschaft & Frieden*, 2014–2.

Friedensgutachten, 2020: *Im Schatten der Pandemie – letzte Chance für Europa*, BICC, HSFK, IFSH, INEF (Bielefeld: transcript).
Fröhlich, C.; Klepp, S. (Eds.), 2020: "Migration and Conflict in a Global Warming Era: A Political Understanding of Climate Change", in: *Social Sciences*, 9,5 (special issue).
Froese, R.; Schilling, J., 2019: "The nexus of climate change, land use, and conflicts", in: *Current Climate Change Reports*, 5: 24–35.
Froese, R.; Scheffran, J.; Schilling, J., 2023: "Transformation durch Eskalation? Klimaprotest zwischen Demokratie und Kriminalisierung", in: *Wissenschaft & Frieden*, 2023/3: 6–9.
Galtung; J., 1995:"Visionen einer friedlichen Welt", in: Senghaas, D. (Ed.): *Den Frieden denken* (Frankfurt: edition suhrkamp): 385–418.
Garcia, H., 2010: *Environmental Geography* (Palm Bay, FL: Apple Academic Press, Inc.)
Ge, Q.; Hao, M.; Ding, F.; Jiang, D.; Scheffran, J.; Helman, D., *et al.*, 2022: "Modelling armed conflict risk under climate change with machine learning and time-series data", in: *Nature Communications*, 13: 2839; at: https://doi.org/10.1038/s41467-022-30356-x.
Germanwatch, 2021: *Global Climate Risk Index 2021* (Bonn: Germanwatch); at: https://www.germanwatch.org/fr/19777.
Gießmann, H.J.; Rinke, B. (Eds.), 2011: *Handbuch Frieden*, (VS Verlag für Sozialwissenschaften, 310–323.
Gioli, G.; Hugo, G.; Manz Costa, M.; Scheffran, J., 2016: "Human Mobility, Climate Adaptation, and Development, Introduction to Special Issue", in: *Migration and Development*: 1–6.
Grêt-Regamey, A.; Huber, S.H:; Huber, R. 2019: "Actors' diversity and the resilience of social-ecological systems to global change", in: *Nature Sustainability*, 2,4: 290–297.
Guterres, A., 2021: "We are losing our suicidal war against nature" (New York: United Nations, SG/SM/20959, 11 October).
Hanson, T.; Brooks, T.M.; Da Fonseca, G.A.B.; Hoffmann, M.; Lamoreux. J.F.; Machlis, G.; Mittermeier, C.G.; Mittermeier, R. A.; Pilgrim, J.D.; 2009: "Warfare in Biodiversity Hotspots", in: *Conservation* Biology, 23, 3: 578–587.
Hanson, T., 2011: "War and Biodiversity Conservation: The Role of Warfare Ecology", in: Machlis, G.E.; Hanson, T.; Špirić, Z.; McKendry, J.E. (Eds.): *Warfare Ecology* (Springer): 125-132.
Hardt, J.N.; Scheffran, J., 2019: "Environmental Peacebuilding and Climate Change: Peace and Conflict Studies at the Edge of Transformation", in: *Policy Brief*, No. 68 (Dec.) (Tokyo: Toda Peace Institute).
Hardt, J.N.; Harrington, C., von Lucke, F., Estève, A., Simpson, N.P. (Eds.), 2023: *Climate Security in the Anthropocene.* The Anthropocene: Politik—Economics—Society—Science (Cham: Springer Nature Switzerland).
Harvey, D., 1990: *The Condition of Postmodernity* (Oxford:Blackwell).
Hegre, H.; Buhaug, H.; Calvin, K. V.; Nordkvelle, J.; Waldhoff, S. T.; Gilmore, E., 2016: "Forecasting civil conflict along the shared socioeconomic pathways", in: *Environmental Research Letters*, 11,5; at: https://doi.org/10.1088/1748-9326/11/5/054002.
Hegre, H., *et al.*, 2019: "ViEWS: a political violence early-warning system", in: *J. Peace Res.* 56: 155–174.
Held, H.; Schellnhuber, H.J., 2004: "Evolution of perturbations in complex systems", in: Steffen. W., *et al.* (Ed.): *Global change and the earth system* (Berlin, Heidelberg: Springer): 145–147.
Herwig, H. H., 1999: "Geopolitik: Haushofer, Hitler and Lebensraum", in: *Journal of Strategic Studies*, 22,2–3: 218–241; at: https://doi.org/10.1080/01402399908437762.
Herwig, H. H., 2016: *The Demon of Geopolitics: How Karl Haushofer Educated Hitler and Hess* (London: Rowman and Littlefield).
Hickel, J.; Kallis; G.; Jackson, T.; O'Neill, D.W.; Schor, J.B.; Steinberger, J.K.; Victor, P.A.; Ürge-Vorsatz, D., 2022: "Degrowth can work—here's how science can help – Wealthy countries can create prosperity while using less materials and energy if they abandon economic growth as an objective", in: *Nature* (12 Dec.).
Higgins, P.; Short, D.; South, N., 2013: "Protecting the planet: a proposal for a law of ecocide", in: *Crime Law and Social Change*, vol. 59, 2.

HIIK, 2023: *Conflict Barometer 2022* (Heidelberg: Heidelberg Institute for International Conflict Research).
Hoare, Q.; Nowell-Smith, G. (Ed.), 1971: *Selections from the Prison Notebooks of Antonio Gramsci* (London: Lawrence & Wishart).
Hoffmann, R.; Dimitrova, A.; Muttarak, R.; Crespo Cuaresma, J.;Peisker, J., 2020: "A Meta-Analysis of Country-Level Studies on Environmental Change and Migration", in: *Nature Climate Change*, 10,14: 904–912.
Holifield, R.; Chakraborty, J.; Walker, G. (Eds), 2018: *The Routledge Handbook of Environmental Justice* (London: Routledge).
Homer-Dixon, T., 1994: "Environmental Scarcities and Violent Conflict: Evidence from Cases", in: *International Security*, vol. 19: 5–40.
Homer-Dixon, T.; Renn, O.; Rockström, J.; Donges, J.; Janzwood, S., 2022: *A call for an international research program on the risk of a global polycrisis*. Technical Paper, (Toronto: Cascade Institute, 2022–3).
Hore, K.; Kelman, I.; Mercer, J.; Gaillard, J.C. (Eds.), 2018: *Climate change and disasters. Handbook of Disaster Research* (Cham: Springer): 145–159.
Howe, B.; Bang, G., 2017: "Nargis and Haiyan: the politics of natural disaster management in Myanmar and the Philippines", in: *Asian Stud. Rev.* 41, 1: 58–78.
Hsiang, S.M.; Burke, M.; Miguel, E., 2013: "Quantifying the Influence of Climate on Human Conflict", in: *Science*, 341: 1212.
Hummel, M.; Scheffran, J.; Simon, H.-R., 2002: *Konfliktfeld Biodiversität* (Münster: agenda).
Huntington, S. P., 1996: *The Clash of Civilizations and the Remaking of World Order* (New York: Simon & Schuster).
IASC, 2008: *Climate Change, Migration and Displacement: Who will be affected?* Working paper, Inter-Agency Standing Committee (31 October).
Ide, T., 2017: "Research methods for exploring the links between climate change and conflict", in: *Wiley Interdisciplinary Reviews: Climate Change*, 8,3; e456, at: https://doi.org/10.1002/wcc.456 Wiley Online Library Web of Science®Google Scholar.
Ide, T., 2018: "Does environmental peacemaking between states work? Insights on cooperative environmental agreements and reconciliation in international rivalries", in: *Journal of Peace Research*, 55,3; 351–365: at: https://doi.org/10.1177/0022343317750216.
Ide, T., 2020: "The dark side of environmental peacebuilding," *World Development*, Elsevier, 127(C).
Ide, T., 2023: *Catastrophes, Confrontations, and Constraints: How Disasters Shape the Dynamics of Armed Conflict* (Cambridge, Mass: MIT Press).
Ide, T.; Bruch, C.; Carius, A., *et al.*, 2021: "The past and future(s) of environmental peacebuilding", in: *International Affairs*, 97,1: 1–16.
Ide T.; Link, P.M.; Scheffran, J.; Schilling, J., 2016: "The Climate-Conflict Nexus: Pathways, Regional Links, and Case Studies", in: Brauch, H. G.; Oswald Spring. Ú; Grin. J.; Scheffran, J. (Eds.): *Handbook on Sustainability Transition and Sustainable Peace*, (Cham: Springer): 285–304.
Ide, T.; Scheffran, J., 2012: *Climate Change: Source of Conflict or Promoter of Cooperation?* Working Paper, CLISEC-26 (Hamburg: Universität Hamburg).
Ide, T.; Scheffran, J., 2014: "On climate, conflict and cumulation: Suggestions for integrative cumulation of knowledge in the research on climate change and violent conflict", in: *Global Change, Peace & Security*, 26,3; 263–279.
IDMC, 2023: *Global Report on Internal Displacement 2023* (Geneva: Internal Monitorin Displacement Centre); at: https://www.internal-displacement.org/global-report/grid2023/.
Ioannides, I., 2022: "What European Union in the "Age of Uncertainty"? Weathering the Geopolitical Storms in a World of Perpetual Crises", in: *Intereconomics*, 57,6: 363–367; at: https://doi.org/10.1007/s10272-022-1088-8.
IOM, 2007: *Migration and the Environment*. Discussion Note (Nov. 1) (Geneva: International Organization for Migration, IOM Council, 94th Session, Doc. MC/INF/288, IOM).

Ionesco, D.; Mokhnacheva, D.; Gemenne, F., 2017: *Atlas der Umweltmigration*, (München: Oekom-Verlag).
IPBES, 2019: *Global assessment report on biodiversity and ecosystem services of the Intergovernmental Science-Policy Platform on Biodiversity and Ecosystem Services*. (Bonn: IPBES secretariat); at: https://doi.org/10.5281/zenodo.3831673.
IPCC, 2014: IPCC, *Climate change 2014: Impacts, adaptation, and vulnerability* (Geneva: IPCC).
IPCC, 2018: *Global Warming of 1.5 °C. An IPCC Special Report on the impacts of global warming of 1.5 °C above pre-industrial levels and related global greenhouse gas emission pathways, in the context of strengthening the global response to the threat of climate change, sustainable development, and efforts to eradicate poverty* (Geneva: IPCC).
IPCC, 2019: *Special Report on the Ocean and Cryosphere in a Changing Climate. Intergovernmental Panel on Climate Change* (Cambridge: Cambridge University Press).
IPCC, 2021: *Climate Change 2021: The Physical Science Basis. Contribution of Working Group I to the Sixth Assessment Report of the Intergovernmental Panel on Climate Change* (Cambridge: Cambridge University Press).
IPCC, 2022a: *Summary for Policymakers Headline Statements*. 6th Assessment Report, WG II (Geneva: Intergovernmental Panel on Climate Change).
IPCC, 2022b: *Climate Change 2022: Impacts, Adaptation and Vulnerability*. 6th Assessment Report, WG II (Geneva: Intergovernmental Panel on Climate Change).
IPCC, 2023: *Climate Change 2023. AR6 Synthesis Report* (Geneva: Intergovernmental Panel on Climate Change).
Jakobeit, C.; Methmann, C., 2012: "Climate Refugees' as Dawning Catastrophe? A Critique of the Dominant Quest for Numbers", in: Scheffran, J.; Brzoska, M.; Brauch, H.G.; Link, P.M.; Schilling, J. (Eds.): *Climate Change, Human Security and Violent Conflict* (Berlin: Springer): 301–314.
Jathe, M.; Scheffran, J., 1995: "Modelling International Security and Stability in a Complex World", in: Tran Thanh, Van J.; Berge, P.; Conte, R.; Dubois, M. (Eds.): *Chaos and Complexity*. Editions Frontieres (city: publisher): 331–332.
Johnke, B.; Scheffran, J.; Soyez, K. (Eds.), 2004: *Abfall, Energie und Klima – Wege und Konzepte für eine integrierte Ressourcenutzung* (Berlin: Erich-Schmidt-Verlag).
Juhola, S.; Filatova, T.; Hochrainer-Stigler, S.; Mechler, R.; Scheffran, J.; Schweizer, P.-J., 2022: "Social Tipping Points and Adaptation Limits in the Context of Systemic Risk: Concepts, models and governance", in: *Frontiers in Climate*, 4:1009234; at: https://doi.org/10.3389/fclim.2022.1009234.
Kamta, F.N.K.; Scheffran, J.; Schilling, J., 2021: "Water resources, forced migration and tensions with host communities in the Nigerian part of the lake chad basin", in: *Resources*, 10: 1–14.
Kaplan, R.D., 2009: "The Revenge of Geography", in: *Foreign Policy*, 172 (May/June): 96–105.
Kelley, C.P.; Mohtadi, S.; Cane, M.A.; Seager, R.-; Kushnir, Y. 2015: "Climate Change in the Fertile Crescent and Implications of the Recent Syrian Drought", in: PNAS, 112,11: 3241–3246.
Kemp, L.; Xu, C.; Depledge, J.; Ebi, K.L.; Gibbins, G.; Kohler, T.A., *et al.*, 2022: "Climate Endgame: Exploring catastrophic climate change scenarios", in: *Proceedings of the National Academy of Sciences*, 119,34: e2108146119; at: https://doi.org/10.1073/pnas.2108146119.
Kim, R. E.; Kotze, L. J., 2021: "Planetary boundaries at the intersection of Earth system law, science and governance: A state-of-the-art review", in: *Review of European Comparative & International Environmental Law*, 30,1, 3–15; at: https://doi.org/10.1111/reel.12383.
Kissinger, H., 2022: "How to avoid another world war", in: *The Spectator*, 17 December.
Klein, N., 2007: *The Shock Doctrine: The Rise of Disaster Capitalism* (New York: Penguin).
Kominek, J.; Scheffran, J. 2012: "Cascading processes and path dependency in social networks", in; Soeffner, H.-G.(Ed.): *Transnationale Vergesellschaftungen* (Wiesbaden: VS Springer).
Koschorke, A., 2023: "Weltordnung: Die Tage des Westens sind gezählt", in: *DIE ZEIT*, 3 Match.
Koubi, V., 2019: "Climate Change and Conflict", in: *Annual Review of Political Science*, 22,1: 343–60.

Kwakkel, J. H.; Walker, W. E.; Haasnoot, M., 2016; "Coping with the wickedness of public policy problems: Approaches for decision making under deep uncertainty", in: *Journal of Water Resources Planning and Management*, 142,3:01816001–1–01816001-5; at: https://doi.org/10.1061/(ASCE)WR.1943%2D5452.0000626.

Lawrence, M.G.; Schäfer, S.; Muri, H.; Scott, V.; Oschlies, A.; Vaughan, N.E., *et al.*, 2018: "Evaluating climate geoengineering proposals in the context of the Paris Agreement temperature goals", in: *Nature Communication*, 9,3734: 1–19.

LeBillon, P.; Duffy, R., 2018: Conflict Ecologies: Connecting Political Ecology and Peace & Conflict Studies, in: *Journal of Political Ecology*, 25,1: 239-260.

Lefebvre, H., 1981 "La présence et l'absence", in: *Revue Philosophique de la France et de l'Etranger* 171 (2):241–242.

Lefebvre, H., 1991: *The production of space* (Oxford, OX, UK: Blackwell).

Lenton, T.; Held, H.; Kriegler, E.; Hall, J. -W.; Lucht, W.; Ramstorf, S.; Schellnhuber, H. J., 2008: "Tipping elements in the Earth's climate system",in: *Proceedings of the National Academy of Science*, 105,6:1786–1793.

Lenton, T.M.; McKay, A.; Loriani, S., et al. (Eds), 2023: The Global Tipping Points Report 2023. (University of Exeter).

Link, P.M.; Brzoska. M.; Maas, A.; Neuneck. G.; Scheffran, J, 2013: "Possible implications of climate engineering for peace and security", in: *Bulletin of the American Meteorological Society*, 94: ES13-ES16.

Link, P. M.; Böhner, J.; Held, H.; Scheffran, J., 2018: "Energy Landscapes", *Bulletin of the American Meteorological Society*, 99, ES71-ES73; at: https://doi.org/10.1175/BAMS-D-17-0295.1.

Longman, 2009: Longman Dictionary of American English (Pearson Education Limited, Essex).

Lüthje, C.; Schäfer, M.; Scheffran, J.; 2011: "Limits to the Anthropocene. What are the challenges and boundaries of science for the post-normal age? EGU General Assembly 2011", in: *Geophysical Research Abstracts*, 13, 2011–11795.

Lutz, D. S., 1999: "Die Europäische Friedens- und Sicherheitsordnung: Vision und Realität. Vortrag im Rahmen des Arbeitskreises Friedens- und Konfliktforschung im Planungsstab des Auswärtigen Amtes in Berlin am 27. September 1999".

Maas, A.; Scheffran, J., 2012: "Climate conflicts 2.0? Climate engineering as a challenge for international peace and security", in: *Sicherheit & Frieden)/Security & Peace*, 30,4:193–200.

Mach, K.J.; Kraan, C.M.; Adger, N.; Buhaug, H.; Burke, M.; Fearon, J.D., *et al.*, 2019: "Climate as a risk factor for armed conflict", in: *Nature*, 571: 193–197.

Mach, K.J.; Adger, W.N.; Buhaug, H.; Burke, M.; Fearon, J.D,; Field, C.B.; Hendrix, C.S.; Kraan, C.M.; Maystadt, J.F.; O'Loughlin. J.; Roessler, P.; Scheffran, J.; Schultz, K.A.; von Uexkull, N., 2020: "Directions for Research on Climate and Conflict", in: Earth's Future, 8,7: 1–7 (e2020EF001532); at: https://doi.org/10.1029/2020EF001532.

Magnus, G., 2009: "Malthus, Marx and the Globalization Debate", in: *The Globalist*, 28 February 2009.

Malthus, T. R., 1798 [1993]: *An Essay on the Principle of Population as It Effects the Future Improvement of Society* (Oxford: Oxford University Press).

Marx, K., 1919: *Theorien über den Mehrwert*. Zweiter Teil, (Stuttgart: Dietz).

Matthew, R. A., 2014: "Integrating Climate Change into Peacebuilding", in: *Climatic Changes* 123,1: 83-93.

Maystadt, J.-F.; Hirvonen, K; Mabiso, A.; Vandercasteelen, J., 2019: "Impacts of hosting forced migrants in poor countries", in: *Annual Review of Resource Economics* 11,1: 439– 459.

McDonald, M., 2013: "Discourses of climate security", in: *Political Geography*, 33: 42–51.

McLeman, R., 2013: *Climate and Human Migration. Past experiences, future challenges* (Cambridge, Cambridge University Press).

McNeill, J.R.; Engelke, P., 2016: *The Great Acceleration: An Environmental History of the Anthropocene since 1945* (Cambridge, Mas: Harvard University Press.

Meadows, D.H.; Meadows, D.; Randers, J.; Behrens, W.W.III, 1972: *The Limits to Growth. A Report for the Club of Rome* (New York: Universe).

Mearsheimer, J. J., 2014: "Why the Ukraine Crisis Is the West's Fault: The Liberal Delusions That Provoked Putin", in: *Foreign Affairs*, (Sep./Oct.): 1–12.
Menon, R., 2023: *An Age of Crises: Prospects for inequality and division*. Pathfinders Report (New York: Center on International Cooperation,Feb. 24).
Menzel, U., 2015: *Die Ordnung der Welt* (Berlin: Suhrkamp).
Merriam, 2009: Merriam-Webster's dictionary and thesaurus, 2007: (Springfield, Mass).
Mesjasz, C., 2010: "Complexity of Social Systems", in: *Acta Physica Polonica A*, 117,4 (April), 706–715.
Meyer, P., 2020: *Arms Control in Outer Space: Mission Impossible or Unrealized Potential?* (city: Canadian Global Affairs Institute, October).
Meyers, R., 2019: "Krieg und Frieden", in: Gießmann, H. J.; Rinke, B. (Eds.), *Handbuch Frieden* (Wiesbaden: VS Springer): 21–50.
Milkoreit, M.; Hodbod, J.; Baggio, J., *et al.*, 2018: "Defining tipping points for social-ecological systems scholarship – an interdisciplinary literature review", in: *Environmental Research Letters*, 13: 1–12.
Mobjörk, M., *et al.*, 2016: *Climate-related security risks. Towards an Integrated Approach* (Stockholm: SIPRI – Stockholm University – Swedish Institute of International Affairs).
Molo, B. 2015: "Resource conflicts", in: Jäger. T, (Ed.): *Handbook of security threats* (Wiesbaden: VS Springer): 33–41.
Mondragón Toledo, G.; Niemann, H.; Scheffran, J.; Wiener, A., 2022: *Conceptualizing Strategic Narratives: The Peace Movement as a Strategic Respondent to COVID-19*. CSS Working Paper Series No 7 (Hamburg:_publisher/institute); at: https://doi.org/10.25592/CSS-WP-007.
Morfi, C.; Nilsson, J.; Hakelius, K.; Karantininis, K., 2021: "Social networks and member participation in cooperative governance", in: *Agribusiness*, 37,2: 264–285; at: https://doi.org/10.1002/agr.21660.
Morin, E.; Kern, A.B., 1999: *Homeland Earth: A Manifesto for the New Millenium. Advances in Systems Theory, Complexity, and the Human Sciences* (Cresskill, N.J: Hampton Press).
Moseley, W.G.; Perramond, E.; Hapke, H.M.; Laris, P., 2014: *An introduction to human-environment geography: local dynamics and global processes* (Oxford, OX, UK: Wiley-Blackwell).
MunichRe, 2021: "Record hurricane season, extreme wildfires – The balance of natural catastrophes in 2020" (Munich: Munich Reinsurance); at: https://www.munichre.com/de.
Münkler, H., 2014: *Der Große Krieg: Die Welt 1914 bis 1918* (Reinbek: Rowohlt).
Murphy, T. W., 2022: "Limits to economic growth", in: *Nature Physics*, 18,8: 844–847; at: https://doi.org/10.1038/s41567-022-01652-6.
Myers, N., 2002: "Environmental refugees: a growing phenomenon of the 21st century", in: *Philosophical Transactions, The Royal Society, Series B*, vol. 357: 609–613.
Nadiruzzaman M.; Scheffran, J.; Shewly, H.J.;Kley, S., 2022 "Conflict-Sensitive Climate Change Adaptation: A Review", in: *Sustainability*: at: https://doi.org/10.3390/su14138060.
Nash, S.L., 2019: *Negotiating Migration in the Context of Climate Change: International Policy and Discourse* (Bristol: Bristol University Press).
Neumann, P.R., 2022: *Die neue Weltunordnung: Wie sich der Westen selbst zerstört*. (Berlin: Rowohlt).
Neuneck, G.; Scheffran, J., 2000a: "Abrüstung am Ende? Zur Kontroverse um die neuen Raketenabwehrpläne der USA", in: *Blätter für deutsche und internationale Politik*, 6/2000: 715–727.
Neuneck, G.; Scheffran, J., 2000b: "Die Grenzen technischer Kriegführung", in: *Spektrum der Wissenschaft*, 1: 90–98.
Oels, A., 2012: "From 'Securitization' of Climate Change to 'Climatization' of the Security Field", in: Scheffran, J.; Brzoska, M.; Brauch, H.G.; Link, P.;Schilling, J. (Eds.): *Climate Change, Human Security and Violent Conflict* (Berlin – Heidelberg: Springer): 185–205.
OPIC, 2022: *For Our Shared Future: Common Security 2022*. (Stockholm: Olof Palme International Center, International Peace Bureau, International Trade Union Confederation).

Ostrom, E., 2000: "Collective Action and the Evolution of Social Norms", in: *Journal of Economic Perspectives*, 14,3: 137–158; at: https://doi.org/10.1257/jep.14.3.137.
Oswald Spring, U. 2023: "Risk Assessment and Management from Climate Change Impacts in Mexico. Gender Perspective in Adaptation and Resilience" in: *Environmental Risk Assessment and Risk Management*, 4,2.
Otto, I.M.; Donges, J.F.; Cremades, R.; Bhowmik, A.; Hewitt, R.J.; Lucht, W., *et al.*, 2020: "Social tipping dynamics for stabilizing Earth's climate by 2050, in: *PNAS*, 117,5: 52354–2365.
OWD, 2023: "Our World in Data: Cumulative CO2 Emissions"; at: https://ourworldindata.org/grapher/cumulative-co-emissions.
Parkinson, S.; Cottrell, L. 2022: "Estimating the Military's Global Greenhouse Gas Emissions", in: *Conflict & Environment Observatory* (CEOBS); Scientists for Global Responsibility; at: https://ceobs.org/.
Pastoors, D.; Drees, L.; Fickel, T.; Scheffran, J. 2022: "'Frieden verbessert das Klima' – Zivile Konfliktbearbeitung als Beitrag zur sozial-ökologischen Transformation, in: *Zeitschrift für Außen- und Sicherheitspolitik,* 23 pp.; at: https://doi.org/10.1007/s12399-022-00911-x.
Pereira, P.; Basic, F.; Bogunovic, I.; Barcelo, D., 2022a: "Russian-Ukrainian war impacts the total environment", in: *Science of the Total Environment*, 837,6, Art.155865; at: https://doi.org/10.1016/j.scitotenv.2022.155865.
Pereira, P.; Zhao, W. W.; Symochko, L.; Inacio, M.; Bogunovic, I.; Barcelo, D., 2022b: "The Russian-Ukrainian armed conflict will push back the sustainable development goals", in: *Geography and Sustainability*, 3,3: 277–287; at: https://doi.org/10.1016/j.geosus.2022.09.003.
Piazza, A., 2021: "Collective Responsibility in the Cooperative Governance of Climate Change", in: *Sustainability*, 13,8: 18, Article 4363; at: https://doi.org/10.3390/su13084363.
Posocco, L.; Watson, I., 2023: "Re-imagining the nation-state: An impetus from the pandemic", in: *Frontiers in Sociology*, 8:1086569; at: https://doi.org/10.3389/fsoc.2023.1086569.
Rajaeifar, M.; Belcher, O.; Parkinson, S.; Neimark, B.; Weir, D.; Ashworth, K., *et al.*, 2022: "Decarbonize the military—mandate emissions reporting", in: *Nature*, 611, 29–32.
Rauschmayer, F., 2013: "Module 10: Designing Participatory Decision Strategies", in: Ring, I.; Kranz, A.; Jepsen, N.; Rauschmayer, F. (Eds.) *Human - Wildlife Conflicts in Europe* (city: Springer): 315-322.
Ratzel, F., 1897 (1923): *Politische Geographie, oder die Geographie der Staaten, des Verkehrs, und des Krieges,* 3rd ed. (Munich – Berlin: Oldenbourg).
Reichstein, M.; Riede, F.; Frank, D., 2021: "More floods, fires and cyclones – plan for domino effects on sustainability goals", in: *Nature*, 592: 347–349.
Reinhard, W., 2016: *Die Unterwerfung der Welt: Globalgeschichte der europäischen Expansion 1415–2015* (München: Beck).
Remoff, H., 2016: "Malthus, Darwin, and the Descent of Economics", in: *American Journal of Economics and Sociology*, 75,4: 862– 903.
Richter, R., 1998: "Umweltflüchtlinge in Afrika", in: Scheffran, J.; Vogt W. (Eds.): *Kampf um die Natur – Umweltzerstörung und die Lösung ökologischer Konflikte*, (Darmstadt: Wissenschaftliche Buchgesellschaft): 42–73.
Rittel, H.; Webber, M., 1973: "Dilemmas in a General Theory of Planning", in: *Policy Sciences*, vol. 4:155–169.
Rockström, J.; Steffen, W.; Noone, K.; Persson, Å.; Chapin, F. S., 2009: "A Safe Operating Space for Humanity", in: *Nature*, 461,7263: 472–475.
Rockström, J.; Gupta, J.; Qin, D., *et al.*, 2023: "Safe and just Earth system boundaries" in: *Nature*; at: https://doi.org/10.1038/s41586-023-06083-8.
Rodriguez Lopez, J.M.; Tielbörger, K.; Claus, C.; Fröhlich, C.; Gramberger, M.; Scheffran, J. 2019: "A Transdisciplinary Approach to Identifying Transboundary Tipping Points in a Contentious Area: Experiences from across the Jordan River Region", in: *Sustainability*, 11: 1184.
Rupke, N.A., 2008: *Alexander Von Humboldt: A Metabiography*. (Chicago: University of Chicago Press).

Rupnik, J., 2015: "Das andere Europa. Die Flüchtlingskrise und der neue Identitätsdiskurs", in: *Lettre international* (Winter): 26–28.
Rüttinger, L.; Smith, D.; Stang, G.; Tänzler, D.; Vivekananda, J., 2015: *A New Climate for Peace: Taking Action on Climate and Fragility Risks* (Berlin: adelphi, *et al.*).
Sachs, J.D., 2020: *The ages of globalization: geography, technology, and institutions* (New York, Columbia University Press).
Sakaguchi, K.; Varughese, A.; Auld, G., 2017: "Climate Wars? A Systematic Review of Empirical Analyses on the Links between Climate Change and Violent Conflict", in: *International Studies*; at: https://doi.org/10.1093/isr/vix022.
Santarius, T.; Scheffran, J.; Tricarico, A., 2012: *North South Transitions to Green Economies.* (Berlin: Heinrich Böll Foundation).
Scartozzi, C., 2020: "Reframing Climate-Induced Socio-Environmental Conflicts. A Systematic Review", in: *International Studies Review*, 0: 1–30.
Schäfer, L., 1995: "Natur", in: Martens, E.; Schnädelbach, H. (Eds.): *Philosophie,* vol. 2 (Reinbek: rororo) 467–506.
Schäfer, M.S.; Scheffran, J.; Penniket, L., 2015 "Securitization of media reporting on climate change? A cross-national analysis in nine countries". In: *Security Dialogue* 471: 1–21.
Scheffer, M., 2009: *Critical Transitions in Nature and Society* (Princeton: Princeton University Press).
Scheffran, J., 1992: "Environmental Applications of Military Information and Communication Technologies", in: Brunn, A.; Baehr, L.; Karpe, H.-J. (Eds.): *Conversion – Opportunities for Development and Environment* (city: Springer, 122–133).
Scheffran, J., 1994: "Kriegs- und Umweltflüchtlinge", in: Böhme, G.; Abindra, R.; Chakraborty, N.; Weiler, F. (Eds.): *Migration und Ausländerfeindlichkeit* (Darmstadt: publisher): 23–38.
Scheffran, J., 1996: "Leben bewahren gegen Wachstum, Macht, Gewalt – Zur Verknüpfung von Frieden und nachhaltiger Entwicklung", in: *W&F*, 3/1996: 5–9.
Scheffran, J.; Jathe, M., 1997: "Kein Frieden mit der Natur ohne Frieden unter den Menschen", in: *Wissenschaft & Frieden*, 1/1997.
Scheffran, J., 1998: "Wege zu einer nachhaltigen Entwicklung des Friedens", in: Scheffran, J.; Vogt, W. R. (Eds.): *Kampf um die Natur—Umweltzerstörung und die Lösung ökologischer Konflikte* (Darmstadt: Wissenschaftliche Buchgesellschaft): 291–301.
Scheffran, J., 2000a: "Zurück zum Kalten Krieg? Russland und der US-Hegemonieanspruch", in: *Wissenschaft & Frieden*, 2/2000.
Scheffran J., 2000b: „Der Konflikt zwischen Natur und Ökonomie aus naturwissenschaftlicher Sicht. Natur und Umwelt", in: Bartmann, H.; John, K.D. (Eds.): *Beiträge zum 9. und 10. Mainzer Umweltsymposium*, in: Chemnitzer Studien, vol. 3 (Düren: Shaker-Verlag): 179–224
Scheffran, J.; Stoll-Kleemann, S., 2003: "Participatory Governance in Environmental Conflict Resolution", in: Deb. K.; Srivastava, L. (Rds.) *Transition Towards Sustainable Development in South Asia* (New Delhi: The Energy and Resources Institute): 307-327.
Scheffran, J., 2006: "Tools in Stakeholder Assessment and Interaction". in: Stoll-Kleemann, S.; Welp, M. (Eds.): *Stakeholder Dialogues in Natural Resources Management: Theory and Practice* (Berlin – Heidelberg: Springer): 153–185.
Scheffran, J., 2008: "The Complexity of Security", in: *Complexity*, 14, 1: 13–21.
Scheffran, J.; Battaglini, A., 2011: "Climate and Conflicts – The security risks of global warming", in: *Regional Environmental Change*, 11 (Suppl. 1): 27–39.
Scheffran, J.; Brzoska, M.; Kominek, J.; Link, P.M.; Schilling, J., 2012a: "Climate Change and Violent Conflict", in: *Science*, 336,6083: 869–871.
Scheffran, J.; Brzoska, M.; Kominek, J.; Link, P.M.; Schilling, J., 2012b: "Disentangling the Climate-conflict Nexus: Empirical and Theoretical Assessment of Vulnerabilities and Pathways", in: *Review of European Studies*, 4,5: 1–13.
Scheffran, J.; Link, M.; Schilling, J., 2012c: "Theories and models of the climate-security link", in: Scheffran, *et al.* (Eds.) *Climate change, human security and violent conflict* (Berlin: Springer): 91–132.

Scheffran, J.; Marmer, E.; Sow. P., 2012d: "Migration as a contribution to resilience and innovation in climate adaptation: social networks and co-development in Northwest Africa", in: *Applied Geography*, 33: 119–127.

Scheffran, J.; Cannaday, T., 2013: "Resistance to climate change policies: the conflict potential of non-fossil energy paths and climate engineering", in: Maas, A., *et al.* (Eds.): *Global environmental change: new drivers for resistance, crime and terrorism?* (Baden-Baden, Nomos).

Scheffran, J.; Ide, T.; Schilling, J., 2014: "Violent Climate or Climate of Violence? Concepts and Relations with Focus on Kenya and Sudan", in: *International Journal of Human Rights*, 18,3: 369–390.

Scheffran, J. 2014: "Der unmögliche Krieg – Jan Bloch und die Mechanik des Ersten Weltkriegs", in: *Wissenschaft & Frieden*, 2/14: 38–42.

Scheffran, J., 2015a: "Technikkonflikte in der vernetzten Welt", in: *Wissenschaft & Frieden*, 2/2015: 6–10.

Scheffran, J., 2015b: "Complexity and Stability in Human-Environment Interaction", in: Kavalski, E. (Ed.): *World Politics at the Edge of Chaos* (New York: SUNY Press): 229–252.

Scheffran, J., 2015c: "Climate Change as a Risk Multiplier in a World of Complex Crises", Working paper presented at the Conference Planetary Security (The Hague: Institute, 2-3 Nov.).

Scheffran, J., 2016a: "Kettenreaktion außer Kontrolle: Vernetzte Technik und das Klima der Komplexität", in: *Blätter für deutsche und internationale Politik*, 3/2016: 101–110.

Scheffran, J., 2016b: "From a Climate of Complexity to Sustainable Peace: Viability Transformations and Adaptive Governance in the Anthropocene", in: Brauch, H. G., *et al.* (Eds.): *Handbook on Sustainability Transition and Sustainable Peace* (Cham: Springer): 305–347.

Scheffran, J.; Burroughs, J.; Leidreiter, A.; van Riet, R.; Ware, A., 2016: *The climate-nuclear nexus* (Hamburg: World Future Council).

Scheffran, J.; Froese. R., 2016 "Enabling Environments for Sustainable Energy Transitions: The Diffusion of Technology, Innovation and Investment in Low-Carbon Societies", in: Brauch, H.G, *et al.* (Eds.): *Handbook on Sustainability Transition and Sustainable Peace* (Cham: Springer International Publishing): 721–756.

Scheffran, J., 2017a: "Der Nexus aus Migration, Klimawandel und Konflikten", in: Scheffran, J. (Ed.): *Migration und Flucht zwischen Klimawandel und Konflikten.* Hamburger Symposium Geographie: 7–40.

Scheffran, J. 2017b: "Complex Crisis Landscapes and Climate Risk Governance: Challenges for European Stability and Transformation", in: Policy Insights, EUC Paper Series, Special Issue – Governing Globalization.

Scheffran, J., 2018a: "Militarisierung oder Zivilisierung? – Ambivalenz der Wissenschaft in der Krise", in: *Wissenschaft & Frieden*, 2/18: 15–20.

Scheffran, J., 2018b: *Biodiversity and Conflict. Supplementary Contribution to IPBES Global Assessment on Biodiversity and Ecosystem Services.* CLISEC Working Paper, University of Hamburg (5 March 2018).

Scheffran, J., 2019a: "Verbrannte Erde: Militär als Verursacher von Umweltschäden und Klimawandel", in: *Friedensforum*, 01/2019: 32–44.

Scheffran, J., 2019b: "Formen hybrider Kriege: Zwischen Ambivalenz und Komplexität" In: *Wissenschaft & Frieden*, 3/2019: 9–12.

Scheffran, J., 2019c: "The entwined Cold War roots of missile defense and climate geoengineering", in: *Bulletin of the Atomic Scientists*, 75,5: 222–228.

Scheffran, J.; Link, P. M.; Schilling, J,, 2019: "Climate and Conflict in Africa", in: *Oxford Research Encyclopedia of Climate Science*; at: https://doi.org/10.1093/acrefore/9780190228620.013.557.

Scheffran, J., 2020a: "Das Anthropozän und seine Grenzen: Überlegungen zu Klimawandel, Nachhaltigkeit und Coronakrise", in: Adloff, F.; Neckel, S., (Eds.): *Gesellschaftstheorie im Anthropozän* (Frankfurt: Campus): 257–279.

Scheffran, J., 2020b: "Climate extremes and conflict dynamics", in: Sillmann, J.; Sippel, S.; Russo, S. (Eds.): *Climate Extremes and Their Implications for Impact and Risk Assessment* (Amsterdam: Elsevier): 293–315.

Scheffran, J., 2020c: "Geopolitik der Energiewende: Infrastrukturen für den nachhaltigen Frieden", in: *Wissenschaft & Frieden*, 4/2020 (November), 38–40.

Scheffran, J., 2020d: "Climate change and weather extremes as risk multipliers: Tipping points, cascading events, and societal instability", Chapter 2, in: Brzoska, M.; Scheffran, J. (Eds.): *Climate Change, Security Risks, and Violent Conflicts. Essays from Integrated Climate Research in Hamburg* (Hamburg: Hamburg University Press): 19–48.

Scheffran. J., 2020e: "Atomwaffen, Umwelt und Klima: Grenzen des fossil-nuklearen Zeitalters", in: *Wissenschaft und Frieden*, 01/2020: 13–16.

Scheffran, J., 2020f: "Weather, War, and Chaos: Richardson's Encounter with Molecules and Nations", in: Gleditsch, N.-P. (Ed.): *Lewis Fry Richardson: His Intellectual Legacy and Influence in the Social Sciences* (Cham: Springer International Publishing): 87–100.

Scheffran, J.; Schürmann; E.; Weber, R., Newfields, T., 2024: "A Viable World in the Anthropocene: Living Together in the Common Home of Planet Earth," *Anthropocene Science* 3: 131–142. https://doi.org/10.1007/s44177-024-00075-7

Scheffran, J., 2021: *"Mythen der etablierten Sicherheitspolitik: 'Der Westen kann die Weltprobleme lösen'", in: Die Friedens-Warte, 3–4/2021; 205–227.*

Scheffran, J., 2022a: "Klimaschutz für den Frieden: Der Ukraine-Krieg und die planetaren Grenzen", in: *Blätter für deutsche und internationale Politik* (April); 113–120.

Scheffran J., 2022b: "Climate change: Human Security Between Conflict and Cooperation", in: *Encyclopedia of Violence, Peace and Conflict*, 3rd Edition (city: Elsevier).

Scheffran, J., 2022c: "Ökozid. Zwischen Klimaklagen und Verbrechen gegen den Frieden", in: *Wissenschaft und Frieden*, 1/22: 38–40.

Scheffran, J., 2022d: "Climate Matching: Verstärkte Nord-Süd-Zusammenarbeit in der Globalen Energiewende", in: Beck, R.-U.; Töpfer, K.; Zahrnt, A. (Eds.): *Flucht: Ursachen bekämpfe Flüchtlinge schützen* (München: Oekom-Verlag): 45–51.

Scheffran, J., 2023a: "Welche Zeitenwende? Globale Risiken, Herausforderungen und Megatrends" in: Staack M. (Ed.): *Einführung in die Internationale Politik – Studienbuch* (Berlin/Boston: Walter de Gruyter): 3–26.

Scheffran, J. 2023b: "Limits to the Anthropocene: geopolitical conflict or cooperative governance?", in: *Frontiers in Political. Science*; at: 5:1190610.https://doi.org/10.3389/fpos.2023.1190610.

Scheffran J. 2023c: "Vom andauernden Krieg zum nachhaltigen Frieden: Aufgaben einer kritischen Friedensforschung – Quo vadis Friedensforschung?", in: *Wissenschaft & Frieden*, 2021/01 (Dossier 96): 21–24.

Schröter, M,; Kraemer, R.; Ceausu, S.; Rusch, G.M., 2017: "Incorporating threat in hotspots and coldspots of biodiversity and ecosystem services", in: *Ambio*, 46: 756–768.

Schilling, J.; Nash, S.L.; Ide, T.; Scheffran, J.; Froese, R.; von Prondzinski, P., 2017 Resilience and environmental security: towards joint application in peacebuilding, in: *Global Change, Peace & Security*, 29,2: 1–21.

Schleussner, C. F.; Donges, J. F.; Donner, R. V., Schellnhuber, H.J., 2016: "Armed-conflict risks enhanced by climate-related disasters in ethnically fractionalized countries", in: *PNAS*, 113,33: 9216–9221.

Scholz, O., 2023: "The Global Zeitenwende How to Avoid a New Cold War in a Multipolar Era", in: *Foreign Affairs*, 102,1: 22–38.

Schröder, U., 2022: "Zeitenwende", in: *IPG-Journal* (15 March).

Selby, J.; Dahi, O.S.; Fröhlich, C.; Hulme, M., 2017: "Climate change and the Syrian civil war revisited", in: *Political Geogr.*, 60: 232–244.

Senghaas, D. (Ed.), 1995: *Den Frieden denken*, (Frankfurt: edition suhrkamp).

Shumilova, O.; Tockner, K.; Sukhodolov, A.; Khilchevskyi, V.; De Meester, L.; Stepanenko, S., *et al.*, 2023: "Impact of the Russia–Ukraine armed conflict on water resources and water infrastructure", in: *Nature Sustainability*, 12; at: https://doi.org/10.1038/s41893-023-01068-x.

SIPRI, 2022: *Trends in World Military expenditure*, 2021. Factsheet (Stockholm: Stockholm International Peace Research Institute).

SIPRI, 2023: *Trends in World Military expenditure*, 2022. Factsheet (Stockholm: Stockholm International Peace Research Institute).

Slettebak, R. 2012 "Don't Blame the Weather! Climate-Related Natural Disasters and Civil Conflict", in: *Journal Article in Journal of Peace Research*, 49, 1: 163–176.

Smith, A., 1776: *An Inquiry into the Nature and Causes of the Wealth of Nations* (London: T. Strahan & T. Cadell in the Strand).

Soja, E.W., 1996: "The extraordinary voyages of Henri Lefebvre; The trialectics of spatiality", in: Soja, E.W.: *Thirdspace. Journeys to Los Angeles and Other Real-and-Imagined Places* (Oxford: Blackwell).

Solomon, N.; Birhane, E.; Gordon, C.; Haile, M.; Taheri, F.; Azadi, H.; Scheffran, J., 2018: "Environmental impacts and causes of conflict in the Horn of Africa: A review", in: *Earth-Science Reviews*, 177: 284–290.

Spangenberg, J. H., 2002: "Environmental space and the prism of sustainability: Frameworks for indicators measuring sustainable development", in: *Ecological Indicators*, 2,3: 295–309; at: https://doi.org/10.1016/S1470-160X(02)00065-1.

Spangenberg, J. H.; Kurz, R., 2023: "Epochal turns. Uncomfortable insights, uncertain outlooks", in: *Sustainable Development*, 16; at: https://doi.org/10.1002/sd.2512.

Stanek, L., 2011: *Henri Lefebvre on Space: Architecture, Urban Research, and the Production of Theory* (Minneapolis: Univ of Minnesota Press).

Steffen, W.; Rockström, J.; Richardson, K.; Lenton, T. M.; Folke, C.; Liverman, D., *et al.*, 2018: "Trajectories of the Earth System in the Anthropocene", in: *Proceedings of the National Academy of Sciences of the United States (PNAS)*, 115,33: 8252–8259.

Stephens, J. C., 2019: "Energy Democracy: Redistributing Power to the People Through Renewable Transformation", in: *Environment*, 61,2: 4–13; at: https://doi.org/10.1080/00139157.2019.1564212.

Sterner, T.;Barbier, E. B.; Bateman, I.; van den Bijgaart, I.; Crepin, A. S.; Edenhofer, O.; *et al.*, 2019: "Policy design for the Anthropocene", in: *Nature Sustainability*, 2,1: 14–21; at: https://doi.org/10.1038/s41893-018-0194-x.

Stogiannos, A., 2018: *The Genesis of Geopolitics and Friedrich Ratzel – Dismissing the Myth of the Ratzelian Geodeterminism* (Cham: Springer).

Stuart-Smith, R.F. *et al.*, 2021: "Increased outburst flood hazard from Lake Palcacocha due to human-induced glacier retreat", in: *Nature Geoscience*, vol. 14:85–90.

Sun, W. W.; Liu, L. J.; Chen, X. J.; Szolnoki, A.; Vasconcelos, V.V., 2021: "Combination of institutional incentives for cooperative governance of risky commons", in: *Iscience*, 24,8; 14, Article 102844; at: https://doi.org/10.1016/j.isci.2021.102844.

Swain, A.; Öjendal, J. (Eds.), 2018: *Routledge Handbook of Environmental Conflict and Peacebuilding* (Abingdon: Routledge).

Taylor, K. B., 2020: "The passing of western civilization", in: *Futures*, 122,9; 102582; at: https://doi.org/10.1016/j.futures.2020.102582.

Theisen, O.M.; Gleditsch, N.P.; Buhaug, H., 2012: "Is climate change a driver of armed conflict?", in: *Climatic Change* 117: 613–625 (2013).

Thonicke, K.; Bahn, M.; Lavorel, S.; Bardgett, R. D.; Erb, K.; Giamberini, M., *et al.*, 2020: "Advancing the understanding of adaptive capacity of social-ecological systems to absorb climate extremes", in: *Earth's Future*, 8: e2019EF001221.

Thrift, N. 2017: "Space", in: Richardson, D.; Castree, N.; Goodchild, M.F.; Kobayashi, A.; Liu, W.; Marston, R.A. (Eds.): *The International Encyclopedia of Geography* (Malden:

Tilley, H., 2014: "Racial Science, Geopolitics, and Empires: Paradoxes of Power", in: *Isis*, 105,4) 773–781; at: https://doi.org/10.1086/679424.

Tollefson, J., 2022: "What the war in Ukraine means for energy, climate and food", in: *Nature*, 604: 232–233.

Tooze, A., 2022: "Welcome to the world of the polycrisis", in: *Financial Times*, 28 Oct.

UN, 1991: *Potential Uses of Military-Related Resources for Protection of the Environment*. Report of the Secretary General (New York: Office for Disarmament Affairs).

UNEP, 2019: *Global Environment Outlook (GEO-6) – Healthy Planet, Healthy People.* (Cambridge: United Nations Environment Programme).
UNEP, 2021: *Making Peace with Nature: A scientific blueprint to tackle the climate, biodiversity and pollution emergencies* (Nairobi: United Nations Environment Programme).
UNHCR, 2023: *Globale Trends 2022* (Geneva: Office of the High Commissioner for Refugees).
UNSG, 2015: *The Challenge of Sustaining Peace. Report of the Advisory Group of Experts for the 2015 Review of the United Nations Peacebuilding Architecture.* United Nations Security Council (New York: UN).
Urdal, H., 2005: "People vs. Malthus: Population Pressure, Environmental Degradation, and Armed Conflict Revisited", in: *Journal of Peace Research,* 42: 417–434; at: https://doi.org/10.1177/0022343305054089.
Urry, J., 2002: *Global Complexity*: (Cambridge: Polity Press).
Urry, J., 2005: "The Complexity Turn", in: *Theory, Culture & Society*, 22,5, 1–14.
van Schaik, L., *et al.*, 2022: *Clean Military Power in the Age of Climate Change.* Expert Group of the International Military Council on Climate and Security, Center for Climate and Security (June).
Van Weezel, S., 2019: "On climate and conflict: Precipitation decline and communal conflict in Ethiopia and Kenya", in: *Journal of Peace Research*, 56,4: 514–528; at: https://doi.org/10.1177/0022343319826409.
Vasconcelos, V. V.; Santos, F. C.; Pacheco, J. M., 2013: "A bottom-up institutional approach to cooperative governance of risky commons", in: *Nature Climate Change*, 3,9, 797–801; at: https://doi.org/10.1038/nclimate1927.
Vasquez, J. A.; Diehl, P. F.; Flint, C.; Scheffran, J.; Chi, S.-H.; Rider, T.J., 2011: "The ConflictSpace of Cataclysm: The International System and the Spread of War 1914–1917", in: *Foreign Policy Analysis*, 7,2; at: https://doi.org/10.1111/j.1743-8594.2011.00128.x.
Vinke, K.; Gabrysch, S.; Paoletti, E.; Rockström, J.; Schellnhuber, H.J., 2020: "Corona and the Climate: A Comparison of Two Emergencies", in: *Global Sustainability,*3: e25; at: https://doi.org/10.1017/sus.2020.20.
Vinke, K.; Rottmann, S.; Gornott, C., *et al.*, 2022: "Is migration an effective adaptation to climate-related agricultural distress in sub-Saharan Africa?", in: *Popul Environ*, 43: 319–345; at: https://doi.org/10.1007/s11111-021-00393-7.
Von Uexkull, N.; Croicu, M.; Fjelde. H.; Buhaug, H., 2016: "Civil conflict sensitivity to growing-season drought", in: *PNAS*, 113,44: 12391–12396.
von Uexkull, N.; Buhaug, H., 2021: "Security implications of climate change: A decade of scientific progress", in: *Journal of Peace Research*, 58,1: 3–17.
von Weizsäcker, E.U.; Wijkman, A., 2018: *Come On: Capitalism, Short-termism, Population and the Destruction of the Planet.* A Report to the Club of Rome (New York: Springer).
Walch, C., 2018: "Disaster risk reduction amidst armed conflict: informal institutions, rebel groups, and wartime political orders", in: *Disasters*, 42,S2: S239–S264.
WBGU, 2007: *Climate Change as a Security Risk* (Berlin: German Advisory Council on Global Change).
WBGU, 2023: *Healthy living on a healthy planet* . (Berlin: Wissenschaftlicher Beirat der Bundesregierung Globale Umweltveränderungen).
Werkner, Z.-J.; Fischer, M. (Eds.). 2019: *Europäische Friedensordnungen und Sicherheitsarchitekturen*, vol. 3 (Wiesbaden: VS Springer).
Werrell, C.E, Femia F (2013) The Arab spring and climate change. A climate and security correlations series. Center for American Progress, Stimson Center, Washington DC.
Westing A., 2011: "Environmental consequences of the second Indochina war: a case study", in: Machlis, G. E.; Hanson, T.; Špiric, Z-; McKendry, J. E: (Eds.): *Warfare Ecology* (Dordrecht: Springer): 11–17.
Whitaker, B.C.G. 1985: *UN Special Rapporteur on Prevention and Punishment of the Crime of Genocide.* E/CN.4/Sub.2/1985/1 4. UN, 2.7.1985.

Witherick, M.; Ross, S.; Small, J., 2001: *Modern Dictionary of Geography* (Oxford University Press).
Witmer, F. D. W.; Linke, A. M.; O'Loughlin, J.; Gettelman, A.; Laing, A., 2017: "Subnational violent conflict forecasts for sub-Saharan Africa, 2015–65, using climate-sensitive models", in: *Journal of Peace Research*, 54,2: 175–192; at: https://doi.org/10.1177/0022343316682064.
Young J.C.; Marzano M.; White R.M.; McCracken D.I.; Redpath S.M.; Carss D.N.; Quine C.P.; Watt A.D., 2010: "The emergence of biodiversity conflicts from biodiversity impacts: characteristics and management strategies", in: *Biodiversity & Conservation*, 19,14: 3973–3990.
Zickgraf, C.; Hut, E.; Castillo, T. (Eds.), 2021: *The State of Environmental Migration 2020 – A Review of 2019*. The Hugo Observatory of the University of Liège (Lièg: Presses Universitaires de Liège); at: https://www.hugo.uliege.be/cms/c_4865980/en/hugo-the-state-of-environmental-migration).
Zierler, D., 2011: *The invention of ecocide* (Atlanta, GA: University of Georgia Press).
Zimmerer, J. (Ed.), 2014: "Climate change, environmental violence and genocide", in: *International Journal of Human Rights* (special issue), 18,2.
Zscheischler, J.; Westra, S.; van den Hurk, B. J. J. M.; Seneviratne, S.I.; Ward. P. J., Pitman, A., *et al.*, 2018: "Future climate risk from compound events", in: *Nat. Clim. Chang.* 8: 469–477.
Zuboff, S., 2019: *The Age of Surveillance Capitalism: The Fight for a Human Future at the New Frontier of Power* (London: Profile Books).

Chapter 9
Governance and Politics in the Anthropocene

John Grin

Abstract This chapter contends that the key factor that has driven humankind and its planet into the Anthropocene is the modernist idea to achieve human progress through *controlling* nature and society on basis of universal principles, without paying attention to or even suppressing local conditions of both 'nature' and 'society'. An alternative orientation for development is to (i) replace the instrumentalisation of nature to serve human needs by mutual alignment of human and social needs with nature through (ii) re-appreciating opportunities for regeneration of energy, resources and the atmosphere. The chapter elaborates this building on two examples: modern, productivist agriculture and associate water management and food consumption practices in Europe and the US; and energy, water and waste disposal systems in American and European cities. These will show how place-based, regenerative development, creating synergy with rather than exploiting nature offers a promising way to remedy society's adverse impacts on the planet during the Anthropocene, and may help improve and maintain a fair and sustainable level of well-being of current and future generations globally. Guiding principles for associate place-based governance include new constitutional understanding of notions like equity and accountability; new finance arrangements and standards that allow for synergy and adaptive governance.

Keywords Place-based · Governance · Modernity · Transition · Food · Water management · Waste disposal · Urban infrastructure

---

John Grin, Professor of Political Science, Amsterdam University Email: j.grin@uva.nl.

J. Grin (✉)
Political Science, Amsterdam University, Amsterdam, Netherlands
e-mail: j.grin@uva.nl

H. G. Brauch (ed.), *Towards Rethinking Politics, Policy and Polity in the Anthropocene*, The Anthropocene: Politik—Economics—Society—Science 35,
https://doi.org/10.1007/978-3-031-71807-6_9

9.1 Introduction: Understanding the Anthropocene as a Perspective on Politics and Governance

Since Paul Crutzen famously exclaimed in February 2000 at a Conference of the *Geosphere–Biosphere Programme* (IGBP) in Cuernavaca, Mexico "This is not the Holocene anymore. We are now in the Anthropocene," this notion has produced a lot of definitional discussion. As yet, this has not led to the formal recognition by academic societies like the *International Union of Geological Sciences* (IUGS) of the Anthropocene as a new geological epoch. Also, while a lot of researchers and professional committees have sought to locate a proper starting point for this new era, these efforts, focusing on stratigraphic markers and other physical indicators, too have not been conclusive.

In this contribution I contend that, at least from the perspective of achieving a better understanding of governance and politics in – and out of! – the Anthropocene, it may help to draw more on the humanities and social sciences. Importantly, the dates that emerge from Brauch (2021: 56–65)'s concise review as the more tenable ones, have one thing in common. They are fundamentally related to Modernity: either marking its take-off (the Columbian Exchange and the Copernican Revolution) or concerning episodes when it got additional momentum, i.e. got more mass and speed, and new direction: the Industrial Revolution, since 1784; the Nuclear Era (1940–1950) and the Great Acceleration after 1950.

This, crucially, is no coincidence. At the core of the Anthropocene concept ever since Crutzen and Stoermer (2000) briefly elaborated the former's exclamation, is the idea that "the term 'anthropocene' for the current geological epoch" was to "emphasize the central role of mankind in geology and ecology." Here of course resonates the core of another idea, Modernity, which may be summarised in two Cartesian axioms:

(i) we can know truth in the form of universal, causal laws, connecting a cause to a later effect;
(ii) we may organise society (including the economy and the polity) such that we may optimally draw on this kind of knowledge's potential to predict, enabling humans to control nature and society, and thus realise progress.

As Ulrich Beck has outlined in his work on risk society and reflexive modernisation, these developments have both led to the depletion of natural resources, pollution and climate change and to institutional arrangements that are not appropriate enable to properly diagnose, let alone remedy these problems. In fact, the unsustainability of current development is the dark backside of what Beck (1997, 2003) calls early modernisation, which sought to realise progress. This yielded blindness and paralysis regarding its side-effects: the particular kind of knowledge as well as the institutional arrangements that promoted and drew upon such knowledge to realise progress, were far from tailored to timely recognising and mitigating these side effects.

Let me emphasise that the key point here is *not* the use of science to understand and improve the world. Rather, the problem is the attempt to *control* nature and society on basis of universal principles, without paying attention to or even suppressing local conditions of both 'nature' and 'society'.[1] Modernity is the idea of not just making use of nature, but controlling it to serve human and social objectives – rather than living with it, aligning out socio-economic practices and nature with each other.

This control paradigm was inspired by prospects offered by the work of modern science's sixteenth and seventeenth centuries pioneers, like Copernicus, Galileo, Newton and Bacon, as well as by Descartes' celebration of that through articulating the general principles cited above. With this embracement of the idea that the human fate would not necessarily be determined by just fate or nature (Toulmin 1990), modernity implied a neglect of the possibility to develop an understanding of nature that would help us to make use of nature without wasting, or destroying it. This neglect has shaped the further development of science, from 'reason' to (instrumental) rationality (Toulmin 2001), which brought about a "logic of homogenisation and virtual elimination of local knowledge" leading (Scott 1998: 302). This modern reliance on universalist principles, equally applied to widely varying contexts, without a place for what Aristotle called mētis – the knowledge of contextual conditions, and the craft to take them into account (Scott 1998: 309–341) – has led to social-economic practices that are fundamentally not sustainable.

Importantly, this choice has not only shaped scientific and social-economic practices, but also governance (tuned to facilitating that kind of practices and their interaction) and politics (informed by the associate rationality to adapt nature and society to serve social progress). This process was driven by what transition studies calls the *co-evolution* (i.e. mutual shaping over time) of dominant problem definitions, practices and structures of science, society, the economy, and the state (Grin et al. 2010: 3, 237–248). The authors just mentioned have each analysed this from a relevant angle. Toulmin has pointed to the reductionism that inhibits a more integral view; Scott has pointed to the risk of authoritarian governance and damage to natural cycles; and Beck has argued that this has led to a risk society with blind eyes and lame limps to the side-effects of early modernisation.

From this viewpoint (made in different ways by adverse authors, including Latour and Haraway), the central role of (hu)mankind in geology and ecology characterising the Anthropocene is *not* understood caused by the interference, as such, of humans with nature, unavoidably increasing with population growth and increase in energy and natural resources intensity of the economy. More fundamentally, it is related to the particular *kind* of interference modern humanity, especially in the Global North, has chosen: dominating over and exploiting nature, all too often depleting mineral and fossil resources; denying or even disrupting the regenerative potential of biotic resources and the composition of the atmosphere – the very processes through which humans characterise the Anthropocene. This is a different understanding than the generic 'blaming' of population growth and welfare growth often seen to be implied in Crutzen's (2019 [2011]: 179) claim that "Humanity will be a major ecological factor for millennia to come. Scientists and engineers face a daunting task: to guide society towards an ecologically sustainable management of the planet in the Anthropocene

[1] This will build on an earlier version of the argument, published in Dutch (Grin 2004, 2007).

era. This requires appropriate human behaviour at all levels ..." Crucially, it focuses attention on the kind of development in the Global North, especially since the start of industrialisation (what Beck called 'early modernisation') rather than pointing at contemporary and future population and welfare growth in the Global South as primary causes.

In addition, the above perspective suggests direction: one in which that mankind (i) replaces the instrumentalisation of nature to serve human needs by mutually alignment of human and social needs with nature through (ii) re-appreciating opportunities for regeneration of energy, resources and the atmosphere. For this approach to the relations between social-economic development and nature I will use as a shorthand throughout this chapter 'sustainable development' – importantly, this implies that we must understand sustainable development as a deep, fundamental change.

In the remainder of this chapter I will illustrate and deepen this understanding through two broad examples (agriculture, water management and their interplay in the rural area; and urban utility functions) how a long standing reliance on the 'control' paradigm adopted to produce welfare, has yielded side effects, and how moving beyond this kind of development implies more place-based approaches which go beyond existing functional differentiations. Admittedly, these examples draw on experience from the Global North. Yet, this is proper given the objective of this paper, as especially these parts of the world are the source of the exploitation of nature that typifies the Anthropocene. Changes in this part of the world are major contribution to global sustainable development, not only intrinsically, but also instrumentally (the development of novel socio-technical practices as well as new global relations, see examples below, may facilitate changes in the Global South) and ethically. In the next section, I will argue that, paradoxically, the increased welfare has not unequivocally led to more experienced life satisfaction. Based on that, I will propose some more generic guiding principles for the shifts in governance needed to manoeuvre society and the economy away from instrumentalising relation to nature characterizing the Anthropocene, and the associate politics.

9.2 A Societal Perspective on Living in and Moving out of the Anthropogenic Condition: Examples

To understand the above general depiction of the Anthropocene and what it would mean to gradually move society away from the Anthropogenic condition, I will now present two brief illustrations. Both cases are broadly situated in the US and Europe, the regions that have contributed most to anthropogenic unsustainabilities. While the discussion of modernisation will abstract from more specific contexts (as modern development tends to do), we will also see that current attempts to deal with their side effects typically re-contextualise practices.

9.2.1 Agriculture, Water Management and Consumption Practices

The development towards the modern, productivist agriculture currently prevalent around the globe (De Schutter 2017; Marsden 2012) took off in the late nineteenth century in the US and Europe, as a response to the agrarian crisis and to the population increase that accompanied industrialisation. Through the 'Green Revolution' of the 1970s, the productivist paradigm has since spread also to the Global South.

Especially in Europe, the pace of modernisation remained moderate, until 1945 when it accelerated after, driven by the need – strongly felt after two world wars and an economic crisis – for sufficient (domestic) food production at affordable prizes as well as by the desire to free labour force for industry, which was supposed to generate the economic growth needed for economic recovery. Modernisation has radically changed the relation between agriculture and nature. Animals and plants were adapted and grown in ways that made them as independent of their natural environment as possible.

As one example, the milk production of cows was optimised through Mendelian crossing, aided by computers, by giving them protein rich power fodder, often produced from maize and soy exported from other continents and by vaccinating them to keep them healthy. Such modernisation made dairy farming more capital and knowledge intensive. The former led to scale enlargement, which went hand in hand with the development of milking machines and robots; for the convenience of the robots, cows were genetically adapted to have 'milkable udders.' The latter led to further specialisation, which – together with global trade effects – led to the end of mixed farming, and further differentiation (between beef farmers and dairy farmers). A similar story can be told about crop production, where 'combinable grain' (i.e. cereal that is equally tall all over the field to help a 'combiner' harvest it without cutting too high, eliminating part of the grain, or too low, adding chaff) was developed; and yield improved and made less dependent on weather, soil and water conditions through artificial fertilizers, optimised to increase productivity much better than animal manure, and smart drainage and water supply systems.

This success had been enabled and promoted by close relations between the primary sector, the knowledge infrastructures and government, networks which generated subsidies, protectionist policies, productivity-improving R&D, supportive water management (controlling the ground water level so as to increase the soil's capacity to carry high cattle densities and machinery while offering sufficient water during the growth season; maintaining proper water quality) and land use policies (to accommodate scale enlargement) and more. Also important for modernisation of production was the development of a matching demand side (Grin 2012): modernisation of food consumption through novelties like new kitchen designs (Oldenziel/ Zachmann 2009) and supermarkets (e.g. Shaw et al. 2004; Scarpellini 2008), developed before World War II in the US and exported to and appropriated by European countries, promoted by Marshal aid and reinforced by trends like a raise of incomes, a shift to less bodily labour and encounters with 'exotic' food through de-colonisation, labour migration and foreign traveling by an increasing share of the population.

It is probably fair to say that what Beck calls the 'politicisation of side effects' took famously off with the publication of Rachel Carson's *Silent Spring* (1962), which put the pesticides on the agenda of countries around the world. In the course of the next two decades a range of other issues would join: manure, animal welfare and animal pandemics which in essence are the implications of the basic achievements of modernisation. Specialisation and global trade led to broken circularity of value chains. One example is that much animal fodder is now imported, often from different continents, which leads to depletion of nutrients and loss of biodiversity in especially Latin America, and excess nutrient concentration in soil and water system ('overfertilisation') in Europe (e.g. Lintsen et al. 2018: ch 23). Intensification of production reinforcing the associate pollution effects and increasing vulnerability to pathogens, spread by global trade; etc. The globalization of food value chains all too often also makes to export from the Global North to the Global South of ecological side-effects or the burden of mitigating them (e.g. Belton et al. 2011 on the implications of fish-labels for communities in Vietnam and Bangladesh).

Similarly, these achievements have undermined the long lasting synergy between agriculture and water management in the rural areas, which is further strained by the consequences of climate change. Excess precipitation cannot be dealt with through the existing approach to drainage, because the consistent lowering of the ground water level has caused land subsidence to the extent that local surface water infrastructures are situated above field level. Moreover, modern agriculture has reduced the soil's water retention capacity, jeopardizing the possibility to make excess precipitation and drought periods balance each other. Finally, farms are a significant share of the so-called 'diffuse sources' that produce water pollution.

As a response to these side-effects the field of agro-ecology (Marsden 2012) is reviving the underlying principles, in a more advanced way (cf. De Schutter 2010; Van der Ploeg et al. 2019; Rossing et al. 2021), "applying ecological concepts and principles to sustainable food systems design and management" "[depending] on natural processes to achieve favorable biological interactions among the different elements of agricultural systems" (Mockshell/Villarino 2019: 65). One core feature is maintaining greater plant and animal diversity above-ground. By promoting this, agroecological approaches stimulate the sub-soil food web diversity, and shift equilibria connected to plant- and animal production and soil water to more benign levels, e.g. increasing the soil's water retention by increasing organic matter in the soil and reducing evapo-transpiration, increasing equilibria of 'good bugs' versus 'bad bugs', etcetera (Veen et al. 2019).

For crop production, this means diversification in time and space (intercropping, crop rotation, strip cropping) and restoring the balance between farming fields and their environment. Regarding animal production, it implies a shift to livestock systems based on 'directive control', of animal behaviour, life conditions and fodder to recursive control, which puts centre stage animal's natural behaviour and natural interaction with the environment as a basis for livestock farming (Bos et al. 2003). Altogether, agriculture will have to be re-embedded in, and contribute to its context.

It will help maintain its own long term viability by maintaining rather than depleting or polluting water, soil and biodiversity. Water and nature management on their turn will continuously seek to align regional water, soil and biodiversity systems to the needs of agriculture and other social functions in the region, enabling multiple value creation. Environmental management functions and farming alike become a matter of adaptive management, responding to each other as well as to external conditions, including, especially weather.

Overall, reducing transcontinental nutrient flows and mitigating the export of side-effects and the burden to deal with them will require regionalization of value chains, or, where this is not deemed desirable, at least systemic provisions to balance flows of nutrients, food products and money with side-effects and efforts.

9.2.2 Energy, Water and Waste Disposal and Urban Practices

Current energy, water and waste disposal systems in American and European cities were born in the second half of the nineteenth and early twentieth centuries, when population growth, industrialisation and the agrarian crisis led to urbanisation. While this intensified existing problems of public hygiene, poverty, bad housing, simultaneously, the science and technology that enabled industrialisation were also seen to imply possibilities to improve not only incomes but also, through economic growth, improved housing, infrastructure and living environment. They could also help remedy these conditions through better infrastructures for energy, water and waste disposal. New social and political movements to deal with these problems found allies in new academic professionals, like engineers and doctors, and their organisations and infrastructures. They pushed for drinking water pipes, electricity and gas networks, sewerage systems and other waste disposal infrastructures that would yield much more control over physical conditions in the living environment (Hughes 1993; Nye 1993; Schot et al. 2010).

Around these physical structures, and enabled by the tacit assumption that fossil fuels could be used without damage and depletion, discursive concepts emerged an institutionalised like 'supply security', 'universal access' and 'affordability for all through network efficiency'. These "structures (…) underlie expectations and assumptions about what is normal and possible" (Nye 1998: 7) and changes in life practices. Lighting yielded opportunities for more family life and home industry in the winters, to public urban life and leisure in the evenings, to around the clock production with work shifts and so on (Schot 1998). Hot water and heated homes enabled showering and other new domestic practices, and set novel standards on what is considered e.g. proper personal hygiene, warm clothing and a comfortable living room (Shove 2003; Shove/Trentman 2018) Electricity and gas enabled changes in the daily lives of especially women (Schwarz Cowan 1985).

Such changes were neither unambiguous nor uncontested; yet they were expressions of the social progress deliberately sought. And, to be sure, achievements like hygiene in the living environment, warm homes and well-cooked, healthy meals have contributed to significant improvements in life expectancy and height. More indirectly, this led to a better educated population, less inequality and more political trust (Van Zanden et al. 2014: 101–138, 249–269); for a detailed analysis of the relation of these outcomes with the development of utility functions for the case of the Netherlands (cf. Lintsen et al. 2018). Yet, after around 1960 – the year around which primary life needs were fulfilled in countries in the global North – this changed in three important respects.

First, the 'expectations and assumptions' to which Nye (and the other authors cited above) refer, together with "abundant and inexpensive energy" (Nye 1998: 187 ff.) and sociological trends like individualisation and the emergence of the *homo ludens*, provided fertile ground to the emergence of 'secondary life needs': wash dyers, dish washers and other novelties that were felt as convenience; larger homes, with separate (heated, if not televised) rooms for all members of the household; audio, video and later also gaming appliances (Nye 1998: 187–215; Lintsen et al. 2018: 355–477). This is one reason why the use of energy has been expanding much faster than GDP growth. Similarly, between 1960 and the mid-1970s, also drinking water use steeply increased. Second, the politicisation of side effects by new social movements and publications like the Club of Rome's (1972) report, together with the oil crises of the 1970s, put energy scarcity, air and water pollution and climate change on the agenda. While this led to more emphasis on energy saving, with some real effect, this was offset by the increase in energy use associated with secondary life needs (Nye 1998: 217–246; Van Zanden et al. 2014: 249–269; Lintsen et al. 2018: 459–482, 524–536). Also, awareness increased that climate change was undermining the tenability of contemporary water management, as during excess rainfalls, sewerage systems flush the water out of cities, so that it cannot be used during drought, when there is scarcity of water and low ground water levels pose threats to the built environment.

After the Paris Climate Agreement on Climate Change (2015; and in some countries, especially Germany, after the 2011 Fukushima nuclear disaster), efforts to drastically reduce the use of fossil fuels started to become more serious; since mid-2022, Putin's malicious actions have added significantly to the sense of urgency. These efforts typically show a replacement of the basic principles underlying the fossil energy system – supply security of energy, which is inexpensive and considered abundant – towards a focus on place-based systems which balance demand and supply, with supply being a combination of decentral provision by actors who often are also consumers ('prosumers') and central supply. Decentral supply typically is not generic but place-based, varying on the condition of the local built environment, the availability of natural resources like sub-soil aquifers and surface waters, and of space for sun and wind and the availability of synergy potential (e.g. industrial excess heat). To be sure, for the foreseeable future, local and regional networks will remain embedded in national and transnational networks to accommodate local/regional imbalances between demand and supply throughout the day and the year. Yet, sufficient production and optimal supply-demand balance on the lower network levels will open up more capacity and require less network capacity at higher levels than solutions which rely more on higher level provisions.

Similarly, to deal with urban water problems, central water purification and removal there is a shift towards decentral retention and re-use provisions (Ricci 2022), as well as to integrating urban water systems and nature within and around cities (e.g. Oral et al. 2020). Regarding solid waste, increasingly cities seek to eliminate waste as much as possible by promoting circularity. And, finally, we see increased attempts to deal with water, waste disposal and energy issues through synergy ('urban nexussing' – Monstadt/Coutard 2019; Grin 2020) between the respective infrastructures and transcending the functional differentiation between these domains that emerged around the sub-problems on which early modernization focussed or even between these domains and food provision and consumption (Covarrubias 2019), bringing together the two examples discussed in this section. Here again, of course, more attention to the local and regional level must be seen not as at odds, but as compatible with embeddedness in higher level systems. Also needed, as the work of Nye, Shove and others show, will be a next round of transformation of lifestyles in everyday hygienic, lighting, sanitation, heating, clothing, and other practices and associate standards and expectations.

Interestingly, this second example on urban utility functions and urban practices leads to similar conclusions as our first case, on agriculture, rural water management and food practices. Here too, we have seen how the Anthropocene condition essentially that science and technology have been used to control the living environment, in ways that exploit nature. Also here, this has shaped both societal structure and everyday practices; and while this remains the case until today, simultaneously the side effects of this development have been politicised. And again, the key implication for science and technology is that, in order to move out of the detriments of the Anthropocene we need a shift away from generic, functionally differentiated solutions which use universal principles to create proper living environments, toward place-based approaches that transcend different domains so as to benefit from specific context to create synergistic solutions to urban needs. And again, *society* faces a need to re-orient itself, structurally (to more place-based, less differentiated structures) in terms of lifestyles and everyday practices, and in terms of the economics of value production, including standards for affordability and efficiency.

9.3 Intermezzo: A Novel Normative Orientation

The problems just discussed were recognised by two Nobel-laureate and economists, Amartya Sen and Joseph Stiglitz, who were charged by then French President Sarkozy to reflect on an alternative measure of societal progress, to replace the Gross Domestic Product. They proposed a dashboard of social values. Drawing on that dashboard a few years later, the OECD showed how after around 1960 GDP growth became much less resource and energy efficient. The increase in pollution, climate change and resource depletion went beyond a direct, linear correlation with GDP, through mechanisms we have just concisely explored for our two examples. This has led to significant reduction of natural capitals.

Importantly, the OECD analysis also leads to a different kind of finding: the more or less linear positive relation between GDP growth and all dimensions of perceived well-being has broken in many OECD countries: homicide rates increased again, trust in institutional dimensions declined (Van Zanden et al. 2014: 247 ff.). In Lintsen et al. (2018), where we analysed the relations between societal practices, outcomes and uses of natural, economic, and social capitals in more detail for the Netherlands, we called this the 'welfare paradox': increasing GDP needs more energy than before, but produces less well-being; and makes social capital decline again. In addition, the financial crises of 1999 and 2008 have shown that money is losing its value, as pensions funds are well able to witness.

In short, the kind of modern development that produced the Anthropocene, had a wider set of repercussions. Contemporary developed societies appear to produce fewer values with their money, produce money in a way that undermines its natural, economic and social capital while satisfactions amongst citizens is not increasing, and even partly declining. But if money production represents less value and depletes capitals, we may wish to shift development ambitions from the GDP towards the ambition to realise the values we nurture, while preserving capital for future generations, through adopting ways of producing and consuming that do not consume capital, but preferably replenish it. Economist Jeroen van den Bergh (2010) has called this orientation 'agnostic growth', or 'a-growth': societies focus on values which they value, realise them in ways that preserve economic, social and natural capital – and will just see whether GDP increases along with it – or not.

Interestingly, economic historian Carlota Perez (2019) has argued out that the transition to an economy based on just that – regeneration through renewable energy and circularity of resource use – also fits the transition from the previous, fossil fuel based, mass production and consumption, high physical mobility era to an information based, networked production, tailored to needs at a lower geographical scale. This shift to a more place based, regenerative economy, including a shift in towards regionalization of value chains, will also pre-empt further bubbles and associate implosion of financial capital, she argues on basis of a historical analysis of the 2008 financial crisis (Perez 2009), which she actually predicted on precisely these grounds (Perez 2002).

In sum, place-based, regenerative development, working with rather than exploiting nature offers not only a promising way to remedy society's adverse impacts on the planet during the Anthropocene. Also, it may maintain conditions for, rather than undermine, a fair and sustainable level of well-being of current and future generations globally.

9.4 Governance, Politics and Strategic Opportunities in the Anthropocene

However promising – ecologically, socially and economically, as we just discussed – the above normative perspective may be in principle, in practice such a fundamental transformation of the relations between society, economy and nature will of course have profound implications – arguably into the general direction implied in current initiatives like those discussed in our examples, also far transcending these in the longer run. Institutional rules and material infrastructures will be needed to de-privilege traditional and naturalise new practices, with due implications for lifestyles of people and economic models of firms. There will be a transformation of how relations, responsibilities and identities are constituted: with a shift from universal to context-specific solutions, constituting ideas like equity will gain a new meaning, and task divisions between actors and societal domains will shift.

Below, I will elaborate the implications for governance, needed to move gradually out of the detriments of Anthropocene society. It is important to realise that such deep change is bound to trigger politics: (dealing with) resistance, re-shaping expectations and relations, defining new principles for justice. The nation-state was modelled to generate societal progress by controlling nature and society on the basis of science and technology (Toulmin 1990; Beck 1997). This has shaped traditional, central government, but also "new waves of governance" (Bevir 2010). As argued elsewhere (Grin et al. 2010), the latter have come with 're-spatialisation' and 're-specialisation', beyond earlier functional differentiations, but making these compatible with the kind of place-based, cross-domain synergistic practices promoting governance needed, is also a highly political process. Therefore, I will end this chapter by a brief reflection on some trends that may be drawn upon in such governance.

9.4.1 Shifts in Governance

Before discussion what shifts in governance may promote and accommodate more place-based, synergistic regenerative strategies for regenerating natural, social and economic capital (Lintsen et al. 2018: 538–567) we must deal with two preliminary issues. First, can we do without some form of global governance? – Crutzen's earlier cited claim refers to "sustainable management of the planet" and appropriate human behaviour at all levels. "I contend that needing governance provisions at the global level is not the same as assuming that these must have, in a top-down way, the primacy."

This can be well illustrated with the case on urban sustainability. While the famous 1992 UN Summit in Rio de Janeiro translated the need to engage cities in global efforts for sustainable development into the 'Local Agenda 21', this did not lead to anything close to 'cities implementing a UN agenda.' In fact, initially the overall effect was mainly limited to cities in which an engaged mayor or another local key player took initiatives and organized themselves into platforms like the US Mayors Network or the so-called C40 network of global cities. In later years, neo-liberalization and the associate concern with the global competitive position of the one city led to "strategic urbanism", in which cities felt that governmental inertia at the national level compelled to address issues like energy security, water resilience and heat stress; this again led to transnational alliances between ('smart') cities, seeking support from each other, and collectively from intergovernmental bodies like the EU, OAS or UN (Bulkeley/Betsill 2013; Bulkeley/Castán Broto 2013). The same happened with the most recent type of local initiatives, undertaken by active citizens, entrepreneurs and citizens, which united themselves in so-called 'transition town movements' or 'city maker movements', which on their turn allied up on the regional and global level (Rumpala 2021). The result has been a 'Multi-Level Governance Type II' (Hooghe/Marks 2003) patchwork of horizontal and vertical networks, in which (institutional) resource distribution and a capacity for 'venue shopping' and other forms of strategic agency may give significant leverage to local and regional actors.

The second preliminary question is, *how* global problems may be resolved through such a web of networks at various levels. As Huntjens and Kemp (2022) point out, referring to transition theory, developments at various loci may reinforce each other through "the co-evolution of developments (…) [i.e.] processes of evolution that are linked with each other." Importantly, this pertains to all kinds of interactions within complex social-ecological systems. Co-evolution between different regional eco-systems may trigger yield ecological change as well as social change at above-regional level and generate horizontal and vertical feedback loops. As Kok et al. (2021) have argued, this may gradually shift power relations so as to promote further change.

Governance thus becomes a matter of mobilizing and orchestrating networks, stimulating changes in governance, R&D, production consumption and so on which correspond to a normative perspective, such as the one discussed in the previous section. Huntjens and Kemp pledge for turning it into a 'Natural Social Contract', and add that to actually realize such a perspective, it must also be translated into more specific principles. Below, I will propose some key principles.

Needed are, first, changes in the polity beyond current arrangements place-based policy making, which are still an addition to, and often been accountable by, 'regular', jurisdictions tailored to generic solutions. It requires a fundamental recognition of place based governance, including a re-definition of constitutional principles like equity and of rules and mechanisms of democratic policy making and accountability.

Second, given the profound changes in everyday practices, novel practices and structures for democratic policy making and accountability also need to be much more sensitive to needs, experiences and feelings in the lifeworld (cp. Eckersley 2004).

Third, it is necessary to develop and sustain new revenue models to realise the synergy sought (Morel et al. 2020; Grin 2020). After all, synergy means: investments, costs and revenues across multiple players and over the short and longer term. It will necessarily affect the practices and institutions of finance.

Relatedly, fourth, there is an urgent need for a new concept of efficiency, beyond the logic of scale of the mass-production and consumption era. Synergistic value propositions imply (also) a logic of synergy: extracting multiple values from one activity. This must be translated into novel standards, not only in the world of finance, but also in governance – think of regulations that seek to ensure lowest tariffs for citizens, of public investment heuristics and so on. But it also has deeper implications for the organisation of societies through de-differentiation and re-differentiation in combinations yet to be discovered.

Fifth, such standards also need to be internalised in research and innovation programmes. This requires a change in both substantive orientation of innovation policies, and in the relations between government, knowledge infrastructure, market and civil society (Cohen 2022).

Sixth, governance arrangements are needed that are able to transform themselves. Especially since the Paris Treaty, radical change is no longer the prerogative of only 'alternative' actors at the fringes of established system, but needs and increasingly and is undertaken by established actors. This implies a dialectic: on the one hand, these actors know the established system well and are well-positioned to adapt it; on the other hand, the latter is complicated by the 'paradox of embedded agency' (Seo/Creed 2002). Drawing on and elaborating basic insights from transition studies (Grin 2020) proposes a combination of new methodical repertoires for planning and design, free spaces ('niches') in and between established organisations and authoritative gremia with the competence to realise structural change.

Last, but not least, in order to deal with the politics which is bound to occur (see above) it also requires strategic action. A key insight here is that while existing power relations and lack of trust may inhibit a transition, it is also true that power relations and trust may gradually become more favourable to the transition (Grin et al. 2010). This has more recently been developed in more accurate accounts of strategies for powering (Hoffman 2016) and trust building (Koole 2023).

9.4.2 Risks and Strategic Opportunities

Such governance and associate democratic reform may be both part of the solution to, and fall prey to contemporary movements towards increasing alienation and associate susceptibility to radical leaders and subversive forces. Already a quarter of a century ago, Cox (1999) warned, using the internet to spread disinformation, could win people's hearts and minds: "Civil society has become the crucial battleground for recovering citizen control of public life," and that the eventual outcome will depend especially on "resurrecting a spirit of association in civil society together with a continuing effort by the organic intellectuals (...) to think through and act towards an alternative social order at local, regional and global levels." Indeed, more recently it has become clear that current malcontent and declining institutional trust and associate susceptibility to populist seductions, are to a significant extent expressions of the current lack of connection between the governing elites and large groups in society.

More specifically, as e.g., Hochschild (2016) has shown, increasingly many people feel 'strangers in their own land.' Their incomes are increasingly uncertain, if not declining; costs for housing and energy are increasing. While they experience swift demographic change, they experience cosmopolitan a dominant culture of 'white wine and cappuccino' which hey detest, and while they suffer from pollution, loss of jobs and other problems due to market behaviour, they also feel abandoned by government, which they feel is being run by a cosmopolitan elite that sees them as bad or backward.

While accounts like these have been seen by some as hyperbolic, the 2022 Ukraine war and energy crisis have underlined these concerns. Even stronger, the fact that the energy crisis centered around prize increases, and manipulable scarcity of fossil energy notwithstanding, it appeared to reinforce resistance against the energy transition as much as it provided additional sense of urgency for it. These events have made clear that there are good reasons to assume that such risks and instabilities will remain for the decades to come.

How to deal with these risks and the politics bound to occur with the transformation described above? First and foremost, it is crucial to design new practices and structural change in such a way that they produce solutions, especially for underprivileged groups. Attention to energy poverty, inclusion of those affected and their needs and lifestyles, and a just transition are quintessential (Williams/Doyon 2019; Sovacool et al. 2021); also essential are actual results and showcasing them.

Second, there are several trends that may be mobilised to legitimate new practices and structures, such as the growing attachment of people to their place and groups and activities there; the growing desire to find pleasure and meaning in work; the growing desire amongst citizens, including entrepreneurs, to re-connect to nature embed the economy more in nature and society; the already mentioned shift from mass production to customisation, and from own business to network economy, signalled by (Perez); rising energy prizes of energy; the increasing awareness under youth of climate change, and the increasing sense in older generations that it is already real enough to be their concern as well.

Yet, again, mobilizing these trends will only be effective if specific changes in practices and structures are indeed meaningfully and demonstrably shaped to fit and accommodate these trends. Ensuring this is a crucial task for those engaged in governance, both because, as we know ever since the work of Dewey (1927), such politics will especially arise around the governance efforts which precisely for that reason need to avoid technocracy and be essentially democratic; and because changes in governance by their nature are changes in power relations, a given that under current social circumstances and democratic risks requires ultimate care and attention.

References

Beck, U., 1997: *The re-invention of politics. Rethinking Modernity in the Global Social Order* (Cambridge: Polity Press).

Belton, B.; Haque, M. M.; Little, D. C.; Sinh, L. X., 2011: "Certifying catfish in Vietnam and Bangladesh: Who will make the grade and will it matter?", in: *Food Policy*, 36, 2: 289–299 at: https://doi.org/10.1016/j.foodpol.2010.11.027.

Bliss, K.; Bickler, C.; Villa, A.; Tippin, L.; Vierweger, A.; Meldrum, J.; Padel, S.; Pearce, B.; Lea, M., 2019: "Getting out of the commodity trap: Enabling diversity through alternative food networks", in: Messéan, A.; Drexler, D.; Heim, I.; Paresys, L.; Stilmant, D.; Willer, H. (Eds.): *First European Conference on Crop Diversification, Budapest, September 18–21, 2019– Book of Abstracts* (Paris - Bundapest: INRA/ÖMKI).

Bos, B. A. J.; Groot Koerkamp, P. W. G.; Groenestein, K., 2003: "A novel design approach for livestock housing based on recursive control—with examples to reduce environmental pollution", in: *Livestock Production Science*, 84: 157–170.

Brauch, H. G. 2021: "Peace ecology in the Anthropocene", in: Oswald Spring, U.; Brauch, H. G. (Eds.): *Decolonising Conflicts, Security, Peace, Gender, Environment and Development in the Anthropocene* (Cham: Springer Nature): 51–186.

Bulkeley H.; Betsill, M. 2013: "Revisiting the urban politics of climate change", in: *Environmental Politics*, 22, 1: 136–154

Bulkeley, H.; Castán Broto, V., 2013: "Government by experiment? Global cities and the governing of climate change", in: *Transactions of the Institute of British Geographers*, 38: 361–375.

Chongtham, I. R.; Gunnarsson, A.; Carton, N.; Steen Jensen, E., 2019: "Lessons from co-innovation process for increasing grain legume production and consumption in Southern Sweden", in: Messéan, A.; Drexler, D.; Heim, I.; Paresys, L.; Stilmant, D.; Willer, H. (Eds.): *First European Conference on Crop Diversification, Budapest, September 18–21, 2019– Book of Abstracts* (Paris - Budapest: INRA/ÖMKI).

Cohen, J. B., 2022: "(Re)searching for the public. Public experimentation with institutional entrepreneurial collectives to open up the research and innovation system from within" (PhD dissertation, Amsterdam: University of Amsterdam); at: https://hdl.handle.net/11245.1/5c6ddb40-25b1-4ba3-834c-0c2086cc94ea.

Covarrubias, M., 2019: "The nexus between water, energy and food in cities: towards conceptualizing socio-material interconnections", in: *Sustainability Science*, 14: 277–287; at: https://doi.org/10.1007/s11625-018-0591-0.

Cox, R. W., 1999: "Civil Society at the Turn of the Millennium: Prospects for an Alternative World Order", in: *Review of International Studies*, 25, 1: 3–28; at: http://www.jstor.org/stable/20097573.

Cretella, A., 2016: "Urban Food Strategies. Exploring Definitions and Diffusion of European Cities' Latest Policy Trend", in: *Metropolitan Ruralities* (Research in Rural Sociology and Development), 23 (Bingley: Emerald Group Publishing Limited): 303–323; at: https://doi.org/10.1108/S1057-192220160000023013.

Crutzen, P. J., 2019 [2011]: "The geology of humanity", in: *The Anthropocene. Key texts of the Nobel prize winner for the new geological age* (Munich: Oekom): 173.

Crutzen, P. J.; Stoermer, E. F., 2000: "The 'Anthropocene'", in: *Global Change Newsletter*, 41 (Stockholm: IGBP).

Davies, A.; Cretella, A.; Franck, V., 2019: "Food Sharing Initiatives and Food Democracy: Practice and Policy in Three European Cities", in: *Politics and Governance*, 7, 4, 8–20; at: https://doi.org/10.17645/pag.v7i4.2090.

Deakin, M.; Diamantini, D.; Borrelli, N. (Eds.), 2016: *The Governance of City Food Systems: Case Studies From Around the World* (Milano: Fondazione Giangiacomo Feltrinelli).

De Schutter, O., 2010: *Report submitted by the Special Rapporteur on the right to food* (New York; UN General Assembly, Human Rights Council. Document A/HRC/16/49).

Dewey, J., 1927: *The Public and Its Problems. An Essay in Political Inquiry* (Athens, Ohio: Alan Swallow Press; New York: Holt).
Eckersley, R., 2004: "The Greening of the Democratic State", in: *The Green State. Rethinking Democracy and Sovereignty* (Cambridge, MA: MIT Press): 139–169.
Grin, J., 2012: "Changing government, kitchens, supermarkets, firms and farms: the governance of transitions between societal practices and supply systems", in: Spaargaren, G.; Loeber, A.; Oosterveer, P. (Eds.), 2012: *Food Practices in Transition. Changing Food Consumption, Retail and Production in the Age of Reflexive Modernity* (New York: Routledge): 35–56.
Grin, J., 2020: "'Doing' system innovations from within the heart of the regime", *in: Journal of Environmental Policy & Planning*, 22, 5: 683–694; at: https://doi.org/10.1080/1523908X.2020.1776099.
Grin, J.; Rotmans, J.; Schot, J., 2010: *Transitions to Sustainable Development. New Directions in the Study of Long-term Structural Change* (New York: Routledge).
Haraway, D. J., 2016: *Staying with the Trouble: Making Kin in the Chthulucene* (Durham, NC: Duke University Press).
Hochschild, A. R., 2016: *Strangers in Their Own Land: Anger and Mourning on the American Right* (New York: The New Press).
Hoffman, J. G., 2016: *From resistance to transformation: Politics of experimentation with new energy systems* (Amsterdam: University of Amsterdam); at: http://hdl.handle.net/11245/1.547364.
Hooghe, L.; Marks, G., 2003: "Unraveling the Central State. But How? Types of Multi-Level Governance", in: *American Political Science Review*, 97, 2: 233–243.
Hughes, T. P., 1993: *Networks of Power. Electrification in Western Society, 1880–1930* (Baltimore, MD: John Hopkins UP).
Huntjens, P.; Kemp, R., 2022: "The Importance of a Natural Social Contract and Co-Evolutionary Governance for Sustainability Transitions", in: *Sustainability*, 14, 5: 2976; at: https://doi.org/10.3390/su14052976.
Kok, K. P. W.; Loeber, A. M. C.; Grin, J., 2021: "Politics of complexity: Conceptualizing agency, power and powering in the transitional dynamics of complex adaptive systems", in: *Research Policy*, 50, 3: 104183; at: https://doi.org/10.1016/j.respol.2020.104183.
Koole, B. P., 2022: "Veganism and plant-based protein crops: Contentious visioning almost obstructing a transition", in: *Environmental Innovation and Societal Transitions*, 42: 88–98; at: https://doi.org/10.1016/j.eist.2021.12.003.
Koole, B. P., 2023: *Trust for Change*, PhD dissertation (Amsterdam: University of Amsterdam).
Latour, B., 2018: *Down to Earth: Politics in the New Climate Regime* (Cambridge: Polity Press).
Lintsen, H. F., V.; Smits, J. P.; Grin, J., 2018: *Well-being, Sustainability and Social Development: The Netherlands 1850–2050* (Cham: Springer).
Marsden, T. K., 2012: "Food systems under pressure: regulatory instabilities and the challenge of sustainable development", in: Spaargaren, G.; Oosterveer, P.; Loeber, A. (Eds.): *Food Practices in Transition: changing food consumption, retail and production in the age of reflexive modernity*. Routledge Studies in Sustainability Transitions, vol. 3 (New York, NY: Routledge): 11.
Marsden, T., 2017: *Agri-food and rural development: sustainable place-making. Contemporary Food Studies: Economy, Culture and Politics* (London - New York: Bloomsbury).
Mawois, M.; Vidal, A.; Revoyron, E. et al., 2019: "Transition to legume-based farming systems requires stable outlets, learning, and peer-networking", in: *Agron. Sustain. Dev.*, 39, 14; at: https://doi.org/10.1007/s13593-019-0559-1.
Meng, X., 2022: "Understanding the effects of site-scale water-sensitive urban design (WSUD) in the urban water cycle: a review", in: *Blue-Green Systems*, 4, 1: 45–57.
Mockshell, J.; Villarino, M. E. J., 2019: "Agroecological Intensification: Potential and Limitations to Achieving Food Security and Sustainability", in: *Encyclopedia of Food Security and Sustainability*, vol. 3 (Amsterdam: Elsevier): 64–70.

Monstadt, J.; Coutard, O., 2019: "Cities in an era of interfacing infrastructures: Politics and spatialities of the urban nexus", in: *Urban Studies*, 56, 11: 2191–2206; at: https://doi.org/10.1177/0042098019833907.

Morel, K.; Revoyron, E.; San Cristobal, M.; Bare, P. V.: 2020: "Innovating within or outside dominant food systems? Different challenges for contrasting crop diversification strategies in Europe", in: *PLoS ONE*, 15, 3: e0229910; at: https://doi.org/10.1371/journal.pone.0229910.

Nye, D. E., 1992: *Electrifying America. Social Meanings of a New Technology, 1880–1940* (Boston, MA: MIT Press).

Nye, D. E., 1999: *Consuming Power. A Social History of American Energies* (Boston, MA: MIT Press).

Oral, H. V.; Pedro C.; Magdalena G. et al., 2020: "A review of nature-based solutions for urban water management in European circular cities: a critical assessment based on case studies and literature", in: *Blue-Green Systems*, 2, 1: 112–136.

Perez, C., 2002: *Technological* Revolutions and *Financial Capital* (Cheltenham: Edward Elgar).

Perez, C., 2009: "The double bubble at the turn of the century: technological roots and structural implications", in: *Cambridge Journal of Economics*, 33 (July): 779–805; at: https://doi.org/10.1093/cje/bep028.

Perez, C., 2019: "Transitioning to smart green growth: lessons from history", in: Fouquet, R. (Ed.): *Handbook on Green Growth* (Cheltenham: Edward Elgar): 447–463.

Ricci, L. 2022. "Integrated Approaches to Ecosystem Services: Linking Culture, Circular Economy and Environment through the Re-Use of Open Spaces and Buildings in Europe", in: *Land*, 11, 8: 1161; at: https://doi.org/10.3390/land11081161.

Rossing, W. A. H.; Albicette, M. M.; Aguerre, V.; Leoni, C.; Ruggia, A.; Dogliotti, S., 2021: "Crafting actionable knowledge on ecological intensification: Lessons from co-innovation approaches in Uruguay and Europe", in: *Agricultural Systems*, 190: 103103; at: https://doi.org/10.1016/j.agsy.2021.103103.

Rumpala, Y., 2021: "'Smart' in another way: the potential of the Fab City approach to reconfigure urban dynamics", in: *Urban Research & Practice*; at: https://doi.org/10.1080/17535069.2021.2009551.

Scarpellini, E., 2008: *Material Nation: A Consumer's History of Modern Italy* (Oxford: Oxford U. P.).

Schot, J. 1998: "The usefulness of evolutionary models for explaining innovation. The case of the Netherlands in the nineteenth century", in: *History and Technology*, 14, 3: 173–200; at: https://doi.org/10.1080/07341519808581928.

Schot, J. W.; Lintsen, H. W.; Rip, A. (Eds.), 2010: *Technology and the making of the Netherlands: The age of contested modernization, 1890–1970* (Cambridge, MA: MIT Press).

Schwarz Cowan, R., 1985: *More Work For Mother: The Ironies Of Household Technology From The Open Hearth To The Microwave* (New York: Basic Books).

Scott, J. C., 1998: *Seeing Like a State. How Certain Schemes to Improve the Human Condition Have Failed* (New Haven, CT - London: Yale U.P.).

Seo, M. G.; Creed, W. D., 2002: "7 Institutional contradictions, praxis, and institutional change: A dialectical perspective", in: *Academy of management review*, 27, 2: 222–247.

Shaw, G.; Curth, L.; Alexander, A., 2004: "Selling Self-Service and the Supermarket: The Americanisation of Food Retailing in Britain, 1945–60", in: *Business History*, 46: 4, 568–582; at: https://doi.org/10.1080/0007679042000231847.

Shove, E., 2003: *Comfort, Cleanliness and Convenience: The Social Organization of Normality* (London, UK: Bloomsbury Publishing PLC).

Shove, E.; Trentman, F. (Eds.), 2018: *Infrastructures in Practice: The Dynamics of Demand in Networked Societies* (London, UK: Routledge).

Sovacool, B. K.: Hess, D. J.; Cantoni, R., 2021: "Energy transitions from the cradle to the grave: a meta-theoretical framework integrating responsible innovation, social practices, and energy justice", in: *Energy Research & Social Science*, 75: 102027.

Toulmin, S., 1990: *Cosmopolis. The Hidden Agenda of Modernity* (Chicago, IL: The University of Chicago Press).
Toulmin, S, 2001: *Back to Reason* (Cambridge, MA: Harvard University Press).
van den Bergh, Jeroen C. J. M., 2010: "Environment versus growth — A criticism of 'degrowth' and a plea for 'a-growth'", in: *Ecol. Econ.*; at: https://doi.org/10.1016/j.ecolecon.2010.09.035.
van der Ploeg, D. J.; et al., 2019: "The economic potential of agroecology: Empirical evidence from Europe", in: *Journal of Rural Studies*, 71: 46–61; at: https://doi.org/10.1016/j.jrurstud.2019.09.003.
van Zanden, J., et al. (Eds.), 2014: *How Was Life?: Global Well-being since 1820* (Paris: OECD); at: https://doi.org/10.1787/9789264214262-en.
Williams, S.; Doyon, A., 2019: "Justice in energy transitions", in: *Environmental Innovation and Societal Transitions*, 31: 144–153.

Chapter 10
Patriarchy Has Produced the Patriacene: Can Gender Challenge Its Persistence?

Úrsula Oswald Spring

Abstract This chapter reviews the evolution of the Patriacene([1] The author thanks an anonymous reviewer for the suggestion of the "Patriacene", which better explains the links between the destruction of society and nature and violent patriarchal behaviour.) over the past 6000 years. The 'system of rule' changed from maternal clans to fratrilineal kinships, god-kings, *pater familias*, colonisation, capitalism, and multinational corporations. Patriarchy is based on violence, wars, and exploitation of humans and nature. Facing these imperialist powers, people in the Global South are helpless to limit the extraction and mistreatments. Thus, women experimented radical alternatives with gift economy, the Aymara's Cosmovision of 'living well' and the Zapatista's 'leading by obeying'. They reversed violent accumulation by understanding that 'there is no better life', promoting subsistence production, inclusive education, traditional health care, and diverse productive activities. Their bottom-up cultural options against the Patriacene are improving well-being, social care, local cohesion, sorority, and ecosystem restoration.

Keywords Bottom-up · Climate change disasters · Cosmovision · Cultural options · Epistemic changes · Global South · Leading by obeying · Living well · Mother earth · Patriacene

Úrsula Oswald Spring is a Research Professor at the National University of Mexico (UNAM) in the Regional Multidisciplinary Research Center (CRIM); Email: uoswald@gmail.com.

Ú. O. Spring (✉)
Regional Multidisciplinary Research Center (CRIM), National University of Mexico (UNAM), Mexico city, Mexico
e-mail: uoswald@gmail.com

H. G. Brauch (ed.), *Towards Rethinking Politics, Policy and Polity in the Anthropocene*, The Anthropocene: Politik—Economics—Society—Science 35,
https://doi.org/10.1007/978-3-031-71807-6_10

10.1 Introduction, Research Questions and Methodology

10.1.1 Structure of the Chapter

This chapter addresses the history of patriarchy and its consolidation during the last 5000 years, producing the destruction of humanity´s relations and the environment. Technological agriculture, cattle raising, urbanisation, population growth, and the high emissions of greenhouse gases (GHG) by the massive use of fossil oil have affected in multiple ways both humankind and nature. Climate change is producing extreme events, acidification of the oceans by absorbing excess GHG from the earth's atmosphere, temperature increase, and sea level rise due to the dramatic loss of sea ice in the Antarctic, Greenland, and the glaciers due to global warming. These global processes have different regional impacts and the Global South is highly affected because these countries are located in the tropics. Crutzen (2002) called this new era of Earth's history the Anthropocene. Depending on different disciplines, until today there is no generally agreed definition of the concept.

The humanities and the social sciences have reacted to the 'Anthropos' in the geosciences and raised new questions about the concept. First, there was the question of when the Anthropocene really began. Later emerged the question of how is it related to patriarchy and which the critics are coming from ecofeminism, Capitalocene, Plantationocene, and Patriacene. An epistemic change centres the analyses of climate change on people living in survival conditions. This bottom-up approach reviews indigenous alternatives related to their Cosmovision of telluric unity. 'Living well' proposed by the Aymara in Bolivia and Peru or 'leading by obeying' by the Zapatistas in Chiapas allows for recovering livelihood, restoring the ecosystem, and achieving food sovereignty. The gift economy, social economy, and women's care economies are additional alternatives for people from the Global South. In the conclusions, the author proposes an alternative and decentralised resource management in the Patriacene from the bottom up. The chapter introduces cultural and gender factors in the analysis to overcome the present patriarchal violence. The interlinks and feedback of socioeconomic, political, cultural, discrimination, and gender processes is a response of people highly affected by climate disasters, which account for limited governmental support. They have learned to survive in increasingly more complex natural conditions, where poverty and extreme events are affecting especially the livelihood of vulnerable women, girls, boys, men, and elderlies. These poor women and men have found in the present monopolised, neoliberal, and patriarchal violent system a possibility to survive. Their bottom-up efforts give them not only force to act but also basic conditions for life within their proper Cosmovision, care, and dignity for the entire community.

10.1.2 Research Questions

Which are historical processes that promoted the transition from maternal clans with sex segregation towards extensive families, where the father is dominant as the backbone of patriarchy. How patriarchal dominance has produced violence, wars, slavery, and the dispossession of women from wealth and power? How could patriarchy promote conquest, primitive accumulation of wealth in small monarchies, and empires, and capitalism controlled by global elites? How could this model of exclusion, wars, and violence stimulate during the last seven decades a globalisation process with a concentration of power in MNE and imperialist governments that resulted in social, race, ethnic, and class divisions? How is therefore patriarchy related to the Patriacene, characterised by the present overexploitation of natural resources; gigantic amounts of garbage, much of it highly toxic; massive use of fossil oil with increasing GHG and climate change impacts? Confronted with this historical destructive process, the chapter asks how is it possible in a decentralised way and from the bottom-up to save humankind, especially vulnerable women and girls, from the destructive Capitalocene, climate disasters, hunger, conflicts, extractivism, and wars.

10.1.3 Introduction

After outlining these research questions, the chapter uses first an anthropological understanding of the evolution of humanity, based on archaeological excavations, myths, narratives, literature, and habits still existing today, to find the way to the origin of patriarchy in the historical evolution of humanity and its consolidation in the modern world. When people settled in more stable conditions, society got organised first in matriarchal clans with sex segregation, where hunters and their food were segregated from women (Reed 1975). The historical development of humanity settled down in small communities, starting with matriarchal clans, to overcome the phase of savagery, where hunters ate flesh from animals and human beings not belonging to their fraternity. The matriarchal clan established clear taboos and rites for sexual relations. To have sex with women men first performed purification rites to be clean from eating meat. The control of the food supply by agriculture and children rising was organised by local women. Both basic needs: food and sex required mechanisms of control or sex taboos, especially to limit cannibalism among the related social women and men groups. During their menstrual period women were not pure and had to wait for seven days to have sex. Women were surrounded, protected, and comforted by other women and during their period they were considered unclean and even other women avoided their contact.

Fratriarchy[1] or fraternal clans helped also to limit further cannibalism through the exchange of gifts with neighbour clans, where reciprocity existed with other women and male communities. Boys were first nourished by mothers, left the matriarchal clan, and passed a rite of passage to integrate with the men's organisation, where they were trained as farmers and progressively less as hunters, due to that agriculture and later cattle raising provided enough food. Several centuries later, women and men engaged in marriage, building their own houses for leaving together, where the children born by the women were provided also by men. In the past, human beings did not have knowledge about the sexual reproduction of children, so processes of couvade and bride price, mostly with payment of cattle, which still exist in multiples cultures today, started to consolidate marriage and father-organised households. These bride prices promoted private propriety and women belonged to the men, who paid for their wives, producing the first indication of patriarchy in traditional society. Irrigation agriculture, cattle rising with multiple by-products (milk, wool, field preparation with animal support), the digging of groundwater wells, technological improvement such as the plough, and commercial interchange with neighbours, produced a surplus and an incipient division of labour.

Women were successively eliminated from their traditional role of social control of food and sex. The primitive accumulation of capital accelerated the process of male leadership, first as father-family and later as a male leader in god-cities, god-states and as monarchs (Oswald-Spring 2016: 170). Slavery eliminated women further from their traditional power structures in the former matriarchal clan together with their loss of control of food, sex, and economic independence. Women and slaves were kept inside their houses for domestic labour and children raising (Fortune 1963). It was forbidden for them to leave their house without male company. They became the physical and social property of the owners of the house. Thus, over several millennia patriarchy evolved as a complex human process from matriarchal clans to a patriarchal family (Reed 1975), later god-cities, monarchies, and empires. The violent process of patriarchy provided the bases for conquest, war, domination, and exploitation, which allowed the emergence of the present capitalist productive system (Mies 1985).

Historically, during the last 6000 years, patriarchy has evolved as a human organized system that has developed from a matriarchal clan to the dominant *pater familia,* later greater social organisations in villages and cities, but always based on violence, discrimination, and exploitation. This chapter addresses the relations of patriarchy as a system based on violence, war, conquest, discrimination and exploitation of humans and nature to concentrate wealth on very limited elites. Through conquests, wars, and invasions empires and dominant nations emerged, using slaves and women for the socioeconomic consolidation and the necessities of the dominant men. The control of territories (colonisation) was crucial to consolidate imperial powers in the past all over the world during the past 500 years, where also capitalism evolved simultaneously. During the last 200 years emerged in the colonies' processes of liberation,

[1] Fratriarchy is the anthropological term for the incipient brotherhood of men, molded by the matriarchal image of the sisterhood of women.

and it was especially in the XX century when countries achieved their independence from their colonial powers, whenever the neo-colonial system allowed the imperialist countries to maintain the exploitation of human labour through the payment of low salaries and the extraction of natural resources.

Based on the same patriarchal models of violence of exploitation, discrimination, and extraction, during the last seven decades emerged a globalisation process, which produced an acceleration of the accumulation of capital and the destruction of natural resources. MNE, supported by imperialist governments, controlled the financial flows and speculation, the productive chains, and promoted a homogenisation of worldwide homogenised consumerism. This present globalisation process after two world wars is also the key motor of global environmental change (Brauch et al. 2009). The increasing use of fossil oil with drastic alteration of the composition of the atmosphere resulted in global warming due to the massive GHG, which produced climate change impacts, especially in the tropics of the Global South. The lack of political and economic will be the dominant force has increased the violent destruction of biodiversity and the exploitation of humans.

Confronted with the increasing destruction of Earth, the political discourse about sustainability (Brundtland Commission 1987) without real outcomes and aggravation of hydrometeorological events (IPCC 2021), together with an environmental (Bailey 2015) and a climate paradox (Bradfield/Vigaud 2017) emerged. Rich and industrialised countries emitted and continue to release 75% of the GHG globally (IPCC 2021) and per capita (see Fig. 10.1), while the Global South suffers from more intensive impacts of climate disasters. This anthropogenic destruction is caused by the imperialist countries in the north, and Crutzen (2002) called this epoch of Earth Anthropocene. In analytical terms, the same mechanisms of patriarchy such as violence, destruction, exploitation, and abuse brought Earth towards potential tipping points (Steffen et al. 2018) and the present era should be called the Patriacene.

Nobel Price Laureate Crutzen (2002: 23) claimed that 'these effects have largely been caused by only 25% of the world population'. Brauch (2022) insisted that there is still no agreement among scientists about this new Earth history. Crutzen/Stoermer (2000: 17) also argued that the geosciences are 'somewhat arbitrary...to assign a specific date to the onset of the 'Anthropocene''. Until present, stratigraphers such as the Anthropocene Working Group (AWG) of the International Commission on Stratigraphy and the International Union of Geological Sciences (Wolfe et al. 2013), have not yet officially approved this new epoch of Earth history.

When reviewing with more detail the emergence and the mechanisms that produced the Anthropocene or Patriacene, there exist abysmal differences among the interpretation. Many researchers do not analyse the androgenic capitalist model of destruction and pollution behind the profits of MNE (Oxfam 2022a, b). The financial elite (The Economists 6 October 2022) care more about the growth rate in the industrialised countries and some social well-being among their people but forget to speak about the costs of disasters in the Global South (ONU 2020). Rhetorically, during the COP 27 in Sharm-el Sheikh, Egypt, Northern countries addressed some of the initial impacts of climate change in their countries, such as the drought and heat waves during 2022, trying to reorganise the supply chain and the energy market

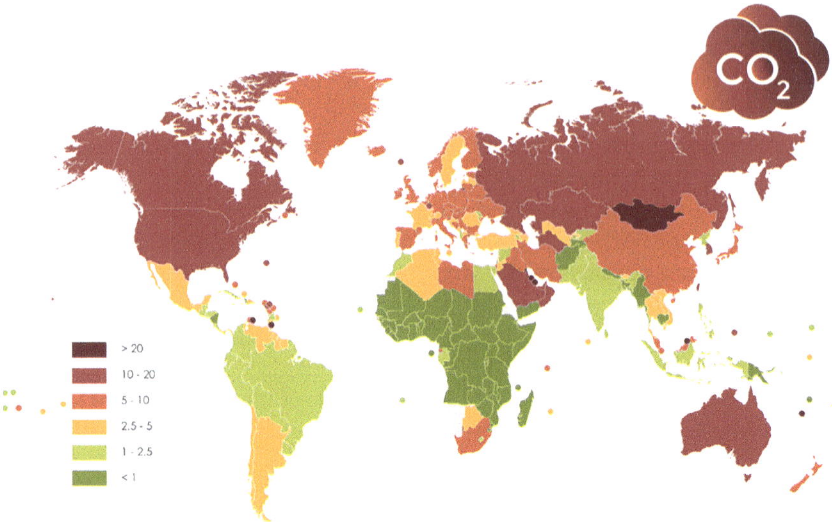

Fig. 10.1 Emissions of CO_{2e} per capita (annual tons during 2019). *Source* Global Carbon Atlas 2021

in favour of their economies (Simionescu et al. 2022), but they have remained unable to negotiate a peace agreement between Ukraine and Russia (Minakov/Kusa 2022). Similar outcomes happened a week later during the G-20 summit in Bali in November 2022.

Further, these global data do not specify the regional and local impacts on vulnerable groups of climate change impacts (Ilesanmi 2018). The analysis of a specific gender perspective is still mostly ignored in the dominant studies of physical, biological, political, economic, and disaster aspects (Senath 2013) of a real Patriacene. The hypothesis beyond this neglected question is related to the dominant patriarchal mindset and worldview of scientists in the north, who prioritise their analysis on the economy, power relations, and disasters occurring in the north (Carbon Brief 2022). Exploitation, modern slavery in sweatshops, gender discrimination, and their exploitation by MNE are mostly silenced (Ahikire 2007). Real commitments by the responsible for the uncontrolled emissions of greenhouse gases (GHG), which have aggravated the impacts of climate change (IPCC 2021), pandemics such as COV-2 SARS (ECLAC 2020), global poverty (Bauer 2015), hunger (FAO 2022), social inequality (Oxfam 2022a, b) are generally ignored. When the pressure in the COP of UNFCCC increases by the affected countries, greenwashing of so-called environmental investments (Kirk 2020) such as *Environment, Society, Governance* (ESG; The Economist July 2022) emerged, whenever most of the participants come from the Global North and their MNE, where often civil organisations and alternative proposals are sometimes repressed with violence (Palik et al. 2019). Business as usual (BAU) continues and in several Southern countries land grabbing and polluting extraction of minerals, the contaminating of water (Wang 2022), soils (Bhatnagar/

Bhatnagar 2022), and air (UNEP 2022a) are rapidly increasing climate risks (IPCC 2022).

Patriarchy has been an evolutionary process starting 5000 to 6000 years ago and the historical evolution of eliminating women from power was complex, and linked to the development of technologies, changes in the social organization, violence, war, processes of acceleration of destruction of natural resources, and exploitation of human beings. Thus, besides the historical process of deconstructing patriarchy, the chapter proposes also an epistemic change to survive in increasingly more complex conditions. Moore et al. (2016) insisted that Earth has reached social and also natural tipping points (Steffen et al. 1998). The present model of exploitation of natural resources is producing the sixth great extinction of life on the planet (Wagler 2011), together with the acidification of the oceans (Colt/Knapp 2016) due to the massive absorption of greenhouse gases. Sea level rise due to the loss of sea and glacier ice (IPCC 2021) is flooding entire islands in the Pacific. Moore et al. argued that the dramatic change on Earth is anthropogenically produced by the present model of production, consumption, and the massive use of fossil energy without real mitigation policies (OPEP 2006).

Facing a survival dilemma (İşeri 2020) by the affected communities, women and men, indigenous and marginal groups started to build resilience from the bottom up to deal with climate impacts. Only 5% of the indigenous people are able to conserve still 80% of the remaining biological diversity (Raygorodetsky 2018). These efforts of multiple ethnic groups worldwide are producing a radical paradigm shift to the challenges of the present capitalist destruction, but also to the occidental approach of Brundtland's sustainability, where the dominant countries have changed nothing. By introducing a care economy with recycling, mostly in the hands of women (ECLAC 2021), drastic changes can be achieved to conserve not only Mother Earth but also vulnerable people. These social groups, basically women, girls, and indigenous are highly affected by the ongoing globalisation process, the massive extraction of minerals (Arach 2018), and the destruction of biodiversity (National Academies of Sciences 2022). The chapter also argues that the time frame is short to change this dominant patriarchal system of production, consumption, and power relations is only about four decades. Will the 6th extinction process force people, including MNE, to return to more sustainable management of natural and human resources, in spite of the natural destruction and high death rates of humans by pandemics and disasters? Is there a way through a women's care economy (ECLAT 2021) or gift economy (Vaughan 1997, 2004) to change in the next four decades the capitalist destructive processes and return toward sustainable management of humans and biological diversity? Without any doubt, this is a global complex enterprise, as 75% of the historical and present GHG originated in industrialised countries, where China has become at present the largest emitter of these toxic gases.

10.1.4 Epistemic Challenges: Towards a Transdisciplinary System Analysis

Transdisciplinary studies unscramble complex problems with multiple feed-backs, changes in external conditions such as climate change, MNE, or an authoritarian government, and directions toward which the processes can be positively oriented. The interdependence of physical and social phenomena, where people, structures, tasks, events, and goals change in time, with regard to resource availability, and in extreme conditions, where only a self-regulating system may prevent people and nature from losing life and well-being. Habermas (2000) distinguished between internal flows of subsystems and external global ones, which generate a highly dynamic structure of permanent adjustment. Prigogine (1983) insisted that self-regulating systems are chaotic and far away from equilibrium. These dissipative structures (Grin et al. 2010) explain the changes in the evolution of humans from matriarchal clans to fratrilineal kinships, which allowed in irrigation cultures changes toward patriarchy with male leaderships protected by a supernatural being, who established god-cities and later by conquest, division of labour, and countries or regions dominated by war gods.

The upcoming systems were highly unstable and therefore leadership related their power to superior religious beings (gods), including new technological innovations from 6000 to 2000 BE (Farrington 1949). For instance, in the Greek cosmovision, a semi-god Zeus could establish the control over earth, sky, and humankind thanks to the control of lightning and thunder (Farrington 1949:18). These changing boundary conditions affected the whole system of the economy and changed power relations. In Rome, the father of a family emerged and the heritage of the first-born son was an adaptive process to limit violence within the family (Benedict 1934). Later, the Catholic Church reinforced the elimination of women from public and financial activities, pointing to their role in the house, and reinforcing the patriarchal domination of men in public, economic, and religious affairs. Wars, conquests, and colonisation allowed Western male leaders and their countries to consolidate the original accumulation of capital. Mies (1985) referred to the exploitation of human beings and natural resources as the key processes of capitalism, producing social stratification, exploitation, violence, and destruction.

After World War II an expansive globalisation process started, led by competition between the USA allied to Europe and the Soviet Union (Gaddis 2014). During and after this war, colonialised countries achieved their independence, while neo-colonial capitalist forces controlled the economy and policy of the former colonies (Bogaerts/Raben 2012). Within the framework of the Bretton Woods Agreements of 1944, MNE developed and flourished, and dominant occidental countries produced an acceleration of production, consumption, and financial capital (Stiglitz 2010). In 1989, the rapid disintegration of the Soviet Union and its countries allowed the USA to remain the sole superpower (Raiklin 1993), while China was emerging as a new economic, political, and military force. In these complex conditions of global environmental change, Crutzen (2003) pointed to a change in Earth's history from the Holocene to the Anthropocene produced by destructive anthropogenic activities of humankind, including natural resources.

These systemic processes not only devastated the existing ecosystem services by overexploitation (Jackson/Jensen 2022) but created also increasing numbers of poor people (Oxfam 2022a, b), new pandemics (COVID-19 Forecasting Team 2022), which are putting humankind as a whole and Mother Earth to the limits of survival. Goals, missions, and assumptions have profoundly changed and the accumulation of capital at any cost, including speculation and illegality, dominated this globalisation process. The following chapter is analysing with greater detail the complex systemic change that occurred during the last 12,000 years of humankind, where patriarchy appeared about 6000 years ago, capitalism 500 years and neoliberalism the last seven decades with the globalisation process (Stiglitz 2007). The systemic analysis of the historical acceleration explains how patriarchy with its violence, exploitation, and discrimination was the origin of the Anthropocene, whenever multiple contradictory processes intervened during thousands of years in the present complex transformation of the Patriocene or Capitalocene.

10.2 From Matriarchal Clans, to Patriarchy and the Patriacene

The following reflections are based on the early history of the human species, bringing light to the hidden view of prehistoric and historic societies. Anthropological studies of primitive societies indicate that society was integrated and got consolidated successively as greater social entities. Later, they established communities, cities, alliances of cities, kingdoms, and more complex relations of human coexistence. Most of these studies, undertaken by male occidental anthropologists, underestimated the history of women and their influence in the evolution and consolidation of the present modern society.

10.2.1 Matriarchal Clans

Morgan (1877) distinguished three important eras of social evolution of human beings starting from 'savagery', towards barbarism up to the present civilization, which have all been related to changes in technology and socioeconomic-productive activities. Savagery was related to hunting and food-gathering a million of years ago[2], when

[2] The oldest hominins are thought to have appeared as early as 7 million B.C.E. The earliest species of the Homo genus appeared around 2 million to 1.5 million B.C.E. Current evidence supports modern Homo sapiens appearing around 190,000 B.E.

homo sapiens emerged together with other prehistoric human beings -today extinct and still present in our DNA-. They slowly evolved and when hunting places were empty, they migrated to new regions, including other continents. Morgan (1877) established that barbarism started with settlements, agriculture, and cattle raising about 8,000 years ago. Manure for soil improvement from domestic animals, wool (sheep, llamas, alpacas, goats), milk (cattle, yaks, goats, sheep, horses, camels), traction (oxen, onagers, donkeys, horses, camels, dogs), and the herding support by dogs improved food and agricultural outputs (Moleón et al. 2014). With irrigation, groundwater from wells, and the evolution of seeds, intensive agriculture started globally (Barker/Goucher 2015). The fertile crescent started to develop about 10,000 years ago emmer wheat, einkorn, barley, flax, chickpea, pea, lentil, and bitter vetch (Brown et al 2011). Asia consolidated the crops of oats, barley, peas, broomcorn millet, rice, tea, rubber, palm oil, coconuts, sugarcane, and jute, where rice got the basic food crop (Fuller 2011). America specialized in potatoes, tomatoes, corn, beans, squash, chili, cacao, peanuts, and vanilla, where the evolution of corn in Central Mexico has undergone a highly complex process from a wild grass called teosinte about 8000 years ago to a corn cob. Further, the transformation of *nixtamalisation* into human food eaten as tortillas was another technological improvement. The expansion of agriculture and cattle raising produced deforestation of native forests, affecting further the hunting of wild animals. In indigenous societies, the traditional system of slash, grave, and burn (*roza, tumba, quema*) allowed the soils to recover during their long period of inactivity, whenever new areas were deforested. With population growth this traditional management of soils and forests disappeared, giving place to agricultural innovation such as the use of ploughs and the fertilisation for more intensive cultivation.

This evolution of agriculture and greater availability of vegetables introduced changes also in the traditional organisation of humankind. Maternal clans were the first form of social organisation that preceded civilised patriarchal society. Morgan (1877) insisted that women settled independently from men and that the matriarchal clan was a nonsexual union, where females possessed maternal functions, and sentiments. They produced vegetable food for their group. There was no direct mother-child caring with exception of breastfeeding, Girls and boys were raised collectively within the maternal clan. The sexual relations of their offspring with their father was unknown and maternal care started with mammal, similar to the primates and at least two third of a female's life was spent nurturing and protecting their infants (Jay 1963). The institution of totemism[3] protected women inside their maternal clans, and the taboo banned women to eat cannibalistic meals. Totemism and taboo avoided cannibalism and prevented incest. Both produced the first food division, where women started much later to eat meat from cattle raising. Tylor (1881) agreed with other anthropologists that the tribal system was based on maternal kinship with egalitarian social and sexual relations. Sexual restrictions were socially and culturally produced (Reed 1975) to limit inbreeding. The behaviour of the mother is a social function

[3] Totemism represents the earliest system of social regulations, based on totem (Separation of plant and animal food) and taboos, which banned sexual relations, cannibalism, or meat.

and not a biological one, where girls and boys were raised inside the matriarchal clans caring about food and regulating sex. Bendict (1934) distinguished between 'my own' matriarchal clan and the outsiders, which often could be attacked by male groups. Later rituals for marriage were developed by male groups, where women intervened as mediators. The first separation of growing boys was eating together with other men (Richards 1932), later young boys were trained as hunters or farmers. They were definitively separated from their mothers and left the matriarchal clan. The separation between women in maternal clans and men groups evolved into new forms of male organisation.

10.2.2 Fratrilineal Kinship

During the Neolithic agricultural revolution, producing grain belts of wheat, oats, barley in the fertile crescent, rice in China and India, and corn in Mesoamerica and South America together with potatoes, cannibalism slowly disappeared or was replaced by ritual cannibalism (see Aztecs). Ritualistic killings were associated with men, later substituted by sacrifices of animals (Frazer 1907–1915). Couvade or making the father was a process of a rite of fathers, where the man acquired the social right to assist his wife in the care and protection of her child (Mead 1935). The couvade is the key explanation of the origin of the father and especially, the patrilineal heritage in patrilocal settlements, and later in patriarchal societies. In traditional social groups, biological fathership did not exist. Paternity began with a social relationship between the women's partner and her child. The emergence of the first name dad is estimated around 1500 BCE, when children in different languages said dad or dada. The biological relationship between a man and his offspring occurred only in the nineteen century and today with DNA analysis, the genetic relationship with the father can be proven.

The rigid separation within the maternal clan was gradually overtaken by greater interlinks between men and women beyond their sexual relations. During this process, a historical change occurred from a sister-brother clan partnership in the maternal clan towards the family of the father, where the social mother was converted into a mother of a family. The myth of Osiris holding up her child Horus is a prefiguration of a sacred mother emerging together with a father family (Budge 1911). Morgan (1877) insisted that in these initial family agreements, sexual intercourse occurred between women and men. The father was socially recognised, whenever this process passed through multiple transitions from matrifamily or matrilineal family to father family. He insisted to distinguish among five phases of family development: (a) the consanguine family, (b) the *punaluan*[4] family, (c) the pairing family, (d) the patriarchal family, and (e) the monogamous family (Morgan 1877: 393–394).

---

[4] *Punalua* from Hawaii is translated as the spouse or intimate companion, equivalent to a pairing family.

Greater settlements demanded more food and natural resources. These demands not only destroyed the local biodiversity and the water for irrigation, especially in the semiarid climate but affected the quality of the rivers and wells for drinking water. More intensive crop systems fertilised with manure produced plagues and fungus. Crop production, exchange, commerce, and financial management emerged and intensified during the past 3000 years (Meillasoux 1977). Through fratriarchy and fatrilineal kinships (Ashley 1958), incest, endogamy, brother organisations, rite of passage of young boys, couvade and bride price, slowly a father-family was building, where private property appeared in the owning of cattle (Ashley 1958). From purchasing marriage to patriarchal power was a small step. Slavery, royal incest (Egypt), father killer (Oedipus), and mother-murder (Orestes) were present in the historical dramas, explaining the difficult process of overcoming social insecurity in maternal relations[5] by moving toward a patriarchal world order. Cleopatra murdered her brother to access the throne, and married two patriarchal men (Julius Caesar and Mark Anthony), but was totally defeated by the new patriarchal forces and committed suicide, while Octavius killed her two sons, who by Roman law could claim their access to the throne.

10.2.3 Evolution and Consolidation of Patriarchy

Within this complex consolidation of social evolution and reorganisation by patriarchy, the chapter revises the origin and root causes of patriarchy. The origins are related to bride prices, private ownership of cattle, and patrilocal settlements, where women together with slaves only lived inside the house. This evolutionary process, reinforced with religious beliefs and legal reinforcement against women, imposed the dominant androgenic economy, It was the out-coming power struggles among male tribes, city-gods, countries, and empires, where violence and social inequality emerged (Mies 1986). Slavery and women inside the house were preconditions for unpaid domestic and sexual services. The use of natural resources for the survival of humanity was never questioned by hunters, irrigated agriculture, or mining. The exploitation of nature beyond its capacity for recovery occurred first by hunting (Haddon 1934) which obliged tribes to migrate to alternative places for survival. Neolithic agriculture started with the settlement of people, and the domestication of plants and animals happened in separate locations on Earth, characterizing the geological era of the Holocene some 11,7000 years ago after the glaciers retreated (Gordon 1935). Women, leaving before often in caves, had early experimented with the growth of cereals to get more food for their children and their recollection was easier. Jared Diamond (1999) postulated that together with the domestication of plants occurred cattle raising, which spread rapidly from the Fertile Crescent to North Africa, Eurasia, and beyond.

[5] Medea a barbarian princess and sorceress, representing their matriarchal heritage, married a Greek hero and committed a blood crime to aid her husband.

Between 6000 to 5000 years ago, in various parts of the world emerged a different social organisation, called patriarchy. Women were displaced from the productive processes and confined with the slaves into their houses. People were led by pater-familias, later god-kings, monarchs and emperors. Conquest and local and regional war emerged. They produced extra incomes, goods, women, and slaves. In America a slave trade organised by Spanish and Portuguese colonisers brought African population to America. Surplus of food and other goods were traded. A division of labour appeared, where men specialised in war, arm production, buildings, and handicrafts. The original accumulation of wealth was concentrated by the male leader of emperor.

Growing populations and more intensive accumulation increased agriculture and cattle raising, at the cost of deforestation and environmental destruction. Plagues obliged scientists to develop chemical pesticides and fertilisers, which affected both humans and nature. Some initial restoration of ecosystem services emerged, granting the resources required for the survival of people (clean water, air, food, and fertile soils). Clean water was crucial after the cholera pandemic appeared in cities. From the XVIII century on, new demands from industrialisation processes emerged, which often highly polluted drainages and drinking water. The intensification of productive processes, population growth, global trade, and from 1970 an intensive use of fossil oil had changed the physical-chemical composition of air, water, and soil, producing climate change. Global warming (IPCC 2022) on Earth produced more and stronger extreme events, such as hurricanes, droughts, and bushfires. During the acceleration of these global production and trading activities, also long-term impacts on biodiversity and ecosystem services emerged (MA 2005). They produced a loss of life quality globally, new plagues in crops, animals, and pandemics in humans (Covid-19). Patriarchy, as a longstanding process, has not only induced changes in the social organisation but also in nature. During the past seven decades of globalisation, this patriarchal management of nature and human beings has accelerated the process of a change of Earth's history from the Holocene to the Anthropocene (Crutzen 2002), better named Patriacene. This acceleration of socio-economic and climate processes is not only related to the geological change in the history of Earth (Crutzen/Störmer 2000; Kragh 2000) but has also generated socio-environmental destruction, resulting in a concentration of wealth (Oxfam 2022a, b) and poverty, especially in the Global South (UNEP 2022a, b). The involvement of MNE and imperialist governments transformed further the economic and power structures, concentrating the wealth in dominant countries and multinational elites (UNEP 2022a, b). Whenever as a long-ongoing process from the early agricultural history to today's climate change, poverty and disasters, patriarchy was always based on violence, conquest, war, discrimination, and exploitation (Reardon 1985).

10.2.4 Patriarchy

Patriarchy, where initially the father controlled the family. It was intimately related to slavery because women and slaves were obliged to stay inside the house and could only leave it with a male escort from the family. The first written evidence of legal patriarchal male management of the household was developed by King Hammurabi of Babylon (1792–1750 BE). Chiselled in stone, it is the oldest and best-conserved document of legal rules in an emerging patriarchal society. As a code, it has reinforced the existing legislation and is regulating marriage, family, propriety rights, orders, wages, professions among men, and the role of slavery. Women were obliged to stand inside their homes and female adultery was punished with the death penalty, while men were free to have several women.

As seen in the former subchapters, patriarchy evolved over thousands of years in traditional societies, where women were controlled initially by the matriarchal clan or a matrifamily, the limit of eating human flesh, and incest by the sexual reproduction process (Reed 1975). After fratilineal kinship and couvade a father-family emerged, where women were leaving with their husbands. Men slowly displaced slowly female privileges of taboos and sex control, and female powers got reduced to care about their offspring. Women were disposed of by property and decision-making, while children were born in fathers' houses, whenever the knowledge about the father's responsibility in the procreation process did not yet exist. The heritage changed from matrilineal to patrilineal, and the first-born son got the proprieties. In the patriarchal family the man, owner, and father were the head of his wife or wives, slaves, and their children. Slavery increased further social discrimination against women. The Spanish colonisation brought African slaves to America, increasing racism, exploitation, and discrimination.

Reardon (1985) stated that war and violence were at the origin of patriarchy, where women lost their public power and were confined within their houses and controlled by men. During the past five centuries of capitalism, wars continued to diversify this dominant Occident androgenic system. Patriarchy promoted new conquests in Africa and Asia. Wars all over the world, exploitation, discrimination, and classism, where not only women were removed from their power, but also other races and indigenous people. Wealth got concentrated in Western male colonizers. During the last seven decades of globalisation, corporate capitalism intensified and accelerated productive-financial processes, the exploitation of humans (Stiglitz 2007), and the destructive extraction of natural resources (Arach 2018). This multinational capitalism consolidated globally the patriarchal system of rule, which is responsible for the exploitation of human force in sweatshops and the destruction of biodiversity. It has produced dramatic impacts on climate change, ecosystem services, and hydrometeorological disasters (hurricanes, floods, and droughts with bushfires). During this acceleration of the last 7 years, the physical-chemical composition of air was so massively altered that the ozone layer was depleted, oceans suffered from acidification, and scientists spoke about the sixth extinction of biodiversity related to the dramatic loss of biological richness (Wagler 2011). Crutzen (2003) analysed these intertwined processes and negative outcomes and suggested a change of the existing history on Earth from the Holocene towards a new one, called the Anthropocene.

After the understanding of the emergency and consolidation of patriarchy, based first on systems of role against cannibalism and incest, Matriarchal clans, paterfamilias, god-kings, written rules of social behaviour, slavery, and the elimination of women from decision-making, colonialism, neo-colonialism, globalisation, and an exploitive and violent economy has produced a change in the history of Earth, the Patriacene. In this epistemic change on Earth, Occident colonised the Global South and later, after emitting massive GHG in their own countries, generated anthropogenic climate change (IPCC 2021, 2022). However, climate impacts are stronger in the tropical regions, where poor and exploited countries were left alone to deal with the increasingly more destructive climate disasters, while Occident produces the high GHG (Mechler et al. 2019).

10.2.5 Anthropocene

The term of the Anthropocene as a geological epoch was first coined by Crutzen/Stoermer (2000) and has since then been used by a part of the scientific community (Crutzen/Brauch 2016) to replace the former Earth epoch which was called the Holocene. This former time represented the current epoch of the Quaternary in the period in Earth's history. The Anthropocene is the result of a significant global impact that human activities exercised on terrestrial ecosystems by clearing forests, and emitting carbon dioxide, nitrogen, and sulphur gases with particles into the atmosphere. The chemical effects of the long-range transport of smoke plumes, and photochemical smog have perturbed the oxidant cycles in the Troposphere. Global warming and hydrometeorological extreme events, acidification of the ocean, the alteration of the nutrient cycles on soil degradation, and sea level rise related to global warming added water from melting ice sheets and glaciers, and from the expansion of seawater due to rising temperatures.

The origin of the beginning and the term Anthropocene is still in dispute. Was it starting in conjunction with migrant hunters who exterminated megafauna in several parts of Earth? Or was the Neolithic Agricultural Revolution about 12,000 years ago and the initiation of irrigation agriculture with plow technology and cattle raising about 6000 to 5000 years ago the origin of the Anthropocene or Patriacene? Was it the drastic change of social organisation with the loss of matriarchal powers and the emergence of the violent patriarchy, which affected Earth in such a way to produce a new era? Dominik et al. (2020) and multiple other researchers, including Paul Crutzen think that the Anthropocene is the outcome of the process of a 'great acceleration' since the 1950s, based on cheap oil and a fast globalization process that has produced global impacts on Earth. Wolfe et al. (2013) stated that the emergence of different definitions of the Anthropocene is not only visible in the sediments, but also in subsoils, water, the air, and in biodiversity. The massive urbanisation process with rural-urban migration (Terrazas/Blitchtein 2022), intensive food production (Payen et al. 2022), industrialisation (Grübler 1995), mining (Arach 2018), and the dramatic increase of global poverty (Oxfam 2022a, b) are integrated processes of the change of Earth era.

Subramanian (2018) linked the recent ongoing process of destruction to the deep roots of the massive extraction of natural resources, industrial processes, consumerism, population growth, and the exploitation of human labour. The social understanding of the global change occurring includes precisely the exploitation of human beings by paying miserable salaries, often without social security in sweatshops in the Global South. Further, a new geological order implies also political threats and how policy is doing in a new era of earth's history. Dalby (2017: 3) asked how to secure 'the conditions for particular forms of collective life', especially when global environmental problems, climate change, sea level rise, human-induced crises such as the crises in Chernobyl, Fukushima, and the war between Russia and Ukraine are affecting globally. The present militarisation of industrialised countries with more than 2 trillion dollars in armed expenditure (SIPRI 2022) is further increasing climate change threats, first by the production of arms and second, by using them and polluting the air, soil, and water.

International organisations, such as the UN with its Security Council (UNSC), were unable to avoid the present war between Russia and Ukraine, which is having global impacts on inflation due to the increase in energy and food prices (Dinucci 2022). There are also other conflicts ongoing in Africa (Muiña/Barros 2021), Asia (Palik et al. 2020), and Latin America (Álvaro 2022). All these struggles are badly managed by UNSC (Daya 2022; Weiss 2022). Further, multiple nuclear essays in 1950 in deserts, underground explosions, and in the Pacific atolls have produced severe alterations in nature and ecosystems. The uncontrolled emissions of greenhouse gases, reaching in August 2022 a level of 418.53 ppm of CO_{2e} (NOAA, 20 November 2022), are aggravating multiple climate change impacts (IPCC 2021), especially in the Global South. They are also increasingly impacting the industrialised countries, such as the extreme temperatures in Spain and Portugal, the severe droughts in California and in the United Kingdom (Kallen/Johnson 2022), and the deadly hurricane Ian in Florida in September 2022.

These extreme events are not only affecting people and their assets but are also destroying biodiversity, and flooding with sea water productive areas. Undoubtedly any armed operation is polluting the environment and putting new pressures on the climate (Dalby et al. 2009), but can also generate long-term impacts, as the defoliant Agent Orange that has caused in Vietnam for Vietnamese, Cambodian, American, Australian, and other soldiers' genetic defects in their offspring (Westing 2013). Whenever it was coined as an earth science concept, the term *anthropos* includes military expenditure since the start of patriarchal conquest at least for the past six thousand years. Therefore, the term Patriacene explain better the past and present geopolitical conditions, the sophistication of warfare and its pollution reinforces the policy of the remaining superpower of the USA and its MNE, which are especially in the Global South extracting minerals and producing by aggravating the environmental and climate conditions. Further, drought and the decline in agricultural production are promoted among vulnerable people forced (Oswald-Spring et al. 2014) and rural-urban migration (Terrazas/Blichtein 2022).

These physical-chemical processes had profound impacts on human beings, especially in the Global South, where temperature rise and socioenvironmental vulnerability (Oswald-Spring 2013) are increasing extreme weather events. However, there have been large changes in global emissions by country and per capita. Asia emits 53% of GHG, where China alone releases 10 billion tons/year and India 2.7 billion tons, followed by the USA with 18%, Europe (27 countries) with 17%, Africa, and Latin America with 3-4% (Our World in Data 2022). The *Emissions Gap Report* (UNEP 2021a) highlighted the fact that global emissions must decline by 45 to 50% by 2030 in order to prevent that temperature rises above 1.5 °C of the global average temperatures since the industrial revolution. Only this reduction would be able to avoid major disasters, especially in countries that have only marginally been responsible for global warming. The following subchapter analyses from a feminist point of view the relation of patriarchy with its violent exploitation of humankind and nature, and the change in Earth's history towards the Patriacene.

10.3 Patriarchy and the Anthropocene from a Feminist Perspective

In this new epoch of Earth's history, multiple complex socio-environmental processes are interlinked and are affecting different genders, races, classes, and ethnic groups. Thus a combined quantitative and qualitative methodology is applied below (Serrano-Oswald 2021) to understand the impact of patriarchy in the global change from the Holocene to the Anthropocene.

10.3.1 Ecofeminism

The feminist critique of the Anthropocene model is linked to the prefix *Anthropos,* which homogenises humanity as a whole and lets without visibility the mechanisms of extraction, exploitation, and discrimination of women, men, and the Global South by MNE and dominant occidental countries. Besides the profound destruction of nature in the world, it has produced inequality, poverty, and discrimination among human beings, where especially women and girls are affected. The concept has homogenised responsibilities and vulnerabilities while reifying the human-nature distinction.

Ecofeminism (D'Eaubonne1974) started to link the exploitation of women with the destruction of nature (Mies 1993). Moore insisted that the Anthropocene is not neutral but produced by the present capitalist system and called this phase of human history as Capitalocene (Moore 2017; Moore et al. 2016). Haraway (2016), a biologist, insisted on the slow ongoing process of humankind evolving from primates, where we contain up to in some cases 99% of their genes. She spoke about an Earth phase of a Plantaciocene and later proposed the term *Chthuluceno* (Haraway 2020), insisting on the historical evolution of animals and humankind.

Warren (1997) distinguished between the oppression of women (a hierarchical system of power and privilege by men) and the domination and destruction of nature (existences without self-understanding). She insisted that capitalism has exercised control even over the reproductive process of nature and humans through genetically modified organisms (Behrokh/Ardekani 2011) or cloning over the natural and female reproductive processes (Mies 1993). Assisted reproduction or surrogacy of bodies has imposed duties and attributions on the ethics of women's care (Saxena et al. 2012). Some ecofeminists, because of the closeness of women to nature, care, and reproduction have fallen into some essentialism, where the feminine is glorified, but underlying the existing vulnerabilities or covering them up (Gaard 2011).

Ecofeminist authors (D'Eaubonne 1974; Mies 1993) have further claimed the origin of the destruction of nature by limited MNE, supported by male financial elites, occidental governments, a global speculative system favouring the accumulation for small financial leaders, the Bretton Wood organisations, tax heavens, and the Wall Street (Stiglitz 2010). 'Governments around the world often patent information about the companies that operate in their country to protect quality and image for domestic and international markets. Conglomerates often have a fear of competitors revealing unjust business World' (Kyove et al. 2021: 216) Most politicians in the North care therefore, about their enterprises, tax incomes, and protect the business of these multinational companies all over the world, without any consideration on the negative impacts elsewhere. MNE are also often influencing with powerful lobbying their government and the Congress (Ülgen 2022).

In the Global South, weak and often corrupt politicians accept the imposed destruction by extraction of their natural resources, extinguishing also their biodiversity, polluting water and soil (Arach 2018), and exploiting humans (Oxfam 2022a, b). They also accepted the payment of miserable salaries for their citizens as an additional stimulus for investments of foreign money thanks to their low wages. Ideologically, religions all over the world have promoted hierarchical obedience to the historical leaderships and to these top MNE leaderships (Barnes 1978), where both women and poor men were dominated and exploited, often in the name of superior beings (Haddon 1934). It is necessary to understand the historical dimension of how the present socioenvironmental crises were produced, and how diverse subjects (gender, class, race, ethnicities) are affected in differentiated ways by these destructive activities. Today, climate disasters are an additional threat for vulnerable people (Oswald-Spring 2013). Understanding the root causes of the present environmental and social disasters, there may be a possibility to respond theoretically, politically, and practically to these complex and interlinked crises (Allen 2011) and revert the ongoing extinction process.

Arons (2020) further asserted the limited understanding of the Anthropocene by geophysicists (Wolfe et al. 2013). In her critical analysis, she insisted that it is necessary to differentiate between the origin of the alteration of Earth by patriarchal behaviour during millennia and especially, the last seven decades of acceleration, where the benefits have been distributed unevenly. Who are the humans who gained during this process of destruction, many of whom are in conflict with the rest of society precisely because of the benefits and costs of the 'Age of Man'.

Feminist Neo-Marxist critics (Stevenson 2020) took such conflicts into fuller account in their argument that a better nomenclature and conceptualisation for this epoch would be to use the term a 'Capitalocene'. This term captures the fact that our ecological crises during the past 70 years of economic acceleration (Peruzzi/Terzi 2021). These destructions have not been precipitated by all humans in some undifferentiated and generalised way (Görg et al. 2020), but more specifically by the global spread of MNE, and North American capitalism (Stiglitz 2010). Precisely, the violent appropriation by very small androgenic elites has first produced patriarchal fathers, god-cities, conquest, colonialism, and today by an intensification of the exploitation of nature and humans an increasing global inequity (Oxfam 2022a, b). These elites with their occidental governments have generated severe socio-economic-ecological injustices globally, especially in the Global South, but are also responsible for the changes in the physical-chemical composition of air, soil, subsoil, and water. These conflicts continue to be at the heart of many human dramas and anthropogenic disasters, both historical and recent.

10.3.2 Capitalocene

Moore (2017: 1) stressed the need for 'historical thinking in coming to grips with capitalism's planetary crises of the twenty-first century'. He insisted that the understanding of the Anthropocene is not neutral 'and do not understand the system of exploitation, profit, power and destruction in the life cycle'. Therefore, Moore proposed the term Capitalocene to overcome the dualism between humans and nature developed by the deep ecologists (Naess 1986), who underestimated the impact of capitalism as a phenomenon of human organisation. Moore (2017) further insisted that the ecological crisis, postulated by many environmentalists did not start with the Industrial Revolution and the invention of the Steam machine by Watts (Abram et al. 2016) but since the agricultural revolution and the first cities. This is precisely also the beginning of patriarchy. As he insisted, the drivers were not steam and coal but imperialism, conquest, violence, profit, and capital accumulation.

The Anthropocene does not separate nature from humans and Moore asked when and where did humankind distinguish from animals and how did humans start using nature for their well-being? During the acceleration of the last seven decades, without doubt, it was the capitalist exploitation and the availability of cheap oil in the fifties that explain the global environmental change (Brauch et al. 2009), including the deep transformations in the power structures and accumulation processes, especially after the rapid disintegration of the Soviet Union (Raiklin 1993). Moore et al. (2016) argued that the changes fit into the processes of power structures and capital accumulation during the colonialisation processes, whenever class, race, ethnicity, and gender discrimination started several thousand years ago, precisely by the patriarchy and produced the Patriacene.

Environmental change is an ongoing process since humans were hunters and gatherers when they destroyed different parts of the world's megafauna and migrated to new hunting regions. The establishment of humans in settlements and the evolution of cattle raising and agriculture with irrigation were push factors towards a new model of social relations, where androgenic militarisation, god-cities, and slavery produced a greater imprint on ecosystem services (MA 2005), which were accelerated by MNE during the past 70 years of natural exploitation. Arons (2020) proposed a type of 'Tragedies of the Capitalocene', insofar as she dramatised stories that traced the dynamics of exploitation of both human and nonhuman resources.

A key limitation of the concept of the Anthropocene is also the link as a new geological epoch with human domination on planet Earth. However, both in academia and in a wider cultural and political context the origin of the concept lacks first the specific social groups affected by the MNE who have caused this change of an epoch. People in the Global South, besides centuries of colonialism, depend for their survival largely on agriculture, even though the process of urbanisation is changing the future also in developing countries. Without a doubt, the last 70 years of intensive use of fossil oil are responsible for the alteration of the physical-chemical composition of the atmosphere and as a result the emergence of global warming. Poor countries are still very marginal in their emissions of GHG (IPCC 2021) but they are directly dependent on favourable climate conditions for their survival (Escudero 2016).

Due to industrial development with high growth rates, China has become the country with the highest GHG emissions, with peaks in deforestation, deviation of rivers, the construction of the biggest dam (Four Gorges), and negative environmental impacts of the Belt & Road initiative (Li et al. 2021) that are all contributing to global warming. India is following this way and will soon be the country with the highest population on Earth, which has produced euphoria and despair. This population is demanding basic services such as food, water, shelter, education, and health, which will negatively affect the availability of natural resources and ecosystem services for healthy living (The Economist 13 May 2022).

Both Moore and Haraway criticised the term *Anthropos* and together with other researchers (IPCC 2021) showed that global capitalism has broken the social metabolism, where nature is no longer able to assimilate the pollution and waste produced by humans in their productive, commercial, and consumptive processes, destroying the natural resources and the capacity of mitigation of ecosystem services (MA 2005). Secondly, we have altered natural resources to such a degree that the environment can no longer produce the necessary and free resources (water, air, soil, subsoil, food [Markandya 2021], fibres, minerals, etc.) within the present level of speed of exploitation and pollution (Arach 2018). Malhi (2017: 79) insisted that the term

Anthropocene often includes an emphasis on (a) the global and pervasive nature of the change; (b) the multifaceted nature of global change beyond just climate change, including biodiversity decline and species mixing across continents, alteration of global biogeochemical cycles, large-scale resource extraction and waste production; (c) the two-way interactions between humans and the rest of the natural world, such that there can be feedbacks at a planetary scale such as climate change; and (d) a sense of a current or imminent fundamental shift in the functioning of our planet as a whole.

10.4 Plantationocene or the Cthulucene

Haraway (2020) goes further and proposes the term *Plantationocene* or the *Chtulucene*, which explains the space-time inexorably diffraction with destructive and reconstructive processes. As a trained biologist Haraway (2016) suggests a multispecies alliance as a way out. Unlike the Anthropocene (or *Capitalocene or Patriacene*), the *Chtulucene* proposed by Haraway would be inhabited by 'multispecies histories and practices in the process of becoming-with, in times that remain at risk, precarious times in which the world is not finished and the sky has not fallen, yet'. She proposed a sort of foundational manifesto for a new civilization in which she hybridises utopia with scientific fact and the account of small experiences that point in the same direction.

Haraway claimed that all these destructive ongoing patriarchal processes are embedded in systems of exploitation, oppression, domination, discrimination, and violence. This patriarchal behaviour has affected and is further destroying the cohabitation of nature and humans, using increasingly more artificial tools for life, food, reproduction, and survival. Haraway is deeply anti-essentialist in blurring the difference between humans and nature with the 'cyborg' or the companion species (Haraway 2016), a typically queer approach. She raises the global causes, vulnerabilities, and responsibilities associated with the Anthropocene and politicises the concept. She unites the feminist visions of environmental justice, health impacts, increased violence with the loss of food sovereignty, obesity, and increasing chronic diseases such as diabetes and cancers.

Haraway (2016) emphasises a 'world-of-life' and peaceful cohabitation of humans and nature with a deep decolonising approach to science (Oswald-Spring/Brauch 2021). Her exceptional vision provided a nexus of biology, economics, culture, feminism, and science fiction. Haraway (2020) is addressing the complex game of how to live well (and die well) in coexistence with a 'planet that is going to shit'. Her approach goes further with the queer-feminist approach, including the past errors during the past six decades of acceleration on Earth with quantum changes (Maldonado 2017). Destructive activities (IPCC 2022) are producing the 6th extinction (Wagler 2011), where she includes not only the specific and global vulnerabilities but proposes a radical change in the understanding of how the Anthropocene and the Capitalocene must be addressed, including the emergence and consolidation of patriarchy. She insisted on alternative concepts such as living well with cyborgs in a multicultural world and without internal colonialism (González Casanova 2003) or the destruction of any living on Earth. People from the Global South, confronted with extreme events, are exploring alternatives for survival.

10.5 Epistemic Changes for the Survival of People in the Global South

After analysing from a feminist critic the emergency of the Patriacene, the key argument of the present chapter is related to the bottom-up alternatives that are emerging among poor people confronted with extreme climate disasters.

10.6 Impacts of Climate Change and the Survival of Poor People

Since November 2022 eight billion people live on Earth, demanding food, water, clean air, transportation, education, life quality, and diversion. Most of these people live in the Global South, where they are highly affected by climate events. Nakate (2021) stated that African communities bear disproportionate consequences of the climate impacts and lack of adaptation (Government of Sierra Leone 2007). As an activist from African nations, representing the Global South, Nakate insisted that Southern voices are not equally being heard as male activists from white nations. This process occurs not only in the different COP of UNFCCC, but also in the World Economic Forum in Davos, in the UN, the UNSC, and in the representation and leadership of the IPCC. She insisted in an Artic Basecamp in Davos during the World Economic Forum in 2022 that:

> People in Uganda, in Africa, and across what's called the Global South are losing their homes, their harvests, their incomes, even their lives, and any hopes of a liveable future right now. This situation is not only terrible, it's also unjust. Although the African continent has just 15% of the world's population, it is responsible for only between 2% and 3% of global energy-related carbon dioxide emissions. The average African's greenhouse gas emissions are a fraction of those of people living in the US, Europe, China, the United Arab Emirates, Australia, or many other countries.

COP 27 in Sharm-el-Sheikh, 2022, in the Egyptian Sinai, was supposed to be an African-oriented conference, preventing further emissions of GHG and giving support to the highly vulnerable countries in the continent. In the decisive sessions, nobody spoke about the devastating drought in Somalia, where from 2011 to 2022 a quarter million people died. In 2022, in Uganda, 2500 people were killed by drought and famine and 8 million people were affected in Ethiopia. Deadly floods in Nigeria produced 600 dead victims since the start of 2022 and 76 victims died, when they were evacuated by boat. Carbon Brief (2022) informed that at least 4000 people have been killed and 19 million were affected since 2022[6]. They are totally abandoned by the industrialized countries, responsible for climate disasters. However, hurricane Ian, which hit Florida and caused 125 lives were reported for weeks in the world press. What are the reasons for the silence of the increasing disasters, deaths, and affectation in the global South?

One sole billionaire emits one million more GEI than a person belonging to 90% of the poor people on the planet. The richest 125 billionaires with 183 MNE control 2.4 trillion dollars. The average of these investments is 3 million tons of CO_2/person, while 90% of poor people emit 2.76 tons/person (Oxfam 2022a). During the COP26 2021 in Glasgow, UK an agreement of 100 billion dollars/year (BD) was negotiated, but only 83 BD were invested. Oxfam (2022a) analysed this amount in detail: 70% came from public funds and were not contributions. Only 24.5 BD were climate finances, the rest came from multilateral banks, often without any discount or special interest rates. During COP27, the supervisory board sent proposals to include agricultural soils, grassland, and various dangerous technologies such as terrestrial and marine geoengineering. For the survival of poor damaged people, affected by these MNE, there is no or only limited support to deal with greater climate change impacts. Further WMO informed during the COP27 that the opportunity to achieve 1.5 °C is almost impossible and with great efforts, the global temperature rise may be controlled by 2 °C.

10.7 From the Bottom-Up

Confronted with these crude realities, the Global South knows that they are alone to deal with catastrophic climate disasters and people have learned to help themselves. As an example, the COVID-19 pandemic gave in Latin America multiple examples of solidarity, caring capacity (Stefanovic 2022), and support in critical situations (Nel-lo et al. 2022) by building social networks. An interesting example was an indigenous community in Morelos, Mexico, who cared about their sick people in a traditional way, giving them food, burying them in their traditions, and after three weeks the entire community had developed a heard immunity to resist the virus without any vaccination or medicine. The number of dead people was limited and lower than in the rest of the state. Asking about their behaviour they insisted that sick people cannot be abandoned. Nobody can have a lockdown when neighbours require food and support, and everybody was working actively to support the sick people. Confronted with these crude realities, the Global South knows that they are alone to deal with catastrophic climate disasters and people have learned to help themselves.

[6] Madagascar and Mozambique have suffered from six severe storms this year, killing at least 890 people, and nearly 2 million people suffered from floods in August and October 2022 in Chad. The temperature reached in Tunisia in July 48 °C, producing extreme wildfires. However, the impacts of African extreme events often go unrecorded, thus the real affected people are much higher. Similar figures were informed by other countries in the Global South (Moreno et al. 2020).

As an example, the COVID-19 pandemic gave in Latin America multiple examples of solidarity, caring capacity (Stefanovic 2022), and support in critical situations (Nello et al. 2022) by building social networks.

An interesting example was an indigenous community in Morelos, Mexico, who cared about their sick people in a traditional way, giving them food, and burying them in their traditions, and after three weeks the entire community had developed herd immunity to resist the virus without any vaccination or medicine. The number of dead people was limited and lower than in the rest of the state. Asking about their behaviour they insisted that sick people cannot be abandoned. Nobody can have a lockdown when neighbours require food and support, and everybody was working actively to support the sick people. They accepted those who have died as part of the cycle of life and death in their cosmovision. Among local communities and in poor city neighbourhood multiple efforts arrived during the pandemic, a flood or an earthquake. The most important objective was to save people, later to disseminate the necessities of the refugees (water, food, medicaments), and to diversify the required social support. A system of self-organization emerged (Grin et al. 2010), sometimes via the internet, and enhanced the initial activities. When finally, the formal authorities arrived, they found a well-established network of collaboration and everybody waited only for financial support from the government, evaluation of the real damages, and the consolidation of refugee camps for longer periods of stay. Often governmental support in remote areas arrives late or never, and people in marginal conditions have learnt to support each other.

The bottom-up analyses on a micro level and generally on specific extreme events (floods, earthquakes or ritual services) are common. It is a response in most countries in the Global South who have learnt that their survival depends in an important way on social networks during an emergency or a disaster. Therefore, strong social webs are established locally to reinforce the personal, religious, spiritual and support processes. Thus, during a disaster, neighbours, friends, and nearby family members are the first to give support to the affected people and save most of the affected people. In indigenous societies in Latin America there exist also clearly established commitments to collective, religious, and communitarian work. In the so-called *mano vuelta or tequio* (the hand turned), family members participate in the assigned collective public works for the improvement of the community, a sacred place or a church festival often mixed with traditional beliefs.[7]

Among neighbours there exists also voluntary support during the building of a house or repairing a roof, which is corresponded, to when the neighbour needs any help. This collaboration is given with pride, and every family participate with free work, sometimes monetary support or offering a religious service during a whole year (*mayordomía*), when a family is elected to care about a specific task. This

[7] An important example is the Virgen of Guadalupe, the 12th of December, celebrated on Tepeyac, a mountain in Mexico City. This place was traditionally used for fertility by goddess *Tonatzin* and represents now a catholic symbol, whenever indigenous and mestizo people go to this place, celebrate this double function in their religious beliefs, called also syncretism.

collaborative behaviour increases social cohesion and confidence among community members. It creates among the participants the feeling of belonging to social groups or communities. Therefore, communities confronted with stronger climate events, have reinforced their traditional links to deal better with catastrophic situations, and especially among indigenous groups, different efforts of care and mutual support have emerged.

10.8 Indigenous Alternatives: Aymara's Living Well and Zapatistas Leading by Obeying

Schaeffer (2022) analysing the artificial border and wall built by the government of Trump, criticises the separation between Tohono O'odham Nation or Fort Yuma Quechan Indian Tribe living in a territory covering Arizona, USA, and Sonora, Mexico. These indigenous groups share a 'land-based ontoepistemologies'. Their 'sacredscience' integrates old-age knowledge with daily practices in a desert region, where the relations with ancestors and their thumbs (on the Mexican side) nourish their cosmovision and their daily way of life was disrupted because their villages were traditionally established on the US side. These entanglements form part of human beings and each indigenous group should have the right to speak to their land and their sacred places in their native language. As a basic human right of these indigenous tribes, today they cannot move freely between their sacred places. Suddenly an artificial wall and borders between the two countries were imposed without any consultation and negotiation among these indigenous groups. This metallic artefact has divided their ancestral land and impedes their crucial way of life, affecting their integration with the territory, the deaths, the costumes, and the cosmovision.

Other ethnic communities or marginal urban areas have understood that decentralized and bottom-up efforts, such as living well (indigenous Aymara in the Andes), *bolsa familia* (Brazil)[8], leading by obeying in the Mayan's Zapatistas shell model in Chiapas (Mexico), but also alternative ways of life have emerged. Gift-giving (Vaughan 1997, 2004) and social economy (Collin 2020) may reinforce the resilience and survival of vulnerable affected people (Oswald-Spring 2013). In all these efforts self-reliant climate change management by vulnerable societies emerged or got consolidated, where solidarity focuses on sorority, fraternity, resilience, and well-being instead of capital accumulation. Confronted with irregular rains, higher temperatures, and lack of water in the complex ecosystem of the Andes (Moreno et al. 2020), the indigenous Aymara in Peru and Bolivia have increased the protection of their traditional way of life. With *sumaj kamaña* (living well) or *suamk kawsay* (good living)

[8] Bolsa Familia is a Brazilian social welfare program introduced by President Ignacio Lula da Silva in 2003 and converted into law in 2004. It is part of the Brazilian federal government's network of social assistance programs. It represents an income transfer program that provides financial assistance to poor families to reduce social inequity in Brazil and allows every family to get enough food.

during the last decade, the indigenous in Bolivia were able to survive a military cue by the local bourgeoisie with military persecution of their local leaders in November 2019. During the legal elections in 2020, the indigenous and multiple other social groups won again the majority of the Congress and the Presidency, while on the 13 of March 2021, the interim Ex-President Jeanine Añez Chávez was arrested, accused of sedition and terrorism during the cue against Morales.

The former indigenous government and the present one were able to reduce radically extreme poverty. The indigenous and allied progressive leaderships have strengthened equity and increased sustainability in their indigenous regions, including multiple women in directive positions (ECLAC 2018). 'Living well' means to increase life quality, which includes social integration, moderation in eating, drinking, dignified working conditions, dancing, and creating happiness by consolidating community cohabitation and limiting conflicts in social and family relations. For Aymara there is no better life, only good life. The respect for nature and the recovery of ecosystem services are crucial for survival in highly complex environmental conditions, where only alternative lifestyles help to understand the complex interrelationship between society and nature. Bolivia designed an alternative constitution, where humans and nature have both the same rights. A model of redistribution within diverse cultural stateshas allowed the poorest indigenous people to improve their life quality with healthy food, higher salaries, and a better health system (ECLAC 2022).

Ceceña (2014:11) insisted that at the centre of their cosmovision is 'the reproduction of life as the means to understand the reproductions process as a whole and as an organizing criterion of economic issues, which could no longer be considered a divided dimension'. Cultural, social, economic, and religious activities are guided by this telluric approach (Arizpe 2004), where integrated HUGE security (Oswald-Spring 2020) orients socio-political, climate and ecological content in their daily life. Diverse Mesoamerican indigenous adopted similar ways of life as the Aymara in the Andes in the mountains of Guatemala and Chiapas. All of them had a limited capacity for adaptation and almost no governmental support either in the prevention or recovery process after an extreme event. The Mayan Zapatistas established after their upraising in 1994 an equilibrium between the requirements of nature and their communities. They recovered former grassland and transformed it into corn and bean fields, where they produced also vegetables. They promoted equal behaviour among women and men in the personal and family spheres. Systematic reforestation with native trees, limited extreme rainfalls during hurricanes. They established rainwater retention in the mountains to get water during the dry season. Indigenous Zapatistas in Chiapas live in peace within their community, whenever they are exposed to a low-ongoing war with the local landlords. In December 2022, 60 paramilitaries attacked Mayan Tzotziles from Las Abejas, killing 18 women, 20 children, and 17 men, while they were prying in a chapel (Hernández 2012). The military was nearby but did not intervene during the massacre, which could have been avoided because the indigenous were not armed.

In political terms and to avoid social stratification among the Zapatistas, the leadership rotates yearly and is without payment. Women, men, young and old, are equally elected as leaders for a year for 'leading by obeying'. The community cares about the well-being of her family during this momentary leadership. Special training is given to young girls and boys in order to qualify them for an alternative peaceful lifestyle with democratic and consensual decision-making processes (Zapatistas from 1994 on). Their participative governance (Gill 2020) has eliminated the exploitation of women by men and equity is a basic principle in decision-making and political leadership (González Casanova 2009). From their upraising on, no feminicide has occurred among their organized *caracoles* (shell models), compared with the rest of Mexico, where daily at least ten women are killed mostly by their intimal partners (INMUJER 2022). Their judicial system is transparent and neighbouring villages, not belonging to the EZLN, ask for judgments thanks to the impartiality of the local judges. Their conflicts are negotiated and their political organisation has evolved thanks to prevention, self-reliance, and care about the most vulnerable, including nature and ecosystem services. Potential offenders are integrated into their village, get councils by elderlies, and collaborate with the community to remedy the caused damage. The whole *caracol* involves in this remediation of conflict management (Menchú 2004). Historical background based on original myths, histories of life, and sustainable management of the environment is also trained in the Zapatista schools.

Twenty years later (2011) the Zapatistas tried to show their democratic process to the whole country with their March Colour of the Earth. But the interests of the oligarchy blocked systematically a greater democratization process and involvement of society until the election in 2018 (Sánchez 2021). In an evaluation of the results, the Zapatistas understood that the other people and communities were not yet ready for a radical change. Within a balance of their own achievements, they mentioned:

> We do not need them to survive. We, who never left even if we were determined to make believe the media of the whole spectrum, resurfaced as indigenous Zapatistas that we are and will be. In these years we have strengthened and we have significantly improved our living conditions. Our standard of living is higher than that of the indigenous communities related to the governments in turn, who receive finical support and waste it on alcohol and useless articles. Our homes have improved without hurting nature by imposing paths that are not sustainable. In our villages, the land that was used to fatten the cattle of farmers and landowners now is for corn, beans, and vegetables that illuminate our tables. (Zapatistas from 1994 on)

There is the last question to be answered: ¿who belongs to the Zapatistas, due to that in multiple communities indigenous are only partially integrated into the EZLN? Without any doubt, the initiators of the movement, who have acquired prestige and consolidated their positions as a local authority in the civil, political, or military spheres, express their leadership. The second groups are the indigenous people who lack in the present Mexican conditions any viable alternatives. They live today in a safer place without violence and internal colonialism, with better medical and educational support and transparency. Whenever they do not have a life of abundance and their precarious economic conditions are only partially mitigated by external aid, they have enough food. Their behaviour represents similar basic principles, such as what Nelson Mandela proposed for South Africa after its independence (Mandela 1995). A third group is the younger generation that was born and raised during the conflict. They have been educated in Zapatista schools and have consolidated an entrenched commitment to the values of the EZLN. Within this sector are also young women who grew up in the ranks of the Zapatists and assumed the discourse of feminine equity and independence. But their family members are often not belonging to the movement. They are pressuring these girls to abandon the EZLN and return to the traditional patriarchal habits of intrafamilial violence. Nevertheless, these young women are solidly trained to understand the lack of alternatives in the traditional way of life, and will not return to the discriminated life observed by their traditional families.

In regional terms and without any doubt, the EZLN has abolished the extractivism of mining, the capitalist agricultural production with transgenic, chemical fertilizers, and pesticides. They eradicated the monopolized distribution and commercialization in the hand of local landlords giving credits with usury interests and underpaying the indigenous handicrafts. The EZLN in their MAREZ (autonomous rebelling municipalities of Zapatistas) and *Caracoles* are also changing the waste generation brought in by the so-called developed countries or cities. Their key message is against the destructive capitalism that started with the colonisation of the region five centuries ago. They are confronted with no governmental support in their daily life, learning to care for others to achieve a dignified livelihood. Their effort in the mountains of Chiapas against exclusive globalization and dignified livelihood did not bring great wealth to their members, but each participant obtains enough food, safe education, a transparent judicial system with conflict prevention, and basic health services, where everybody is treated with dignity and respect.

In June 2021, five women and two men crossed on a raft for 47 days in the Atlantic and arrived in the Azores Islands in Portugal. Their trip represented a global struggle against capitalism also in Europe. It symbolized also 'a reverse crossing' making visible the former culture of conquest, colonialism, and exploitation, imposed 500 years by the Spaniards and still in the minds of many people (Zapatistas 2020).

10.8.1 Women's Care or Gift Economy

Taking up the initial research questions, patriarchy was described as a violent system with the exploitation of humankind and natural resources that has produced climate change (IPCC 2021), greater inequity (Oxfam 2022a, b), discrimination (World Bank 2022), wars all over the world (Reardon 1985), and invisibility of women in policy and economy (Ilesanmi 2018). After more than 500 years of colonialism and seven decades of neoliberal capitalism with an acceleration of the global economy (Fraser 2005), including the destruction of nature (Núñez et al. 2003), climate change (IPCC 2021), and the new era of Earth, the Patriacene, the Global South is highly affected

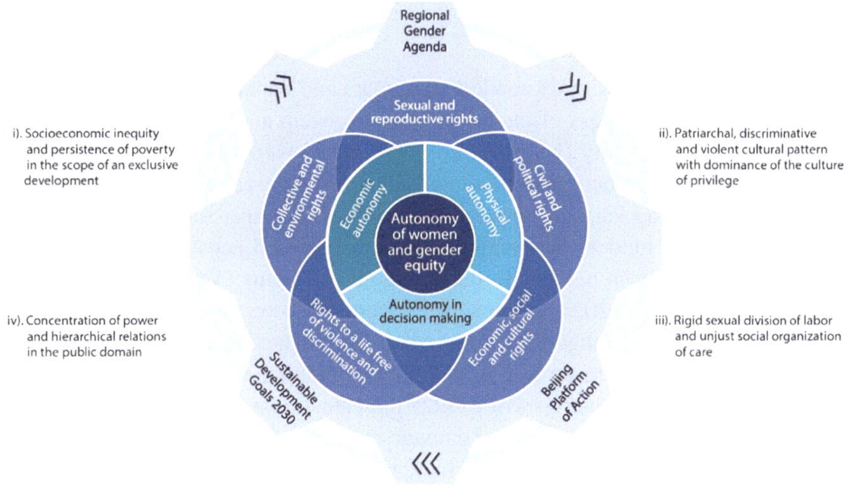

Fig. 10.2 Women's care economy. *Source* ECLAC (2021: 3) (translated by the author)

(Carbon Brief 2022). Poor indigenous, women, girls (Ilesanmi 2018), and men are searching for radical alternatives to the ongoing socioenvironmental destructive processes. ECLAC (2021) proposed a women's care economy for all discriminated people globally, including poor people in the Global North (Fig. 10.2).

This model empowers women and men, especially marginal ones, with physical, economic, political, and decision-making potential. Particularly, sexual and reproductive rights for any woman on earth are granted (Fraser 2008) and gender violence is reduced, especially the eradication of female genital mutilation (Chappell et al. 2021). A life without violence generates creativity and promotes drastic changes toward resilience in times of increasing climate change threats. These rights were developed in the past by women and men in the Beijing Platform for Action, reinforced by the UN Sustainable Development Goals of the Agenda 2030, and by Belem do Parra in Latin America. National and local policies of equity in policy and executive jobs, reinforcement of laws against gender violence, and equal salaries for similar work are basic political tools to limit patriarchal and discriminative behaviour against women, girls, the elderly, and youth. The limits to achieving these rights are not legally limited, but the obstacles are deeply embedded in the cultural androgenic understanding of life and the management of nature. They are reproduced inside the families by women and men, reinforced by all religions, and socially accepted mostly without punishments. Thus, in today's economic, social, and cultural crises, an alternative paradigm of women's and men's care economy is crucial to increase the recognition (Fraser 2005) and the empowerment of women, girls (OECD 2022), and poor people in the Global South.

10.9 Conclusions

Within this indigenous telluric knowledge, Aymara and Zapatistas communities have started a restoration process of the destroyed ecosystem and the ecological conservation of their biodiversity (Núñez et al. 2003). They were able to grant sufficient food for all their members and promoted an alternative educational system. For their participants, the learning process is clear: for marginal indigenous societies another alternative model is emerging, where nature and ecosystem services are understood as gifts from nature, similar to the gift economy (Vaughan 1997; 2004) or as recently ECLAC (2021) promoted the women's economy of care. Increasing climate change threats oblige these alternative approaches. In different parts of the world are emerging bottom-up activities in favour of human welfare and the recovery of Mother Earth. As the outcomes of COP27 indicate, financial tycoons are unable to pay the 100 billion dollars for loss and damage of the poor affected countries (Muiña/ Barros 2021), knowing that they have increased their wealth during the pandemic by trillions of dollars (Forbes 2021). Paying decent salaries to workers with social security occurs in industrialised countries, but not in the sweatshops of the Global South.

Limiting the exploitation of natural resources by open pit mining and regional and local restorations of ecosystems are out of the MNE's agenda, where only profit matters. They have forgotten the free ecosystem services such as clean water, fresh air, healthy soils, and biodiversity, which were basic demands by the civil society in Sharm-el-Sheikh. As these multinationals are destroying Mother Earth from the top down, only for enriching small elites. They challenge the bases of survival, while an increasing number of humans are exploring alternative ways. Only through a culturally diverse bottom-up care approach, humankind as a whole and the environmental biodiversity have a chance to survive. These alternatives may help highly vulnerable people to deal with increasingly more complex climate conditions. Globally, the lack of GHG reduction is also responsible for the upcoming threats of new pandemics (UN Women 2021), encapsulated in thick glaciers. There are also unknown emerging risks (WEF 2021), which may involve the disappearance of part or the entire humanity on Earth. To limit this catastrophic outcome, the Global South is reacting. Poor people account for abundant natural resources and traditional medical knowledge (Ahmad et al. 2021). Community-based collaboration can provide people with healthy food (Lowder et al. 2021). They are experimenting with radical changes in the traditional meritocratic educational system that has historically affected women (Powell 2016). Within an understanding of care or gift economy, an active participation of all women, girls, elderly, and men, based on HUGE security with sustainable peace (Oswald-Spring 2020), the planet may get some limited chances for specific regions. The rest of the military-financial-multinational-research policy-guided world will turn into irreversible tipping points (Steffen et al. 2018) with increasing more financial crises and collapses of the occidental welfare system.

The Global South, abandoned and exploited is exploring diverse alternatives with examples such as the women's care economy (ECLAC 2021), gift economy (Vaughan 1997, 2004), and sororities (Zaiser 2019). The remaining timeframe designed by IPCC (2021), the Emission Gap Report (UNEP 2021a), and the Carbon Brief (2022) are very short. They have insisted on multiple serious unreported disasters and deaths in the Global South, especially in Africa. Thus, the changes must occur in less than four decades. Modifications in the patriarchal dominant mindset of multinational male-led enterprises is almost impossible in this short time. However, also these extremely rich men have no better alternative than living on our common Earth. Our planetary system has the potential to recover when the remaining biodiversity is protected and their exploitative behaviour is abolished. As an example of why indigenous societies and women are crucial in this regard: five percent of indigenous people are protecting 80 per cent of the remaining biodiversity without any financial support (Raygorodetsky 2018). Thus, within a bottom-up approach by poor people, it is possible and urgent to act in the Patriacene before irreversible processes with unpredictable negative loops may exterminate numerous groups of people and valuable ecosystems.

References

Abram, N.; McGregor, H.; Gergis, J.; Phipps, S., 2016: "The Industrial Revolution kick-started global warming much earlier than we realized"; at: https://iceds.anu.edu.au/news-events/news/industrial-revolution-kick-started-global-warming-much-earlier-we-realised.

Ahikire, J., 2007: "Localised or localising democracy: Gender and the politics of decentralisation in contemporary Uganda" (Kampala: Fountain Publishers).

Ahmad, K.; Ahmad, M.; Huber, F.; Weckerle, C., 2021: "Traditional medicinal knowledge and practices among the tribal communities of Thakht-e-Sulaiman Hills, Pakistan", in: BMC Complementary Medicine and Therapies, 21,230; at: https://doi.org/10.1186/s12906-021-034 03-1.

Allen, A., 2011. "Feminist Perspectives on Power", Stanford Encyclopedia of Philosophy, at: http://plato.stanford.edu/entries/feminist-power/.

Álvaro García Linera, J. L., 2022: *La política como disputa de las esperanzas* (Buenos Aires: CLACSO).

Arizpe, L., 2004: *Los retos culturales de México* (Mexico City: Miguel Ángel Porrúa - CRIM/UNAM).

Arons, W., 2020: "Tragedies of the Capitalocene", in: Journal of Contemporary Drama; at: https://doi.org/10.1515/jcde-2020-0003.

Ashley, M., 1958: Man: His First Million Years (New York: New American Library).

Bailey, D., 2015: "The Environmental Paradox of the Welfare State: The Dynamics of Sustainability", in: *New Political Economy*, 20,6: 739-811.

Barker, G.; Goucher, C., 2015 (Eds.): *A World with Agriculture 12,000 BCE-500BCE* (Cambridge: Cambridge University Press).

Barnes, D. F., 1978: "Charisma and Religious Leadership: An Historical Analysis", in: *Journal for the Scientific Study of Religion*, 17,1: 1-18.

Bauer, J., 2015: *Climate Change, Human Systems, and Policy. Global Warming, Poverty, and Ethical Issues*, vol. 1 (Paris: Encyclopedia of Life Support Systems).

Behrokh, M. M.; Ardekani, A. M., 2011: "Genetically Modified Foods and Social Concerns", in: *Avicenna J. Med Biotechnol*, 3,3: 109-117.
Bendict, R., 1934: *Patterns of Culture* (New York: New American Library).
Bhatnagar, A.; Bhatnagar, R., 2022: *Desertification and Land Degradation. Concept to Combating* (Boca Raton: CRC Press).
Boas, F., 1940: Race, *Language and Culture* (New York: Free Press).
Bogaerts, E.; Raben, R., 2012 (Eds.): *Beyond Empire and Nation: The Decolonization of African and Asian societies, 1930s-1970s* (Boston: Brill).
Bosch, F.; Guerra Rondón, L., 2022: "Geopolítica de la transición energética: ¿qué asociaciones UE-ALC?", in: *Jornada de Estudio. Geopolítica de la transición energética* (Paris: 10 June).
Bradfield, L.; Vigaud, N., 2017: "Unraveling East Africa's Climate Paradox: Patterns and Mechanisms", in: Simon Wang, S.-Y.; Yoon, J.-Ho; Funk, C. C.; Gillies, R. R. (Eds.): *Climate Extremes. Patterns and Mechanisms* (New Jersey: Wiley and The American Geophysical Union); at: https://doi.org/10.1002/9781119068020; chap. 16.
Brauch, H. G.; Oswald-Spring, Ú.; Grin, J. et al. (Eds.), 2009: *Facing Global Environmental Change. Environmental, Human, Energy, Food, Health and Water Security Concepts* (Berlin: Springer).
Brown, T.; Jones, M.; Wayne, P.; *et al.*, 2011: The complex origins of domesticated crops in the Fertile Crescent (New York: University Books).
Brundtland Commission, [World Commission on Environment and Development], 1987; *Our Common Future* (Oxford: Oxford University Press).
Carbon Brief, 2022: "Analysis: Africa's unreported extreme weather in 2022 and climate change", at: https://www.carbonbrief.org/analysis-africas-unreported-extreme-weather-in-2022-and-climate-change/%23:%7E:text=Carbon%20Brief%92s%20analysis%20of%20disaster,since%20the%20start%20of%202022.
Ceceña, A. E., 2014: "Del desarrollo al 'vivir bien': la subversión epistémica", in: Girón, A. (Ed.): *Del "vivir bien" al "buen vivir", entre la economía feminista, la filantropía y la migración: hacia la búsqueda de alternativas.* Mexico City: IIEc-UNAM: 11-22.
Chappell, A. G.; Sood, R.; Hu, A., Folsom, S. M.; Wafford, Q. E.; Sumanas, W.; Bowen, D. K.; Post, L. A.; Jordan; Sumanas. W., 2021: "Surgical management of female genital mutilation-related morbility: A scoping review", in: *Journal of Plastic, Reconstructive & Aesthetic Surgery*, 74, 10 (October): 2467-2478.
Childe, G. V., 1936: *Man Makes Himself* (London: Watts & Co).
Civilization Symposium Series (New York: McGraw Hill).
Collin, L., 2020: "Economía local y diversa, una opción resiliente, sustentable de trabajo digno", in: Oswald-Spring, Ú. et al. (Eds.): *Transformando al mundo y a Méx*ico (Cuernavaca: CRIM-UNAM): 117–140.
Colt, S.; Knapp, G., 2016: "Economic Effects of an Acidification Catastrophe", in: *American Economic Review*, 106,5: 615-619.
COVID-19 Forecasting Team, 2022: "Variation in the COVID-19 infection–fatality ratio by age, time, and geography during the pre-vaccine era: a systematic analysis", in: *Lancet*, 399: 1469–1488; at. https://doi.org/10.1016/S0140-6736(21)02867-1.
Crawley, E., 1902: *The Mystic Rose* (2 vols.) (New York: Meridian).
Crutzen, P. J., 2002: "Geology of Mankind", in: *Natur*e, 415: 6867: 23.
Crutzen, P. J.; Brauch, H. G., (Eds.), 2016: *Paul J. Crutzen: A Pioneer on Atmospheric Chemistry and Climate Change in the Anthropocene* (Cham: Springer International Publishing).
Crutzen, P. J.; Stoermer, E. F., 2000: "The 'Anthropocene'", in: *Global Change Newsletter*, 41: 17.
D'Eaubonne, F., 1974: *Le Féminisme ou la Mort* (Paris: Pierre Horay).
Dalby, S., 2017: "Anthropocene Formations: Environmental Security, Geopolitics and Disaster", in: *Theory, Culture, and Society*, 34, 2–3; at: https://doi.org/10.1177/0263276415598629.
Dalby, S.; Brauch, H. G.; Oswald-Spring, Ú. 2008: "Towards a Fourth Phase of Environmental Security": in: Brauch, H. G. et al. (Eds.); *Facing Global Environmental Change: Environmental, Human, Energy, Food, Health and Water Security Concepts* (Heidelberg: Springer): 787–796.

Daya, A., 2022: "A Crisis of Consent in UN Peace Operations"; at: https://reliefweb.int/report/democratic-republic-congo/crisis-consent-un-peace-operations.
De Sena, A., 2014 (Ed.): *Las políticas hechas cuerpo y lo social devenida emoción: lecturas sociológicas de las políticas sociales* (Buenos Aires: UBA).
Diamond, J., 1999: *Guns, Germs, and Steel: The Fates of Human Societies* (New York: Norton & Co)
Dinucci, M., 2022: "Ucrania: todo estaba escrito en el plan de la RAND Corporation"; at: https://cronicon.net/wp/ucrania-todo-estaba-escrito-en-el-plan-de-la-%E2%80%8Erand-corporation/.
ECLAC, 2018: *Los pueblos indígenas en América* (Abya Ayala) (Santiago: ECLAC).
ECLAC, 2020: *Facing the challenge of COVID-19* (Santiago: ECLAC).
ECLAC, 2021: *Economía del Cuidado* (Santiago: ECLAC).
ECLAC, 2022: *Los Impactos Socioeconómicos de la Pandemia COVID-19 en América Latina y El Caribe* (Santiago: ECLAC).
Escudero, J. C., 2016: "Estrategias de supervivencia y adaptación", in: *Red Sociales, Revista del Departamento de Ciencias Sociales*, 4, 1: 04–06.
FAO, 2022: *Food Outlook. Biannual Report on Global Food Markets* (Rome: FAO).
Farrington, B., 1949: *Greek Science*, 2 vols. (Harmondsworth: Penguin).
Forbes 2021: "How Much Money America's Billionaires Have Made During the Covid-19 Pandemic" (30 April); at: https://www.forbes.com/sites/chasewithorn/2021/04/30/ameri-can-billionaires-have-gotten-12-trillion-richer-during-the-pandemic/?sh=73811122f557.
Fortune, R. F., 1932: *Sorcerers of Dobu: The Social Anthropology of the Doby Islanders of the Western Pacific* (New York: Dutton).
Fraser, N., 2005: "Reframing Justice in a Globalizing World", in: New Left Review, 36: 1- 19.
Fraser, N., 2008: "Social Justice in the Age of Identity Politics: Redistribution, recognition and participation", in: Henderson, G. L.; Waterstone, M. (Eds.): *Geographic Thought: A Praxis Perspective* (London: Taylor & Francis): 72–89.
Frazer, J. G., 1907–1915. *The Golden Bough*, 12 vol. (London: Macmillan).
Fuller, D., 2011: "Pathways to Asian Civilizations: Tracing the Origins and Spread of Rice and Rice Cultures", in: *Rice*, vol. 4: 78-92.
Gaard, G., 2011: "Ecofeminism Revisited: Rejecting Essentialism and Re-Placing Species in a Material Feminist Environmentalism", in: *Feminist Formations*, 23, 2 (Summer): 26–53.
Gaddis, J. L., 2014: *The Cold War: A New History* (Harmondsworth: Penguin Books).
García Linera, A.; Luis, J., 2022: *La política como disputa de las esperanzas* (Buenos Aires: CLACSO).
Gill, S., 2020: "Global Governance 'As it was, is, and Ought to B' - A Critical Reflection", in: *Foro Internacional*, 40, 4: 1261–1293; at: https://doi.org/10.24201/fi.v60i4.2800.
Girón, A. (Ed.): Del "vivir bien" al "buen vivir", entre la economía feminista, la filantropía y la migración: hacia la búsqueda de alternativas (Mexico City: IIEc-UNAM): 11–22.
Global Carbon Atlas, 2021: https://www.fs.usda.gov/ccrc/tool/global-carbon-atlas-.
Global CCS Institute, 2019; *Bioenergy and Carbon Capture and Storage*; at: https://www.globalccsinstitute.com/wp-content/uploads/2019/03/BECCS-Perspective_FINAL_PDF.pdf.
González Casanova, P. 2003: *Colonialismo interno. Una redefinición* (Mexico City: IIS-UNAM).
González Casanova, P., 2009: *Zapatismo Movimientos sociales. Movimientos indígenas. Movimiento Zapatista EZLN* (Buenos Aires: CLACSO).
Görg, C.; Plank, C.; Wiedenhofer, D.; *et al.*, 2020: "Scrutinizing the Great Acceleration: The Anthropocene and its analytic challenges for social-ecological transformations", in: *The Anthropocene Review*, 7,1: 42–61.
Government of Sierra Leone, 2007: *National Adaptation programme of Action (NAPA)*; at: https://unfccc.int/resource/docs/napa/sle01.pdf.
Grin, J.; Rotmanns, J.; Schot, J., 2010: *Transitions to Sustainable Development. New Directions in the Study of Long Term Transformative Change* (New York - London: Routledge).

Grübler, A. 1995. "Industrialization as a Historical Phenomenon", in: Socolow, R.; Andrews, C.; Berkhout, F.; Thomas, V. (Eds.): *Industrial Ecology and Global Change* (Cambridge: University Press): 43–68.

Habermas, J, 2000: *La constelación posnacional: ensayos políticos* (Barcelona: Paidós).

Haddon, A. C. 1934: *History of Anthropology* (London: Watts).

Hamilton, E., 1940: *Mythology* (New York: New American Library).

Haraway, D. J., 2016: *A Cyborg Manifesto Science, Technology, and Socialist-Feminism in the Late Twentieth Century* (Minneapolis: University of Minnesota Press. ProQuest Ebook Central).

Haraway, D. J., 2020: *Seguir con el problema: Generar parentesco en el Chthuluceno* (El origen del mundo nº 1), (Bilbao: Consonni).

Hernández Navarro, L., 2012: "Acteal: impunidad y memoria", in: *El Cotidiano*, 172 ((Mexico City: UAM Azcapozalco, March-April): 99–115.

Hernando, R. C., 2022: "Multiple headwinds derail recovery", in: *APEC Regional Trends Analysis* (February).

Forbes 2021: "How Much Money America's Billionaires Have Made During the Covid-19 Pandemic" (30 April).

IGAD-ICPAC, 2022; *Agriculture Hotspots* (August); at: Agriculture_Hotspots_August_2022_9iN1Txx.pdf (icpac.net).

Ilesanmi, O.O., 2018. "The Invisibility of Women's Organizarion in Decision Making Processes in Africa and Its Implications", in: *Front. Sociol*; at: https://doi.org/10.3389/fsoc.2018.00038

INEGI, 2022: *Encuesta Nacional de Ingresos y Gastos de los Hogares Estacional 2020 Primer trimestre* (Aguascalientes: INEGI).

INMUJERES, 2022: El INMUJERES en 2021 (Mexico City: INMUJERES).

IPCC, 2021: Sixth Assessment Report, Climate Change 2021: The Physical Science Basis, the Working Group I (Geneva: IPCC).

IPCC, 2023: *AR Sixth Synthesis Report, Climate Change 2023* (Geneva: IPCC); at: https://www.ipcc.ch/report/sixth-assessment-report-cycle/.

İşeri, E., 2020: *From the security dilemma to the human-centered survival dilemma in the Eastern Mediterranean: The environmental security approach* (Istanbul: H. Böll Foundation).

Jackson, W.; Jensen, R.; 2022: "The politics of overconsumption and getting the right scale"; at: https://www.commondreams.org/views/2022/09/14/politics-overconsumption-and-getting-scale-right.

Jay, P., 1963: "The Female Primate", in: Seymour, M.; Faber, M.; Wilson, R. (Eds.): *Man and Civilization Symposium Series*. New York: McGraw Hill.

Kallen, J.; Johnson, S., 2022: "Europe was once green and water-rich. Now, it's more and more like California"; at: https://www.latimes.com/world-nation/story/2022-09-04/drought-europe-climate-change.

Kirk, R., 2020: "Green Capitalism – Greenwashing over Sustainability?"; at: https://89initiative.com/green-capitalism-greenwashing-over-sustainability/.

Kragh, H., 2000: "Max Planck: the reluctant revolutionary", in; *Physic World*, 13, 12: 31–36; at: https://doi.org/10.1088/2058-7058/13/12/34.

Kyove, J.; Streltsova, K.; Odibo, U.; Cirella, G.T., 2021: "Globalization Impact on Multinational Enterprises", in: *World*, vol. 2: 216–230; at: https://doi.org/10.3390/world2020014.

Li, J. V.; Cheong, T. S.; Xunpeng, S., 2021: "A Counterfactual Baseline for Assessing Future Environmental Impact: A Case Study of the Belt and Road Initiative, in: *Frontiers in Environmental Science*, 9, 724095 (October); at: https://doi.org/10.3389/fenvs.2021.724095.

Lowder, S. K.; Sánchez, M. V.; Bertini, R., 2021. "Which farms feed the world and has farmland become more concentrated?", in: *World Dev.*: 142:105455; https://doi.org/10.1016/j.worlddev.2021.105455.

MA, 2005: Millennium Ecosystem Assessment. Ecosystems and the Human Well-Being (Washington: Island Press).

Maldonado, C. E., 2017: "¿Ciencias sociales cuánticas?", in: *Le Monde Diplomatique*, 165 (April): 34-35.

Malhi, Y., 2017: "The Concept of the Anthropocene", in: *Annu. Rev. Environ. Resour*, 42: 77–104.
Mandela, N., 1995: *Long Walk to Freedom: The Autobiography of Nelson Mandela* (London: Abacus).
Markandya, A. et al., 2021: "Covid, the Environment and Food Systems: Contain, Cope and Rebuild Better", in: *Frontiers in Environmental Science*, 9, 674432; at: https://doi.org/10.3389/fenvs.2021.674432.
Markandya, A.; Salcone, J.; Hussain, S.; Mueller, A.; Thambi, S., 2021: "Covid, the Environment and Food Systems: Contain, Cope and Rebuild Better", in: *Frontiers in Environmental Science*; 9, 674432; at: https://doi.org/10.3389/fenvs.2021.674432.
Mead, M., 1935: *Sex and Temperament in Three Primitive Societies* (New York: Morrow).
Mechler, R.; Bouwer, L.M.; Schinko, T.; Surminski, S.; Linnerooth-Bayer, J., 2019: *Loss and Damage from Climate Change. Concepts, Methods and Policy Options* (Cham: Springer Nature).
Meillasoux, C., 1977: "Mujeres, graneros y capital", in: *Nueva Antropología*, 2, 8 (April).
Menchú, R., 2004: "Culturas Indígenas, Cosmovisión y Futuro", in: Oswald-Spring, Ú. (Ed.): *Resolución noviolenta en sociedades indígenas y minorías* (Mexico City: Coltax): 49–62.
Mies, M., 1985: *Patriarchy & Accumulation on a World Scale. Women in the International Division of Labour* (London: Zed Books).
Mies, M., 1993: *Ecofeminism* (Blackpoint: Fernwood Publications).
Minakov, M.; Kusa, I., 2022: "Ukraine-Russia Negotiations: What's Possible?"; at: https://www.wilsoncenter.org/blog-post/ukraine-russia-negotiations-whats-possible.
Moleón, M., Sánchez-Zapata, J. A.; Margalida, A.; Carrete, M.; Owen-Smith, N.; Donázar, J. A., 2014: "Humans and Scavengers: The Evolution of Interactions and Ecosystem Services", in:_ *Bioscience*, 64, 5 (May): 394–403; at: https://doi.org/10.1093/biosci/biu034.
Moore, J.W.; Parenti, M.; Christ, E. et al., 2016: *Anthropocene or Capitalocene? Nature, History, and the Crisis of Capitalism* (Oakland: PM Press).
Moore, Jason W. 2017: "The Capitalocene, Part I: on the nature and origins of our ecological crisis", in: *The Journal of Peasant Studies*; at: https://doi.org/10.1080/03066150.2016.1235036.
Moreno, J. M.; Laguna, C.; Barrios, V. et al., 2020: *RIOCCCADAPT Report* (Madrid: McGraw-Hill).
Morgan, L., 1877: *Ancient Society* (Chicago: Kerr).
Muiña, F.; Barros, H., 2021: "Etiopía: entre la sangre y las urnas", in: OPINIÓN; at: https://nuso.org/articulo/etiopia-urnas-y-sangre/.
Naess, A., 1986: "The Deep Ecological Movement: Some Philosophical Aspects", in: *Philosophical Inquiry*, vol. 8: 1-14.
Nakate, V., 2021: A Bigger Picture: My Fight to Bring a New African Voice to the Climate Crisis (Boston: Houghton Mifflin Harcourt).
National Academies *of Sciences, Engineering, and Medicine*, 2022: Biodiversity at Risk: Today's Choices Matter (Washington: The National Academies Press).
Nello, O.; Blanco, I.; Goma, R. (Eds.), 2022: *El apoyo mutuo en tiempos de crisis. La solidaridad ciudadana durante la pandemia Covid-19* (Buenos Aire: CLACSO).
Núñez, I.; González-Gaudiano, E.; Barahona, A., 2003: "Biodiversity: History and context of a concept", in: *Interciencia*, 287, 7: 387–393+426.
OECD, 2022: "How fragile contexts affect the well-being and potential of women and girls"; in: OECD (Ed.): *OECD Development Co-operation Directorate* (Paris: OECD).
ONU, 2020: *América Latina y el Caribe: La segunda región más propensa a los desastres* (New York: UN); at: https://news.un.org/es/story/2020/01/1467501.
OPEP, 2006: "The In-Salah CCS experience Sonatrach, Algeria", in: *The First International Conference on the Clean Development Mechanism* (Riyadh, Saudi Arabia: 19–21 September).
Oswald-Spring, Ú., 2013: "Dual vulnerability among female household heads", in: *Acta Colombiana de Psicología*, 16,2: 19-30.
Oswald-Spring, Ú., 2016: "Development with Sustainable-Engendered Peace: A Challenge during the Anthropocene", in: Brauch, H. G.; *et al.* (Eds.): *Sustainability Transition and Sustainable Peace Handbook* (Cham: Springer International: 161–186)

Oswald-Spring, Ú., 2020: *Earth at Risk in the 21st Century. Rethinking Peace, Environment, Gender, and Human Water, Health, Food, Energy Security, and Migration* (Cham: Springer Nature).
Oswald-Spring, Ú.; Brauch, H. G., (Eds.), 2021: *Decolonising Conflicts, Security, Peace, Gender, Environment and Development in the Anthropocene* (Cham: Springer).
Oswald-Spring, Ú.; Serrano-Oswald; S. E.; Estrada, A. et al., 2014: *Vulnerabilidad Social y Género entre Migrantes Ambientales* (Cuernavaca: CRIM-DGAPA-UNAM).
Our World in Data, 2022: "Who emits the most CO2 today?".
Oxfam, 2022a: Inequality kills. The unparalleled action needed to combat unprecedented inequality in the wake of COVID-19 (London: Oxfam).
Oxfam, 2022b: Carbon Billionaires. The investment emissions of the world's richest people (London: Oxfam).
Palik, J.; Rustad, S. A.; Methi, F., 2020: *Conflict Trends in Asia, 1989–2019*, PRIO Paper (Oslo: PRIO).
Payen, F. T.; Evans, D.L.; Falagán, N. et al., 2022: "How much food can we grow in urban areas? Food production and crop yields of urban agriculture: A meta-analysis", in: *Earth's Future*; at: https://doi.org/10.1029/2022EF002748.
Pearsall, D. M., 2015: "Early agriculture in the Americas", in: Barker, G: Goucher, C. (Eds.). A *World with Agriculture 12,000 BCE-500BCE* (Cambridge: Cambridge University Press).
Peruzzi, M.; Terzi, A., 2021: "Accelerating Economic Growth: The Science beneath the Art", in: *Economic Modelling*, vol. 103.
Powell, S., 2016: *Gender Equality and Meritocracy. Contradictory discourses in the academy* (Uppsala: Swedish University of Agricultural Science).
Prigogine, I., 1983: "La evolución de la complejidad y las leyes de la naturaleza", in: Lazlo, F.; Biermann, J. (Eds.): *Goals in a Global Community* (New York: Pergamon Press): 221–304.
Raiklin, E., 1993: "The Disintegration of the Soviet Union", in: *Intern. Journal of Social Economics*, 20, 3–4: 2–132; at: https://doi.org/10.1108/EUM0000000000522.
Raygorodetsky, G., 2018. "Indigenous peoples defend Earth's biodiversity—but they're in danger", in: National Geographic (16 November); at. https://www.nationalgeographic.com/environment/2018/11/can-indigenous-land-stewardship-protect-biodiversity-/.
Reardon, Betty, 1985; *Sexism and the War System* (New York: Syracuse University Press).
Reed, Evelyn, 1975: *Women's Evolution. From matriarchal clan to patriarchal family* (New York: Pathfinder Press).
Richards, A.I., 1932: *Hunger and Work in a Savage Tribe: A Functional Study of Nutrition among the southern Bantu* (Cleveland: World).
Sánchez, M., 2021: "México. 20 años de la Marcha del Color de la Tierra, que marcó 'un antes y un después' en el zapatismo"; at: https://www.resumenlatinoamericano.org/2021/02/07/mexico-20-anos-de-la-marcha-del-color-de-la-tierra-que-marco-un-antes-y-un-despues-en-el-zapatismo/.
Saxena, P.; Mishra, A.; Malik, S., 2012: "Surrogacy: Ethical and Legal Issue", in: *Indian J. Community Med.*, 37,4: 211-213.
Schaeffer, F. A., 2022: *Unsettled Borders: The Militarized Science of Surveillance on Sacred Indigenous Land* (Durham: Duke University Press).
Senath, E. Rt., 2013: "Gender, Governance and Participatory Development in Uganda: Lessons for the Local Government", in: Prah, M. (Ed.): *Insights into Gender Equality, Equality and Power Relations in Sub-Saharan Africa* (Addis Ababa: Fountain Publishers): 23–41.
Serrano-Oswald, S. E., 2021 (Ed.): *Diseño de una metodología triangulada de indicadores cualitativos y cuantitativos, que evalúe la prevalencia de la violencia política por razón de género en México y su impacto en el ejercicio de los derechos políticos de las candidatas a puestos de elección popular en el proceso electoral federal 2020-2021* (Cuernavaca: CRIM-UNAM-INE).
Simionescu, M.; Schneider, N.; Gavurova, B., 2022: "Decarbonized Energies and the Wealth of Three European Nations: A Comparative Nexus Study Using Granger and Toda-Yamamoto Approaches", in: *Front. Environ. Sci.*, 9: 817982; at: https://doi.org/10.3389/fenvs.2021.817982.
SIPRI, 2022: "World military expenditure passes $2 trillion for first time".

Stefanovic, A. F. (Ed.), 2022: *Caring in times of COVID-19: a global study on the impact of the pandemic on care work and gender equality* (Santiago: ECLAC).
Steffen, W. et al., 1998. "Trajectories of the Earth System in the Anthropocene", in: *PNAS*, 115, 33: 8252–8259; at. https://doi.org/10.1073/pnas.1810141115.
Stevenson, N., 2020: "Critical theory in the Anthropocene: Marcuse, Marxism and ecology", in: *European Journal of Social Theory*, 242: 136843102096272; at: https://doi.org/10.1177/1368431020962726.
Stiglitz, J., 2007: *Globalisation and its Discontent* (New York: W.W. Norton).
Stiglitz, J., 2010: *Freefall: America, Free Markets, and the Sinking of the World Economy* (New York: W.W. Norton).
Subramanian, K.R., 2018. "The Crisis of Consumption of Natural Resources", in: *International Journal of Recent Innovations in Academic Research*, 8,2: 2635-3040.
Terrazas, J.; Blitchtein, D., 2022: "Rural–urban migration as a factor associated with physical and sexual intimate partner violence Peru 2015–2017: a secondary analysis of a national study", in: *BMC Women's Health*, 22, 67; at: https://bmcwomenshealth.biomedcentral.com/articles/10.1186/s12905-022-01648-7.
The Economist, 2022: "A broken idea", in: *The Economist*, Special Report: ESG investing. (23 July).
The Economist, 2022a: "A new macroeconomic era is emerging. What will it look like? A great rebalancing between governments and central banks is under way", in: *The Economist*, 6 October.
The Economist, 2022b: "The Indian economy is being rewired. The opportunity is immense and so are the stakes". in: *The Economist* (13 May, Special Issue); at: https://www.economist.com/leaders/2022/05/13/the-indian-economy-is-being-rewired-the-opportunity-is-immense.
Tylor, E.B., 1881: *Anthropology* (Ann Arbor: University of Michigan Press).
Ülgen, S. (Ed.), 2022: *Rewiring Globalization* (Brussel: Carnegie Endowment for International Peace).
UN Women, 2021: Beyond COVID-19: *A feminist plan for sustainability and social justice* (New York: UN Women).
UNEP, 2021a: *Climate action holds key to tackling global conflict* (Nairobi: UNEP).
UNEP, 2021b: *The Emissions Gap Report* (Nairobi: UNEP).
UNEP, 2022a: *Human Security Report 2022* (New York: UNEP).
UNEP, 2022b: *The Closing Window Climate crisis calls for rapid transformation of societies. The Emissions Gap Report 2022* (Nairobi: UNEP).
UNEP, 2022c: "Pollution Action Note – Data you need to know", at: https://www.unep.org/interactive/air-pollution-note/?gclid=CjwKCAiAjs2bBhACEiwALTBWZS150zYdwEw4I70t00D5G1SUPE8Uu3PIjFY1UhdjoWYiH0sOQdFPKhoCnA0QAvD_BwE.
Vaughan, G., 1997: *For-Giving: A Feminist Criticisms of Exchange* (Austin: Plain View Press).
Vaughan, G., 2004: *The Gift; Il Dono*. New Series 8. (Meltemi: University of Bari).
Wagler, R., 2011: "The Anthropocene Mass Extinction: An Emerging Curriculum Theme for Science Educators", in: *The American Biology Teacher*, 73, 2: 78–83; at: https://doi.org/10.1525/abt.2011.73.2.5. S2CID 86352610.
Wang, X.C., 2022: "Safe water reuse through a quasi-natural water cycle", in: *Water Reuse*; at: http://iwaponline.com/jwrd/article-pdf/doi/https://doi.org/10.2166/wrd.2022.039/1079986/jwrd2022039.pdf.
Warren, K.J. (Ed.), 1997: *Ecofeminism. Women, Culture Nature* (Bloomington: Indiana University).
WEF [World Economic Forum], 2021: *The Global Risk Report 2021* (Davos: WEF).
Weiss, Thomas, 2022: "UN Security Council is powerless to help Ukraine – but it's working as designed to prevent World War III".
Westing, A.H., 2013: *Texts on Environmental and Comprehensive Security* (Cham: Springer).
Wolfe, A.P. et al., 2013: "Stratigraphic expressions of the Holocene–Anthropocene transition revealed in sediments from remote lakes", in: *Earth-Science Reviews*, 116 (January): 17-34.
World Bank, 2022: *Nearly 2.4 Billion Women Globally Don´t Have Same Economic Rights as Men* (Washington: World Bank).

Zaiser, H., 2019: *Sororidad 2.0. La lucha feminista y sus nuevas dimensiones en la actualidad mexicana* (Mexico City: F. Heinrich Böll).
Zapatistas, 1994–2022: "EZLN: Ejército Zapatista de Liberación Nacional"; at: http://www.ezln.org.mx/.
Zapatistas, 2020: "Una Montaña en Alta Mar" (5 October); at: http://enlacezapatista.ezln.org.mx/2020/10/05/sexta-parte-una-montana-en-alta-mar/.

Conclusions and Outlook: The Anthropocene in Political and Social Science: Integrated Concepts in Geography, Global History and Planetary Politics

Hans Günter Brauch[1]

Key Messages of the Ten Chapters in This Book[2]

Crutzen's concept of the 'Anthropocene', which he coined in 2000, and his thesis of [hu]mankind as a "geological (f)actor" (2002) has so far hardly been addressed by policy-makers, and there are only a few references in policy fields and by political institutions. The Anthropocene has not yet reached the mainstream of Political Science and International Relations research. Social scientists noted this discussion, that occurred first in the research networks of global environmental change and Earth systems science, as well as in the geological research communities—and has been debated nearly exclusively by natural scientists.

Two decades later the concept was taken up in three policy reports in Germany by the German Advisory Council on Global Change (WBGU), in the EU by the European Environment Agency (EEA) and in the UN system by the UNDP. Thus, the Anthropocene has become an item in environment and development policy. But the concept has not been addressed by top national, supranational and international policymakers and decision-making bodies. In Germany the concept was explored by the (WBGU) in its 2020 flagship report *Rethinking Land in the Anthropocene: from Separation to*

[1] Hans Günter Brauch has been chairman of the non-profit scientific society Peace Research and European Security Studies (AFES-PRESS) since 1987 and since May 2020 chairman of the board of the Hans Günter Brauch Foundation for Peace and Ecology in the Anthropocene (HGB Foundation).

[2] The author is grateful to Simon Dalby (ret. from Wilfried Laurier Univ., Waterloo, Canada), Czeslaw Mesjasz (Cracow Economic Univ., Poland), and Georg Simonis (ret. from FU Hagen, Germany) for comments and suggestions.

H. G. Brauch (ed.), *Towards Rethinking Politics, Policy and Polity in the Anthropocene*,
The Anthropocene: Politik—Economics—Society—Science 35,
https://doi.org/10.1007/978-3-031-71807-6

Integration[3] and in the United Nations in the UNDP's Human Development Report 2020 on *The Next Frontier: Human Development and the Anthropocene.*[4] The EEA addressed it in March 2023 in a briefing on *Exiting the Anthropocene? Exploring fundamental change in our relationship with nature*[5]:

> The ways in which societies, institutions and citizens relate to and value nature have played a key role in the interconnected biodiversity, climate change, natural resource and health crises we face. This briefing explores how to reframe the relationships between humans and nature. It examines how holistically understanding humans' deep interconnection with other life forms and ecosystems could lead to new motivations to protect nature and accelerate the societal transformation we need to live well within the limits of the planet.

This book, with 10 chapters, offers a wide diversity of disciplines, scientific approaches and themes addressed by its nine authors. Their key messages reflect their individual perspectives on the Anthropocene. They contribute to an agenda-setting and a continued need to rethink the impact of the Anthropocene on a reconceptualising of politics, policy and polity. These chapters address a variety of themes and methods from different disciplinary perspectives, moving "towards rethinking politics, policy and polity in the Anthropocene". However, they are not systematically assessing whether a rethinking in Political Science and International Relations can be observed in the Anthropocene since the middle of the twentieth century, nor do they try to reconceptualise these terms in light of the Anthropocene epoch of Earth history and whether living in the Anthropocene has resulted in a new 'second *Sattelzeit*' (Geulen 2010) that suggests new basic concepts and revised meanings of these basic concepts. Both have been suggested by a few authors in the history of science, while other scholars and social thinkers proposed that humanity needs a "new enlightenment" period.

The first two chapters by *Hans Günter Brauch*—who studied political science, modern history and international law and specialised as a social scientist in international relations—focused on: "Towards Rethinking Politics, Policy and Polity in the Anthropocene: Dual Context and Threat to Survival in the Peace and Security versus The Anthropocene Narrative" and on "The Anthropocene and International Political Order: Towards an Integrated Analysis".

Brauch discussed the Anthropocene from a dual perspective of both international relations as well as peace and security studies and from ecology, trying to combine Crutzen's "geological time" with Braudel's socio-historical time of the *longue durée.* By focusing on changes in international order he tried to build bridges between the

---

[3] WBGU, 2020: *Rethinking Land in the Anthropocene: from Separation to Integration* (Berlin: WBGU); at: https://www.wbgu.de/fileadmin/user_upload/wbgu/publikationen/hauptgutachten/hg2020/pdf/WBGU_HG2020_en.pdf.

[4] UNDP, 2018: *UNDP Human Development Report 2020: The Next Frontier: Human Development and the Anthropocene* (New York: UNDP); at: https://digitallibrary.un.org/record/3895090?ln=en&v=pdf or https://hdr.undp.org/content/human-development-report-2020

[5] EEA, 2023: Briefing: *Exiting the Anthropocene? Exploring fundamental change in our relationship with nature* (Copenhagen: EEA); at: https://www.eea.europa.eu/publications/exiting-the-anthropocene.

"peace and security narrative" in the context of the social sciences and the "Anthropocene narrative" based on knowledge from global environmental change and Earth systems science on the other. Thereby he aimed at a combined "peace (and) ecology" approach focusing on two existential anthropogenic threats posed by nuclear weapons and human action and agency as a "geological actor" resulting in manifold societal outcomes of human activities on the Earth system through extreme weather events, sea level rise, temperature rise and change in precipitation, as well as in societal and (geo)political) tipping points.

While from the peace and security perspective the threat is posed by the other (opponent, enemy), from an ecological perspective we (humankind) are the key threat to the survival of our species. While the first perspective has been legitimised from a Hobbesian worldview by the development of threat with military power through armaments and weapons of mass destruction and by conflict resolution and peace building with cooperative diplomatic means, the alternative ecological approach requires global cooperation in containing the physical effects of global environmental change (by means of adaptation, mitigation, resilience) and in addressing the threats posed by the burning of fossil fuels and by biodiversity loss.

Hans Günter Brauch's key messages in Chap. 1 on "Towards Rethinking Politics, Policy and Polity in the Anthropocene: Dual Contextual Change in the Peace and Security versus The Anthropocene Narrative" addressed the linkages between GHG emissions and three stages of post-World War II international order in a bipolar (Cold War, 1947–1989), unipolar (1990–2022) and an emerging multipolar (since 2022) global policy context in which two fundamentally different anthropogenic existential dangers have existed: weapons of mass destruction in the nuclear age and environmental survival threats posed by extreme weather events due to global climate change and potential geopolitical tipping points.[6]

In the Anthropocene the necessary rethinking of *Politik* (*politique, la Politica*) and of 'politics', 'policy' and 'polity' should move in the second, post-1945, *Sattelzeit* to a reconceptualisation of these key concepts in both policymaking and in Political Science and International Relations, where policy problems should be addressed simultaneously from a peace and security perspectives and from ecological concerns. Thus, developing a peace ecology will become an intellectual challenge and ecological peace policies a crucial political task. This will gradually require a decoupling from national sovereignty and an increased emphasis on planetary boundaries.

In Chap. 2 on "The Anthropocene and International Political Order: Towards an Integrated Analysis", Hans Günter Brauch concluded that

> International politics and the study of international relations have reached a turning point with the collapse of a cooperative security and peace system in Europe and with a possible transition towards an authoritarian century (Menzel 2023) with a possible return to ColdWar-type Hobbesian geopolitical rationales and mind-sets which may also reduce the willingness to move ahead with cooperative policies involving all major GHG-emitting countries. The US strategy to persuade and force its partners to reduce their economic and political cooperation with China may result in changes in supply lines. US domestic politics has been, is and

[6] The conclusions in the 10 chapters were provided by the respective authors and slightly adapted for this text summary.

> will remain fragile and relatively unpredictable with regards to climate policies and its policies towards China and Russia. Hobbesian geopolitical mind-sets could distort the global priority goals to be achieved in the realm of cooperative global climate policy. In the worst case the prediction of the Stern Report (2006) that the costs for the violent impacts of the physical consequences on humankind may rise for the next generations will come true. This necessitates research on planning for the outcome of the present wars in Europe and the Middle East, including diplomatic initiatives of nations and mediations among the warring parties and their supporters.

The third chapter, by *Czeslaw Mesjasz* who has a background in physics and management and works at the Cracow University of Economics, on "The Anthropocene and Complexity", offered "a Survey of Ideas":

> Environmental studies, including the Anthropocene, have been influenced by complexity-related ideas, e.g., equilibrium, feedback, emergence, self-organisation, chaos, the edge of chaos. This refers to mathematical and simulation models, and complexity-related analogies and metaphors. They help study the Anthropocene and the relationships between human systems and nature. Future studies of complexity-related ideas in research and policy-oriented interpretations of the Anthropocene should focus on elaborating more advanced inter- and multidisciplinary research methodologies. More profound complexity-related ideas in developing policy recommendations for a new approach are proposed in this book, i.e., a comprehensive idea of global peaceful cooperation interrelated with effective dealing with fundamental environmental challenges in the Anthropocene.

In the fourth chapter the human geographer *Simon Dalby* addressed "Politics and Geology in Anthropocene Geopolitics". The concept of the Anthropocene provides a mechanism for critiquing the implicit contextualisations which limit contemporary social sciences' ability to grapple with contemporary global perils. It also provides the opportunity to tackle the mutual silence across the disciplines, especially across the natural and human sciences. Focusing on the key point that the industrial production decisions taken by the rich and powerful among humanity are shaping the future of the technosphere, and as such the increasingly artificial configuration of the biosphere too, the Anthropocene also shows the limitations of traditional environmentalism, dependent on formulations of a nature separate from humanity and its production processes. A "planet politics" that takes the Anthropocene recontexualisation seriously, both for scholarly analysis and political action, offers the possibilities of overcoming the epistemological gaps between the human and natural sciences that so constrain contemporary thinking.

"Thinking Politically about the Anthropocene" was discussed in Chap. 5 by *Richard Ned Lebow*, who concluded:

> Professionally led by organisations with widespread credibility, there is no reason for those of us interested in the Anthropocene to attempt to duplicate their efforts. Rather, we should devote our efforts to reordering the problems our several disciplines currently prioritise, developing concepts germane to the study of the interaction of environment and politics, and above all, thinking about how we can bring the resulting research findings to the attention of relevant political actors and audiences.

Dahlia Simangan's comprehensive review in Chap. 6 provided "An International Relations Perspective" on "Peace and Security Discourses on the Anthropocene." While security is more prominent in the literature than peace, the Anthropocene extends the analysis of these concepts beyond the traditional state-centric and human-centric subjects in IR. Conventional IR topics and human-centric concerns are now being examined vis-à-vis the expanded notions of time and space in the Anthropocene, while still reflecting the problem-solving aspect of IR. IR will benefit from more context-specific analysis from other regions or locales to better situate the principles of plurality and justice, and from more studies on other environmental manifestations of the Anthropocene, such as biodiversity loss. A deeper and sustained conversation between IR and other disciplines is imperative in making sense of the philosophical, ontological, and epistemological questions about peace and security in the Anthropocene.

In a case study, "The Anthropocene and German Political Science", in Chap. 7 *Georg Simonis* addressed the question of why the concept and hypothesis of the Anthropocene has elicited little response in German political science. From the margins of the discipline, the notion of a new age in which anthropogenically generated technologies and their forms of use threaten the conditions of human life is only slowly diffusing into the core of political science. The explanatory approach referred to Niklas Luhmann's systems theory. However, there are individual political scientists, especially from the field of environmental politics, who have made important contributions to the political science discussion of the Anthropocene. Oriented towards policy fields, the chapter provided an overview of relevant contributions of German Anthropocene research, mainly in the fields of governance research and international relations theory. As a result of its thematically structured analysis, the chapter also offered an overview of focal points of Anthropocene research in political science.

In Chap. 8, *Jürgen Scheffran*, a trained physicist who was a professor of Integrative Geography at Hamburg University, offered conclusions and outlooks for future research and political challenges in "Planetary Boundaries, Polycrisis and Politics in the Anthropocene: Climate Pathways, Tipping Cascades and Transition to Sustainable Peace in Integrative Geography". Integrative Geography provides a framework for studying complex human-nature interactions of environmental, economic, social, political and scientific-technical limits, crises and conflicts of the Anthropocene. In a multi-polar world order, challenges and opportunities include transformation of fossil capitalism, power shifts in the North-South relationship, and the influence of social networks and civil society on democracy. Policies, politics and polities are to address complexity challenges of compound effects, tipping points and cascading risks, moving from negative to positive tipping cascades that foster synergies between a social-ecological and a conflict transformation. Governance of a sustainability transition combines technical and societal innovations steering expansive human development to orderly and stabilising pathways. The available environmental space is to be used in a planned and sensible way to guarantee an acceptable life for all inhabitants in the common house of Planet Earth, adapting human needs to the available possibilities and reorganising human coexistence in harmony with nature. The cohabitation of different world orders to address these problems is more promising than further geopolitical power struggles, while crisis and conflict management should protect the foundations for sustainable problem-solving in the context of transformation to a viable world. Peace research can help understand the links between environmental degradation and violence, and to proactively develop integrated concepts and synergies for sustainable peace.

The third trained physicist, who holds a professorship in Political Science at Amsterdam University, *John Grin,* dealt in Chap. 9 with "Governance and Politics in the Anthropocene". He argued that the core of the problem of the Anthropocene is the idea central to Modernity: controlling nature, based on modern science's universalist laws that enable prediction—and hence control. To eventually overcome the Anthropocene condition, humans need to learn to align fulfilment of their needs with nature, through regenerative, place-based practices of production and consumption. The governance of such a re-orientation requires greater sensitivity of democratic and policy-making procedures to the life worlds and physical conditions in a specific place, novel standards for shaping solutions, and a reflexive capacity to envisage and pursue changes in the practices of consumption, production, innovation and governance itself—as well as of the structural (institutional, discursive and material) contexts in which they are embedded. Transition studies suggest guidelines for strategies and opportunities to achieve this.

In Chap. 10, the Mexican ecologist and feminist Úrsula Oswald-Spring criticised the neutral perspective of the Anthropocene concept, because it is missing the main causes of the exploitation of the environment, women, indigenous, and vulnerable people. She introduces an alternative concept called the 'Patriacene', analysing patriarchy as the key cause of the historical-colonial empires and the present transnational military-financial-political-scientific complex, both based on war, violence, and exploitation of nature and humans. Patriarchy maintains the present system of wealth concentration and socioenvironmental destruction, while only the care of vulnerable human beings and nature will be able to ensure the survival of Mother Earth and humankind.

Rethinking of Concepts in Other Scientific Disciplines During the Anthropocene: History of Science and History

In the humanities, especially in the research field of History of Science, a rethinking has started on "the Anthropocene" (Verburgt/de Waal 2022; Renn 2022b; Pesa 2022; Simon/Thomas 2022; Coen/Jonsson 2022) and on "Science for the Anthropocene" (Renn 2020, 2022a).

Verburgt and de Waal (2022: 366) present "the Anthropocene as both a challenge and an opportunity, opening up new avenues of research and widening spatial, temporal, and topical boundaries while giving rise to urgent problems, dilemmas, and choices that will need to be faced" by putting "these challenges and opportunities on the history of science's agenda, and, thereby, [starting] a discipline-wide dialogue". They argue that the "Anthropocene problematises history as a concept that encompasses past, present, and future in a temporal whole", shakes "the foundations of history as an institutionalised and professionalised scholarly practice of writing about and understanding the past" and that the Anthropocene invites historians "to reflect critically on the (a)historicity of the notion itself and on its role as a dominant way—at once scientific, political, and existential—of historicising our times" Verburgt/de Waal 2022: 367).

For Renn (2022b: 377), the "'Anthropocene' entails different historical horizons banded together—in particular *longue durèe* transformations [e.g. of international order, hgb] and the now-accelerating environmental and socio-epistemic changes". Renn further argues "that we need a new transdisciplinary, transformative science in order to understand the techno-Earth System from an integrative perspective". From the perspective of an emerging "geoanthropology"[7] he concludes that the

> Anthropocene as a *new geological epoch forces us to think geological and historical time together as a new form of planetary time* [emphasis added]. Natural history and human history thus become entangled in new ways. Realizing that the Anthropocene needs to be addressed as an urgent crisis, we have to integrate physical, historical, and evolutionary time in order to identify the critical time zones marked by the intrinsic time-behaviour of [a] complex techno-Earth-System. And finally, taking the Anthropocene as an opportunity to establish a sustainable relation between our species and the planet requires us to integrate the social time of human histories with the anthropological time dimension that belongs to each of them, thus giving rise to a multi-fingered species time, a reference point for any kind of history suitable as a resource for the alliances we need to build in order to cope with the challenges of the Anthropocene. (Renn 2022b: 383)

The Anthropocene has not only provoked such a discourse in the research area of the History of Science but also among an increasing number of historians from many other research areas, among them Dipesh Chakrabarty (2021, 2022: 264), who have contributed to a rethinking of the different measurements of time that are associated with Earth and world history between an anthropogenic and a planet-centred way of thinking (Zalasiewicz 2017).

[7] See: Max-Planck-Institute on Geoanthropology in Jena; at: https://www.mpg.de/9347805/geoanthropology. "Research at the Max Planck Institute of Geoanthropology focuses on the interrelationships between the geosphere and human-made systems. One of the central topics is human-ecosystem dynamics, for which data and expertise from climate research, biodiversity research and the social sciences are brought together. Inter- and transdisciplinary research projects also deal with urbanisation, the global food system and global material, energy and information flows. The key questions range from the deep past to the distant future and include the question of how humanity has driven the emergence of the Anthropocene and can still positively influence its course. The Max Planck Institute of Geoanthropology emerged from the Max Planck Institute for the Science of Human History, which was founded in 2014. The Senate of the Max Planck Society decided to rename it in June 2022."

The debate has also been taken up by political geographers, such as Simon Dalby (2020, 2022, 2023) in his book *Anthropocene Geopolitics* and in several journal articles (Dalby 1998, 2002a, b, c, 2008, 2009, 2013, 2016; Butler 2021), and by Jürgen Scheffran who has contributed to an approach aiming at an integrative geography (see their chapters in this volume).

Rethinking the Anthropocene Concepts in Political Science and International Relations

A process of rethinking the Anthropocene concept has also been underway in the social sciences and humanities. In the research programme on the History of Science and in History several authors contributed to a rethinking of "history in the Anthropocene". Several authors suggested that geological time and the historical time of the *longue durée'* should be thought together in the Anthropocene as "planetary time". However, these emerging discourses have not addressed the impact on *Politik* and its subconcepts of politics, policy and polity.

From a perspective of the "history of concepts", Christian Geulen (2010) has suggested a reconceptualisation of basic concepts for a second *Sattelzeit* (Geulen)—inspired by the concept Reinhart Koselleck applied to the period from 1750 to 1850 that partly overlapped with the era of enlightenment in philosophical and societal discourses. However, these debates on a reconceptualisation of modern views and of a second *Sattelzeit* have so far been totally unrelated to the discourses on the Anthropocene.

From a variety of publications in the natural and social sciences, and from the humanities and cultural studies, Dürbeck (2018) has distinguished five narratives on the Anthropocene: (1) the narrative focusing on catastrophes or apocalypses; (2) a court narrative, (3) the narrative of the "great transformation", (4) the (bio) technological narrative and (5) an interdependence narrative.

Outlook: The Anthropocene and the Social Sciences

The scientific literature on the Anthropocene—including books—has continued to spread into other disciplines in the social sciences and humanities (Chakrabarty 2021), among them Anthropology (Arizpe 2019; Antweiler 2022, 2023a, b), Geography (Dalby 2000) and Sociology (Latour/Schultz 2022a, b Schroer 2022), as well as Philosophy (Raffnsøe 2016) and in other languages, especially in French, German, Spanish et al. (Brauch 2021a).

Nevertheless, most mainstream authors in Political Science and International Relations have largely ignored this body of already available literature and have not yet participated in this emerging discourse, nor have they recognised the importance of the human intervention in the Earth system and the role of humanity as a geological 'actor' in the Anthropocene since the start of the nuclear era in 1945 and the great accelaration by 1950. Most political scientists and International Relations specialists have so far restrained themselves from thinking of "geological and historical time together as a new form of planetary time" (Renn 2022b).

A wider perspective that transgresses disciplinary boundaries is still often lacking, while the urgency to address the Anthropocene debate and agenda and to respond to the Anthropocene narrative has become obvious due to a rapid increase in the observable physical impacts of anthropogenic climate change. Impacts of extreme weather events (such as storms, hurricanes, typhoons, droughts and forest fires) have increased in environmental hotspot regions like the Mediterranean, North and Central America and in South, Southeast and Eastern Asia.

The prevailing mindset of policymakers and the worldviews of most social scientists are still focused on the Hobbesian security narrative and its policy agenda, and this trend has further been strengthened by the Russian war of agression against Ukraine and the new war between Hamas and Israel since 7 October 2023, that has legitimated a significant increase in military spending (at least in Europe, North America, the Middle East and Southeast and East Asia) since 2022 (SIPRI Yearbook 2023).

The ten chapters of this book have partly challenged the prevailing narrow orthodoxy and encouraged those who take the need for a broader holistic assessment seriously and contribute to the development of methodologies and approaches to think about both narratives of security and peace studies on the one hand and ecology concerns in environmental policy research on the other, and in Earth Systems Studies focusing on issues of global environmental change issues together,

This requires intensified efforts in areas such as "environmental security" triggered by the studies of Bächler,[8] Homer-Dixon,[9] and Dalby[10] and in "environmental cooperation" or "environmental peacemaking" developed by Conca[11] since the 1990s and promoted by the Environmental Peacebuilding Association[12] since 2018, and in "peace ecology"[13] or the ecology of peace. All of these concepts differ with regard to their specific research focus, their methological approach and the policy-relevant goals of the authors who have coined or developed this literature. The Anthropocene as a historical and political context and turning-point has been addressed by Dalby (2020) and this author elsewhere (Brauch 2021a).

---

[8] See Bächler (1990, 1994, 1998, 1999a, b; Bächler/Spillmann et al. 2002a, b).

[9] See Homer-Dixon (1991, 1994, 1995, 1996, 1999, 2000; Homer-Dixon/Blitt 1998; Homer-Dixon/Deligiannis 2009).

[10] See Dalby 1998, 2000, 2002a, b, c, 2008, 2009, 2013, 2016, 2020; Dalby/Brauch/Oswald Spring 2009).

[11] See Conca (1994, 2002) and Conca/Carius/Dabelko (2005).

[12] See the publication of the team of authors that were instrumental for the establishment of the *Environmental Peacebuilding Association* since 2018.

[13] See the debate on peace and ecology linkages and on peace ecology by (Kyrou 2007a, b, c; Amster 2014; Brauch 2020).

This intellectual bridgebuilding is the key goal of this author in establishing an *International Science Prize on Peace and Ecology in the Anthropocene.*[14] This is funded by his small foundation, established on 8 May 2020, the 75th anniversary of the end of World War II and the liberation from Fascism, about ten weeks before (according to the AWG) the Anthropocene started with the first test of a nuclear weapon on 16 July 1945 in Alamogordo in New Mexico in the USA.

Research on the Anthropocene is addressed so far in five academic journals: *The Anthropocene* (published by Elsevier, since 2013), *Anthropocene Science* (published by Springer Nature), the *Anthropocene Review* (Sage, since 2014), *Elementa: Science of the Anthropocene* (published by Sage), and the *Anthropocene Magazine* (from Future Earth).

As at September 2023 several book series have emerged or been announced: *The Anthropocene: Politik—Economics—Society—Science* (since 2016) and the *Anthropocene—Humanities and Social Sciences* series (since 2023), both published by Springer Nature; *Routledge Research in the Anthropocene* (published by the Taylor & Francis Group); *The Anthropocene Epoch* (with texts by the biologist E.O. Wilson); *AnthropoScene: The SLSA Book Series* (published by Penn State University Press in collaboration with the Society for Literature, Science and the Arts); and the forthcoming *African and Asian Anthropocene: Studies in the Environmental Humanities* (published by Brill publishers in Leiden).

Furthermore, at least two reference books have been published so far: The *Encyclopedia of the Anthropocene* (in five volumes coedited by D. A. Dellasala and M. I. Goldstein, published in 2018 by Elsevier) and the *Handbook of the Anthropocene—Humans between Heritage and Future* (co-edited by N. Wallenhorst and C. Wulf, published by Springer Nature in 2023).

This book aims to put the "Anthropocene concept" and the emerging "Anthropocene narrative" in many disciplines in the social sciences and humanities on the agenda of Political Science and International Relations. The importance of holistic approaches was also stressed in a report of the German Academy of Science Leopoldina (2022: 81–87) on *Earth System Science: Discovery, Diagnosis, and Solutions in Times of Global Change*, that discusses the contextualisation of Earth Systems Science in the Anthropocene (2022: 48–53) and raises two key questions: firstly, "What are the critical thresholds of key interconnected tipping elements of Earth's climate?"

> At our current state of knowledge, keeping global warming as close as possible to 1.5 °C seems to be the only option to minimise its adverse consequences. Yet, the coupled nature of the Earth System, with the possibility of cascading effects, can cause further tipping dynamics and makes self-propagating 'run-away' processes a real risk. How such cascades

[14] See "Award of the First International Science Prize 2023 of the Hans Günter Brauch Foundation (HGBF) for Peace and Ecology in the Anthropocene on the Theme: Climate Change and Conflicts", at: http://hgb-stiftung.org/html/Award_winners.html (1 November 2023). The first winner of the international science prize 2023 is Tobias Ide for his book: *Catastrophes, Confrontations and Constraints – How Disasters Shape the Dynamics of Armed Conflict* published by MIT Press.

> function and at what levels of Anthropocene forcing they may be triggered is one of the largest gaps in our knowledge. ...

The Report of Leopoldina also asks, "How can Earth System stewardship avoid the risks?"

> The relationship between the degree of human forcing and the plasticity and buffering capacity of many of Earth's subsystems is poorly understood. A fundamental question concerns the level at which social dynamics at a given time and location result in reaching critical thresholds in closely connected social-environmental systems elsewhere. An example is provided by the intensity of international food trade that is mainly controlled by a few key actors such as multinational companies and financial investors in industrial countries. This high-level control may weaken local farming and dietary diversity and make societies more dependent on a limited number of globally traded food items. This in turn contributes to less resilient and more unhealthy dietary practices that may be fragile in the face of climate and water-related extremes like droughts, floods, and freshwater over-use. Finally, the impacts will cascade across markets and regions, ultimately hitting local farming communities hard. Climate change and resulting effects on food prices and security are also connected to human migration and political conflict.
>
> The challenges that arise from such cascading thresholds affecting interconnected social-environmental systems across scales are (1) developing and reconciling sustainable socio-environmental and institutional behaviour and decision-making as outlined in the UN sustainable development goals, (2) establishing a global sustainable governance system for social-environmental risk assessment and management, and (3) avoiding disruptive environmental changes that lead to crossing critical thresholds by employing technological and socio-environmental innovations. All these steps must be accomplished in a just and equitable fashion. Meeting these challenges will allow humankind to assume a role of stewardship of the Earth System, ensuring a sustainable trajectory within planetary boundaries.

To answer these key questions, and obtain a holistic understanding and a sound diagnosis of the social-environmental aspects of the Earth System, three main courses of action are urgently needed [in addressing] "The Grand Challenges of Earth System Science" according to the Report by Leopoldina: (1) To properly assess biogeophysical risks, we need to determine the tipping points of subsystems and the danger of crossing them; (2) Two approaches to mitigate global warming are currently envisaged: radiation management and negative CO_2 emission technologies; and (3) Biogeophysical risks and tipping points that are mediated by the biosphere are highly interwoven with societal tipping points. From the perspective of the Report of Leopoldina there is an urgent need to include the human domain as a major planetary system in holistic studies of the Earth System.

These three social-environmental aspects of the Earth System stressed by the Report of Leopoldina also call for an "Academic Education for Earth System Research: Challenges and Future Needs" from schools to universities, which "may offer master's degrees in Earth system science with in-depth training in one of three directions: (1) in the STEM disciplines (Science, Technology, Engineering, and Mathematics); (2) in Data Science; (3) in communication, social, psychological, political, and social sciences." The proposal of the Leopoldina Report for its implementation in school and university education requires multi- and interdisciplinary research "that combines curricula in the geological, natural and the social sciences". This would also include a bridgebuilding between the peace and security narrative and the Earth System Science narrative in the Anthropocene epoch of Earth history (or the Anthropocene narrative).

From 12 to 13 December 2023 the newly established Max Planck Institute (MPI) for Geoanthropology in Jena (Germany) organised a workshop on "Bridging the Gap—Integrating Historical Knowledge and Complex Human-Environmental Systems Modelling".[15]

Both the Leopoldina Report of 2022 and the workshop of the MPI for Geoanthropology in December 2023 are at the forefront of a bridgebuilding of scientists in the natural and social sciences and the humanities within the framework of the Anthropocene. For such a bridgebuilding to be possible, a rethinking of politics, policy and polity is essential, for example by separating the thinking on *Politik* from national sovereignty and moving towards governance at different levels. So far, the mainstream research in Political Science and International Relations can contribute little to "bridging the gap" between knowledge produced in Political Science and International Relations and the knowledge needed for "Complex Human-Environmental Systems Modelling".

In addition, these much broader holistic multidisciplinary (Prigogine/Stengers 1984, 1997), interdisciplinary (Piaget) and transformative (USNSF 2007; WBGU 2011; Cramer 2023) approaches (Oswald Spring/Brauch/Scheffran 2016) have not yet reached the mainstream discourse in the social and political sciences. Thinking about Crutzen's geological time and Braudel's three historical times of structures, conjunctions and events together with agency, actors and governance (e.g., Grin, Biermann, Simonis et al.) requires also a fundamental rethinking of the basic concepts of politics, policy and polity during the second *Sattelzeit* and in the Anthropocene epoch of Earth history since 1945.

What impact may have the vote of the SQS on 4 March 2024 leaked to the New York Times? In the Introduction this author has reviewed and updated the controversial debate within the geological community whether the geological community has definitely rejected the Anthropocene as a new Epoch of Earth History. However, even if the Epoch of the Holocene has not yet ended and the Anthropocene Epoch was not officially been approved by geologists this vote has no impact on the debate within the social and policy science communities.

[15] See Max Planck Institute for Geoanthropology: "Bridging the Gap – Integrating Historical Knowledge and Complex Human-Environmental Systems Modelling", at: https://www.shh.mpg.de/2362899/integrating-historical-knowledge-and-complex-human-environmental-systems-modelling1.html (31 October 2023). This workshop explored "research strategies that can generate actionable knowledge to inform policies addressing the escalating climate crisis".

References

Amster, R., 2014: *Peace Ecology* (Boulder, CO: Paradigm).

Antweiler, C., 2022: *Anthropologie im Anthropozän – Theoriebausteine für das 21. Jahrhundert* (Darmstadt: WBG Academic).

Antweiler, C., 2023a: "Apocalypse", in: Wallenhorst, N.; Wulf, C. (Eds.), 2023: *Handbook of the Anthropocene – Humans between Heritage and Future* (Cham: Springer Nature): 1473–1476.

Antweiler, C., 2023b: "Society", in: Wallenhorst, N.; Wulf, C. (Eds.), 2023: *Handbook of the Anthropocene – Humans between Heritage and Future* (Cham: Springer Nature): 1305–1308.

Arizpe Schlosser, L., 2019: *Culture, International Transactions and the Anthropocene* (Cham: Springer- Springer International Publishing).

Bächler, G., 1990: *Ökologische Sicherheit und Konflikt.* Arbeitspapiere der Schweizerischen Friedensstiftung, No. 5 (Bern: Schweizerische Friedensstiftung).

Bächler, G., 1994: *Umweltflüchtlinge: Das Konfliktpotential von Morgen* (Münster: Agenda).

Bächler, G., 1998: "Why Environmental Transformation Causes Violence", in: *Environmental Change and Security Report*, No. 4: 24–44.

Bächler, G., 1999a: "Environmental Degradation in the South as a Cause of Armed Conflict", in: Carius, A.; Lietzmann, K. (Eds.): *Environmental Change and Security: A European Perspective* (Berlin: Springer): 107–130.

Bächler, G., 1999b: *Violence Through Environmental Discrimination – Causes: Rwanda Arena, and Conflict Model* (Dordrecht – Boston – London: Kluwer Academic Publishers).

Bächler, G.; Böge, V.; Klötzli, S.; Libiszewski, S.; Spillmann, K. R. (Eds.), 1996: *Kriegsursache Umweltzerstörung. Ökologische Konflikte in der Dritten Welt und Wege ihrer friedlichen Bearbeitung*, vol. I (Chur-Zürich: Rüegger).

Bächler, G.; Spillmann, K. R. (Eds.), 1996: *Kriegsursache Umweltzerstörung. Regional- und Länderstudien von Projektmitarbeitern/Environmental Degradation as a Cause of War: Regional and Country Studies of Research Fellows*, vol. II (Chur: Rüegger).

Bächler, G.; Spillmann, K. R.; Suliman, M. (Eds.), 2002: *Transformation of Resource Conflicts: Approach and Instruments* (Bern – Berlin: Peter Lang).

Benner, S.; Lax, G.; Crutzen, P. J.; Pöschl, U.; Lelieveld, J.; Brauch, H. G. (Eds.): *Paul J. Crutzen and the Anthropocene: A New Epoch in Earth's History* (Cham: Springer Nature Switzerland).

Brauch, H. G., 2008: "Introduction: Globalization and Environmental Challenges: Reconceptualizing Security in the 21st Century", in: Brauch, H. G.; Oswald Spring, Ú.; Mesjasz, C.; Grin, J.; Dunay, P; Behera, N. B.; Chourou, B.; Kameri-Mbote, P.; Liotta, P. H. (Eds.): *Globalization and Environmental Challenges – Reconceptualizing Security in the 21st Century* (Berlin: Springer-Verlag): 27–43.

Brauch, H. G., 2016a: "Historical Times and Turning Points in a Turbulent Century: 1914, 1945, 1989 and 2014?", in: Brauch, H. G.; Oswald Spring, Ú.; Bennett, J.; Serrano Oswald, S. E. (Eds.): *Addressing Global Environmental Challenges from a Peace Ecology Perspective* (Cham – Heidelberg – New York – Dordrecht – London: Springer International Publishers): 11–54.

Brauch, H. G., 2016b: "Conceptualizing Sustainable Peace in the Anthropocene: A Challenge and Task for an Emerging Political Geoecology and Peace Ecology", in: Brauch, H. G.; Oswald Spring, Ú.; Grin, J.; Scheffran, J. (Eds.): *Handbook on Sustainability Transition and Sustainable Peace* (Cham – Heidelberg – New York – Dordrecht – London: Springer): 187–236.

Brauch, H. G., 2021a: "The Anthropocene concept in the natural and social sciences, the humanities and law – a bibliometric analysis and a qualitative interpretation (2000–2020), in: Benner, S.; Lax, G.; Crutzen, P. J.; Pöschl, U.; Lelieveld, J.; Brauch, H. G. (Eds.): *Paul J. Crutzen and the Anthropocene: A New Epoch in Earth's History* (Cham: Springer Nature Switzerland): 289–438.

Brauch, H. G., 2021b: "Peace Ecology in the Anthropocene", in: Oswald Spring, Ú.; Brauch, H. G (Eds.): *Decolonising Conflicts, Security, Peace, Gender, Environment and Development in the Anthropocene* (Cham: Springer Nature Switzerland): 51–186.

Brauch, H. G.; Oswald Spring, Ú.; Grin, J.; Scheffran, J. (Eds.), 2016: *Handbook on Sustainability Transition and Sustainable Peace (Cham:* Springer International Publishing).

Butler, D. T., 2021: "The Anthropocene: A Special Issue", in: *Annals of the American Association of Geographers*, 111,3: 633–637.
Charakbarty, D., 2021: *The Climate of History in a Planetary Age* (Chicago: The University of Chicago Press).
Charakbarty, D., 2022: *Das Klima der Geschichte im planetarischen Zeitalter* (Berlin: Suhrkamp).
Coen, D. R.; Jonsson, F. A., 2022: "Between History and Earth System Science", in: *Isis*, 113,2 (June): 407–416.
Conca, K., 1994: "In the Name of Sustainability: Peace Studies and Environmental Discourse", in: Käkonen, J. (Ed.): *Green Security or Militarized Environment* (Dartmouth: Aldershot): 7–24.
Conca, K., 2002: "The Case for Environmental Peacemaking", in: Conca, K.; Dabelko, G. (Eds.): *Environmental Peacemaking* (Baltimore: Johns Hopkins University Press): 1–22.
Conca, K.; Carius, A.; Dabelko, G. D., 2005: "Building Peace through Environmental Cooperation", in: The Worldwatch Institute (Ed.): *State of the World 2005: Redefining Global Security* (New York – London: W.W. Norton): 144–157.
Cramer, P., 2023: "Transformative science for a changing world – Inaugural Speech by the President of the Max Planck Society Professor Dr. Patrick Cramer on the occasion of the Annual General Meeting 22.06.2023, Lokhalle Göttigen"; at: https://www.mpg.de/20613384/20230622-rede-cramer-inaugural-speech.pdf (4 November 2023).
Crutzen, P. J., 2002: "Geology of mankind – The Anthropocene", in: *Nature*, 415: 23.
Dalby, S., 1998: "Ecological Metaphors of Security: World Politics in the Biosphere", in: *Alternatives*, 23,3: 291–319.
Dalby, S., 2000: "Geopolitics and Ecology: Rethinking the Contexts of Environmental Security", in: Lowi, M. R.; Shaw, B. R. (Eds.): *Environment and Security. Discourses and Practices* (Basingstoke – London: Macmillan; New York: St. Martin's Press): 84–100.
Dalby, S., 2002a: "Environmental Change and Human Security", in: *Canadian Journal of Policy Research*, 3,2: 71–79.
Dalby, S., 2002b: "Security and Ecology in the Age of Globalization", in: Woodrow Wilson International Center for Scholars (Ed.): *Environmental Change & Security Project Report*, Issue No. 8 (Summer): 95–108.
Dalby, S., 2002c: *Environmental Security* (Minneapolis: University of Minneapolis Press).
Dalby, S., 2008: "Security and Environment Linkages Revisited", in: Brauch, H. G.; Oswald Spring, Ú.; Mesjasz, C.; Grin, J.; Dunay, P.; Behera, N. C.; Chourou, B.; Kameri-Mbote, P.; Liotta, P. H. (Eds.): *Globalization and Environmental Challenges: Reconceptualizing Security in the 21st Century* (Berlin – Heidelberg: Springer-Verlag): 165–172.
Dalby, S., 2009: *Security and Environmental Change* (Cambridge: Polity).
Dalby, S., 2013: "Peace in the Anthropocene" in: *Peace Review*, 25,4 (October).
Dalby, S., 2016: "Framing the Anthropocene: The Good, the Bad and the Ugly", in: *Anthropocene Review*, 3,1: 33–51.
Dalby, S., 2020: *Anthropocene Geopolitics: Globalization, Security, Sustainability* (Ottawa: University of Ottawa Press).
Dalby, S.; Brauch, H. G.; Oswald Spring, Ú., 2009: "Environmental Security Concepts Revisited During the First Three Phases (1983–2006)", in: Brauch, H. G.; Oswald Spring, Ú.; Grin, J.; Mesjasz, C.; Kameri-Mbote, P.; Behera, N. C.; Chourou, B.; Krummenacher, H. (Eds.), 2009: *Facing Global Environmental Change: Environmental, Human, Energy, Food, Health and Water Security Concepts* (Berlin – Heidelberg – New York: Springer-Verlag): 781–790.
Décultot, E.; Fulda, D. (Eds.), 2016: *Die Vielfalt der Sattelzeit. Strukturen und Tendenzen des Historischen Erzählens um 1800 im Deutsch-Französischen Vergleich Veranstaltung 2013 Halle (Saale)* (Berlin-Boston: Walter de Gruyter).
Dellasala, D. A.; Goldstein, M. I. 2018: *Encyclopedia of the Anthropocene* (Oxford: Elsevier).
Deutsche Gesellschaft Club of Rome, 2017: "Wir brauchen eine Neue Aufklärung"; at: https://clubofrome.de/neue-aufklarung.
Dürbeck, G., 2018: "Narrative des Anthropozän – Systematisierung eines interdisziplinären Diskurses", in: *Kulturwissenschaftliche Zeitschrift*, 2/2018 (i.p.).

EEA, 2023: Briefing: *Exiting the Anthropocene? Exploring fundamental change in our relationship with nature* (Copenhagen: EEA); at: https://www.eea.europa.eu/publications/exiting-the-anthropocene.

Environmental Peacebuilding Association, at: https://www.environmentalpeacebuilding.org/about/.

Gabriel, M.; Horn, C.; Katsman, A.; Krull, W.; Lippold, A. L.; Pelluchon, C.; Venzke, I., 2022: *Towards a New Enlightenment – The Case for Future-Oriented Humanities* (Bielefeld: transcript); at: https://www.transcript-verlag.de/shopMedia/openaccess/pdf/oa9783839465707.pdf.

Geulen, Christian, 2010: "Plädoyer für eine Geschichte der Grundbegriffe des 20. Jahrhunderts", in: *Zeithistorische Forschungen/Studies in Contemporary History*, 7: (Göttingen: Vandenhoeck & Ruprecht): 79–97.

Homer-Dixon, T. F., 1991: "On the Threshold: Environmental Changes as Causes of Acute Conflict", in: *International Security*, 16,2 (Fall): 76–116.

Homer-Dixon, T. F., 1994: "Environmental Scarcities and Violent Conflict: Evidence From Cases," in: *International Security*, 19,1 (Summer): 5–40.

Homer-Dixon, T. F., 1995: "Correspondence: Environment and Security", in: *International Security*, 20,3: 189–194.

Homer-Dixon, T. F., 1996: "Strategies for Studying Causation in Complex Ecological-Political Systems", in: *The Journal of Environment and Development*, 5,2 (June): 132–148.

Homer-Dixon, T. F., 1999: *Environment, Scarcity, and Violence* (Princeton, NJ: Princeton University Press).

Homer-Dixon, T. F., 2000, 2001: *The Ingenuity Gap: How Can We Solve the Problems of the Future?* (New York – Toronto: Alfred A. Knopf).

Homer-Dixon, T. F.; Blitt, J. (Eds.), 1998: *Ecoviolence: Links among Environment, Population, and Security* (Lanham, Md.: Rowman and Littlefield).

Homer-Dixon, T. F.; Deligiannis, T., 2009: "Environmental Scarcities and Civil Violence", in: Brauch, H. G.; Oswald Spring, Ú.; Grin, J.; Mesjasz, C.; Kameri-Mbote, P.; Behera, N. C.; Chourou, B.; Krummenacher, H. (Eds.): *Facing Global Environmental Change: Environmental, Human, Energy, Food, Health and Water Security Concepts* (Berlin – Heidelberg – New York: Springer-Verlag): 309–324.

Ide, T., 2023: *Catastrophes, Confrontations and Constraints – How Disasters Shape the Dynamics of Armed Conflict* (Cambridge Mass.: MIT Press).

Kyrou, C., 2007a: "Peace Ecology: An Emerging Paradigm in Peace Studies", in: *The International Journal of Peace Studies*, 12,2.

Kyrou, C., 2007b: "Peace Ecology: Theoretical and Methodological Reflections", Paper presented at conference on Cutting Edge Theories and Recent Developments in Conflict Resolution, Syracuse, NY, 27–28 September.

Kyrou, C., 2007c: "Peace Ecology: Evaluative and Methodological Reflections", Paper presented at the annual meeting of the ISA 48th Annual Convention, Chicago, 28 February 2007.

Latour, B.; Schultz, N., 2022a: *Zur Entstehung einer ökologischen Klasse: Ein Memorandum* (Berlin: Suhrkamp).

Latour, B.; Schultz, N., 2022b: *On the Emergence of an Ecological Class: A Memo* (Cambridge: Polity).

Leopoldina, Nationale Akademie der Wissenschaften, 2022: *Report on Tomorrow's Science: Earth System Science – Discovery, Diagnosis, and Solutions in Times of Global Change* (Halle: Leopoldina)

Menzel, U., 2023: *Wendepunkte – Am Übergang zum autoritären Jahrhundert* (Berlin: Suhrkamp)

Nolte, N., 2010: „Vom Fortschreiben und Umschreiben der Begriffe. Kommentar zu Christian Geulen", in: *Zeithistorische Forschungen/Studies in Contemporary History*, Online-Edition, 7,1: 98–103; at: https://zeithistorische-forschungen.de/1-2010/4616, https://doi.org/10.14765/zzf.dok-1789 (24 September 2023).

Oswald Spring, U.; Brauch, H. G.; Scheffran, J. 2016: "Sustainability Transition with Sustainable Peace: Key Messages and Scientific Outlook", in: Brauch, H. G.; Oswald Spring, U.; Grin, J.;

Scheffran, J. (Eds.): *Handbook on Sustainability Transition and Sustainable Peace* (Cham – Heidelberg – New York – Dordrecht – London: Springer): 887–928.
Pesa, I., 2022: "A Planetary Anthropocene? Views from Africa", in: *Isis*, 113,2 (June): 386–395.
Prigogine, I.; Stengers, I., 1984: *Order out of Chaos: Man's New Dialogue with Nature* (London: Flamingo).
Prigogine, I.; Stengers, I., 1997: *The End of Certainty* (Glencoe: The Free Press).
Renn, J., 2020: *The Evolution of Knowledge: Rethinking Science for the Anthropocene* (Princeton: Princeton University Press).
Renn, J., 2022a: *Die Evolution des Wissens: Eine Neubestimmung der Wissenschaft für das Anthropozän* (Berlin: Suhrkamp).
Renn, J., 2022b: "From the History of Science to Geoanthropology", in: *Isis*, 113,2 (June): 377–385.
Schroer, Markus, 2022: *Geosoziologie. Die Erde als Raum des Lebens* (Berlin: Suhrkamp).
Simon, Z. B.; Thomas. J. A., 2022: "Earth System Science, Anthropocene Historiography, and Three Forms of Human Agency", in: *Isis*, 113,2 (June): 396–406.
SIPRI, 2023: *SIPRI Yearbook 2003: Armaments, Disarmament and International Security* (Oxford: Oxford University Press).
UNDP, 2018: *UNDP Human Development Report 2020: The Next Frontier: Human Development and the Anthropocene* (New York: UNDP); at: https://hdr.undp.org/content/human-development-report-2020.
U.S. National Science Foundation [USNSF], 2007: *Enhancing Support of Transformative Research at the National Science Foundation* (Washington: NSF).
Verburgt, L.M.; de Waal, E., 2022: "Rethinking History of Science in the Anthropocene", in: *Isis*, 113,2 (June): 366–376.
Wallenhorst, N.; Wulf, C. (Eds.), 2023: *Handbook of the Anthropocene – Humans between Heritage and Future* (Cham: Springer Nature).
WBGU, 2011: *World in Transition—A Social Contract for Sustainability* (Berlin: German Advisory Council on Global Change, July 2011 (Berlin: WBGU).
WBGU, 2020: *Rethinking Land in the Anthropocene: from Separation to Integration* (Berlin: WBGU); at: https://www.wbgu.de/en/publications/publication/landshift.
Weizsäcker, E. U. v.; Wijkman, A., 2017: *Wir sind dran. Club of Rome: Der große Bericht: Was wir ändern müssen, wenn wir bleiben wollen. Eine neue Aufklärung für eine volle Welt* (Gütersloh: Gütersloher Verlagshaus).
Witze, A., 2023: "This quiet lake could mark the start of a new Anthropocene epoch – The dawn of a new geological epoch is recorded in the contaminated sediment at the bottom of Crawford Lake in Canada", in: *Nature*, 12 July 2023.
Zalasiewicz, J. 2017: "The Extraordinary Strata of the Anthropocene", in: Oppermann, S.; Iovino, S. (Eds.): *Environmental Humanities. Voices from the Anthropocene* (London: Rowman & Littlefield Publishers): 124.

Photo Gallery: Brainstorming During the AFES-PRESS Workshop with Nobel Laureate Paul J. Crutzen in Mosbach on 31 May 2017 in the Editor's Private Library

See Photos 1, 2, 3, 4, 5, 6, 7, 8, 9, 10, 11, 12, 13, 14, 15, 16, 17, 18, 19, 20, 21, 22, 23, 24, 25, 26 and 27.

Photo 1 Nobel Laureate Prof. Dr. Paul J. Crutzen during the Workshop in Mosbach on "'Politik' in and for the Anthropocene". *Source* © Hans Günter Brauch

H. G. Brauch (ed.), *Towards Rethinking Politics, Policy and Polity in the Anthropocene*,
The Anthropocene: Politik—Economics—Society—Science 35,
https://doi.org/10.1007/978-3-031-71807-6

Photo 2 Nobel Laureate Prof. Dr. Paul J. Crutzen during the Workshop in Mosbach on "'Politik' in and for the Anthropocene". *Source* © Hans Günter Brauch

Photo 3 Discussions in the Workshop on "Politik in and for the Anthropocene" with Prof. Dr. Paul J. Crutzen in the private library of Dr. Hans Günter Brauch in Mosbach, Baden-Württemberg, Germany on 31 May 2017. *Source* © Hans Günter Brauch

Photo 4 Discussions in the Workshop on "Politik in and for the Anthropocene" with Prof. Dr. Paul J. Crutzen in the private library of Dr. Hans Günter Brauch in Mosbach, Baden-Württemberg, Germany on 31 May 2017. *Source* © Hans Günter Brauch

Photo 5 Mr. Michael Headon (Wales, UK), Prof. Dr. Klaus Hüfner, Nuclear physicist, Heidelberg University, and Prof. Dr. John Grin, Department of Political Science, University of Amsterdam. *Source* © Hans Günter Brauch

Photo 6 Ass. Prof. Dr. Serrena Erendira Serrano Oswald, Prof. Dr. Úrsula Oswald Spring (speaking), Prof. Dr. Lourdes Arizpe Schlosser (representing CRIM, UNAM, Cuernavaca, Mexico), Prof. Dr. Paul J. Crutzen (Max-Planck-Institute for Chemistry, Mainz), Astrid Kaltenbach (personal secretary of Prof. Crutzen), Dr. Lidia Mesjasz (Cracow, Poland) and PD Dr. Hans Günter Brauch. *Source* © Hans Günter Brauch

Photo 7 Prof. Dr. Paul J. Crutzen with Hans Günter Brauch in the Workshop on "Politik in and for the Anthropocene" in the private library of the editor in Mosbach, Baden-Württemberg, Germany on 31 May 2017. *Source* © Hans Günter Brauch

Photo 8 Prof. Dr. Paul J. Crutzen, Nobel laureate of chemistry, Mainz with the honorary deputy Mayor of Mosbach, Mr. Georg Nelius, in the private library of Dr. Hans Günter Brauch in Mosbach, Baden-Württemberg, Germany on 31 May 2017 on the occasion of the signing of Prof. Crutzen into the Golden Book of Mosbach. *Source* © Hans Günter Brauch

Photo 9 Workshop on 31 May 2017 on "'Politik' in and for the Anthropocene" in the private library of the editor of this book in Mosbach, Germany with Prof. Dr. Paul J. Crutzen (MPI for chemistry, Mainz), PD Dr. Hans Günter Brauch (political scientist, FU University of Berlin (ret.), Mosbach), Prof. Dr. Jörg Hüfner (theoretical physicist, Heidelberg University), Prof. Dr. John Grin, Amsterdam University). *Source* Photo © Hans Günter Brauch

Photo 10 Workshop on 31 May 2017 on "'Politik' in and for the Anthropocene" during lunch in the living room of the editor of this book in Mosbach, Germany showing Prof. Dr. Lourdes Arizpe Schlosser (UNAM, CRIM, Cuernavaca, Mexico), Ass. Prof. Serrena Erendira Serrano Oswald (UNAM, CRIM, Cuernavaca, Mexico), Dr. Lidia Mesjasz (Cracow Economic University), Prof. Dr. Paul J. Crutzen (MPI for Chemistry, Mainz), Prof. Dr. John Grin, Amsterdam University) and Prof. Dr. Jürgen Scheffran (Department of Geography, Hamburg University. *Source* Photo © Hans Günter Brauch

Photo 11 Workshop on 31 May 2017 on "'Politik' in and for the Anthropocene" during lunch in the library of the editor in Mosbach, Germany showing Prof. Dr. Ursula Oswald Spring and Prof. Dr. Lourdes Arizpe Schlosser (UNAM, CRIM, Cuernavaca, Mexico), Prof. Dr. Paul J. Crutzen (MPI for Chemistry, Mainz), PD Dr. Hans Günter Brauch (political scientist, FU University of Berlin (ret.), Mosbach). *Source* Photo © Hans Günter Brauch

Photo 12 Workshop in Dr. Hans Günter Brauch's private library in Mosbach with Prof. Dr. Úrsula Oswald Spring and Prof. Dr. Lourdes Arizpe Schlosser (UNAM, CRIM, Cuernavaca, Mexico). *Source* Photo © Hans Günter Brauch

Photo 13 Workshop in H. G. Brauch's library in Mosbach with Prof. Dr. Jürgen Scheffran, Prof. Dr. Paul J. Crutzen (MPI for Chemistry, Mainz), Prof. Dr. Jörg Hüfner (theoretical physicist, Heidelberg University), PD Dr. Hans Günter Brauch (political scientist, FU University of Berlin (ret.), Mosbach), Jochen Herkert (Rector, Nicolais-Kistner Grammar School, Mosbach) and Prof. Dr. John Grin (Amsterdam University). *Source* Photo © Hans Günter Brauch

Photo 14 Workshop in H. G. Brauch's library in Mosbach with Ms. Astrid Kaltenbach, personal secretary of Prof. P.J. Crutzen, Prof. Dr. Paul J. Crutzen (MPI for Chemistry, Mainz), and Ph.D. Candidate A. S. M. Mostafizur Rahman (Heidelberg University, Institute for South Asia Studies). *Source* Photo © Hans Günter Brauch

Photo 15 Workshop in H. G. Brauch's library in Mosbach with the three trained physicists below H. G. Brauch's survival hexagon with Prof. Dr. John Grin (The Netherlands), Prof. Dr. Czeslaw Mesjasz (Poland), Prof. Dr. Jürgen Scheffran (Germany) and philosopher Dr. Frederik Arends (Leiden, The Netherlands). *Source* Photo © Hans Günter Brauch

Photo 16 Workshop in H. G. Brauch's library in Mosbach with philosopher Dr. Frederik Arends (Leiden, The Netherlands), Ph.D. Candidate A. S. M. Mostafizur Rahman (Heidelberg University), Mr. Rainer Fischbach (Berlin) and Ass. Prof. Dr. Serena Erendira Serrano Oswald. *Source* Photo © Hans Günter Brauch

Photo 17 During the discussions in the the Workshop on "Politik in and for the Anthropocene" with Prof. Dr. Paul J. Crutzen in H. G. Brauch's library in Mosbach on 31 May 2017 during. *Source* Photo © Michael Headon, Wales, UK

Photo 18 During the discussions in the the Workshop on "Politik in and for the Anthropocene" with Prof. Dr. Paul J. Crutzen in H. G. Brauch's library in Mosbach on 31 May 2017 during. *Source* Photo © Michael Headon, Wales, UK

Photo 19 During the discussions in the Workshop on "Politik in and for the Anthropocene" with Prof. Dr. Paul J. Crutzen in H. G. Brauch's library in Mosbach on 31 May 2017 during. *Source* Photo © Astrid Kaltenbach, MPIC in Mainz

Photo 20 During the discussions in the Workshop on "Politik in and for the Anthropocene" with Prof. Dr. Paul J. Crutzen in H. G. Brauch's library in Mosbach on 31 May 2017 during. *Source* Photo © Astrid Kaltenbach, MPIC in Mainz

Photo 21 Nobel Laureate Prof. Dr. Paul J. Crutzen, Max-Planck-Institute for Chemistry in Mainz, visits other parts of the private scientific library of PD Dr. Hans Günter Brauch during the workshop on "Politik in and for the Anthropocene". *Source* Photo © Astrid Kaltenbach

Photo 22 Hans Günter Brauch shows hos private scientific library to Nobel Laureate Prof. Dr. Paul J. Crutzen, on 31 May 2017. *Source* Photo © Astrid Kaltenbach

Photo 23 Prof. Dr. Paul J. Crutzen signs the Golden Book of the town of Mosbach

Photo 24 Prof. Paul J. Crutzen signed the Golden Book of the Town Mosbach with Georg Nelius (deputy mayor of Mosbach), Ursula Oswald Spring and Hans Günter Brauch watching

Photo 25 Nobel laureate in chemistry of 1995 Prof. Paul J. Crutzen, Max-Planck-Institute of Chemistry signed the Golden Book of the Town Mosbach with the participants of the workshop and of the signing ceremony which is posted online at: http://www.afes-press.de/html/video/08_Video_31-5-17_Crutzen_Golden_Book.m4v

Photo 26 The head of the public library of Mosbach (Mediathek); Mr. Ph.D. Candidate shows to the honourable deputy mayor, Mr. Georg Nelius the Golden Book of the town of Mosbach on 31 May 2017

Photo 27 Workshop in front of the house of the editor with wine grapes in Mosbach with PD Dr. Hans Günter Brauch (political scientist, FU University of Berlin, ret.), Prof. Dr. Paul J. Crutzen (MPI for Chemistry, Mainz), and Ph.D. Candidate A. S. M. Mostafizur Rahman (Heidelberg University). *Source* Photo © Hans Günter Brauch. Most photos were taken by Ph.D. Candidate Mr. A. S. M. Mostafizur Rahman, Bangladesh (Heidelberg University)

On the Max Planck Institute for Chemistry

Aiming at an integral scientific understanding of chemical processes in the Earth System from molecular to global scales

Current research at the Max Planck Institute for Chemistry in Mainz aims at an integral understanding of chemical processes in the Earth system, particularly in the atmosphere and biosphere. Investigations address a wide range of interactions between air, water, soil, life and climate in the course of Earth history up to today's human-driven epoch, the *Anthropocene*: see at: https://www.mpic.de/3865097/the-anthropocene.

Research at the Max Planck Institute for Chemistry has been at the forefront of science throughout its existence. Since the Institute's foundation in 1912, three of its directors were awarded with the Nobel Prize for Chemistry: Richard Willstätter in 1915 for the revelation of the structure of chlorophyll and other plant pigments, Otto Hahn in 1944 for the discovery of nuclear fission, and Paul Crutzen in 1995 for the elucidation of atmospheric ozone chemistry.

The research departments and focal points of the Institute have gone through a history of change and scientific evolution. What began in 1912 with classical organic, inorganic and physical chemistry at the Kaiser Wilhelm Institute for Chemistry in Berlin evolved into radiochemistry and nuclear physics in the 1930s, leading to the discovery of nuclear fission by Otto Hahn, Lise Meitner, Fritz Strassmann, and Otto Robert Frisch. As the Institute was severely damaged towards the end of World War II it was moved to the Swabian Alps in today's Baden-Wuerttemberg. There the chemists worked provisionally from 1944 to 1949 until the Institute moved a second time to the campus of the newly founded Johannes Gutenberg University in Mainz. At the same time it was integrated into the Max Planck Society, the successor of

H. G. Brauch (ed.), *Towards Rethinking Politics, Policy and Polity in the Anthropocene*,
The Anthropocene: Politik—Economics—Society—Science 35,
https://doi.org/10.1007/978-3-031-71807-6

the Kaiser Wilhelm Society, Institute for Chemistry. Since 1959 the Institute also carries the name "Otto Hahn Institute" in honor of its previous director and the first president of the Max Planck Society. In the 1960s and 1970s the Institute's research portfolio was extended from Physical Chemistry, Nuclear Physics and Mass Spectrometry to Cosmochemistry, Isotope Cosmology and Air Chemistry. Meteorites and moon dust samples were studied and the interplay of atmospheric gases, particles and meteorology were investigated. In the 1980s new departments for Geochemistry and Biogeochemistry were founded, in 2001 the Particle Chemistry department was established jointly with the Institute for Atmospheric Physics at the Johannes Gutenberg University of Mainz. In 2012 the Multiphase Chemistry Department and in 2015 the Climate Geochemistry Department was founded. The newest department is the Aerosol Chemistry Department, founded in 2024.

Nowadays, the research focus of the Max Planck Institute for Chemistry is on Earth System science, in particular on the chemical processes occurring in the atmosphere and their interactions with the biosphere and oceans. It also includes the influence of humans, as unprecedented urbanisation and industrialisation in the past centuries have changed the course of natural processes on our planet, in an epoch now known as the Anthropocene.

Currently, the institute employs some 300 staff in five departments (Aerosol Chemistry, at https://www.mpic.de/5587552/aerosol-chemistry, Atmospheric Chemistry, at: https://www.mpic.de/3537962/Atmospheric_Chemistry, Climate Geochemistry at: https://www.mpic.de/3669258/climategeochemistry, Multiphase Chemistry, at: https://www.mpic.de/3537984/Multiphase_Chemistry and Particle Chemistry, at: https://www.mpic.de/3537995/Particle_Chemistry) and additional research groups. Scientists conduct laboratory experiments, collect samples and record measurement data during field campaigns utilizing airplanes, ships, and vehicles. The practical work is complemented with mathematical models that simulate chemical, physical, and biological processes from molecular to global scales. A description of current research topics is given in the Institute Reports https://www.mpic.de/3916170/scientific-report.

The Nobel Prize in Chemistry in 1995

The Royal Swedish Academy of Sciences has decided to award the 1995 Nobel Prize in Chemistry to Professor Paul Crutzen, Max Planck Institute for Chemistry, Mainz, Germany (Dutch citizen), Professor Mario Molina, Department of Earth, Atmospheric and Planetary Sciences and Department of Chemistry, MIT, Cambridge, MA, USA and Professor F. Sherwood Rowland, Department of Chemistry, University of California, Irvine, CA, USA *for their work in atmospheric chemistry, particularly concerning the formation and decomposition of ozone.*

Paul J. Crutzen
Prize share: 1/3

Mario J. Molina
Prize share: 1/3

F. Sherwood Rowland
Prize share: 1/3

- **Paul Crutzen** was born in 1933 in Amsterdam and deceased in 2021 in Mainz. Dutch citizen. Doctor's degree in meteorology, Stockholm University, 1973. Member of the Royal Swedish Academy of Sciences, the Royal Swedish Academy of Engineering Sciences and Academia Europaea. Former Address: Professor **Paul Crutzen**, Max Planck Institute for Chemistry, P.O. Box 3060, D-55020 Mainz, Germany.

H. G. Brauch (ed.), *Towards Rethinking Politics, Policy and Polity in the Anthropocene*, The Anthropocene: Politik—Economics—Society—Science 35,
https://doi.org/10.1007/978-3-031-71807-6

- **Mario Molina** was born in 1943 in Mexico City, Mexico, and deceased in 2020. Ph.D. in physical chemistry, University of California, Berkeley. Member of the US National Academy of Sciences. Former Address: Professor **Mario Molina**, Department of Earth, Atmospheric and Planetary Sciences MIT 54 – 1312, Cambridge MA 02139, USA
- **F. Sherwood Rowland** was born in Delaware, Ohio, USA, 1927 and deceased in 2012. Doctor's degree in chemistry, University of Chicago, 1952. Member of the American Academy of Arts and Sciences and of the US National Academy of Sciences, where he was Foreign Secretary.

Source: "Press Release: The 1995 Nobel Prize in Chemistry". *Nobelprize.org*. Nobel Media AB 2014. Web. 3 Jan 2015. http://www.nobelprize.org/nobel_prizes/chemistry/laureates/1995/press.html and at: https://www.nobelprize.org/prizes/chemistry/1995/press-release/ and for the Award Ceremony Speech by Professor Ingmar Grenthe of the Royal Swedish Academy of Sciences is at: http://www.nobelprize.org/nobel_prizes/chemistry/laureates/1995/presentation-speech.html. Interview with *Paul Crutzen* by Astrid Gräslund at the meeting of Nobel Laureates in Lindau, Germany, June 2000. Paul Crutzen talks about family background, early education and interest in natural science; his work in the Institute of Meteorology in Stockholm (5:02); his discovery (6:55); the ozone layer (15:42); the Greenhouse Effect (19:10); ozone holes (23:43); and the consequences of a 'Nuclear winter' (27:09). See a Video of the Interview, 34 min., at: http://www.vega.org.uk/video/programme/111.

Paul Crutzen Who Coined the Term Anthropocene

The research of Paul J. Crutzen has been mainly concerned with the role of chemistry in climate and biogeochemistry, and in particular the photochemistry of ozone in the stratosphere and troposphere. In 1970 he hypothesised that natural ozone production by the action of solar ultraviolet radiation on molecular oxygen (O_2) is mainly balanced by destruction processes, involving NO and NO_2 as catalysts. These catalysts in turn result from the oxidation of N_2O, a product of microbiological nitrogen conversion in soils and waters. He and Prof. Harold Johnston of the University of California, Berkeley, pointed out that NO emissions from large fleets of supersonic aircraft could cause substantial ozone losses in the stratosphere.

In the years 1972–1974 Crutzen proposed that NO and NO_2 could catalyse ozone production in the background troposphere by reactions occurring in the CO and CH_4 oxidation chains. Additional photochemical reactions leading to ozone loss were likewise identified. These gross ozone production and destruction terms are each substantially larger than the downward flux of ozone from the stratosphere, which until then had been considered the main source of tropospheric ozone.

In 1979–1980 Crutzen and co-workers drew attention to the great importance of the tropics in atmospheric chemistry. In particular, some measurement campaigns in Brazil clearly showed that biomass burning in the tropics was a major source of air pollutants, on a par with, or larger than, industrial pollution in the developed world.

In 1982 Crutzen, together with Prof. John Birks of the University of Colorado, drew attention to the risk of darkness and strong cooling at the earth surface as a consequence of heavy smoke production by extensive fires in a nuclear war ('nuclear winter'). This study and additional studies by R. Turco, B. Toon, T. Ackerman, J. Pollack and C. Sagan and by the Scientific Committee on Problems of the Environment (SCOPE), to which Crutzen contributed, showed that more people could die from the indirect consequences of a nuclear war than by the direct impacts of the nuclear explosions.

In 1986, together with Dr. F. Arnold of the Max Planck Institute of Nuclear Physics in Heidelberg, Crutzen showed that nitric acid and water vapour could co-condense in the stratosphere at higher temperatures than required for water ice formation, providing a significant part of a chain

H. G. Brauch (ed.), *Towards Rethinking Politics, Policy and Polity in the Anthropocene*, The Anthropocene: Politik—Economics—Society—Science 35,
https://doi.org/10.1007/978-3-031-71807-6

of events leading to rapid ozone depletion at high latitudes during late winter and spring (the so-called Antarctic 'ozone hole').

His most recent research is concerned with the role of clouds in atmospheric chemistry as well as photochemical reactions taking place in the marine boundary layer, involving catalysis by halogen gases produced by marine organisms. Also, his last research deals with the chemical and climatic effects of the heavy air pollution which is found over Asia and other regions in the developing world: the so-called ABC (Atmospheric Brown Clouds) phenomenon.

Prof. Paul J. Crutzen on the eve of his 80th birthday on 2 December 2013 at the symposium The Anthropocene organised in his honour by the Max Planck Institute for Chemistry in Mainz. Photo by Carsten Costard who granted permission.

Hans Günter Brauch Foundation for Peace and Ecology in the Anthropocene

The Hans Günter Brauch Foundation for Peace and Ecology in the Anthropocene (HGBS) is a foundation under civil law with legal capacity which has its registered office in 74821 Mosbach in Baden-Württemberg.

Aims and Purposes of This Foundation

The purpose of the Foundation is to promote science and research, education and upbringing, international peace and environmental protection, in particular peace research, ecology and the interrelationships of peace and ecology (peace ecology) in the Anthropocene, the new epoch in the history of the Earth.

The Purpose of the Foundation Shall Be Realised Through

(a) Promoting local competitions on questions of the transition to sustainability and peace at general-education grammar schools in the Neckar-Odenwald County, which will be awarded annually to each school from 2022 onwards.
(b) An international science prize for peace and ecology has annually been awarded since 2023.

H. G. Brauch (ed.), *Towards Rethinking Politics, Policy and Polity in the Anthropocene*, The Anthropocene: Politik—Economics—Society—Science 35,
https://doi.org/10.1007/978-3-031-71807-6

(c) Maintenance of a foundation house with archives, library and the foundation office (after the founder's death).
(d) Conducting scientific events, research projects and their publication on peace research and ecology (currently not possible).

Members of the Governing Board

- Dr. Hans Günter Brauch (1947), Mosbach, Germany, Chairman of the Board Political Scientist,
- Prof. Dr. Úrsula Oswald Spring (1946), Mexico, Switzerland), Deputy Board Chair, UNAM, CRIM, Cuernavaca, Mexico,
- Dr. Hans Friedrich Happes, M.A. (1953), Deputy Board Chair, Former headmaster, Nicolaus-Kistner Grammar School in Mosbach (Ret.),
- Georg Nelius (1949), Member of the Board, former SPD member of the State Parliament (2007–2021), Deputy Mayor of Mosbach, Chair of the SPD Faction in the Town Council. Member of the County council. Neckar-Odenwald Council.

International Science Prize for Peace and Ecology in the Anthropocene

Starting in July 2023, an international science prize has been awarded annually by the *Hans Günter Brauch Foundation for Peace and Ecology in the Anthropocene* (HGBF) in Mosbach, Baden-Württemberg in South Germany that combines:

- *Peace and Ecology as* two distinct research programmes in the social sciences that evolved since the 1950s and 1970s and that have coexisted without a joint research perspective.
- *Peace Research* emerged between World War I and World War II in the idealist or Kantian tradition of international relations that offered critiques of the new realist tradition in international law and politics that prevailed after World War II.
- *Peace Research* started in North America and Scandinavia and spread in the 1970s to other parts of the world after the *International Peace Research Association* (IPRA) had been established in 1964.
- The '*ecology concept*' was coined in the 1860s by Ernst Häckel, a German biologist. Since World War II the concept widened as 'deep ecology', 'human ecology', 'political ecology', 'social ecology', 'geoecology' and 'ecofeminism'.
- In political science and international relations *ecological perspectives, environmental studies* started since the 1960s and *research on global environmental change* followed in the 1980s.
- *Peace Research and Environmental Studies* have coexisted as two research programmes with little conceptual, theoretical and empirical interchange.

- *Paul J. Crutzen*, Nobel Laureate in Chemistry, coined the *Anthropocene* in February 2000 as a *New Epoch in Earth's History*.
- *The Anthropocene Working Group (AWG)* proposed in 2019 as its start the years 1945 (first nuclear weapon test in Alomogordo) and 1950 when the great acceleration started.

The *Hans Günter Brauch Foundation* (HGBF) *on Peace and Ecology in the Anthropocene* (PEA) was established on 8 May 2020 on the occasion of the 75th anniversary of the end of World War II and of the liberation from Fascism to combine a joint peace research and ecology perspective on the new challenges facing humankind's survival in the Anthropocene.

The *HGBF on Peace and Ecology in the Anthropocene* supports *globally scientific research* and *locally education* in secondary schools in the county where the founder was born and has lived. The HGBF funds *local educational prizes* for grammar or high school students and globally this *international science prize*. Both Prizes will be awarded jointly to enhance the exchange of scientific results to secondary schools and the cooperation between scientists and pupils.

Between peace and security issues on the one hand and environment and sustainable development problems on the other, various relationships and linkages have been addressed in the social science literature since the 1960s in the Anthropocene Epoch:

- early conceptual debate on linkages between peace and the environment (since 1980s);
- conceptual discussion on environmental and ecological security (since 1989);
- research on the impacts of weapons and wars on the environment (Westing et al.);
- impact of environmental degradation and stress on environmental conflict (Baechler, Homer-Dixon);
- environmental peace-making (Conca) and role of the environment in post-conflict peacebuilding;
- emerging discourse and policy debate on climate change and conflicts (since 1989);
- early approaches to peace ecology and their shortcomings (Kyrou, Amster, Oswald/Brauch).
- theoretical and empirical work on sustainability transition, transformation of the economy, society and political realm and climate neutrality.

These manifold research themes have not yet resulted in a peace ecology research programme in the Anthropocene Epoch. The theme of *the First International Science Prize of the HGB-Foundation* in 2023 was: Climate Change and Conflicts in 2023 and the first recipient is Prof. Dr. Tobias Ide (Assoc. Prof. Murdoch University, Perth (Australia) who also teaches at the University of Hiroshima (Japan) and at the TH in Braunschweig (Germany). International Recognitions were awarded to Dr. Tom Deligiannis, Laurier University (Canada) and Dr. Richard Marcantonio, Notre Dame University (USA).

Chairman of the Board of Trustees (*Kuratorium*) of the HGB-Foundation: Lord Mayor of the town of Mosbach, Julian Stipp. Mosbach is since 2021 a member of "Mayors for Peace".

See at: http://hgb-stiftung.org/index.html and http://hgb-stiftung.de/index.html.

The theme of the *Second International Science Prize of the HGB-Foundation* in 2024 was: "Ecocide: Impacts of Wars and/or Climate Change on Food Security since 1945 that was awarded to Assoc. Prof. Graham Thaler (Oxford University) for his book on *Saving a Rainforest and Losing the World - Conservation and Displacement in the Global Tropics* (Yale University Press, 2024) and to Dr. Victor Tsilonis for his book on *The Jurisdiction of the International Criminal Court* (Springer Nature Switzerland, 2024). The award ceremony and an interview is at: http://hgb-stiftung.org/html/Award_winners_2024.html.

The theme for the third call in 2025 is: "Impacts of Wars and Global Environmental and Climate Change on Food Insecurity, Hunger and Migration" and the prize will be awarded on 9 October 2025 in Mosbach in South Germany. The deadline for submissions is 31 May 2025. Details on the call are at: http://hgb-stiftung.org/html/Science_Prize_2025.html and on the theme at: http://hgb-stiftung.org/html/Science_Prize_Wars_GEC_Food_Insecurity_Migration.html.

About the Editor

Hans Günter Brauch

Hans Günter Brauch (Germany), Dr., was until 2012 Adj. Prof. (Privatdozent) at the Faculty of Political and Social Sciences, Free University of Berlin; since 1987 he has been chairman of *Peace Research and European Security Studies* (AFES-PRESS). He is editor of: *Hexagon Book Series on Human and Environmental Security and Peace*; *SpringerBriefs on Environment, Security, Development and Peace* (ESDP); *SpringerBriefs on Pioneers in Science and Practice*; *Pioneers in Arts, Humanities, Science, Engineering, Practice*; and of *The Anthropocene: Politik – Economics – Society – Science* with Springer International Publishing, He was a guest professor of international relations at the universities of Frankfurt am Main, Leipzig, Greifswald, and Erfurt, a research associate at Heidelberg and Stuttgart, and a research fellow at Harvard and Stanford Universities. In the autumn and winter of 2013/2014 he was a guest professor at Chulalongkorn University in Bangkok. He has been a member of the IPRA Council (1996–2000 and 2016–2018) and of its Executive Committee (2016–2018), Chairman of IPRA's Subcommission on Statutes (2016–2018), Co-convenor of IPRA's *Ecology and Peace Commission* (EPC) from 2012 to 2018.

He has published on security, armaments, climate, energy, migration, and Mediterranean issues in English and German, and his works have been translated into Chinese, Spanish, Greek, French, Danish, Finnish, Russian, Japanese, Portuguese, Serbo-Croatian, and Turkish. Recent books in English: (co-edited with Liotta, Marquina, Rogers, Selim): *Security and Environment in the Mediterranean: Conceptualising Security and Environmental Conflicts*, 2003; (co-edited with Oswald Spring, Mesjasz, Grin, Dunay, Chadha Behera, Chourou, Kameri-Mbote, Liotta): *Globalisation and Environmental Challenges: Reconceptualizing Security in the 21st Century, 2008;* (co-edited with Oswald Spring, Grin, Mesjasz, Kameri-Mbote, Chadha Behera, Chourou, Krummenacher): *Facing Global Environmental Change: Environmental, Human, Energy, Food, Health and Water Security Concepts* (2009); (co-edited with Oswald Spring): *Reconceptualizar la Seguridad en el Siglo XXI* (2009); (co-edited with Oswald Spring, Mesjasz, Grin, Kameri-Mbote, Chourou, Dunay, Birkmann): *Coping with Global Environmental Change, Disasters and Security – Threats, Challenges, Vulnerabilities and Risks* (2011); (co-edited with Scheffran, Brzoska, Link, Schilling, 2012): *Climate Change, Human Security and Violent Conflict: Challenges for*

H. G. Brauch (ed.), *Towards Rethinking Politics, Policy and Polity in the Anthropocene*, The Anthropocene: Politik—Economics—Society—Science 35,
https://doi.org/10.1007/978-3-031-71807-6

Societal Stability; (co-edited with Oswald Spring, Grin, Scheffran, 2016): *Handbook on Sustainability Transition and Sustainable Peace;* (co-edited with Oswald Spring, Bennett and Serrano Oswald, 2016): *Addressing Global Environ-mental Challenges from a Peace Ecology Perspective*; (co-edited with Oswald Spring, Bennett and Serrano Oswald, 2016): *Regional Ecological Challenges for Peace in Africa, the Middle East, Latin America and Asia Pacific*; co-edited with Oswald Spring, Collins and Serrano Oswald, 2018: *Climate Change, Disasters, Sustainability Transition and Peace in the Anthropocene*; 2021: co-edited with Oswald Spring: *Decolonising Conflicts, Security, Peace, Gender, Environment and Development in the Anthropocene*; 2021: co-edited with Benner, Lax, Crutzen, Pöschl, Lelieveld: *Paul J. Crutzen and the Anthropocene: a new Epoch in Earth's History*; 2024: co-edited with D. J. Dunn and P. Burton: *John W. Burton: A Pioneer in International Relations, Peace Theory, World Society, Human Needs, Conflict Prevention, Management and Resolution.*

Address: PD Dr. Hans Günter Brauch, Alte Bergsteige 47, 74821 Mosbach, Germany
Email: brauch@afes-press.de
Website: http://www.afes-press.de and http://www.afes-press-books.de/

Hans Günter Brauch is the Sole Editor of Two Book Series of these Volumes on 'Pioneers' in Peace Research

Pioneers in Science and Practice (PSP)

Detailed websites are online by clicking on the '*book titles*' to the *AFES-PRESS-Books websites* and by clicking on 'Springer' to the book website of *Springer Nature* where the books may be purchased and information on accesses of the book and chapters can be found, including specific information on citation.

Editor of these Volumes on Pioneers in Peace Research

Pioneers in Science and Practice (PSP)

1: Arthur H. Westing: *Arthur H. Westing: Pioneer on the Environmental Impact of War* (2013)
5: Johan Galtung; Dietrich Fischer: *Johan Galtung: Pioneer of Peace Research* (2013)
6: Dieter Senghaas: *Dieter Senghaas: Pioneer of Peace and Development Research* (2013)
7: Chadwick F. Alger: *Chadwick F. Alger: Pioneer in the Study of the Political Process and on NGO Participation in the United Nations* (2013)
8: Chadwick F. Alger: *The UN System and Cities in Global Governance* (2013)
9: Chadwick F. Alger: *Peace Research and Peacebuilding* (2014)
13: Arthur H. Westing:*Texts on Environmental and Comprehensive Security* (2014)
19: Hans Günter Brauch; Teri Grimwood (Eds.): *Jonathan Dean: Pioneer in Détente in Europe, Global Cooperative Security Arms Control and Disarmament* (2015)

H. G. Brauch (ed.), *Towards Rethinking Politics, Policy and Polity in the Anthropocene*,
The Anthropocene: Politik—Economics—Society—Science 35,
https://doi.org/10.1007/978-3-031-71807-6

21: Ulrich Bartosch (Ed.): *Carl Friedrich von Weizsäcker: Pioneer of Physics, Philosophy, Religion, Politics and Peace Research* (2014)
25: Ulrich Bartosch (Ed.): *Carl Friedrich von Weizsäcker: Major Texts on Politics and Peace Research* (2015)
26: Betty A. Reardon; *Dale Snauwaert: Betty A. Reardon: A Pioneer in Education for Peace and Human Rights* (2015)
27: Betty A. Reardon: *Betty A. Reardon: Key Texts in Gender and Peace* (2015)
29: Nils Petter Gleditsch: *Nils Petter Gleditsch: Pioneer in the Analysis of War and Peace* (2016)
30: Morton Deutsch; Peter T. Coleman (Eds.): *Morton Deutsch: A Pioneer in Developing Peace Psychology* (2015)
31: Morton Deutsch; Peter T. Coleman: *Morton Deutsch (Eds.): Major Texts on Peace Psychology* (2016)
32: Marek Thee; Nils Petter Gleditsch; Stein Tønnesson; Marta Bivand Erdal (Eds.): *Marek Thee: My Story: A Journey through the 20th Century* (2013)
34: Harvey Starr (Ed.): *Bruce M. Russett: Pioneer in the Scientific and Normative Study of War, Peace, and Policy* (2014)
41: Kalevi Holsti (Ed.): Kalevi Holsti: *A Pioneer in International Relations Theory, Foreign Policy Analysis, History of International Order, and Security Studies* (2016)
42: Kalevi Holsti (Ed.): Kalevi Holsti: *Major Texts on War, Peace, and International Order* (2016)
50: Paul J. Crutzen, Hans Günter Brauch (Eds.): *Paul J. Crutzen: A Pioneer on Atmospheric Chemistry and Climate Change in the Anthropocene* (2016)

Pioneers in Arts, Humanities, Science, Engineering and Practice (PAHSEP)

1: Louis Kriesberg: *Louis Kriesberg: Pioneer in Peace and Constructive Conflict Resolution Studies* (2016)
6: J. Russell Boulding (Ed.): *Elise Boulding: A Pioneer in Peace Research, Peacemaking, Feminism and the Family: From a Quaker Perspective* (2017)
7: J. Russell Boulding (Ed.): *Elise Boulding: Writings on Peace Research, Peacemaking and the Future* (2017)
8: J. Russell Boulding (Ed.): *Elise Boulding: Writings on Feminism, the Family and Quakerism* (2017)
9: J. Russell Boulding (Ed.): *Elise Boulding: Autobiographical Writings and Selections from Unpublished Journals and Letters* (2017)
10: Patricia M. Shields (Ed.): *Jane Addams: Progressive Pioneer of Peace, Philosophy, Sociology, Social Work and Public Administration* (2017)
13: Herbert C. Kelman, USA; Ronald J. Fisher (Eds.): *Herbert C. Kelman: Pioneer in the Social Psychology of Conflict Analysis and Resolution* (2017)

14: Ronald J. Fisher: *Ronald J. Fisher: A North American Pioneer in Interactive Conflict Resolution* (2017)
17: Úrsula Oswald Spring: *Úrsula Oswald Spring: A Pioneer on Gender, Ecology, Peace, Security, and Development* (2018)
18: Úrsula Oswald Spring: *Earth at Risk in the 21st Century: Rethinking Peace, Environment, Gender, and Human, Water, Health, Food, Energy Security, and Migration* (2018)
20: Dale T. Snauwaert (Ed.): *Exploring Betty A. Reardon's Perspective on Peace Education: Looking Back, Looking Forward* (2019)
21: Paul Rogers: *Paul Rogers: A Pioneer in Critical Security Analysis and Public Engagement* (2019)
22: Galia Golan: *Galia Golan: An Academic Pioneer on the Soviet Union, Peace and Conflict Studies, and a Peace and Feminist Activist* (2019)
23: *I. William Zartman: I. William Zartman: A Pioneer in Conflict Management and Area Studies: Essays on Contention and Governance* (2019)
24: Luc Reychler, Arnim Langer (Eds.): *Luc Reychler: A Pioneer in Sustainable Peacebuilding Architecture* (2020)
25: Charles Lewis Taylor, Bruce M. Russett (Eds.): *Karl W. Deutsch: Pioneer in the Theory of International Relations* (2020)
26: Nathan C. Funk; Meena Sharify-Funk: *Abdul Aziz Said: A Pioneer in Peace, Intercultural Dialogue, and Cooperative Global Politics* (2022)
27: Nils Petter Gledisch (Ed.): *Lewis F. Richardson: His Intellectual Legacy and Influence in the Social Sciences* (2020)
28: Raimo Väyrynen: *Raimo Väyrynen: A Pioneer in International Relations, Scholarship and Policy-Making* (2021)
29: Harvey Starr: *Pioneer in the Study of Conflict Processes and International Relations* (2020)
30: Peter Wallensteen: *Peter Wallensteen: A Pioneer in Making Peace Researchable* (2022)
31: Peter Wallensteen; Armend Bekaj (Eds.): *Alva Myrdal: A Pioneer in Nuclear Disarmament* (2022)
32: Mary Kaldor, Joshua Kaldor-Robinson, Peter Pringle, Henrietta Wilson, Ghionis Alexander, Guthrie Richard (Eds.): *Julian P. Perry Robinson: Pioneer in the Prohibition of Chemical and Biological Warfare* (2025)
33: David J. Dunn, Hans Günter Brauch, Patricia Burton (Eds.): *John W. Burton: A Pioneer in Conflict Prevention, Management and Resolution Key Contributions on International Relations, Peace Theory, World Society, and Human Needs* (2024)
34: Kathleen Vogel, Nicole Ball, Milton Leitenberg (Eds.): *Milton Leitenberg: A Pioneer in Peace Research and Arms Control* (2025)
35: Roger Boyd, Simon Dalby (Eds.): *Simon Dalby A Pioneer in International Relations - Key Contributions On Critical Geopolitics, Environmental Security and the Anthropocene* (2025)
36: Dieter Senghaas: *Dieter Senghaas: Hearing Peace: Music, Sound and Notes in Peace Education* (2025)

The Anthropocene: Politik – Economics – Society – Science (APESS)

1: Susanne Benner, Gregor Lax, Paul J. Crutzen, Ulrich Pöschl, Jos Lelieveld, Hans Günter Brauch (Eds.): *Paul J. Crutzen and the Anthropocene: A New Epoch in Earth's History* (2022)
2: *Maja Göpel: The Great Mindshift: How a New Economic Paradigm and Sustainability Transformations go Hand in Hand (2016)*
3: *Audley Genus (Editor): Sustainable Consumption: Design, Innovation and Practice (2016)*
4: *Hans Günter Brauch, Úrsula Oswald Spring, Juliet Bennett, Serena Eréndira Serrano Oswald (Eds.): Addressing Global Environmental Challenges from a Peace Ecology Perspective (2017)*
5: *Úrsula Oswald Spring, Hans Günter Brauch, Serena Eréndira Serrano Oswald, Juliet Bennett (Eds.): Regional Ecological Challenges for Peace in Africa, the Middle East, Latin America and Asia Pacific (2017)*
6: *M. Laura Vazquez Maggio: Mobility Patterns and Experiences of the Middle Classes in a Globalizing Age: The Case of Mexican Migrants in Australia* (2017)
7: *Tamer M. Qarmout: Delivering Aid Without Government: International Aid and Civil Society Engagement in the Recovery and Reconstruction of the Gaza Strip (2017)*
8: *David Curran: More than Fighting for Peace? The Role of Conflict Resolution in Training Programmes for Military Peacekeepers (2017)*
9: *Heather Devere - Kelli Te Maihāroa - John Synott (Eds.): Peacebuilding and the Rights of Indigenous Peoples Experiences and Strategies for the 21st Century (2017)*
10: Michael Durant Thomas: *The Securitization of Climate Change: Australian and United States' Military Responses (2003 - 2013)*
11: Zerrin Savaşan: *Paris Climate Agreement: A Deal For Better Compliance? Lessons Learned from the Compliance Mechanisms of the Kyoto and Montreal Protocols* (2017)
12: Lydia Wanja Gitau: *Trauma-sensitivity and Peacebuilding. Considering the Case of South Sudanese Refugees in Kakuma Refugee Camp* (2018)
13: Eleonore Emkic: *Reconciliation and Education in Bosnia and Herzegovina. From Segregation to Sustainable Peace* (2018)
14: Mohan Jyoti Dutta: *Imagining India in Discourse: Meaning, Power, Structure* (2017)
15: Sabit Erşahin, Selim Kapur, Erhan Akça, Ayten Namlı, Hakkı Emrah Erdoğan (Eds.): *Carbon Management, Technologies, and Trends in Mediterranean Ecosystems* (2017)
16: Abubakar Ayesah Uy: Peacebuilding and Sustainable Human Development. The Pursuit of the Bangsamoro Right to Self-Determination (2019)
17: Lourdes Arizpe Schlosser (Ed.): *Culture, International Transactions and the Anthropocene* (2019)

18: Tsugihiro Watanabe, Selim Kapur, Mehmet Aydın, Rıza Kanber, Erhan Akça (Eds.): *Climate Change Impacts on Basin Agro-ecosystems* (2019)
19: Dumisani Ngwenya: *Healing the Wounds of Gukurahundi in Zimbabwe. A Participatory Action Research Project* (2018)
20: Klaus von Beyme: *From Post-Democracy to Neo-Democracy* (2018)
21: Imtiaz Ahmed: *South Asian Rivers. A Framework for Cooperation* (2018)
22: Sidonia Angom: *Women in Peacemaking and Peacebuilding in Northern Uganda* (2018)
23: Ma. Luisa Marván, Esperanza López-Vázquez (Eds.): *Preventing Health and Environmental Risks in Latin America* (2018)
24: Úrsula Oswald Spring, Serena Eréndira Serrano Oswald (Eds.): *Risks, Violence, Security and Peace in Latin America. 40 Years of the Latin American Council of Peace Research (CLAIP)* (2018)
25: Hans Günter Brauch, Úrsula Oswald Spring, Andrew E. Collins, Serena Eréndira Serrano Oswald (Eds.): *Climate Change, Disasters, Sustainability Transition and Peace in the Anthropocene* (2018)
26: Nayak, Amar KJR (Ed.): *Transition Strategies for Sustainable Community - Systems. Design and Systems Perspectives* (2019)
27: Carl Middleton, Vanessa Lamb (Eds.): *Knowing the Salween River: Resource Politics of a Contested Transboundary River* (2019)
28: Saleemul Huq, Jeffrey Chow, Adrian Fenton, Clare Stott, Julia Taub, Helena Wright (Eds.): *Confronting Climate Change in Bangladesh. Policy Strategies for Adaptation and Resilience* (2019)
29: Luis-Alberto Padilla: *Sustainable Development in the Anthropocene. Towards a New Holistic and Cosmopolitan Paradigm* (2021)
30: Úrsula Oswald Spring, Hans Günter Brauch (Eds.): *Decolonising Conflicts, Security, Peace, Gender, Environment and Development in the Anthropocene* (2021)
31: Zerrin Savaşan, Vakur Sümer (Eds.): *Environmental Law and Policies in Turkey* (2020)
32: Clarilza Prado de Sousa, Serena Eréndira Serrano Oswald (Eds.): *Social Representations for the Anthropocene: Latin American Perspectives* (2021)
33: Judith Nora Hardt, Cameron Harrington, Nick Simpson, Franziskus von Lucke, Adrien Estève (Eds.): *Climate Security in the Anthropocene - Exploring the Approaches of United Nations Security Council Member-States* (2023)
34: Jean Chrysostome K. Kiyala - Geoff Harris (Eds.): *Civil Society and Peacebuilding in Sub-Saharan Africa in the Anthropocene: An Overview* (2021)
35: Hans Günter Brauch (Ed.): *Towards Rethinking Politics, Policy and Polity in the Anthropocene* (2025)
36: Hussein Solomon, Jude Cocodia (Ed.): *African Security in the Anthropocene* (2023)
37: Jean Chrysostome K. Kiyala, Norman Chivasa (Ed.): *Climate Change and Socio-political Violence in Sub-Saharan Africa in the Anthropocene - Perspectives from Peace Ecology and Sustainable Development* (2024)

38: Amar K. J. R. Nayak (Ed.): *Regenerative Ecosystems in the Anthropocene - A Transdisciplinary Ecosystemic Framework for Regenerativeness* (2024)
39: Antonina Ivanova Boncheva, José Ernesto Rangel Delgado (Eds.): *Transition to a Safe Anthropocene in the Asia-Pacific - Sustainability, Climate Action, and Green Technology* (2025)

Hexagon Series on Human, Environmental Security and Peace (HESP)

For individual titles see at:
AFES-PRESS: http://www.afes-press-books.de/html/hexagon.htm
Springer: https://www.springer.com/series/8090

SpringerBriefs in Environment, Security, Development and Peace (ESDP)

Sub Book Series

ASEAN Studies (ESDP-AS)
Mediterranean Studies (ESDP-MeS)
Migration Studies (ESDP-MiS)
Peace and Security Studies (ESDP-PSS)
Sustainable Development and Sustainability Transition Studies (ESDP-SDST)

For individual titles see at:
AFES-PRESS: http://www.afes-press-books.de/html/SpringerBriefs_ESDP.htm
Springer: https://www.springer.com/series/10357

Stand Alone Book (SAB) Volumes http://www.afes-press-books.de/html/SpringerBriefs_ESDP.htm

For individual titles see at:
AFES-PRESS: http://www.afes-press-books.de/html/SAB.htm
Springer: All volumes listed under AFES-PRESS are linked to the individual Springer Book pages on the web.

About the Contributors

Simon Dalby is Professor emeritus at Wilfrid Laurier University, Waterloo, Ontario, a Senior Fellow at the Centre for International Governance Innovation and Senior Research Fellow at the University of Victoria Centre for Global Studies. He was educated at Trinity College Dublin, the University of Victoria and holds a Ph.D. from Simon Fraser University. Prior to joining Wilfrid Laurier University he was Professor of Geography, Environmental Studies and Political Economy at Carleton University. He has served as co-editor of *Geography Compass* and *Geopolitics* journals, as the sustainability theme lead for the Borders in Globalization research program, and from 2012 to 2018 he was CIGI Chair in the Political Economy of Climate Change at the Balsillie School of International Affairs. He is coeditor of *Reframing Climate Change* (Routledge 2016) and *Achieving the Sustainable Development* Goals (Routledge 2019) and author of *Anthropocene Geopolitics* (University of Ottawa Press 2020), *Rethinking Environmental Security* (Edward Elgar 2022) and *Pyromania: Fire and Geopolitics in a Climate Disrupted World* (Agenda 2023), Roger Boyd, Simon Dalby (Eds.): *Simon Dalby A Pioneer in International Relations - Key Contributions On Critical Geopolitics, Environmental Security and the Anthropocene* (2025)

Address: Prof. Dr. Simon Dalby, 245 Spinnaker Drive, Mayne Island, B.C. V0N2J2 Canada

Email: sdalby@gmail.com

Website: https://www.balsillieschool.ca/people/simon-dalby/

John Grin (M.Sc. in physics, 1986; Ph.D. in technology assessment, 1990) is full professor of public policy and governance at the Department of Political Science at the University of Amsterdam, which he joined after post-doc positions at Princeton University and at VU University, Amsterdam. His research focuses on the design and governance of transitions and associate policies, especially in the domains of agrifood, urban utility systems, water management and health care. He co-established and co-led a major network (KSI, 2005–2010) to develop transition studies, leading

H. G. Brauch (ed.), *Towards Rethinking Politics, Policy and Polity in the Anthropocene*, The Anthropocene: Politik—Economics—Society—Science 35,
https://doi.org/10.1007/978-3-031-71807-6

to several hundreds of scholarly publications, practice experiments, a book series and a new journal as well as the International Sustainability Transition Network. He is member of the Editorial Board, *J Environmental Policy and Planning* and co-chair of AFES PRESS. Grin has been director (2006–2010) and interim director (March-September, 2022) of the Amsterdam Institute of Social Science Research, and founder and co-director (2010–2022) of its programme group Transnational Configurations, Conflicts and Governance. He was Chief Science Officer at the Dutch Ministry of Agriculture, Food and Nature and is currently member of the Acceleration Dialogue of the Dutch National Climate Platform.

Address: Prof Dr. John Grin, Department of Political Science, University of Amsterdam Postbus 15578, 1001 NB Amsterdam, Netherlands

E-mail: j.grin@uva.nl

Website: www.uva.nl/profiel/j.grin

Richard Ned Lebow is Professor Emeritus of International Political Theory in the War Studies Department of King's College London, having recently retired after fifty-four years as a university professor. He is Honorary Fellow of Pembroke College, University of Cambridge, and James O. Freedman Presidential Professor Emeritus at Dartmouth College. He is a Fellow of the British Academy and a member of The Atheneum. His most recent books are *The Quest for Knowledge in International Relations: How Do We Know?* (Cambridge 2021), and coauthored with Feng Zhang, *Justice East and West and International Order* (Oxford, 2022). In 2022, he also published *Rough Waters and Other Stories* (Ethics International, 2021) and *Obsession* (Pegasus, 2020), a classic English murder mystery. Lebow is also an avid long-distance runner, back-country skier, and tennis player.

Address: Prof. Richard Ned Lebow, Ph.D., Department of War Studies, King's College London, London WC2R 2LS, UK

Email: Richard.Lebow@kcl.ac.uk

Websites: https://www.kcl.ac.uk/people/richard-ned-lebow; http://afes-press-books.de/html/SpringerBriefs_PSP_Lebow.htm.

Czeslaw Mesjasz (Poland) is Associate Professor at the Cracow University of Economics, Physics (M.Sc.), Jagiellonian University, Cracow, Poland, Organisation and Management (MA), Cracow University of Economics, Cracow. Ph.D. and Habilitation from the Cracow University of Economics. His research interests include management theory, strategic management, project management, conflict management and negotiation – management and international relations, applications of systems approach in social sciences (management, security), econophysics – complexity of social systems (management systems), theoretical foundations of corporate governance; studies of relations between complexity and environment, complexity and ignorance in the studies of society. Author of about 300 publications and unpublished works in Polish and English. Works translated into Chinese,

Danish, Spanish, and Turkish. Member of several academic institutions: Polish Economic Association; International Peace Research Association (IPRA) – from 1990 until 1994, convenor of IPRA Security and Disarmament Commission; International Studies Association (ISA); Academy of Management; European Academy of Management; International Sociological Association – Research Committee RC 51, (Sociocybernetics); Econophysics Section of the Polish Physical Society; member of the Board of Arbeitsgruppe Friedensforschung und Europäische Sicherheitspolitik (AFES-PRESS), Mosbach, until 2022; Co-operation with the Millennium Project of the American Council for the United Nations University since 1998.

Address: Assoc. Prof. Dr. Habil. Czeslaw Mesjasz, Cracow University of Economics, 31-510 Kraków, ul. Rakowicka 27, Poland

Email: mesjaszc@uek.krakow.pl

Prof. Dr. Úrsula Oswald-Spring (Mexico) is a full-time researcher at the Regional Multidisciplinary Research Centre of the National Autonomous University of Mexico (CRIM-UNAM). She has studied medicine, psychology, philosophy, modern languages, anthropology, and ecology in Madagascar, Paris, Zurich, and México and has a Ph.D. from the University of Zurich. She is the first MRF-Chair on Social Vulnerability at the United National University Institute for Environment and Human Security (UNU-EHS); a lead author of the IPCC and task force of IIASA. She was elected in 2016 in Sierra Leone as Secretary General and in 1998 in South Africa as President of the International Peace Research Association. Between 2002–2006, she was also General Secretary of the Latin-American Council for Peace Research (CLAIP), where she is now Founder Honorary President. She was the first General Attorney (an ombudswoman for the environment) and Minister of Environmental Development in Morelos. She has written 70 books, more than 422 scientific articles and book chapters. She got several prizes for her academic work: Sor Juana Inés de la Cruz; the Ecological Merit of Tlaxcala; the Medal of General Emiliano Zapata; Women of the Year 2000; the Fourth Decade of Development of UN; Academic Women from UNAM, and the Merit of 50 years of Academic Work. She has worked with women, peasants, and environmental organisations.

Address: Prof. Dr. Úrsula Oswald-Spring, Rio Privada 1, Vista Hermosa, Cuernavaca, Morelos, Mexico

Email: uoswald@gmail.com

Website: http://www.afes-press.de/html/download_oswald.html

Jürgen Scheffran (Germany) is Professor emeritus of Geography at the University of Hamburg and Founder of the Research Group *Climate Change and Security* (CLISEC) in the *Center for Earth System Research and Sustainability* (CEN) and the *Climate Excellence Clusters* CliSAP/CLICCS. After the study and doctorate in physics, he worked in interdisciplinary research groups in environmental science, peace and conflict research at the universities of Marburg, Darmstadt, Paris and Illinois, as well as the *Potsdam Institute for Climate Impact Research* (PIK). In addition to collaboration in science and peace networks, he has been involved in projects for the United Nations, the Office of Technology Assessment of the German Bundestag and in the Expert Commission on the Causes of Forced Migration of the German government. Main research interests are: climate security, environmental migration and resource conflicts; nexus water-food-energy and urban-rural-coastal interactions; sustainability, complex systems and models; technology assessment, international security and arms control. He co-authored several hundred articles and dozens of books, including *Climate Change, Security Risks, and Violent Conflicts* (Hamburg University Press 2020); *Agent-based Modeling of Environmental Conflict and Cooperation* (Routledge 2019); *Handbook of Sustainability Transitions and Sustainable Peace* (Springer 2016); *Climate Change, Human Security and Violent Conflict* (Springer 2012*); Handbook of Bioenergy Economics and Policy* (Springer 2010); *Biofuels from Agricultural Wastes and Byproduc*ts (Wiley/Blackwell, 2010); *Advanced Methods for Decision Making and Risk Management in Sustainability Science* (Nova Science 2007).

Address: Prof. Dr. Jürgen Scheffran, Institute of Geography, CEN, Grindelberg 7, 20144 Hamburg, Germany

Email: juergen.scheffran@uni-hamburg.de

Website: www.clisec.uni-hamburg.de

Dahlia Simangan is Associate Professor at Hiroshima University's Center for Peaceful and Sustainable Futures (CEPEAS), The IDEC Institute. She contributes to the research activities of the university's *Network for Education and Research on Peace and Sustainability* (NERPS) and teaching activities of the International Peace and Co-existence Program of the Graduate School of Humanities and Social Sciences. Prior to that, she was a Kanagawa University *Japan Society for the Promotion of Science* (JSPS) postdoctoral fellow for research in Japan, nominated by the *United Nations University – Centre for Policy Research* (UNU-CPR) in Tokyo. She holds a Ph.D. in International, Political and Strategic Studies from the Australian National University. She is the author of *International Peacebuilding and Local Involvement: A Liberal Renaissance?* (London: Routledge, 2019) and a member of *Planet Politics Institute.*

Address: Prof. Dr. Dahlia Simangan, IDEC Building, Hiroshima University, 1-5-1 Kagamiyama, Higashi-Hiroshima, Hiroshima, 739-8529 Japan

Email: simangan@hiroshima-u.ac.jp

Website: https://dahliasimangan.com/

Georg Simonis (Germany) born in 1943, studied political science in Berlin (Otto Suhr Institute of the Free University), doctorate and habilitation in political science at the University of Constance. After working at the Universities of Darmstadt and Duisburg, he became managing director of the NRW research programme "Mensch und Technik- Sozialverträgliche Technikgestaltung" (1985–1992). From

1990 to 1992 Director at the Institute for Work and Technology of the Science Centre on North Rhine-Westfalia (NRW) in Gelsenkirchen for the area of technology diffusion. Since 1991 Professor of International Politics at the Institute for Political Science at the Fernuniversität in Hagen. Founding director of the interdisciplinary postgraduate programme in environmental sciences (2000–2008). After retirement, teaching positions at the Universities of Osnabrück, Wuppertal, Herdecke/Witten, Fernuniversität in Hagen. Numerous publications, most recently: *Global Governance. Entstehung – Institutionen – Analyse* (Wiesbaden: Springer VS): 2022.

Address: Prof. Dr. Georg Simonis, FernUniversität in Hagen, Institut für Politikwissenschaft, 58084 Hagen, Visitors'address: Gebäude 1, KSW, Bauteil C, Universitätsstr. 33, 58097 Hagen

Email: Georg.Simonis@fernuni-hagen.de

Website: https://www.fernuni-hagen.de/polis/lg2/team/georg.simonis.shtml

The Neckar-Odenwald District Rich in Culture and in Peace with Nature

NECKAR-ODENWALD KREIS

The Neckar-Odenwald District is located in the north of the state of Baden-Württemberg and borders with counties in the state of Hesse and Bavaria. Geographically it combines these subregions: the Odenwald forest with its highest mountain (Katzenbuckel (cat's humb, 626,8 m) in the North, the Neckar Valley, the Small Odenwald and Kraichgau and in the East the Madonnenland and Bauland. Geologically and language wise the district is diverse. Its towns are historically a part of Franconia while the Neckar valley were a part of the Electoral Palatinate (Kurpfalz).

In Peace with Nature

The district attracts visitors with its enchanting half-timbered towns, castles, palaces, an adventurous stalactite cave and numerous certified cycling and hiking trails. An ideal holiday destination to enjoy yourself while actively discovering nature and culture. Nestled in the Neckartal-Odenwald Nature Park and the UNESCO Bergstraße-Odenwald Geopark, the region is home to many geological highlights, such as the Katzenbuckel, the Margareten and Wolfsschlucht gorges and the Eberstadt stalactite cave.

The Roman Museum in Osterburken, the Odenwald Open-Air Museum and many other well curated museums inform about the history of the district and its inhabitants, e.g. the UNESCO World Heritage Site *Upper Germanic-Raetian Limes*, the *Odenwald Limes* and the famous knight Götz von Berlichingen. The *Neckarsteig*, certified as a 'Quality Trail Hikeable Germany', leads 128 km from Heidelberg to

H. G. Brauch (ed.), *Towards Rethinking Politics, Policy and Polity in the Anthropocene*,
The Anthropocene: Politik—Economics—Society—Science 35,
https://doi.org/10.1007/978-3-031-71807-6

Bad Wimpfen and was honoured as 'Germany's most beautiful hiking trail 2018'. The district is a true Eldorado for cyclists. The Neckar Valley Cycle Path and the Odenwald Madonna Cycle Path were praised as quality cycle routes.

Innovative, Globally Active Industry aiming at Sustainable Employment

The Neckar-Odenwald district as a business location offers a total of around 40,000 jobs. The German Armed Forces are one of the largest employers in the Neckar-Odenwald district:

- Johannes-Diakonie Mosbach, with about 2600 employees, cares for and supports around 3200 people with disabilities at more than 30 locations in communities in Baden.
- Since 1 January 2005, the Neckar-Odenwald district has around 900 employees.
- The Braun plant in Walldürn is the production site for the company's entire shaver division and employs around 1100 people.

Other global companies such as Getrag (gear manufacturing), Honeywell International (valves) and Fibro (standard parts) have sites in the Neckar-Odenwald district. Medium-sized companies, which are often world market leaders in their field, are well represented: Companies such as AZO GmbH (automated feeding systems for handling raw materials) and Bleichert (conveyor systems), Maschinenfabrik Gustav Eirich and Grammer Interior Components GmbH, Mosca (packaging machines), KWM Weishaar (sheet metal processing for the printing machine and computer industry, air conditioning and clean room technology, automotive industry, machine and rail vehicle construction) and MAG Hüller Hille as well as Scheuermann & Heilig GmbH (precision parts in the areas of spring, stamping, bending and assembly technology) are also among the largest employers in the district, each with several hundred employees.

The Neckar-Odenwald district has been officially certified as a Fairtrade district since 24 September 2020. Fair trade complements regional marketing, which is an integral part of the 'Fair and Regional' campaign. This promotes both fairly traded and regional products accordingly. In this way, the district assumes social responsibility and positions itself as an innovative and cosmopolitan district that sets an example and provides impetus for the topic of sustainable procurement. Over 100 co-operation partners are active in the Fairtrade district as part of the 'Fair and Regional' campaign with numerous activities throughout the year.

Rich in History, Art and Culture

The Neckar-Odenwald district is criss-crossed by two Roman Limes sites—the Neckar-Odenwald Limes and the Upper Germanic-Raetian Limes. Both run from north to south and both have well-restored remains of forts, watchtowers, redoubts, settlements and Roman roads. The Neckar-Odenwald Limes lies to the west and cuts through the Neckartal-Odenwald Nature Park from Mudau-Schloßau to Mosbach-Bergfeld. The Upper Germanic-Raetian Limes touches the Neckar-Odenwald district north-east of Walldürn and extends in the Neckar-Odenwald district as far as Osterburken. It was declared a UNESCO World Heritage Site in 2005. The Limes nature trail north of Walldürn and the Osterburken Roman Museum are a branch of the Baden-Württemberg Archaeological State Museum.

Castles and Palaces

The Castle Road, one of the most traditional holiday routes in Germany, leads through the Neckar Valley past the following castles and palaces in the Neckar-Odenwald district: Zwingenberg Castle (Zwingenberg), - Guttenberg Castle (Haßmersheim-Neckarmühlbach) and Hornberg Castle (Neckarzimmern). Around 40 museums, heritage centres and other collections exist in the Neckar-Odenwald district to preserve its cultural heritage.

Zwingenberg Castle and Castle Music Festival

Hornberg Castle of Knight Götz von Berlichingen

The concentration camp memorial in Mosbach-Neckarelz in the Comenius-Special School, together with the 'Goldfish' history trail, provides impressions of the everyday life of the concentration camp prisoners and forced labourers who were deployed in the underground armaments production at Neckarelz. Several schools have joint educational projects with the concentration camp memorial.

Mosbach an Emerging University Town

Mosbach is a town in the north of Baden-Württemberg, Germany. It is the seat of the Neckar-Odenwald district and has a population of approximately 25,000 distributed in six boroughs. Located about 35 km east of Heidelberg, it is situated south of the Odenwald mountains. The town is part of the conservation area Neckartal-Odenwald Nature Park and the UNESCO Bergstrasse-Odenwald Nature Park. Mosbach is a member of the "Mayors of peace".

Mosbach is twinned with: Budapest II, Hungary; Château-Thierry, France; Finike, Turkey; Lymington, England, United Kingdom and Rosolina, Italy. Historic sites include: the historic town centre with the pedestrian area and timber-framed houses, such as: the Palm House built in 1610, which is the town's emblem, the Salzhaus, which is the oldest timber-framed house, the old town hall with tower, the former collegiate church, now a parish church, of which the nave is used by the Protestants, and the chancel by the Roman Catholics, the *Tempelhaus* in Neckarelz, which has the character of both a castle and a church. Mosbach lies on two *heritage* routes: the *Burgenstraße* ("*Castle Road*"), linking many historic castles, the *Deutsche Fachwerkstraße* ("*German Half-timbered Road*"), joining the locations of many of the best German half-timbered buildings.

H. G. Brauch (ed.), *Towards Rethinking Politics, Policy and Polity in the Anthropocene*, The Anthropocene: Politik—Economics—Society—Science 35,
https://doi.org/10.1007/978-3-031-71807-6

Institutions of Higher Education

Mosbach is administrative centre of the Neckar-Odenwald District. It has several schools of Higher Education (Nicolaus-Kistner, Auguste Pattberg grammar schools, and an economic and technological and other specialised high schools and of a Vocational and Middle school.

Since 1981 Mosbach is one of nine locations of the *Baden-Württemberg Cooperative State University* (DHBW) in the federal state of Baden-Württemberg in south-western Germany. More than 3,600 students are enrolled in 33 undergraduate study degrees. In addition to their studies, the students work at one of the more than 1,100 companies, which cooperate with the university as corporate partners. This is the core of the *DHBW cooperative study concept*. The DHBW Mosbach offers a wide range of departments in its *School of Business* and the *School of Engineering*. Many of these departments have subdivisions within their units, such as the department of Business Studies offering spezialisations in International Business, Banking, and Health Care Management, amongst others.

See more here: https://www.mosbach.de/startseite and on the DHBW at: https://www.mosbach.dhbw.de/fileadmin/user_upload/dhbw/ressorts/auslandsamt/international_englisch/about_us/Image_Brochure_DHBW_Mosbach_2019.pdf; and here: https://www.facebook.com/mosbach.dhbw.de/.

Index

H. G. Brauch (ed.), *Towards Rethinking Politics, Policy and Polity in the Anthropocene*, The Anthropocene: Politik—Economics—Society—Science 35,
https://doi.org/10.1007/978-3-031-71807-6

G

H

I

J

K

N

O

P

The manufacturer's authorised representative in the EU is Springer Nature Customer Service Centre GmbH, Europaplatz 3, 69115 Heidelberg, Germany. If you have any concerns regarding our products, please contact ProductSafety@springernature.com

Printed and bound by CPI Group (UK) Ltd, Croydon, CR0 4YY
01/10/2026
02225955-0008